Bio-Based Polymers and Composites

Bio-Based Polymers and Composites

Richard P. Wool

Xiuzhi Susan Sun

ELSEVIER
ACADEMIC PRESS

AMSTERDAM • BOSTON • HEIDELBERG • LONDON • NEW YORK
OXFORD • PARIS • SAN DIEGO SAN FRANCISCO • SINGAPORE
SYDNEY • TOKYO

Elsevier Academic Press
30 Corporate Drive, Suite 400, Burlington, MA 01803, USA
525 B Street, Suite 1900, San Diego, California 92101–4495, USA
84 Theobald's Road, London WC1X 8RR, UK

This book is printed on acid-free paper. ∞

Library of Congress Cataloging-in-Publication Data
Wool, Richard P.
Bio-based polymers and composites / Richard P. Wool, Xiuzhi Susan Sun.—1st ed.
p. cm.
Includes bibliographical references and index.
ISBN 0-12-763952-7 (alk. paper)
1. Plant polymers. 2. Biopolymers. 3. Plants—Composition I. Sun, Xiuzhi Susan. II. Title.
QK898.P76W66 2005
620.1′92—dc22 2005013910

British Library Cataloguing in Publication Data
A catalogue record for this book is available from the British Library

For all information on all Elsevier Academic Press publications visit our Web site at www.books.elsevier.com

Printed in the United States of America
05 06 07 08 09 9 8 7 6 5 4 3 2 1

Dedicated with respect to

Those people devoted to the research, development, and marketing of bio-based materials and technology

Those people who pioneer green chemistry and engineering

Those people who protect the environment and ecosystem

Those people who promote humanity's sustainable development

Special people in our lives:

Deborah, Sorcha, Meghan, and Breeda
Dong Hai, Felix, and George

Contents

PREFACE

The conversion of biomass to useful materials such as polymers and composites has considerable economic and environmental value, particularly in times of global warming and diminishing petroleum oil reserves. This book provides considerable detail on the chemistry, physics, and engineering development of plastics, adhesives, composites, foams, elastomers, and coatings from plant co-products such as oils, starch, proteins, lignin, and cellulose, as well as animal processing waste such as chicken feathers and many other natural products. Recent advances in genetic engineering, composite science, and natural fiber development offer significant opportunities for new, improved green materials from renewable resources that are optionally recyclable, biocompatible, and biodegradable, thereby enhancing global sustainability. The development of bio-based materials is consistent with the principles of Green Chemistry and Engineering, which pertain to the design, commercialization, and use of processes and products that are technically and economically feasible while minimizing the generation of pollution at the source and the risk to human health and the environment. Utilization of the (free) energy of sunlight to grow materials by photosynthesis helps remove global warming gases such as CO_2 and reduces the use of fossil fuels. The bio-based materials technologies presented herein are also highly integrated into the biofuel refining process and can utilize their byproducts such as glycerol, protein, fibers, and lignin from biodiesel and ethanol processing lines.

This book will appeal to a broad range of scientists in several fields, including chemistry, biochemistry, physics, materials science, engineering, agriculture, and biotechnology-related disciplines. By selecting the fatty acid distribution function of plant oils, and the amino acid sequence of proteins, we show how the molecular connectivity can be controlled through chemical functionalization to produce linear, branched, or cross-linked polymers with useful thermal and mechanical properties. These bio-based (Green) materials can be used as adhesives, foams, films, rigid or flexible plastics, coatings, elastomers, rubbers, composite resins, carbon nanotubes dispersants, and nanoclay exfoliants. When such bio-based resins are combined with natural fibers (plant and poultry) starch and lignin, new low-cost composites are produced that are economical in many high-volume applications. These high-performance composites can be used in construction, furniture, hurricane-resistant housing, agricultural equipment, automotive sheet-molding compounds, civil and rail infrastructures, marine applications, electronic materials, and sports equipment.

Chapter 1 provides an overview of biomass resources and the basics of plant-derived polymers and sustainability issues. Beginners and experts in the field will find this a useful overview and introduction to our biomass feedstock processing, along with the development of plant materials synthesis and formation, especially with protein, starch polymers, and oils (Chapters 2 and 3). In Chapter 2, there is also a discussion on what it takes to produce genetically improved plant materials. Chapter 3 introduces the current technology for plant polymers, oil extraction, and refining. Chemists will be quite interested in Chapter 4, which shows many of the chemical pathways used to make polymers and composite resins from plant oils. Since all contain high fractions of triglycerides, the resulting polymers are also biocompatible and some biodegradable. The composites manufacturing community will be interested in Chapter 5, where we describe how the first John Deere tractors and Harvesters were made from soy oil resins. The new bio-based foams shown in this chapter can also be used in foam-core composite applications such as windmill blades, housing insulation, and car seats. The physics and fracture mechanics community will find the discussion on fracture in Chapter 6 to contain significant new percolation theories and ideas on structure-property relations of bio-based materials and polymers. Chapter 7 applies the fundamentals of the latter chapter to the important issue of selecting the fatty acid distribution function to design the molecular architecture and tailor the properties of the polymers—a key issue for genetic engineers and natural oil refiners. Pressure sensitive adhesives (PSA), elastomers, and coatings from genetically engineered plant oils are presented in Chapter 8. These soft materials, along with the other materials, are potentially new multibillion-dollar businesses. The uniqueness of these materials compared to their petroleum-derived equivalents is that in addition to being inexpensive they are also biocompatible and in some cases biodegradable. Most of them can also be made from byproducts of the biodiesel process, which is a most fortuitous dual utilization of the new biodiesel plants springing up around the world. It makes more environmental and economic sense to take a gallon of fatty acids (biodiesel) and make 8–10 lbs of valuable materials (at \$1–2/lb), rather than burning it in a fuel-inefficient sports utility vehicle (at \$2.50/gal). Chapter 9 discusses soy protein polymer structure, curing behavior, and engineering properties. The information is necessary to develop adhesives, coatings, films, and gels from soy protein polymers. Chapter 10 shows how the complexity of the soy proteins can be harnessed to produce valuable adhesives with desirable performances for wood particleboard, plywood, wood veneer, strawboard, fiberboard packaging, children's glue, color paints, and casting material for the foundry industry. The protein field is highly complex and is a great research area for future work. Biodegradable thermoplastics derived from poly(lactic acid) (PLA) and starch are thoroughly addressed in Chapter 11, including starch structure, compatibility of PLA and starch, thermal and mechanical properties, physical aging, and plasticization. The last five chapters deal with specialized developments and applications of bio-based polymers and composites such as computer circuit boards from chicken feathers (Chapter 12), potentially adding new meaning to "Farmer in the Dell," hurricane-resistant housing and tsunami emergency shelters (Chapter 13), carbon nanotube composites (Chapter 14), nanoclay composites, which could be a route to new self-healing materials (Chapter 15), and lignin-based polymers and composites, which highlight new discoveries with this fairly intractable waste material (Chapter 16).

Global warming problems will not be resolved in a sustainable manner by utilization of biomass for biofuels, such as bioethanol and biodiesel, although their use will help reduce the need for imported oil and provide economic support in the agricultural community. Global warming and the rapidly expanding energy needs (three times today's energy consumption) of developing nations, with an estimated 9–10 billion people by 2050, can be met in a sustainable manner only by solar, fusion, and nuclear energy in combination with bioenergy. With the anticipated demise of the oil industry by 2050 and the rapid acceleration of global warming effects through fossil fuel consumption, the development of alternative routes to making environmentally friendly bio-based materials should be most welcomed by the present and next generation of scientists and engineers.

ACKNOWLEDGMENTS

The authors are particularly grateful to their students, postdocs, research associates, university and industrial colleagues for their support and hard work in bringing this research to fruition. Dr. Mark Paster of the U.S. Department of Energy (DOE) played an important role in mentoring the collaboration between Kansas State University and the University of Delaware and financially supporting this work. RPW appreciates the financial support of his Affordable Composites from Renewable Resources (ACRES) group from DOE, EPA, NSF, USB, UD, and Cara Plastics Inc. Pre-submission editorial assistance from David Banks and Jue Lu is much appreciated by RPW. XSS appreciates the financial support of her Bio-Materials & Technology Lab from the U.S. Department of Energy, the U.S. Department of Agriculture, the Consortium of Plant Biotechnology Research, Environmental Protection Agency, NSF, Kansas Soybean Commission, Kansas Wheat Commission, Kansas Technology Enterprise Corporation, industrial sectors, and Kansas State University. We both appreciate the much-tried patience of the Editorial staff at Elsevier.

About the Authors

Richard Wool is a Professor of Chemical Engineering and Director of the Affordable Composites from Renewable Resources (ACRES) Program at the University of Delaware. He was Director of the Center for Composite Materials at the University of Delaware and Professor of Materials Science and Engineering at the University of Illinois at Urbana–Champaign (1977–1995). He received his B.S. degree in Chemistry from University College Cork, Ireland (1970) and Ph.D. in Materials Science and Engineering from the University of Utah (1974). He held Visiting Professorships at Trinity College Dublin (Physics), École Polytechnique Paris (Condensed Matter Physics), and Milano (Natta Institute). He is author of the book *Polymer Interfaces: Structure and Strength*, Hanser/Gardner (1995), has edited several books, and has published over 100 archival papers.

Professor Wool is a Fellow of the American Physical Society, member of the American Chemical Society, Materials Research Society, American Institute of Chemical Engineers, Society of Plastic Engineers, Neutron Scattering Society, a founding member of the Bio/Environmentally Degradable Plastics Society, and was Chairman of the ASTM Committee for Biodegradable Plastics. He is a consultant for Nike (Oregon), DuPont (Delaware), Raytheon Missile Defense Systems (Arizona), and Tetra Pak (Italy). He is the founder and President of Cara Plastics Inc., which built the first John Deere tractors, round hay balers, and Harvester composite parts from soybeans, and developed the biodegradable plastic industry in China. He teaches Green Engineering at the University of Delaware and has interests in polymer physics, interfaces, fracture, fractals, bio-based materials, and environmental issues. He holds several patents in the field of bio-based polymers and composites.

Xiuzhi Susan Sun is a Professor in the Department of Grain Science and Industry at Kansas State University and is Director of the Bio-Materials & Technology Laboratory. She received her Ph.D. in Biological & Agriculture Engineering (1993) from the University of Illinois at Urbana–Champaign, and did her postdoctoral training at Texas A&M University. She specializes in biological materials science and engineering, focusing on utilization of renewable plant materials for industrial products, especially for bio-based adhesive, resins, composites, and structured protein polymers. Her research interests also include thermal and rheological behavior, and structure and functional properties of plant-related polymeric materials and ingredients. She is the author of 100+ peer-reviewed journal articles and patents and is the Associate Editor of the *Journal of Cereal Chemistry*. She has been an invited speaker for many scientific symposia and international conferences. Dr. Sun is a member of the American Association of Cereal Chemists, the BioEnvironmental Polymer Society, the American Chemical Society, the American Society of Biological and Agricultural Engineers, and The Scientific Research Society, Sigma Xi. She is the USDA National Research Initiative Technical Panel Manager of Biobased Products and Bioenergy for 2004 and 2005. She regularly participates in national strategic research planning workshops and program review panels in bio-based materials and bioenergy for the USDA, DOE, EPA, and NSF.

1

Overview of Plant Polymers: Resources, Demands, and Sustainability

Xiuzhi Susan Sun

Advances in petroleum-based fuels and polymers have benefited mankind in numerous ways. Petroleum-based plastics can be disposable and highly durable, depending on their composition and specific application. However, petroleum resources are finite, and prices are likely to continue to rise in the future. In addition, global warming, caused in part by carbon dioxide released by the process of fossil fuel combustion, has become an increasingly important problem, and the disposal of items made of petroleum-based plastics, such as fast-food utensils, packaging containers, and trash bags, also creates an environmental problem. Petroleum-based or synthetic solvents and chemicals are also contributing to poor air quality. It is necessary to find new ways to secure sustainable world development. Renewable biomaterials that can be used for both bioenergy and bioproducts are a possible alternative to petroleum-based and synthetic products.

Agriculture offers a broad range of commodities, including forest, plant/crop, farm, and marine animals, that have many uses. Plant-based materials have been used traditionally for food and feed and are increasingly being used in pharmaceuticals and nutraceuticals. Industrial use of agricultural commodities for fuels and consumer products began in the 1920s, but they were soon replaced by petroleum-based chemicals after World War II because of petrochemicals' low cost and durability. This chapter focuses on bio-based polymers derived from plant-based renewable resources, their market potential, and the sustainability of the agriculture industry of the future.

The three major plant-based polymers are protein, oil, and carbohydrates. Starch and cellulose, also called polysaccharides, are the main naturally occurring polymers in the large carbohydrate family. Agricultural fiber is also a member of the carbohydrate family. Natural fiber such as flax, hemp, straw, kenaf, jute, and cellulose consists mainly of cellulose, hemicellulose, and lignin, but is usually listed as a material when used as a fiber in composites, as discussed in Chapters 5 and 13.

Corn, soybean, wheat, and sorghum are the four major crops grown in the United States (Table 1.1), with total annual production of about 400 million metric tons (800 billion pounds) in the year 2000. Annually, 10–15% of these grains are used for food, 40–50% for feeds, and the rest could be for various industrial uses. Based on U.S. Department of Agriculture statistics, the total land used for crops is about 455 million acres, which is about 20% of the total usable land (Figure 1.1) [1]. Including other crops, such as rice, barley, peanuts, and canola, the United States has the potential to produce about 550 million metric tons of grains and legumes. At least 150 million metric tons of grains and legumes are available for nonfood industrial uses. In general, seeds make up about 45–52% of the dry mass of a plant. This means that there is the potential to produce about 400 million metric dry tons of cellulosic sugar-based biomass (agriculture fiber residues) annually in the United States alone based on the total production of corn, soybean, wheat, and sorghum. Including other crops, plants, and forest products, the total annual U.S. production of cellulosic sugar-based biomass could be about 800 million dry tons.

1.1 PLANT PROTEINS

Plant proteins are amino acid polymers derived mainly from oilseeds (i.e., soybeans) and grains (i.e., wheat and corn) and are usually produced as by-products

TABLE 1.1 Production of selected grains and legumes (million metric tons).

	Wheat	Soybean	Corn	Sorghum
World production	578	172	585	55
United States	60 (2nd)	75 (1st)	253 (1st)	12 (1st)
Other countries	99.6 (1st)	37 (2nd)	106 (2nd)	9 (2nd)
	China	Brazil	China	India
	37 (3rd)	15.4 (4th)	40 (3rd)	2.8 (6th)
	France	China	India	China

Sources: From Ref. [31] and USDA World Agriculture Production, July 27, 2001.

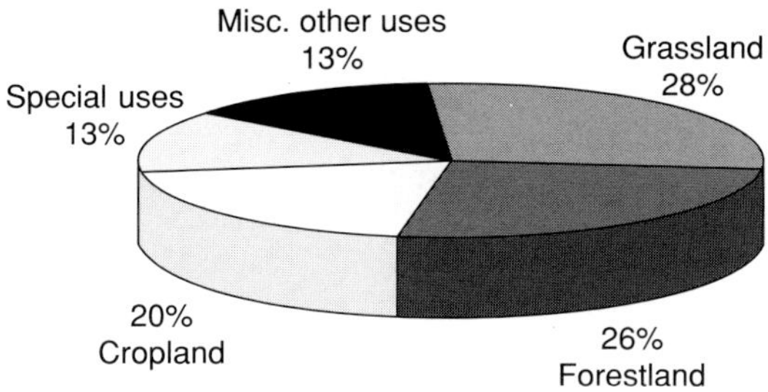

FIGURE 1.1 Land use and distribution. Total useful land in the United States is about 2.3 billion acres.

of processing oils and starches (Table 1.2). The potential U.S. protein production is about 120 billion pounds of soybean meal containing about 50% protein, about 20 billion pounds of wheat gluten containing about 70% protein, and about 40 billion pounds of corn gluten containing about 65% protein. Of the corn protein, about 30% is a functional protein called *corn zein protein* [2]. Plant proteins are widely used as major ingredients for food, feed, pharmaceuticals, nutraceuticals, paper coating, textile sizing, and, increasingly, adhesives. Plant proteins are complex macromolecules that contain a number of chemically linked amino acid monomers, which together form polypeptide chains, constituting the primary structure. The helix and sheet patterns of the polypeptide chains are called secondary structures. A number of side chains are connected to the amino acid monomers. These side chains and attached groups interact with each other, mainly through hydrogen and disulfide bonds, to form tertiary or quaternary structures. These proteins often have large molecular weights, in the range of 100,000–600,000 Dalton (Da) (Dalton = grams per mole), which makes them suitable for polymers and adhesives.

Proteins can be modified by physical, chemical, and enzymatic methods. Modification results in structural or conformational changes from the native structure without alteration of the amino acid sequence. Modifications that change the secondary, tertiary, or quaternary structure of a protein molecule are referred to as *denaturation* modifications [3]. The compact protein structure becomes unfolded during denaturation, which is accompanied by the breaking and reforming of the intermolecular and intramolecular interactions [4].

Physical modification methods mainly involve heat [5] and pressure [6] treatments. Heat provides the protein with sufficient thermal energy to break hydrophobic interactions and disassociate the subunits [5]. The disassociation and unfolding expose the hydrophobic groups previously enclosed within the contact area between subunits or on the interior of the folded molecules. For example, soybean protein disassociates and coagulates at high pressure and exhibits large hydrophobic regions and high viscosity [6]. This

TABLE 1.2 Average composition of cereal grains and oil seeds (% dry weight basis).

Cereal Grains	Protein	Fat	Starch	Fiber	Ash	Source
Wheat	12.2	1.9	71.9	1.9	1.7	[45]
Rye	11.6	1.7	71.9	1.9	2.0	[45]
Barley	10.9	2.3	73.5	4.3	2.4	[45]
Oats	11.3	5.8	55.5	10.9	3.2	[45]
Maize	10.2	4.6	79.5	2.3	1.3	[45]
Millet	10.3	4.5	58.9	8.7	4.7	[45]
Sorghum	11.0	3.5	65.0	4.9	2.6	[45]
Rice	8.1	1.2	75.8	0.5	1.4	[45]
Oil Seeds						
Soybean	51–70[a]	18–26	—	6.5	3.7–7.4	[47]
Rapseed	36–44[a]	38–50	—	12–18	7.4–8.8	[47]
Sunflowers	20.8	54.8	18.4	2.1	3.4	[47]
Peanuts	30	50	14	2.9	3.1	[47]
Canola	22.0	41.0	22	10.0	5.0	[46]
Caster bean	12–16	45–50	3–7	23–27	2	[47]
Cottonseed	22	19.5	35	19.0	4.5	[46]
Copra	4.6–8.0	68–79	17.4–21	4.6–7.7	2.4–3.7	[47]
Safflower	21	41.0	14.5	19.0	4.5	[46]
Linseed	22–26	41.5–45.5	27–31	5.5–9.7	4.3–2.7	[47]
Sesame	20	52	23	—	5.6	[47]

Sources: From Refs. [45], [46], and [47].
[a]Oil-free basis.

process is covered in more detail in Chapter 10 in the discussion of the development of soy protein adhesives and composites.

Chemical modification methods may cause alteration of the functional properties, which are related closely to protein size, structure conformation, and the level and distribution of ionic charges. Furthermore, chemical treatments could cause reactions between functional groups, resulting in either adding a new functional group or removing a component from the protein. Chemical modification methods include acetylation, succinylation, phosphorylation, limited hydrolysis, and specific amide bond hydrolysis. Acetylation is the reaction between a protein amino, or a hydroxyl group, and the carboxyl group of an acetylating agent. The acetylation reaction can modify the surface hydrophobicity of a protein [7]. Succinylation converts the cationic amino groups in the protein to an anionic residue, which increases the net negative charge, resulting in an increase in hydrophobicity under specific succinylating conditions [8]. This treatment also increases the viscosity [9]. Phosphorylation is another effective method to increase negative charges, thereby affecting gel-forming ability and cross-linking [10]. Gel-forming ability can also be increased by alkylation treatment [8]. Chemical hydrolysis is one of the most popular methods for protein modifications by acid-based agents. For example, peptide bonds on either side of aspartic acid can be cleaved at a higher rate than other peptide bonds during mild acid hydrolysis [11]. The hydrophobicity of a protein greatly increases under specific conditions of mild acid hydrolysis [12, 13].

1.2 PLANT OILS

Plant oils, such as soy oil, corn oil, and flax oil, can be derived from many crops (Table 1.2). The United States has the potential to produce about 30 billion pounds of soy oil, 25 billion pounds of corn oil, and many billion pounds of oils from other oilseeds as listed in Table 1.2. Plant oils are triglycerides and contain various fatty acids. Soybean, a major oil plant, contains about 20% oil. Soy oil is inexpensive in the United States, selling at about $0.20/lb. Refined soy oil contains more than 99% triglycerides and about eight major fatty acids, including linoleic, oleic, linolenic, palmitic, and stearic acids (see Table 4.1 in Chapter 4) [14]. These fatty acids differ in chain length, composition, distribution, and location. Some are saturated, and some are unsaturated, which results in differences in the physical and chemical properties of the oil. Control of the fatty acid distribution function is essential to optimize polymer properties, as discussed in Chapters 4 through 8. Such plant oil can be physically treated and chemically modified to meet specific industrial applications [15].

Adhesives and resins can be derived from bio-based oils using similar synthetic techniques to those used with petroleum polymers. Many active sites from the triglycerides, such as double bonds, allylic carbons, and ester groups, can be used to introduce polymerizable groups. Wool and coworkers [16] prepared soy oil-based resins by functionalizing the triglycerides (Chapter 4). This was accomplished by attaching polymerizable chemical groups, such as maleinates and acrylic acid, or by converting the unsaturated sites to epoxies or hydroxyl functionalities, making the triglycerides capable of polymerizing via ring-opening, free-radical, or polycondensation reactions.

The second method of producing resins from oil is to reduce the triglycerides into monoglycerides. Polymerizable groups, such as maleate half esters, can be attached to the monoglycerides, allowing them to polymerize through free-radical polymerization [17].

The third method is to functionalize the unsaturated sites and reduce the triglycerides to monoglycerides, which can form monomers by reacting with maleic anhydride, allowing polymerization by free-radical polymerization [18, 19]. Such reactions produce bio-based polymers that have properties and costs comparable to those of petrochemical-based adhesives and composite resins. These processes will be discussed in more detail in later chapters.

1.3 PLANT STARCHES

Starch is a carbohydrate polymer that can be purified from various sources with environmentally sound processes and green engineering (see Section 11.1 in Chapter 11 for structures). Corn, wheat, sorghum, and potato are all major resources, containing about 70–80% starches (Table 1.2). The potential U.S. starch production is about 455 billion pounds each year from wheat, corn, and sorghum. However, only 5 billion pounds of starch are produced annually in the United States, mainly from corn. These starches have been used in the food industries, as well as in the paper and other nonfood industries. This number is expected to increase to about 10 billion pounds in the near future with the development of biopolymers, such as poly(lactic acid) (PLA), as substitutes for petroleum-based plastics [20], as discussed in Chapter 11. Ethanol production from starch as a liquid fuel substitute will also increase until new hydrogen- and methanol-based fuels become viable in the next 10–20 years.

Starch is a polysaccharide of repeating glucose monomers and is a mixture of two polymers: linear amylose linked by α-1,4-bonds and branched amylopectin linked by α-1,6-bonds. At a given molecular weight, amylose swells to a much larger volume in solution than amylopectin [21], but the more amorphous amylopectin absorbs more water than amylose at elevated temperatures [22]. Linear amylose polymers can also align their chains faster than branched amylopectin polymers. The branched amylopectin can have an infinite variety of structures, depending on the frequency of branching and the length of the branched chains. Different physical properties are associated with these various structures. These molecules can be cross-linked by themselves, or with other multifunctional reagents. As the cross-linking increases, the cross-linked polymer becomes less water soluble (Chapter 11).

Many modified starches are produced by chemical substitution of the hydroxyl groups attached to the starch molecules. The type of modification, degree of substitution, and modification conditions will greatly affect the characteristics of the final modified starch and, consequently, product quality. Four major starch modification methods have been used [23]: (1) pregelatinization, such as disintegration of the crystalline starch granules by heat, high pH, or shear force, to obtain water-soluble amorphous products; (2) degradation of starch by acids or enzymes to reduce the molecular weight, resulting in a lower viscosity; (3) chemical substitution by either esterification with acid anhydrides or by etherification with epoxide compounds; and (4) cross-bonding with bifunctional esterifying or etherifying reagents to increase the starch molecular weight, resulting in a higher viscosity.

1.4 AGRICULTURAL FIBERS AND CELLULOSE

Agricultural fibers include crop residuals, such as straw, stems, hulls, and milling by-products (e.g., brans) from wheat, corn, soybean, sorghum, oat, barley, rice, and other crops. The major chemical composition of these fibers is similar to wood fibers and includes cellulose, lignin, and pentosan. Wood fiber contains about 40–45% cellulose, 26–34% lignin, and 7–14% pentosan. In comparison, wheat straw contains about 29–35% cellulose, 16–21% lignin, and 26–32% pentosan [24]. Wheat straw also contains about 0.6%–3.6% protein [25]. Other cereal straws, such as rice, barley, oat, and rye, have chemical compositions similar to that of wheat straw [26]. Large quantities of agricultural fibers are available for biofuels and bioproducts. For example, about 400 million metric tons (800 billion pounds) of dry crop residues are available, based on the grain production of corn, soybean, wheat, and sorghum at a straw-to-seed ratio of from 45 to 52% [27–31]. Among these residues, about 150 billion pounds are wheat straw [32]. Wheat straw is usually used for fuel, manure, cattle feed, mulch, and bedding materials for animals [33]. Particleboard can be prepared using wheat straw [34–36], sunflower stalks [37], rice straw, cotton stalks, sugar cane bagasse, flax [38], maize husks, and maize cobs [33].

Natural fibers can be used for composites (see Chapters 5, 10, and 13) as harvested, or they can be used as raw materials for cellulose production. Cellulose can be modified into cellulose esters, such as cellulose acetate, cellulose acetate propionate, and butyrate, which are currently used as major components of thermoplastics. Cellulose, a major component of natural fibers, occurs in nature largely in crystalline forms made up of partially aligned or oriented linear polymer chains, consisting of up to 10,000 β-1, 4-linked anhydroglucose units. Cellulose chains are compacted aggregates, held together by hydrogen bonds forming a three-dimensional structure, which imparts mechanical strength to cellulose and contributes to its biodegradation and acid hydrolysis [39]. Hemicellulose is mainly composed of β-1, 4-linked D-xylopyranoyl units with side chains of various lengths containing L-arabinose, D-glucuronic acid, or its 4-*O*-methyl ether, D-galactose, and possibly D-glucose [40]. Lignin is mainly made up of phenlypropane units. Lignin is encrusted in the cell wall and partly covalently bonded with hemicellulose polysaccharides. Lignin is often a by-product of cellulose or paper pulping manufacture. It is inexpensive and mainly used for fuel and reformed composite materials [41]. Lignin may also have potential use in adhesives. It can be functionalized (Chapter 16) to make it soluble in composite resins and be used as a comonomer and interfacial agent for natural fibers and soy-based resin composites.

1.5 MARKET POTENTIAL FOR PLANT POLYMERS

Materials and composites used for construction, automobile parts, furniture, packaging, utensils, printing, coatings, and textile sizing represent a large market (about $100 billion) that includes a broad variety of products, such as adhesives, resins, plastics, binders, fibers, paints, inks, additives, and solvents. For example, about 20 billion pounds of adhesives are used annually in the United States. Among those adhesives, about 8 billion pounds are formaldehyde-based adhesives, 3.5 billion pounds are thermoset- and thermoplastic-based adhesives, 7.5 billion pounds are latex based, 0.5 billion pounds are isocyanate based, and the rest are various adhesives with unique applications. The latex-based adhesives are mainly used for packaging, coating, labeling, ink, paints, office glues, furniture, furnishings, and similar uses. The formaldehyde-based adhesives primarily include urea formaldehyde and phenol formaldehyde adhesives, which are mainly used for plywood, particleboard, medium-density fiberboard, and oriented strand board for construction and furniture. Generally, the adhesive is about 5–20 wt% of a wood-based composite material used in construction, with the rest of the composite comprised mainly of fiber materials. With an average of 10% adhesive used in such composites, the total annual fiber demand is about 150 billion pounds.

The demands for thermoplastic resin are another indicator of market potential. Narayan [20] did a search in 1994 and found that about 54.2 billion pounds of thermoplastic resins and 11.1 billion pounds of styrene-based latex resins were produced in 1992 in the United States. These resins are used mainly in packaging, construction, furniture, and adhesives (Table 1.3). About 22 billion pounds of plastic waste was discharged in 1992 with an annual rate of increase of 5.9% [42]. This figure is expected to reach 42 billion pounds by 2007. Based on U.S. Environmental Protection Agency (EPA) statistics, about 10 million pounds of plastic wastes are produced aboard

TABLE 1.3 Thermoplastic resin uses and distributions.

Thermoplastic Resins	Amount (Billions of Pounds)	Styrene-Based Latex	Amount (Millions of Pounds)
Packing	18.2	Adhesives, inks, and coatings	461
Building and construction	7.6	Furniture and furnishings	369
Electrical and electronic	2.6	All other	313
Furniture and furnishings	2.4		
Consumer and institutional	5.9		
Industrial	0.6		
Adhesives, inks, and coatings	1.2		
Transportation	2.5		
Exports	6.8		
All other	6.6		

Source: Facts and Figures of the U.S. Plastics Industry, Society of the Plastic Industry, 1993.

government ships [20]. These wastes can be used as an indicator for market potential for both bio-based and biodegradable materials.

An example of disposable items produced from thermoplastics is given in Figure 1.2. These thermoplastics are commonly used for packaging containers, films, closures, foams, cutlery, utensils, loose fill, and other applications. Many other single-use or short-term-use items, such as diapers, personal and feminine hygiene products, masks, gowns, gloves, and even computer hardware and television frames, all have market potential for bio-based materials.

1.6 SUSTAINABLE AGRICULTURE INDUSTRY OF THE FUTURE

Durability, compatibility, affordability, and sustainability are the challenges of converting renewable resources into industrial materials. Sustainable development provides growth of both ecological integrity and social equity to meet basic human needs through viable economic development over time. When a new material is designed and manufactured, one consideration should be sustainability, including resource availability, land use, biodiversity, environmental impact, energy efficiency, soil conservation, and the impact on the social community. Besides a favorable life cycle analysis, research and development of bio-based products should consider the limits that will maintain sustainable development. The design of bio-based materials should favor increased materials supplements, optimized land use, improved plant biodiversity, minimized environmental pollution, and improved energy efficiency, while at the same time meeting consumer demands. The principles of green engineering discussed in the Preface are a useful guide for the design of new green materials derived from biomass.

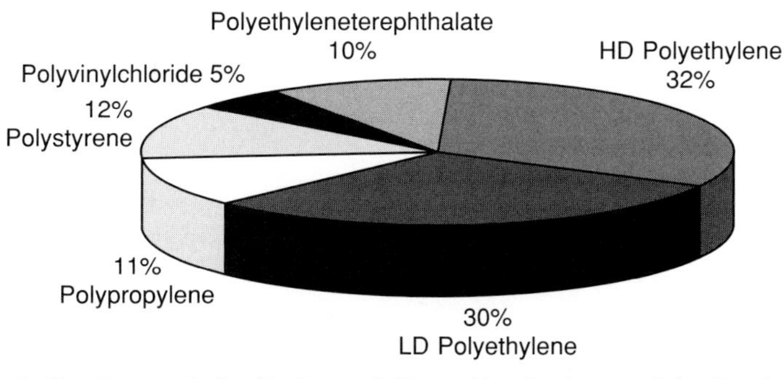

FIGURE 1.2 Uses and distributions of disposable plastic materials. Total disposable plastics is about 13,655 million pounds. (*Source: Facts and Figures of the U.S. Plastics Industry,* Society of the Plastic Industry, 1993.)

About 467 million dry tons of biomass are available for energy use, including energy crops (switch grass, hybrid poplar, and willow), forest residues, mill residues, sludge, biogas, and other wastes [43]. In addition, about 550 million dry tons of crop residues are produced annually in the United States, based on total grain and legume production [31]. Some of these residues need to be returned to fields to maintain soil quality (such as soil carbon balance), and some are used for manure or animal bedding, but approximately 70% of these crop residues may be available for energy uses. Burning of residual natural fibers is now forbidden in most Western countries and their utilization in materials as proposed herein has a double environmental benefit.

The total amount of energy consumption in the United States is about 100 quadrillion Btu annually [44]. About 40% of the Btu comes from petroleum oils. The total estimated cellulosic sugar-based biomass available for biofuel is about 467 million dry tons in addition to 385 million dry tons of crop residues. Based on current technology, biomass materials would contribute about 10–15% of the total energy annually used in the United States [43]. To make sugar-based cellulosic biomass economically viable for energy, advanced technology is being developed to convert these biomasses into biofuels at higher efficiency. In addition, plant production needs to be increased at least three- to fourfold during the next 40 years to meet national biofuel needs. It makes excellent environmental sense to utilize grains and waste agriculture fibers for materials and fuels that otherwise would be derived from petroleum. However, such energy and material conversions should be done in a sustainable green engineering manner such that a gallon of ethanol fuel does not require a gallon of petroleum to produce.

The total estimated market potential for bioproducts could be about 160 billion pounds (about 80 million metric tons). There are about 250 million metric tons of grains and legumes potentially available in the United States annually for industrial products. Polymers from grains and legumes require much less energy to convert into useful materials for some, but not all, bioproducts. Protein, oil, carbohydrates, and/or their monomers, including amino acids, fatty acids, sugars, and phenolics are all important platforms as coproducts of a feedstock system and meet the large demands for bioproducts, including adhesives, resins, composites, plastics, lubricants, coatings, solvents, inks, paints, and many other chemicals (Figure 1.3). Plant materials can rarely be used as they are, but they can be converted into functional polymers for consumer products after bioconversions, reactions, and modifications with physical, chemical, enzymatic, and genetic approaches.

Plant material structures are genetically controlled, which consequently affects product performance. Plant materials are studied in this book in relation to their product performance. Proteins are complex macromolecules that contain a number of amino acid monomers linked by amide bonds. The sequences of these amino acids and composition determine protein structure,

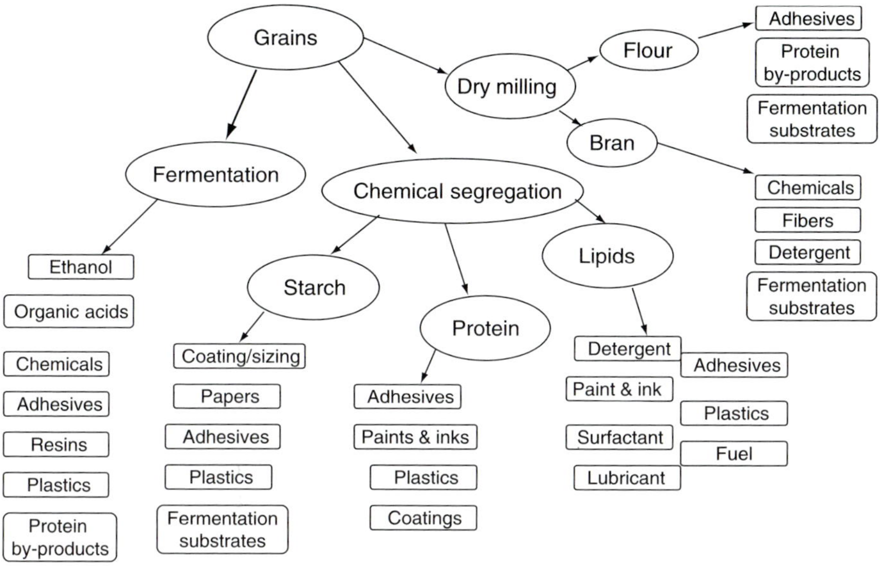

FIGURE 1.3 Diagram of possible industrial products from biorefinery process of grains and legumes. Application potentials are beyond those listed in the diagram.

functional groups, and conformational structures that affect both processing and end product quality. The triglyceride oil molecular structure is essentially that of a three-arm star where the length of the arms, the degree of unsaturation, and the fatty acids' content and distribution are the important structural variables for product quality. Advanced technologies, such as biopolymer simulation and modeling, surface structure analysis, chemical structure analysis, synthesis, thermal phase transitions, and rheological behavior analysis, should be used to obtain the information required to better understand bio-based polymers.

Research and development for a sustainable agricultural industry for plant-based materials and composites include five major units: plant science, production, bioprocessing, utilization, and products designed to meet society's demands. Based on several road maps developed by federal funding agencies for bio-based materials research, we summarize the critical research needs and directions as follows: Research efforts in plant science should focus on genomics, enzymes, metabolism, and bioinformatics. This allows for a better understanding of gene regulations, plant metabolic pathways, carbon flow, functional genomics, molecular evolution, and protein/oil/carbohydrate formation, which helps in developing tools and technologies for functional gene markers, gene switching, gene screening and sequencing, and gene manipulations. Research efforts in production focus on plant and grain quality consistency, unit costs, yield, infrastructure, and designed plants. It is important to produce components with favorable traits, improve yields, understand interactions of genotypes with environment, control bio-based polymer and compound quality, develop harvesting technologies, and use land economically. For bio-based polymer and materials science and engineering, attention should be given to bio-based polymer chemistry, reactions and modification pathways, processing technologies, enzyme metabolism for bioconversion, bioseparation, molecular structure and functional performance, scale-up, economics, and infrastructure. Understanding these areas will allow for development of new technologies for cost-effective conversion of plant materials into functional industrial materials. Plant materials utilization focuses on market/function identification, bioproduct designs, new bio-based materials development, performance definition, life cycle analysis and cost-value analysis, material standards improvement, new analytical method development, infrastructure and distribution system, and the main driver, economics.

1.7 CONCLUSION

Plant protein, oil, starch, and cellulosic materials are all important platforms for bioproduct applications. Lignin from cellulosic-based biomass should also be utilized for biofuels and bioproducts. Agricultural commodities typically cannot be used as they appear in nature. They need to be converted into functional polymers and materials by various technologies including chemical reactions, fermentation, and modifications. Research efforts need to focus on interdisciplinary approaches that integrate plant science, production, processing, and utilization. Integrated research teams in the areas of materials science and engineering, plant science, biochemistry/chemistry, and economics should be assembled in collaboration with universities, institutions, national laboratories, and industries to achieve what we need in this and coming centuries.

ACKNOWLEDGMENT

The authors thank Dr. Forrest Chumley for his thorough review of Chapter 1.

REFERENCES

1. Vesterby, M.; Krupa, K. S. Major Uses of Land in the United States, *Statistical Bulletin No. 973,* U.S. Department of Agriculture, Washington, DC; 2001.
2. Shukla, R.; Cheryan, M. *Industrial Crops and Products* 2001, **13,** 171–192.
3. Hettiarachchy, N. S.; Kalapathy, U; Myers, D. J. *J. Am. Oil Chem. Soc.* 1995, **72**(12), 1461–1464.

4. Careri, G.; Giansanti, A.; Gratten, E. *Biopolymers* 1979, **18,** 1187–1203.
5. Niwae, E.; Wang, T.; Kanoh, S.; *et al. Bull. Japanese Soc. Sci. Fisheries* 1988, **54**(10), 1851.
6. Kajiyama, N.; Isobe, S.; Uemura, K.; *et al. Int. J. Food Sci. Technol.* 1995, **30**(2), 147.
7. Kim, S. H.; Rhee, J. S. *J. Food Biochem.* 1989, **13**(3), 187.
8. Kim, S. H.; Kinsella, J. E. *Cereal Chem.* 1986, **63,** 342.
9. Kim, S. H.; Kinsella, J. E. *J. Food Sci.* 1987, **52**(5), 1341.
10. Frederick, F. S. *Biochem. Food Proteins,* Hudson, B. J., Ed.; Elsevier Science Publishers, Ireland, Ltd.; 1992.
11. Han, K. K. R.; Biserte, G. *Int. J. Biochem.* 1983, **15,** 875.
12. Matsudomi, N.; Sasaki, T.; Kato, A.; *et al. Agric. Biol. Chem.* 1985, **49**(5), 1251.
13. Wagner, J. R.; Gueguen, J. *J. Agr. and Food. Chem.* 1995, **43**(8) 1993–2000.
14. Liu, K. *Soybeans: Chemistry, Technology, and Utilization,* International Thomson Publishing, New York; 1997.
15. Wool, R. P.; Kusefoglu, S. H.; Palmese, G. R.; *et al.*, U.S. Patent 6,121,398; 2001.
16. Wool, R. P.; Kusefoglu, S. H.; Palmese, G. R.; *et al.*, U.S. Patent 6,121,398; 2000.
17. Tracker, D. J.; Borden, G. W.; Smith, O. W. U.S. Patent 3,979,270; 1976.
18. Tracker, D. J.; Borden, G. W.; Smith, O. W. U.S. Patent 3,931,075; 1976.
19. Wool, R. P.; Khot, S. N.; LaScala, J. J.; *et al.*, *Affordable Composites and Plastics from Renewable Resources; Part I: Synthesis of Monomers and Polymers,* 177–204; *Part 2: Manufacture of Composites,* 205–224, *Advancing Sustainability through Green Chemistry and Engineering*, R. L. Lankey and P. T. Anastas, Eds., ACS Symposium Series 823, 2002.
20. Narayan, R., Polymeric materials from agriculture feedstocks, In *Polymers from Agricultural Coproducts,* Fishman *et al.*, Eds., ACS Symposium Series 575; 1993.
21. Thomas, D. J.; Atwell, W. A. *Starches: Practical Guides for the Food Industry,* American Association of Cereal Chemists, St. Paul, MN; 1997.
22. Tester, R. F.; Morrison, W. R. *Cereal Chem.* 1990, **67**(6), 551.
23. Whistler, R. L.; BeMiller, J. N. *Carbohydrate Chemistry for Food Scientists,* American Association of Cereal Chemists, St. Paul, MN; 1999.
24. Karr, G. S. Acetylation of Ground Wheat Straw for the Production of Strawboard, *M.S. Thesis,* Kansas State University, Manhattan; 1998.
25. Atchison, J. E.; McGovern, J. N. History of paper and the importance of agro-based plant fibers. In *Pulp and Paper Manufacture,* Vol. 3, Secondary Fibers and Agro-Based Pulping, Hamilton, F.; Leopold, B., Eds.; TAPPI Press, Atlanta; 1983.
26. Rowell, R. M. Opportunities for Lignocellulosic Materials and Composites. *ACS Symposium Series 476*; American Chemical Society, Washington, DC; 1992, pp. 12–27.
27. Sharma, R. C.; Smith, E. L. *Crop Sci.* 1986, **26,** 1147–1150.
28. King, K. M.; Greer, D. H. *Agron. J.* 1986, **78,** 515–521.
29. Rajcan, I.; Tollenaar, M. *Field Crops Res.* 1999, **60,** 245–253.
30. Schapaugh, W. T.; Wilcox, J. R. *Crop Sci.* 1980, **20,** 529–533.
31. *World Agricultural Supply and Demand Estimates,* World Agricultural Outlook Board, U.S. Department of Agriculture, Washington, DC; most recent edition: WASDE–399, June 11, 2003.
32. Zucaro, J.; Reen, R. The Second Forest: Filling the Wood Source Gap While Creating the Environmental Performance Board of the 21st Century. Developing Composites from Wheat Straw. In *Proc. Washington State Univ. Int. Particleboard/Composite Materials 29th Symp.;* 1995, pp. 225–231.
33. Sampathrajan, A.; Vijayaraghavan, N. C.; Swaminathan, K. R. *Bioresour. Technol.* 1992, **40**(3), 249–251.
34. Han, G.; Zhang, C.; Zhang, D.; *et al. J. Wood Sci.* 1998, **44**(4), 282–286.
35. Karr, G.; Sun, X. *Indust. Crops Products* 2000, **12,** 19–24.
36. Mo, X.; Hu, J.; Sun, X.; *et al. Indust. Crops Products* 2001, **14,** 1–9.
37. Khristova, P.; Yossifov, N.; Gabir, S.; *et al. Cellul. Chem. Technol.* 1998, **32**(3–4), 327–337.

38. Heslop, G. In *Proc. Washington State Univ. Int. Particleboard/Composite Materials 31st Symp.;* 1997, pp. 109–113.
39. Theander, D. Review of Straw Carbohydrate Research. In *New Approaches to Research on Cereal Carbohydrates,* Hill, R. D.; Munck, L., Eds.; Elsevier Science Publishers, Amsterdam; 1985, p. 217.
40. Lawther, J. M.; Sun, R. C.; Banks, W. B. *J. Agric. Food Chem.* 1995, **43,** 667.
41. McCarthy, J. L.; Islam, A. In *Historical, Biological, and Materials Perspectives,* Glasser, W. G.; Northey, R. A.; Schultz, R. A., Eds.; American Chemical Society, Washington, DC; 2000, Chap. 1, pp. 2–99.
42. Anonymous, *Plastics Eng.* 1994, **50,** 34.
43. Biobased Products and Bioenergy Road Map, Draft, July 18, 2001, United States, Department of Energy.
44. *Annual Energy Review,* Energy Information Administration, Washington, DC; 2001; http://www.eia.doe.gov/emeu/aer/pdf/03842001.pdf.
45. Lasztity, R. *The Chemistry of Cereal Proteins,* 2nd ed., CRC Press, Boca Raton, FL; 1995.
46. Lusas, E. W. Oilseeds and Oil-Bearing Materials. In *Handbook of Cereal Science and Technology*, 2nd ed., Kulp, K.; Ponte, Jr., G., Eds.; Marcel Dekker, New York; 2000, pp. 297–362.
47. Salunkhe, D. K.; Chavan, J. K.; Adsule, R. N.; *et al., World Oilseeds Chemistry, Technology, and Utilization,* Van Nostrand Reinhold, New York; 1991.

2

Plant Materials Formation and Growth

Xiuzhi Susan Sun

The protein, carbohydrate, and oil-based polymers mentioned in Chapter 1 are synthesized in plants through the utilization of sunlight for energy and carbon sources such as CO_2 in the air. The plants' utilization of global warming gases for molecular building blocks and sunlight as a free source of energy attests to the environmentally friendly nature of bio-based materials. Photosynthesis is an essential procedure for polymer formation. The unique properties of chlorophyll enable green plant cells to receive the radiant energy of sunlight. With the aid of ferredoxin, cytochromes, and other compounds in the stacks of membranes, the light reactions of photosynthesis produce nicotinamide adenine dinucleotide phosphate (NADPH), adenosine triphosphate (ATP), and molecular oxygen. Photosynthesis produces two major compounds: fructose diphosphate and hexoses, which are essential for protein, lipid, and carbohydrate synthesis.

The goal of this chapter is to provide a general picture about how polymers form in plants and what major variables control the quality and growth of these polymers. This chapter is designed for scientists who undertake materials science and engineering with a focus on plant polymers for industrial uses. The information and data presented in this chapter are mainly from the books *Biology* [1], *Biochemistry & Molecular Biology of Plants* [2], and *Molecular Biotechnology* [3]. People who are interested in learning more about plant polymer formation and growth should read these or other related books.

2.1 PLANT MATERIAL SYNTHESIS

2.1.1 BASIC TERMS AND THEIR FUNCTIONS

Amino Acid

An amino acid is an organic compound containing an amino group ($-NH_2$) and a carboxyl group ($-COOH$), as listed in Table 2.1. Amino acids are monomers of protein molecules linked by amide bonds ($-NHCO-$) to form a peptide chain. This is the same linkage used in polyamides or nylon. For example, alanine (Table 2.1) has the structure $NH2-CHCH_3-COOH$.

Proteins

Proteins are large macromolecules containing carbon, hydrogen, oxygen, nitrogen, and usually sulfur and phosphorus (Figure 2.1). Proteins are

TABLE 2.1 Amino acids and their codons.[a]

Amino Acid	Three-Letter Code	One-Letter Code	Codons
Alanine	Ala	A	GCC, GCU, GCG, GCA
Arginine	Arg	R	CGC, CGG, CGU, CGA, AGA, AGG
Asparagine	Asn	N	AAU, AAC
Aspartic acid	Asp	D	GAU, GAC
Cysteine	Cys	C	UGU, UGC
Glutamic acid	Glu	E	GAA, GAG
Glutamine	Gln	Q	CAA, CAG
Glycine	Gly	G	GGU, GGC, GGA, GGG
Histidine	His	H	CAU, CAC
Isoleucine	Ile	I	AUU, AUC, AUA
Leucine	Leu	L	UUA, UUG, CUA, CUG, CUU, CUC
Lysine	Lys	K	AAA, AAG
Methionine	Met	M	AUG
Phenylalanine	Phe	F	UUC, UUU
Proline	Pro	P	CCU, CCC, CCA, CCG
Serine	Ser	S	UCU, UCC, UCA, UCG, AGU, AGC
Threonine	Thr	T	ACU, ACC, ACA, ACG
Tyrosine	Tyr	Y	UAU, UAC
Tryptophan	Trp	W	UGG
Valine	Val	V	GUU, GUC, GUA, GUG
"Stop"	—	—	UAA, UAG, UGA

Source: Regenerated from Figure 9.4 of Ref. [2].

[a]The name of each amino acid (e.g., alanine) can be abbreviated by using a three-letter code (Ala) or a one-letter code (A). The genetic code relates codons, that is, three nucleotide sequences of mRNA, to the amino acids they specify.

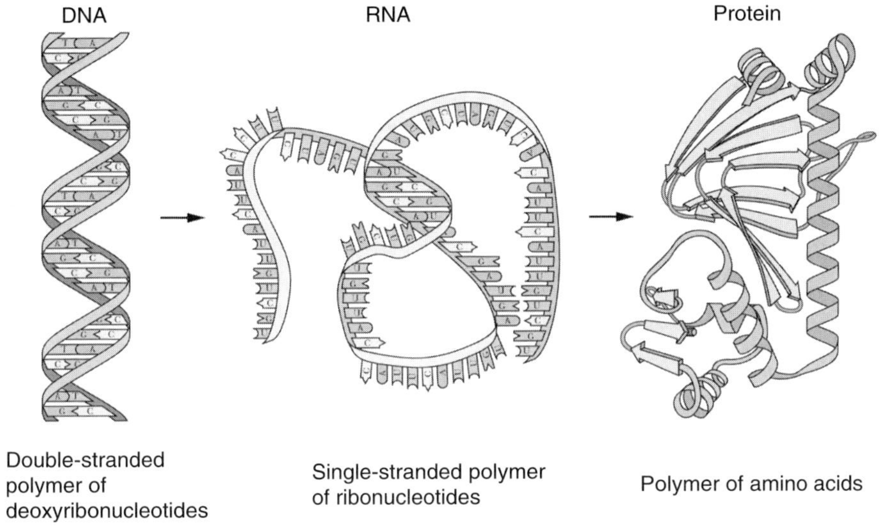

FIGURE 2.1 For genes that yield protein products, genetic information flows from double-stranded DNA to single-stranded RNA to protein [2].

composed of amino acid chains linked by amide bonds to form a three-dimensional complex structure. Proteins are the principal compounds in all cells. Some proteins are bioactive, and are also called *enzymes,* which are responsible for the cell life cycle and the metabolism and synthesis of other compounds, such as lipids and carbohydrates. Some proteins are not bioactive and are called *storage proteins;* they are very stable until needed in seed germination. Storage proteins have many physical and chemical properties and are excellent bio-based polymers for many applications.

DNA

DNA, which stands for deoxyribonucleic acid, is present in chromosomes and contains genetic information coded in specific sequences of its constituent nucleotides. DNA is a polymer with a double-strand molecular structure (Figure 2.1). The nucleotides on one strand pair with those on the other strand. Three major components of DNA are sugar, base, and phosphate groups. DNA is a directional molecule with a free phosphate group at one end and a free hydroxyl group at the other end. DNA is very stable and carries all the necessary genetic information for cell growth, development, structure, and reproduction. DNA is responsible for ensuring that progeny cells contain the same information as the parent and for accommodating the changes and adaptations in evolution.

Gene

A gene is defined biochemically as that segment of DNA (or in a few cases RNA) that encodes the information required to produce functional biological products.

RNA

RNA, or ribonucleic acid, is a nucleic acid containing sugar ribose and is present in both the nucleus and the cytoplasm. RNA is the primary product of gene expression and has an important role in protein synthesis and other cellular functions (Figure 2.1). The several types of RNA vary in their function and size. Three major types of RNA include ribosomal RNA (rRNA), messenger RNA (mRNA), and transfer RNA (tRNA). The ribosomal rRNA forms complex three-dimensional structures that combine with polypeptides to create ribosomes, which are the organelles responsible for protein synthesis. The mRNA carries instructions that dictate the amino acid sequences of proteins. The ribosomes serve as a platform for decoding the mRNA. The transfer tRNA plays the role of an adapter to translate the codons of mRNA into particular amino acids.

Translation

Translation is the mechanism by which specialized riboprotein complexes convert the mRNA sequence into corresponding sequences of amino acids linked by peptide bonds in order to form a polypeptide chain (Figure 2.1).

2.1.2 PROTEIN SYNTHESIS

Protein synthesis is essential to cell function. The life cycle of a protein is illustrated in Figure 2.2 starting with DNA transcription and moving to mRNA translation, to protein maturation, to protein function, and finally to protein degradation. As mentioned, plants utilize sunlight as energy for synthesis, which is called photosynthesis. Although protein photosynthesis in plants has many similarities to protein synthesis in animals and other organisms, the photosynthetic complexes serve as examples of multiple-subunit structures and processes that are unique to plant cells. Protein synthesis occurs in three subcellular compartments, namely, cytoplasm, plastids, and mitochondria (Figure 2.3). Each of them contains different protein synthetic machinery. About 75% of the protein is synthesized in the cytoplasm and 20% in the chloroplast, whereas only a few proteins are synthesized in the mitochondria. The proteins in each unit are synthesized by distinct mechanisms. Therefore, plant cells contain three types of ribosomes, three groups of tRNAs, and three sets of factors for protein synthesis.

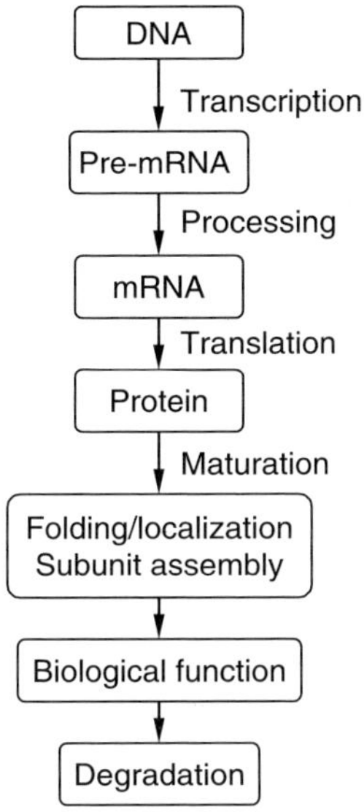

FIGURE 2.2 Flowchart of the life cycle of a protein from synthesis to degradation. (*Source:* Adapted from Figure 9.1 of Ref. [2].)

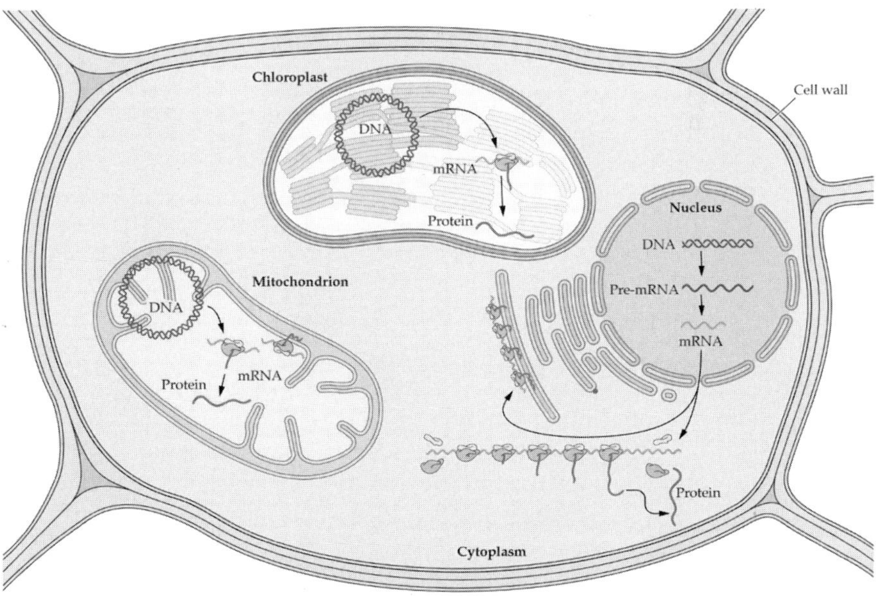

FIGURE 2.3 A typical plant cell synthesizes proteins in three distinct compartments: the cytosol, the plastids, and the mitochondria. Translation of mRNAs transcribed in the nucleus occurs in the cytosol. In contrast, both transcription and translation of plastid and mitochondrial mRNA take place within those organelles [2].

In protein synthesis, at a given DNA transcription from parents, pre-mRNA and mature mRNA are produced based on the information from DNA. The mRNA is then translated to a protein with a specific amino acid sequence as shown earlier in Figure 2.1. Three nucleotides (also called codons or the trinucleotide sequence) are needed to translate into one amino acid at a time (Table 2.1). Any mistake during translation could cause a cell not to function. The tRNA is responsible for correctly linking amino acids to mRNA. The tRNA can recognize different codons, can differentiate more than one codon, and should be able to tolerate a mismatch to avoid mistranslation from mRNA to amino acids. In Table 2.1, note that more than one codon code exists for some amino acids. For example, the UGU and UGC codons are both for the amino acid cysteine. In this case, tRNA would allow for wobble base pairing in the system. The wobble pairing would not affect protein structure and quality but may influence the yield of a particular protein.

The ribosomes in each compartment act as catalysts, accelerating the formation of peptide bonds between amino acid residues. Like many other polymerization reactions, protein synthesis contains three phases: initiation, polymerization, and termination. Initiation of a protein synthesis is a complex process. A start site on the mRNA is selected to establish the reading frame. The tRNA, which is charged with methionine amino acid, interacts with the AUG codon on the mRNA, so that every nascent polypeptide has an N-terminal methionine. The small subunit of ribosomes has to identify the correct AUG codon at which to begin reading the mRNA. Generally, a simple sequence, such as . . . AUGG . . . , is enough for such initiation. Then, the large ribosomal subunit binds to the small subunit and holds the mRNA and Met-tRNA in the correct orientation. Polymerization adds amino acid residues to the growing polypeptide chain. Three important sites include the peptide-tRNA-binding site (P site), the aminoacyl-tRNA-binding site (A site), and the exit site (E site). These three sites are used sequentially as the polypeptide chain is synthesized. Termination of protein synthesis occurs at a specific signal in the mRNA. The polypeptide chain polymerization process ceases when a ribosome reaches one of three stop signs (codons) on the mRNA. These codons are UAA, UAG, and UGA. After proteins reach their appropriate subcellular location in the cell, proteins undergo final grooming and optimization through the removal of some unnecessary subunits, such as formyl groups used in the initiation stage, signal sequences, or some segments of the original polypeptides. This process is called post-translational modification.

Proteins have to fold into a three-dimensional structure before their biological functions can occur. During or after the translation process, the linear polypeptide chain rearranges to yield the three-dimensional conformation of the protein. The mechanism of this process still needs further study. Based on a model study, a linear chain protein first folds into secondary structures, such as α-helices and β-sheets. These secondary structures align with each other and interact with each other in a complicated manner, resulting in a final three-dimensional structure. One important protein in the protein folding process is called a *chaperone,* and it can facilitate protein folding and inhibit any incorrect formation by preventing protein from incorrectly interacting within a polypeptide, between polypeptides, or between polypeptides and other macromolecules. Chaperones increase the yield of functional proteins.

Most proteins in all mature cells are storage proteins, which are stable three-dimensional complex macromolecules. During the life cycle, these storage proteins become degraded upon germination, which is a very important step in protein synthesis in a plant. One function of protein degradation is to remove abnormal proteins and eliminate molecules that are no longer needed. Abnormal proteins may result from errors in previous protein synthesis or folding, spontaneous denaturation, disease, stress, or oxidative damage. If these abnormal proteins are not removed, they may form large insoluble aggregates and eventually poison the cell. Another function of protein degradation is to promote the accumulation of oligomeric protein complexes and to ensure optimum ratios of various enzymes. Protein degradation plays a very important role in regulating many of the biological processes that ensure protein synthesis will occur correctly in a new plant. Upon degradation, the storage proteins are the main sources for amino acids in new protein synthesis.

These storage proteins can be isolated through a process that occurs before degradation, which is discussed in Chapter 3. The isolated protein polymers can be used for food, feed, and industrial products (see Chapters 9 and 10).

2.1.3 PLANT OIL SYNTHESIS

Plant oil is one type of lipid, stored in an organelle in the form of triglycerides, during oilseed development. A lipid is a molecule with diverse structural groups that are hydrophobic. Lipids contain a large variety of fatty acids, pigments, and secondary compounds that are metabolically unrelated to fatty acid metabolism. Different plant species contain lipids with different fatty acid composition and distribution. Lipids play various roles in plants, including protein modification, photoprotection, membrane damage protection, signal transduction, waterproofing, and surface protection. Lipids help form a hydrophobic biological membrane that separates cells from their surroundings and keeps chloroplasts, mitochondria, and cytoplasm apart, thus preventing or regulating diffusion of species in and out of the cells.

This section will focus on triglyceride (also called triacylglycerol) synthesis. Plant oil is a mixture of various triglycerides (Chapter 4). One glycerol is attached to three different fatty acids to form a triglyceride. Glycerolipid and fatty acids are synthesized in the oilseed simultaneously during seed development, before forming diacylglycerol and subsequently triacylglycerols.

Plant oils contain about 15 different fatty acids. Some of them are saturated, including lauric, palmitic, stearic, arachidic, behenic, and lignoceric acids; some are unsaturated, including oleic, petroselenic, linolenic, α-linolenic, γ-linoleic, roughanic, and erucic acids; and two of them are unusual fatty acids, ricinoleic and vernolic acids (for structures, see Chapter 4). These fatty acids have chain lengths of 12 to 24 carbons, with one to three double bonds for unsaturated fatty acids. These fatty acids are synthesized in organelles, called plastids, and originate from a photosynthetic bacterial symbiote. The mechanism of fatty acid photosynthesis is similar to the mechanism for fatty acid production in bacteria.

The enzymes in fatty acid biosynthesis are acetyl-CoA carboxylase (CoA means coenzyme A) and fatty acid synthase. The fatty acid synthase refers to several individual enzymes. One of them is acyl-carrier protein (ACP), which is essential in fatty acid synthesis. Figure 2.4 presents the overview flowchart of fatty acid biosynthesis. The synthesis of a fatty acid is initiated by the ATP-dependent carboxylation of acetyl-CoA, catalyzed in a two-step reaction by acetyl-CoA carboxylase to form malonyl-CoA. The formation of malonyl-CoA is the first step in fatty acid synthesis. The malonyl group is then transferred to ACP.

Subsequently, a condensation reaction occurs because of the decarboxylation of the malonyl moiety. A carbon–carbon bond forms between C-1 of an acetate "primer" and C-2 of the malonyl group on ACP in the condensation process. Then a fully reduced acyl-ACP is obtained through a sequence of reactions—beginning with reduction, followed by dehydration, and then reduction again—to complete a two-carbon synthesis cycle. In the reduction steps, much reducing power is used, because NADPH is generated from the photo system. Each two-carbon addition cycle includes two reduction steps. For a typical 18-carbon fatty acid, eight cycles are needed, and 16 NADPH molecules are consumed. In the first cycle, the condensation reaction is catalyzed by ketoacyl-ACP synthase (KAS) III. For the next six cycles, the condensation reaction is catalyzed by KAS I, and finally, the conversion of 16-carbon to 18-carbon is catalyzed by KAS II.

The synthesis of saturated fatty acids, as shown in Figure 2.4, generally stops at 16 or 18 carbons by one of several reactions. Hydrolysis of the acyl moiety from ACP by a thioesterase enzyme is one of the most common reactions in terminating fatty acid synthesis to produce matured fatty acids. This process transfers the acyl moiety directly onto glycerolipids. Shorter or longer saturated and/or unsaturated fatty acids usually occur after desaturation, and elongation involves enzymes called transferase and desaturase. The matured fatty acids are transferred out of the plastid by an unknown mechanism. Various thioesterase enzymes, which are responsible for different fatty acid syntheses, exist in different plants. For example, one thioesterase called

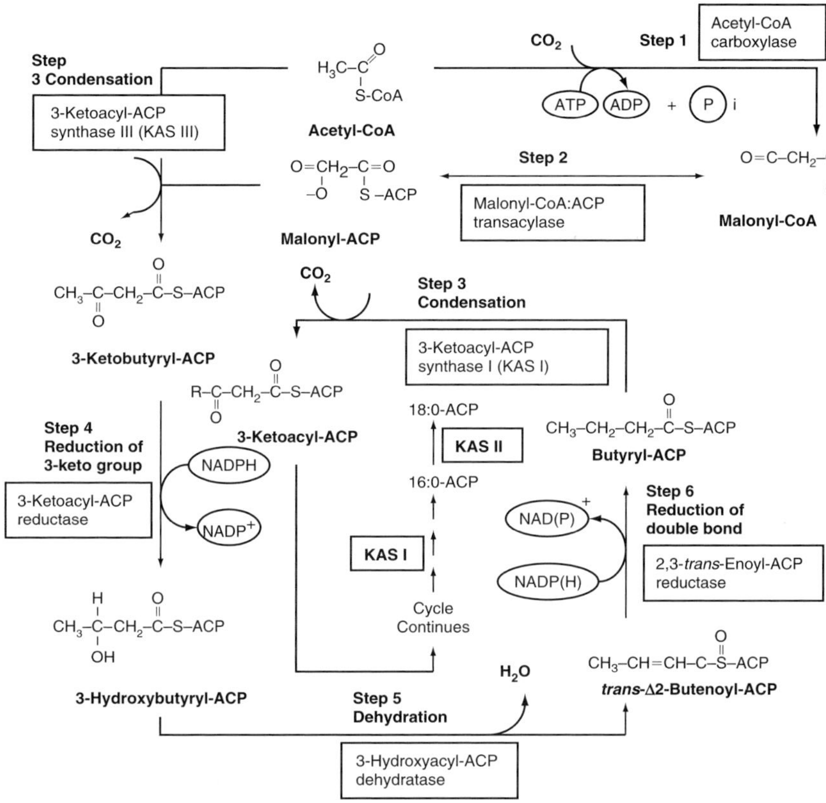

FIGURE 2.4 Flowchart for fatty acid photosynthesis. Fatty acids grow by addition of 2-carbon units. Synthesis of 16-carbon fatty acids requires that the cycle be repeated seven times. During the first turn of the cycle, the condensation reaction (step 3) is catalyzed by ketoacyl-ACP synthase (KAS) III. For the next six turns of the cycle, the condensation reaction is catalyzed by isoform I of KAS. Finally, KAS II is used during the conversion of 16:0 to 18:0. (*Source:* Adapted from Ref. [2].)

FatA is responsible for the formation of 18-carbons with one degree of double bond unsaturation, and another class called FatB is responsible for the formation of the shorter chain, saturated acyl-ACP, either 10-carbon or 12-carbon fatty acids. This explains why different oilseed plants contain oils with varied fatty acid compositions and distributions. These variations significantly influence the final properties of the acids as a material; the cariations are controlled by DNA gene formation. In Chapter 8, we show how oils with a very high oleic (one double bond per fatty acid) content are essential for the synthesis of pressure-sensitive adhesives (PSA), elastomers, and coatings, whereas in Chapter 7, we show that more highly unsaturated oils (such as linseed oil) are most useful for highly cross-linked composite resins.

Glycerolipid synthesis is a complex and highly organized interaction inside and out of the plastid organelle. Inside the plastid, the acyl-ACP is condensed with glycerol 3-phosphate by a soluble enzyme called G3P acyltransferase. The condensed product is then converted to phosphatidate, which is the key element for diacylglycerol synthesis. The mechanism of glycerolipid synthesis outside the plastid is similar to that inside the plastid except that acyl-CoA is used as a substrate. Lipids move to other organelles, or are transferred into the inner membranes by an unknown mechanism. Diacylglycerol is produced in the endosperm by the dephosphorylation of the phosphatidate by the enzyme phosphatidate phosphohydrolase, which occurs in the inner chloroplast envelope membrane, microsomes, and soluble fractions.

Triacylglycerol synthesis involves acyltransferase enzyme and acyl-exchange reactions, which move fatty acids between pools of membrane and storage lipids (Figure 2.5). Steps 1 to 5 of Figure 2.5 describe fatty acid synthesis in the plastid. All fatty acids used for triacylglycerol synthesis come from the acyl-CoA pool. Saturated 16-carbon-ACP (16:0-ACP) and 18:1-ACP are the major products of fatty acid synthesis and 18:0-ACP desaturase activity in the plastid in oilseeds. These products are used for phosphatidylcholine synthesis, which is further used in the synthesis of unsaturated fatty acids, including 18:1, 18:2, and 18:3. These fatty acids also exchange with the acyl-CoA pools and can be modified into longer chain fatty acids, such as 20:1-CoA and 22:1-CoA, by the elongation process. In many oilseeds, the phosphatidylcholine is a direct precursor of unsaturated species of diacylglycerols used for triacylglycerol synthesis. Synthesis of diacylglycerol also uses the components of the acyl-CoA pool. The triacylglycerol can then be synthesized by final acylation of diacylglycerol by the enzymes acyl-CoA:1 and 2-diacylglycerol *O*-acyltransferase. Triacylglycerol accumulation forms oil bodies in the oilseeds, in a droplet surrounded by a monolayer of phospholipids.

2.1.4 CARBOHYDRATE SYNTHESIS

Polysaccharides are major components of carbohydrates such as cellulose and starch. As mentioned before, two major products of photosynthesis are fructose diphosphate and hexoses. The hexose pool contains glucose 6-phosphate, glucose 1-phosphate, and fructose 6-phosphate. Carbon can enter or leave the hexose pool because of enzyme activities during starch and sucrose degradation and synthesis. Glucose 1-phosphate is converted into uridine diphosphate glucose (UDP-glucose) by the enzyme UDP-glucose pyrophosphorylase. Sucrose may be synthesized in cytosol from UDP-glucose and fructose 6-phosphate by the enzyme sucrose-phosphate synthase. Sucrose is a major product of photosynthesis in green leaves and serves as the main long-distance transport component in most plants.

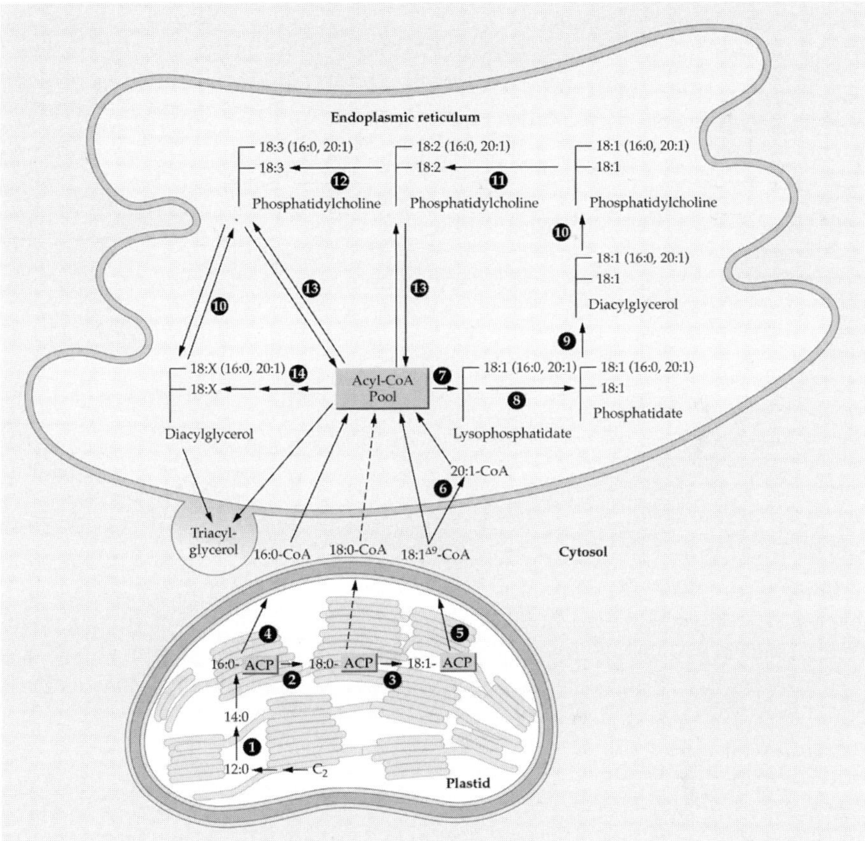

FIGURE 2.5 Abbreviated scheme for the reactions of triacylglycerol synthesis in oilseeds. The enzyme-catalyzed steps are indicated by numbers and involve the following enzymes: (1) KAS I-dependent and KAS III-dependent FAS; (2) KAS II-dependent FAS; (3) stearoyl-ACP desaturase; (4) palmitoyl-ACP thioesterase; (5) oleoyl-ACP thioesterase; (6) oleate elongase; (7) acyl-CoA; glycerol-3-phosphate acyltransferase; (8) acyl-CoA:lysophosphatidate acyl-transferase; (9) phosphatidate phosphatase; (10) CDP-choline:diacylglycerol choline phosphotransferase; (11) oleate desaturase, *FAD2*; (12) linoleate desaturase, *FAD3*; (13) acyl-CoA: *sn*-1 acyllysophosphatidylcholine acyltransferase; (14) same as in (7), (8), and (9), but any fatty acids used are from the acyl-CoA pool [2].

Starch, a polymer of glucose linked by α-bonds, is synthesized and stored in the plastids. Starch synthesis is initiated in adenine diphosphate glucose (ADP-glucose) by the enzyme ADP-glucose pyrophosphorylase (Figure 2.6). The ADP-glucose may be synthesized in the cytosol and imported into the plastids as a substrate for starch synthesis. Amylose and amylopectin are two major polymers of starch. Amylose is a linear chain, and amylopectin is a branched chain. Individual glucose is added to the linear amylose chain at the

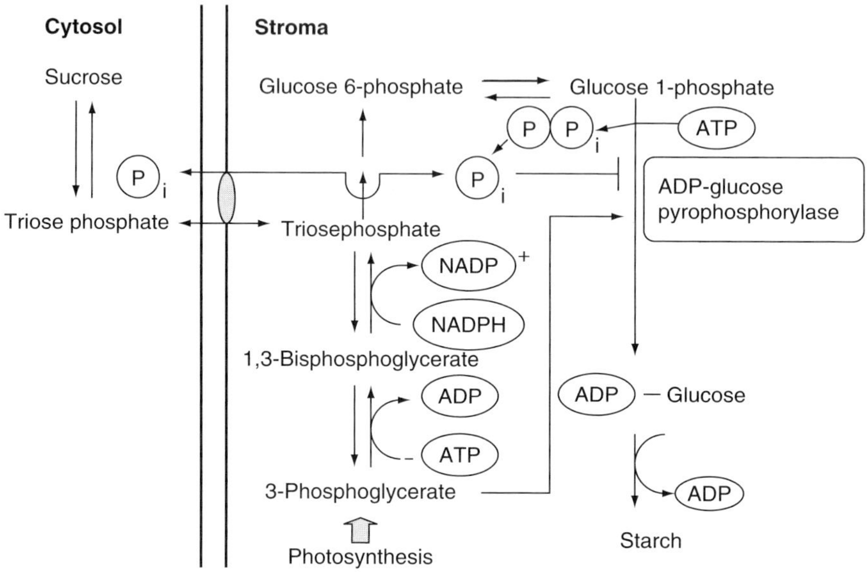

FIGURE 2.6 Flowchart for starch synthesis in chloroplasts. When 3-phosphoglycerate is abundant, starch synthesis is activated. Inorganic phosphate, an indicator of the status of the triose phosphate pool, inhibits starch synthesis. (*Source:* Adapted from Figure 13.16 of Ref. [2].)

nonreducing end by the enzyme starch synthase, and added to the branched amylopectin chain at the reducing end by the enzyme amylo-(1-4),(1-6)-transglycosylase. The synthesized starch is then stored in the endosperm of matured seeds and can be readily isolated using environmentally friendly biorefinery processes for food and industrial uses.

During seed germination, sucrose can be hydrolyzed to free hexoses or converted to UDP-glucose and fructose; this is the sucrose degradation process that involves the enzymes sucrose synthase and invertase. Sucrose degradation can generate substrates for cell wall synthesis. Cell walls contain two major components: polysaccharides and lignin. The main compositions of polysaccharides are composed of cellulose and hemicellulose, which are polymers of glucose linked by β-bonds. Cellulose, which is a molecular first cousin of starch, is synthesized by the cellulose synthase complex, an enzyme associated with the plasma membrane, which uses the UDP-glucose as a substrate.

During seed germination, starch can be hydrolyzed into glucose by amylases, which can combine with debranching enzymes to synthesize new starch during plant development. All the genes stored in the seed produce the correct enzymes for the next generation of starch synthesis. Figure 2.7 presents the diagram of starch degradation during seed germination. A similar

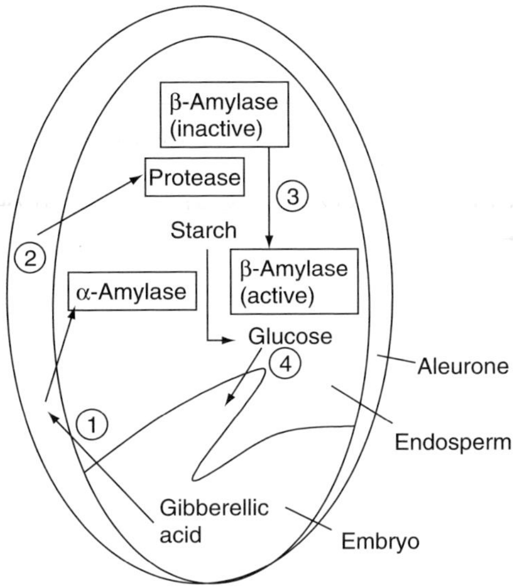

FIGURE 2.7 Role of gibberellic acid (GA) in mobilizing the carbohydrate reserves of germinating cereal seeds. The breakdown of starch in the endosperm of monocots is triggered by the release of gibberellins from the embryo. (*Source:* Adapted from Figure 13.27 of Ref. [2].)

mechanism hydrolyzes starch into glucose in many starch fermentation processes for creating chemicals, products, and biofuels.

The mechanisms of carbon flow and carbohydrate synthesis are still unclear and need further exploration.

2.2 PLANT GROWTH

Many variables influence plant development, including light, temperature, water, and nutrients. Protein synthesis is a key part of plant development because all enzymes responsible for bio-based polymer synthesis, such as storage proteins, oil, starch, and celluloses, are synthesized during protein synthesis. Protein synthesis is significantly affected by many physiological and environmental variables. Protein synthesis can be diminished by stresses such as anaerobiosis, heat shock, or viral infection. During seed germination, protein synthesis is tightly regulated by light and hormones. Different plants require different weather conditions and nutrients, which is evident by the fact that some plants can only grow in certain climates or locations. An optimum environment should be provided for plants for high-quality yield and polymer production.

A plant begins with a seed. A typical seed contains an embryo and an energy body. Figure 2.8 shows the structure of a soybean seed and corn

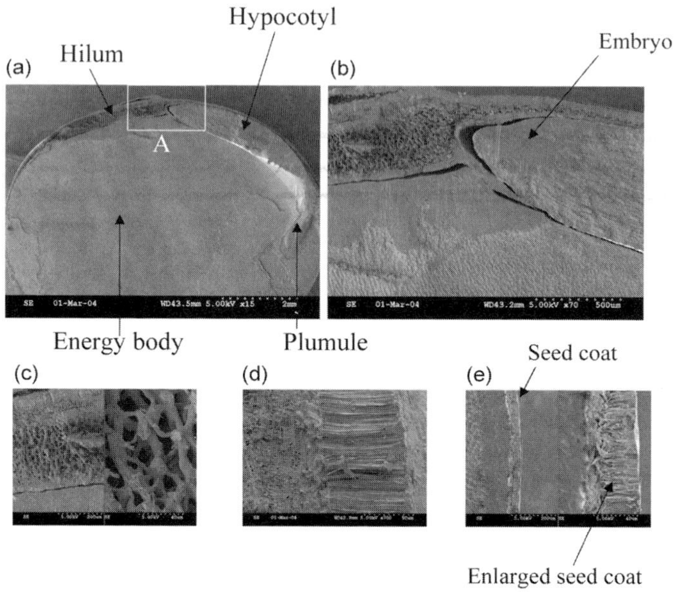

FIGURE 2.8A A mature soybean seed: *a,* the cross section of a soybean seed; energy body includes storage proteins, oil bodies, and carbohydrate; *b* is the enlarged image of the part marked "A" on image a; *c* is the enlarged image of hilum structure; *d* is the enlarged image of hilum skin; and *e* is the enlarged image of seed coat. (*Source:* Courtesy of S. Zutara and X. Sun.)

kernel [4]. Plant growth includes seed germination, organic compound synthesis, cellular respiration, skeletal system establishment, turgor pressure formation, plant digestion, plant circulation, plant hormone production, plant coordination, and impulse transmission. We discussed organic compound (polymer) synthesis in Section 2.1. In this section, we briefly discuss what a plant needs to be assured of healthy growth and successful polymer synthesis.

Cellular respiration is responsible for taking in oxygen, removing carbon dioxide, and transferring energy from glucose and other substrates to ATP and other forms of biologically useful energy. When a plant is illuminated, as in a greenhouse, the rate of photosynthesis is 10 to 30 times higher than that for cellular respiration. Plant leaves contain enough chlorophyll for photosynthesis. The skeletal structure of a plant, the straw and stems, supports the leaves. Cell walls stretch and are supported by the osmotic pressure generated by the solution in the cell sap containing sugar, salts, and other organic molecules. This pressure is called *turgor,* and it is important for cell growth. During plant growth, the concentration of solute in the cell sap increases, resulting in more water diffusion into the cell to increase turgor pressure. If the salt concentration outside the cell sap is higher than inside, water in the

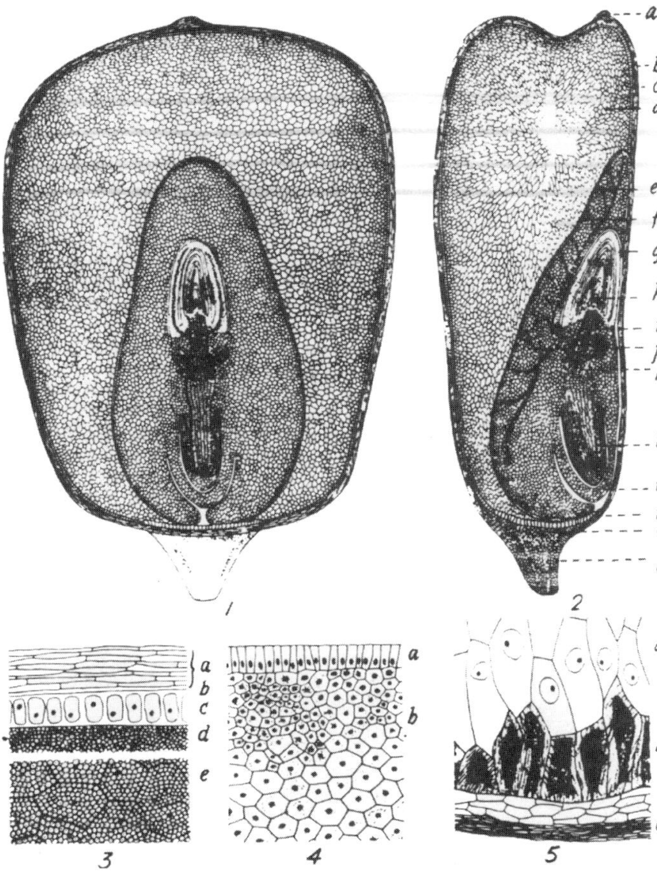

FIGURE 2.8B A mature corn kernel. 1 and 2, vertical sections in two planes of a mature kernel of dent corn, showing arrangement of organs and tissues: *a,* silk scar; *b,* pericarp; *c,* aleurone; *d,* endosperm; *e,* scutellum; *f,* glandular layer of scutellum; *g,* coleoptile; *h,* plumule with stem and leaves; *i,* first internode; *j,* lateral seminal root; *k,* scutellar node; *l,* primary root; *m,* coleorhiza; *n,* basal conducting cells; *o,* brown abscission layer; *p,* pedicel or flower stalk (×7). 3, Enlarged section through pericarp and endosperm: *a,* pericarp; *b,* nucellar membrane; *c,* aleurone; *d,* marginal cells of endosperm; *e,* interior cells of endosperm (×70). 4, Enlarged section of scutellum: *a,* glandular layer; *b,* interior cells (×70). 5, Vertical section of the basal region of endosperm: *a,* ordinary endosperm cells; *b,* thick-walled conducting cells of endosperm; *c,* abscission layer (×350). (*Source:* Reprinted from Ref. [4].)

cell sap will move out, causing cell shrinking and wilting. This process is called *plasmolysis.* Plants do not have a digestion system like that of animals; instead plants takes nutrients through cell walls and stems where substrates are synthesized during plant development.

The seed provides initial energy and enzymes. Once the skeletal structure forms and the leaf/root systems are established, nutrients are first provided by the sunlight and nutrients from the soil. These can then be converted into various nutrients and substrates for synthesis of a variety of compounds. Because plants have no mechanical hearts and no blood vessels, circulation instead involves the xylem and phloem systems. The xylem transports water and minerals from the roots up the stems to leaves, and phloem transports nutrients manufactured in the leaves down the stems for storage and use in the stems and roots. The material in the xylem and phloem is called sap and is somewhat analogous to the blood plasma of humans and animals. The sap is a complex mixture of substances, both organic and inorganic compounds. About 98% of the sap is water; the 2% solid portion contains salts, amino acids, hormones, enzymes, and other proteins. The content of the sap varies from plant to plant, and from one part of a plant to another, and from season to season. A major difference between a plant and an animal is that a plant produces almost no waste. Plants recycle constituent compounds. Plants have no nervous system and no sense organs, but like all living beings, plants can transmit excitation, although very slowly. Plants also produce hormones in actively growing tissue at the tips of stems and roots. These hormones have many different types of effects on metabolism and cell divisions, including stimulating individual cell growth and cell division in the cambium, initiating new root formation and seed development, and inhibiting lateral buds and abscission formation.

2.3 TRANSGENIC PLANTS

Conventional plant breeding has a long history of improving plant cultivars and influencing the farming world by increasing yield, improving disease resistance, and improving environmental tolerance. Advances in biotechnology allow us to better understand living plants, making it possible to genetically modify certain plants. Specific genes can be manipulated to produce grains containing desirable polymers and compounds with favored properties. Genetic engineering can improve and accelerate plant breeding. Creating transgenic plants involves several major procedures, including gene cloning, gene expression, and cloning DNA vectors, as well as those steps described in polymer photosynthesis. Because plants can be genetically modified to yield polymers with desirable properties, it is very possible to utilize plants as vehicles for the production of polymers with specific functionality for industrial uses.

Soybean has been traditionally bred for many years to alter its oil and protein composition. Crude protein content can now range from 30% to 50% with crude oil ranging from 2% to 30%. Soy protein and soy oil content are negatively correlated, which means that both protein and oil cannot increase at the same time [5]. Burton [6] pointed out that the oil content was negatively correlated with the total soybean yield, and often the oil content is in the range of 22–24%. It is a challenge then to improve both soybean protein and oil content. However, soybean garners attention in genetic studies because of its economic importance and low production costs. In 1999, about 22 million hectares of land were used for genetically modified soybeans [7].

As mentioned before, oil contains various fatty acids that significantly affect oil extraction, processing, and utilization. Fatty acids can be genetically modified as well. Research efforts have focused on improving oil nutritional quality, oxidative stability in processing and storage, and oil properties. Oils with low linolenic acid are available to stabilize oil flavors; oils with high oleic acids improve oxidative stability; and oils with high stearic acid improve production of shortening and margarine by replacing hydrogenated oils [8].

Genes can also be introduced into soybean to produce oil that contains particular fatty acids to meet customers' demands. How oil structure and fatty acids affect bio-based products, such as adhesives, resins, plastics, coating, and other products, is not yet fully understood, but considerable progress has been made. Understanding structure and functionality is essential to improve soy oil quality for industrial uses. In an example, Wool and his coworkers found that the degree of unsaturation is positively correlated to oil resin strength. This research will be discussed fully in Chapters 7 and 8.

Similarly, soy protein quality can be genetically improved. Soy protein contains about 18 main amino acids that significantly affect protein quality and properties. Research efforts focus on changing the ratio of glycinin and conglycinin components, and on specific amino acids, such as lysine and methionine that contribute to nutrition. Soybean varieties containing only glycinin or conglycinin as their major proteins content are available at the laboratory scale and are being studied by scientists at DuPont and Monsanto in the United States. The relationship between protein structure and the performance of various industrial products needs to be identified. Sun and her coworkers studied soy protein fractions and chemically synthesized peptides for adhesive performance, which will be discussed in detail in Chapter 10, but other research is very limited. More research is necessary to study the structure and functionality of bio-based polymers. Such research would be useful to plant scientists involved in genetically improving soybean or other plants.

Another crop for which genetic modification has proven successful is corn. Genetically modified corn hybrids have improved insect resistance, herbicide tolerance, disease resistance, male sterility in hybrid seed production, and yield [7]. Corn yield has increased from 2970 to 3618 kg/acre in the past 20 years. Approximately 60% of the corn produced per year is used for feed, with the result that much research has focused on modifying corn nutritional value. Animal meat quality and production can be improved by improving corn protein quality and the amino acids with nutritional functions. High-oil corn is now commercially available to increase the profits of feedstock industries. High-amylose corn and waxy corn are also commercially available for special applications, such as food additives, syrups, sweeteners, paper coating, and adhesives.

Corn is also used as a biofactory for producing high-value end use products, such as vaccines, therapeutic proteins, feed enzymes, and specialty chemicals, all at a much lower cost than cell culture technologies. This concept is also used to produce polyesters from corn. However, the cost of the polyester extracted from the corn is 10 times higher than petroleum-based polyesters because of the complicated extraction procedures. The cost of manufacturing polyesters from plants will eventually be reduced.

Many genetically modified wheat and sorghum varieties using both traditional breeding and biotechnology are commercially available, but mainly for improving yield, disease and insect resistance, and environmental tolerance [9, 10]. Some genetically modified wheat varieties have higher protein content, improved protein quality, and improved eating qualities. Sorghum, a crop high in starch, receives the least attention in genetics because of its low starch extraction, bioconversion rate, and low digestibility. However, sorghum may be a good crop for industrial uses because it needs much less water and fertilizer than soybean, corn, and wheat, and can grow well in tropical and dry lands. Research is under way to identify useful genes to improve sorghum as an industrial crop for starch extraction, bioconversion into energy and bioproducts, and as a biofactory for chemicals.

REFERENCES

1. Villee, C. A. *Biology,* W. B. Saunders, Philadelphia; 1977.
2. Buchanan, B. B.; Gruissem, W.; Jones, R. L. *Biochemistry & Molecular Biology of Plants,* American Society of Plant Physiologists, Rockville, MD; 2000.
3. Glick, B. R.; Pasternak, J. J. *Molecular Biotechnology Principles & Applications of Recombinant DNA,* ASM Press, Washington, DC; 1994.
4. Benson, G. D.; Pearce, R. B. In *Corn: Chemistry and Technology,* Watson, S. A.; Ramstad, P. E., Eds.; American Association of Cereal Chemists, St. Paul, MN; 1987, Chap. 1, pp. 1–28.
5. Schapaugh, W., Kansas State University, Department of Agronomy, Manhattan, KS; Personal communication.
6. Burton, J. W. In *Soybeans: Improvement, Production, and Uses,* Wilcox, J. R., Ed.; American Society of Agronomy, Madison, WI; 1987, pp. 211–247.
7. Armstrong, C. L.; Spencer, T. M.; Stephens, M. A., et al. In *Transgenic Cereals,* Brien, L. O.; Henry, R. J., Eds.; American Association of Cereal Chemists, St. Paul, MN; 2000.
8. Liu, K. S. *Soybeans Chemistry, Technologies, and Utilization,* ITP International Thomson Publishing, Washington, DC; 1997.
9. Anderson, O. D.; Blechl, A. E., In *Transgenic Cereals,* Brien, L. O.; Henry, R. J., Eds., American Association of Cereal Chemists, St. Paul, MN; 2000.
10. Godwin, I. D.; Gray, S. J. 2000, In *Transgenic Cereals,* Brien, L. O.; Henry, R. J., Eds.; American Association of Cereal Chemists, St. Paul, MN; 2000.

3

Isolation and Processing of Plant Materials

Xiuzhi Susan Sun

Biorefining can be defined as a process that converts whole grains and legumes or fiber residues into useful fractions, chemicals, and polymers by physical, chemical, or enzymatic methods or by microorganism conversion technologies. Protein, oil, lignin, and carbohydrate are major materials that can be found in the grains and legumes presented in Table 1.2 in Chapter 1. Based on a particular plant's nature, one can always find one of these materials in any kind of grain or legume. These materials are often coproducts of each other in a plant. Biorefinery technologies for manufacturing these polymers and materials are generally similar, but they do vary depending on differences in composition and structure from one plant to another. This chapter focuses on the concepts and technologies used in biorefining to isolate protein, starch, and oil-based polymers out of grain and legumes, but our discussion is limited to bioconversion and to the energy and mass balance of a biorefinery system.

3.1 Oil Extraction and Refining

3.1.1 OIL EXTRACTION

Two major approaches are commonly used for oil extraction: mechanical extraction and solvent extraction. Mechanical extraction has a long history, and it is the simplest and safest technology for oil extraction. Figure 3.1

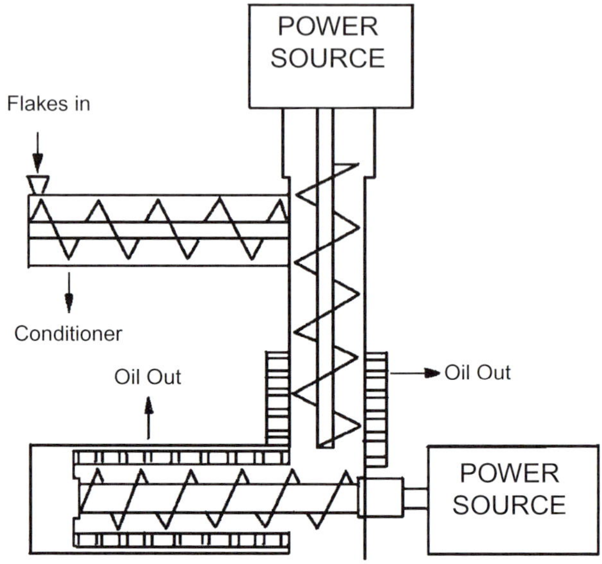

FIGURE 3.1 A mechanical oil extractor with two expellers and a conditioner. (Adapted from Ref. [8].)

presents a typical mechanical extractor for oil extraction [1]. The principle of mechanical oil extraction is to apply mechanical forces to rupture the beans, cells, and oil bodies, and then squeeze out the oil. Heat is usually generated in the barrel due to intensive shearing, which may cause protein denaturation. The main advantages of mechanical extraction are low initial costs and no environmental issues in that the process does not use hazardous solvents. The disadvantage of mechanical extraction is the low oil recovery rate, which means that the oil content left in the soybean meal after extraction is high, often 7% or higher. Mechanical extraction technology is still popular worldwide for small or medium-sized companies, where the soy meals are often used for animal feeds.

In contrast, the solvent method allows higher oil recovery and low oil residue in the soybean meals, but requires a larger investment than does mechanical extraction. It also has environmental issues related to the use of hazardous solvents. However, solvent oil extraction is still the main technology used in large feedstock companies in the United States. Hexane is a commonly used solvent for oil extraction. The main principle of solvent extraction is that the solvent diffuses into oil bodies and solubilizes the oil; the oil is then carried out of the cell when the solvent diffuses out. Therefore, the diffusion rate is critical for the efficiency of the oil extraction process. The thickness and size of the oilseed flakes are important factors in improving the diffusion rate. Thinner flakes may increase the diffusion rate, but may produce more fine particles in flaking. The solvent flows over these fine particles instead of diffusing through them, which reduces oil recovery.

Soybean contains about 40% protein and 20% oil and is considered a main source of proteins and oils. In addition, soybean growth requires fixed nutrients and less water, and produces fewer agriculture residues than other crops. Soy oil extraction and refining technologies at both the laboratory and commercial scales are well developed. Laboratory scale often involves two major different settings depending on project tasks. To study the relationships between structures/genomic/composition of oils or soybean and their functional properties, small amounts of oil samples are often obtained using a soxhlet oil extraction method with petroleum ester solvent. To improve or develop processing technologies, laboratory-scale oil extraction equipment is often used with adjustable processing parameters, and then scaled up into pilot scale and commercial scales. In this section, the soxhlet method and commercial-scale processing technologies of soybeans are introduced. The methods and technologies of oil extraction from other oilseeds are similar.

Soxhlet Method

The sample size for the soxhlet method is from 20 to 40 grams in powder form with particle sizes smaller than 1 mm. Traditionally, the sample is usually wrapped with filtration paper and exposed to circulating petroleum solvent for many hours (Figure 3.2). The circulating solvent carries the oil away from the sample. Oil concentration in the solvent increases with extraction time, flowing in an exponential pattern. An oil sample is obtained by evaporating the solvent. This method also is often used for determining the oil content of a material. For oil content measurement, one should follow the American Oil Chemists' Society (AOCS) standard method. The traditional soxhlet method for oil extraction takes as long as 12 hours. A new technique has recently been developed that utilizes the soxhlet principle, which reduces the extraction time to 30 min [2]. An example of soybean oil extraction is given next.

About 30 grams of soybeans are ground using a laboratory mill to a particle size of less than 1 mm. Immediately after grounding, the sample is transferred quantitatively into thimbles, and the thimbles are loaded with the ground sample, which is then weighed. The oil is then extracted from the ground sample using a Soxtec HT2 1045 system, provided by Foss Tecator with petroleum ether, at a temperature of from 40 °C to 60 °C. The oil collection cup volume is about 50–70 mL, and the stove temperature is from 98 °C to 120 °C. The extraction takes a total of about 45 min, with 15 min of boiling and 30 min of rinsing. The oil extracted is then collected in amber glass bottles and stored at 2 °C for further analysis and uses.

Industrial Oil Extraction

In this section, we continue to use soybean as an example. Several main procedures for soy oil processing are shown in Figure 3.3. Plant tissues,

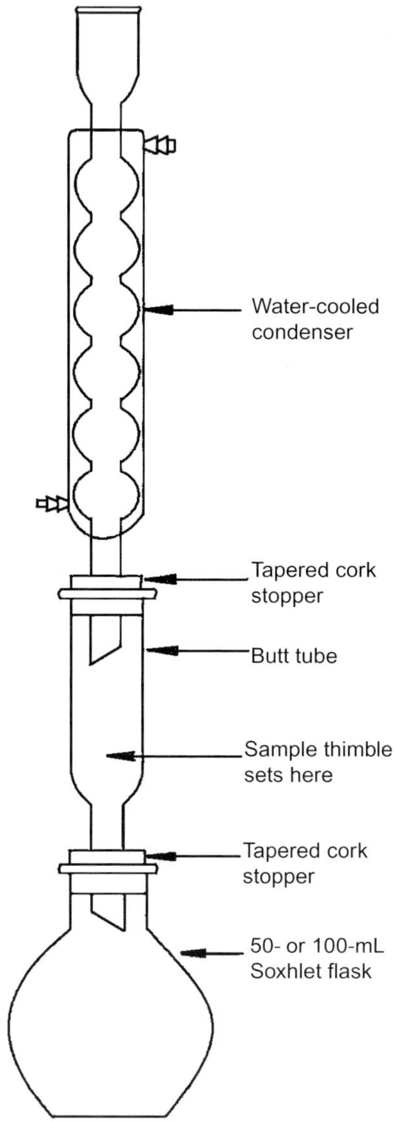

FIGURE 3.2 Laboratory oil extraction of soxhlet apparatus.

foreign materials, pebbles, dust, etc., are removed during the cleaning step. The drying/tempering step ensures that the soybean has a moisture content of about 10%, which is the optimum moisture content for easily removing the soybean hull. This procedure often takes 1–20 days depending on the initial moisture content of the beans. The beans are then cracked into four to six

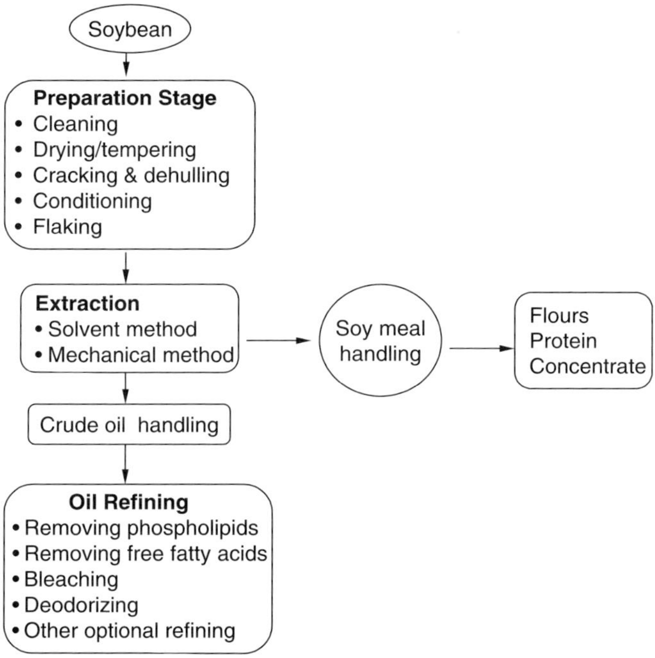

FIGURE 3.3 Flowchart of industrial soybean oil extraction process.

small pieces, and simultaneously the hull is removed from the small pieces using a counter-rotating cracking roller machine with a capacity of 500–600 tons/day. Screening is often used to separate large grits and fine soy flours from the cracking process. The cracked beans are then conditioned at 65–70°C in the presence of steam to plasticize the beans, preparing them for flaking. The cracked bean pieces need to be pressed into flakes with a thickness of 2.5–3.7 mm. Flaking is an important procedure prior to solvent oil extraction. Some cells containing oil bodies are ruptured during flaking, which allows oil to flow out and also reduces solvent penetration distance into the oil bodies.

The thickness and oil extraction yield for soybean can be expressed by Eq. (3.1), which was developed in 1955 by Orthmer and Agarwal [3]:

$$-dC/dt = kF^{-3.97}C^{3.5}, \tag{3.1}$$

where C is the oil concentration of the flake, t is the time, F is the thickness of the flake, and k is the constant. Flake thickness and oil concentration in oil bodies are very important in soybean oil extraction. The oil extraction rate is rapidly reduced in the first 10 min due to the reduction of oil residue in oil bodies.

Moisture content is also another important factor affecting oil extraction efficiency. The solvent diffusion coefficient decreases at about 0.4×10^{-9} cm^2/sec for every 1% increase in moisture contents ranging from 10% to 22% [4]. As discussed in Chapter 2, oil is a mixture of various triacylglycerols (trigylcerides) attached with three different fatty acids. The composition and distribution of the fatty acids are factors that influence the oil extraction process. Karnofsky [5] reported that phospholipid content increased with extraction time. It is necessary to study further the behaviors of oil composition and structure in the oil extraction process; such study would help to improve extraction efficiency or develop new technologies. The composition of oilseed can be improved or modified through breeding or genetic engineering, which could also improve extraction efficiency.

Two major pieces of solvent oil extraction equipment include the rotary extractor and the loop extractor. The rotary extractor is also called a deep-bed extractor. Soy flakes are added to the rotating bins, and each bin is a cylindrical rotocel. The flakes are held in an upper chamber; the solvent circulates the flakes and then drains out. After extraction, the soy flakes are discharged as soy meals. The rotary extractor can handle thick soy flakes because of constant contact between the flakes and the solvent. The loop-style extractor (Figure 3.4) circulates the flakes in both vertical and horizontal routes. The solvent flows in a countercurrent direction to the flakes. In

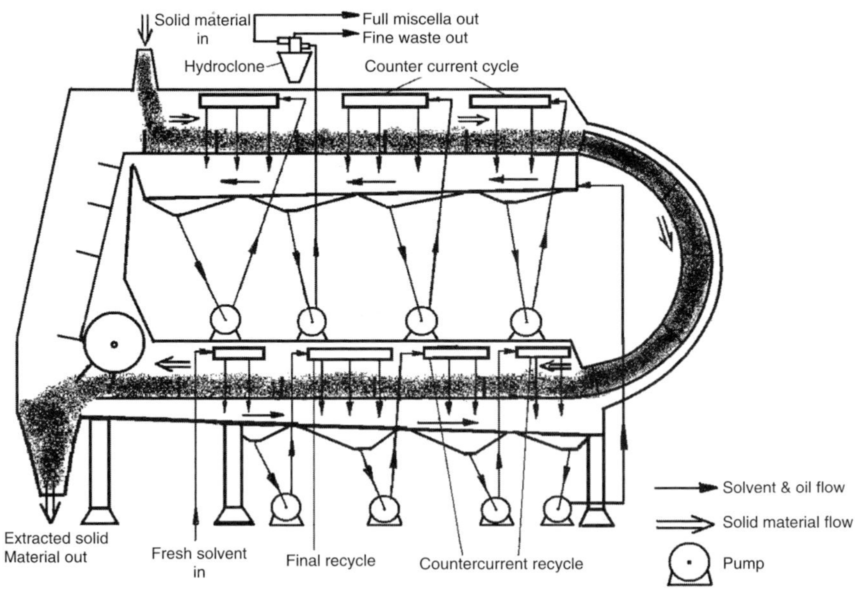

FIGURE 3.4 Countercurrent loop solvent oil extractor. (*Source:* Redrawn from the original picture provided by Crown Iron Works Co., Minneapolis, MN.)

contrast, the thickness of flakes for a loop extractor is much thinner than that used for a rotary extractor because the countercurrent flow results in a shorter contact duration between flakes and solvent.

The oil needs to be separated from the solvent. In addition, the used hexane needs to be recycled according to environmental regulations. The hexane content in oil should be less than 1000 ppm. Crude oil is easily oxidized due to moisture, heat, air, light, etc. As a result, crude oil is often stored in stainless steel tanks with appropriate agitation preventing settling of gummy substances. The moisture content of the crude oil should be as low as 0.2% [6].

3.1.2 OIL REFINING

Oil refining refers to any postextraction treatment that is utilized to obtain high-quality oil for various uses. Crude oil contains many impurities including phosphatides, free fatty acids, gummy substances, color bodies, tocopherols, sterols, hydrocarbons, ketones, and aldehydes. These impurities are removed in the refining process. Typically, oil refining is used to remove phospholipids, free fatty acids, and odors and to improve oil color. Oil refining involves many thermodynamic processes such as mass and heat transportation phenomena, where a food engineer or chemical engineer can help to improve oil quality and processing efficiency.

Crude oil contains phospholipids at the level of about 500–900 ppm, which causes browning, off-flavor formation, and gum settling at the bottom of the container during storage and long-distance transportation [7]. This gum is very difficult to remove. Oils are required to contain phospholipids of less than 0.02%. The procedure of removing phospholipids in industrial oil refining is referred to as *degumming*. A commonly used method to remove phospholipids is to add warm water into the crude oil at 70°C (Figure 3.5). The polar phospholipids are attracted in the water phase, and are then separated from oil by centrifuge based on their difference in density. The degummed oil is then dried by vacuum dryer and cooled to room temperature for storage. The phospholipid separated from the crude oil is also called lecithin, which is a useful additive for food and animal feed. The temperature of the warm water needs to be the same as the oil temperature to avoid emulsion formation. The amount of warm water added to the crude oil is usually about 75% of the phospholipid content. Insufficient water can result in a dark and viscous lecithin product, and excess water can cause poor water–oil phase separation, resulting in low oil recovery. Some insoluble phospholipid residues cannot be removed by using a water method. The residue is often removed using an acid degumming method instead. The reaction time is only about a few minutes, but causes a dark color in lecithin.

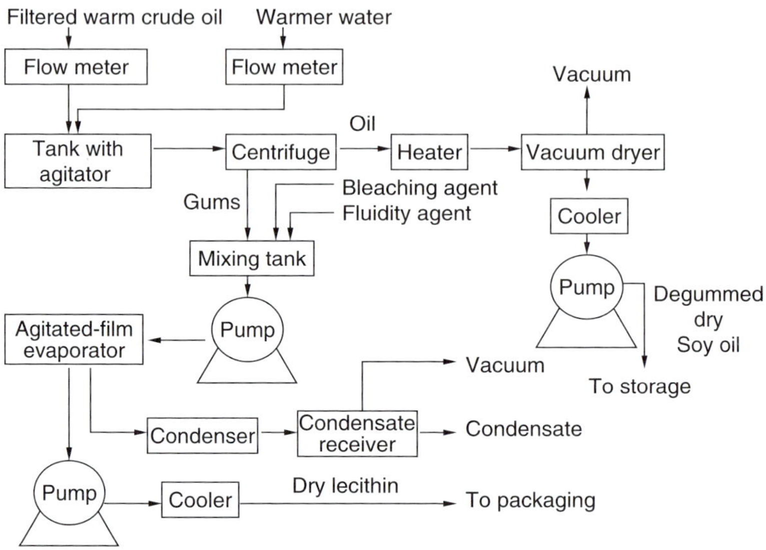

FIGURE 3.5 Degumming process to remove phospholipids from crude soy oil, and processing of soy lecithin, a by-product of degumming. (Adapted from Ref. [8].)

Crude oil also contains about 0.3–0.7% free fatty acids that reduce the smoking point of the oil and increase foaming on heating [8]. The free fatty acids need to be reduced to about 0.05% or less, a process often referred to as *neutralization* in oil refining. Several methods are available to remove these free fatty acids. An alkali method is used most often in industry. Sodium hydroxide is often added to a pretreated crude or degummed oil with phosphoric acid. The product of the sodium salt of the fatty acids, such as soaps, can be removed by centrifugation. Some triglycerides are lost during neutralization, along with some phospholipid residues, pigments, and insolubles with the soaps. Any soap residues in the refined oil can be removed by a hot-water washing method. Other alkali-based chemicals can also be used for this process.

For food uses, degummed and neutralized oil is often bleached to improve oil appearance (that is, to obtain a light color). Many residues, such as phospholipids, soaps, and free fatty acids, from the degumming and neutralization processes can be further removed during the bleaching process. Bleaching is an adsorption process, and the impurities and residues are removed by a powdered adsorbent containing silica gel that binds phospholipids, free fatty acids, and soap residues. Odor is another quality factor of oil for food uses. Therefore, a deodorization process is required when refining food-grade oils.

Refining procedures for industrial-grade oils such as resins, adhesives, lubricants, and coatings can be simplified based on a specific application, or they can be integrated with the reactions and treatment processes required for bioproduct manufacturing. Optimum integrated processing systems for industrial-grade oils are under development and should be commercialized along with plant oil-based bioproducts.

3.2 STARCH WET MILLING

Many starch-rich crops contain about 70% starch (see Table 1.2), mainly in the endosperm. Starch granules are wrapped in the protein matrix. The protein matrix has to be opened in order to free the starch granules from the protein matrix. Typical procedures for starch extraction include germ removal, fiber removal, protein matrix removal, and final starch purification (Figure 3.6). Starch extraction requires large amounts of water, which is why starch extraction is also referred to as *wet milling*. Commercial corn starch milling is an example of this process. The laboratory procedure of starch extraction is similar to the industry procedure, but it uses a laboratory scale for the mills, screen, and centrifuge, and it processes in batch style.

3.2.1 PROTEIN MATRIX DISPERSION

Corn has been used for starch products for many years because corn germ and pericarp are easy to remove. Corn oil and protein are recovered from the

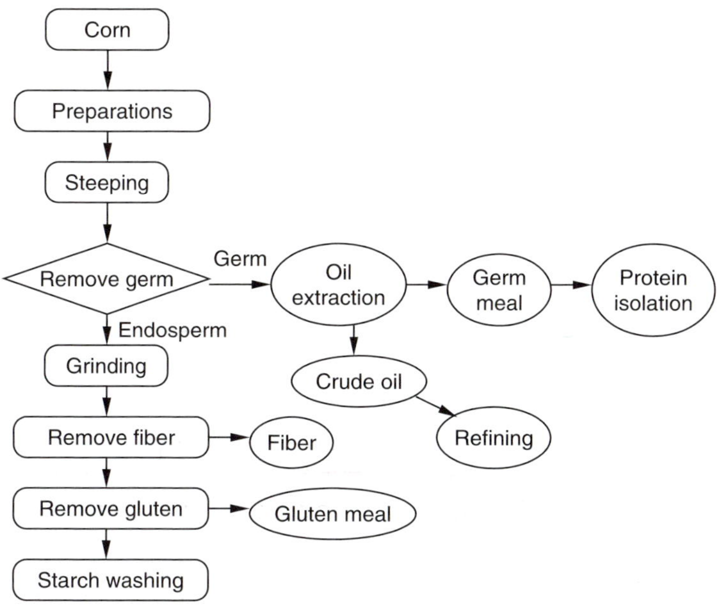

FIGURE 3.6 Flowchart of corn biorefinery system.

germ as coproducts of corn starch. Steeping is the first step of corn wet milling. The main purpose of steeping is to open the protein matrix and free starch granules for extraction. Another function of steeping is to allow easy removal of the germ from the corn.

Commercial corn, in kernel form, is loaded into a steeping tank after cleaning and general preparation. Steeping can take as long as 36–48 hours. Steeping is a diffusion, chemical, and biological process involving water diffusion, lactic acid bacteria, and yeast growth. The steeping water contains about 0.1–0.2% sulfur dioxide, which provides the combination of germicidal, softening, and protein matrix dispersing characteristics that is essential for high starch yields. A higher level of sulfur dioxide would promote faster protein disintegration, but the pH of the steeping water can be too low that it can interfere with the lactic fermentation. Lactic acid is produced during steeping, which can inhibit yeast growth and other undesirable organisms. The pH of the steeping water is in the range of 2.5–4.0. The temperature of the steeping water is 49–53 °C. A higher temperature would increase the water diffusion rate into the corn, but a temperature of 53 °C is the highest limit that lactic acid bacteria can tolerate, and 49 °C is the lowest limit for minimizing yeast growth. Besides this, a steeping temperature above 53 °C may cause starch gelatinization and protein denaturation, which would reduce the starch yield and change the starch characterization.

The steeping water flows in a countercurrent direction to the corn. Part of the steeping water is recycled in the steeping process and fresh water, with sulfur dioxide, needs to be added. The steeped corn contains about 45% water after leaving the steeping tank. For a continuous steeping process, multiple steeping tanks are usually used. The flow rate and steeping time are used to determine the number of steeping tanks.

3.2.2 GERM SEPARATION

After removing bulk water by draining, the steeped corns are opened by being torn up with a specially designed machine that leaves the germ intact for later separation. The major parts of the machine are two plates with teeth; one plate is fixed and the other rotates. Because of the high oil and protein content, the germ has a higher shearing strength than the endosperm after such long steeping. The force applied by the machine is just enough to tear open the corn endosperm so that the germ can be detached from the endosperm without damage. The germ is then separated from the endosperm fragments based on density difference using hydrocyclones. The flow diagram in Figure 3.7 illustrates the germ separation process. To ensure complete separation, the procedure is repeated using secondary germ cyclones.

The hydrocyclone is a major separation unit that is used several times during the starch wet milling process, for instance, for fiber and starch

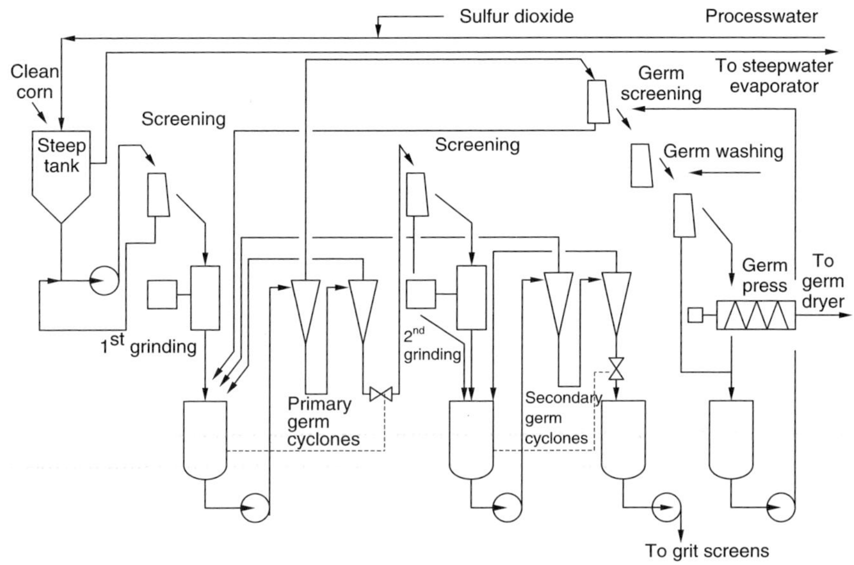

FIGURE 3.7 Flow diagram of process to remove corn germ from corn kernel (35,000 bushel/day). (Adapted, courtesy of Steve Eckhoff.)

separations. The principle of a cyclone is based on the difference of the density of two matters in a manner similar to that of a centrifuge (Figure 3.8). The mixture is fed at the top of the cyclone through a tangential inlet, at a specific speed, depending on the nature of the materials. The matter with higher density moves toward the cyclone wall because of the rapid swirling and is then discharged from the underflow outlet. The matter with lower density moves upward due to the vortex created in the cyclone because of the rapid swirling. The dimension of the cyclones, the angle of the tangential inlet, the mixture flow rate, and the speed are determined mainly by the characteristics of the materials and separation quality and quantities.

The germ from the overflow of the cyclone is washed at least three times as it passes through gravity-fed screens with a 1-mm aperture (Figure 3.7). Any germs with heavier density left in the endosperm slurry are recovered using the secondary cyclones. The parameters of the secondary cyclones are different from those of the primary cyclones, which allows for maximum recovery of germs from endosperm slurry.

3.2.3 FIBER SEPARATION

After the germ is removed, the endosperm slurry is passed over a 50-μm screen to isolate the fine particles of starch released from the protein matrix and some fine particles of proteins and fibers. The coarse materials are milled

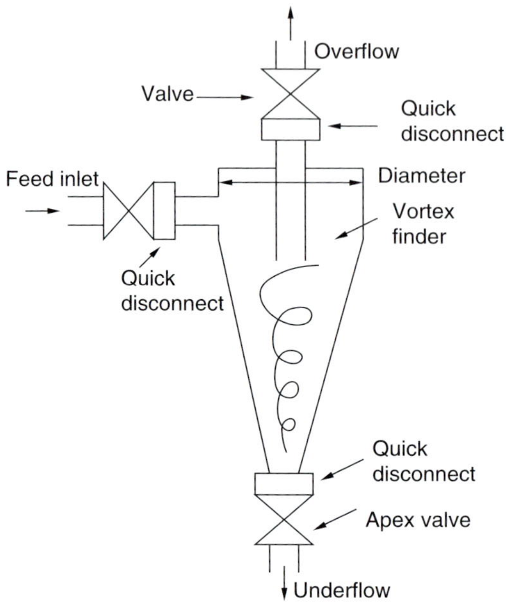

FIGURE 3.8 Diagram of hydrocyclone working principle. (Adapted, courtesy of Steve Eckhoff.)

again to liberate as much of the starch, still attached to the protein matrix or hull fragments, as possible. The fibrous hull is not easily fragmented and is washed off on a series of screens (Figure 3.9). The endosperm slurry is fed to the sixth screen. The washed fiber passes from the sixth screen to the dewatering and the drying equipment. The damaged protein matrix contains a certain amount of fibers and is referred to as *gluten* in starch in the wet milling industry. The starch and gluten from the fiber-washing screen travel back to the first screen and discharge from the starch separation process. Screen 1 has a 50-μm aperture to minimize fiber content in starch and gluten slurry, and the remaining screens have 70-μm apertures to ensure process capacity.

3.2.4 STARCH SEPARATION

The starch and gluten slurry discharged from the fiber-washing screen, screen 1, shown in Figure 3.9 is fed into the starch separation system (Figure 3.10). A series of cyclones, similar to germ cyclones but smaller in diameter, is used to separate starch from gluten. Before entering the cyclone system, any sand or large particles that may block cyclone nozzles are removed from the slurry mixture. The slurry moisture content is also reduced using a thickener to ensure efficient separation. The principle of the thickener is based on centrifuge theory.

The main separation of starch from gluten takes place at the primary separator. This is a nozzle-bowl centrifuge. The density of the starch is

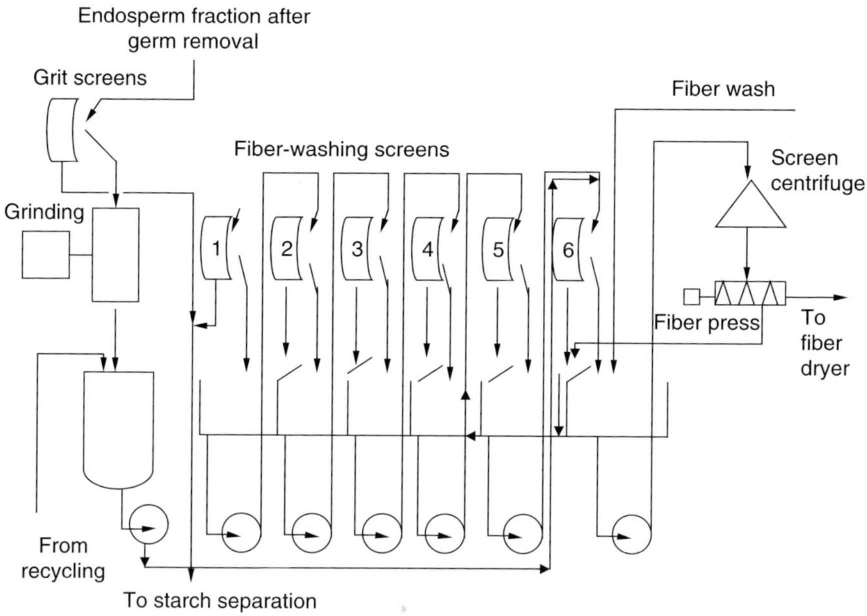

FIGURE 3.9 Flow diagram of fiber separation from corn starch and protein. (Adapted, courtesy of Steve Eckhoff.)

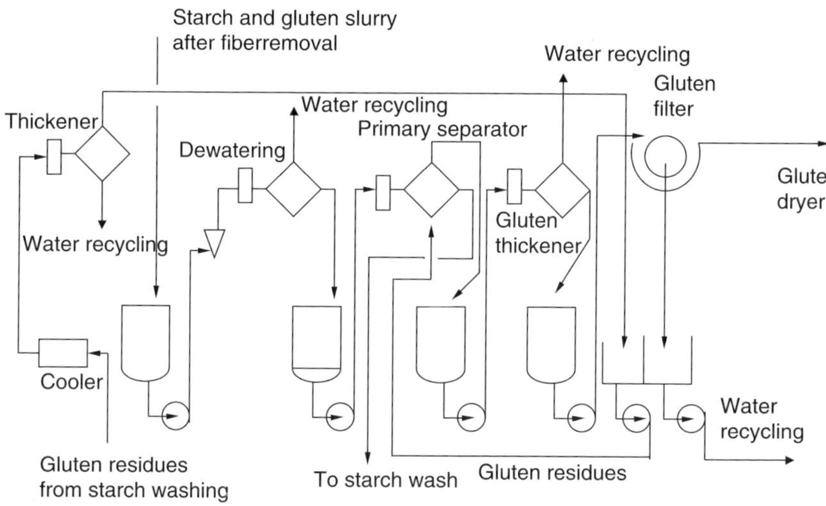

FIGURE 3.10 Flow diagram of separation of corn protein from starch. (Adapted, courtesy of Steve Eckhoff.)

greater than that of the gluten, so that the slurry containing the starch is discharged as overflow, and the gluten is discharged as underflow. A portion of the underflow is recycled to the feed inlet to ensure the separation is thorough and to keep the proper density at the nozzles. Some machines may have an internal built-in recycler. A primary separator with a 91-cm bowl, rotating at about 3000 rpm, normally has a capacity of 1016 tons per day. The quality of the separation is dependent on proper steeping and other factors. The process can be monitored in the field by spinning samples of the various streams in tubes in a laboratory centrifuge. In addition, the cyclone system needs to be cleaned and maintained at regular intervals.

Starch from the primary separator contains about 2–4% gluten. Further purification is achieved by multistage countercurrent washing. Some systems use additional nozzle centrifuges or a combination of centrifuges and hydrocyclones. The most common washing system uses a number of series of hydrocyclones with small diameters. It is very common to use up to a 15-cyclone series, with each series containing many cyclones in parallel housed in just one machine. Wash water is introduced in the final washing series, and impurities containing some starch are removed during this long series of washing. The wash water discharged from the series is recycled as steeping water.

The cyclones for starch washing are very small, so they are referred to as *cyclonettes.* About 10 mm in diameter and about 152 mm in overall length, they are usually made of high-strength nylon. Clamshell-style houses are designed for high-pressure feeds ranging from 7 to 9 kg/cm^2. Such houses commonly hold several hundred cyclonettes. Because of their small dimensions, the cyclones in the washing system are often designed with a self-cleaning strainer that has a 0.8-mm aperture in the feed inlet connected to the system and 20 mesh pot strainers with an opening of 0.84 mm between each series. The washing system is controlled by regulating the flow of starch from the underflow of the final series and the flow of wash water. The overflow from the first series must be adjusted from time to time during the process to maintain the balance.

Parameters of the washing system include pressure, temperature, density distribution, feed flow rate, feed speed, and the number of cyclonette series. All of these parameters have to be controlled to determine starch quality and washing efficiency. The number of cyclonette series is often determined by a calculation based on the processing parameters. The quality of the wash water is very important because the water is the carrier of starch granules during the entire separation process. It has been found that impurities in water can easily cause contamination and affect starch quality, consequently influencing the quality of the end product. The starch slurry discharged from the washing system is then sent for drying or to a modification process according to end use requirements.

3.2.5 STARCH PROCESS FROM OTHER GRAINS

Sorghum is one of the grains that contains high starch (see Table 1.2). The United States produces about 12 million tons of sorghum each year, and 98% of this sorghum is consumed as animal feed. Corn is the most popular raw material for starch processing because the starch yield from sorghum is low due to grain structure issues. First, the protein matrix in sorghum is tightly attached to the starch granules. The steeping process used for corn steeping cannot completely disperse the protein matrix, which results in a low starch recovery. In fact, this is also the reason for low recovery of products related to bioconversion processes. Second, the germ of a sorghum kernel is very small compared to the corn germ, which makes it very difficult to separate it from the sorghum kernel without causing damage. The germ is usually included in sorghum starch processing. Third, unlike corn, the sorghum pericarp is weaker, meaning that it is easily shattered during processing. The coproducts from corn germ and pericard contribute to industry profits; therefore, the industry has more experience with corn than sorghum. In addition, sorghum hybrids have been selected based on agronomic and animal feed values disregarding starch process potentials.

The procedures of sorghum starch extraction are similar to that of corn except that the germ separation step is voided. Also, steeping time is shorter for sorghum than for corn because of the smaller kernel size, resulting in less water usage. Another major advantage of sorghum is that sorghum can survive in very dry conditions and needs much less water for growth compared to corn. As biotechnology advances, it may be possible to improve the protein matrix properties of sorghum by gene manipulation that would be favorable for starch and other bioconversion processes.

Wheat is another starch-rich grain (see Table 1.2). The United States produces about 60 million tons of wheat per year, and 50% of this is used to produce wheat flour by dry milling. Wheat starch is popular in Europe. However, wheat starch is produced in the United States as a by-product of wheat protein. Like sorghum, wheat also has a very small germ so that it is difficult to separate it from the endosperm. The endosperm is tightly attached to the pericarp and also difficult to separate, resulting in low recovery. Wheat protein has a special character that forms a continuous water-insoluble matrix after interacting with water. Because of this characteristic, wheat starch processing is very different from corn and sorghum processing.

The wheat starch process includes wheat flour dry milling, dough formation, starch washing, and starch drying. Wheat is often tempered to a certain moisture content and then dry milled to produce wheat flour with about a 70% recovery rate. Flour dough is made by adding water to form a continuous protein matrix. Then the starch is washed out of the protein matrix by soaking the dough in excess water. The starch can be purified by passing the starch slurry through a series of cyclonettes, as described for corn starch washing. The protein matrix, after starch washing, is in skeleton form and dewatered. The dewatered protein matrix has a strong viscoelastic property like rubber and is ready for further processing.

Starch from other grains, such as rice, is usually produced for special uses because of specific properties, such as small granule size, special gelatinization and rheological behaviors, and other properties.

3.3 PROTEIN ISOLATION

As mentioned in Chapter 2, a plant contains two types of proteins: bioactive proteins and storage proteins. Bioactive proteins are mainly the various enzymes required for the germination, growth, and synthesis of plant polymers. Storage proteins are the main sources of amino acids for new protein synthesis during growth. The storage proteins can be isolated from carbohydrates and lipids as industrial polymers. Many methods for protein isolation on a laboratory scale utilize characteristics of proteins, such as surface structure and charges, solubility in water and various solvents, density, molecular weight, and chemical compositions. The methods introduced in this chapter are all commonly used procedures.

3.3.1 SOY PROTEINS

As shown in Table 1.2, soybean contains about 40% protein and is, therefore, a major source of storage proteins. Soy protein is often isolated from the soy meals after oil extraction and is also called a coproduct of soy oil. The soy meal contains about 50% protein, 30–35% carbohydrates, and less than 1% lipids. The remaining matter is mainly ash and moisture. The soy meal from soy oil extraction contains about a 30–35% hexane residual. Therefore, the first step of soy protein processing is to remove the hexane from the soy meal. To avoid heat damage, the hexane residual is removed by superheated hexane vapor in a desolventizing system. This system can be either a container or a conveyer. For the container system, the superheated hexane vapor is introduced into an agitating container that holds the defatted soy flakes. For the conveyer system, the superheated hexane vapor and flakes come into contact in a conveying tube. This contact time is relatively short compared to that of the container system. For food uses, the desolventized soy meal has to be deodorized. Heat treatment is often applied to feed applications for nutritional purposes. Soy proteins are denatured upon heat treatment, having a lower protein dispensability index (PDI) than those that have not been heat treated. Soy meal with a high PDI is often preferred for industry uses.

The desolventized soy meal is processed into several products (Figure 3.11), such as soy grits, soy flour, soy concentrates, and soy protein isolates, depending on end uses and demands. Soy grits and soy flour are obtained by grinding the soy meal followed by screen separation. Particle sizes for grits range from 10 to 80 mesh, and from 100 mesh and above for soy flour. Soy protein concentrates, containing about 70% protein, can be obtained by removing soluble carbohydrate components. Three methods are commonly used for soy protein concentrate processing: acid leaching, aqueous ethanol extraction, and moist heat methods. For the acid leaching method, the soy meal is dissolved in water at a solid content of about 5–10%. The pH of the slurry is adjusted to an isoelectric point of protein, which is about 4.5. Protein at the isoelectric point precipitates, becomes water insoluble, and precipitates once it is centrifuged. The separated protein is then neutralized to a pH value of 7.0 and spray-dried into powder. For the alcohol extraction, alcohol-soluble carbohydrates can be removed from the soy meal using aqueous alcohol flowing concurrently in a column. The soy concentrates, containing proteins and some alcohol-insoluble polysaccharides, are desolventized and dried for future uses. The soy concentrates produced using an alcohol method have low nitrogen solubility due to denaturation by alcohol [8]. For the moist heat process, the soy meals are heated in the presence of

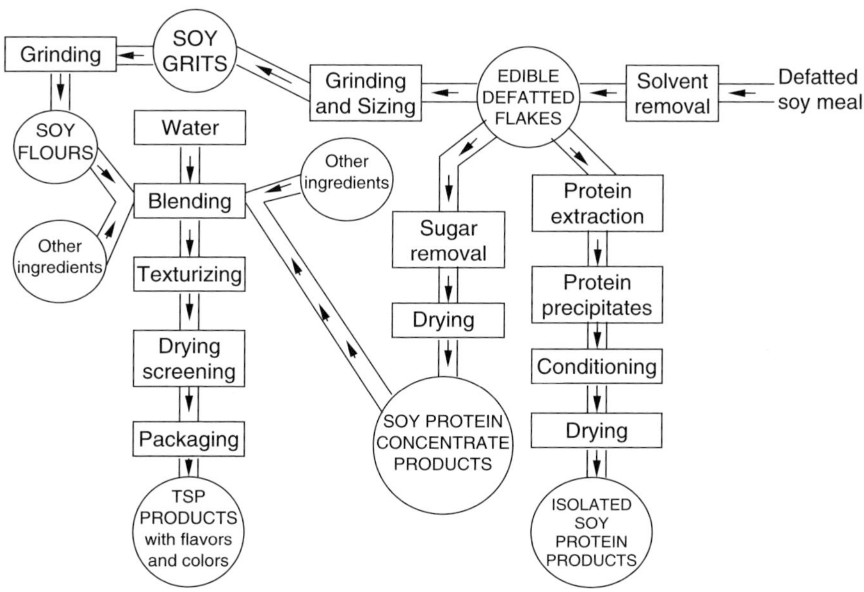

FIGURE 3.11 Flowchart of soy meal processing into various soy protein-based products. (*Source:* Adapted from Ref. [8].)

moisture to allow protein to denature, resulting in water-insoluble proteins. Water-soluble carbohydrates are then removed by water leaching through the soy meal.

Soy protein isolate (SPI) is produced from the soy meal by various processing technologies, depending on end uses and property specifications. The SPI protein content should be 90% or above on a dry base. Traditionally, soy protein isolate is extracted from the soy flour using a mild alkali precipitation method in water. Figure 3.12 shows a typical laboratory procedure for SPI extraction. Soy flour is dissolved in water at about 5–10% solid content, and stirred until complete dispersion. The pH of the slurry is adjusted to about 8.5 with 2N sodium hydroxide, allowing higher protein solubility. The water-insoluble carbohydrates and other particles are removed by centrifuge. The pH of the supernatant, containing proteins, is adjusted to an isoelectric point to obtain minimum protein solubility, and then centrifuged to separate protein from water and other particles. The precipitated SPI is washed and dried, or neutralized to a pH of 7.6, and then dried to powder. Commercial processing of SPI follows a similar procedure. The SPI can be modified during isolation or after drying, for various applications, which will be discussed in detail in Chapters 9 and 10.

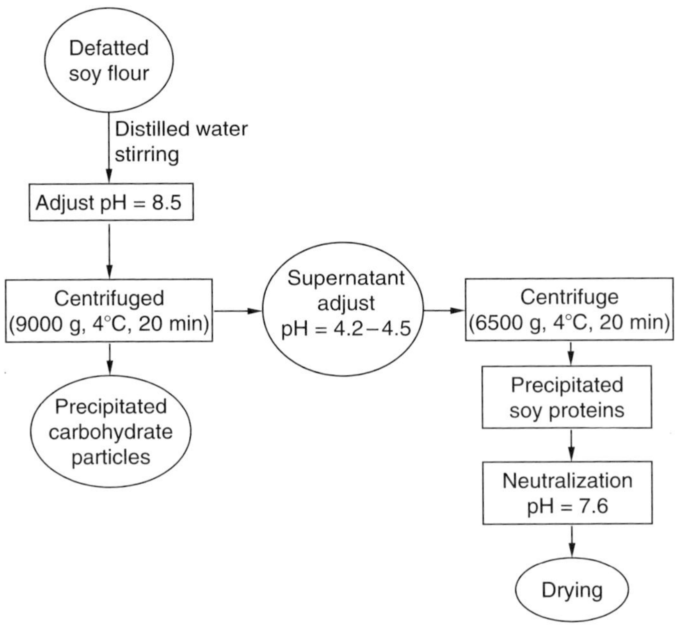

FIGURE 3.12 Flowchart of procedures for isolating soy protein from defatted soy flour.

3.3.2 SOY PROTEIN FRACTIONS

Soy storage proteins are globulins consisting of many subproteins. Two major subproteins are glycinin (11S) and conglycinin (7S), counting for about 80% of total soy proteins. Glycinin and conglycinin can be isolated from soy protein isolate, or soy flour, by utilizing the solubility property of proteins and carbohydrates, according to the laboratory procedure described in Figure 3.13 [9]. Soy flour is dissolved in water at a solid content of 4–10%, and stirred until completely dispersed. Water-insoluble carbohydrates and other particles are removed by centrifuge after adjusting the pH of the slurry to 7.5 in order to obtain higher protein solubility. The pH of the supernatant is adjusted to 6.4, which is the isoelectric point of glycinin protein, and then stored for 24 hr to allow better separation. It is then centrifuged. The precipitated glycinin protein is neutralized to pH 7.6 by 2N sodium hydroxide and dried to powder. The supernatant, containing conglycinin, is diluted, then adjusted to a pH of about 4.8, which is the isoelectric point of conglycinin, and centrifuged. The precipitated conglycinin is neutralized and dried to powder. The commercial procedure of producing glycinin and conglycinin is similar to this laboratory procedure.

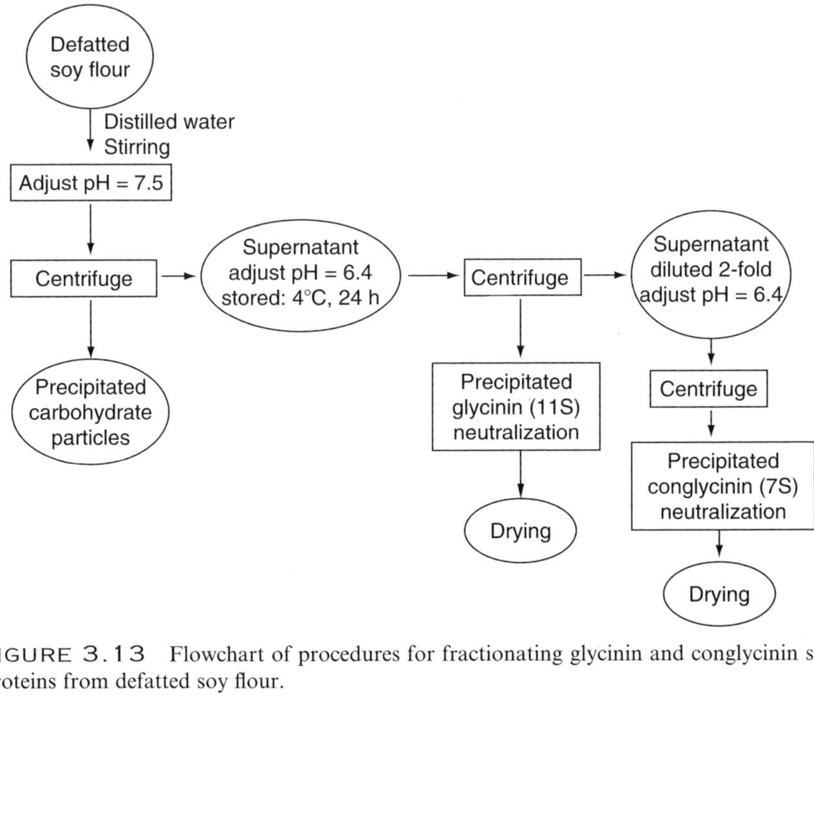

FIGURE 3.13 Flowchart of procedures for fractionating glycinin and conglycinin subproteins from defatted soy flour.

Glycinin and conglycinin also consist of many subproteins. Glycinin has two major subproteins called basic and acidic polypeptides. Isolation of basic and acidic polypeptides follows the procedures described by Demodaran and Kinsella [10] with modification. Glycinin is dissolved in distilled water at a solid content of about 0.5% in 30 mM Tris buffer (pH 8.0) containing 15 mM mercaptoethanol. The solution is heated to 90°C for 30 min, followed by centrifugation of 10,000 *g* at 4°C. The precipitated matter is washed twice using a Tris buffer (pH 8.0), then dissolved in distilled water, and then freeze-dried and collected as basic polypeptides. The supernatant is filtered through a 0.45-mm membrane and freeze-dried. The powder is dissolved in water and dialysis against water and then freeze-dried as acidic subunits.

Major subproteins from conglycinin include α-, β-, and γ-conglycinins. Isolation of the subproteins from conglycinin follows the procedure proposed by Iwabuchi and Yamauchi [11] and Thanh and Shibasaki [12]. Conglycinin powder is first purified by dissolving it in a pH 7.6 potassium phosphate buffer (2.6 mM KH_2PO_4, 32.5 mM K_2HPO_4, 0.4 M NaCl, 10 mM 2-mercaptoethanol) at a conglycinin solid content of 3%, and then ammonium sulfate is added to the protein solution to 75% saturation. The precipitation is removed by centrifuge, and the supernatant is further adjusted with ammonium sulfate to a saturation of 90%. After centrifugation, the precipitation is collected and dissolved in the same phosphate buffer and dialysis against water and then freeze-dried. Isolation of fractions from conglycinin can be accomplished by a chromatographic method [13]. A 2.5- × 100-cm column is packed with gel DEAE-Sephadex A-50, swollen in 19 mM phosphate buffer (16 mM K_2HPO_4, 3 mM K_2HPO_4, 0.2 M NaCl, 10 mM 2-mercaptoethanol, pH 7.8). The conglycinin is dissolved in the 19 mM phosphate buffer in the column. The column is then washed with the buffer. Chromatography elution is performed with a linear concentration gradient in a NaCl concentration, from 0.2 to 0.4 M (2 L each) at a flow rate of 0.5 mL/min. About six fractions are collected, concentrated, and freeze-dried as subproteins of conglycinin.

3.3.3 CORN PROTEINS

Corn contains about 6–12% protein (see Table 1.2). Corn protein is obtained primarily from corn germ and corn gluten from starch processing. The oil of the germ can be extracted using the technologies described in Section 3.1, such as the hexane solvent method. The germ meal is traditionally further processed for animal feeds. The gluten from the endosperm is called corn gluten meal. The corn gluten is lighter than starch, and so the overflow stream from the hydrocyclone contains gluten (Figure 3.10). The gluten is then dewatered using either a centrifuge or a rotary drum filter. The latter is commonly used in corn gluten processing (Figure 3.14) because of its high efficiency and low cost, with a processing capacity of about 500 ton/day.

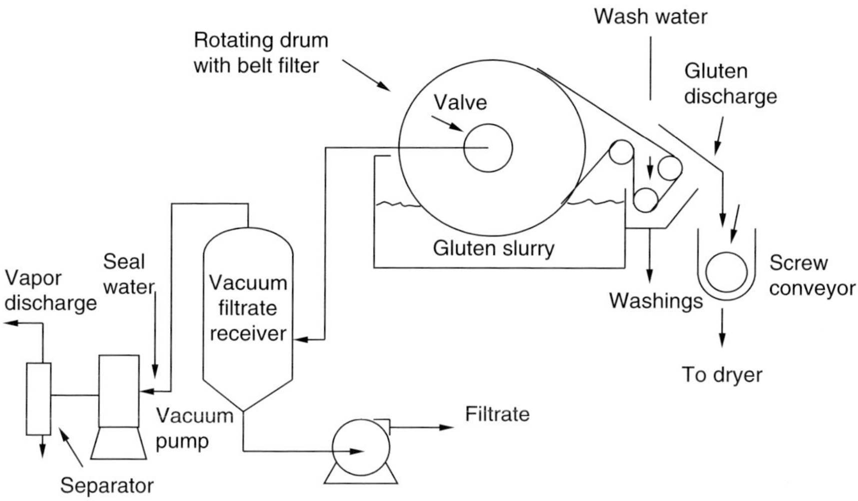

FIGURE 3.14 Flow diagram of corn gluten processing.

A belt covered on the drum serves as a filtering surface. The drum dips into the container with gluten slurry. A vacuum is applied to build up a cake that is sucked free of surplus moisture and then discharged when the belt is pulled away from the drum. The belt is then washed clean with washing water for the next dip. The dewatered gluten is then dried for further processing.

Corn gluten meal contains about 60% protein, and about 50% of this protein is a functional prolamine protein, called zein, which can be isolated from the corn gluten meal as an industrial polymer. Zein is insoluble in water but becomes soluble in the presence of alcohol, a high concentration of urea or alkali [14, 15]. Ethanol is the most commonly used solvent for zein extraction. Corn gluten meal is dissolved in 93% ethanol, at a high pH, and an elevated temperature ranging from 50° to 60°C for about 0.5–2 hr. The zein can be extracted by filtration or centrifuge. The precipitated zein is often vacuum dried and ground into powder for future uses.

3.3.4 WHEAT PROTEINS

Wheat contains about 7–14% storage protein (see Table 1.2). The by-product of wheat starch processing is often referred as to wheat gluten and contains about 70% protein. As mentioned in the discussion of wheat starch extraction, wheat gluten can be separated from starch by a dough washing method. After starch extraction, the gluten is a continuous matrix. Because of its viscoelastic behavior, like rubber, the gluten is palletized and dried using belt drying technology. The dried pallets are ground into powder as commercial gluten product. High-pressure expansion technology is also used by industries to prepare gluten pallets with loose structure and low density, which reduces drying and post-grinding energies.

Glutenin and gliadin are two major subproteins of wheat protein that are not water soluble. Gliadin is soluble in alcohol and usually extracted using a 70% alcohol solution, whereas glutenin can be extracted using a 0.3-M acetic acid solution [16]. The procedures of extracting gliadins and glutenins can start with wheat flour [17]. Wheat flour is dissolved in a solution containing 0.4 mol/L NaCl salt and 0.067 mol/L $HKNaPO_4$ with a pH of 7.6, stirred for about 10 min at room temperature, and centrifuged. Supernatant 1 is discharged. Sediment 1 is dissolved in a 50% (w/v) 2-propanol solution, stirred for about 10 min, and centrifuged. Supernatant 2 contains mainly gliadins, which are precipitated by adding 1% (w/v) dithioerythritol and 0.08 mol/L Tris/HCl with a pH of 8.0. Sediment 2 is redissolved in the 2-propanol solution, stirred for 20 min at an elevated temperature of 60 °C, and centrifuged. The glutenin can then be isolated by precipitating supernatant 3 and adding the dithioerythritol and Tris/HCl with a pH of 8.0. Other laboratory procedures can also be found in the literature [18–20]. Soluble proteins primarily contain albumins and globulins that can be extracted using a salt solution [16]. Many subproteins, of both insoluble and soluble protein fractions, can be further isolated using column chromatography methods in the laboratory for polymer study purposes.

Starch and Protein Polymer Drying

Commercial starch and protein are in powder forms and drying is a necessary procedure in processing. Drying involves temperature, flow rates, drying rates, and drying media, which can significantly alter the properties of the final starch and protein products [21, 22]. Commonly used drying technologies for protein and starch include freeze drying, spray drying, belt drying, and flash drying. The selection of drying technology depends on the properties required for end uses and also costs. Freeze drying is relatively expensive compared to other drying technologies, and it is often used at the laboratory scale to avoid structure damage. Spray drying is the most commonly used technology in the feedstock industry for starch and protein drying. Starch or protein dried using spray drying has a fluffy surface structure, low density, high solubility, and low viscosity. These properties can be altered within a limited range by simulating spray drying parameters, such as air temperature and speed, slurry flow rate, nozzle spray speed, solid concentration, and particle travel distance. Polymers dried using belt drying often have low solubility, high density, high viscosity, and a hard surface structure. Proteins or modified proteins can be very viscous and are often palletized for belt drying. The dried pallets are then ground into powder from 100 mesh US sieves (150 μm) and above to the required particle size. The properties of starch and protein dried using flash drying are in between those achieved by spray and belt drying.

REFERENCES

1. Norris, F. A. Extraction of Fats and Oils. In *Baileys Industrial Oil and Fats Products,* Vol. II, Swern, D., Ed.; Wiley, New York; 1982.
2. Randall, E. L. *Extractor Assembly,* U.S. Patent 3,798,133; 1974.
3. Orthmer, D. F.; Agarwal, J. C. *Chem. Engineering Prog.* 1955, **51,** 372–378.
4. Fan, H. P.; Morris, J. C.; Wakeman H. *Ind. Eng. Chem. Soc.* 1948, **60,** 203.
5. Karnofsky, G. *J. Am. Oil Chem. Soc.* 1949, **26,** 564–569.
6. Erickson, D. R., Ed. *Practical Handbook of Soybean Processing and Utilization*, American Oil Chemists' Society Press, Champaign, IL; 1995.
7. Wiedermann, L. H. J. *Am. Oil Chem. Soc.* 1981, **58,** 159–166.
8. Liu, K. S. *Soybean Chemistry, Technology, and Utilization,* Chapman & Hall, New York; 1997.
9. Thanh, V. H.; Shibasaki, K. *J. Agric. Food Chem.* 1976, **24**(6), 1117–1121.
10. Damodaran, S.; Kinsella, J. E. *J. Agric. Food Chem.* 1982, **30,** 812–817.
11. Iwabuchi, S.; Yamauchi, F. *J. Agric. Food Chem.* 1987, **35,** 200–205.
12. Thanh, V. H.; Shibasaki, K. *Biochim. Biophys. Acta* 1976, **439,** 326–338.
13. Iwabuchi, S.; Yamauchi, F. *J. Agric. Food Chem.*1987, **35,** 205–209.
14. Wilson, C. M. Proteins of Kernel. In *Corn Chemistry and Technology,* Ramstad, P. E.; Watson, S. A., Eds.; American Association of Cereal Chemists, St. Paul, MN; 1999.
15. Shukla, R.; Cheryan, M. *Zein: The Industrial Protein from Corn, Industrial Crops and Products*, 2001, **13,** 171–192.
16. Wrigley, C. W.; Bietz, J. A. In *Wheat Chemistry and Technology*, American Association of Cereal Chemists, St. Paul, MN; 1988, 195–276.
17. Wieser, H., In *Gluten 96: Proceedings of the Sixth International Gluten Workshop,* Wrigley, C. W., Ed.; Cereal Chemistry Division, Royal Australian Chemical Institute, North Melbourne, Australia; 1996.
18. Beckwith, A. C.; Nielsen, H. C.; Wall, J. S.; *et al., Cereal Chem.* 1966, **43,** 14–28.
19. Fu, B. X.; Sapirstein, H. D. *Cereal Chem.* 1996, **73**(1), 143–152.
20. Bean, S. R.; Lyne, R. K.; Tilley, K. A.; *et al. Cereal Chem.*, 1998, **75**(3), 374–379.
21. Baker, C. G. J., *Industrial Drying of Foods,* Blackie Academic & Professional, New York; 1997.
22. Gould, W. A. *Unit Operations for the Food Industries,* CTI Publications, Timonium, MD; 1996.

4

POLYMERS AND COMPOSITE RESINS FROM PLANT OILS

RICHARD P. WOOL

Recent advances in genetic engineering, composite science, and natural fiber development offer significant opportunities for developing new, improved materials from renewable resources that can biodegrade or be recycled, enhancing global sustainability. A wide range of high-performance, low-cost materials can be made using plant oils, natural fibers, and lignin. By selecting the fatty acid distribution function of plant oils via computer simulation and the molecular connectivity, we can control chemical functionalization and molecular architecture to produce linear, branched, or cross-linked polymers. These materials can be used as pressure-sensitive adhesives, elastomers, rubbers, and composite resins. This chapter describes the chemical pathways that were used to modify plant oils and allow them to react with each other and various comonomers to form materials with useful properties.

4.1 INTRODUCTION

Polymers and polymeric composite materials have extensive applications in the aerospace, automotive, marine, infrastructure, military, sports, and industrial fields. These lightweight materials exhibit excellent mechanical properties, high corrosion resistance, dimensional stability, and low assembly costs. Traditionally, polymers and polymeric composites have been derived from petroleum; however, as the applications for polymeric materials increase, finding alternative sources of these materials has become critical. In recent years, the Affordable Composites from Renewable Sources (ACRES) program at the University of Delaware has developed a broad range of chemical routes to use natural triglyceride oils to make polymers and composite materials [1, 2]. These materials have economic and environmental advantages that make them attractive alternatives to petroleum-based materials.

Natural oils, which can be derived from both plant and animal sources, are abundant in most parts of the world, making them an ideal alternative to chemical feedstocks. These oils are predominantly made up of triglyceride molecules, which have the structure shown in Figure 4.1. Triglycerides are composed of three fatty acids joined at a glycerol juncture. The most common oils contain fatty acids that vary from 14 to 22 carbons in length with 0 to 3 double bonds per fatty acid. Table 4.1 shows the fatty acid distributions of several common oils [3]. Exotic oils are composed of fatty acids with other types of functionalities, such as epoxies, hydroxyls, cyclic groups, and furanoid groups [4]. Because of the many different fatty acids present, on a molecular level these oils are composed of many different types of triglycerides with numerous levels of unsaturation. With newly developed genetic engineering techniques, the variation in unsaturation can be controlled in plants such as soybean, flax, and corn; however, some oils are better suited to polymer resin development.

Besides applications in the foods industry, triglyceride oils have been used extensively to produce coatings, inks, plasticizers, lubricants, and agrochemicals [5–11]. In the polymers field, the use of these oils as toughening agents was investigated. Barrett [12] has reviewed an extensive amount of work on using these oils to produce interpenetrating networks (IPNs). It was found that IPNs formed by triglycerides could increase the toughness and fracture resistance in conventional thermoset polymers [13–16; see also 17–25]. In these works, the functional triglyceride was a minor component in the polymer matrix acting solely as a modifier to improve the physical properties of the main matrix. Consequently, the triglyceride-based materials were low-molecular-weight, lightly cross-linked materials incapable of displaying the necessary rigidity and strength required for structural applications by themselves.

4.2 SYNTHETIC PATHWAYS FOR TRIGLYCERIDE-BASED MONOMERS

Triglycerides contain active sites amenable to chemical reaction: the double bond, the allylic carbons, the ester group, and the carbons alpha to

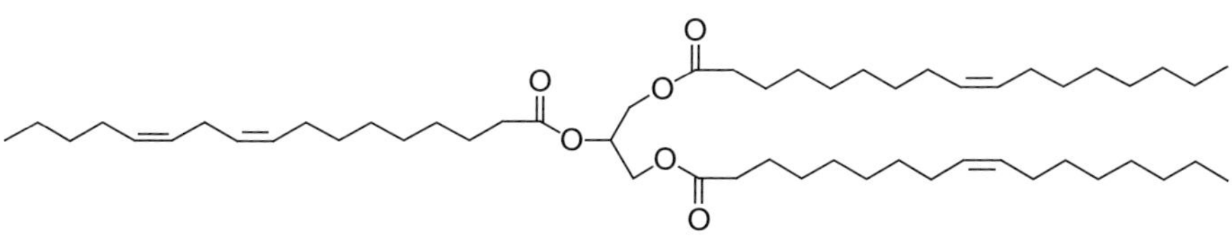

FIGURE 4.1 Triglyceride molecule, the major component of natural oils.

TABLE 4.1 Fatty acid distribution in various plant oils.

Fatty acid	# C: # DB[a]	Canola	Corn	Cottonseed	Linseed	Olive	Palm	Rapeseed	Soybean	High oleic[b]
Myristic	14:0	0.1	0.1	0.7	0.0	0.0	1.0	0.1	0.1	0.0
Myristoleic	14:1	0.0	0.0	0.0	0.0	0.0	0.0	0.0	0.0	0.0
Palmitic	16:0	4.1	10.9	21.6	5.5	13.7	44.4	3.0	11.0	6.4
Palmitoleic	16:1	0.3	0.2	0.6	0.0	1.2	0.2	0.2	0.1	0.1
Margaric	17:0	0.1	0.1	0.1	0.0	0.0	0.1	0.0	0.0	0.0
Margaroleic	17:1	0.0	0.0	0.1	0.0	0.0	0.0	0.0	0.0	0.0
Stearic	18:0	1.8	2.0	2.6	3.5	2.5	4.1	1.0	4.0	3.1
Oleic	18:1	60.9	25.4	18.6	19.1	71.1	39.3	13.2	23.4	82.6
Linoleic	18:2	21.0	59.6	54.4	15.3	10.0	10.0	13.2	53.2	2.3
Linolenic	18:3	8.8	1.2	0.7	56.6	0.6	0.4	9.0	7.8	3.7
Arachidic	20:0	0.7	0.4	0.3	0.0	0.9	0.3	0.5	0.3	0.2
Gadoleic	20:1	1.0	0.0	0.0	0.0	0.0	0.0	9.0	0.0	0.4
Eicosadienoic	20:2	0.0	0.0	0.0	0.0	0.0	0.0	0.7	0.0	0.0
Behenic	22:0	0.3	0.1	0.2	0.0	0.0	0.1	0.5	0.1	0.3
Erucic	22:1	0.7	0.0	0.0	0.0	0.0	0.0	49.2	0.0	0.1
Lignoceric	24:0	0.2	0.0	0.0	0.0	0.0	0.0	1.2	0.0	0.0
Average # DB/triglyceride		3.9	4.5	3.9	6.6	2.8	1.8	3.8	4.6	3.0

[a]The number of carbon atoms and C=C double bonds per fatty acid. For example, 18:2 means that the fatty acid is 18 carbon atoms long and contains 2 C=C double bonds.
[b]Genetically engineered high oleic acid content soybean oil (DuPont).

the ester group. These active sites can be used to introduce polymerizable groups on the triglyceride using the same techniques applied in the synthesis of petrochemical-based polymers. The key step is to reach a higher level of M_W and cross-link density, as well as to incorporate chemical functionalities known to impart stiffness in a polymer network (e.g., aromatic or cyclic structures). Figure 4.2 illustrates several synthetic pathways that accomplish this [1].

In structures **5, 6, 7, 8,** and **11,** the double bonds of the triglyceride are used to functionalize the triglyceride with polymerizable chemical groups. From the natural triglyceride, it is possible to attach maleates (**5**) [6–11] or to convert the unsaturation to epoxy (**7**) [26–28] or hydroxyl functionalities (**8**) [29, 30]. Such transformations make the triglyceride capable of reaction via ring-opening or polycondensation polymerization. These particular chemical pathways are also accessible via natural epoxy and hydroxyl functional triglycerides [12, 14–16]. It is also possible to attach vinyl functionalities to the epoxy and hydroxyl functional triglycerides. Reaction of the epoxy functional triglyceride with acrylic acid incorporates acrylates onto the triglyceride (**6**), while reaction of the hydroxylated triglyceride with maleic anhydride incorporates maleate half-esters and esters onto the triglyceride (**11**). These monomers can then be blended with a reactive diluent, similar to most conventional vinyl ester resins and cured by free-radical polymerization.

The second method for synthesizing monomers from triglycerides is to convert the triglyceride to monoglycerides through a glycerolysis (**3A**) reaction or an amidation reaction (**2, 3B**) [31–36]. Monoglycerides are used in surface coatings, commonly referred to as alkyd resins, because of their low cost and versatility [32]. In those applications, the double bonds of the monoglyceride are reacted to form the coating. However, monoglycerides also can react through the alcohol groups via polycondensation reactions with a comonomer, such as a diacid, epoxy, or anhydride. Alternatively, maleate half esters can be attached to these monoglycerides (**9**) allowing them to undergo free-radical polymerization.

The third method is to functionalize the unsaturation sites as well as reduce the triglyceride into monoglycerides. This can be accomplished by glycerolysis of an unsaturated triglyceride, followed by hydroxylation or by glycerolysis of a hydroxy functional triglyceride. The resulting monomer can then be reacted with maleic anhydride, forming a monomer capable of polymerization by the free-radical mechanism [1].

Although the structure of triglycerides is complex in nature, it is possible to characterize some aspects of it using proton nuclear magnetic spectroscopy (^{1}H NMR) and Fourier transform infrared (FTIR) spectroscopy. A typical ^{1}H NMR spectrum of soybean oil is shown in Figure 4.3, with peak

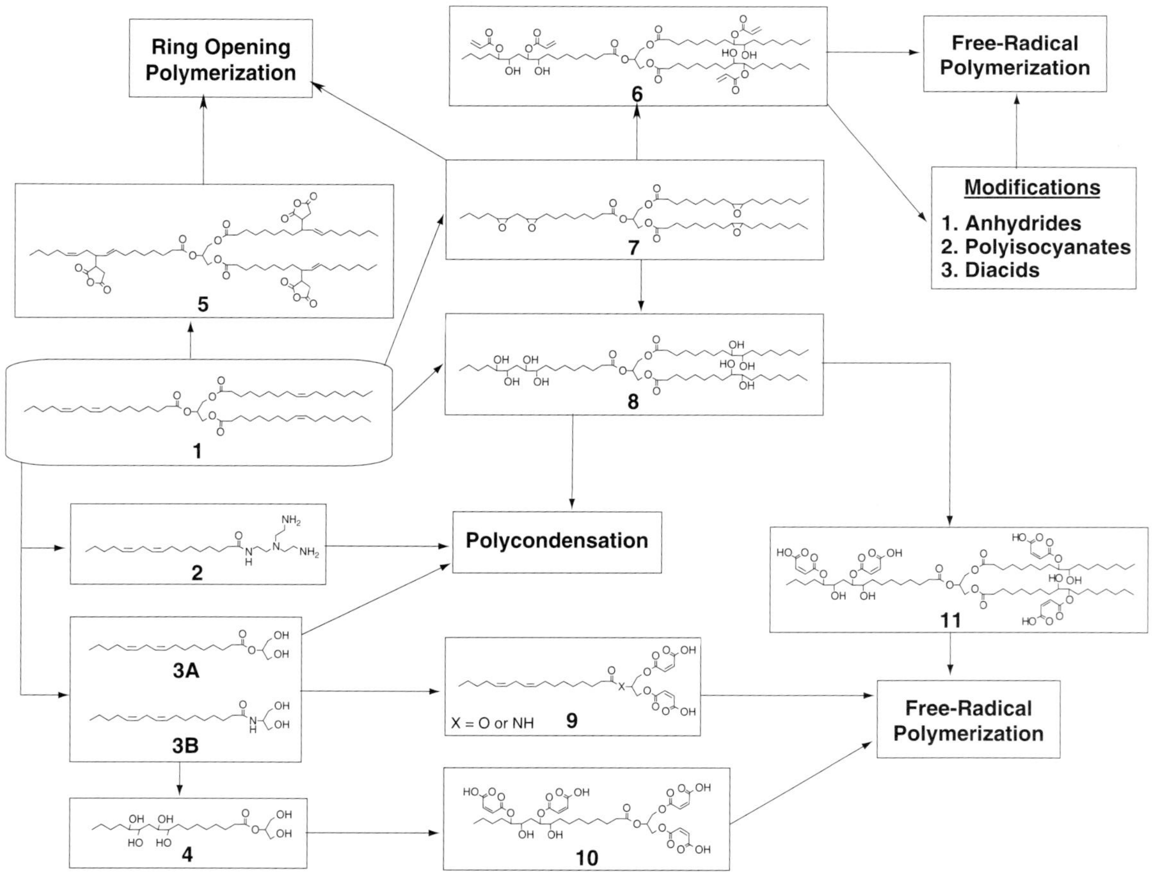

FIGURE 4.2 Chemical pathways leading to polymers from triglyceride molecules [1].

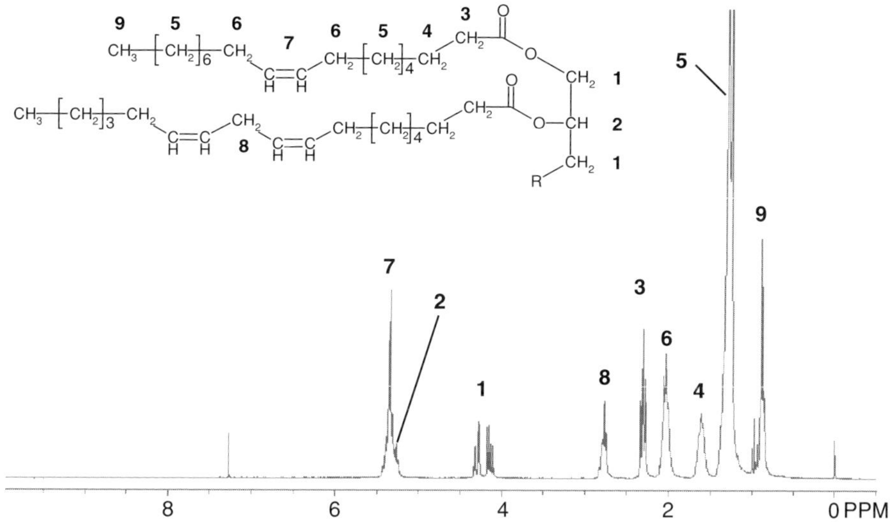

FIGURE 4.3 ^{1}H NMR spectrum of soybean oil ($CDCl_3$). R represents a third fatty acid.

assignments. The two sets of peaks at 4.0 to 4.4 ppm are produced by the four glycerol methylene protons per triglyceride [4]. The triplet set of peaks at 2.3 ppm is produced by the six protons in the alpha position relative to the carbonyl groups. The peak at 0.9 ppm is produced by the nine methyl protons per triglyceride at the end of each fatty acid chain. These three groups of peaks provide a standard by which other peaks can be used to quantitatively characterize functional groups in the triglyceride.

In this work, we focus on three triglyceride monomers, shown in Figure 4.4, which have been found to be promising candidates for use in the composites and engineering plastics fields. They are acrylated epoxidized soybean oil (AESO), the maleinized soybean oil monoglyceride (SOMG/MA), and maleinized hydroxylated soybean oil (HSO/MA). These monomers, when used as a major component of a molding resin, have shown properties comparable to conventional polymers and composites, and these properties will be presented. In addition, they can be used as a matrix in synthetic and natural fiber-reinforced composites.

4.2.1 ACRYLATED EPOXIDIZED SOYBEAN OIL

Acrylated epoxidized oils (Figure 4.4) are synthesized from the reaction of acrylic acid with epoxidized triglycerides. Epoxidized triglycerides can be found in natural oils, such as vernonia plant oil, or can be synthesized from more common unsaturated oils, such as soybean oil or linseed oil, by a standard epoxidation reaction [37]. The natural epoxy oil, vernonia oil, has

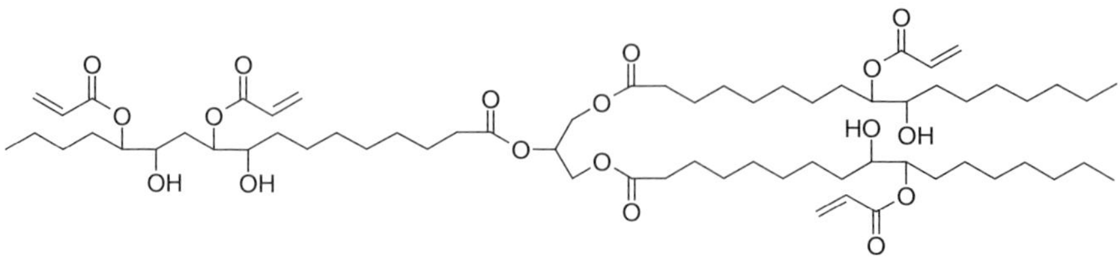

Acrylated Epoxidized Soybean Oil (AESO)

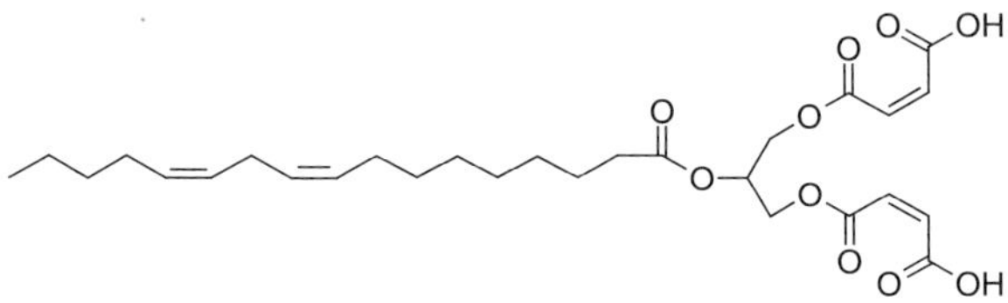

Maleinized Soybean Oil Monoglyceride (SOMG/MA)

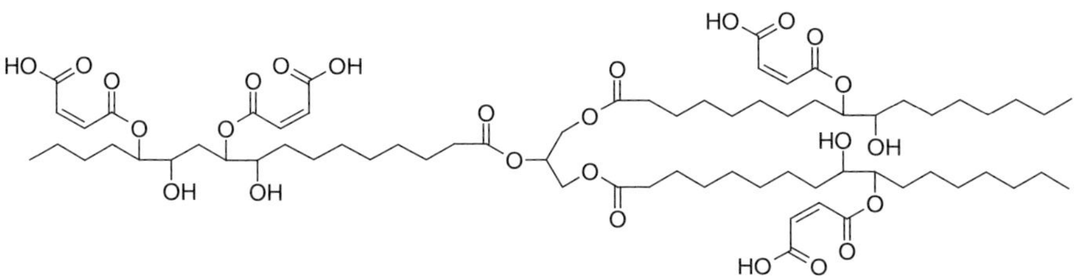

Maleinized Hydroxylated Soybean Oil (HSO/MA)

FIGURE 4.4 Triglyceride-based monomers.

a functionality of 2.8 epoxy rings per triglyceride [0]. Epoxidized soybean oil is commercially available and is generally sold with a functionality of 4.1–4.6 epoxy rings per triglyceride, which can be identified via ^{1}H NMR [20, 38]. Epoxidized linseed oil is also commercially available when higher epoxy content is required. Predominantly, these oils are used as alternative plasticizers in polyvinyl chloride in place of phthalates [39–41], but their use as a toughening agent also was explored [20, 23–25, 42]. With the addition of acrylates, the triglyceride can be reacted via additional polymerization. AESO was used extensively in surface coatings and is commercially manufactured in forms such as Ebecryl 860 [7, 43, 44]. Urethane and amine derivatives of AESO have also been developed for coating and ink applications [8, 9, 45].

The reaction of acrylic acid with epoxidized soybean oil occurs through a standard substitution reaction and was found to have first-order dependence with respect to epoxy concentration and second-order dependence with respect to acrylic acid concentration [46]. However, epoxidized oleic methyl ester was found to display second-order dependence on both epoxy and acrylic acid concentrations [47]. Although the reaction of epoxidized soybean oil with acrylic acid is partially catalyzed by the acrylic acid, the use of additional catalysts is common. Tertiary amines, such as *N,N*-dimethyl aniline, triethylamine, and 1,4-diazobicyclo[2.2.2]octane, are commonly used [38, 48]. Additionally, more selective organometallic catalysts have been developed that reduce the amount of epoxy homopolymerization [49, 50].

AESO can be blended with a reactive diluent, such as styrene, to improve its processability and to control the polymer properties to reach a range acceptable for structural applications. By varying the amount of styrene, it is possible to produce polymers with different moduli and glass transition temperatures. Polymer properties can also be controlled by changing the M_W of the monomer or the functionality of the acrylated triglyceride. Consequently, a range of properties, and therefore applications, can be found. Subsequent to the acrylation reaction, the triglyceride contains both residual amounts of unreacted epoxy rings as well as newly formed hydroxyl groups, both of which can be used to further modify the triglyceride by reaction with a number of chemical species, such as diacids, diamines, anhydrides, and isocyanates. The approach presented here is to oligomerize the triglycerides with reagents that have chemical structures conducive to stiffening the polymer, such as cyclic or aromatic groups. Reacting AESO with cyclohexane dicarboxylic anhydride (Figure 4.5A) forms oligomers, increasing the entanglement density as well as introducing stiff cyclic rings to the structure. Reaction of the AESO with maleic acid (Figure 4.5B) also forms oligomers and introduces more double bonds. Although it is desirable to maximize the conversion of hydroxyls or epoxies, the viscosity increases dramatically at high levels of conversion. Eventually, this can lead to gelation, so the reaction must be carefully monitored. After oligomerization, the modified AESO resin can be blended with styrene and cured in the same manner as the unmodified AESO resin.

4.2.2 MALEINIZED SOYBEAN OIL MONOGLYCERIDE

Maleinized soybean oil monoglyceride (Figure 4.4) is synthesized from the triglyceride oil in two steps [33]. The first is a standard glycerolysis reaction to convert the triglycerides into monoglycerides by reacting triglycerides with glycerol; see [31]. The product is generally a mixture of mono- and diglycerides, as illustrated in Figure 4.6. Using excess glycerol can aid in conversion. Additionally, the reaction can be run in solvent or in the presence of an emulsifier catalyst [34]. Once the reaction is completed, it is possible to

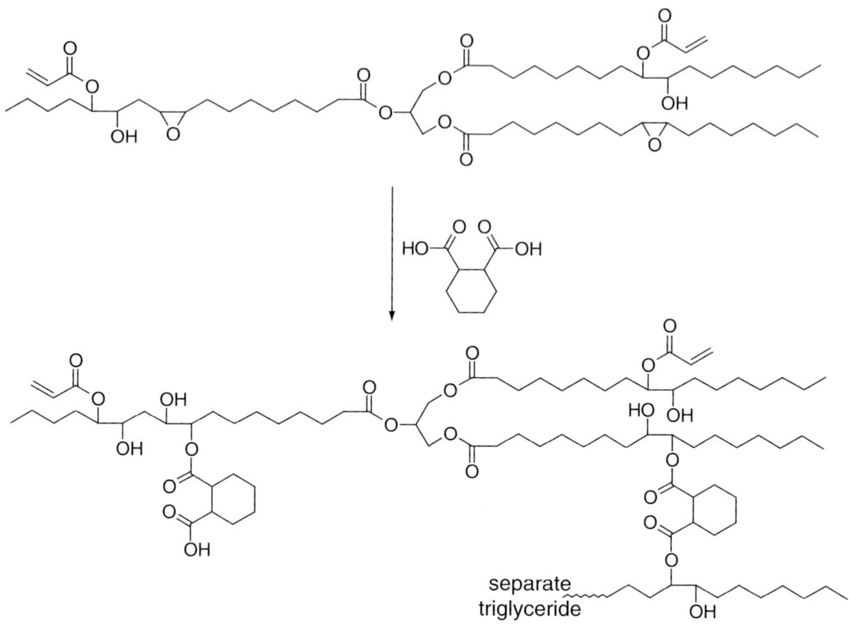

FIGURE 4.5A Modification of AESO by reaction with cyclohexane dicarboxylic anhydride.

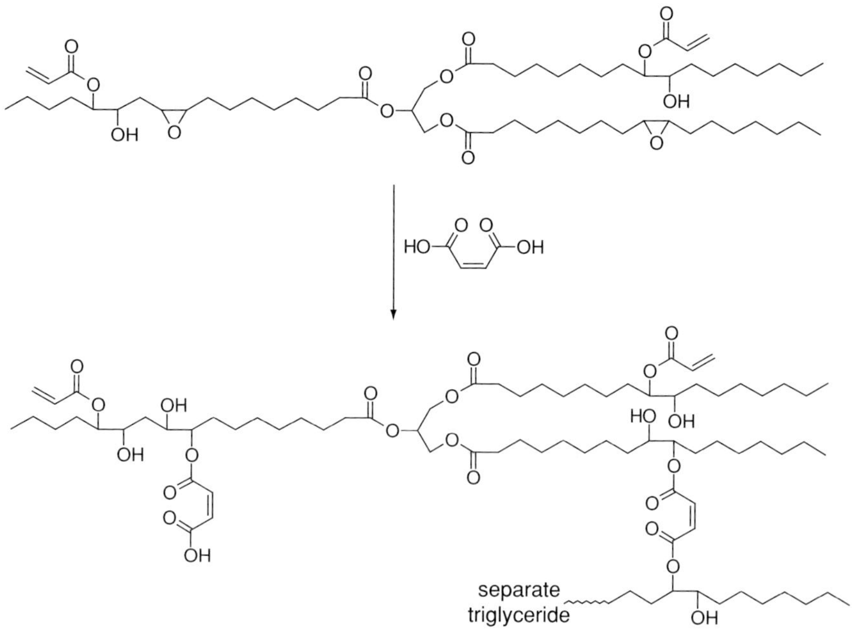

FIGURE 4.5B Modification of AESO by reaction with maleic acid.

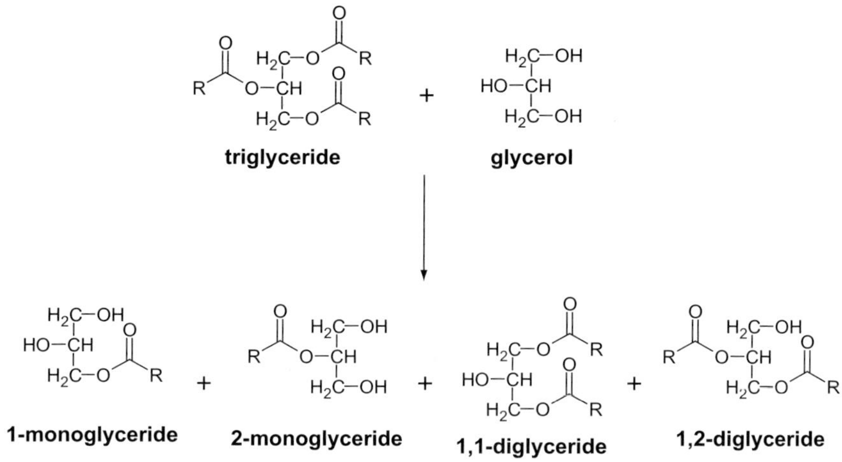

FIGURE 4.6 Glycerolysis of triglycerides to form mixtures of monoglycerides and diglycerides.

separate a portion of the unreacted glycerol by cooling the product rapidly [33]. The presence of glycerol is not detrimental to the end polymer, because it can be reacted with maleic anhydride in the same manner as the monoglycerides and incorporated into the end polymer network.

The maleinization of the soybean oil monoglyceride (SOMG) mixture at temperatures below 100 °C produces monoglyceride, diglyceride, and glycerol maleate half-esters. This reaction makes no attempt to produce a polyester, and the half-ester formation is expected to proceed at low temperatures in the presence of either acid or base catalysts without any by-products. A good indication of the success of this reaction is to follow the signal intensity ratio of maleate vinyl protons to fatty acid vinyl protons (N_M/N_{FA}) in the ^{1}H-NMR spectrum. The use of 2-methylimidazole and triphenyl antimony as catalysts was shown to be successful when conducting the reaction at temperatures of 80–100 °C with a 3:2 weight ratio of glycerides to maleic anhydride ($N_M/N_{FA} = 0.85$) [33, 51]. Once these maleates have been added, the monoglycerides can react via addition polymerization. Because maleates are relatively unreactive with each other, the addition of styrene increases the polymerization conversion and imparts rigidity to the matrix.

To increase the glass transition temperature (T_g) and modulus of the SOMG/MA polymer, more rigid diols can be added during the maleinization reaction. Two such diols are neopentyl glycol (NPG) and bisphenol A (BPA), which may increase the rigidity of the end polymer network. Although their addition to the maleinization mixture will reduce the renewable resource content of the final resin, they should result in higher T_g values for the end polymer. The synthesis of maleate half-esters of organic polyols, including NPG and BPA and the cross-linking of the resulting maleate half-esters with a vinyl monomer such as styrene, has been reported [52, 53]. The literature also abounds with examples of unsaturated polyesters prepared from NPG and maleic anhydride with other polyols and diacids [54–57]. However, the copolymers of NPG and BPA bis maleate half-esters with SOMG maleate half-esters are new.

Here we present the properties of the SOMG/MA polymer as well as the effect of adding NPG and BPA on the mechanical properties of the final polymers. For this purpose, mixtures of SOMG/NPG and SOMG/BPA, prepared at the same weight ratio, were maleinized, and the copolymers of the resulting maleates with styrene were analyzed for their mechanical properties and compared to those of SOMG maleates.

4.2.3 MALEINIZED HYDROXYLATED OIL

Maleinized hydroxylated oil (HO/MA) is synthesized in a manner similar to both the AESO and the SOMG/MA monomers. The double bonds of an unsaturated oil are used to attach the polymerizable groups by converting the double bonds of the triglyceride to hydroxyl groups. The hydroxyls can then be used to attach maleates. As shown in Figure 4.2, there are two routes to synthesize the hydroxylated triglyceride. The first is through an epoxidized intermediate. By reacting the epoxidized triglyceride with an acid, the epoxies can be easily converted to hydroxyl groups [29, 58]. Alternatively, the hydroxylated oil can be synthesized directly from the unsaturated oil, as described in [0]. After hydroxylation, the oil can be reacted with maleic anhydride to functionalize the triglyceride with maleate half-esters. A molar ratio of 4:1 anhydride to triglyceride was used in all cases, and the reaction was catalyzed with *N,N*-dimethylbenzylamine. Once the maleinization reaction is finished, the monomer resin can be blended with styrene similar to the other resins presented here.

4.2.4 SOPERMA: SOYBEAN OIL PENTAERYTHRITOL GLYCERIDE MALEATES

In the preceding sections, we reported the preparation of soybean oil monoglyceride maleates (SOMGMA) by E. Can [51]. Soybean oil was reduced to monoglycerides through a glycerolysis reaction, and the glycerolysis product, which was a mixture of mono- and diglycerides as well as unreacted triglycerides and free glycerol, was reacted with maleic anhydride to convert the free hydroxyls to maleate half-esters, thus allowing them to free radically polymerize. The use of pentaerythritol instead of glycerol in the same synthetic route, as shown in Figure 4.7, offers certain advantages, such as the presence of more hydroxyl groups and, therefore, more reactive sites for malination, which should result in a higher cross-link density for the resulting polymers. There are no standard conditions for the alcoholysis of the

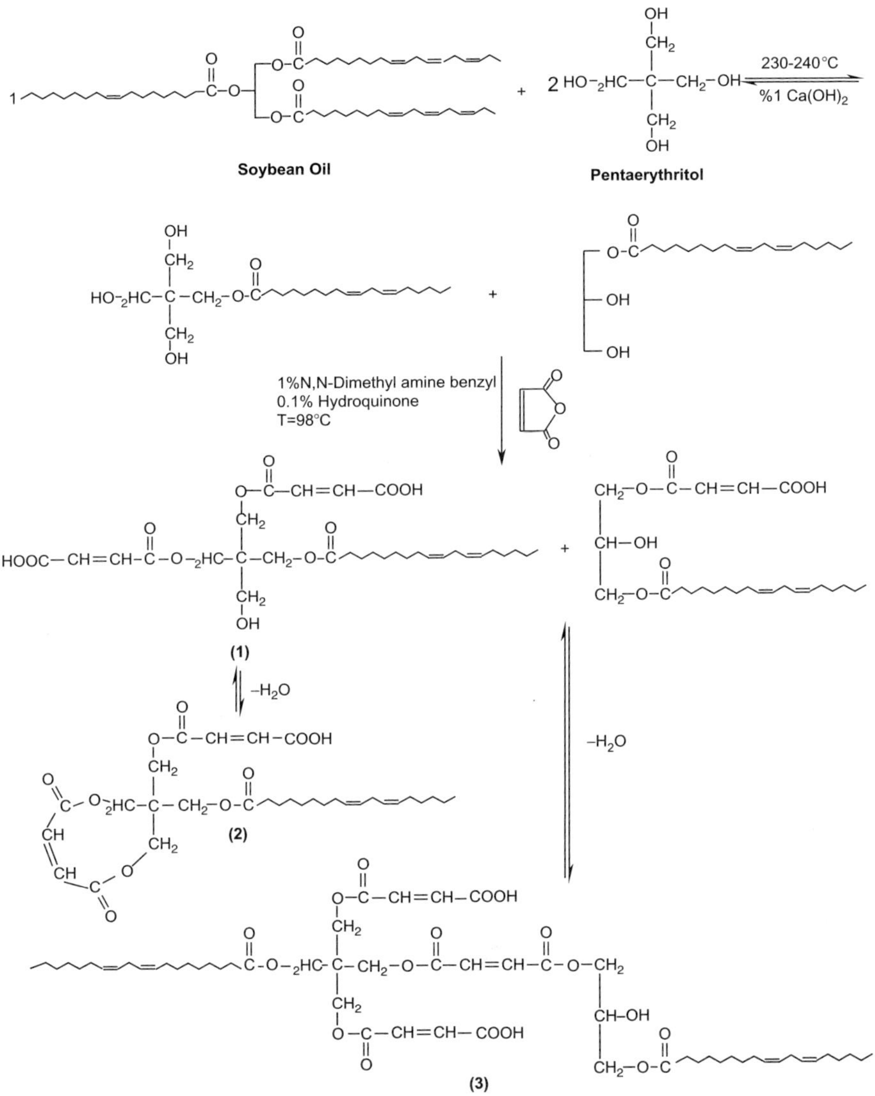

FIGURE 4.7 The reaction scheme for soybean oil pentaerythritol alcoholysis and malination reaction (SOPERMA).

triglyceride oils with pentaerythritol; the reactant molar ratios and reaction conditions change according to the end use of the product. We used soybean pentaerythritol molar ratios of SO:PER = 1:2 and SO:PER = 1:3, which should give mixtures of monoglycerides and pentaerythritol monoesters as the main products. Higher amounts of pentaerythritol were avoided because this would further decrease the triglyceride content of the formulation. The idealized reaction scheme for the soybean oil pentaerythritol alcoholysis reaction is shown in Figure 4.7.

The soybean oil pentaerythritol alcoholysis reactions were carried out for the first time by E. Can [65, 66] at 230–240 °C for 0.5, 2, and 5.5 h. A reaction temperature of 180–190 °C was also employed. $Ca(OH)_2$ was used as a catalyst at a concentration of 1% of the total weight of the oil and the polyol. $Ca(OH)_2$ forms soap with the free fatty acids in the oil and promotes the reaction at least in part by increasing the solubility of pentaerythritol in the oil. $Ca(OH)_2$ was reported to be an effective catalyst for the glycerolysis of soybean oil; it increases the monoglyceride yield and reduces the triglyceride content of the glycerolysis product. The amount of reactants and catalysts used for the soybean oil pentaerythritol alcoholysis reactions and the malination reactions, at different mole ratios, are shown in Table 4.2.

For the malination of the alcoholysis products, maleic anhydride was in a 1:1 molar ratio with the number of hydroxyls on pentaerythritol used in the alcoholysis reaction, thus molar ratios of SO:PER:MA = 1:2:8 and 1:3:12 were employed. A molar ratio of SO:PER:MA = 1:3:7.62 was also used to reduce the unreacted maleic anhydride content of the latter reaction. A reaction temperature of 95–100 °C was used, because when lower reaction temperatures were used, for example, 60 °C, the reaction rate was significantly lower and therefore not preferred. Temperatures above 100 °C led to gelation of the product due to polyesterification of both the maleate half-esters and maleic anhydride with the free hydroxyls and were therefore avoided. Table 4.2 shows the amounts of the reactants used for the malination reactions done at different SO:PER:MA molar ratios. The product at room temperature was a light brown solid. The SO:PER:MA(1:3:12) and SO:PER:MA(1:2:8) products were prepared in a similar manner by changing the molar ratios of the reactants in the formulation.

TABLE 4.2 The amount of ingredients used in the alcoholysis of soybean oil with pentaerythritol and the resulting malination reactions.

	SO:PER:MA (1:2:8)		SO:PER:MA (1:3:12)		SO:PER:MA (1:3:7.62)	
Reactants	Weight of Reactant (g)	Moles of Reactant	Weight of Reactant (g)	Moles of Reactant	Weight of Reactant (g)	Moles of Reactant
Soybean oil	5	0.0057	5	0.0057	5	0.0057
Pentaerythritol	1.554	0.01143	2.331	0.01714	2.331	0.01714
$Ca(OH)_2$	0.0655	—	0.0733	—	0.0733	—
Maleic anhydride	4.479	0.04571	6.720	0.06857	4.267	0.0435
N,N-Dimethylbenzylamine	0.1103	—	0.1405	—	0.116	—
Hydroquinone	0.011	—	0.014	—	0.0116	—

4.2.5 SOGLYME: SOYBEAN OIL MONOGLYCERIDE METHACRYLATES

Soybean oil monoglyceride methacrylates (SOGLYME) were synthesized by E. Can [65, 66]). Soybean oil monoglyceride methacrylates were prepared in a two-step process. First soybean oil was glycerolized in the presence of $Ca(OH)_2$ as a catalyst at 230–240 °C for 5 h. The glycerolysis of soybean oil under these conditions gives a product with an equilibrium mixture containing the monoglycerides, diglycerides, and the two starting materials (E. Can [51]). The glycerolysis product was then reacted with methacrylic anhydride at 55 °C to form the methacrylate esters of the glycerides and methacrylic acid. Pyridine, which is an effective catalyst in the reaction of methacrylic anhydride with alcohols, was used as the catalyst. Hydroquinone was used to inhibit the radical polymerization of the reactive methacrylate esters. The idealized reaction schemes for the glycerolysis and methacrylation are shown in Figure 4.8.

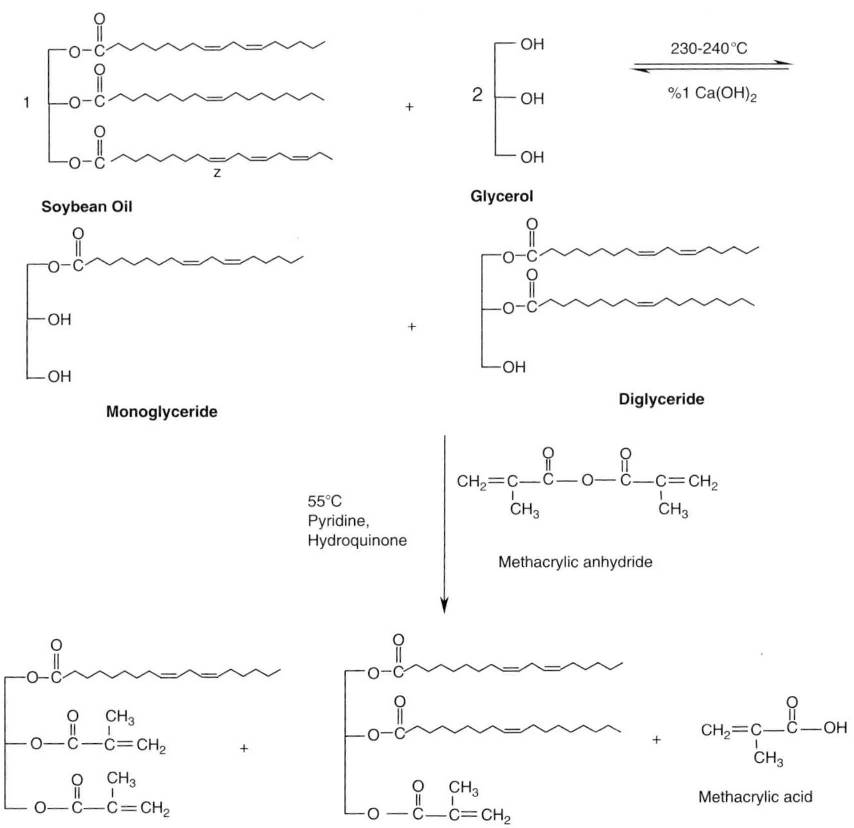

FIGURE 4.8 The reaction scheme for soybean oil glycerolysis and methacrylation reactions (SOGLYCMA).

The glycerolysis of soybean oil was carried out in a 1:2.4 molar ratio of soybean oil to glycerol. For the following methacrylation reaction, methacrylic anhydride was in a 1:1 molar ratio with the number of hydroxyls on glycerol used in the glycerolysis reaction; thus the molar ratio of soybean oil, glycerol, and methacrylic anhydride was SO:GLYC:ME = 1:2.4:7.2. The amounts of reactants and catalysts used for the soybean oil glycerolysis and methacrylation reactions are shown in Table 4.3. The resulting SOGLYME product was a light yellow liquid.

4.2.6 COPERMA: CASTOR OIL PENTAERYTHRITOL GLYCERIDE MALEATES

Castor oil is not commonly used in alkyd resin formulations and there are few reports on the alcoholysis of castor oil triglycerides. For the preparation of castor oil-based monomers, castor oil was first alcoholized with glycerol, pentaerythritol, and an aromatic diol; BPA propoxylate and the alcoholysis products were then malinated, as shown in Figure 4.9 (Can [66]). Bisphenol A propoxylate was used specifically to introduce the rigid aromatic rings onto the triglyceride structure. The maleate esters of castor oil alcoholysis products have never been synthesized before; thus the castor oil-based monomers presented here are totally new resins. The alcoholysis reactions of castor oil were carried out for 2 h at 230–240 °C in the presence of $Ca(OH)_2$ as catalyst, similar to the soybean oil alcoholysis reactions. The malination reactions were carried out for 5 h at 98 °C to ensure the completeness of the malination of the secondary hydroxyls of castor oil. *N,N*-Dimethylbenzylamine, which is reported to be an effective catalyst for the malination of hydroxylated oils, was used as a catalyst. Castor oil was also directly malinated to see the effect of the alcoholysis step on the mechanical properties of the resulting polymers.

The molar ratio of castor oil to maleic anhydride was 1:3 for malination of castor oil; therefore, the reaction was carried out in an excess of maleic anhydride assuming that 1 mol castor oil contains 2.7 mol of hydroxyls.

TABLE 4.3 Reactant amounts used in soybean oil glycerolysis and methacrylation reactions (SO:GLYC:ME = 1:2.4:7.2).

Component	Weight (g)	Moles
Soybean oil	119.93	0.1371
Glycerol	30.30	0.329
Calcium hydroxide	0.7514	—
Methacrylic anhydride	152.14	0.987
Pyridine	3.024	—
Hydroquinone	0.302	—

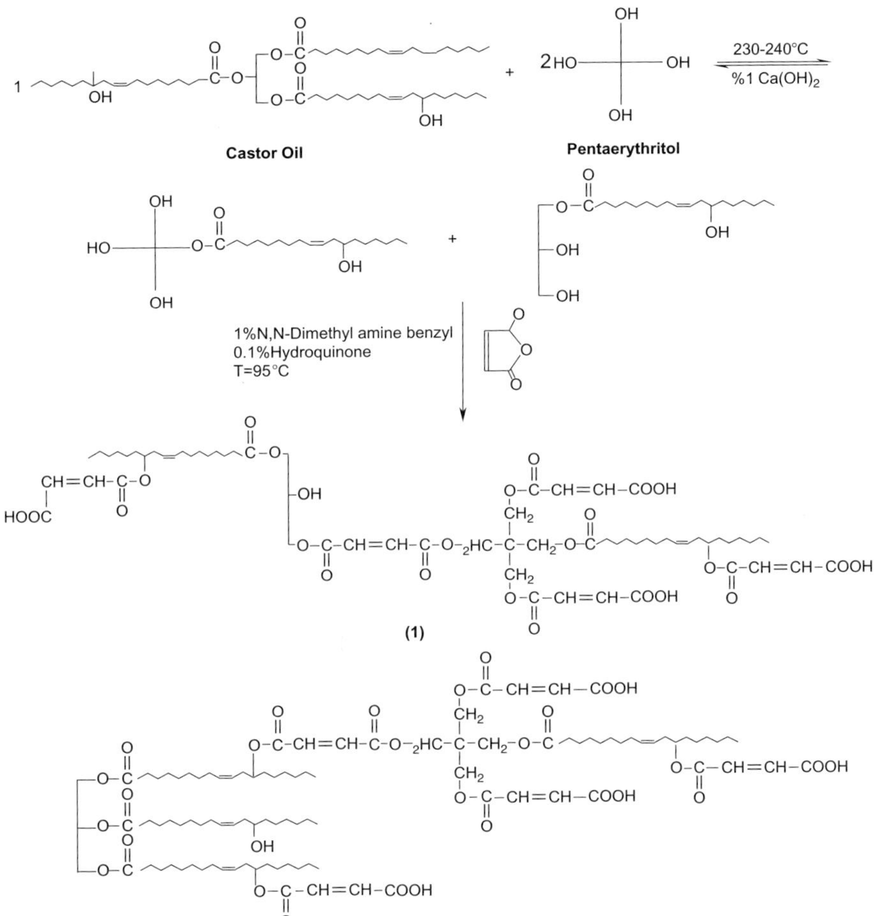

FIGURE 4.9 The reaction scheme for castor oil pentaerythritol alcoholysis and malination reactions (COPERMA).

The reactants used in this reaction as well as their mole numbers and masses are given in Table 4.4. The COPERMA product was a light brown solid.

4.2.7 COGLYMA: CASTOR OIL MONOGLYCERIDE MALEATES

The castor oil glycerolysis reaction was carried out in a molar ratio of CO:GLY = 1:2.2 using reactants shown in Table 4.5. The reaction mixture was heated to 230–240 °C and agitated under N_2 atmosphere for 2 h at this temperature. The reaction product at room temperature was a light brown liquid. The idealized structures of both the reactants and products for the

TABLE 4.4 Reactant amounts used in castor oil pentaerythritol alcoholysis and malination reactions (CO:PER:MA = 1:2:10.7).

Component	Weight (g)	Moles
Castor oil	120	0.13
Pentaerythritol	35.33	0.26
Calcium hydroxide	0.778	
Maleic anhydride	178.99	1.826
N,N-Dimethylbenzylamine	3.35	
Hydroquinone	0.335	

TABLE 4.5 Reactant amounts used in castor oil glycerolysis and malination reactions (CO:GLY:MA = 1:2.2:9.3).

Component	Weight (g)	Moles
Castor oil	100	0.108
Glycerol	21.9	0.238
Calcium hydroxide	1.22	—
Maleic anhydride	98.08	1.001
N,N-Dimethylbenzylamine	2.20	—
Hydroquinone	0.220	—

glycerolysis reaction are shown in Figure 4.10. Both unreacted castor oil and excess glycerol exist as by-products in the reaction.

For the malination of the castor oil glycerolysis product, maleic anhydride was used in a molar ratio sufficient to malinate the hydroxyls of both castor oil and the glycerol used in the glycerolysis reaction. Thus the molar ratio of castor oil (CO), glycerol (GLY), and maleic anhydride (MA) was CO:GLY:MA = 1:2.2:9.3. The castor oil glycerolysis product as prepared above was heated to 90 °C with mechanical stirring, then the specified amounts of maleic anhydride and hydroquinone were added. The mixture was stirred at this temperature until the maleic anhydride melted and mixed with the castor oil glycerolysis product. *N,N*-Dimethylbenzylamine was added and the reaction mixture was heated to 98 °C. The mixture was agitated at this temperature for 5 h. The reaction product at room temperature was a light brown solid.

4.2.8 COBPPRMA: CASTOR OIL BISPHENOL A PROPOXYLATE GLYCERIDE MALEATES

The castor oil (CO), pentaerythritol (PER) alcoholysis reaction was carried out in a molar ratio of CO:PER = 1:2 (E. Can [66]). The reaction product at room temperature was a light brown liquid. The idealized structures of both

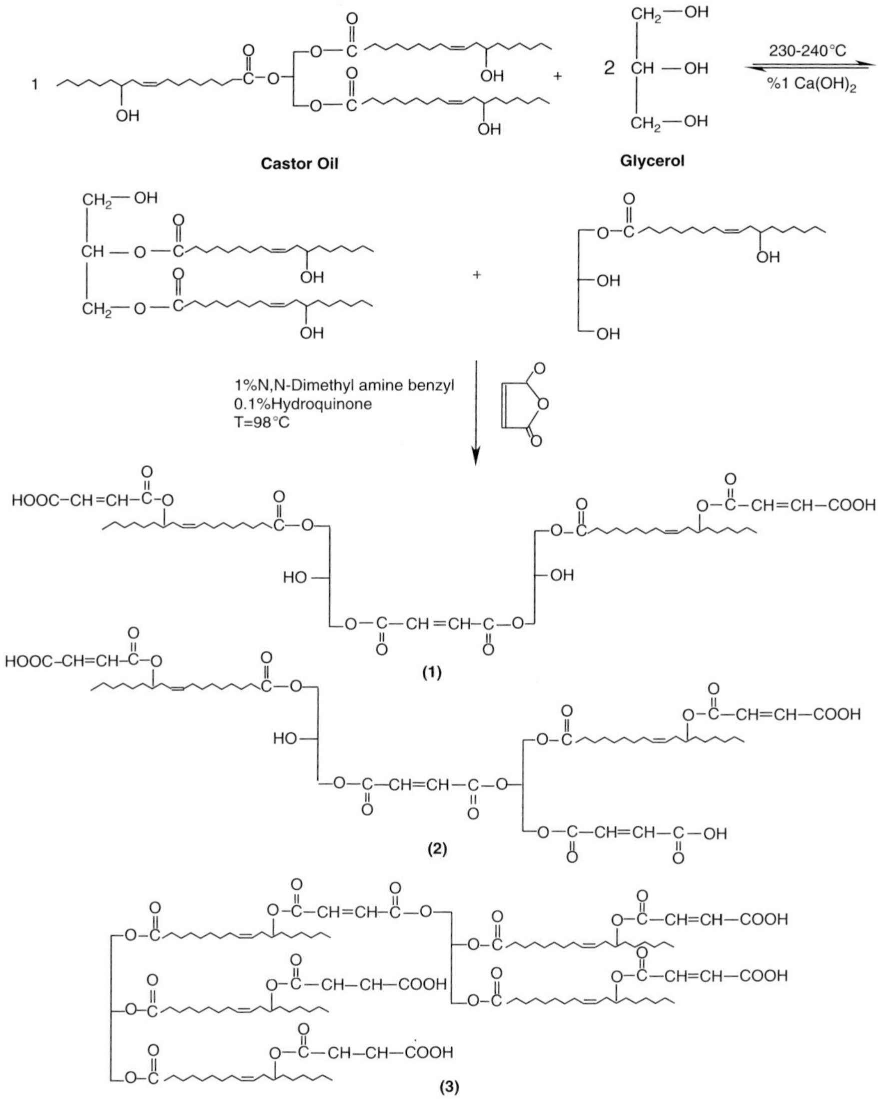

FIGURE 4.10 The reaction scheme for castor oil glycerolysis and malination reactions (COGLYMA).

the reactants and products for the alcoholysis reaction are shown in Figure 4.11. Both unreacted castor oil and excess glycerol exist as by-products in the reaction. For the malination of the castor oil pentaerythritol alcoholysis product, maleic anhydride was used in a molar ratio to malinate the hydroxyls of both castor oil and the pentaerythritol used in the alcoholysis reaction.

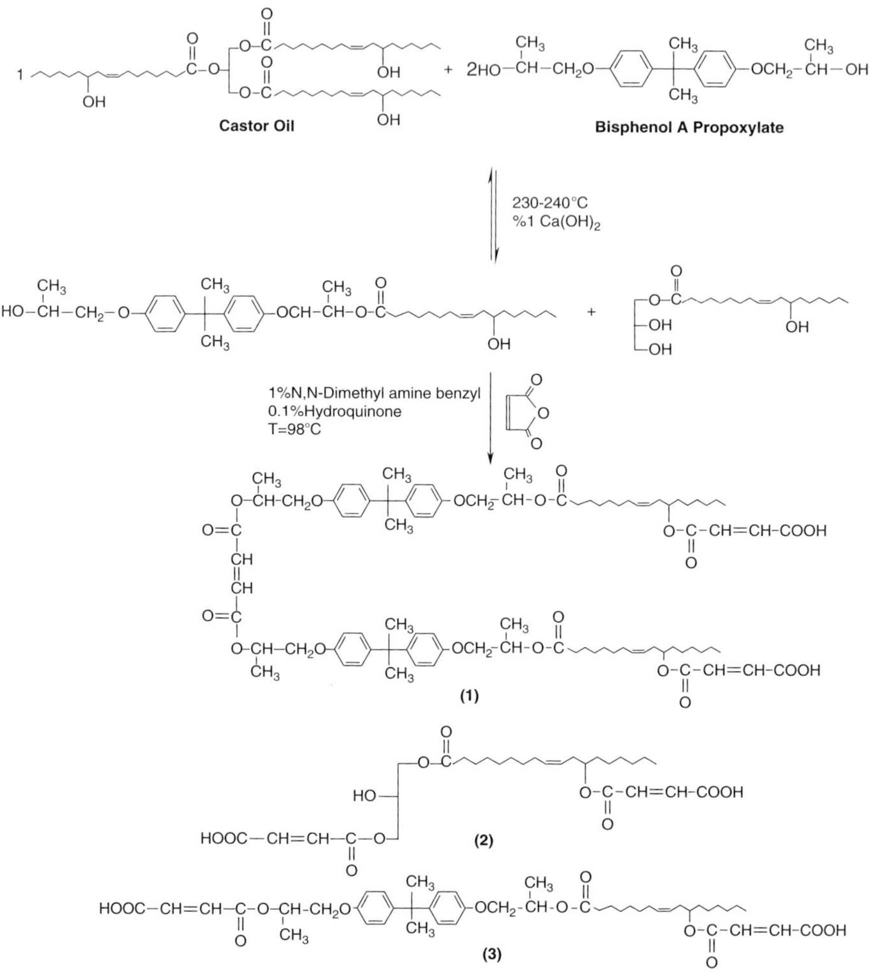

FIGURE 4.11 The reaction scheme for castor oil bisphenol A propoxylate alcoholysis and malination reactions (COBPAPRMA).

Thus, the molar ratio of castor oil (CO), pentaerythritol (PER), and maleic anhydride (MA) was CO:PER:MA = 1:2:10.7. The malinated product at room temperature was a light brown solid.

4.2.9 COMA: CASTOR OIL MALEATES

COMA consists of the castor oil (CO) which was directly maleated on the 3 hydroxyl groups using maleic anhydride (MA). The ratios of reactants used are shown in Table 4.6 (E. Can [66]).

TABLE 4.6 Reactant amounts used in malination of castor oil (CO:MA = 1:3).

Component	Weight (g)	Moles
Castor oil	100	0.108
Maleic anhydride	31.82	0.325
N,N-Dimethylbenzylamine	1.32	—
Hydroquinone	0.132	—

4.3 POLYMERS FROM PLANT OILS

4.3.1 ACRYLATED EPOXIDIZED SOYBEAN OIL POLYMERS

Acrylated epoxidized soybean oil (AESO), shown in Figure 4.4, was examined for its ability to produce high T_g and high modulus polymers. A commercial form of AESO, Ebecryl 860, was blended with various amounts of styrene to determine the effect of blending on mechanical and dynamic mechanical properties. The AESO used had an average functionality of approximately 3 acrylates per triglyceride as determined by ^{1}H NMR [38]. The optimal number of acrylates per triglyceride to obtain maximum stiffness and strength is about 5 acrylic acid groups per triglyceride, as discussed in Chapter 8. An example ^{1}H NMR spectrum of AESO is shown in Figure 4.12. Similar to soybean oil, the triplet peak at 2.3 ppm can be used as a basis for the protons to present alpha to the carbonyls in the triglyceride. The three peaks in the range of 5.8 to 6.5 ppm represent the three protons of the acrylate group.

Styrene monomer was blended with the AESO along with a free-radical initiator, 2,5-dimethyl-2,5-di(2-ethylhexanoyl peroxy) hexane. The addition of styrene to any type of unsaturated polyester is common practice in the composite liquid molding resin field. Its low cost and low viscosity improve the price and processability of the resin. For triglyceride-based polymers, the styrene also imparts a rigidity that the triglyceride does not naturally possess. The amount of initiator used was 1.5 wt% of the total resin weight (AESO plus styrene). For tensile testing of the polymers, samples were prepared in accordance with ASTM D 638. The resin was cured at 60 °C for 12 h, followed by 125 °C for 1.5 h. Samples for dynamic mechanical analysis (DMA) testing were prepared by pouring resin into a rubber gasket between two metal plates covered with aluminum foil. Samples were cured at 65 °C for 1.5 h and postcured at 125 °C for 1.5 h.

4.3.2 SYNTHESIS OF MODIFIED ACRYLATED EPOXIDIZED SOYBEAN OIL POLYMERS

To improve the properties of the AESO-based resins, modified forms of the AESO were synthesized. These modifications involved partially reacting

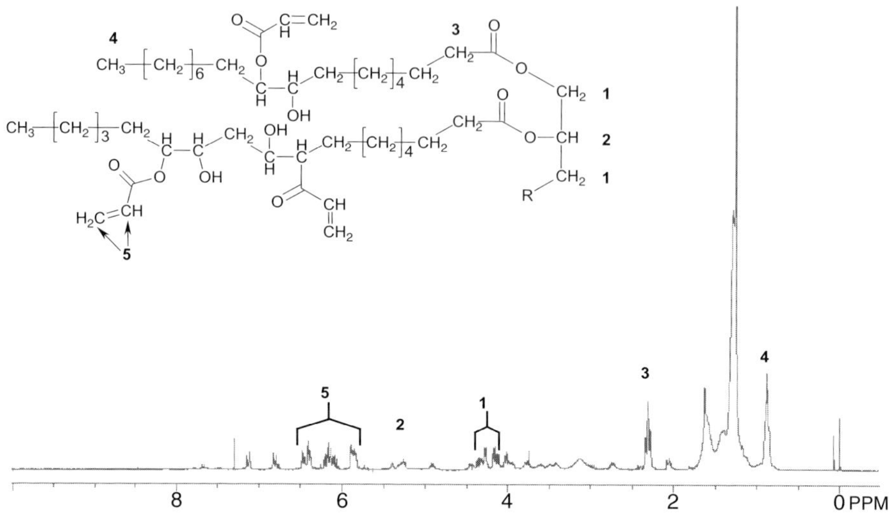

FIGURE 4.12 ^{1}H NMR spectrum for acrylated epoxidized soybean oil (Ebecryl 860, UCB Chemicals Co.).

epoxidized soybean oil with acrylic acid and reacting the remaining epoxies with anhydrides or diacids. A more detailed explanation of the synthesis of partially acrylated epoxidized soybean oil can be found in other sources [38]. In summary, a mixture of epoxidized soybean oil was mixed with a stoichiometric amount of acrylic acid (about 1500 g ESO to 460 g acrylic acid). Hydroquinone was added as a free-radical inhibitor in the amount of 0.07 wt% of the total reactants' weight, as well as 1,4-diazobicyclo[2.2.2]octane to act as a catalyst in the amount of 0.1 wt% of the total reactants' weight. This was reacted at 95 °C for about 11 h, after which it was allowed to cool to room temperature. The resulting product had approximately 1.7 acrylates/triglyceride and 0.4 residual epoxy/triglyceride according to ^{1}H NMR. The remaining 2.3 epoxies were lost to epoxy homopolymerization [38].

The first modification was the reaction of AESO with cyclohexane dicarboxylic anhydride (CDCA), as illustrated earlier in Figure 4.5A. In a typical reaction, the synthesized AESO was reacted with 7.4% of its weight in CDCA and 0.1% of its weight in 2-methyl imidazole, which catalyzes the reaction [38]. After reacting at 110 °C for about 3 h, the majority of the anhydride and epoxy groups was consumed, as indicated by FTIR spectroscopy. The second modification was the reaction of AESO with maleic acid (Figure 4.5B). This was accomplished by reacting the synthesized AESO with 11% of its weight in maleic acid [38]. The reaction was held at approximately 80 °C for 4 h, during which consumption of the epoxies was again confirmed by FTIR spectroscopy.

The modified resins were then blended with styrene and initiator in the amounts of 66 wt% modified AESO, 33 wt% styrene, and 1 wt% 2,5-dimethyl-2,5-di(2-ethylhexanoyl peroxy) hexane initiator. After curing at 65 °C for 1.5 h and postcuring at 125 °C for 1.5 h, the polymers' dynamic mechanical properties were analyzed and compared to the unmodified AESO resin.

4.3.3 MALEINIZED SOYBEAN OIL MONOGLYCERIDE RESIN SYNTHESIS

The maleinized soybean oil monoglyceride (Figure 4.4) was synthesized by breaking the triglycerides into monoglyceride and then functionalizing the alcohol groups with maleic anhydride. The glycerolysis reaction was done by heating the triglycerides in the presence of glycerol and a catalyst. In a typical reaction, glycerol was heated at 220–230 °C for 2 h under an N_2 atmosphere to distill off any water present [33]. The amount of soybean oil reacted with the glycerol was 4 g soybean oil to 1 g glycerol, a molar ratio of 4.75 mol glycerol to 1 mol triglyceride. The soybean oil was added in five portions to the glycerol, each portion 1 h apart. With the first portion, commercial soap was added in the amount of 1% of the total oil amount to act as an emulsifier and catalyst. The solution was heated at 230 °C under N_2 while being stirred. After 5.5 h, the reaction was immediately cooled to room temperature with an ice bath, causing glycerol to separate from the mixture. On removal of this layer, approximately 90% of the reaction solution, consisting of glycerides and glycerol, was recovered.

Maleinization of the mixture was accomplished by heating 60 g of glyceride/glycerol mixture to about 80 °C while being stirred. Maleic anhydride was then added in the amount of 40 g. As the anhydride melted, 0.6 g triphenyl antimony was added as a catalyst along with 0.01 g hydroquinone. The reaction was complete after 5.5 h, according to FTIR and ^1H NMR, resulting in a mixture of maleinized glycerides and glycerol (SOMG/MA) [33].

4.3.4 MALEINIZED SOYBEAN OIL MONOGLYCERIDE/NEOPENTYL GLYCOL RESIN SYNTHESIS

Modifying the procedure given in [53], SOMG/NPG/MA resin was synthesized as follows [51]. Forty-five grams of SOMG was placed into a 250-mL round-bottom flask equipped with a temperature controller and a magnetic stirrer and then heated to 125 °C. Fifteen grams of NPG (0.144 mol) was then added to SOMG, and as the NPG melted, 58.3 g maleic anhydride was added. As the three compounds formed a homogenous solution, 0.06 g triphenyl antimony catalyst and 0.015 g hydroquinone were added. The solution was stirred for 6.5 h at 120 °C. ^1H NMR analysis of the product showed the formation of both the SOMG and NPG maleate and later fumarate vinyl groups. The product was a light yellow viscous liquid at room temperature.

4.3.5 MALEINIZED SOYBEAN OIL MONOGLYCERIDE/BISPHENOL A RESIN SYNTHESIS

The preparation of maleates of BPA and ethylene and propylene oxide adducts of BPA was reported in [52]. For this work, SOMG and BPA were maleinized as a mixture [51]. Forty-five grams of SOMG was placed into a 250-mL round-bottom flask equipped with a temperature controller and a magnetic stirrer and heated to 125 °C. Fifteen grams of BPA (0.0657 mol) was added to the SOMG, and as BPA dissolved, 42.88 g maleic anhydride (0.4375 mol) was added. As the three compounds formed a homogenous solution, 0.6 g triphenyl antimony and 0.01 g hydroquionone were also added. The solution was then stirred for 9 h at 125 °C until maleic anhydride consumption was completed. The ^{1}H NMR analysis of the product showed the formation of both the SOMG and BPA maleate and later fumarate vinyl groups. The reaction product was an orange-colored viscous liquid (98 g) at room temperature.

4.3.6 COPOLYMERIZATION OF THE MALEATES WITH STYRENE

The copolymerization of SOMG/MA, SOMG/NPG/MA, and SOMG/BPA/MA with styrene were all run under the same conditions for comparison of the mechanical properties of the resulting polymers. For this purpose a certain weight ratio of the maleate mixture was mixed with 35% of its own weight of styrene in a closed vial. All of the maleate products were found to be soluble in styrene. *tert*-Butyl peroxy benzoate radical initiator, 2% by weight of the total mixture, was added. Nitrogen gas sparging and vacuum degassing were carried out for 5 min. The solution was then transferred to a rectangular rubber gasket mold sandwiched between two steel plates. The resin-filled mold was heated to 120 °C at a rate of 5 °C/min and was cured at this temperature for 3.5 h. It was then postcured at 150 °C for 1 h. Samples were clear, homogeneous, and free of voids or gas bubbles. The polymer samples were polished and prepared for DMA, which was conducted in a three-point bending geometry on a Rheometrics Solids Analyzer II. The temperature was ramped from 30° to 200 °C at a rate of 5 °C/min, with a frequency of 1 Hz and strain of 0.01%.

4.3.7 MALEINATED HYDROXYLATED OIL POLYMER SYNTHESIS

The HO/MA shown in Figure 4.4 uses the unsaturation of the triglyceride to incorporate polymerizable groups. This monomer was used in a series of experiments to understand how triglyceride structure can affect the synthesis and dynamic mechanical properties of the end polymer [59]. Olive oil, cotton-seed oil, soybean oil, safflower oil, linseed oil, triolein, and a genetically engineered high oleic soybean oil were converted into HO/MA resins. The levels of unsaturation for these oils are shown in Table 4.1. The fatty acid chain lengths for all of these oils are between 17.5 and 18 carbons, making the unsaturation level essentially the only difference among oils.

Hydroxylation was done by stirring the oil (~100 g) vigorously in the presence of formic acid (150 mL) and 30% (aq) hydrogen peroxide (55 mL) at 25 °C [30, 59]. The reaction time was 18 h to reach a maximum conversion of double bonds. Formic acid, peroxide, and water were then removed from the hydroxylated oil by dissolving the reaction mixture in diethyl ether and washing multiple times with water and then saturated (aq) sodium bicarbonate until a neutral pH was reached. The solution was then washed with saturated (aq) sodium chloride and dried over sodium sulfate. Finally, the ether was evaporated off under vacuum. The extent of hydroxylation can be characterized by ^{1}H NMR. An example ^{1}H NMR spectrum is presented in Figure 4.13 with corresponding peak assignments [59]. The extent of hydroxylation has a linear dependence on the level of unsaturation. Generally, for every double bond present on the triglyceride, an average of 1.6 hydroxyls can be added [59].

The purified hydroxylated oil was reacted with maleic anhydride in a ratio of 1 mol triglyceride to 4 mol anhydride. The hydroxylated oil was heated to a temperature of about 80 °C, and finely ground maleic anhydride was then added. Upon dissolving of the anhydride, *N,N*-dimethylbenzylamine was added to catalyze the reaction. The reaction was continued for 3 h, and the extent of maleinization was determined by ^{1}H NMR. An illustrative ^{1}H NMR is shown in Figure 4.14 [59]. Under these reaction conditions, the

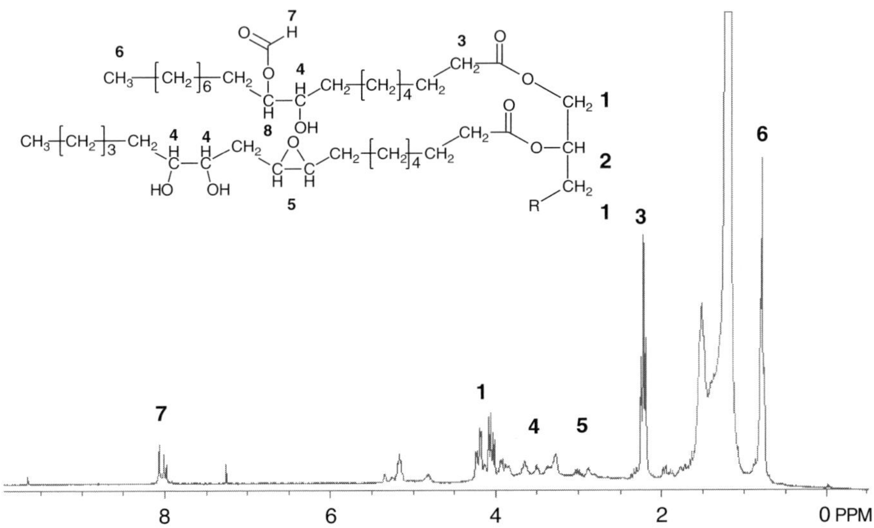

FIGURE 4.13 ^{1}H NMR of hydroxylated soybean oil. Treating the oil with formic acid and hydrogen peroxide results in conversion of the double bonds to hydroxy groups.

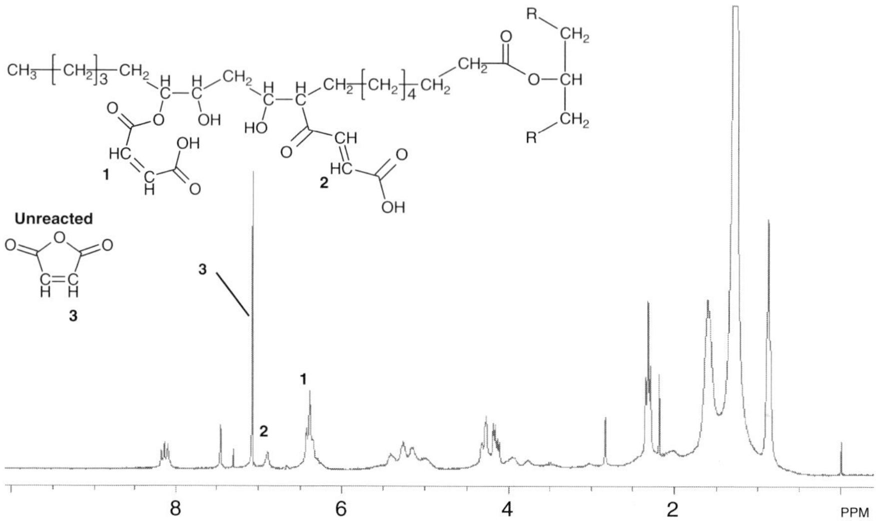

FIGURE 4.14 ^{1}H NMR of maleinized hydroxylated soybean oil. Peaks 1 and 2 represent the maleate half-esters and fumarate half-esters, respectively. Peak 3 represents unreacted maleic anhydride.

extent of functionalization plateaus in the range of 2.1 to 2.8 maleates/triglyceride for all oils [51]. Approximately 20–25% of the maleates attached to the triglycerides isomerize to form fumarate groups (*trans* confirmation). Unreacted maleic anhydride remained in the resin and was polymerized during the cure reaction. The HO/MA resins were then dissolved in styrene in a molar ratio of 7:1 styrene to HO/MA. Resins were cured using 2,5-dimethyl-2,5-di-(2-ethylhexanoylperoxy)hexane at 65 °C for 1.5 h and post-cured at 120 °C for 1 h. DMA was conducted in a three-point bending geometry on a Rheometrics Solids Analyzer II. Temperature was ramped from 30° to 175 °C at a rate of 5 °C/min, with a frequency of 1 Hz and strain of 0.01%.

4.3.8 SOPERMA AND COPERMA POLYMER SYNTHESIS

The general-purpose unsaturated polyester (UP) resin is a linear polymer with the number average molecular weights in the range of 1200–3000 g/mol. Depending on the chemical composition and molecular weight, they can be viscous liquids or solids. The plant oil-based resins we prepared are not linear polymers but similar mixtures of monomers or oligomers with different molecular weights. The number and weight average molecular weights of different species in each of these plant oil-based resins were in the range of ~300–2000 g/mol as presented in Section 4.2. With the exception of the COMA and SOGLYME resins, which were liquid at room temperature, all the malinated glyceride-based resins we prepared were pastelike solids at room temperature. The melting points of these resins were in the range of 60–70 °C, similar to the general-purpose UP resins whose melting points are in the range of 60–77 °C.

To prepare the styrenated plant oil-based resins, which were solid at room temperature, the resin was first heated in an oil bath above its melting point (~70 °C), and then the necessary amount of styrene was added and the mixture was agitated in the oil bath at a temperature of 80 °C until styrene and the resin became totally mixed. The resin was then processed at room temperature for polymerization. For the resins that were liquid at room temperature, styrene was added, mixed, and processed at room temperature. Apart from the SOGLYME resin (methacrylated soybean oil monoglycerides), all the styrenated resins showed phase separation in the microscale. Figure 4.15 shows the optical microscopic pictures of the SOPERMA-styrene mixtures at decreasing SOPERMA concentrations at 500× magnification. The SOPERMA forms nonuniform droplets in the continuous styrene matrix in the 1- to 5-μm size range. As can be seen in Figure 4.15, these droplets become less dense and more uniform in size as the concentration of SOPERMA decreases from 80 to 20 wt%. Similarly, COPERMA, COGLYMA, and COB-PAPRMA resins containing 33 wt% styrene were found to be incompatible with styrene and exhibited phase separation.

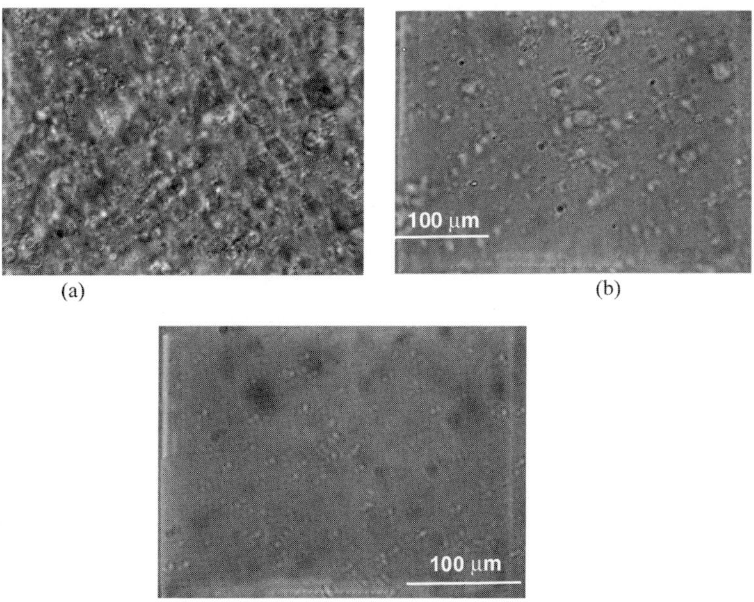

(a) (b) (c)

FIGURE 4.15 Optical microscopic pictures of the SOPERMA-styrene mixtures (500×): (a) 80 wt%, (b) 60 wt%, and (c) 20 wt% SOPERMA.

The general-purpose UP resins based on propylene glycol, phthalic, and maleic anhydride are miscible with styrene. The molecular weights of the general-purpose UP resins have a number average molecular weight of 900 g/mol and a weight average molecular weight of 2400 g with a polydispersity of 2.7. The acid number of these polyesters is around 50 mg KOH/g. The incompatibility of the malinated plant oil-based resins in styrene can be attributed to both the abundance of the acid groups of the maleate half-esters and the presence of high-molecular-weight species in these resins. The acid number of all the malinated glyceride-based resins was found to be above 200 mg KOH/g. This value is much higher than that of the general-purpose UP resins. The molecular species present in the malinated glyceride-based resins are highly polar compared to styrene, and the strong interactions of these molecules via hydrogen bonding between the acid groups result in the insolubility of the malinated glycerides in styrene. Thus, the SOGLYME resin based on methacrylated glycerides, both with the acrylic acid by-product and without acrylic acid, showed no phase separation in styrene. The methacrylate half-esters as shown earlier in Figure 4.8 do not carry acid functionality, and oligomer formation during the methacrylation reaction cannot occur. Additionally, these molecules are less polar than maleate half-esters and they cannot interact via hydrogen bonding, which favors their compatibility with styrene. The SOPERMA resin was also found to be insoluble in more polar monomers such as acrylic acid and acrylonitrile. The immiscibility of the SOPERMA resin in these solvents shows that the molecular weight and strong interactions between the malinated glyceride molecules play an important role in the immiscibility of these molecules in polar solvents.

A liquid molding resin should have properties within a certain operating range to be successfully used in molding processes. Possibly the most stringent requirement is the resin's viscosity, which must range between 200 and 1000 cP. At viscosities lower than 200 cP, air pockets will remain in the mold after injection. At viscosities greater than 1000 cP, voids may occur in the part, the time required for injection increases, and problems with fiber wetting can arise during composite preparation.

The surface free energy of a liquid, also referred to as *surface tension*, determines most of the surface and interfacial properties such as wetting, adhesion, and adsorption. Surface tension results from an imbalance of molecular forces in a liquid. At the surface of the liquid, the liquid molecules are attracted to each other and exert a net force, pulling themselves together. High surface tension values mean that the molecules tend to interact strongly; thus, polar materials show high surface energy values. The surface energy of a liquid molding resin may be especially important for reinforcement of the resins by fibers. The wetting of a fiber with a liquid resin can be judged by the difference between the surface energies of the fiber and the resin. For the most desirable condition, proper wetting and spreading in resin transfer molding processes, the surface energy of the fiber should be high, whereas the surface energy of the resin should be low. Table 4.7 shows the specific gravity, viscosity (η), and surface energy values for the plant oil-based resins and general-purpose UP resins containing 33 wt% styrene.

As can be seen from Table 4.7, apart from the COMA and the SOGLYME resins, these resins show viscosities that are in a range that is suitable for liquid molding processes. Although the COMA and SOGLYME resins show low viscosities, the viscosities of these resins can easily be increased by decreasing the amount of the styrene diluent. It was found by Can [66] that as the weight fraction of styrene was increased, the viscosity decreased in an exponential manner. This result is very desirable since it shows that a small amount of comonomer can be used to make these resin systems much easier to process.

The surface energy values of these resins show values closer to those of UP resins, and are significantly lower than those of vinyl ester resins (32–34 mN/m), which may have difficulty in wetting fiber substrates. Among the malinated resins, the surface free energy value is highest for the COPERMA resin and lowest for the COMA resin in proportion to the maleate content of the resins. A higher maleate content should result in a higher polarity and a higher surface free energy. The SOGLYME resin shows the lowest surface free energy among all the resins. The methacrylated glycerides do not carry an acid functional group as malinated glycerides and, therefore, are less polar than the corresponding maleates.

The curing of UP resin is accomplished via free-radical cross-linking polymerization between the UP molecules and styrene. The UP molecules are the cross-linkers, while the styrene acts as an agent to cross-link the adjacent polyester molecules. Similarly, in the plant oil-based resin systems, the functionalized glycerides act as the cross-linker units, and styrene is the agent that links the adjacent glyceride molecules. Styrene is the most commonly used vinyl monomer in unsaturated polyester resins due to its low viscosity, low cost, and reactivity with the unsaturated sites of polyesters. The

TABLE 4.7 The specific gravity, viscosity (η), and surface energy values for the plant oil-based resins (33 wt% styrene) and general-purpose UP resins.

Resin	Specific Gravity (g/mL)	Viscosity, η (cP)	Surface Energy (mN/m)
SOPERMA	0.94	343	27.38
COPERMA	1.06	363	28.84
COGLYMA	1.04	213	26.36
COBPAPRMA	0.98	183	27.2
COMA	0.75	92	26.02
SOGLYME	0.87	51	22.7
GP-UP	1.14	200–2000	24–30

unsaturation present on the UP backbone is very sluggish in homopolymerization. The reactivity ratio of styrene and maleic/fumaric acid esters is about 0, indicating that this system has a tendency to form alternating copolymers. Because the malinated plant oil-based resins were found to be insoluble in styrene at room temperature, it was especially important for us to determine the conversion of polymerization for the maleate and styrene monomers.

For the determination of the conversion and kinetics of the polymerization, the styrenated SOPERMA, COMA, and COPERMA resins were prepared as described earlier. The styrene concentration was 33 wt% for each resin. *tert*-Butyl peroxybenzoate was used as the initiator. The initiator concentration was 2 wt% for the SOPERMA and COMA resins and 1.5 wt% for the COPERMA resin. The curing of all resins was carried out at 120 °C for 3 h for comparison of conversion and rate of polymerization in different resins. The SOPERMA and COPERMA resins were also postcured at 150° and 160 °C, respectively, for 1 h after 2 h at 120 °C. Although all the resins showed similar conversion versus time profiles during the isothermal cure, the final conversion values were different for each resin.

Table 4.8 lists the total conversion (α) and the conversion of maleates (α_{ma}) and styrene (α_{st}) for the COMA, SOPERMA, and COPERMA resins (33 wt% styrene) at the end of 3 h at 120 °C. As can be seen, the final maleate conversion decreases as the maleate content of the resin increases from COMA to COPERMA. The styrene conversion and thus the total conversion are also considerably lower for the SOPERMA and COPERMA resins than for the COMA resin. It is expected that the molecular mobility of the resin decreases as the cross-link density increases, resulting in lower total conversion. As a result, styrene monomer as well as some resin can be trapped in the network and cannot participate in polymerization. However, although the COPERMA resin has a higher maleate content and, therefore, a higher cross-link density than the SOPERMA resin, the styrene and thus the final conversions for the SOPERMA and COPERMA resins do not show a significant difference. The maleate conversion was higher than styrene conversion for all of the resins.

TABLE 4.8 Total conversion (α) and the conversion of maleates (α_{ma}) and styrene (α_{st}) for the COMA, SOPERMA, and COPERMA resins (33 wt% styrene) at the end of 3 h at 120 °C.

Resin	α_{ma}	α_{st}	α
COMA	0.998	0.921	0.967
SOPERMA	0.979	0.828	0.886
COPERMA	0.952	0.835	0.885

The total ultimate conversion of unsaturated polyesters ranges from 0.75 to 0.9 and increases with increasing temperatures. Similarly, the final conversions for the isothermal cure of the plant oil-based resins were lower than the complete conversion. During an isothermal cure, when the increasing glass transition temperature of the resin reaches the reaction temperature and the material evolves from the rubbery state to the glassy state, the rate of propagation becomes diffusion controlled. This process, referred to as *vitrification,* may virtually terminate the polymerization, limiting the conversion that can be reached isothermally. This was true in particular for the SOPERMA and COPERMA resins, which possess glass transition temperatures (T_g) when fully cured [T_g(SOPERMA) = 139 °C and T_g(COPERMA) = 146 °C) that are higher than the cure temperature. Thus a postcure was necessary for these systems to increase the conversion.

4.4 PROPERTIES OF PLANT OIL RESINS

4.4.1 VISCOELASTIC AND MECHANICAL PROPERTIES OF AESO-STYRENE POLYMERS

The storage moduli, E', of the AESO-styrene neat polymers at various temperatures and compositions are shown in Figure 4.16. At room tempera-

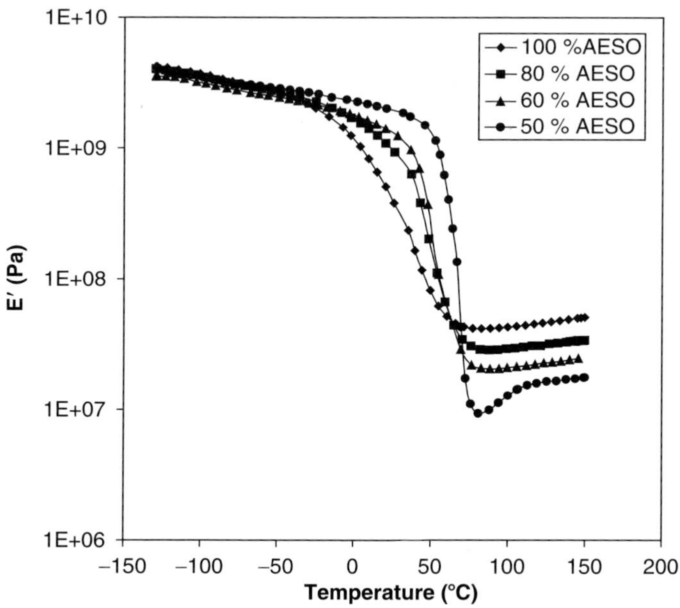

FIGURE 4.16 Storage modulus (E') of AESO-styrene copolymer as a function of temperature.

ture the polymers display moduli proportional to the amount of styrene present, which is expected from the tensile properties presented earlier. Additionally, at room temperature all of the polymers are in the transition phase from the glassy region to the rubbery plateau. Even at temperatures as low as −130 °C, it does not appear that these polymers have reached a characteristic glassy plateau. At extremely low temperatures, all compositions exhibit essentially equal moduli of about 4 GPa. At higher temperatures, the compositions show moduli inversely proportional to the amount of styrene present. According to rubber elasticity theory [60], the lower styrene content polymers have a higher cross-link density, as observed in Figure 4.16.

The T_g is often designated by either the temperature at which the dynamic loss modulus E'' value is at a peak or the temperature at which the loss tangent tan δ exhibits a peak [61]. As shown in Figure 4.17, all of the AESO-styrene copolymers exhibit two peaks in E''. A minor relaxation occurs in the range of −85° to −95 °C, showing little dependence on composition. The much larger relaxation, corresponding to the T_g, occurs in the range of −10° to 60 °C and also becomes sharper in nature with the addition of styrene. These peaks are shown in the tan δ graph in Figure 4.18. The temperature at which these peaks occur exhibits a linear dependency on composition, increasing with the amount of styrene present in the system, as illustrated by Figure 4.19.

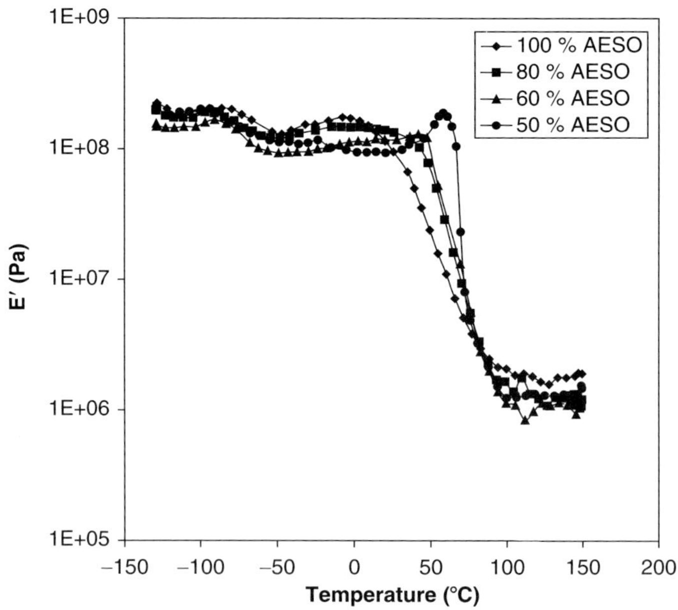

FIGURE 4.17 Loss modulus (E'') of AESO-styrene copolymer as a function of temperature.

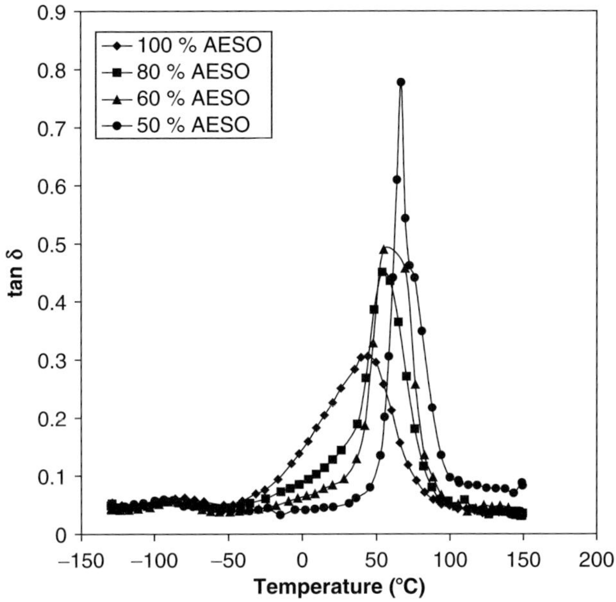

FIGURE 4.18 Damping peak (tan δ) of AESO-styrene copolymer as a function of temperature.

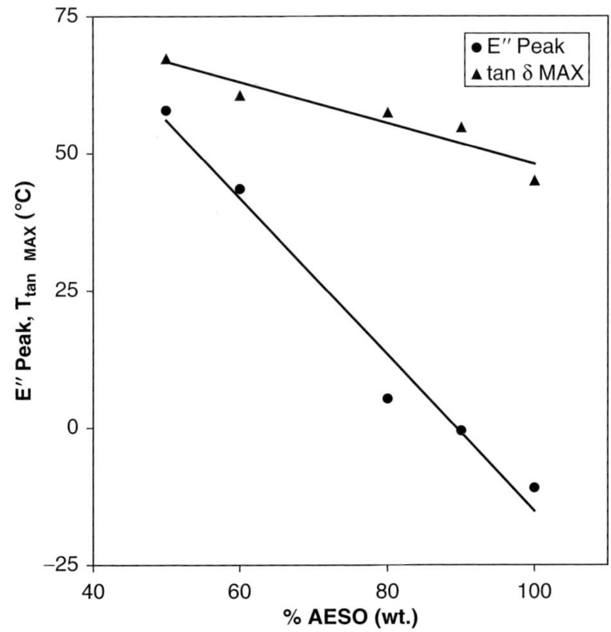

FIGURE 4.19 The E'' and tan δ peak temperatures of various compositions of AESO-styrene polymer.

The dynamic mechanical behavior just discussed is a combination of three factors: cross-link density, copolymer effects, and plasticization. As the amount of AESO increases, so does the number of multifunctional monomers. Therefore, the overall cross-link density will be greater with increasing amounts of AESO, as supported by the high-temperature moduli shown in Figure 4.16. Increasing the cross-link density slows the transition in E' from glassy to rubbery behavior. Additionally, the tan δ peak broadens and decreases in height [61]. The copolymer effect occurs frequently when there are differences in the reactivity or structure of the different monomers. If one monomer is more reactive, it is depleted faster, causing polymer formed later in the reaction to be composed mostly of the slower reacting monomer. This causes heterogeneity in the composition of the total polymer. If these monomers differ from each other in their physical properties, such as very different T_g's, a general broadening of the glass–rubber transition is frequently observed, due to this gradient [61].

The other factor in the dynamic mechanical behavior, plasticization, is due to the molecular nature of the triglyceride. The starting soybean oil contains fatty acids that are completely saturated and cannot be functionalized with acrylates. Therefore, these fatty acids act in the same manner as a plasticizer, introducing free volume and enabling the network to deform more easily. The addition of even small amounts of plasticizer to polymers was known to drastically broaden the transition from glassy to rubbery behavior and reduce the overall modulus [61]. This plasticizer effect presents an issue that may be inherent to all natural triglyceride-based polymers that use the double bonds to add functional groups. However, with advances in genetic engineering, it may be possible to reduce this trend by reducing the amount of saturated fatty acids present, thus sharpening the glass–rubber transition. This issue is addressed later in the properties of HO/MA polymers produced from genetically engineered high oleic content oil and synthetic triolein oil. The existence of some saturated fatty acids, though, can contribute to improved toughness and ballistic impact resistance [62].

4.4.2 TENSILE PROPERTIES OF AESO-STYRENE POLYMERS

The tensile moduli of three AESO-styrene copolymers at room temperature are shown in Figure 4.20. The pure AESO polymer has a modulus of about 440 MPa. At a styrene content of 40 wt% the modulus increases fourfold to 1.6 GPa. In this region, the dependency on composition appears to be fairly linear. The ultimate tensile strengths of these materials, as shown in Figure 4.21, also show linear behavior. The pure AESO samples exhibited strengths of approximately 6 MPa, whereas the polymers with 40 wt% styrene show much higher strengths of approximately 21 MPa. Therefore, it is apparent that the addition of styrene drastically improves the properties of the end resin.

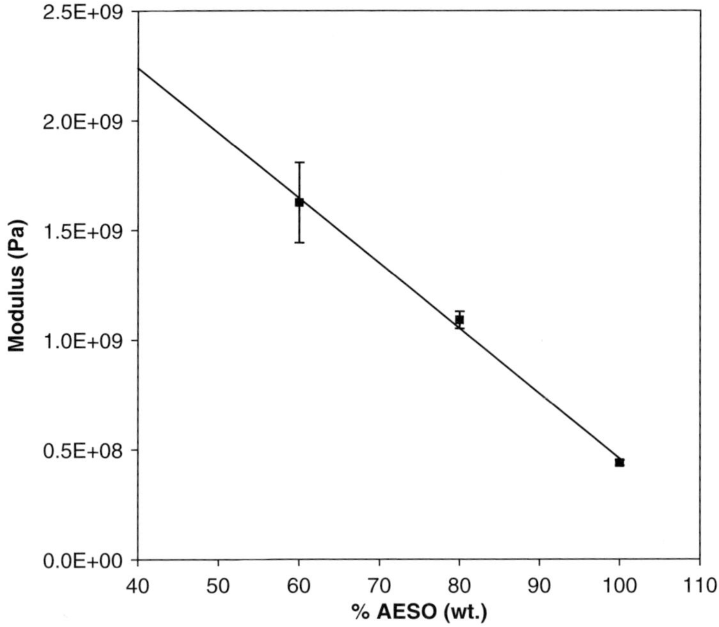

FIGURE 4.20 Tensile modulus of AESO-styrene copolymers.

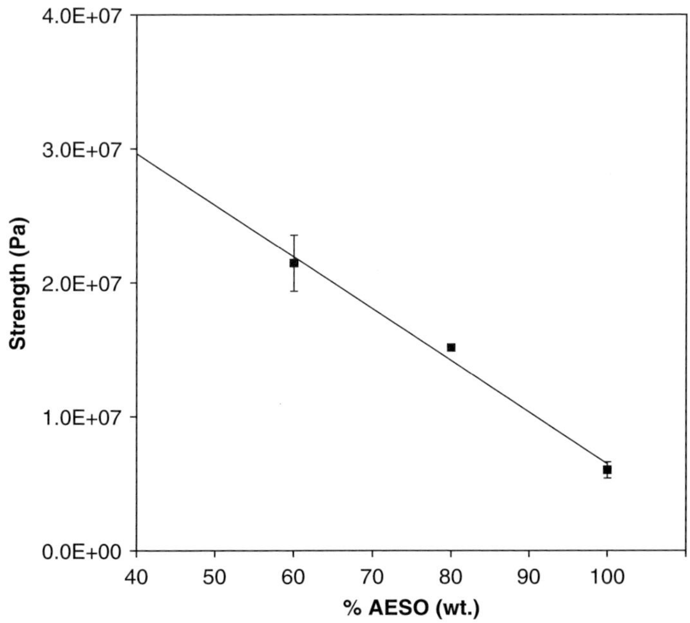

FIGURE 4.21 Ultimate tensile strength of AESO-styrene copolymers.

4.4.3 DYNAMIC MECHANICAL BEHAVIOR OF MODIFIED AESO RESINS

The dynamic mechanical properties of the AESO polymers modified by cyclohexane dicarboxylic anhydride (CDCA) and maleic acid were better than the unmodified polymers. As shown in Figure 4.22, the storage modulus at room temperature increases with both of these modifications. The storage modulus of the unmodified AESO resin at room temperature is 1.3 GPa, whereas the cyclohexane dicarboxylic anhydride modification increases the modulus to 1.6 GPa. The maleic acid modification provides the most improvement, raising the storage modulus to 1.9 GPa. The T_g, as indicated by the peak in tan δ, does not show any large increase from the anhydride modification, as shown in Figure 4.23. However, the maleic acid modification shifts the tan δ peak by almost 40 °C, showing a peak at 105 °C. The increased broadness of the peak can be attributed to increased cross-link density.

4.4.4 SOMG/MA POLYMER PROPERTIES

As seen in Figure 4.24, the tan δ peak for the SOMG/MA polymer occurs at around 133 °C, and the polymer has an E' value of approximately 0.92 GPa at room temperature. It is apparent that the glass transition is rather broad due to the broad molecular weight distribution of the SOMG maleates. The distribution of soybean oil monoglyceride monomaleates, monoglyceride

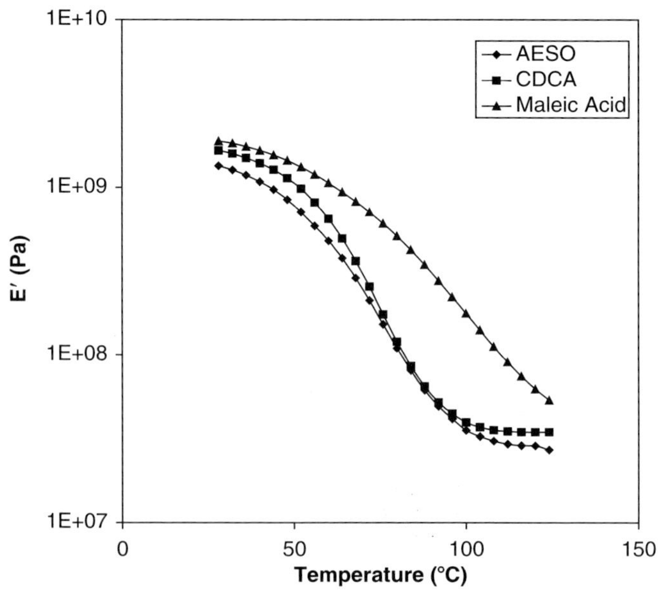

FIGURE 4.22 Storage modulus (E') of modified AESO resins as a function of temperature.

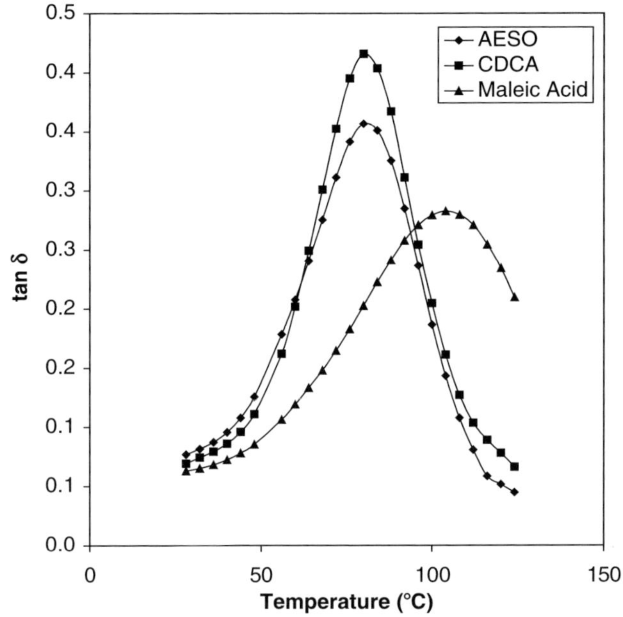

FIGURE 4.23 Damping peak (tan δ) of modified AESO resins as a function of temperature. Peaks in tan δ were found at 81 °C (CDCA modified) and 105 °C (maleic acid modified) compared to 79 °C for the synthesized AESO.

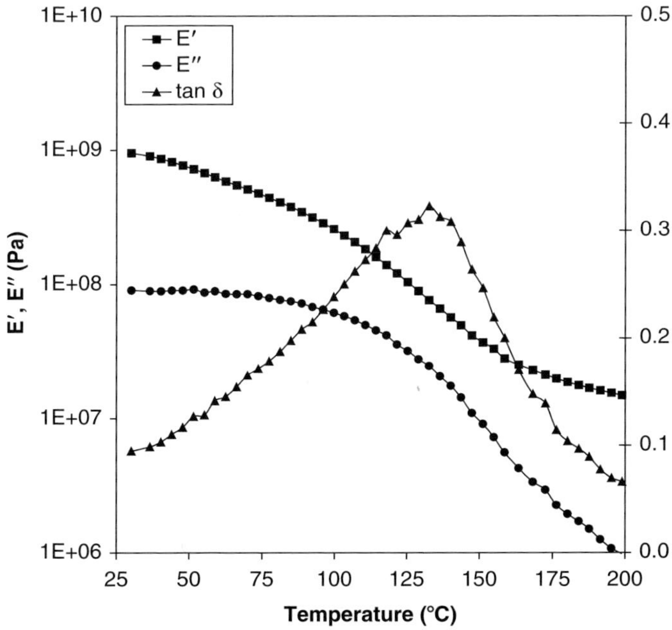

FIGURE 4.24 Dynamic mechanical behavior for SOMG/MA polymer.

bismaleates, diglyceride monomaleates, and glycerol tris maleates was confirmed by mass spectral analysis, which was reported in a previous publication [63]. The tensile tests performed on the copolymers of SOMG maleates with styrene showed a tensile strength of 29.36 MPa and a tensile modulus of 0.84 GPa as calculated from the force displacement graph.

4.4.5 SOMG/NPG MALEATES (SOMG/NPG/MA)

The DMA of SOMG/NPG/MA polymer showed a tan δ peak at approximately 145 °C and an E' value of 2 GPa at room temperature. The 12 °C increase in the T_g and the considerable increase in the modulus of the copolymers of SOMG/NPG maleates with styrene compared to that of the SOMG maleates can be attributed to the replacement of the flexible fatty acid chains by the rigid methyl groups of NPG. The overall dynamic mechanical behavior of the SOMG/NPG/MA polymer was very similar to that of the SOMG/MA shown in Figure 4.24. Despite the higher T_g and modulus, there remained a broad glass transition. The tensile strength of the SOMG/NPG/MA polymer was found to be 15.65 MPa, whereas the tensile modulus was found to be 1.49 GPa.

Maleinized pure NPG polymerized with styrene (NPG/MA) was prepared to compare its properties with the SOMG/NPG/MA polymer [51]. DMA analysis of the NPG/MA showed a tan δ peak at approximately 103 °C and an E' value of about 2.27 GPa at 35 °C. The high T_g value observed for the SOMG/NPG/MA system (~145 °C) is attributed to a synergistic effect of both the NPG and SOMG together since the T_g value observed for the NPG/MA system (~103 °C) is much lower. This is probably due to the incorporation of the fatty acid unsaturation into the polymer in the SOMG/NPG/MA system. On the other hand, the comparatively higher E' value observed for the NPG maleates explains the increase in the E' observed for the SOMG/NPG/MA system compared to that of the SOMG/MA system. The decrease in tensile strength of the SOMG/NPG/MA system compared to that of SOMG/MA may be attributed to a broader molecular weight distribution of this system compared to that of the SOMG maleates.

4.4.6 SOMG/BPA MALEATES (SOMG/BPA/MA)

The DMA of this polymer showed a tan δ peak at around 131 °C and an E' value of 1.34 GPa at 35 °C. The introduction of the rigid benzene ring on the polymer backbone considerably increased the modulus of the final polymer compared to that of the SOMG maleates. The T_g of this polymer, however, was not very different from that of the SOMG maleates (133 °C). This was attributed to a lower yield in the maleinization of the BPA, as determined from ^{1}H NMR data [51]. Like the SOMG/NPG/MA polymer, the SOMG/BPA/MA displayed the characteristic gradual glass transition shown in Figure 4.24.

4.4.7 HO/MA DYNAMIC MECHANICAL POLYMER PROPERTIES

The dynamic mechanical properties of the HO/MA polymers were found to be better than those of the AESO polymers. Little variation was seen between the polymers made from the different oils. At room temperature, the E' for all of the oils existed between 1.45 and 1.55 GPa, showing no dependence on saturation level. The dynamic mechanical behavior was similar between the different oils, with the typical behavior being shown in Figure 4.25. The temperatures at which a maximum in tan δ was exhibited ranged from 107° to 116°C, which are all substantially higher than the AESO base resin. These properties are fairly close to those shown by conventional petroleum-based polymers. However, the distinctive triglyceride behavior still exists, in that the glass transitions are extremely broad and that, even at room temperature, the materials are not completely in a glassy state. Again this is probably due to the saturated fatty acids of the triglycerides that act as a plasticizer.

Although the extent of maleinization was approximately the same from oil to oil, it is possible to see how the slight differences affect the T_g. In Figure 4.26,

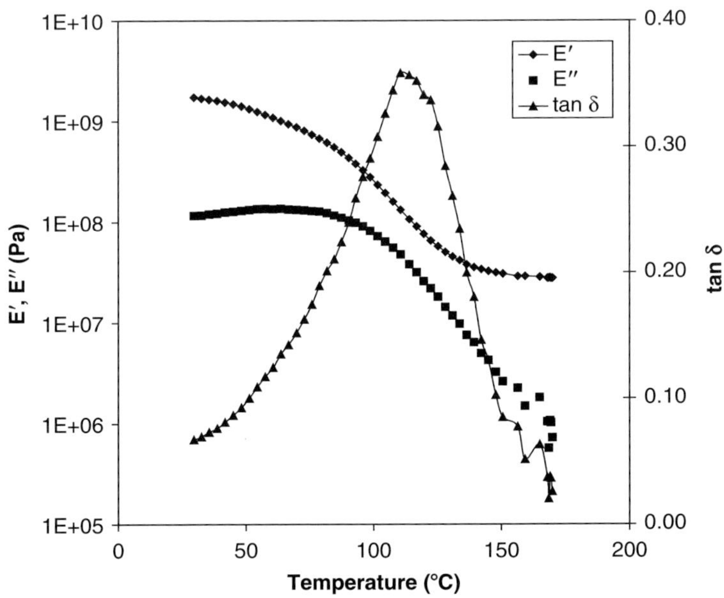

FIGURE 4.25 Representative dynamic mechanical behavior for HO/MA polymers.

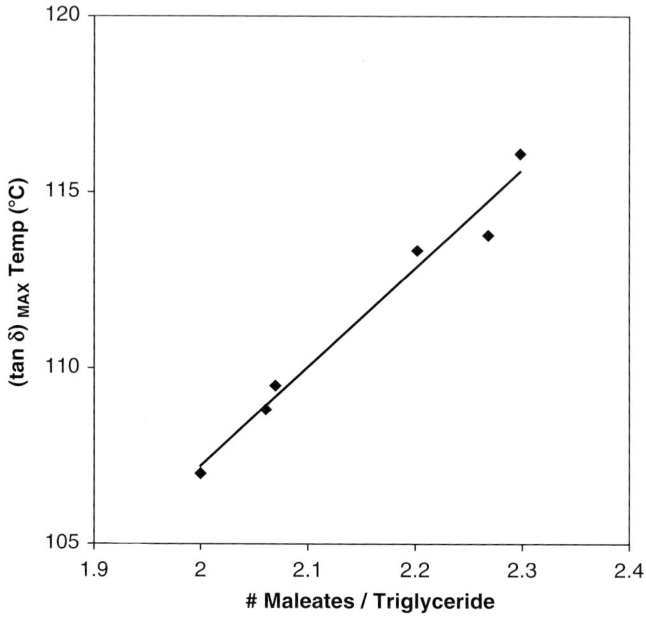

FIGURE 4.26 Peak in tan δ as a function of maleate functionality.

the tan δ peak temperature was plotted as a function of maleate functionality. Within this range, the behavior is linear, suggesting that if higher levels of functionalization are able to be reached, the properties should improve accordingly [64]. However, it is expected that past a certain extent of maleate functionality, the tan δ peak temperature dependence will plateau. Work is currently being pursued to test the limits of this behavior.

It was previously stated that the broadness in the glass transition may be inherent to all triglyceride-based polymers. However, work with genetically engineered oil and synthetic oil has shown that it is possible to reduce this characteristic. The genetically engineered high oleic soybean oil has an average functionality of three double bonds/triglyceride and has the fatty acid distribution shown in Table 4.1. The maleinized form of this oil had a functionality of two maleates/triglyceride. The properties of polymers from this material were compared to polymers from triolein oil, which is monodisperse, consisting only of oleic fatty acid esters (18 carbons long and one double bond). The maleinized triolein oil had a functionality of 2.1 maleates/triglyceride. Thus, the only difference between the two oils is the fatty acid distribution of the high oleic oil versus the monodisperse triolein oil.

The dynamic mechanical properties of polymers made from these oils are shown in Figure 4.27. The T_g of these two polymers does not seem to differ much, judging from either their tan δ peak or the inflection in the E'.

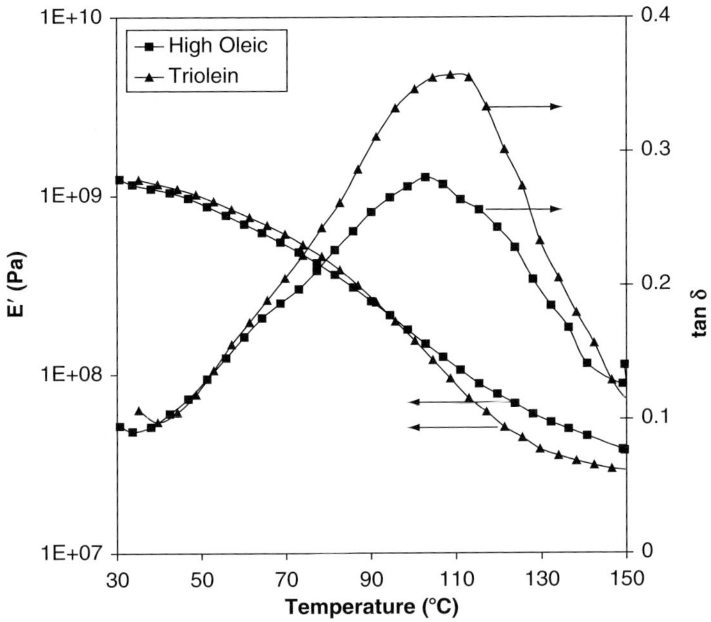

FIGURE 4.27 Dynamic mechanical properties of polymers made from maleinized hydroxylated high oleic oil and triolein oil. The monodisperse triolein displays a sharper transition from the glassy region to the rubbery region.

However, the broadness of the transitions does differ. It is apparent that the triolein polymer has a sharper E' transition from the glassy region to the rubbery region. This is evident also in the tan δ peak, which has a higher peak height. The transition is not yet as sharp as petroleum-based polymers. This is probably caused by the triolein monomer having a functionality of only two maleates/triglyceride. Consequently, there is still a plasticizer effect present, but this effect may be reduced by controlling the reaction conditions to reach higher conversions.

4.4.8 SOPERMA POLYMER PROPERTIES

The typical DMA behavior of the SOPERMA-styrene polymer (40 wt% styrene) is shown in Figure 4.28, where we can see that, at room temperature, these polymers are already in transition from the glassy region to the rubbery plateau. Most thermoset polymers show a distinct glassy region in which the modulus is independent of temperature. This is not observed for the SOPERMA-styrene polymers. The SOPERMA-styrene polymers show a very broad transition from the glassy to the rubbery state. Because of this broad transition, these polymers do not show a clear peak in the loss modulus E''. Thus, the tan δ curve is also very broad. The broad transition observed for the SOPERMA-styrene polymers is a result of two major effects. One

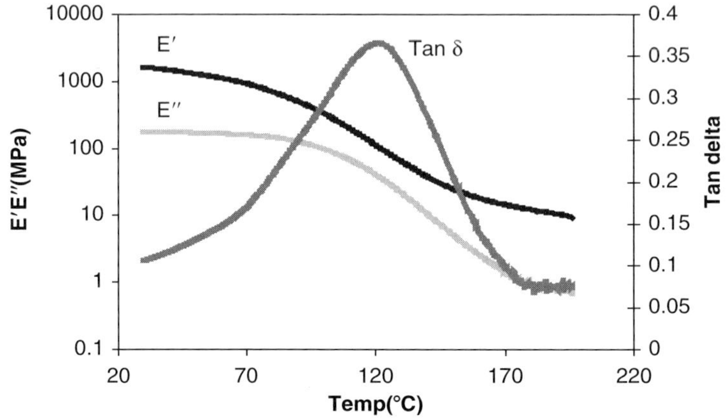

FIGURE 4.28 Typical DMA behavior of the SOPERMA-styrene polymer (40 wt% styrene).

major effect is the phase separation, which results in higher T_g SOPERMA-rich and lower T_g styrene-rich regions in the polymer matrix. Another effect that may result in a broad glass transition is the plasticizing effect of the fatty acids present in the SOPERMA monomer which are not functionalized. The transition from the glassy to the rubbery state broadens significantly with the addition of small amounts of plasticizers to polymers.

Figures 4.29(a) and (b) show the flexural modulus and flexural strength of the SOPERMA-styrene polymers as a function of styrene concentration. As can be seen both the flexural modulus and flexural strength of the polymers increase with increasing concentrations of styrene despite the decrease in cross-link density, ν. Thus the rigid aromatic structure of the styrene monomer as compared to the SOPERMA monomer with flexible fatty acid chains dominates the effect of cross-link density. The linear dependence of flexural modulus on styrene concentration above 30 wt% styrene follows Eq. (4.1):

$$E_f = 0.0278(W_{\text{styrene}}) + 0.2643, \tag{4.1}$$

where W_{styrene} presents the weight percentage of styrene. This correlation predicts the flexural modulus of 100% polystyrene as 3.04 GPa. The flexural modulus values of standard polystyrenes of different grades are in the range of 2.9–3.8 GPa. In the same manner, the dependence of flexural strength on styrene concentration follows Eq. (4.2):

$$S_f = 0.6159(W_{\text{styrene}}) + 27.696, \tag{4.2}$$

which predicts the flexural strength of the 100% polystyrene as 89.29 MPa. The strength values of standard polystyrene samples are in the range of 70–100 MPa.

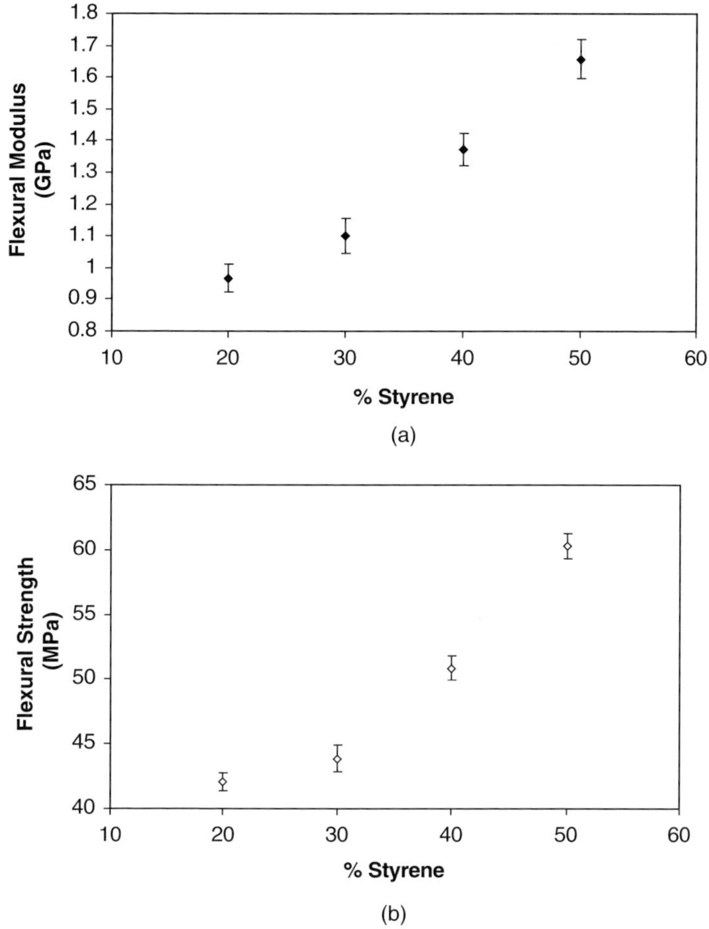

FIGURE 4.29 The change of (a) flexural modulus and (b) flexural strength of SOPERMA-styrene polymers at increasing styrene weight percentages.

The addition of butyrated kraft lignin to SOPERMA, as discussed for soy-based resins in Chapter 16, had a large effect on the polymer properties. The T_g's (as determined from the tan δ maximum) of the SOPERMA-styrene polymers as a function of lignin concentration are shown in Figure 4.30(a). As can be seen, there is a slow increase in T_g at low concentrations and then at 5 wt% lignin the T_g increases significantly. The T_g of the SOPERMA-lignin composite should be influenced by both the cross-link density of the system as well as the higher T_g of kraft lignin (167 °C). The cross-link densities, as determined using the modulus in the rubbery region, are shown in Figure 4.30(b), where we see that the cross-link density of the network increases with lignin until 5 wt% and then starts to decrease again. The increase in cross-link

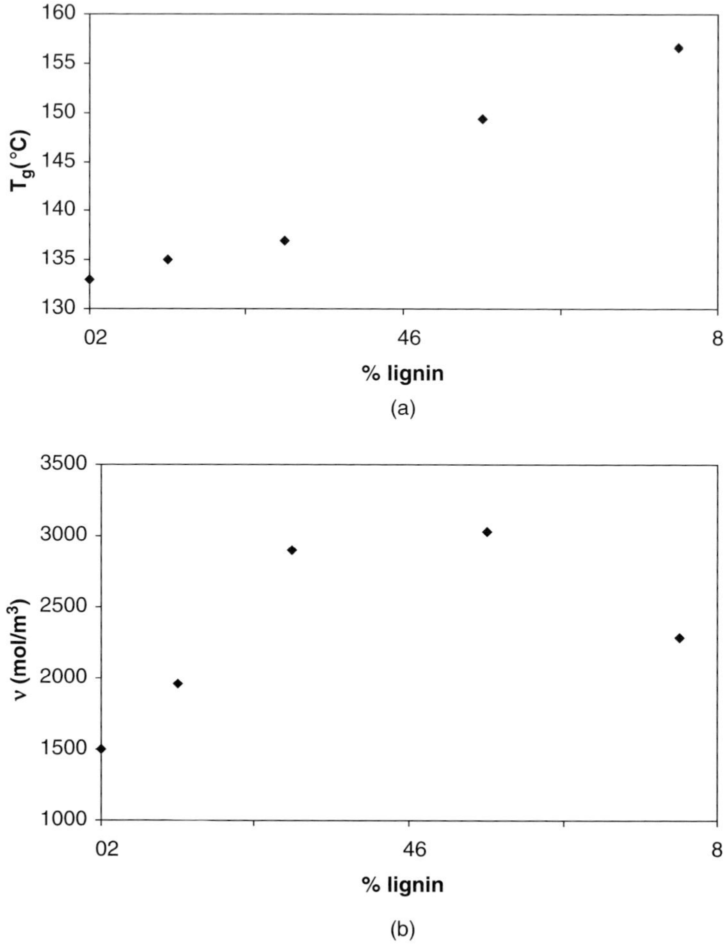

FIGURE 4.30 (a) Glass transition temperature, T_g, and (b) cross-link density (ν) variation of the SOPERMA-styrene polymers as a function of butyrated lignin content.

density with lignin addition may be attributed to specific interactions between the polymer matrix and the lignin molecule. The carboxylic acid groups of the SOPERMA monomer may interact with the available hydroxyl groups or thiol groups of the lignin molecule. Additionally, lignin may have effects on the kinetics of polymerization of both the SOPERMA and styrene monomer due to its inhibition effect on radical polymerization, which may affect the cross-link density. However, more work needs to be done to evaluate this effect. The cross-link density levels off at 5 wt% load and starts to decrease for higher concentrations. Thus, at this point the cross-link density must decrease due to the increase in the volume fraction of lignin, which cannot apparently interact with the matrix any more. The significant increase in T_g at

5 and 7.5 wt% lignin shows that the T_g of the system approaches the T_g of the kraft lignin (167 °C) component at these high concentrations.

Figure 4.31 shows that both the flexural modulus and flexural strength of the SOPERMA-styrene polymers increase continuously with increasing lignin content of the resin. Because butyrated kraft lignin is dissolved in the polymer matrix, it should undergo the same strain as the polymer matrix. Thus, the modulus of the composite should increase with the introduction of the rigid aromatic structure of lignin to the system. Additionally, the cross-link density increase with lignin addition is also expected to increase the modulus. On the other hand, the increase in flexural strength with increasing

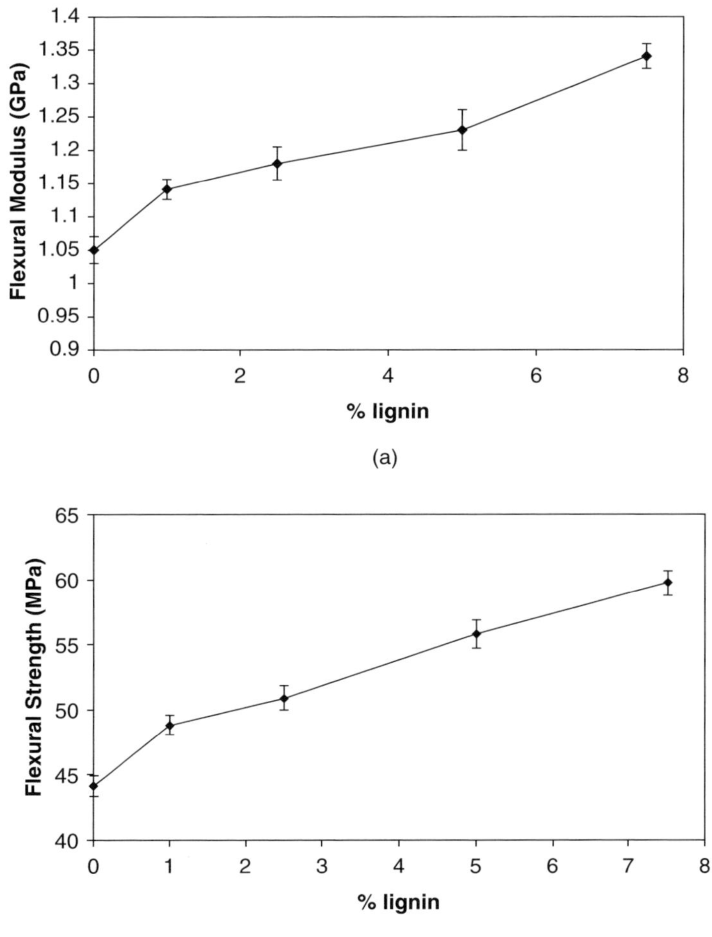

FIGURE 4.31 The change of (a) flexural modulus and (b) flexural strength of SOPERMA-styrene polymers at increasing butyrated lignin content.

lignin content may be attributed to both the increase in modulus with lignin addition and an increase in cross-link density up to 5 wt% lignin load.

4.4.9 SOGLYME POLYMER PROPERTIES

As discussed in Section 4.3.8, the soybean oil monoglyceride methacrylates (SOGLYME) were prepared by methacrylation of the soybean oil glycerolysis product by methacrylic anhydride, as shown earlier in Figure 4.8. The methacrylated glycerides did not show phase separation in styrene. The crude methacrylated soybean oil monoglycerides contain methacrylic acid as a by-product. Methacrylic acid is itself a reactive diluent and acts as a comonomer in the system. Thus, we will examine the mechanical properties of the polymers prepared from this crude product (SOGLYME-MEA) as well as the polymers prepared, using styrene as the third comonomer (SOGLYME-MEA-ST). We will also look at the properties of the polymers prepared from the acid-extracted product with styrene (SOGLYCME-ST).

Figures 4.32(a) and (b) show the storage modulus (E'), loss modulus (E''), and tan δ values as a function of temperature for the polymers prepared from the crude methacrylated soybean oil monoglycerides that contain methacrylic acid as a by-product (SOGLYME-MEA) and the styrenated resin (SOGLYME-MEA-ST) (33 wt%), respectively. As can be seen, these polymers also show a broad transition from the glassy state to the rubbery state, similar to the SOPERMA-styrene polymers. The polymers prepared from these resins did not show phase separation, which means that the phase separation observed in the SOPERMA-styrene polymers is not the only factor responsible for the broad transition observed in these triglyceride-based polymers. The plasticization effect of the long flexible fatty acid chains present in the cross-linked monomer has a significant effect on the observed behavior. Additionally, as can be seen, the tan δ peak is broader for the SOGLYME-MEA polymer compared to the SOGLYME-MEA-ST polymer showing that the higher cross-link density of the former polymer also has a significant effect on broadening the glass transition.

The E' values, T_g values (tan δ maxima) as determined from DMA, flexural moduli, and flexural strengths of the SOGLYME-MEA, SOGLYME-MEA-ST (33 wt% styrene), and SOGLYME-ST (33 wt% styrene) polymers are listed in Table 4.9. As can be seen, the SOGLYME-MEA-ST polymer has the highest modulus and strength followed by the SOGLYME-MEA polymer. The T_g's of these two polymers do not show a significant difference. The SOGLYME-ST polymer, on the other hand, exhibits considerably lower modulus, strength, and T_g values compared to the other two polymers. The properties of the individual monomers in these polymer systems as well as the cross-link density are both detrimental to the mechanical properties. The cross-link densities as determined by using the modulus values in the rubbery region of these polymers are shown in Table 4.10.

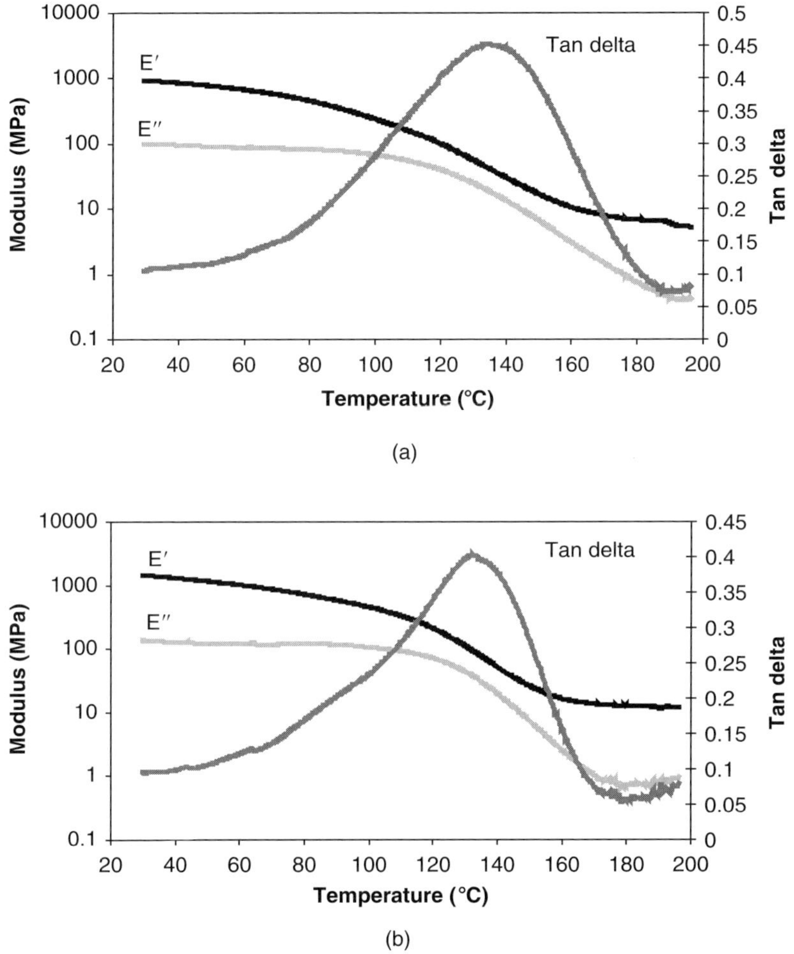

FIGURE 4.32 Storage modulus (E'), loss modulus (E''), and tan δ values as a function of temperature for (a) SOGLYME-MEA and (b) SOGLYME-MEA-ST polymers.

TABLE 4.9 The 30°C E' values, T_g values (tan δ maxima), flexural modulus, and flexural strength of the SOGLYME-MEA, SOGLYME-MEA-ST (33 wt% styrene), and SOGLYME-ST (33 wt% styrene) polymers.

Resin	E' (30°C) (GPa)	T_g (°C)	Flexural Modulus (GPa)	Flexural Strength (MPa)
SOGLYME-MEA	0.86	134.7	0.80	26.5
SOGLYME-MEA-ST	1.15	132.6	1.04	49.0
SOGLYME-ST	0.23	65.5	0.26	4.0

TABLE 4.10 The cross-link densities (ν) of the SOGLYME polymers.

Resin	ν (mol/m^3)
SOGLYME-MEA	701
SOGLYME-MEA-ST	609
SOGLYME-ST	520

The cross-link densities of the SOGLYME-MEA and SOGLYME-MEA-ST polymers are in the neighborhood of the SOPERMA-styrene polymer at 50 wt% styrene concentration. The mechanical properties observed for these two polymers, however, are much lower than those observed for the SOPERMA-styrene polymers. The SOPERMA-styrene polymer at 50 wt% styrene has a flexural modulus of 1.65 GPa and a flexural strength of 62 MPa. This fact clearly shows that the maleate comonomer compared to the methacrylates, and the styrene comonomer compared to the methacrylic acid, bring more rigidity and strength to the triglyceride-based polymers. The cross-link density of the SOGLYME-MEA-ST polymer is lower than that of the SOGLYME-MEA polymer, as expected due to the increase in comonomer content. The SOGLYME-MEA-ST polymer, despite its lower cross-link density, still shows superior properties compared to those of the SOGLYME-MEA polymer due to the presence of the rigid styrene molecules in the polymer matrix. The significantly lower modulus, strength, and T_g observed for the SOGLYME-ST polymer is unexpected, especially when considering the properties of the SOGLYME-MEA polymer, and can only be attributed to its lower cross-link density compared to the other polymers. Chapter 6 will show that the fracture strength σ of all bio-based polymers depends on modulus and cross-link density ν, as $\sigma \sim [E\nu]^{1/2}$.

4.5 CASTOR OIL-BASED POLYMER PROPERTIES

The basic fatty acid constituent of castor oil is ricinoleic acid, which is a hydroxy monounsaturated fatty acid (12-hydroxy *cis* 9-octodecenoic acid) (~87%). Castor oil was thus first alcoholized with pentaerythritol, glycerol, and an aromatic diol BPA propoxylate, and the alcoholysis products were then malinated with maleic anhydride, introducing maleate functionality to both the polyol hydroxyls and fatty acid hydroxyls. The resulting resins were labeled COPERMA, COGLYMA, and COBPAPRMA, respectively. The reaction schemes and the structures of the final malinated products for the COGLYMA, COPERMA, and COBPAPRMA products were shown earlier in Figures 4.9, 4.10, and 4.11, respectively. Additionally, castor oil was directly malinated and castor oil maleates (COMA) were also prepared. In this section, we introduce the properties of these castor oil-based polymers and analyze their properties with reference to the network structure. The effect of styrene concentration on mechanical properties of the COPERMA-styrene polymers is also explored and compared to the observed trend for the SOPERMA-styrene polymers.

4.5.1 EFFECT OF STYRENE CONCENTRATION ON THE COPERMA-STYRENE POLYMER PROPERTIES

Figure 4.33 shows the storage modulus values, E', of the COPERMA-styrene polymers as a function of temperature at increasing styrene concentrations as determined from DMA. As can be seen, the room temperature modulus values increase significantly, going from 20 wt% styrene to 30 wt% styrene. The increase in styrene concentration has a much less pronounced effect on the modulus at higher concentrations.

The changes in flexural modulus and flexural strength of the COPERMA-styrene polymers at increasing styrene concentrations are shown in Figures 4.34(a) and (b), respectively. The increase in flexural modulus and strength with increasing styrene concentrations follows a trend similar to that observed for the storage modulus. A significant increase in both the modulus and strength is observed while going from 20 to 30 wt% styrene, but this increase levels off rapidly at higher concentrations. As discussed in

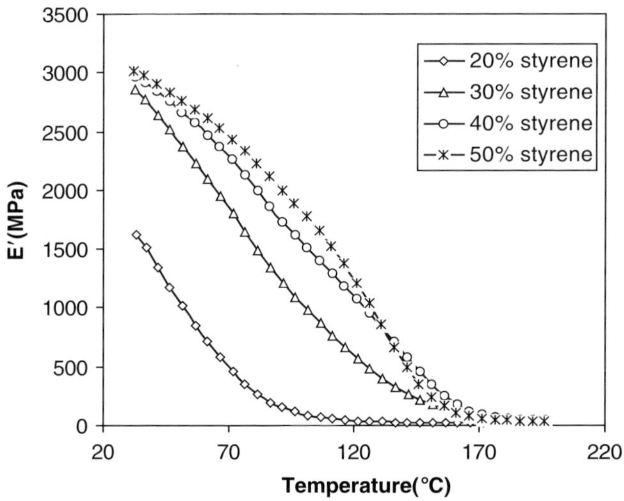

FIGURE 4.33 The change of storage modulus (E') values with temperature for the COPERMA polymers at increasing styrene concentrations.

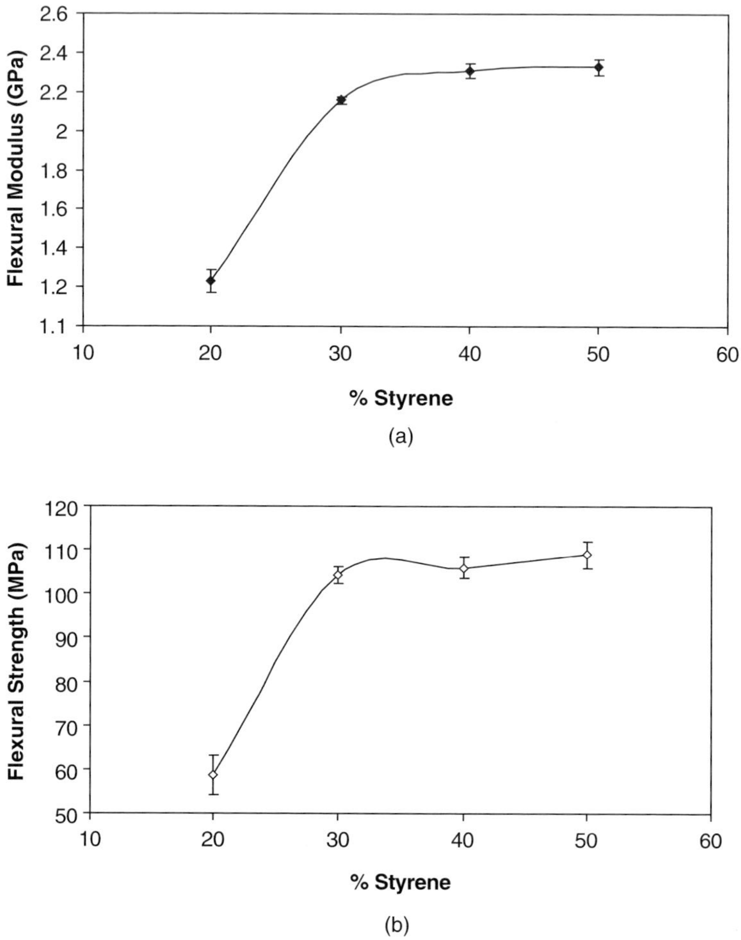

FIGURE 4.34 The change in (a) flexural modulus and (b) flexural strength of the COPERMA polymers with increasing concentrations of styrene.

Section 4.4, the SOPERMA-styrene polymers showed a continuous increase in both the modulus and strength, with styrene content in similar concentrations.

To explain the difference in the effect of styrene concentration on the mechanical properties of these two polymers, it is useful to determine the cross-link densities of the COPERMA-styrene polymers. Figure 4.35 shows the cross-link densities (ν) of the COPERMA-styrene polymers at increasing styrene concentrations as determined from the rubbery modulus by DMA. The cross-link densities (ν) of the SOPERMA-styrene polymers at the same styrene concentrations are also shown in the same figure, for comparison. For the COPERMA resin, which has a much higher maleate content per

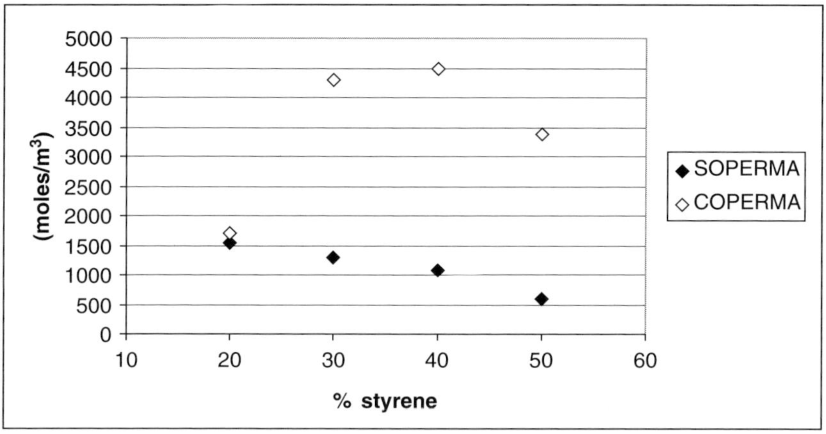

FIGURE 4.35 The cross-link densities of the SOPERMA and COPERMA polymers at increasing styrene concentrations.

triglyceride than the SOPERMA resin, at 20 wt% styrene, the styrene concentration is too low to incorporate all of the maleates into polymerization, since the maleate half-esters do not homopolymerize. The molar ratio of styrene double bonds to maleate double bonds for the 20 wt% styrene COPERMA resin is approximately 0.8, as determined from ^{1}H NMR analysis. Thus, the 20 wt% styrene COPERMA polymer has the lowest cross-link density. At 30 and 40 wt% styrene, the molar ratio of styrene double bonds to maleate double bonds is 1.3 and 1.8, respectively, thus a significant increase in cross-link density is observed at 30 wt%, and at 40 wt% the cross-link density reaches its optimum value where all the available reactive groups of the COPERMA monomer can react with styrene. At higher concentrations, the cross-link density starts to decrease again since the added styrene increases the length of the segments between the cross-links. Thus, the modulus and strength of the COPERMA polymers at 20 wt% styrene are especially low and show a big increase at 30 and 40 wt% styrene due to the significant increase in cross-link density. After this point, the styrene content does not seem to have a significant effect on the modulus and strength. This behavior is different from that of the SOPERMA polymers, which showed a continuous increase in modulus and strength in the same styrene concentrations, despite the decrease in cross-link density. The introduction of the rigid aromatic rings of the styrene comonomer into the SOPERMA monomer with long flexible fatty acid chains results in a higher net increase in both the modulus and strength of the network than that observed for the COPERMA polymers because the fatty acids present in the COPERMA monomer are malinated and therefore incorporated into the network.

The tan δ curves for the COPERMA-styrene polymers at increasing styrene concentrations, shown in Figure 4.36, also reflect the trend observed in cross-link density. As can be seen, the 20 wt% styrene polymer with the lowest cross-link density shows the tan δ maximum at the lowest temperature and therefore has the lowest T_g. The tan δ maximum shifts to higher temperatures with increases up to 40 wt% styrene due to the increase in cross-link density. After this point the increase in styrene concentration decreases the cross-link density and the T_g starts to decrease again. Thus the highest T_g is observed with 40 wt% styrene at 156 °C, for the COPERMA-styrene polymers.

4.5.2 COMPARISON OF COPERMA AND SOPERMA-STYRENE POLYMER MECHANICAL PROPERTIES

As can be seen in Figure 4.35, the COPERMA-styrene polymers show significantly higher cross-link densities than those of the SOPERMA-styrene polymers, especially at 30 wt% and higher weight percentages of styrene. Table 4.11 shows the properties of the 30 wt% styrene SOPERMA and COPERMA polymers for a direct comparison. It can be seen that the modulus value nearly doubles and the flexural strength shows even a larger increase with the change from soybean oil to castor oil in the formulation. The glass transition temperature of the COPERMA polymer is about 9 °C higher than that of the SOPERMA polymer. The incorporation of the fatty acid chains into the polymerization both increases the cross-link density and

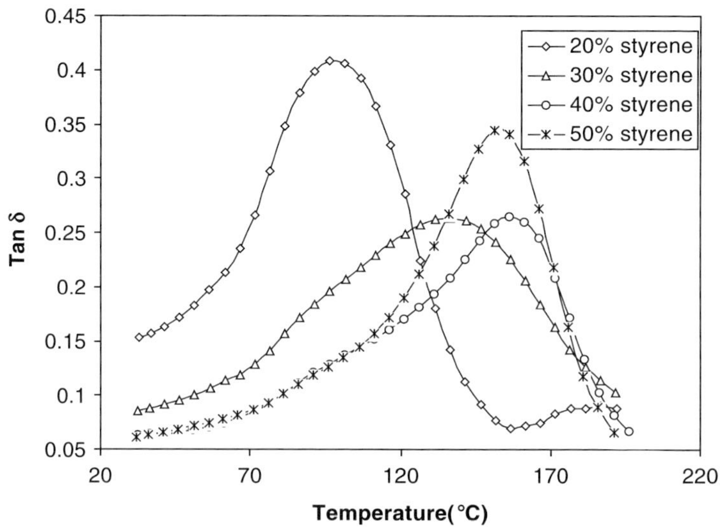

FIGURE 4.36 The change of tan δ values with temperature for the COPERMA polymers at increasing styrene concentrations.

TABLE 4.11 The mechanical properties of COPERMA and SOPERMA polymers at 30 wt% styrene.

Property	COPERMA (30% styrene)	SOPERMA (30% styrene)
Flexural strength (MPa)	104.23	43.86
Flexural modulus (GPa)	1.95	1.10
T_g (°C)	144	135
Storage modulus (GPa)	2.88	1.24

reduces the plasticization effect of the fatty acid chains in the COPERMA polymer, which in turn leads to a considerable increase in modulus, strength, and T_g compared to those properties of the SOPERMA polymer.

For the COPERMA resin, which has the highest maleate content among all the castor oil-based resins, the styrene concentration should be more than 30 wt% percent to fully incorporate all of the maleates during polymerization. This level should be lower for the other resins, which show lower maleate contents than the COPERMA resin. Thus we used 33 wt% styrene concentration for the preparation of other castor oil-based polymers, which keeps the renewable content of the resin within a reasonable range and also gives us a chance to compare the properties of these materials to the commercially available unsaturated polyesters that use similar formulations.

The ratio of the flexural strength of COPERMA/SOPERMA is $\sigma_1/\sigma_2 = 104.2/43.9 = 2.4$. If we apply the square root law for strength (Chapter 6), where

$$\sigma_1/\sigma_2 = [E_1 \nu_1 / E_2 \nu_2]^{1/2},$$

then we can readily compare theory with experiment. Using $E_1 = 1.95$ GPa and $E_2 = 1.1$ GPa (Table 4.11), $\nu_1 = 4300\ \text{mol/m}^3$ and $\nu_2 = 1300\ \text{mol/m}^3$ (Figure 4.35 at 30% styrene), then the predicted ratio for strength is $\sigma_1/\sigma_2 = 2.4$, which is in excellent accord with the experimental ratio.

4.5.3 THERMOMECHANICAL PROPERTIES OF CASTOR OIL-BASED POLYMERS

All of the malinated castor oil-styrene–based polymers exhibited broad transition profiles from the glassy state to the rubbery state, similar to the SOPERMA-styrene polymers. The broad transitions observed were attributed to the phase separation observed on the microscale and the high crosslink density exhibited by these polymers. The plasticization effect of the fatty acids should have a less pronounced effect for the castor oil-based polymers because the hydroxy fatty acids that constitute the majority of the fatty acids (87%) in castor oil were malinated and therefore connected to the network structure. Figure 4.37 shows the tan δ curves of the castor oil-based polymers as determined by DMA.

As discussed above, the damping peak position is a sensitive indicator of cross-linking. As the cross-link density increases, the tan δ maximum shifts to higher temperatures, the peak broadens, and a decrease in the tan δ value is observed. The cross-link densities of the malinated castor oil-based polymers (33 wt% styrene) determined by using the value of E in the rubbery region of the polymers as determined by DMA are shown in Table 4.12. As can be seen, the COPERMA polymer has the highest cross-link density, followed by COGLYMA, COMA, and COBPAPRMA polymers. The COBPAPRMA polymer shows the lowest cross-link density, despite the higher maleate content than both COMA and COGLYMA which can be attributed to the bulkiness of the Bisphenol A propoxylate moiety. As can be seen in Figure 4.37, the COPERMA polymer with the highest cross-link density exhibits the highest T_g and shows the broadest peak, with the lowest tan δ value. The COBPAPRMA, which has the lowest cross-link density, shows the highest tan δ values as expected; however, its tan δ max temperature is about 14°C higher than that of the COMA polymer. The higher T_g observed for the COBPAPRMA may be explained by the presence of the rigid aromatic backbone of BPA propoxylate, as compared to the aliphatic fatty acid backbone of the COMA polymer. Thus, the monomer chemical structural influence dominates the cross-link density effect on T_g.

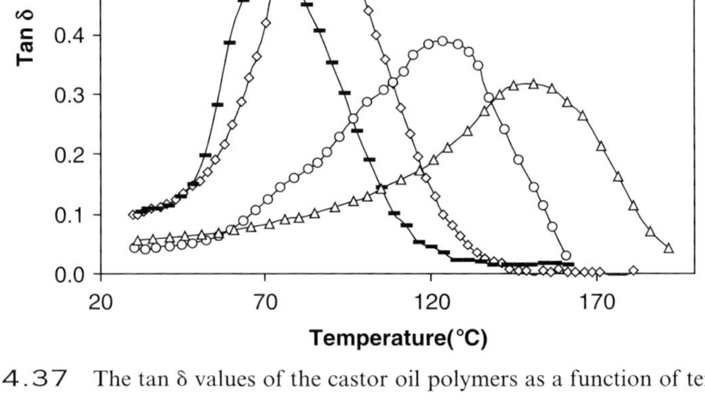

FIGURE 4.37 The tan δ values of the castor oil polymers as a function of temperature.

TABLE 4.12 Cross-link densities of castor oil polymers.

Resin type	Mc (g/mol)	V (mol/m^3)
COPERMA	255	4310
COGLYMA	441	2494
COMA	732	1511
COBPAPRMA	783	1418

4.5.4 MECHANICAL PROPERTIES OF CASTOR OIL-BASED POLYMERS

The storage modulus values at 30 °C and the T_g's as determined from DMA, as well as the flexural modulus, flexural strength, and the surface hardness values of the castor oil polymers are given in Table 4.13. The styrene content of each resin was 33 wt%. The mechanical property hardness is the ability of the material to resist indentation, scratching, abrasion, cutting, and penetration. This property may be important for structural materials that require a high resistance to indentation or abrasion. The hardness of a polymer reflects such other qualities as resilience, durability, uniformity, strength, and abrasion resistance. As can be seen in Table 4.13, the surface hardness of the castor oil-based polymers changes proportionately with the strength of the polymers.

The observed mechanical properties of the castor oil-based polymers can be explained both in terms of the cross-link density and the chemical structures of the polyols used. The COPERMA polymer, which has the highest cross-link density, shows superior properties to the other castor oil-based polymers. The COPERMA polymer with its T_g around 150 °C and flexural modulus of 2.2 GPa and flexural strength of 105 MPa exhibits the highest T_g and strength obtained from any triglyceride-based thermoset resins. The COBPAPRMA polymer's modulus, strength, and surface hardness values are higher than those of both COMA and COGLYMA polymers and ap-

TABLE 4.13 The mechanical properties of castor oil polymers.

Resin type	E' (30 °C) (GPa)	T_g (°C)	Flexural Modulus (GPa)	Flexural Strength (MPa)	Surface Hardness (D)
COPERMA (1:2:10.7)	2.94	150	2.21	104.60	89.3
COGLYMA (1:2.2:9.2)	2.40	124	1.76	78.89	86.1
COBPAPRMA (1:2:6.7)	2.69	86	2.17	83.20	88.5
COMA (1:3)	1.15	72	0.78	32.83	77.1
Ortho-UP	—	135–204	3.45	80	—
Iso-UP	—	135–204	3.59	100	—

proach those of COPERMA polymer, although its cross-link density is slightly lower than these two polymers. The aromatic structure of the BPA propoxylate moiety brings both rigidity and strength to the polymer network. Thus, this resin shows both high modulus and strength with a reasonable T_g, despite its lower maleate content, which is beneficial for the formulation since it decreases the nonrenewable content of the polymer. The COMA polymer shows the lowest modulus, strength, surface hardness, and T_g values due to its low cross-link density and also shows that a multifunctional unit at the center of the triglyceride monomer structure is essential for improved properties for these polymers.

Figure 4.38 gives a comparison between the mechanical properties of these bio-based resins and the properties of two of the most commonly used UP resins: Orthophthalic (Ortho-UP) and Isophthalic (Iso-UP) UP resins. As can be seen, the properties of castor oil-based polymers are in a comparable range with those of the commercially successful UP resins. As shown in this chapter and in more detail in Chapter 16, the properties of these bio-based resins can be significantly improved by the addition of lignin, which introduces the required aromatic groups for high stiffness and high T_g polymers and further increases their bio-based content with low-cost renewable material.

4.6 SUMMARY OF PLANT OIL-BASED POLYMER PROPERTIES

Triglyceride oils are an abundant natural resource that has yet to be fully exploited as a source for polymers and composites. The different chemical functionalities allow the triglyceride to be converted to several promising monomers. When blended with comonomers, these monomers form polymers with a wide range of physical properties. They exhibit flexural strength up to 100 MPa and tensile moduli in the 1–3 GPa range, with T_g ranging from 70 to 150 °C, depending on the particular monomer and the resin composition. DMA shows that the transition from glassy to rubbery behavior is extremely broad for these polymers as a result of the triglyceride molecules acting both as cross-linkers and as plasticizers in the system. Saturated fatty acid chains are unable to attach to the polymer network, causing relaxations to occur in the network. However, this transition can be sharpened by reducing the saturation content, as demonstrated with the genetically engineered oil and pure triolein oil.

This area of research sets a foundation from which completely new materials can be produced with novel properties. Work continues on optimizing the properties of these green materials and understanding the fundamental issues that affect them. We can use computer simulation to optimize the choice of the fatty acid distribution function (Chapter 6) and determine the resulting architecture and mechanical properties for the particular chemical pathways

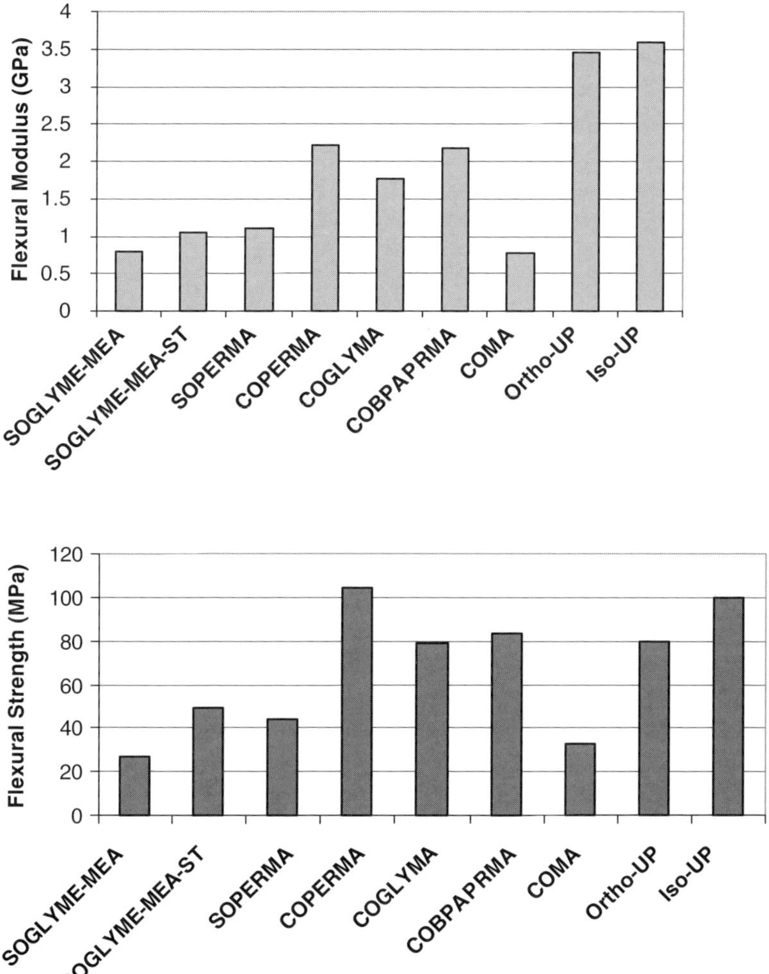

FIGURE 4.38 A property comparison of soybean oil and castor oil-based polymers with commercial petroleum-based Ortho-UP and Iso-UP resins.

shown in Figure 4.2. Use of a computer significantly reduces the number of chemical trials required for a system with a large number of degrees of freedom and suggests the optimal oil most suited to a particular type of resin. In this manner, more renewable resources can be used to meet the material demands of many industries.

REFERENCES

1. Wool, R. P.; Kusefoglu, S. H.; Palmese, G. R.; *et al.* U.S. Patent 6,121,398; 2000.
2. Wool, R. P. *Chemtech.* 1999, **29,** 44.

3. Liu, K. *Soybeans: Chemistry, Technology, and Utilization,* Chapman and Hall, New York; 1997.
4. Gunstone, F. *Fatty Acid and Lipid Chemistry,* Blackie Academic and Professional, New York; 1996.
5. Cunningham A.; Yapp, A. U.S. Patent 3,827,993; 1974.
6. Bussell, G. W. U.S. Patent 3,855,163; 1974.
7. Hodakowski, L. E.; Osborn, C. L.; Harris, E. B. U.S. Patent 4,119,640; 1975.
8. Trecker, D. J.; Borden, G. W.; Smith, O. W. U.S. Patent 3,979,270; 1976.
9. Trecker, D. J.; Borden, G. W.; Smith, O. W. U.S. Patent 3,931,075; 1976.
10. Salunkhe, D. K.; Chavan, J. K.; Adsule, R. N.; Kadam, S. S. *World Oilseeds: Chemistry, Technology, and Utilization,* Van Nostrand Reinhold, New York; 1992.
11. Force, C. G.; Starr, F. S. U.S. Patent 4,740,367; 1988.
12. Barrett, L. W.; Sperling, L. H.; Murphy, C. J. *J. Am. Oil Chem. Soc.* 1993, **70,** 523.
13. Qureshi, S.; Manson, J. A.; Sperling, L. H.; Murphy, C. J. In *Polymer Applications of Renewable-Resource Materials*; Carraher, C. E., Sperling, L. H., Eds.; Plenum Press, New York; 1983.
14. Devia, N.; Manson, J. A.; Sperling, L. H.; Conde, A. *Polym. Eng. & Sci.* 1979, **19,** 878.
15. Devia, N.; Manson, J. A.; Sperling, L. H.; Conde, A. *Polym. Eng. & Sci.* 1979, **19,** 869.
16. Devia, N.; Manson, J. A.; Sperling, L. H.; Conde, A. *Macromolecules* 1979, **12,** 360.
17. Sperling, L. H.; Carraher, C. E.; Qureshi, S. P.; *et al.* In *Polymers from Biotechnology,* Gebelein, C. G., Ed.; Plenum Press, New York; 1991.
18. Sperling, L. H.; Manson, J. A.; Linne, M. A. *J. Polym. Mater.* 1984, **1,** 54.
19. Sperling, L. H.; Manson, J. A. *J. Am. Oil Chem. Soc.* 1983, **60,** 1887.
20. Fernandez, A. M.; Murphy, C. J.; DeCosta, M. T.; *et al.* In *Polymer Applications of Renewable-Resource Materials,* Carraher C. E.; Sperling, L. H., Eds.; Plenum Press, New York; 1983.
21. Sperling, L. H.; Manson, J. A.; Qureshi, S. A.; Fernandez, A. M. *Ind. Eng. Chem.* 1981, **20,** 163.
22. Yenwo, G. M.; Manson, J. A.; Pulido, J.; *et al. J. Appl. Polym. Sci.* 1977, **21,** 1531.
23. Frischinger; I.; Dirlikov, S. *Polymer Comm.* 1991, **32,** 536.
24. Frischinger; I.; Dirlikov, S. In *Interpenetrating Polymer Networks,* Advances in Chemistry Series 239; Sperling, L. H., Kempner, D., Utracki, L., Eds., American Chemical Society, Washington, DC; 1994, p. 517.
25. Rosch, J.; Mulhaupt, R. *Polymer Bull.* 1993, **31,** 679.
26. Meffert, A.; Kluth, H.; Denmark Patent 4,886,893; 1989.
27. Rangarajan, B.; Havey, A.; Grulke, E. A.; Culnan, P. D. *J. Am. Oil Chem. Soc.* 1995, **72,** 1161.
28. Zaher, F. A.; El-Malla, M. H.; El-Hefnawy, M. M. *J. Am. Oil Chem. Soc.* 1989, **66,** 698.
29. Friedman, A.; Polovsky, S. B.; Pavlichko, J. P.; Moral, L. S. U.S. Patent 5,576,027; 1996.
30. Swern, D.; Billen, G. N.; Findley, T. W.; Scanlan, J. T. *J. Am. Chem. Soc.* 1945, **67,** 786.
31. Sonntag, N. O. V. *J. Am. Oil Chem. Soc.* 1982, **59,** 795.
32. Solomon, D. H. *The Chemistry of Organic Film Formers,* Wiley, New York; 1967.
33. Can, E. *M.S. Thesis,* Bogazici University, Turkey; 1999.
34. Bailey, A. E. In *Bailey's Industrial Oil and Fat Products,* Swern, D., Ed.; Wiley, New York; 1985.
35. Hellsten, M.; Harwigsson, I.; Brink, C. U.S. Patent 5,911,236; 1999.
36. Cain, F. W.; Kuin, A. J.; Cynthia, P. A.; Quinlan, P. T. U.S. Patent 5,912,042; 1995.
37. Eckwert, K.; Jeromin, L.; Meffert, A.; *et al.* B. U.S. Patent 4,647,678; 1987.
38. Khot, S. N. *M.S. Thesis,* University of Delaware; 1998.
39. Wypych, *J. Polyvinyl Chloride Stabilization,* Elsevier, Amsterdam; 1986.
40. Sears, J. K.; Darby, J. R. *The Technology of Plasticizers,* Wiley & Sons, New York; 1982.
41. Carlson, K. D.; Chang, S. P. *J. Am. Oil Chem. Soc.* 1985, **62,** 934.

42. Raghavachar, R.; Letasi, R. J.; Kola, P. V.; *et al. J. Am. Oil Chem. Soc.* 1999, **76,** 511.
43. Pashley, R. M.; Senden, T. J.; Morris, R. A.; *et al.* U.S. Patent 5,360,880; 1994.
44. Likavec, W. R.; Bradley, C. R. U.S. Patent 5,866,628; 1999.
45. Bordon, G. W.; Smith, O. W.; Trecker, D. J. U.S. Patent 4,025,477; 1974.
46. La Scala, J. J.; Wool, R. P. *J. Am. Oil Chem. Soc.,* 79, **59,** 2002.
47. S. P. Bunker, *M.S. Thesis,* University of Delaware; 2000.
48. Chu, T. J.; Niou, D. Y. *J. Chin. Inst. Chem. Eng.* 1989, **20,** 1.
49. Betts, A. T. U.S. Patent 3,867,354; 1975.
50. Mitch, E. L.; Kaplan, S. L. In *Proceedings 33rd Annual SPE Technical Conference,* Atlanta; 1975.
51. Can, E.; Kusefoglu, S.; Wool, R. P.; *J. Appl. Polym. Sci.*, 2001, **69,** 81.
52. Gardner, H. C.; Cotter, R. J. European Patent 20,945; 1981.
53. Thomas, P.; Mayer, J. U.S. Patent 3,784,586; 1974.
54. Lee, S. H.; Park, T. W.; Lee, S. O.; *Polymer (Korea),* 1999, **23,** 493.
55. Shione, H.; Yamada, J. Japanese Patent 11,147,222; 1999.
56. Hasegawa, H. Japanese Patent 11,240,014; 1999.
57. Johnson, L. K.; Sade, W. T. *J. Coat. Tech.* 1993, **65,** 19.
58. Solomons, T. W. G. *Organic Chemistry,* Wiley, New York; 1992.
59. La Scala, J.; Wool, R. P. *Polymer,* 46 (2005) 61-69.
60. Flory, P. *J. Principles of Polymer Chemistry,* Cornell University, Ithaca, NY; 1975.
61. Nielsen, L. E.; Landel, R. F. *Mechanical Properties of Polymers and Composites*, Marcel Dekker, New York; 1994.
62. Wool, R. P.; Khot, S. N. In *Proceedings ACUN-2,* Sydney; 2000.
63. Can, E.; Kusefoglu, S. H.; Wool, R. P. *J. Appl. Polym. Sci.* 2001, **82,** 703.
64. Khot, S. N.; La Scala, J. J.; Can, E., *et al. J Appl. Polym. Sci.*, 2001, **82,** 703.
65. Can, E. *Ph.D. Thesis,* University of Delaware; 2004.
66. Can, E.; Wool, R. P., Paper submitted for publication (2005).

5

Composites and Foams from Plant Oil-Based Resins

Richard P. Wool

When bio-based resins derived from natural oils (described in Chapter 4) are combined with natural fibers (flax, hemp, jute, cellulose, switch grass, chicken feathers), glass fibers, graphite fibers, and lignin, new low-cost composites are produced that are economical in many high-volume applications. These composites are used in agricultural equipment, automotive sheet molding compounds, civil infrastructures, marine applications, housing, and the construction industry. Examples are given for the synthesis, manufacture, and properties of plant-based resins and glass, flax, lignin, and hemp composites, including sheet molding compound. This chapter deals primarily with highly cross-linked triglyceride-based resins suited for liquid molding applications with fibers to form composites. Linear polymers suited to elastomers, pressure-sensitive adhesives, and coatings requiring molecular weights of the order of 10^6 g/mol, are treated in Chapter 8. Such polymers are made primarily with the individual fatty acids with a preference for high oleic (1 C=C per FA).

Henry Ford made the first auto parts in 1938 from soybean proteins with fiberglass reinforcement, as shown in Figure 5.1. This figure shows Mr. Ford performing an impact test on the hood of his car with the blunt end of an ax. Mr. Ford and Thomas Edison were committed to developing bio-based materials to reduce pollution in their mass production of products. However, global wars intervened and the rise of the petroleum-based plastic industry rapidly replaced the more environmentally friendly materials envisaged by these industrial giants. However, 60 years later with significant improvements

FIGURE 5.1 The first auto part made by Henry Ford from soybean proteins with fiberglass reinforcement in 1938. (From the collections of the Henry Ford Museum and Greenfield Village.)

in agricultural biotech, genetic engineering, crop production, polymer science, and the composite engineering field, the new field of bio-based materials is blossoming, spurred on by global environmental problems and petroleum feedstock finite resources. Such advances are evidenced by our ability to make agricultural equipment parts from soybeans as shown in Figure 5.2, bulletproof parts for armored vehicles, hurricane-resistant housing (Chapter 13), high-speed computer chips (Chapter 13), and other products. This chapter provides background on the development of the first generation of bio-based composites, such as those used in John Deere tractors, round hay balers, and combine harvesters. Improvements on the plant oil-based resins for composite applications are discussed in Chapter 7.

5.1 TRIGLYCERIDE-BASED COMPOSITE MATERIALS

All of the resins presented here are suitable for use as a matrix in a composite material. Their low viscosity (200–1000 cP) and method of curing (free radical; similar to all unsaturated polyesters) make them ideal candidates for use in conventional resin transfer molding (RTM) processes. Most polymer matrix composites are made by embedding strong fibers such as carbon, aramid, glass, or natural fibers in a polymer matrix. The high strength and modulus of the embedded fibers impart strength and rigidity to the material that surpass that of the neat polymer [1]. In recent years natural fibers have attracted attention as potential reinforcements due to the high cost of synthetic fibers. These cellulose-based fibers offer the advantages of biodegradability, low density, nonabrasive nature, and low cost.

Depending on their origin, natural fibers can be grouped into seed, bast, leaf, and fruit qualities. Bast and leaf quality fibers are the most commonly

FIGURE 5.2 The first John Deere round hay baler panel made with soybean oil.

used in composite applications. Examples of bast fibers include hemp, jute, flax, ramie, and kenaf. Leaf fibers include sisal and banana leaf fibers. Properties for these fibers include excellent tensile strength and modulus, high durability, low bulk density, good moldability, and recyclability. These natural fibers have an advantage over glass fibers in that they are less expensive, abundantly available from renewable resources, and have a high specific strength. While high-performance carbon fibers remain superior to natural fibers in high-end applications, natural fibers have comparable properties to glass fibers in high-volume applications [2]. The properties of flax, jute, sisal, and hemp fibers are shown in Table 5.1 and are compared to the commonly used E-glass fiber [3]. Notably, flax fiber has a modulus higher than that of E-glass. Flax is also less dense and, hence, produces a lighter composite with good mechanical properties.

Numerous studies on the properties of natural fiber composites using jute [4–8], banana [9], agave [9], hemp [9, 10], flax [10–12], bamboo [13], pineapple [14], and rubber wood [15] have appeared in the literature. For certain applications, natural fiber composites such as those made from flax or hemp fiber are not sufficient because of the low strength of these fibers. However, combining natural fibers with stronger synthetic fibers like glass could offer an optimum balance between performance and cost. These "hybrid" composites, which use two different types of fibers, have been examined in such forms as jute/glass hybrids with epoxy and polyester matrix materials [16, 17].

TABLE 5.1 Properties of natural and E-glass fibers (3).

	Density (kg/m^3)	Tensile Modulus (GPa)	Tensile Strength (GPa)
Flax	1500	100	1.1
Jute	1450	2.5–13	0.46–0.53
Sisal	1450	9.4–15.8	0.57–0.64
Hemp	1480	—	0.69
E-glass	2540	76	1.5

In previous works, the natural fibers were combined with petroleum-derived matrix resins. The resins presented here offer the unique potential of combining natural fibers with resins based on natural renewable resources. Here we present the properties of glass-reinforced composite materials made from AESO resin as well as all-natural fiber composite materials reinforced by flax and hemp fibers [18]. Additionally, we review the properties of hybrid composites manufactured from AESO-based resins reinforced with flax and glass fibers [19].

5.2 MANUFACTURING OF GLASS FIBER-REINFORCED COMPOSITES

The properties of glass fiber-reinforced composites made from the AESO and HSO/MA polymers were examined [20]. AESO, 1500 g, was mixed with 750 g styrene, 113 g divinyl benzene, 18 g cobalt naphtholate activator, and 68 g Trigonox 239 free-radical initiator. The resin was infused into a glass fiber (50% volume fraction) preform using Seeman's composite resin injection molding process. The composite was cured for 12 h at room temperature and postcured at 150 °C for 2 h. The properties were compared with those of a commercial high-performance vinyl ester resin containing the same fiber volume fraction and manufactured under the same conditions.

The HSO/MA composites were made using a resin composed of 100 g HSO/MA monomer synthesized in the manner previously mentioned, 45 g styrene, 5 g divinyl benzene, and 2.25 g 2,5-dimethyl-2,5-di-(2-ethylhexanoyl-peroxy)hexane as an initiator. The mixture was injected into a resin transfer mold containing two mats of glass fiber. The mold was heated at 65 °C for 1.5 h and was postcured at 1205 °C for 1 h. Properties were compared with those of the vinyl ester resin Dow Derakane 411C50 prepared under the same conditions. The tensile, flexural, and compressive properties for both composites were measured per ASTM D 3039, ASTM D 790, and ASTM D 3410, respectively.

5.2.1 MANUFACTURING OF FLAX AND HEMP COMPOSITES

More details on the method used to manufacture the flax and hemp composites can be found elsewhere [18, 21]. Composites were manufactured using Durafiber Grade 2 flax fiber and hemp fibers within the AESO-based polymer. The hemp was obtained in the form of a nonwoven mat with an oriented fiber direction.

The resin used in these all-natural composites was prepared by blending AESO, styrene, and divinyl benzene in the ratio 100:45:5 by weight. To initiate the free-radical polymerization reaction, 1.5% of the total resin weight of 2,5-dimethyl-2,5-di-(2-ethylhexanoylperoxy)hexane was added. The composite was manufactured using an RTM process. For the flax composites, flax fibers were uniformly arranged in the mold in a random mat. Hemp composites were made with two layers of the oriented hemp mats placed with their orientation perpendicular to each other. Resin was injected into the mold, and the composite was cured at 90 °C for 1 h, followed by a postcure at 110°C for 1 h. The tensile and flexural properties were measured per ASTM D 3039 and ASTM D 790, respectively.

5.2.2 MANUFACTURING OF GLASS/FLAX HYBRID COMPOSITES

Hybrid composites were manufactured symmetrically and asymmetrically using E-glass woven fiber and Durafiber Grade 2 flax fibers in an AESO-based polymer [19]. In the symmetric composites, a layer of flax fiber was sandwiched between two layers of glass fiber. The asymmetric composites were produced by uniformly arranging the flax fibers at the bottom of the mold in a random mat and then placing two layers of woven glass fabric on top.

The resin for these studies was prepared by mixing AESO with 50 parts per hundred by weight (phr) styrene, 4.5 phr Trigonox 239A free-radical initiator, and 1.2 phr cobalt naphthalate activator. After injecting the resin into the mold, the composite was cured overnight at 2.1 MPa and room temperature. The composite was postcured at 110 °C for 2 h, after which the composite panels had fiber weight fractions in the range of 0.31 to 0.40. Composites with glass/flax ratios of 100/0, 80/20, 60/40, 40/60 and 0/100 were made. The tensile, compression, and flexural properties were measured per ASTM D 3039, ASTM D 3410, and ASTM D 790, respectively. Impact tests were conducted per ASTM D 3763.

5.3 COMPOSITE PROPERTIES

5.3.1 AESO AND HSO/MA GLASS FIBER COMPOSITES

The properties of the glass fiber-reinforced AESO composites are shown in Table 5.2 [20]. The physical properties are very close to those of the Dow

TABLE 5.2 Properties of glass fiber reinforced AESO-based polymer and Dow PC100 vinyl ester polymer.

	Testing Direction	Tensile Strength (MPa)	Tensile Modulus (GPa)	Compressive Strength (MPa)	Compressive Modulus (GPa)
AESO	0°	463.2	24.8	302.6	24.8
Dow PC100	0°	458.4	23.8	420.5	23.4
AESO	90°	321.9	20.7	180.6	20.7
Dow PC100	90°	324.0	17.6	339.1	17.9

PC100 resin. The tensile strength, tensile modulus, and compressive modulus all are similar to the properties of the vinyl ester resin. The only apparent shortcoming is in the compressive strength of the AESO resin. This can be attributed to the lower strength of the AESO neat polymer. However, according to DMA, the AESO composite still displayed a T_g close to that of the neat polymer at approximately 80 °C. This is much lower than the T_g of the vinyl ester polymer, which was found to be about 128 °C.

The HSO/MA composite properties were found to be even more successful at replicating the properties of a vinyl ester composite, as shown in Table 5.3. The flexural modulus and compressive strength for the HSO/MA composite were of same magnitude as the vinyl ester composite, while the flexural strength was found to be slightly lower. Additionally, the T_g of the HSO/MA composite was found to be approximately 128 °C, which equals that found for the vinyl ester composite.

These results indicate that although the properties of the neat soy oil based polymers are less than those of the vinyl ester polymers, the composite material properties are very similar. In tensile deformation, the fiber reinforcement is able to support the majority of the load leading to an acceptable modulus and strength. The area that needs improvement is compression deformation, where the polymer bears the majority of the stress.

5.3.2 FLAX COMPOSITE PROPERTIES

Figure 5.3 shows the tensile and flexural strength as a function of fiber content for the Durafiber Grade 2 flax composites. The tensile strength of the

TABLE 5.3 Properties of HSO/MA-based polymer and Dow 411C50 vinyl ester composites.

	Flexural Modulus (GPa)	Flexural Strength (MPa)	Compressive Strength (MPa)
HSO/MA	34.5	669	200
Dow DK 411C50	35.8	813	290

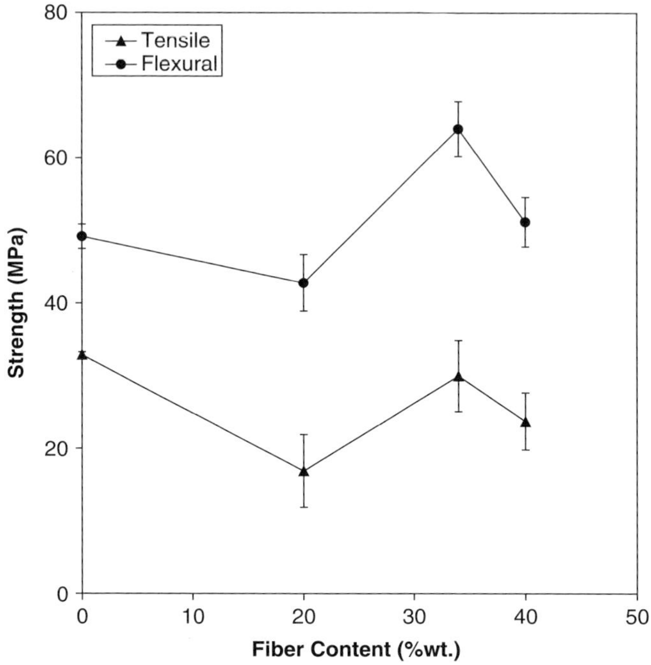

FIGURE 5.3 Strength dependence on composition for flax (Durafiber Grade 2)-reinforced AESO polymer.

AESO/flax fiber composite was found to have a maximum value of 30 MPa at 34% fiber content, which is comparable to the tensile strength of the AESO neat resin (~30 MPa). The flexural strength showed a similar trend, exhibiting a maximum value at approximately 34% fiber content. The flexural moduli of these materials behaved similarly showing a maximum at 34% fiber content, while the tensile moduli increased with fiber content as illustrated in Figure 5.4. Other researchers have noticed this optimization phenomenon, which was explained in terms of increasing fiber–fiber interactions as the fiber content increases [2]. This reduces the level of fiber–matrix interaction, thereby weakening the composite. Percolation theory has also been used to explain this effect [22].

5.3.3 HEMP COMPOSITE PROPERTIES

Composites made of 20 wt% hemp fiber were found to display a tensile strength and modulus of 35 MPa and 4.4 GPa, respectively. The flexural properties of the composites were found to be anisotropic. Samples tested with the side of the composite in contact with the upper surface of the mold

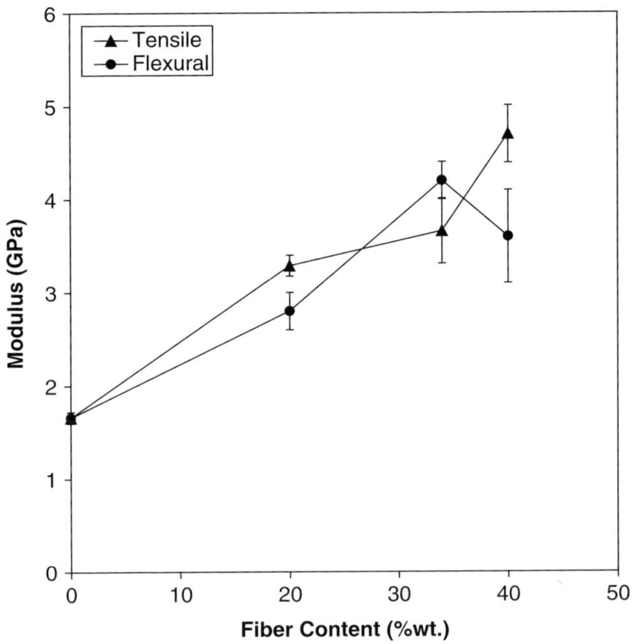

FIGURE 5.4 Modulus dependence on composition for flax (Durafiber Grade 2)-reinforced AESO polymer.

during the curing process display a yield strength of 35.7 ± 5.9 MPa and a modulus of 2.6 ± 0.2 GPa. Samples tested with the side in contact with the lower surface of the mold during cure have a yield strength of 51.3 ± 2.7 MPa and a modulus of 2.7 ± 0.2 GPa. Thus, the modulus does not appear to be affected by orientation, while the yield strength has a 44% increase. This increase in strength is substantial and can be attributed to the orientation of the fibers in the mat.

The mechanical properties of the all-natural composites are comparable to the properties shown by wood. For example, a typical hard wood has a tensile modulus of about 10 GPa, with a fracture stress of about 30 MPa when the stress is exerted parallel to the fiber axis and about 3 MPa when the stress is exerted normal to the fiber axis. The considerable advantage of the all-natural composites is that the unidirectional high-strength properties of wood can be obtained in all directions for the randomly oriented fiber composite. In addition, the ease of manufacturing complex shapes via normal composite liquid molding operations provides a significant cost advantage for these materials. The ACRES group is currently designing roofs for houses using these materials, which are expected to be more hurricane resistant than current designs.

5.3.4 HYBRID NATURAL-GLASS FIBER COMPOSITES

The tensile modulus, tensile strength, and compressive strength of the glass/flax hybrid composites for different glass/flax ratios and composite constructions are shown in Table 5.4. As can be expected, these properties all increase with increasing glass fiber content. The 100% flax fiber-reinforced materials show a tensile strength and modulus of 26.1 ± 1.7 MPa and 1.9 ± 0.1 GPa, respectively. At the other extreme, the 100% glass fiber-reinforced materials show a tensile strength and modulus of 128.8 ± 1.1 MPa and 5.2 ± 0.1 GPa, respectively.

As shown in Table 5.5, the asymmetric composites have tensile moduli similar to the symmetric composites' moduli. However, the tensile and compression strengths of the asymmetric composites were noticeably less than those of the symmetric composites. This is due to the different modes of failure exhibited by the two types of composites. The symmetric composites undergo tensile failure at the peak load, while the asymmetric composites fail by shear delamination at the glass/flax interface due to the difference in the tensile moduli of the two fiber types.

The flexural properties of the glass/flax hybrid composites are shown in Table 5.6. The flexural modulus and strength for the glass fiber composite are much higher than those for the flax fiber composite due to the higher

TABLE 5.4 Properties of symmetric glass/flax hybrid composites.

	Weight Fractions				
Glass/Flax Ratio	Glass	Flax	Tensile Modulus (GPa)	Tensile Strength (MPa)	Compression Strength (MPa)
100/0	0.35	0.00	5.2 ± 0.1	128.8 ± 1.1	89.8 ± 3.2
80/20	0.25	0.06	3.5 ± 0.1	123.3 ± 1.2	71.6 ± 2.6
60/40	0.23	0.16	3.2 ± 0.1	109.1 ± 1.0	62.3 ± 3.1
40/60	0.16	0.24	2.9 ± 0.2	82.6 ± 1.4	33.6 ± 0.8
0/100	0.00	0.31	1.9 ± 0.1	26.1 ± 1.7	18.5 ± 2.4

TABLE 5.5 Properties of asymmetric glass/flax hybrid composites.

	Weight Fractions				
Glass/Flax Ratio	Glass	Flax	Tensile Modulus (GPa)	Tensile Strength (MPa)	Compression Strength (MPa)
80/20	0.25	0.06	3.4 ± 0.1	111.7 ± 2.1	65.3 ± 4.8
60/40	0.24	0.16	3.1 ± 0.1	90.6 ± 2.4	46.2 ± 0.6
40/60	0.16	0.25	2.7 ± 0.3	68.9 ± 2.1	30.1 ± 2.2

TABLE 5.6 Flexural properties and energy absorption on impact of glass/flax hybrid composites.

Glass/Flax Ratio	Weight Fractions: Glass	Weight Fractions: Flax	Composite Construction	Loading/ Impact Face	Flexural Modulus (GPa)	Flexural Strength (MPa)	Energy Absorbed (J)
100/0	0.35	0	—	—	9.0 ± 0.2	205.5 ± 4.5	16.5 ± 0.2
80/20	0.25	0.06	Symmetric	—	6.9 ± 0.2	130.3 ± 3.0	17.7 ± 1.9
	0.25	0.06	Asymmetric	Glass	6.3 ± 0.3	87.8 ± 3.9	13.3 ± 0.3
			Asymmetric	Flax	5.0 ± 0.1	189.0 ± 8.5	25.8 ± 1.1
60/40	0.23	0.16	Symmetric	—	6.0 ± 0.2	115.3 ± 2.5	18.0 ± 0.3
	0.24	0.16	Asymmetric	Glass	4.0 ± 0.3	80.1 ± 0.7	14.7 ± 0.3
			Asymmetric	Flax	4.7 ± 0.3	146.9 ± 5.5	27.6 ± 2.6
40/60	0.16	0.24	Symmetric	—	5.8 ± 0.5	83.3 ± 5.4	18.5 ± 0.2
	0.16	0.25	Asymmetric	Glass	3.8 ± 0.1	73.2 ± 7.5	15.1 ± 0.3
			Asymmetric	Flax	3.3 ± 0.4	111.1 ± 9.5	28.7 ± 1.2
0/100	0	0.31	—	—	3.8 ± 0.2	61.0 ± 3.4	1.4 ± 0.2

modulus and strength of glass fibers. The 100% flax-reinforced composites display a flexural strength and modulus of 61.0 ± 3.4 MPa and 3.8 ± 0.2 GPa, respectively. The 100% glass fiber-reinforced composites have a flexural strength and modulus of 205.5 ± 4.5 MPa and 9.0 ± 0.2 GPa, respectively. Additionally, there is an obvious anisotropy in the behavior of the asymmetric composites depending on the surface that bears the load. The maximum flexural strengths occurred when the flax surface bore the exerted load or impact. In such an orientation, the glass fibers bear a tensile load from the bending of the sample.

The impact energy of the hybrid composites ranged from 13.3 ± 0.3 to 28.7 ± 1.2 J. The maximum impact energy absorbed (28.7 ± 1.2 J) was shown by the asymmetric 40/60 glass/flax ratio composite when the flax surface was the load-bearing face. The energy absorption by the symmetric hybrid composites seems to be only marginally higher than that of the 100% glass fiber composite, a difference made even more insignificant when considering the standard deviations.

5.4 SHEET MOLDING COMPOUND

Unsaturated polyesters synthesized from petroleum-based chemicals have enjoyed long-term leadership in polymeric composites since 1941. The versatility and low cost make these resins very popular in applications in construction, transportation, electric, and electronic industries. One of the major applications of unsaturated polyester is its use in sheet molding compound

(SMC) [23, 24]. SMC is one of the major polymer composites used in the automobile industry due to its light weight, high strength, dimensional stability, and very good surface quality. In Chapter 4, we reported on the synthesis and characterization of new thermosetting resins [25–32] from soybean oil and SMC applications by J. Lu [31]. Figure 5.5 shows a scheme to prepare SMC resins from soybean oil. The functionalized soybean oil, which include maleated hydroxylated soybean oil (MHSO) and maleated acrylated epoxidized soybean oil (MAESO), when combined with styrene, can form the rigid polymers by free-radical polymerization.

Typical compositions of SMC are shown in Table 5.7. It is very common in SMC applications to thicken the materials before molding for easy handling and good fiber-carrying capability. The high viscosity reduces the segregation of reinforcement during molding and polymerization shrinkage [23]. The most common thickeners for unsaturated polyesters are alkaline earth metal oxides or hydroxides and diisocyanate compounds. Although using

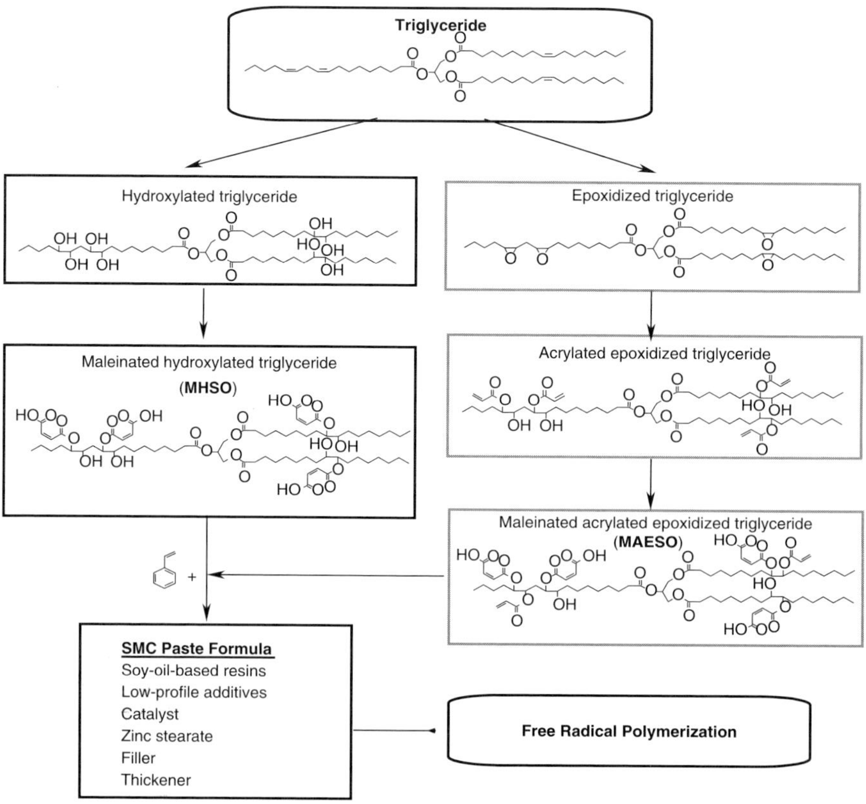

FIGURE 5.5 A scheme to prepare SMC from soybean oil.

TABLE 5.7 Sheet molding compound.

Component	Weight Percent Ranges
Styrene	5–14%
Soy resin	10–12%
Styrene/soy ratio	0:1–1:1
Glass	25–35%
Glass fiber sizing	–
Natural fibers	0–35%
Natural fiber sizing	–
Calcium carbonate	35–45%
Low profile additive	3–4%
Initiator	1.00%
Divalent cation	0.5–1.0
Zinc stearate	1.00%
Other additives	0–10%

diisocyanate compounds as thickener can result in a highly stable viscosity, the formation of covalent urethane bonds that are stable at the molding temperatures causes the reduced sensitivity to heat. The triglyceride-based monomers have been tested to develop thickening with diisocyanate because these molecules possess hydroxyl groups [33], but this method was not considered in this work. For the thickening process using alkaline earth metal ions, magnesium oxide (MgO) is a popular choice because of its low cost and high reactivity. An ideal viscosity profile during SMC processing is that, first, the initial viscosity of the compound should be low enough to permit a good fiber wet-out, and during thickening, the viscosity should increase quickly to reach a moldable viscosity to reduce the storage cost. After that, the viscosity should remain stable to keep a long shelf life. During the molding process, when SMC sheet is placed in a heated mold, with a temperature and pressure increase, the viscosity should decrease in a second to keep a good mold flow, otherwise, the mold may not be completely filled or an unusual air trap may develop. The mechanism of thickening behavior for unsaturated polyester (UP) was extensively investigated in the past and it is always a subject of much controversy [34–39]. A two-stage reaction theory [Figure 5.6(a)] postulated that the high viscosity results from the formation of high-molecular-weight products as a result of reaction of MgO with dicarboxylic acid. At the same time, the basic magnesium salt can interact with carbonyl or hydroxyl oxygens of UP to form complexation, which results in a coordinated three-dimensional (3-D) network, as shown in Figure 5.6(b). In this section, the thickening behavior of these new polymers using MgO paste is examined using three different ratios of MA to triglyceride. These ratios correspond to 1:2 (MAESO2), 1:3 (MAESO3), and 1:4 (MHSO) triglycerides: MA molar ratios.

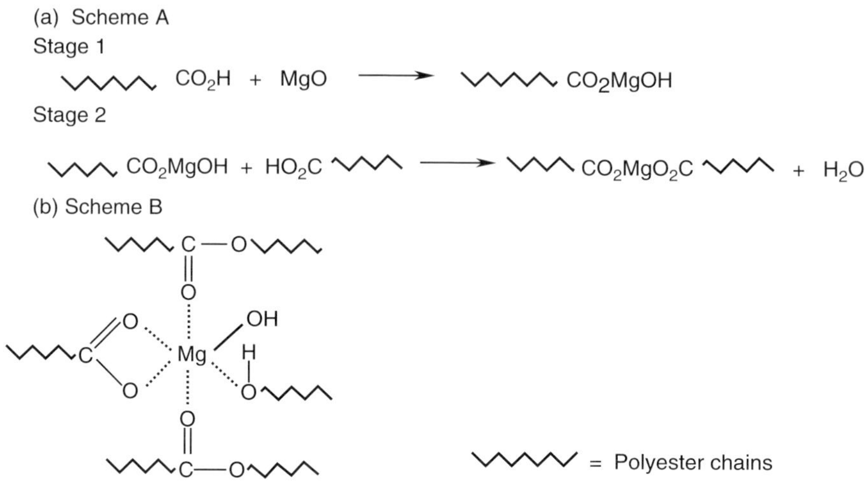

FIGURE 5.6 Thickening mechanism of unsaturated polyester.

5.4.1 SMC THICKENING BEHAVIOR

The thickening process is an essential step in SMC applications. The carboxylic acid groups in unsaturated polyester are able to react with magnesium ion, which causes at least a 1000-fold increase in viscosity in 2–3 days. Figure 5.7 shows the viscosity changes of MAESO2 and MHSO systems when thickened with 1.5 wt% MgO paste. The starting viscosity for both resins is approximately 1200 cP. It takes less than 40 h for the viscosity of both resins to reach more than 10^6 cP, which is a common moldable viscosity range during the SMC process. After that, the viscosity fluctuates a little, which may result from the humidity change in the environment, because the water content can affect the thickening behavior [40]. Compared to the thickening behavior of the commercial UPs, the triglyceride-based resins need a smaller amount of thickener and less time to achieve the same saturated viscosity. This is possibly due to the distribution of carboxylic acid groups on fatty acid backbones. Some triglyceride molecules may have more than two acid groups, which results in the formation of a cross-linked network. Another possible reason is that triglyceride monomers have unreacted hydroxyl groups (approximately 1.2 hydroxyl groups/triglyceride), which can coordinate with the magnesium basic salt to form a 3-D network as shown earlier in Figure 5.6(b).

The links between magnesium ion and carbonyl oxygen are weak, and they break up at high temperature. To understand the viscosity changes during heating, the thickened sheet was placed in a 150°C silicon oil bath and the viscosity of the sheet was followed by the Brookfield viscometer. Figure 5.8 shows the viscosity and temperature changes versus time during heating.

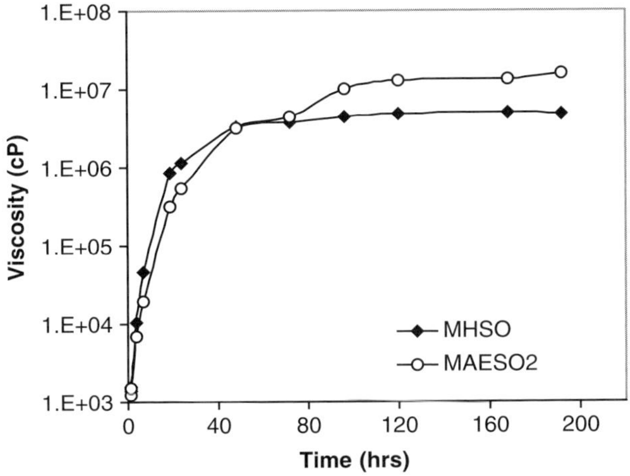

FIGURE 5.7 Viscosity changes during the maturation process of triglyceride-based resins with 1.5 wt% MgO.

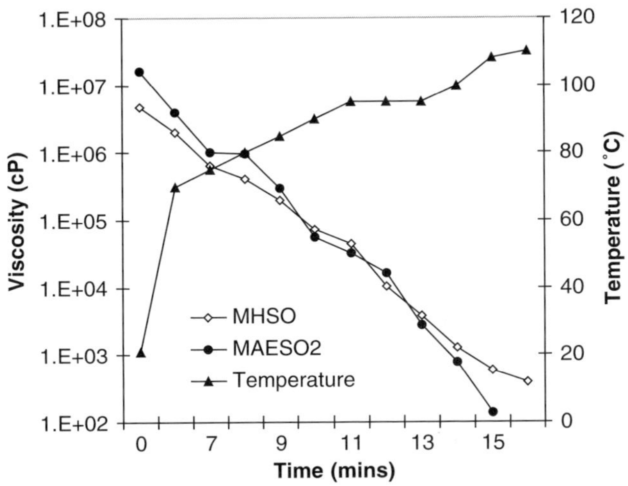

FIGURE 5.8 Viscosity changes during heating.

With the temperature of the resins increasing, the viscosity decreases dramatically from 10^7 cP to the initial value of unthickened resins, which means that all of the thickening bonds are broken; this entire process takes about 15 min. In the real case, the viscosity may not drop that much as the curing reaction starts.

The effect of molecular structure and the amount of MgO on the thickening behavior were also examined. Figure 5.9 shows that MAESO3 initially has a faster viscosity increase than MAESO2, but finally they reach a similar viscosity. The possible reason is that MAESO3 has more carboxylic

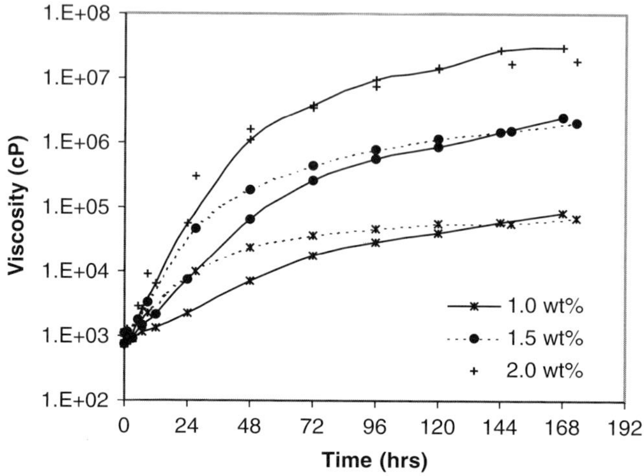

FIGURE 5.9 Viscosity changes using different amount of thickener: .... MAESO3, —MAESO2.

acid groups, which make the condensation reaction much easier. Many other factors could affect the thickening process, such as the content of water and initial molecular weight. Figure 5.9 also shows that, with more thickener, the faster the viscosity rise and higher the final viscosity. One weight percent MgO based on the total weight of the resin is not enough for these triglyceride-based resins to reach a moldable viscosity, but more than 2 wt% MgO may not be necessary for the thickening process. The compound viscosity design can be achieved by combining the resins and different amount of thickeners.

5.4.2 THERMAL AND MECHANICAL PROPERTIES OF SMC

Figure 5.10 shows the temperature dependence of the loss factor (tan δ) for maleated soybean oil compared to acrylated soybean oil. The glass transition temperatures for these new SMC polymers are in the range of 100–115 °C. Apparently, with the modification of MA, the single transition of the curves shifts to high temperatures, and the intensity of tan δ decreases with increasing molar ratio of MA to AESO, which indicates an increased cross-link density and lower toughness. The glass transition temperature of these polymers is very important because the T_g determines an upper bound for the temperature at which the polymers can be used. As shown in Figure 5.11, the glass transition temperature of the new SMC polymers exhibits a linear fit with the cross-link density, which is following the theories of Wool and Fox and Loshaek discussed in Chapter 7:

$$T_g = T_{gL} + K_{FL} \cdot v \tag{5.1}$$

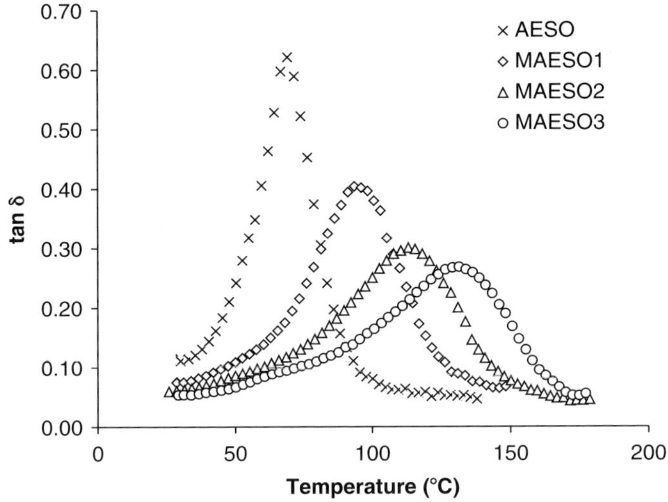

FIGURE 5.10 Temperature dependence of the loss factor tan δ for MAESO polymers at different molar ratios of MA to AESO.

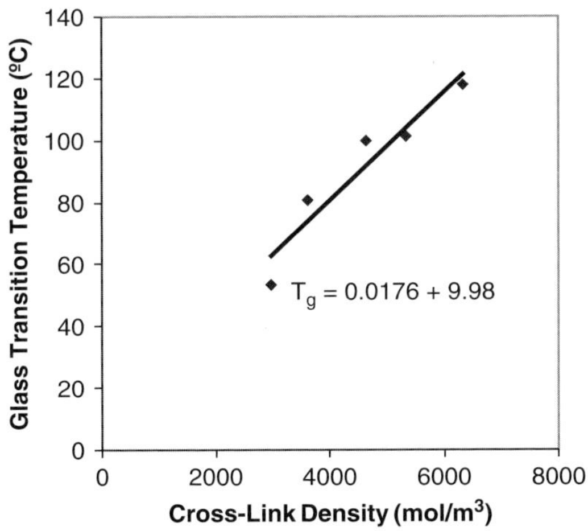

FIGURE 5.11 The glass transition temperature of triglyceride-based polymers follows a linear fit with the cross-link density.

where T_{gL} is the glass transition temperature for an infinite straight-chain polymer, K_{FL} is a universal constant, and ν is the cross-link density. The linear fit of the experimental data gives a value of 9.98 °C and $0.0176\,m^3/mol$ for T_{gL} and K_{FL}, respectively.

Figure 5.12 shows the mechanical behavior of the triglyceride-based polymers from flexural tests. Basically they all show a typical deformation of brittle plastics in terms of the stress and strain. Beyond the yield point, the deformation of MHSO and MAESO1 polymers ceases to be elastic, but the failure of MAESO2 and MAESO3 occurs at a strain of less than 7.0%. Table 5.8 shows that these polymers have a flexural strength in a range of 60–90 MPa and modulus in a range of 2.4–3.0 GPa. Figure 5.13 shows the tensile stress–strain behavior of these new polymers. Again, they show a typical deformation of brittle plastics with tensile strength in a range of 27–44 MPa, and tensile modulus in a range of 1.6–2.5 GPa. The Poisson ratio is approximately 0.4, which is in the range of typical plastics (Table 5.8). The Poisson ratio slightly decreases with increasing maleic anhydride modification. We can see that the mechanical properties of these new polymers are comparable to the commercial resins, such as ortho-unsaturated polyester, and iso-unsaturated polyester, which are used in the SMC industry. Also, the advantage

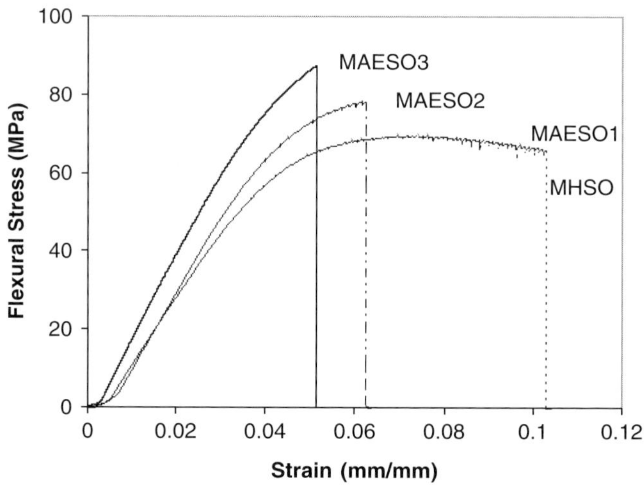

FIGURE 5.12 Flexural behavior of the triglyceride-based polymers.

TABLE 5.8 Mechanical properties of triglyceride-based polymers.

Polymer sample	AESO	MAESO1	MAESO2	MAESO3	MHSO4
Flexural Strength (MPa)	54.84 ± 1.53	69.55 ± 0.72	77.06 ± 1.50	87.24 ± 5.24	61.42 ± 0.75
Flexural modulus (GPa)	1.90 ± 0.05	2.47 ± 0.12	2.68 ± 0.09	3.03 ± 0.28	2.46 ± 0.19
Tensile strength (MPa)	21	29.52 ± 2.66	39.70 ± 3.27	44.08 ± 3.03	27.04 ± 5.71
Tensile modulus (GPa)	1.63	1.81 ± 0.08	2.18 ± 0.13	2.47 ± 0.0.26	1.61 ± 0.05
Elongation		0.040	0.052	0.0439	0.038
Poisson Ratio	0.43	0.412	0.394	0.392	0.394

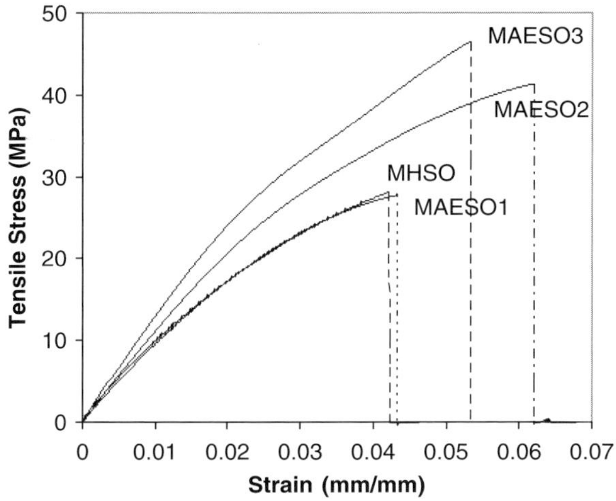

FIGURE 5.13 Tensile stress–strain curves for the triglyceride-based polymers.

gained here is in the amount of renewable material in the resins, which contains up to 50 wt% soybean oil.

It is obvious that with more MA modification, both flexural and tensile properties increase as a result of increasing the molar ratio of MA to AESO. The modification with MA increases the functional groups on the triglyceride molecules, resulting in an increase in cross-link density, which increases the mechanical properties. Figure 5.14 shows the effect of the cross-link density on the tensile strength of AESO-based polymers. The cross-link density was calculated based on the kinetic theory from the rubbery modulus at temperatures well above T_g [31]. Initially, the tensile strength increases rapidly with increasing cross-link density, and then it becomes insensitive to the cross-linking. The tensile strength is expected to increase with cross-linking because weak van der Waals bonds are replaced by strong covalent bonds, but the high cross-link density results in the development of submicroscopic cracks from the internal stress as the mobility of the molecular segments decreases [41]. A prediction relating the molecular structure to the properties of these new polymers would be valuable. Indeed, by knowing the triglyceride molecular structure, La Scala et al. [42, 43] showed that the use of a simple model, such as vector percolation (details in Chapter 7), can predict the mechanical properties of these new polymers. Percolation theory relates the polymer properties to the difference between the level of perfection of the network, p, and the percolation threshold, p_c [44–46]. The critical stress, σ_c, required to break a network is

$$\sigma_c = (E\nu D_0[p - p_c])^{1/2}, \tag{5.2}$$

where E is the Young's modulus, D_0 is the C−C bond rupture energy, and ν is the cross-link density. Because $[p - p_c]$ is essentially constant, then we have the simple

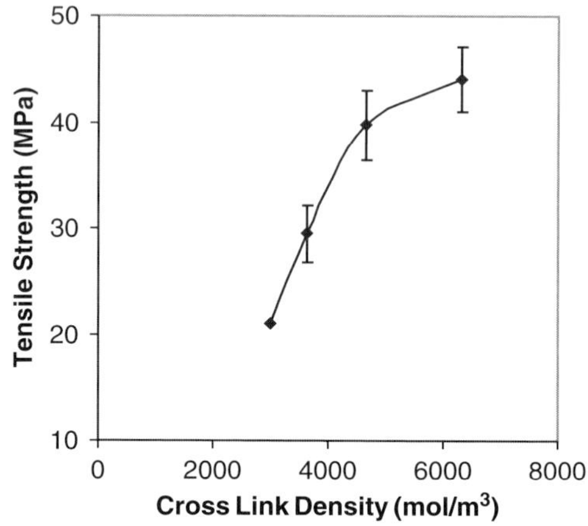

FIGURE 5.14 Effect of cross-link density on the tensile strength of triglyceride-based polymers.

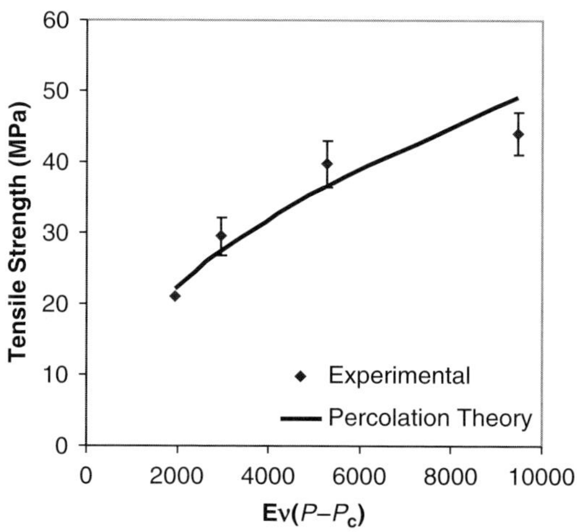

FIGURE 5.15 Comparison of polymer strength with vector percolation behavior prediction.

relation for the fracture stress as $\sigma \sim [E\nu]^{1/2}$ [47]. Figure 5.15 shows a comparison of the percolation theory with the tensile strength data. We can see that the percolation theory fits the experimental data very well. From these results, the modulus and strength of triglyceride-based polymers can be increased by simply increasing the chemical functionalities on the triglycerides. This can be done by using highly unsaturated oils with high linolenic fatty acid content, or by additional chemical functionalization as shown earlier in Figure 5.5.

The fracture toughness, which can be expressed by the critical stress intensity factor K_{1c} and critical strain energy release rate G_{1c}, quantify the resistance of a polymer to initiate and propagate cracks. For these triglyceride-based polymers, with the increase in the cross-link density, a transition from ductility to brittleness happens due to the restriction of the molecular mobility. Figure 5.16 shows the microstructure of the fracture surfaces near the notch for AESO- and MAESO2-based polymers. The fracture surface of the AESO-based polymer is very rough, whereas the MAESO2-based polymer has a smooth mirror-like surface, which indicates a brittle structure. The relationship between fracture toughness and cross-link density is shown in Figures 5.17 and Figure 5.18. Critical stress intensity factor K_{1c} varies from 0.4 to 1.5 MPa · m$^{1/2}$ and G_{1c} ranges from 100 to 1500 J/m^2.

The size of the plastic zone, r_p, can be calculated from [48]

$$r_p = \frac{1}{2\pi}\left(\frac{K_{1C}}{\sigma_y}\right)^2 \tag{5.3}$$

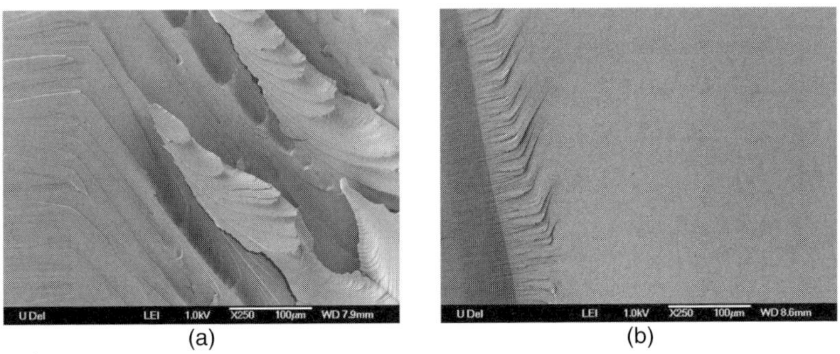

FIGURE 5.16 SEM photographs for the fracture surfaces near the notch for triglyceride-based polymers: (a) AESO polymer and (b) MAESO polymer.

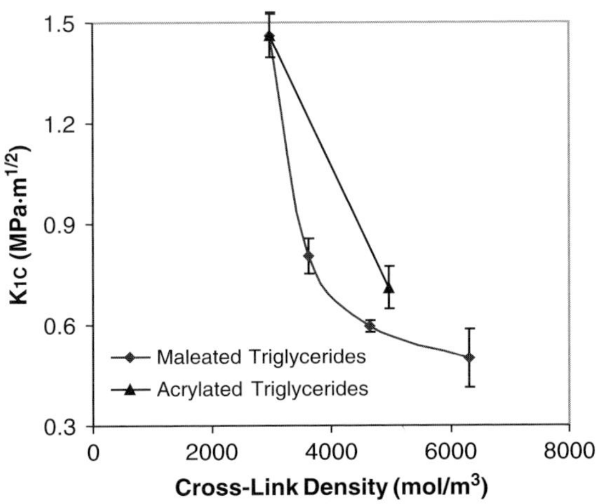

FIGURE 5.17 Effect of cross-link density on the critical stress intensity factor, K_{1c}.

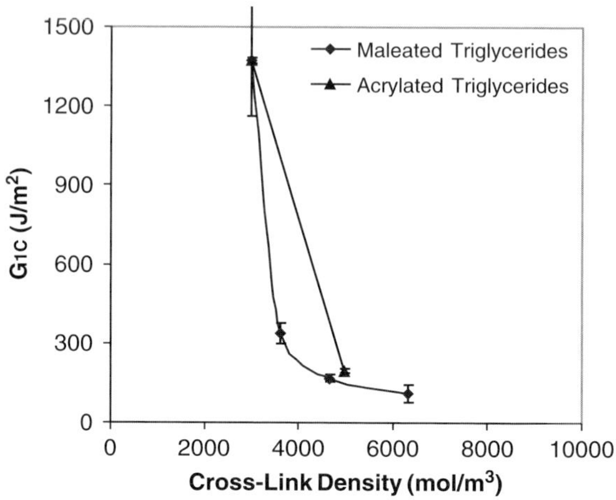

FIGURE 5.18 Effect of cross-link density on the critical strain energy release rate, G_{1c}.

where σ_y is the yield stress. Both Fischer [49] and Bos and Nusselder [50] have shown that the size of the plastic zone is directly proportional to the cross-link density of epoxy resins. Figure 5.19 is a plot of the size of the plastic zone against the cross-link density of the polymers. The size of the plastic zone for AESO polymer is 772 μm; it significantly decreases for MAESO polymers, for which it is in the range of 20–110. Again, the size of the plastic zone for these triglyceride-based polymers does not show a first-order dependence on the cross-link density.

5.4.3 SUMMARY OF SMC

Figure 5.20 shows a John Deere harvester rear wall prototype made by the ACRES group. The exterior body panels of these harvesters are now being made commercially with soy-based SMC. This work provided experimental evidence that these resins can be used in SMC applications for the automotive, trucking, and agricultural industries and for many other applications. When these resins were thickened with MgO paste, the viscosity increased quickly during the maturation process and kept a stable value during room temperature storage. The thickening behavior is affected by the monomer structure and the amount of thickener. The flexural strength and moduli of these polymers varied from 61 to 87 MPa and from 2.4 to 3.0 GPa, respectively. The tensile strength and moduli varied from 27 to 44 MPa and from 1.6 to 2.5 GPa, respectively. The elongation at break was approximately 5.0%. The properties of these polymers are obviously related to the amount of the functional groups on the fatty acid backbone, and they can be predicted using percolation theory. The fracture toughness of these new polymers decreased with increasing cross-link density, similar to all composite resins. The new polymers possessed mechanical properties comparable to those of the com-

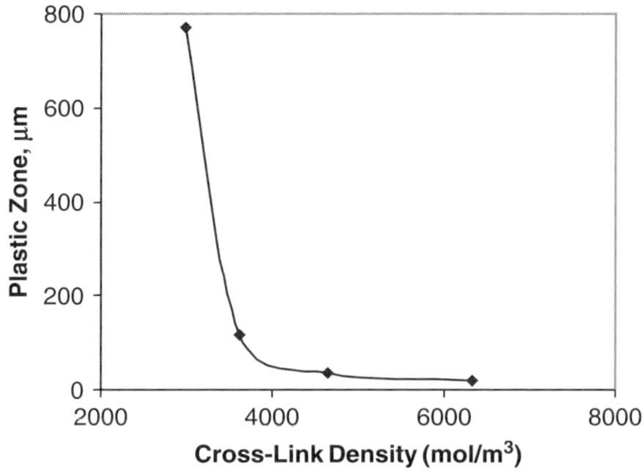

FIGURE 5.19 The size of plastic zone versus the cross-link density for the triglyceride-based polymers.

FIGURE 5.20 The first SMC prototype panel made for John Deere harvesters.

mercially available unsaturated polyesters that are commonly used in SMC applications.

5.5 BIO-BASED POLYMERIC FOAMS

Polymeric foams are complex gas/solid structured materials, consisting of a multitude of gas cells inside a solidified polymer matrix. This two-phase architecture presents numerous physical and mechanical advantages over simple polymers, such as a higher weight-to-strength ratio, added flexibility, lower thermal and electrical conductivities, and better shock absorption and sound-dampening properties [51, 52]. Polymer foaming is a mature industry; the total U.S. foamed plastics demand is projected to grow nearly 3% annually to 8 billion pounds in 2005 [53] and is valued at 18 billion. Densities of solid polymeric foams typically range from 1.6 to 960 kg/m^3, according to the needs of a wide range of applications that include furniture, construction, transportation (high-density foams), cushioning, packaging (flexible foams), insulation, and filtration (low-density foams) [51, 52].

To date, some polymeric foams have been produced out of soybean oil polyurethanes and out of starch. Starch-based plastics are used in specific industrial applications where biodegradability is required. In 1999, the market for starch-based biopolymers was estimated at about 20,000 ton/year, with a strong incidence of soluble foams for packaging and films [54]. Polyurethanes (PUs) are formed from the reaction of a diol or triol with a diisocyanate, in 1:1 proportions (Figure 5.21). Many different plant oil triglycerides have successfully been functionalized for the production of polyols used for PU plastics or foams (Figure 5.22). In January 2004, the United Soybean Board estimated that 400 million pounds of soybean oil are used annually in synthesizing polyols for the production of PU foams [55]. Bio-based PU and PU foams display mechanical and thermal properties comparable to those of petroleum-based ones [56]. In addition, soy-based polyols are cheaper than petroleum-based polyols [55]. However, because of the 1:1 stoichiometry of the polyurethane reaction, plant-based PU foams still use a large proportion of chemicals produced by the petroleum industry (diisocyanates).

The ACRES group has designed several new thermosetting polymers from plant oils (Chapter 4) [20]. Initial investigations are performed on process parameters and additives for the foaming of AESO with pressurized carbon dioxide (CO_2) into foams. Rigid polymers such as cured AESO generally form closed-cell materials (e.g., polyurethanes, epoxy resins, silicones, polyvinylchlorides, or polystyrene foams). The expected resulting AESO/CO_2

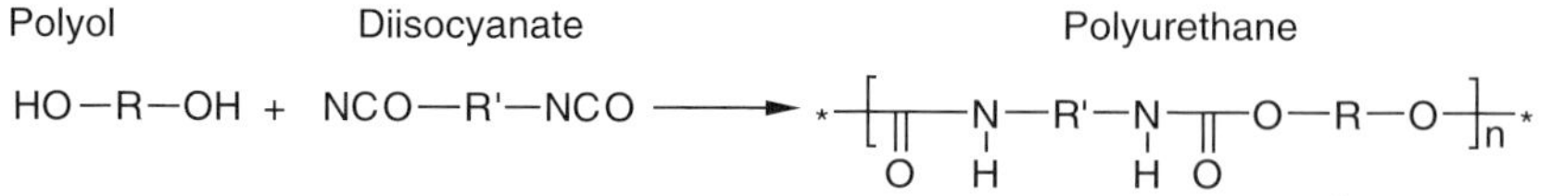

FIGURE 5.21 General polyurethane formation reaction.

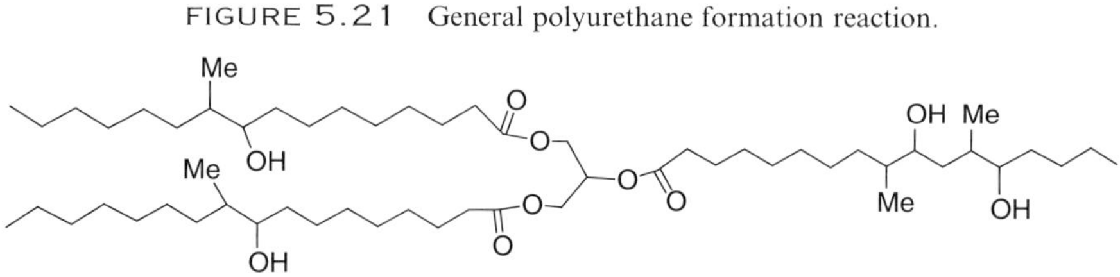

FIGURE 5.22 Structure of a polyol made from soybean oil triglyceride. The triglyceride is first epoxidized with hydrogen peroxide, then the epoxy rings are opened with methanol [56, 61].

cured foams would have a higher bio-based content than already produced soybean-based polyurethane foams, and would be stronger and less readily biodegradable than starch foams. Potential applications include its inclusion in a hurricane-resistant roofing structure designed by O'Donnell et al. [57] (Chapter 13), windmill blades; interior design panels; emergency housing for tsunami, flood, earthquake, and hurricane survivors; building insulation; and tissue scaffolds.

During polymerization, AESO forms a gel due to its high cross-linking potential; on average, 6.8 cross-links per monomer can theoretically be created [42]. The results are rigid thermosetting resins. AESO and its derivatives have been found to exhibit tensile moduli of around 1 GPa and glass transition temperatures in the range of 70–150 °C. Polymeric foams using these plant oil-based resins should produce materials strong enough to be used in structural applications.

5.5.1 AESO/CO_2 FOAMS

The process presented in Figure 5.23 was inspired from the works of Wei [58] and Mohamed [59] on the high-pressure foaming of polymer-fiber composites in a Parr reactor. A 400-mL Parr high-pressure reactor is connected to a pressurized CO_2 tank. Carbon dioxide was chosen as the blowing agent to be used in the high-pressure reactor because it is inexpensive, nontoxic, nonreactive, and environmentally benign. It is already used in polymer foaming processes as a replacement for HCFC gases [52] because of its high solubility in organic solvents (~2.5 mL/g at 25 °C in benzene, toluene, or heptane; even higher in methanol and acetone). The amount of CO_2 that dissolves into the monomer mixture (at saturation) is a function of the temperature and pressure. Raising the temperature in the reactor causes the pressure to increase linearly. For gases with a critical point lower than $T_c = 140°C$ and $P_c = 100$ bars, the supercritical state can be reached (CO_2: $T_c = 31.04$ °C, $P_c = 73.8$ bars), where the gas's solubility into AESO becomes theoretically infinite. The foam's density depends on the amount of blowing agent dissolved; hence, lightweight foams may be obtained.

The polymerization begins as soon as the thermal initiator decomposes at the initiation temperature. The formation of gas cells is triggered by opening the release valve on top of the reactor. Either part or all of the pressurized gas is vented. The pressure drop is used as the driving force causing the expansion of the foam. When the solubility of the gas is reduced, excess gas vaporizes and gas cells grow according to Eq. (5.4):

$$\Delta P = \gamma / r, \tag{5.4}$$

where ΔP is the difference between the pressure in the gas cell and the pressure in the liquid matrix, γ is the surface tension, and r is the gas cell's radius [51].

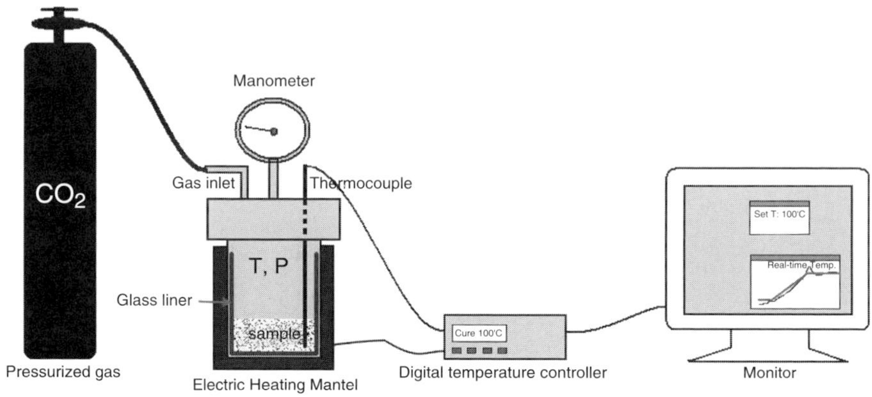

FIGURE 5.23 High-pressure thermostatic reactor setup [60].

For a thermosetting polymer, foaming will be successful only if cell formation is triggered just before the gel point. If triggered too early, the foam will slowly collapse before the matrix solidifies, under the influence of cell coarsening and coalescence and monomer drainage due to gravity. Therefore, it is vital to study the stability of the liquid foam before polymerization. Increasing the stability of the liquid foam before polymerization consists of slowing down the foam collapse, through action on the process parameters T and P, and with the choice of efficient additives, such as a nucleating agent and surfactant.

Figure 5.24 presents foam density measurements as a function of T_R and P_R. As a general trend, high temperature and high pressure (70 °C, 57–77 bars) result in higher density, and the lowest density is obtained at room temperature. An increasing density trend can be seen in the diagonal of the graph, from low T_R and P_R to high T_R and P_R. This shows a stronger influence of temperature on the solubility of CO_2 in AESO than pressure. Using these results, we can tune the density of a poured foam to a desired value by keeping the pressure higher than 40 bars and adjusting the temperature of the system to vary the solubility of carbon dioxide in AESO. The solubility of CO_2, S_{CO2} was determined by L. Bonnaille as S_{CO_2}(25 °C, 1 bar) ~ 0.002 g/g AESO, S_{CO_2}(25 °C, 45 bars) ~ 0.0077 g/g AESO [60].

We can use the solubility values to estimate the original cell size in the CO_2/AESO foams formed from $T_R = 25$ °C and $P_R = 45$ bars. The original foam density is 0.25 g/mL. If we neglect the mass of the gas, we have

$$\rho_{\text{foam}} = \frac{m_{\text{AESO}}}{V_{\text{gas}} + V_{\text{AESO}}} \tag{5.5}$$

and

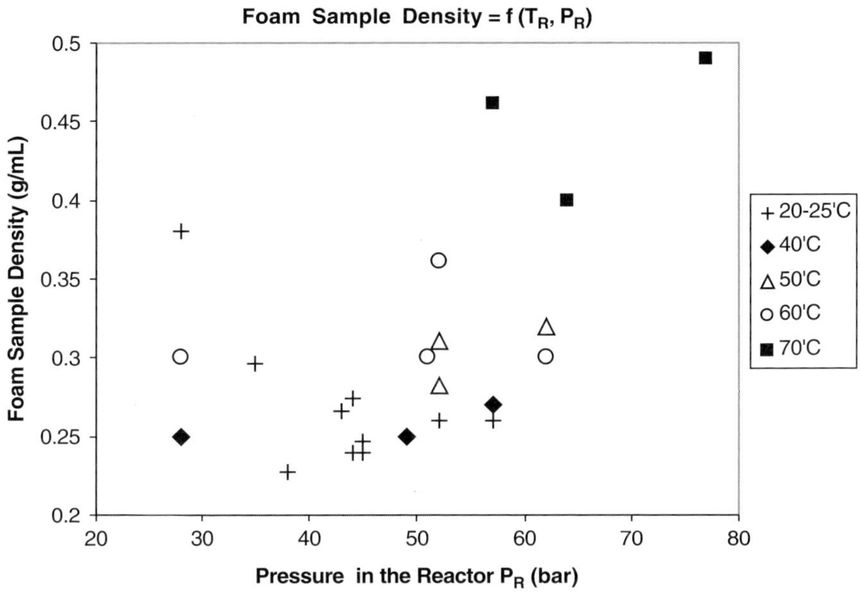

FIGURE 5.24 Density of foam samples extracted from the high-pressure thermostatic reactor with the pouring process. Samples were kept at T_R and P_R for at least 12 h.

$$V_{\mathrm{AESO}} = \frac{m_{\mathrm{AESO}}}{\rho_{\mathrm{AESO}}}. \tag{5.6}$$

The density of AESO is ~1.05 g/mL from volumetric measurement at T_{room}. Using Eqs. (5.5) and (5.6), the volume of gas in the foam V_{gas} can be calculated. Next, from the ideal gas law valid at low pressure, we have

$$V_{\mathrm{gas}} P_{\mathrm{gas}} = nRT, \tag{5.7}$$

where $T = T_{\mathrm{room}}$, P_{gas} is the pressure of the gas in the foam cells, and n is the number of moles of gas that participated in the foam, that is, the total amount of gas dissolved at T_R and P_R, minus the equilibrium solubility at T_{room} and P_{atm}:

$$n = \frac{S_{\mathrm{CO_2}}(T_{\mathrm{room}},45\ \mathrm{bars}) - S_{\mathrm{CO_2}}(T_{\mathrm{room}},1\ \mathrm{bar})}{M_{\mathrm{CO_2}}}, \tag{5.8}$$

where $M_{\mathrm{CO_2}}$ is the molecular weight of carbon dioxide (44 g/mol). Next, P_{gas} is inserted in Eq. (5.4), where $\Delta P = P_{\mathrm{gas}} - P_{\mathrm{atm}}$. Finally, the cell's radius r can be estimated from the experimental measurement of the surface tension of the AESO/CO_2 mixture.

Figure 5.25 presents the density profiles as a function of time of three samples pressurized at T_{room} and $P_R = 28$, 38, and 44 bars, foamed at atmospheric pressure, and allowed to collapse at room temperature. The initial

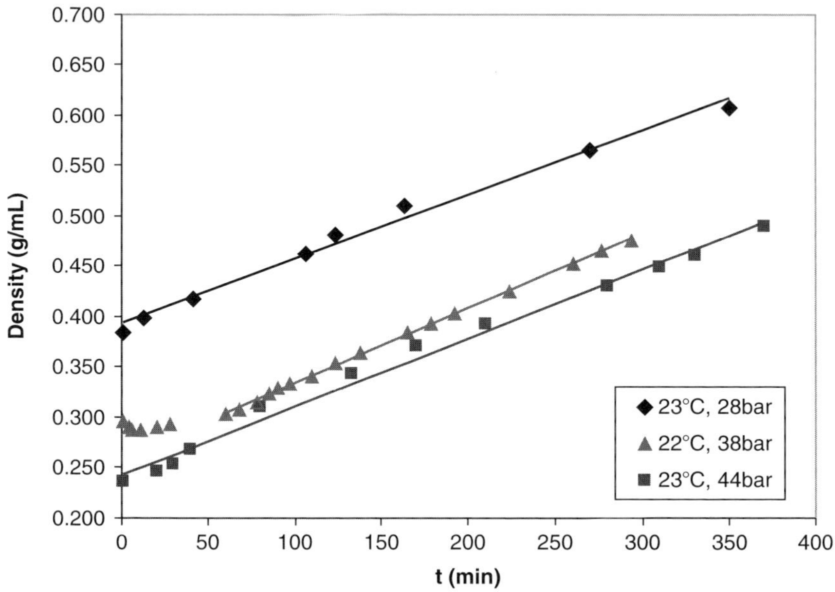

FIGURE 5.25 Density change of three samples foamed by pouring, after pressurization with CO_2 at $T_R = 22°$ or 23 °C and $P_R = 28$, 38, or 44 bars.

density is seen to decrease with higher reactor pressure, but the slope of the three linear density profiles is constant with a collapse rate of $\sim 7.10^{-4}$ g/(mL · min). These data suggest that at $t = 0$, all three foams have similar cell sizes. The number of cells in the foam shown in Figure 5.26 contained in a square of 200 × 200 pixels was counted and converted to the volumetric cell density N_c:

$$N_c = \left[\frac{\text{cell number}}{(200\text{ pixels})^2} . \text{scale}^2\right]^{3/2} \text{ in cells/cm}^3. \tag{5.9}$$

The volumetric cell density as a function of time is presented in Figure 5.27. The cell volumetric density is seen to decrease proportionally to the inverse of the square root of time:

$$N_c \approx \frac{\alpha}{\sqrt{(t - t')}}, \tag{5.10}$$

where α is a constant and $(t - t')$ is the time elapsed since the beginning of linear foam collapse. The mean volume-equivalent diameter d_V of a cell is calculated from the following relation:

$$\frac{4}{3}\pi d_V^3 = \frac{V_{\text{gas}}(t)}{N_c(t)}, \tag{5.11}$$

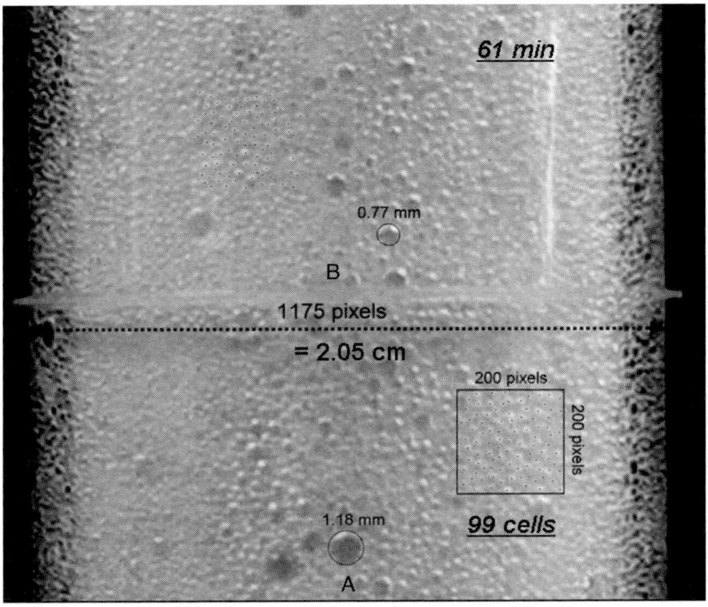

FIGURE 5.26 Digital picture of the central section of a foam inside a graduated cylinder, 61 min into foam collapse; the number of cells per unit area is counted to calculate the average cell size as a function of time. Two large cells, A and B, are highlighted for observations of individual cell growth.

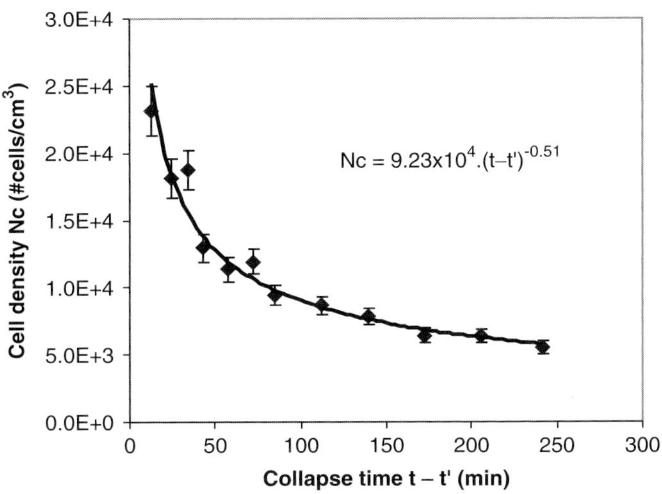

FIGURE 5.27 Volumetric cell density N_c as a function of time during foam collapse.

where $N_c(t)$ is already known from Eq. (5.10) and $V_{gas}(t)$ can be obtained from the foam density profile as a function of time:

$$\rho_{foam}(t) = \frac{m_{AESO}}{V_{gas}(t) + \frac{m_{AESO}}{\rho_{AESO}}}. \tag{5.12}$$

As expected, the average cell size increases with time. The cells' volume-equivalent diameter appears to follow a power law of the collapsing time: $d_V = 0.13(t - t')^{0.14}$ in the foam example considered.

5.5.2 FOAM NUCLEATING AGENTS

The introduction of solid particles in the AESO mixture provides a solid–liquid interface with a lower local surface tension than in the bulk, where bubbles from dissolved gas can nucleate easily. The best nucleating agent needs to (a) produce the lowest local surface tension; that is, have little affinity with AESO; (b) offer the greatest surface area possible; that is, be finely ground; and (c) stay homogeneously suspended in the mixture; that is, have a density close to that of the monomer. In light of these considerations, four candidates were tested:

- Starch powder is a ground biomaterial that would add to the bio-content of the foam. Its density is close to that of AESO.
- Loose keratin fibers from chicken feathers. These mix very well with AESO and would not only permit heterogeneous nucleation, but also provide some structural support to the foam.
- Metallic powders are commonly used as industrial nucleating agents. Aluminum powder is the lightest.
- Cobalt dust shows an extremely poor wetting compatibility with AESO. Plus, the tiny particles offer great surface area.

In Figure 5.28, the foam rise starts between −10 and −24″Hg, and is fastest with the keratin fibers; that is, at −25″Hg, the foam density is 0.62 g/mL with keratin fibers, whereas the plain AESO sample has barely started to expand. At −25″Hg, the aluminum and cobalt powders have equally good performance, with a foam density of 0.82 g/mL, and the starch-based mixture comes third with only 5% expansion ($d = 1$ g/mL).

The superiority of keratin fibers as a nucleating agent appears in the foam longevity also: The AESO/keratin fiber foam was still standing and growing at more than 95% vacuum after all of the other samples collapsed from excessive cell sizes relative to the viscoelastic properties. Table 5.9 gives a summary of the performance of the four nucleating agents tested. Starch powder is an efficient nucleating agent, because it lowers the amount of vacuum needed to start foaming, and improves by 20% the minimum density that can be reached through the creation of denser, smaller air cells than pure

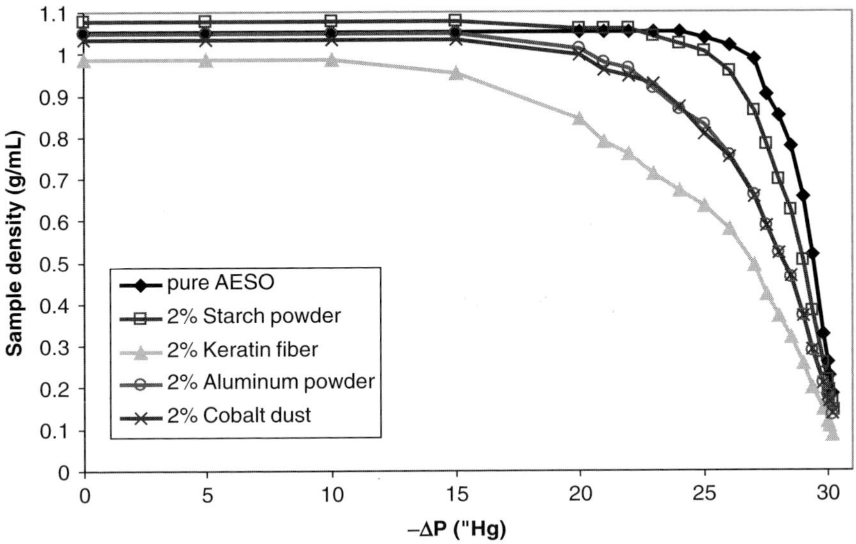

FIGURE 5.28 Foam density profiles as a function of vacuum applied, at 20 °C, for AESO + 2% of different nucleating agents.

TABLE 5.9 Foam rise in vacuum oven at 20°C. Test of 2 wt.% of different nucleating agents into AESO.

Nucleating Agent	Foam Expansion Begins	Density at Foam Collapse
Pure AESO	−24″Hg	0.178
2% Starch powder	−22″Hg	0.140
2% Ground keratin fibers	−10″Hg	0.084
2% Aluminum powder	−15″Hg	0.118
2% Cobalt dust	−15″Hg	0.119

AESO. The two metallic powders have equivalent quality, and allow the formation of foam much earlier, with a decrease in minimum density of 33% compared to the pure AESO foam. Finally, ground keratin fibers appear to be excellent foam enhancers. They trigger the foam expansion at a small pressure drop and can lead to a minimum foam density one-half that of pure AESO. Small keratin fibers are believed to position themselves into the cell ribs and contribute to the strength and stability of the foam.

5.5.3 STRUCTURE OF CURED FOAMS

Figure 5.29 shows the structure of three cured foams prepared under different processing conditions. The foam in Figure 5.29(a) consists of 97 wt% AESO cured with 2 wt% Esperox 28 initiator using a 1 wt% cobalt

(a)

FIGURE 5.29 Structure of three cured foams prepared under three different conditions as described in the text.

catalyst solution containing 6% cobalt ions. The sample was extracted from the reactor at T_{room} and $P = 52$ bar. The foam then passed through a heater at 100 °C and exited at $P = 1$ atm with $T \sim 45$ °C and density $\rho = 0.42$ g/mL. The cure started within 10 min after extraction from the reactor and the foam was placed in a vacuum oven at 16 Hg (~50% vacuum) at 45 °C. Due to the advanced cure, only limited expansion occurred, which produced a foam with a final density of $\rho = 0.35$ g/mL with cell sizes ranging from 0.2 to 0.5 mm. As can be seen from the vertical cut through the sample of Figure 5.29(a), a mixture of open (dark) and closed cells (light) was formed.

Figure 5.29(b) shows a foam (horizontal cut) prepared first with 100 wt% pure AESO, which was blown from the reactor at 100 °C and $P_{CO_2} = 90$ bar. The foam was extracted into a nitrogen atmosphere at 1 bar and $T = 47$ °C. The initiator-catalyst package, which was the same as that of Figure 5.29(a), was then whipped mechanically into the foam. It sat for 15 min before a 50% vacuum was applied at 45 °C for extra foam expansion. It was then postcured at 140 °C for 12 h. A mixture of open and closed foam cells was obtained with a density of $\rho = 0.33$ g/mL and a cell size range from 0.2 to 1.0 mm. In this case, we obtained larger cells but thicker cell walls at the same density, compared to the foam shown in Figure 5.29(a).

Figure 5.29(c) shows a view of the surface of a 0.6-mm-thick foam that was cured by photoinitiation. The foam consisted of 96.5 wt% AESO, 2 wt% styrene, 1% Esperox 28, and 0.5 wt% UV photoinitiator. The liquids were placed in a reactor at T_{room} and a CO_2 pressure of $P = 52$ bar. The foam was extracted at T_{room} and 1 bar and placed between two glass sheets and cured for 20 min of exposure on both sides to an 800-W UV light, followed by

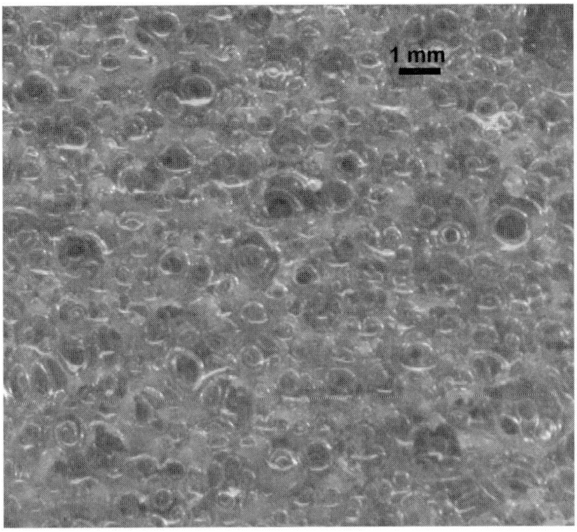

(b)

FIGURE 5.29 (Continued)

(c)

FIGURE 5.29 (Continued)

postcure at 140 °C for 12 h. The resulting cured foam had a density of ρ = 0.25 g/mL with cell sizes in the range of 0.1–0.3 mm and consisting mostly of closed cells. These foams were semirigid, had excellent structural integrity, and contained the highest fraction of bio-based materials compared to all other PU-type foams.

5.5.4 SUMMARY OF BIO-BASED FOAMS

Bio-based foams are a rapidly growing industry, with a large contribution from plant oil-based polystyrenes in the construction area, and from starch foams in packaging applications. The petroleum foaming industry is faced with several issues, and bio-based foams have shown properties making them fit to replace some petroleum-based foams on the market. However, a resilient thermosetting foam system with a bio-based content higher than 50% is still to be designed. In this work, we implemented a high-pressure, thermostatic process using carbon dioxide as a blowing agent to foam cross-linked soybean oil polymers. We examined the compatibility of carbon dioxide with the acrylated epoxidized soybean oil monomer (AESO), and the dynamics of the CO_2/AESO-monomer foams. Additives, such as nucleating agents (starch powder, ground keratin fibers, etc.) and surfactants, proved useful for the creation of foams with a large concentration of small cells and a low bulk density. These foams cam be readily polymerized into soft or rigid strong, structural foams with a higher bio-based content than is currently available commercially. Since these foams are derived from triglycerides, they can also be expected to have application as biomaterials requiring compatibility with tissue, such as tissue scaffolds, artificial skin, and wound healing materials. The bio-compatibility of these triglyceride-based foams with tissue has been demonstrated by C. M. Klapperich (Boston University) and R. P. Wool (unpublished work 2005).

5.6 SUMMARY OF BIO-BASED COMPOSITES

The resins derived from plant oils are suitable for use in many molding processes to produce composite materials. At low glass fiber content (35 wt%), composites produced from AESO by RTM displayed a tensile modulus of 5.2 GPa and a flexural modulus of 9 GPa. They also exhibit a tensile strength of 129 MPa and flexural strength of 206 MPa. At higher fiber contents (50 wt%), composites produced from AESO displayed tensile and compression moduli of 24.8 GPa each. The tensile and compressive strengths were found to be 463.2 and 302.6 MPa, respectively. Besides glass fibers, natural fibers such as flax and hemp can be used in the composite materials. Hemp composites of 20% fiber content display a tensile strength of 35 MPa and a tensile modulus of 4.4 GPa. The flexural modulus was approximately

2.6 GPa and flexural strength in the range of 35.7–51.3 MPa, depending on the test conditions. The flax composite materials also have tensile and flexural strengths in the ranges of 20–30 MPa and 45–65 MPa, respectively. The properties exhibited by both the natural and synthetic fiber-reinforced composites can be combined through the production of "hybrid" composites. These materials combine the low cost of natural fibers with the high performance of synthetic fibers, resulting in properties spanning a wide range.

REFERENCES

1. McCrum, N. G.; Buckley, C. P.; Bucknall, C. B. *Principles of Polymer Engineering*, Oxford University Press, New York; 1997.
2. Devi, L. U.; Bhagawan, S. S.; Thomas, S. *J. Appl. Polym. Sci.* 1997, **64,** 1739.
3. Bledzki, A. K.; Reihmane, S.; Gassan, J. *J. Appl. Polym. Sci.*, 1996, **59,** 1329.
4. Saha, A. K.; Das, S.; Bhatta, D.; *et al. J. Appl. Polym. Sci.* 1999, **71,** 1505.
5. Ghosh, P.; Ganguly, P. K. *Plast. Rub. Comp. Proc. Appl.* 1993, **20,** 171.
6. Gassan, J.; Bledzki, A. K. *Polym. Comp.* 1997, **18,** 179.
7. Gowda, T. M.; Naidu, A. C. B.; Rajput, C. *Comp. Part A: Appl. Sci. Manuf.* 1999, **30,** 277.
8. Shalash, R. J. A.; Khayat, S. M.; Sarah, E. A. *J. Petrol. Res.* 1989, **8,** 215.
9. Mishra, S.; Naik, J. B. *J. Appl. Polym. Sci.* 1998, **68,** 1417.
10. Hargitai, H.; Czvikovszky, T.; Gaal, J.; *et al.* In *Proceedings of the First Conference on Mechanical Engineering*, Budapest, 1998.
11. Hornsby, P. R.; Hinrichsen, E.; Tarverdi, K. *J. Mater. Sci.* 1997, **32,** 443.
12. Mieck, K. P.; Luetzkendorf, R.; Reussmann, T. *Polym. Comp.* 1996, **17,** 873.
13. Chen, X.; Guo, Q.; Mi, Y. *J. Appl. Polym. Sci.*, 1998, **69,** 1891.
14. George, J.; Sreekala, M. S.; Thomas, S.; *et al. J. Reinf. Plast. Comp.* 1998, **17,** 651.
15. Rozman, H. D.; Kon, B. K.; Abusamah, A.; *et al. J. Appl. Polym. Sci.* 1998, **69,** 1993.
16. Mohan, R.; Kishore R. M.; Shridhar, M. K.; *et al. J. Mater. Sci. Lett.* 1983, **2,** 99.
17. Shah, A. N.; Lakkad, S. C. *Fiber Sci. Tech.* 1981, **15,** 41.
18. Williams, G. I.; Wool, R. P. *J. Appl. Comp. Materials.* 2000, **7,** 421.
19. Morye, S. S.; Wool, R. P. *Polym. Comp.* (in press 2005).
20. Wool, R. P.; Kusefoglu, S. H.; Palmese, G. R.; *et al.* U.S. Patent 6,121,398; 2000.
21. Williams, G. I. *M.S. Thesis,* University of Delaware, 1999.
22. Wool, R. P.; Khot, S. N. In *Proceedings ACUN-2,* University of New South Wales, Sydney, Australia 2000.
23. Kia, H.G., Ed. *Sheet Molding Compounds Science and Technology,* Hanser/Gardner Publications, Inc., Cincinnati; 1993.
24. Melby, E. G.; Castro, J. M. In *Comprehensive Polymer Science,* Allen, G.; Bevington, J. C., Eds.; Pergamon Press, Oxford; 1989, pp. 51–109.
25. Can, E.; Kusefoglu, S.; Wool, R. P. *J. Appl. Polym. Sci.* 2001, **81**(1), 69–77.
26. Can, E.; Kusefoglu, S.; Wool, R. P. *J. Appl. Polym. Sci.* 2002, **83**(5), 972–980.
27. Guo, A.; Demydov, D.; Zhang, W.; *et al. J. Polym. Environ.* 2002, **10**(1–2), 49–52.
28. Khot, S. N.; Lascala, J. J.; Can, E.; *et al. J. Appl. Polym. Sci.* 2001, **82**(3), 703–723.
29. Li, F. K.; Larock, R. C. *J. Appl. Polym. Sci.* 2001, **80**(4), 658–670.
30. Li, F. K., Larock, R. C. *J. Polym. Sci. B: Polym. Phys.* 2001, **39**(1), 60–77.
31. Lu, J.; Khot, S. N.; Wool, R. P. *Polymer*, 46(1), 71–80 (2005).
32. Wool, R. P.; Lu, J.; Khot, S. N. Sheet Molding Compound Resins from Plant Oils. U.S. patent pending (Approved 2005).
33. Lo, K. Application of Isocyanates in the Thickening of Soy Oil Resins for SMC Application. *Undergraduate Thesis,* University of Delaware, 2000.

34. Burns, R.; Gandhi, K. S.; Hankin, A. G.; *et al. Plastics & Polymers* 1975, **43**(168), 228–235.
35. Gandhi, K. S.; Burns, R. *J. Polym. Sci. A: Polym. Chem.* 1976, **14**(4), 793–811.
36. Rao K.B.; Gandhi, K. S. *J. Polym. Sci. A: Polym. Chem.* 1985, **23**(8), 2133–2150.
37. Rao K.B.; Gandhi, K. S. *J. Polym. Sci. A: Polym. Chem.* 1985, **23**(8), 2305–2317.
38. Vancsó-Szmercsányi, I.; Szilágyi, A. *J. Polym. Sci. A: Polym. Chem.* 1974, **12,** 2155.
39. Vancsó-Szmercsányi, I.; Kalló, A. *J. Polym. Sci. A: Polym. Chem.* 1982, **20,** 639.
40. Alve, F. *J. Polym. Sci.* A 1971, **9,** 2233.
41. Nielsen, L. E. *J. Macromolec. Sci.: Rev. Macromolec. Chem.*1969, **C3**(1), 69.
42. Lascala, J. J. *Ph.D. Dissertation,* Department of Chemical Engineering, University of Delaware; 2002.
43. Lascala, J. J.; Wool, R. P. *Polymer*, 2005, **46**(1), 61–69.
44. Feng, S.; Sen, P. N. *Phys. Rev. Lett.* 1984, **52**(3), 216–219.
45. Kantor, Y.; Webman, I. *Phys. Rev. Lett.* 1984, **52**(21), 1891–1894.
46. Wool, R. P. *Polymer Interfaces, Structure and Strength,* Hanser Publishers, New York; 1995.
47. Lu, J.; Wool, R. P., *Polymer* (in press 2005).
48. Williams, J. G. *Fracture Mechanics of Polymers.* Halsted Press, New York; 1984.
49. Fischer, M. *Adv. Polym. Sci.* 1992, **100,** 313–355.
50. Bos, H. L.; Nusselder, J. J. H. *Polymer* 1994, **5**(13), 2793–2799.
51. Klempner, D.; Frisch, K. C. *Handbook of Polymeric Foams and Foam Technology,* Oxford University Press; 1991.
52. Landrock, A. H. *Handbook of Plastic Foams: Types, Properties, Manufacture and Applications,* Noyes Publications; 1995.
53. Freedonia Group; Market Report # R154-449; June 2001; *Mindbranch.com.*
54. Bastioli, C. *Starch-Starke* 2001, **53,** 351–355.
55. USB. United Soybean Board; January 2004. http://www.unitedsoybean.org/newuses/
56. Guo, A.; Javni, I.; Petrovic, Z. *J. Appl. Polym. Sci.* 2000, **77,** 467–473.
57. O'Donnell, A.; Dweib, M. A.; Wool, R. P. *Compos. Sci. Technol.* 2004, **64,** 1135–1145.
58. Wei, X. *Ph.D. Thesis,* Department of Fiber and Polymer Science, North Carolina State University, Raleigh; 2000.
59. Mohamed, T. S. *M.S. Thesis,* Department of Textile Engineering, North Carolina State University, Raleigh; 2002.
60. Bonnaille, L. *M.S. Thesis,* Department of Chemical Engineering, University of Delaware; 2004.
61. Zlatanic, A.; Lava, C.; Zhang, W.; *et al. J. Polym. Sci. B: Polym. Phys.* 2004, **42,** 809–819.

6

FUNDAMENTALS OF FRACTURE IN BIO-BASED POLYMERS

RICHARD P. WOOL

The design of bio-based polymers is strongly influenced by the need to achieve useful mechanical, thermal, electrical, rheological, and other physical properties, such as environmental degradability or resistance to degradability. Properties such as mechanical strength, stiffness, fatigue resistance, heat distortion temperatures, chemical resistance, and processability are related to molecular architectures and chemical composition. We know how to design the molecular architecture to obtain certain classes of polymers. For example, thermoplastics such as PLA and PHBV, which are processed by melt injection molding (cups, spoons), blow molding (bottles), and extrusion (fibers, films), should be linear polymers with molecular weights M^* about 8 times their critical entanglement molecular weight M_c. But what determines M_c and why should $M^* = 8M_c$? How does the strength depend on M_c and M^* and are there penalties to pay for improper design, for example, $M = 5M_c$ instead of $8M_c$?

Polymer coatings such as paint should be made as latex particles of linear polymer chains dispersed in an aqueous (preferable) solvent. As the solvent evaporates, the latex particles coalesce, and the polymer chains need to interdiffuse a distance equal to their radius of gyration to obtain good film strength. Additional reactions, such as photo curing or oxidation to promote light cross-linking, may be needed to improve the creep resistance of the coating. To make pressure-sensitive adhesives (PSAs), we also need linear polymer chains with a certain degree of branching and a very low glass transition temperature T_g. But how should the adhesion of a PSA to a solid substrate be controlled by stickier groups so that strong adhesive rather than weak cohesive failure occurs during detachment? This is an issue for postage stamps, labels, and sticky tape. To make elastomers, we need lightly cross-linked polymers, which are best derived from linear polymers, which can be lightly cross-linked and are capable of reversible deformations to about 500% strain. The elastomeric network should not contain defects. Such elastomers should also be capable of being compounded with fillers such as carbon black to make rubber materials such as auto tires.

To make highly rigid, high-performance composite resins, we need a highly cross-linked molecular structure that is defect free and a high T_g. How does the fracture stress depend on the cross-link density and how should we design fatty acid distribution functions and chemical functionalization to achieve optimal properties? In addition to fracture of polymers in the bulk or virgin state, a number of important polymer interface issues must be addressed by researchers of bio-based products. These involve welding of thermoplastics, lamination of composites, coalescence of latex particles in coatings and elastomers, blends of incompatible polymers, reinforcement of incompatible interfaces with compatibilizers, polymer solid interface such as fiber-reinforced composites, and adhesion of polymers to substrates. A useful review of polymer interfaces is provided in [1]. These issues are critical for the successful advance of new bio-based plastics, adhesives, and resins and are discussed in Chapters 6 through 8 using theoretical and experimental examples. Computer simulations are used to make predictions for thermosets derived from plant oils and explore the utility of the fundamental theories.

Chapter 6 provides the reader with the basic theory of strength of polymers and interfaces with general examples for linear and cross-linked polymers such as thermoplastics, elastomers, and cross-linked materials. Chapter 7 deals primarily with highly cross-linked triglycerides derived from functionalized oils with optimized fatty acid distribution functions to make high-performance, high-T_g, composite resins; and Chapter 8 examines the field of low-T_g PSAs such as that used in cellophane tape, duct tape, and postage stamps. Readers with a background in physics, materials science, engineering, and chemistry should find this chapter informative and enabling in terms of the understanding and design of new bio-based materials.

6.1 FRACTURE OF POLYMERS: FUNDAMENTAL THEORY

A useful approach to evaluating the fracture energy G_{1c} of a polymer or interface of A/B polymers is represented in Figure 6.1 [1, 2]. Typically, a crack propagates through the interface region preceded by a deformation zone at the crack tip. For cohesive failure, the fracture energy can be determined by the *J*-integral method, as described by Hutchinson *et al.* [3–5], where G_{1c} is the integral of the traction stresses with crack opening

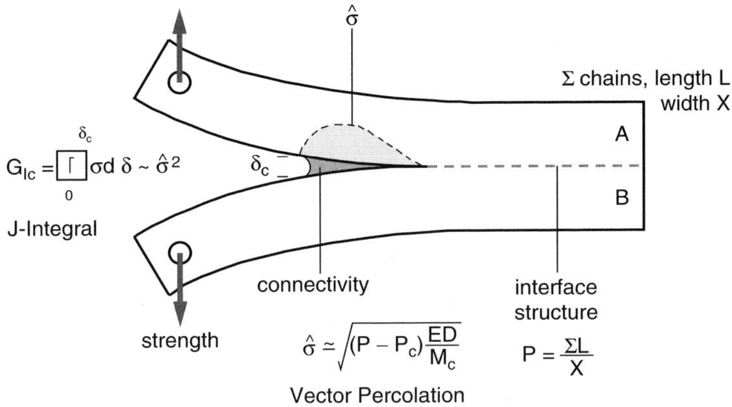

FIGURE 6.1 The microscopic entanglement structure, for example, at an interface or in the bulk, is related to the measured macroscopic fracture energy G_{1c} via the RP theory of breaking connectivity in the embedded plastic zone (EPZ) at the crack tip. The RP theory determines $\sigma_{\max}$ in the EPZ, which is related to G_{1c} via Hutchinson's J-integral theory. The percolation parameter p is related to the interface molecular structure via $p \sim \Sigma L/X$, where Σ is the number of chains of length L in an interface of width X.

displacements δ, in the cohesive zone, following yielding at a local yield or craze stress σ_Y. The cohesive zone at the crack tip breaks down by a vector, or rigidity percolation process [6–11], as described herein, at a maximum stress value, $\sigma_m > \sigma_Y$. Typical ratios of σ_m/σ_Y are about 4–10 [3]. Both σ_m and δ are rate dependent and, in the simplest case, the fracture energy is determined by

$$G_{1c} = \sigma_m \delta_m, \tag{6.1}$$

where δ_m is the critical crack opening displacement. Both σ_m and δ_m depend on the zone structure and the microscopic deformation mechanisms controlling the percolation fracture process via disentanglement and bond rupture.

The strength of a polymer is associated with molecular connectivity, which can be short range or long range in nature. Short-range connectivity is associated with primary and secondary bonds; long-range connectivity is associated with entanglements in amorphous polymers and crystals in semicrystalline polymers. Fracture involves the breaking of this connected structure and can be a very complex process. To simplify the fracture process, we consider the connected structure in terms of a simple three-dimensional lattice, which could represent the entanglement network in amorphous polymers, the cross-linked network of thermosets, or the crystal lattice of semicrystalline polymers.

Consider the lattice shown in Figure 6.2. The transmission of forces through a lattice as a function of the fraction p of bonds in the lattice was analyzed by Kantor and Webman [6], Feng *et al.* [7, 8], Thorpe *et al.* [8, 9],

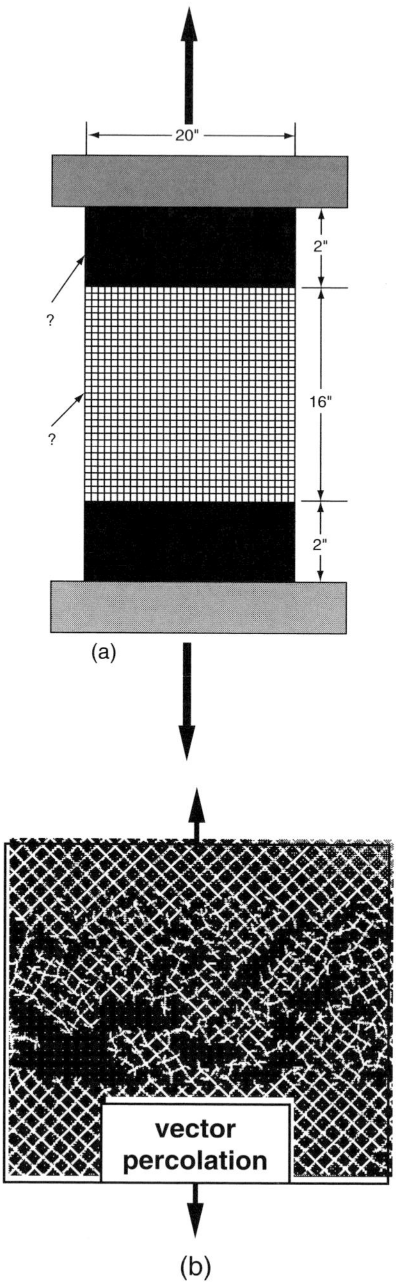

FIGURE 6.2 The role of percolation in the random fracture of bonds in a model net at constant strain [1]. (a) The net of modulus E is stressed in uniaxial tension to a stress σ. (b) Random fracture events in the net result in a percolating system near the fracture threshold and a very broad distribution of stress on the bonds.

and others [1, 10, 11]. DeGennes [12] first suggested that conductivity, or scalar percolation, could also be used to quantize the modulus of elasticity E of randomly connected networks, such as gels. Analyses based on the Born model of the microscopic elasticity of a lattice [13] gave results for the elasticity that resembled conductivity percolation when shear terms were neglected in the Hamiltonian for the elastic energy, as

$$E \sim [p - p_c]^t, \tag{6.2}$$

where p is the occupied fraction of lattice bonds, p_c is the percolation threshold, and the conductivity exponent, $t \approx 1$–2. However, when shear terms dominated the elasticity, a new form of elasticity began to emerge that potentially belonged to a new universality class than conductivity percolation. Kantor and Webman [6] reformulated the Hamiltonian for the elastic energy, accounting for both tensile bond stretching and angle bending between the fractal, tortuously connected "strings" of connected bonds remaining in the lattice near p_c, and the macroscopic elasticity became:

$$E \sim [p - p_c]^\tau, \tag{6.3}$$

where τ is the vector percolation exponent, which is larger than the conductivity (or scalar) percolation exponent t in Eq. (6.2). The vector percolation threshold p_c in Eq. (6.3) can also be greater than the scalar percolation threshold, which is due to the "sloppiness" of the lattice near p_c, thus permitting, for example, the transmission of electrons through the weakly connected fractal structure, but not the sensible transmission of vectors.

The mean size of the average cluster near p_c, namely, the correlation length ξ, is given by [14]

$$\xi \sim [p - p_c]^{-\nu}, \tag{6.4}$$

where ν is the critical exponent for scalar percolation. The relation between the dimensionality d of the lattice, the critical exponent ν, and the vector percolation exponent τ, as derived by Kantor and Webman [6], is

$$\tau = d\nu + 1. \tag{6.5}$$

For example, in $2d$ (e.g., graphene sheets or single-wall carbon nanotubes), using percolation values recommended by Stauffer [14], $\nu \approx 1.3$ and $\tau \approx 3.6$, whereas, in $3d$, $\nu \approx 0.85$ and $\tau \approx 3.55$. The values for the conductivity exponent t in Eq. (6.2) are $t \approx 1.2$ in $2d$, and $t \approx 2$ in $3d$. These are sufficiently smaller than τ values in Eq. (6.3), which, as noted by Kantor and Webman, is evidence that vector percolation belongs to a different universality class than scalar percolation.

The vector percolation process addresses several important points. First, consider a $2d$ lattice near the percolation threshold p_c, as shown in Figure 6.2. Due to the random fractal connectivity of the lattice, the stress distribution $\phi(\sigma)$ in the bonds becomes highly nonuniform such that some bonds are highly stressed, while others bear little stress. The existence of highly stressed bonds is a prelude to molecular fracture and parallels the "hot bonds" in conductivity percolation, where hot bonds arise from high current density in some individual bonds near the percolation threshold. The hot bonds overheat like electrical fuses in the high current density and break.

The concept of mechanical "hot bonds" is relevant to fracture of polymers in general and is the basis for understanding why materials fracture at macroscopic stresses, which are orders of magnitude less than the molecular fracture stresses. When polymers such as polypropylene and polyethylene are subjected to uniform tensile stresses, it was shown using infrared and Raman spectroscopy that the molecular stress distribution can be quite broad, even though the applied stress is well below the macroscopic fracture stress [15, 16]. The development of the molecular stress distribution $\phi(\sigma)$ is due to the inherent sloppiness of the lattice. Thus, in the *J*-integral fracture mechanics model, the maximum fracture stress near the crack tip σ_m, described in Figure 6.1 and Eq. (6.1), remains closer to the yield stress than to the much higher molecular fracture stress.

Another point of interest is that only a fraction $[p - p_c]$ of the bonds needs to be fractured before complete failure occurs in a $2d$ or $3d$ network. Thus, in a deformation zone at a crack tip, the crack advances through the zone by breaking a fraction $[p - p_c]$ of bonds or fibrils in parts of a craze network. The broken bonds do not lie on the same plane, as is often assumed intuitively, but are distributed over the deformation zone volume. The deformation zone near the fracture is best described as a volume of material preceding the crack tip that contains a considerable number of defects.

An important corollary to the existence of the threshold p_c is that when $p < p_c$, the lattice connectivity is broken and no significant strength exists beyond that of nonbonded potentials and van der Waals interactions. Thus, the molecular lengths ($L \sim M$) must be long enough, the areal density of chains Σ at the interface must be great enough, and the number of entanglements in the lattice N, at an interdiffusion distance X or interface width w, has to exceed the percolation threshold before strength develops. This means that an initial investment is needed before strength develops, such that when $G_{1c} \sim [p - p_c]$, there exists corresponding critical parameters such as M_c, L_c, Σ_c, X_c, N_c, w_c, etc., which are all related to each other through the percolation parameter p [1, 2].

To convert these percolation concepts into quantitative fracture terms, consider the vector percolation experiment shown in Figure 6.2, applied to any $3d$ lattice in general with tensile modulus E. The Hamiltonian for the stored elastic energy can be formulated using the Born and Huang [13] or the Kantor and Webman [6] approach for specific lattices, or using the more simple engineering strain energy density approach as follows. The stored elastic strain energy density U (energy per unit volume) in the lattice due to an applied uniaxial stress σ is determined by

$$U = \sigma^2/2E. \tag{6.6}$$

The modulus E as a function of the bond fraction p near p_c is derived from Eq. (6.3) in its normalized form as

$$E = E_0[p - p_c]^{\tau}/[1 - p_c]^{\tau}, \tag{6.7}$$

in which E_0 is the apparent modulus for the perfect lattice when $p = 1$.

The stored strain energy can also be determined for the general case of multiaxial stresses [1] and lattices of varying crystal structure and anisotropy. The latter could be important at interfaces where mode mixing can occur, or for the fracture of rubber, where U is a function of the three orthogonal stretch ratios λ_1, λ_2, and λ_3. For example, in the Mooney equation [17], the finite deformation strain energy density function is given by

$$U(\lambda_1, \lambda_2, \lambda_3) = C_1(\lambda_1^2 + \lambda_2^2 + \lambda_3^2 - 3) + C_2(\lambda_1^{-2} + \lambda_2^{-2} + \lambda_3^{-2} - 3), \tag{6.8}$$

in which C_1 and C_2 are constants. In the ideal theory of rubber elasticity proposed by Flory [18], $C_2 = 0$ and using the constant volume assumption, $\lambda_1\lambda_2\lambda_3 = 1$, Eq. (6.8) becomes

$$U(\lambda) = C_1[\lambda^2 + 2/\lambda - 3], \tag{6.9}$$

in which the constant C_1 is related to the cross-link density ν and modulus E of the rubber. When appropriate, these and other strain energy functions can be used in the energy balance analysis described later.

The stored strain energy dissipation per unit volume U_f to fracture a network consisting of ν bonds per unit volume is

$$U_f = \nu D_0[p - p_c], \tag{6.10}$$

where D_0 is the bond fracture energy, and $[p - p_c]$ is the percolation fraction of bonds that must be broken to cause fracture in the network. In this approach, the strain energy U is first stored in the net, and we inquire if this energy is sufficient to break $\nu[p - p_c]$ bonds per unit volume when it releases at a critical strain energy density $U^* = \sigma^{*2}/2E$, such that at the critical condition,

$$U^* \geq U_f. \tag{6.11}$$

It is important to note that we assume the random fracture approximation (RFA) is applicable. This assumption has certain implications, the most important of which is that it bypasses the real evolutionary details of the highly complex process of the lattice bond stress distribution $\phi(\sigma)$, creating bond rupture events, redistribution of $\phi(\sigma)$, microvoid formation, coalescence, propagation, and, finally, macroscopic failure. The fractal nature of the distributed damage clusters is evident (Figure 6.2) and the RFA, while providing a facile solution to an extremely complex process, remains a physically realistic and useful assumption. For highly cross-linked polymers, the bond fracture energy D_0 may be reduced by a factor f to $D_0/(1+f)$ due to free-radical propagation reactions creating f additional ruptured bonds per mechanical scission event. For now, we will assume that $f = 0$.

Substituting for U^* and U_f in Eq. (6.11) and solving for the critical stress σ^*, we obtain the *Net solution* for the critical fracture stress as

$$\sigma^* = \{2E\nu D_0[p - p_c]\}^{1/2}. \quad (6.12)$$

This equation predicts that the fracture stress increases with the square root of the bond density ν. The percolation parameter p is, in effect, the normalized bond density such that for a perfect net without defects, $p = 1$, and for a net that is damaged or contains missing bonds, then $p < 1$. Obviously, as p approaches p_c, the fracture stress decreases toward zero and we have a very fragile material. This fracture relation could therefore be used to evaluate durability, fatigue damage accumulation, healing processes, or retention strength of a material by tracking damage through a single parameter p. For thermosets, p is related to the extent of reaction of the cross-link groups and this could be critical in the fiber–matrix interface of composites [19]. Note that the Net solution refers to the stress required to cause fracture in a unit volume of the net in uniaxial tension. When applied to interfaces, we let the volume of material or Net contain the interface such that we can calculate σ^* with a knowledge of p based on a local normalized entanglement density [1, 2]. In all applications of the RP model, the stressed state is the reference state used to assess percolation and connectivity. This will become more apparent when we examine disentanglement, for example, where an unraveling or disentanglement process in the stretched state breaks the connectivity. The next section contains several applications of the RP percolation model of polymer fracture.

6.2 APPLICATIONS OF FRACTURE THEORY

6.2.1 FRACTURE OF ENTANGLED POLYMERS

Entangled linear polymers in bulk, both amorphous and semicrystalline, form "sloppy" nets of irregular entanglement lengths, whose average length is determined by the familiar entanglement molecular weight M_e or by the critical entanglement molecular weight M_c, which is about twice M_e. Critical entanglement molecular weight M_c represents a segment of an entangled chain that is long enough to form a bridge or loop of three crossings (3P) through a plane in the melt. An entangled net forms when the number of chains ($\Sigma \sim M^{-1/2}$) intersecting the plane equals the number of bridges [20]. Thus, when $\Sigma = 3P$, $M_c = 9(\Sigma/P)^2 M$.

This description of connectivity, which is based on a percolation concept of entanglements [20], was examined by Uhlherr *et al.* [21] using a computer simulation of linear polyethylene in the melt and found to be accurate. By sampling the amorphous structure, they found that the average segment length that intersected a randomly placed plane three times was equivalent to the critical entanglement molecular weight M_c. Thus, the bridge with three crossings is the basic mesh element of the network capable of transmitting vectors and defines the number of bonds that have to be broken to reduce a high-molecular-weight net to the critically connected net by either bond rupture or disentanglement.

For semicrystalline polymers, the net becomes quite complex due to the microcrystalline structure and orientation functions of crystals and amorphous regions. However, it is well known that the amorphous regions play a critical role in the fracture process, as reviewed by Kausch [22]. When a tensile stress σ is applied to the polymer, due to the irregularity of the network, hot bonds break at molecular stresses, which are typically two orders of magnitude greater than the applied macroscopic stress σ. Rupture of the hot bonds occurs randomly in the net and they accumulate and connect in a percolation fashion, as discussed in the last section. As the bonds break (we assume one hot bond per entanglement length), the stored energy U in the net is consumed and eventually approaches zero at the vector percolation threshold, p_c. Macroscopic fracture occurs when the stored energy is released by percolating random microscopic fracture events, implied schematically in Figure 6.2.

The Net solution for the fracture stress [Eq. (6.12)] can be further simplified by the following assumptions: (1) For very high molecular weights, typically $M > 8M_c$, $p \approx 1$ (no initial chain end effects on the entanglement density); (2) $p_c \approx 1/2$ and, hence, $2(p - p_c) \approx 1$; (3) the stress in the craze deformation zone is higher than the macroscopic stress due to the area reduction at high draw ratio λ, which gives $\sigma^* = \sigma_c/\lambda$, where σ^* is the applied stress at the craze/glass fibril interface and λ is the fibril draw ratio; (4) for most drawing processes, $\lambda \approx 4$; and (5) the entanglement density $\nu = \rho/M_e$. Making these substitutions in Eq. (6.12), the estimation for the fracture stress of high-molecular-weight polymers undergoing vector percolation via random bond fracture is

$$\sigma^* \approx [ED_0\rho/16M_e]^{1/2}. \tag{6.13}$$

This equation contains constants with well-known values and can therefore be used to predict the fracture stress of most polymers. For example, the bond dissociation energy D_0 is about 80 kcal/mol (335 kJ/mol) for C–C bonds. Thus, for many polymers below T_g, using $E \approx 2\,\text{GPa}$, $\rho \approx 1.2\,\text{g/cc}$, and $M_c \approx 2M_e$, we can express the critical fracture stress in terms of M_c as

$$\sigma^* = \sigma_o/M_e^{1/2}, \tag{6.14}$$

where the constant $\sigma_o \approx 4.6\,\text{GPa (Daltons)}^{1/2}$. On this basis, poly(lactic acid) (PLA) with a modulus $E \approx 1\,\text{GPa}$ and $M_c \approx 12{,}000\,\text{Da}$ (reported by J. Dorgan *et al.* [22a])

would be expected to have a fracture stress of $\sigma^* \approx 60\,\text{MPa}$. This compares with $E = 61$ MPa reported by Sun *et al.* in Chapter 11. However, starch with a very high entanglement molecular weight $M_e \approx 100{,}000$ (reported by C. J. Carriere [22b]) is expected to be relatively weak with $\sigma^* \approx 14\,\text{MPa}$, assuming one can reach molecular weights of the order of 10^6 Da. If the molecular weight of starch were less, say, $M = 400{,}000$ Da, then one would obtain $\sigma^* \approx 3\,\text{MPa}$. Thus starch, while being an abundant, low-cost bio-based polymer with excellent biodegradability characteristics, is fundamentally very weak. However, its high modulus can still provide useful reinforcement in polymers allowing applications such as packaging foam material and an additive to promote biodegradability and biodisintegration in polymer composites, as discussed in Chapter 11.

Values for the fracture stress as a function of M_c using Eq. (6.14) are listed in Table 6.1 and compared with tensile fracture data reported by several investigators. Clearly, there is a very strong effect of the entanglement density $\nu \sim 1/M_c$ such that $\sigma \sim [E\nu]^{1/2}$, and polymers with very high M_c values, such as starch and polystyrene, will be expected to be very brittle compared to those with low M_c values such as those observed for polyethylene and polycarbonate.

To explore the general utility of this fracture relation for amorphous and semicrystalline polymers, we examine the following work: Vincent [25] analyzed the tensile fracture stress σ of a broad range of polymers as a function of the number of backbone bonds per cross-sectional area (Ω) and found a nearly linear relation, $\sigma \sim \Omega$, as shown in Figure 6.3. The bond areal density Ω, is related to M_c via the theory of entanglements for random walk chains [20], via $\Omega \sim M_c^{-1/2}$:

$$\Omega = \{5.56(C_\infty j/M_0)^{1/2} b\rho N_a\}/M_c^{1/2} \tag{6.15}$$

TABLE 6.1 Comparison of RP theory and experimental fracture stress.

Polymer	M_c (g/mol)	σ (Theory) (10^4 MPa/$M_c^{1/2}$)	σ (Experiment) (MPa)	T (°C)	Reference
PE	4,000	158	160	−196	25
PP	7,000	119	98	−120	25
PVC	11,000	95	142	−180	25
PMMA	18,400	74	68	−60	25
PC	4,800	144	145	−140	25
			120	20	24
PS	31,000	57	56	20	23
PTFE	13,200	87	117	−196	25
PLA	12,000	65 ($E = 1$ GPa)	64	20	
Starch	200,000	22	—	20	

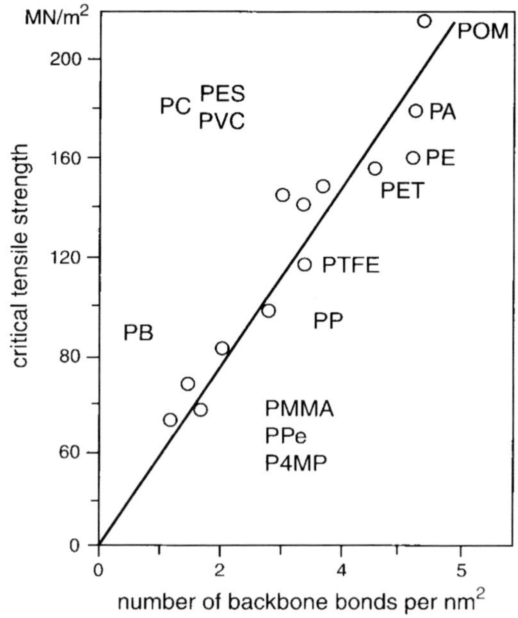

FIGURE 6.3 Tensile strength σ versus number of backbone bonds per monomer $\Omega = 1/a$, reported for a range of polymers by Vincent [25]. The solid line is the theoretical line for the vector percolation analysis of strength discussed herein.

or

$$\Omega = \beta / M_c^{1/2}. \tag{6.16}$$

Here the parameter β is defined by

$$\beta = [5.56(C_\infty j/M_0)^{1/2} b \rho N_a], \tag{6.17}$$

in which $C_\infty, j, M_0, b,$ and N_a are the characteristic ratio, number of backbone bonds per monomer, monomer molecular weight, bond length, and Avogadro's number, respectively. Combining Eq. (6.14) with Eqs. (6.15) through (6.17), we obtain the relation for the fracture stress as a function of Ω:

$$\sigma = (\sigma_0/\beta)\Omega. \tag{6.18}$$

Thus, if the ratio σ_0/β is constant, then the linear behavior $\sigma \sim \Omega$, as shown in Figure 6.3, is consistent with the Net solution. Consider the β factor in Eq. (6.17): For many polymers, the characteristic ratio C_∞ is in the range of 7–10, the ratio M_0/j is the molecular weight per backbone bond (~30–50 g/mol) and will not vary extensively, $b = 1.54A$ is constant, and the density is about 1 g/cc, such that the parameter β is approximately constant. Thus, since σ_0 in Eq. (6.14) is also fairly constant, Eq. (6.18) is considered to describe the observed fracture behavior shown in Figure 6.3. However, due to the preceding assumptions, the data are not expected to fall exactly on the straight fit line due to some differences in σ_0 and β for each polymer. As a specific test case, consider polypropylene (PP), where, from Figure 6.3, $\sigma = 98\,\text{MPa}$ and $\Omega = 3$. For PP, $C_\infty = 5.8, j = 2$, $M_0 = 42\,\text{g/mol}$, $b = 1.54A$, $\rho = 1.0\,\text{g/cc}$, and $M_c = 7000$ g/mol, and we obtain $\beta = 270\,\text{Da}^{1/2}/\text{nm}^2$. Hence, we predict that $\Omega = 3.2$, which compares reasonably with the actual value, $\Omega = 3$ in Figure 6.3. The fracture stress for PP as predicted from Eq. (6.14) is $\sigma^* = 119\,\text{MPa}$, which is of the same order of magnitude as that reported by Vincent ($\sim 98\,\text{MPa}$). The slope of the line in Figure 6.3 is $\sigma_0/\beta = 36.8\,\text{MPa/nm}^2$, as obtained by Vincent from a linear regression analysis, which compares with $\sigma_0/\beta = 37\,\text{MPa/nm}^2$ for PP. Deviations from the average linear relation $\sigma \sim \Omega$ are predicted by the theory for polymers with known constants σ_0/β, and we conclude that the percolation theory of fracture of polymers is in good accord with a broad range of fracture data.

The RP fracture model solves the problem of why we observe relatively low macroscopic fracture stresses, while individual bonds are breaking at much higher "hot-bond" stresses. This is also true near crack tips where the maximum stress, which could be infinite due to the stress singularity, is in fact quite low, as noted by Hutchinson [3]. The "hot-bond" fracture stress σ_h, is determined from the anharmonic nature of the C–C bond using the Morse potential energy function [1]:

$$\sigma_h = D_0 a_m/(2a\cos\theta), \tag{6.19}$$

where $a_m = 1.99/\text{Å}$ is the Morse anharmonicity parameter, a is the cross-sectional area of the molecule, and $\cos\theta$ is the angle between the bond and the applied stress (typically, $\cos\theta \approx 1$). Using values for polystyrene with $a = 111\,\text{Å}^2$ and $D_0 = 80\,\text{kcal/mol}$, we obtain the hot-bond fracture stress, $\sigma_h \approx 5\,\text{GPa}$, which contrasts with the macroscopic fracture stress $\sigma^* \approx 50\,\text{MPa}$. Thus, the applied macroscopic stress creates a broad stress distribution with molecular stresses that are up to 100 times greater than the applied stress. As bonds break, the stress distribution $\phi(\sigma)$ rearranges, the strain energy decreases and moves around, vectors propagate along remaining connected pathways, and microvoids form, coalesce, and eventually lead to microscopic crack propagation through the fractal percolation clusters.

6.2.2 FRACTURE BY DISENTANGLEMENT

Fracture by disentanglement is considered to proceed by the mechanism shown in Figure 6.4, where we depict the response of an (average) entangled chain (with polymer concentration ϕ) to a constant (step function) draw ratio λ as follows: (A) The average entangled chain with $M/M_c > 1$ is uniaxially

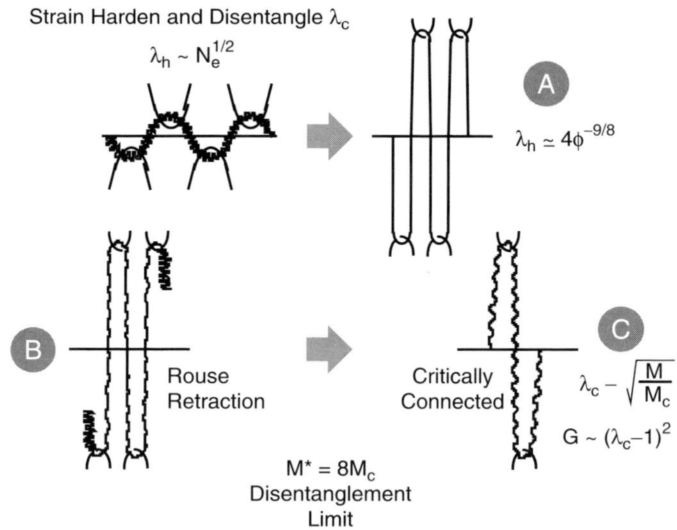

FIGURE 6.4 Disentanglement mechanism. (A) Tightened slack between entanglements. (B) Retraction and disentanglement by Rouse relaxation. (C) Critically connected state at draw ratio λ_c, where each chain crosses the plane three times. The symbol ϕ represents the polymer concentration.

deformed to a constant draw ratio λ. The extension is accommodated by extending the random walk (slack) between entanglements, such that the end-to-end vector R_e between entanglements behaves affinely as $R_e(\lambda) = \lambda R_e$. (B) Rouse-like dynamics causes a retraction of the extended chain primitive path length $L(\lambda)$ and the stored strain energy begins to release. The retraction process will be rate sensitive. As the chain shortens toward its equilibrium path length, it begins to lose entanglements and becomes critically connected at $L_c \approx 2\lambda R_e$. The time dependence of the retraction process can be approximated as a simple exponential such that the stressed fraction of the primitive path $L(t)$ as a function of time is $L(t)/L(\lambda) \approx \exp -t/\tau_{\mathrm{RO}}$, where τ_{RO} is the Rouse relaxation time of the chain. (C) When the chain retracts to a critical length $L_c = 2R_e\lambda$, then each chain possesses one bridge ($P = 3$ crossings) and the network becomes critically connected, as described in [20]. This state corresponds to the failure time of the entanglement network, τ_f, and is determined by

$$\tau_f \sim \tau_{\mathrm{RO}} \ln M/M_c. \tag{6.20}$$

When $M \approx M_c$, disentanglement is nearly instantaneous, but approaches τ_{RO} when $M \approx 8M_c$, which is the strain-hardened ($\lambda \approx 4$) upper bound for chain pullout without bond rupture [1]. Thus, bond rupture would be necessary to complete the fracture process and the value of $M^* \approx 8M_c$ sets an upper limit for fracture by disentanglement or chain pullout. A corollary to this mechanism is that at deformation rates faster than the disentanglement times, or when disentanglement cannot occur, the mechanism changes to bond rupture.

The percolation parameters $[p - p_c]$ associated with the disentanglement process are derived as follows: p is the normalized entanglement density, which is defined as

$$p = g(\lambda)N_\nu/\nu, \tag{6.21}$$

where $g(\lambda)$ is the number of entanglements per chain, N_ν is the number of chains per unit volume, and ν is the entanglement density of the perfect net with infinite molecular weight. We define $g(\lambda)$ as

$$g(\lambda) = [M/M_e(\lambda)] - 1. \tag{6.22}$$

The chain ends effectively contribute to the loss of one entanglement. Because $N_\nu = \rho/M$ and $\nu = \rho/M_e$, we then have

$$p = \frac{\{[M/M_e(\lambda)] - 1\}\rho/M,}{\rho/M_e(\lambda)} \tag{6.23}$$

such that

$$p = [1 - M_e(\lambda)/M], \tag{6.24}$$

where $M_e(\lambda)$ is the stretch-dependent M_e value after it has relaxed to position C, as shown in Figure 6.4. The value of $M_e(\lambda)$ is given by the following approximation:

$$M_e(\lambda) = \lambda^2 M_e \tag{6.25}$$

and $M_e(\lambda)$ increases between entanglement points due to the retraction process at constant λ. A more detailed treatment of disentanglement would account for the orientation function of the entanglements and lateral contraction, as discussed elsewhere [1]. Substituting for $M_e(\lambda)$ in Eq. (6.24), we have the following relation for p:

$$p = 1 - \lambda^2 M_e/M. \tag{6.26}$$

An important consequence of the latter equation is that when $\lambda = 1$, there exists a critical value of molecular weight $M = M_c$ for which $p = p_c$ and we obtain the relation between M_e and M_c as

$$M_c = \frac{M_e}{1 - p_c} \tag{6.27}$$

Since $p_c \approx 1/2$, we note that $M_c \approx 2M_e$, as commonly observed. Also, M_e is determined from the onset of the rubbery plateau by dynamic mechanical spectroscopy and M_c is determined at the onset of the highly entangled zero-shear viscosity law, $\eta \sim M^{3.4}$. This provides a new interpretation of the critical entanglement molecular weight M_c as the molecular weight at which entanglement percolation occurs with the onset of long-range connectivity. Concomitantly, the dynamics changes from single chain, Rouse-like behavior, to that of chains significantly impeded by others, as in Reptation. It also represents the transition from the Nail (weak fracture) [1, 26] to the Net (strong fracture) solution, and marks the onset of significant strength development via the formation of stable, strong, oriented fibrillar material in the deformation zones preceding the crack advance.

When $M > M_c$, we obtain the critical draw ratio for fracture λ_c from Eqs. (6.26) and (6.27):

$$\lambda_c \approx (M/M_c)^{1/2}. \tag{6.28}$$

The maximum molecular weight M^* at which disentanglement can occur is determined when strain hardening occurs at $\lambda_c \approx 4$ such that $M^* \approx 8M_c$. Note that Eq. (6.28) does not have the orientation correction factor [1] of order $2^{1/2}$, which gives a factor of 8 rather than 16, when $\lambda = 4$. Thus, fracture by disentanglement occurs by first straining the chains to a critical draw ratio λ_c and storing mechanical energy of order $G \sim (\lambda_c - 1)^2$. The stretched chains then relax by Rouse-like retraction and disentangle, when the energy released is sufficient to relax them to the critically connected state corresponding to the percolation threshold, p_c. This leads to an obvious dependence of the fracture energy on molecular weight, as described below.

6.2.3 MOLECULAR WEIGHT DEPENDENCE OF FRACTURE ENERGY

The molecular weight dependence of fracture during disentanglement is considerable and varies by several orders of magnitude in the range M_c to $M^*(\sim 8M_c)$ [1]. From Eq. (6.28), the critical draw ratio is $\lambda_c \sim (M/M_c)^{1/2}$, and given that $G \sim (\lambda_c - 1)^2$ in the simple elastic approximation, the molecular weight dependence of fracture behaves approximately as

$$G_{1c} \sim [(M/M_c)^{1/2} - 1]^2 \tag{6.29}$$

or

$$G_{1c} \sim M[1 - (M_c/M)^{1/2}]^2, \tag{6.30}$$

when M is in the range $M_c \leq M \leq 8M_c$. At $M = M^* = 8M_c, G^* \sim 0.42M^*$ such that Eq. (6.30) gives the following ratio:

$$G_{1c}/G^* = 0.3M/M_c[1 - (M_c/M)^{1/2}]^2. \tag{6.31}$$

When $M < M_c$, the role of entanglements is no longer present and Eq. (6.31) cannot be used since $G_{1c} = 0$ at M_c. However, the Nail solution applies for weak interfaces and the chain segments simply pull out at fracture such that $(G_{1c} - G_0) \sim M$, where G_0 is the surface energy term, and we obtain [1]

$$(G_{1c} - G_0)/(G_c - G_0) = M/M_c, \tag{6.32}$$

where G_c is the fracture energy due to pullout at M_c. Typically, the surface energy term G_0 is of order $0.1\,\mathrm{J/m^2}$ and the value of G_c at M_c is about 1–5 $\mathrm{J/m^2}$, which, if compared to typical G_{1c} values for entangled polymers ($\sim 1000\,\mathrm{J/m^2}$), is smaller by several orders of magnitude.

The critical stress intensity factor K_{1c} is related to G_{1c} by the linear elastic fracture mechanics approximation $K_{1c} = (EG_{1c})^{1/2}$ and is derived from Eq. (6.29) as

$$K_{1c} \sim (M^{1/2} - M_c^{1/2}). \tag{6.33}$$

A plot of K_{1c} versus $M^{1/2}$ should be linear with intercept at $M_c^{1/2}$. Alternatively, at $M^* = 8M_c, K^* \sim (M^{*^{1/2}} - M_c^{1/2}) \sim 1.83$, such that the ratio of K_{1c} values at $M_c < M < M^*$ is

$$K_{1c}/K^* = 0.55[(M/M_c)^{1/2} - 1]. \tag{6.34}$$

At high rates of strain compared to $1/\tau$, at the inverse disentanglement time [Eq. (6.20)], or when disentanglement cannot occur ($M > M^*$), bond rupture occurs randomly in the network and the percolation parameter p becomes dominated by chain ends. In this case, the entanglement molecular weight M_e does not depend on strain and Eq. (6.24) gives

$$p = 1 - M_e/M. \tag{6.35}$$

Since $G_{1c} \sim [p - p_c]$, and $p_c = 1 - M_e/M_c$, we obtain

$$G_{1c} = G^*[1 - M_c/M], \tag{6.36}$$

where G^* is the plateau fracture energy at high molecular weight. The latter equation is identical to the empirical relation for the molecular weight dependence of fracture suggested by P. J. Flory, who coincidentally developed the first percolation theory applied to polymer gelation [18].

Figure 6.5 and Table 6.2 show the molecular weight dependence of the fracture energy for polystyrene in which the three solutions are represented for the three pertinent regions involving pullout, disentanglement, and bond rupture mechanisms. When $M > M_c$, the Nail solution of Eq. (6.32) applies to very fragile glasses and $G_{1c} = 0.18 + 4 \times 10^{-5}M$. When $M_c < M < M^*$, disentanglement predominates and Eq. (6.31) applies such that little bond rupture occurs and a linear regression of the data gives $G_{1c} = [0.1132M^{1/2} - 22.81]^2$ with best fit parameters, $M_c = 40{,}600\,\mathrm{g/mol}$, and $G^*(8M_c) = 1734\,\mathrm{J/m^2}$. When $M > M^*$, bond rupture dominates and Eq. (6.36) applies such that the best fit parameters are $G^*(\infty) = 1413\,\mathrm{J/m^2}$, and $M_c = 70{,}771\,\mathrm{g/mol}$.

The following examples illustrate the use of the above equations.

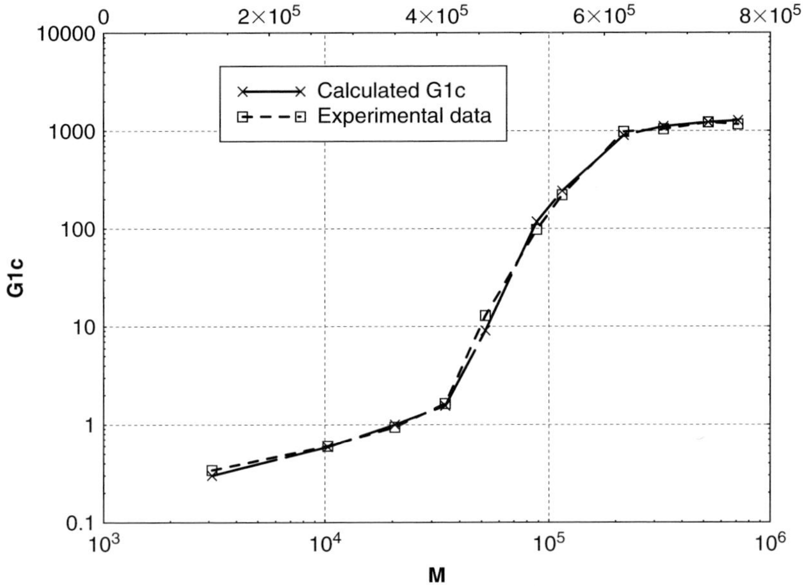

FIGURE 6.5 Experimental and theoretical values of fracture energy G_{1c}(J/m^2) versus molecular weight (g/mol) for polystyrene in the three molecular weight ranges $M < M_c$, $M < 8M_c$, and $M < \infty$, where $M_c = 30{,}000$ and $\log M_c = 4.48$.

TABLE 6.2 Experimental and RP theory results for G_{1c} versus M data for polystyrene.

Experimental					Calculated Results		
M	G_{1c}	log(M)	log (G_{1c})	G_{1c}, Eq. (6.32)	G_{1c}, Eq. (6.31)	G_{1c}, Eq. (6.36)	$Rog(G_{1c})$
3100	0.34	3.491362	−0.46852	0.3007			−0.52187
10300	0.6	4.012837	−0.22185	0.5887			−0.23011
20500	0.94	4.311754	−0.02687	0.9967			−0.00144
34500	1.64	4.537819	0.214844	1.5567			0.192205
52050	12.9	4.716421	1.11059		9.10216237		0.959145
88340	97.6	4.946157	1.98945		117.426605		2.069767
114921	220	5.060399	2.342423		242.29413		2.384343
217600	980.8	5.337659	2.99158		899.766037		2.95413
330070	1040.5	5.518606	3.017242			1110.034	3.045336
524220	1220.66	5.719514	3.086595			1222.24	3.087157
713950	1160.45	5.853668	3.064626			1272.934	3.104806

Example 6.1. The Vanlung-Jones company made a PLA plastic batch by fermentation of corn starch with a molecular weight $M = 48{,}000$ Da. The following questions can be asked:

- What are the M_c, M_e, and M^* values for PLA?
- Estimate the optimal fracture stress σ^* of PLA at M^*.
- What is the fracture stress σ of this PLA with $M = 48{,}000$ Da?

Solution 6.1(a). The M_c value for a random walk chain in the melt and its equivalent amorphous state in the semicrystalline state for linear polymer chains is given for 3/1 helical polymers such as PP, PLA, PVC, PMMA, etc., by R. P. Wool as

$$M_c \approx 61 C_\infty M_j, \tag{6.1.1}$$

where C_∞ is the characteristic ratio and M_j is the molecular weight per repeat unit of the backbone. The front factor 61 is precise for 3/1 helical polymers; polymers with other helicities will have other similar values as reported by Wool [1]; M_j is related to the monomer repeat unit molecular weight via

$$M_j = M_0/j, \tag{6.1.2}$$

where j is the number of backbone atoms per monomer, such that for PLA, $M_0 = 72$ Da, $j = 3$, and thus $M_j = 24$ Da. Because $C_\infty = 7$[1] for PLA, $M^* \approx 8M_c$, and $M_e \approx M_c/2$, then the answers follow as $M_c \approx 10,200$ Da, $M_e = 5100$ Da, and $M^* \approx 82{,}000$ Daltons.

Solution 6.1(b). Using the relation for the fracture stress,

$$\sigma^* = [ED_0\rho/16M_e]^{1/2}, \tag{6.1.3}$$

in which $E \approx 1$ GPa, $\rho \approx 1$ g/cc, $D_0 = 80$ kcal/mol $= 334.4$ J/mol for a C–C bond and $M_e = 5100$ Da, then $\sigma^* \approx 64$ MPa. This compares with experimental values of 61 MPa reported in Chapter 11.

Solution 6.1(c). Using $\sigma/\sigma^* = 0.55[(M/M_c)^{1/2} - 1]$, with $M_c = 10,200$ Da, $\sigma^* = 64$ MPa, then the fracture stress with $M = 48{,}000$ Da is $\sigma = 41$ MPa.

6.2.4 FRACTURE OF AN IDEAL RUBBER

Rubber is one of the first natural polymers used by humankind. Consider an ideal rubber whose strain energy function $U(\lambda)$ is given by Eq. (6.9). Realistically, this function is defined for small λ values since strain hardening near $\lambda_{sh} \approx 4$ is not considered. However, we can inquire as to the requirements to fracture a rubber material at $\lambda < \lambda_{sh}$. The constant C_1 is related to the modulus E, via $C_1 = E/2$ and E is determined from Flory's entropy elasticity

[1] The $C_\infty = 7$ value for PLA was reported by J. Dorgan at the AICHE Annual *Mtg*, October 2003 with experimental M_c values of $M_c \approx 10{,}000$ Da.

theory as [18] follows:

$$E = \nu RT, \tag{6.37}$$

where ν is the cross-link density of the rubber. The strain energy density function for simple uniaxial deformation becomes

$$U(\lambda) = 0.5\,\nu RT(\lambda^2 + 2\lambda^{-1} - 3). \tag{6.38}$$

The percolation dissipation fracture energy is

$$U_f = \nu D_0[p - p_c]. \tag{6.39}$$

Equating $U(\lambda)$ and U_f, we obtain the critical λ_c value as

$$\lambda_c{}^2 + 2\lambda_c^{-1} = 3 + 2D_0[p - p_c]/RT. \tag{6.40}$$

Notice that the entanglement density ν canceled out, which is consistent with critical draw ratios being independent of M_e. There are three solutions to this cubical equation, one trivial and two useful. The first approximate solution for λ_c in tension in Eq. (6.40) involves neglecting the $2/\lambda$ term compared to the λ^2 term such that when $\lambda^2 \gg 2/\lambda$, the solution for positive tension becomes

$$\lambda_c \approx \{3 + 2D_0[p - p_c]/RT\}^{1/2}. \tag{6.41}$$

For example, if we have a perfect net, $p = 1$, $p_c \approx 1/2$, $D_0 = 80\,\text{kcal/mol}$, $R = 1.986\,\text{cal/mol K}$, and $T = 300\,\text{K}$, which gives $2D_0[p - p_c]/RT = 134$, such that the critical value in Eq. (6.41) is $\lambda_c = 11.6$. Thus, since the critical value is much greater than the strain hardening value of $\lambda = 4$, we can see that a typical ideal rubber material will not break without first strain hardening.

The second solution to Eq. (6.40) involves compression with $\lambda < 1$, which is equivalent to equibiaxial extension in the other two directions, and if we make the assumption that $\lambda^2 \ll 1/\lambda$, then we obtain the critical draw ratio in compression as follows:

$$\lambda_c \approx 2/[3 + 2D_0/RT(p - p_c)]. \tag{6.42}$$

If as before, we let the term $2D_0[p - p_c]/RT = 134$, then $\lambda_c \approx 0.007$, such that from the constant volume assumption ($\lambda_1\lambda_2\lambda_3 = 1$), the equibiaxial draw ratios are about $\lambda_2 \approx 11.7$, which again is unrealistic. Thus, we can conclude that an ideal rubber with a perfect network cannot be readily broken in tension or compression.

Examining Eqs. (6.40) and (6.41), it can be seen that to break an ideal rubber without strain hardening at $\lambda_c < 4$, one or more of the following events needs to occur:

- *P decreases to* p_c: This can occur, by either defect creation during synthesis, bond rupture during fatigue, biodegradation, or photodegradation. This is how a pencil eraser works with fragile rubber networks, where $p \approx p_c$. Also, the lifetime of a rubber subject to fatigue or oxidative aging with bond cleavage would be influenced. To achieve values of $\lambda_c \approx 4$, using Eq. (6.40), the term $[p - p_c]$ should decrease from its initial value of 0.5 to about $[p - p_c] = 0.05$.
- *The bond energy* D_0 *decreases:* This can occur with labile bonds, such as weak divalent cation linkages. This could be a method of designing recyclable elastomers, which would become thermoplastic at a certain temperature. At room temperature, the bond energy would have to decrease by a factor of about 10 to $D_0 = 8\,\text{kcal/mol}$ in order to achieve the critical draw ratio of $\lambda_c \approx 4$.
- *The modulus E increases:* The rubber modulus $E \sim T$ could be increased such that when $\lambda_c = 4$, Eq. (6.40) would be satisfied for the perfect rubber net with $T \approx 3000\,\text{K}$, which corresponds to a 10-fold increase in E above that of room temperature. Of course, these temperatures are unrealistic, but the 10-fold increase in modulus is readily obtained during strain hardening, and that is the typical mechanism of fracture for rubber. We have all experienced this effect when attempting to break a rubber band. This phenomenon is not unique to rubber, but occurs with all highly entangled polymers, and it explains why crazes need to strain harden at local temperatures near their glass transition temperature during fracture of glassy polymers. The crazes typically do not fracture at their strain hardening draw ratio ($\lambda \approx 4$) but continue to bear increasing loads in the deformation zone (Figure 6.1) before they fracture [27–30].

In evaluating the fracture of rubber, we have given an example for an ideal rubber using the strain energy term $U(\lambda)$ in Eq. (6.9). However, many other strain energy density functions can be used in a similar manner, and with different constraints on the λ values, as dictated by the sample geometry and triaxial loading conditions, including compression ($\lambda < 1$). The stress resulting from fracture at these draw ratios is determined from $\sigma = dU(\lambda)/d\lambda$. Using Eqs. (6.37) and (6.38) for the ideal rubber, we obtain the uniaxial fracture stress, $\sigma^* = E(\lambda_c - 1/\lambda_c)$, in which the critical value of λ_c is determined by Eq. (6.40). The maximum fracture stress that one can obtain at $\lambda_c \approx 4$ is $\sigma^* = 3.75E$, before strain hardening occurs and the elasticity mechanism changes from entropic to enthalpic, where the internal energy of the molecules changes through bond and valence angle deformation.

If we use the simple elastic strain energy function for rubber, where $U = \sigma^2/2E$, we can examine the role of modulus E and cross-link density

ν more directly. Since the entropic modulus $E = \nu RT$ and $p \approx 1$, $[p - p_c] \approx 1/2$, we obtain from Eq. (6.12):

$$\sigma^* = \nu (D_0 RT)^{1/2}, \tag{6.43}$$

in which $\sigma^* \sim \nu$ is linearly proportional to the cross-link density at constant T. Because $E \sim \nu$, Eq. (6.43) can also be stated as follows:

$$\sigma^* = E[D_0/RT]^{1/2}. \tag{6.44}$$

For example, at $T = 300\,\mathrm{K}$, and using $D_0 = 335\,\mathrm{kJ/mol}$, the latter equation predicts that $\sigma \approx 0.36E$. Bueche and Berry [31] analyzed the fracture of silicone rubber materials as a function of their modulus E. Based on the Griffith theory of fracture, they expected to find that $\sigma \sim E^{1/2}$. Instead they found that the tensile stress was linearly proportional to modulus with a slope of 0.46, which is in close agreement with Eq. (6.44) where the term $[D_0/RT]^{1/2} \approx 0.36$.

Since the fracture energy $G \sim \sigma^2/E$, then we expect from the latter two relations that the dependence of G on cross-link density is given by

$$G \sim \nu. \tag{6.45}$$

This simple linear relation applies when viscoelastic effects are negligible at low deformation rates. Chang *et al.* [32, 33] examined the effect of peel rate on the fracture energy of lightly cross-linked polybutadiene at temperatures in the range of -40° to $130\,^\circ\mathrm{C}$. At very low deformation rates, they found from the time–temperature master plots that the constant minimum fracture energy G_0 in peel was linearly proportional to ν as follows:

$$G_0 = 60\,\nu\,\mathrm{J/m^2}(\mathrm{m^3}/10^{26}). \tag{6.46}$$

The ν values used by Chang *et al.* gave G values in the range of $1-60\mathrm{J/m^2}$ at low deformation rates; at high rates or low temperature, G increased to about $G_\infty \approx 3000\,\mathrm{J/m^2}$ due to bulk viscoelastic energy dissipation.

6.2.5 FRACTURE OF THERMOSETS

For highly cross-linked thermoset polymers, such as epoxies, polyurethanes, unsaturated polyesters, vinyl esters, and functionalized triglyceride networks, the percolation parameter $p \approx 1$, and using $[p - p_c] \approx 1/2$, Eq. (6.12) for the tensile strength becomes

$$\sigma \approx [E\nu D_0]^{1/2}. \tag{6.47}$$

Because D_0 is constant, the fundamental relation for thermosets becomes

$$\sigma \sim [E\nu]^{1/2}. \tag{6.48}$$

The latter relation was explored by LaScala *et al.* [34, 35] and Lu *et al.* [36, 37] and found to be descriptive of the tensile properties of highly cross-linked triglyceride thermosets. A plot of σ versus $(E\nu)^{1/2}$ was linear but the slope indicated that D_0 was reduced by a factor $f = 1000$, possibly due to the free-radical propagation reactions discussed by Kausch [22]. This being the case, the strength could be increased by the presence of free-radical traps, such as hydroquinone or lignin. Computer simulations by Lorenz *et al.* [38] on the fracture of cross-linked thermosets as a function of ν also found that $\sigma \sim \nu^{1/2}$.

For composite materials, the role of ν in the fiber–matrix interface becomes critical, especially since $\nu \sim X$, where X is the degree of reaction of the C=C double bonds (unsaturated polyesters) or for the polyfunctional linkages (epoxies, urethanes) used to create the cross-linked network. Typically, X is reduced near the solid surface of the fiber and this can lead to facile fracture in the interface. The cross-link density as a function of the extent of reaction typically behaves as [19]

$$\nu = \nu_\infty[X - X_c], \tag{6.49}$$

where X_c is the extent of reaction at the gelation threshold, and ν_∞ occurs at $X = 1$ for the fully reacted system. The governing relation for the strength of composites with interface failure will be determined as a function of X by

$$\sigma/\sigma_\infty = [X - X_c]^{1/2}/[1 - X_c]^{1/2}, \tag{6.50}$$

where σ_∞ is the virgin strength determined at $X = 1$. Thus, incomplete reaction ($X < 1$) at some location in the thermoset matrix, typically near the fiber surface, due to either stoichiometric imbalance or steric effects, can significantly weaken the material.

While the fracture stress of thermosets increases as $\sigma \sim \nu^{1/2}$, the fracture energy G_{1c} of thermosets will be expected to decrease with increasing ν due to a loss of ductility. Since $G_{1c} = \sigma\delta$, where δ is the critical crack opening displacement, there will be a competition between σ increasing with ν while δ decreases with ν. If $\delta \sim R_e\lambda$, where R_e is the radius of gyration of the cross-link length and λ is the draw ratio of the cross-link, then since $R_e \sim M_e^{1/2} \sim \nu^{-1/2}$ and $\lambda \sim M_e^{1/2} \sim \nu^{-1/2}$, we expect that $\delta \sim \nu^{-1}$, which reflects the reduced ductility of thermosets with increasing ν values. Lu and Wool [36] also found that the plastic zone at a crack tip in thermosets decreases considerably with increasing ν. Combining relations for σ and δ, the fracture energy for thermosets is predicted to behave as follows:

$$G_{1c} \sim \nu^{-1/2}. \tag{6.51}$$

This is consistent with experiments of Lemay *et al.* [39] and Lu *et al.* [36, 37]. However, other studies by Levita *et al.* [40] and Pearson and Yee [41] show that $G_{1c} \sim \nu^{-1}$. With very highly cross-linked systems, the random walk relation for $R_e \sim M_e^{1/2}$ may no longer be valid. If the conformation of the cross-link changes to a self-avoiding walk where $R \sim M_e^{3/5}$, then $G_{1c} \sim \nu^{-0.7}$. In the extreme case where $R_e \sim M_e$ for highly extended rigid rod-like cross-links, then we predict that $G_{1c} \sim \nu^{-3/2}$. Thus, the exact relation for G_{1c} versus ν depends on the microscopic details and the exponent is expected to vary between $-1/2$ and $-3/2$.

6.2.6 FRACTURE OF CARBON NANOTUBES

Soybean oil and functionalized oils (Chapter 4) behave as solvents for carbon nanotubes and will be discussed in more detail in Chapter 14. The tensile strength σ of a single-wall carbon nanotube (SWNT), as shown in Figure 6.6 [42], can be determined from the RP theory of fracture as follows. Consider a graphene sheet, which can be rolled into a nanotube of diameter d. The sheet has a width πd, thickness h, and unit length L such that its volume is πdhL. The areal density of bonds in the sheet is $\nu = j/b^2$, where b is the C–C bond length and $j = 1.15$ is the hexagonal lattice factor. The critical force F_c applied to an area πdh necessary to break the sheet in uniaxial tension is determined from Eq. (6.12) to be

$$F_c = [2E\nu D_0(p - p_c)\pi^2 d^2 h]^{1/2}. \tag{6.52}$$

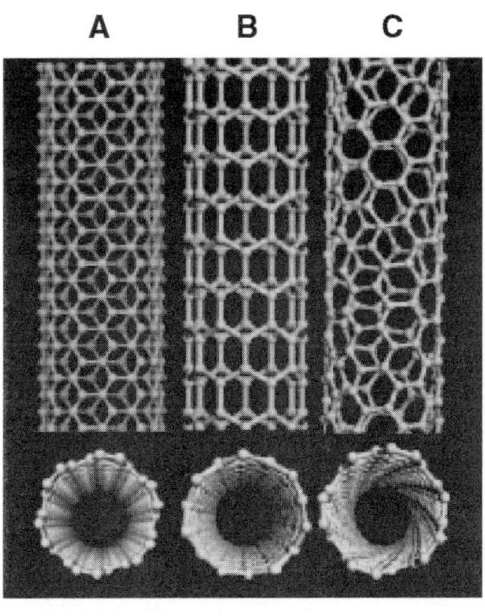

FIGURE 6.6 Schematic illustrations of the structures of (A) armchair, (B) zigzag, and (C) chiral SWNTs. Projections normal to the tube axis are shown at the top and perspective views along the tube axis are on the bottom. The areal density of bonds is $\nu = 1.15/b^2$, where $b = 0.142$ nm is the bond length. (*Source:* Courtesy of R. Baughman *et al.* [42].)

If we now wrap the sheet into a SWNT of diameter d and cross-sectional area $A = \pi d^2/4$, the applied macroscopic nominal stress $\sigma = F_c/A$, which is necessary to reach the critical force, is determined:

$$\sigma = [32E\nu D_0(p - p_c)h/d^2]^{1/2}. \quad (6.53)$$

Substituting for the bond density $\nu = j/b^2$ and sheet thickness $h \approx b/2$, we obtain the critical fracture stress for a SWNT as

$$\sigma = [16ED_0(p - p_c)j/bd^2]^{1/2}. \quad (6.54)$$

For example, using the values of modulus $E \approx 1\,\text{TPa}$, hexagonal lattice parameter $j = 1.15$, C=C bond energy $D_0 = 518\,\text{kJ/mol}$, bond length $b = 0.142\,\text{nm}$, $h = 0.071\,\text{nm}$, lattice degree of perfection $p \approx 1$ (no defects), and percolation threshold in two dimensions $p_c \approx 0.6$, the fracture stress of a defect-free SWNT as a function of tube diameter d (nm) is given by

$$\sigma(d) = 211\,\text{GPa} \cdot \text{nm}/d \quad (6.55)$$

The critical force F_c to fracture a SWNT is determined from Eq. (6.52) using the same values:

$$F_c(d) = 166d\,\text{nN/nm}. \quad (6.56)$$

Thus, if $d \approx 2\,\text{nm}$, $\sigma \approx 100\,\text{GPa}$ when $p = 1$ and $F_c = 332\,\text{nN}$. However, for nanotubes with defects, such as those that occur during the chemical vapor deposition (CVD) process [43], then $p < 1$ and the fracture stress can decrease significantly as p approaches p_c. For example, if the SWNT contains 10% defects at any location along the tube, then the stress decreases to 180 GPa and $F_c = 287\,\text{nN}$. For multiple-wall nanotubes (MWNTs) whose strength is determined by the outer layer, when the diameter increases to $d = 10\,\text{nm}$, the nominal fracture stress is reduced to about 21 GPa and $F_c = 1660\,\text{nN}$. These results are consistent with data on the fracture of SWNTs and MWNTs that were obtained by many investigators and reviewed by Thostenson *et al.* [43].

6.3 MICROSCOPIC TO MACROSCOPIC FRACTURE RELATIONS

6.3.1 RELATION OF RP MODEL TO *J*-INTEGRAL FRACTURE MECHANICS

Using the Hutchinson [3–5] theory of fracture for embedded process zones (EPZs), such as crazes at crack tips, the fracture energy is approximated using the *J*-integral approach as shown earlier in Figure 6.1. The EPZ model addresses small-scale yielding at the crack tip and describes the work of separation as a function of deformation rate, yield stress, and the stress integral as a function of displacement δ, in the zone. The vector percolation model describes the maximum stresses attainable in the deformation zone. However, the crack opening displacement δ is determined by the drawability or ductility of the entangled matrix, as described by Kramer *et al.* [44, 45]. This process is exceedingly complex, and for polymers, it involves the non-Newtonian flow and plastic deformation of the strain-hardened, craze-like material in the zone. The craze fibrillar structure evolves from a combination of Saffmann-Taylor meniscus "finger" instability, combined with cavitation processes, depending on the rate and the molecular weight, as discussed in [1]. Most materials exhibit the craze-like fibrillar structure under dilatational plane-strain conditions. The microstructure of crazes formed from entangled amorphous polymers was well described by Kramer and Berger [45] and Kambour [46]. Thus, the deformation zone initiates when the stress field at the crack tip exceeds the yield stress, σ_y, and propagates with increasing traction stresses σ, up to the maximum stress σ^*. Bjerke and Lambros [47] have shown that the local temperature approaches the glass transition temperature in the craze zone for dynamically propagating cracks in PMMA, thus enhancing the disentanglement process at the crack tip. As the zone breaks down, the stresses decrease near the crack tip (as p approaches p_c) but the displacements continue to increase (modulus E approaches zero) and finally the crack advances at δ_c. The traction stresses have a displacement function $\sigma(\delta)$, as shown in Figure 6.7, such that the fracture energy G_{1c} is the integral of $\sigma(\delta)$ over the crack opening displacement range, $\delta = 0$, to $\delta = \delta_c$. Solutions to the EPZ integrals are complex with regard to rate and the nonlinear constitutive character of the drawing process, and are usually obtained by computer simulation [3–5].

To provide linkage of the vector percolation description of σ^* with fracture experiments involving G_{1c} measurements, we approximate $\sigma(\delta)$ as a simple box function with maximum value σ^*, such that G_{1c} is obtained as

$$G_{1c} = \sigma^* \delta_c. \tag{6.57}$$

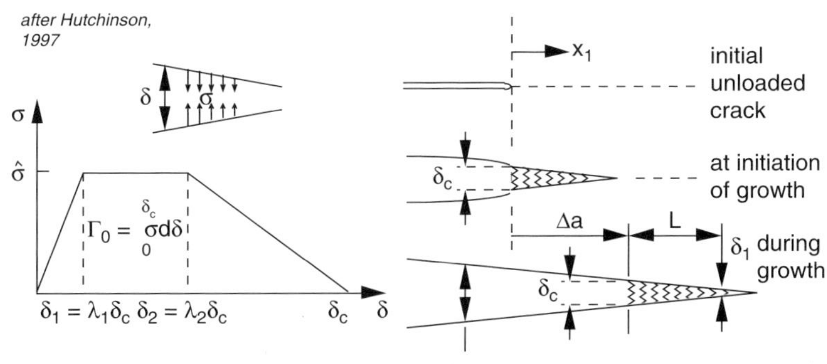

FIGURE 6.7 Fracture energy determination using the *J*-integral theory [3–5].

For example, if $G_{1c} = 1000\,\text{J/m}^2$ and the critical stress $\sigma^* = 50\,\text{MPa}$, then from Eq. (6.57), we obtain $\delta_c = 20\,\mu\text{m}$, which is considerably larger than typical chain dimensions where $R_g \approx 0.01\,\mu\text{m}$. In the traditional Dugdale model [48, 49], $\sigma^* = \sigma_y$ and the familiar result, $G_{1c} = \sigma_y\delta_c$, is obtained. However, in the EPZ model, σ^* exceeds σ_y and typically σ^*/σ_y is in the 4–8 range, and this ratio is rate dependent. Consistent with the previous assumptions on the size of the deformation zone relative to the crack length, we make the assumption that the critical crack opening displacement δ_c is proportional to the maximum stresses σ^* in the deformation zone, via

$$\delta_c = \psi\sigma^*. \tag{6.58}$$

The proportionality constant ψ is associated with the plastic drawing process. In effect, the highly complex non-Newtonian plastic zone drawing process is viewed as a pseudoelastic (nonreversible) process, such that the higher the σ^* value, the higher the δ_c values. There is considerable evidence for this behavior for linear amorphous polymers [1]. Substituting for δ_c in Eq. (6.57) we obtain

$$G_{1c} = \sigma^{*2}\psi, \tag{6.59}$$

where σ^* is given by the VP model, via Eq. (6.8), and ψ is a constant to be determined. For example, if $G_{1c} \approx 1000\,\text{J/m}^2$ and $\sigma^* \approx 50\,\text{MPa}$, then $\psi \approx 0.4\,\mu\text{M/MPa}$. The latter ψ value should depend on the exact non-Newtonian flow constitutive details of the glass-to-craze zone drawing process, temperature, and rate of deformation. Thus, the final equation for the fracture energy of polymers with craze deformation zones is given by

$$G_{1c} = 2\psi(p - p_c)^{\tau+1}E_0D_0\rho/(\lambda^2 M_e)(1 - p_c)^{\tau}. \tag{6.60}$$

When $G_{1c} \sim \sigma^{*2}$, the latter equation predicts for high molecular weight ($M > 8M_c$ and $p \approx 1$) glassy polymers, so that

$$G_{1c}{}^* \approx G_0/M_c, \tag{6.61}$$

where $G_0 \approx 3 \cdot 10^4\,\text{J/m}^2 \cdot \text{Da}$. This is essentially constant for linear glassy polymers with comparable modulus and density. The latter relation highlights the role of M_c in fracture processes, where the higher entanglement density nets require larger fracture energies. Thus, we would expect polycarbonate to have a fracture energy about 5–6 times that of polystyrene, or $G_{1c} \approx 6\,\text{GPa}$, consistent with experiment [24].

6.3.2 RELATIONSHIP TO THE GRIFFITH THEORY OF FRACTURE

The Griffith theory of fracture is the most classical and useful approach in linear elastic fracture mechanics (LEFM) and involves a material with a central crack of length $2a$, or a single edge notch of length a. When external tensile loads are applied, the critical fracture stress (in plane stress) is given by

$$\sigma = [EG_{1c}/\pi a]^{1/2}, \tag{6.62}$$

where E is Young's modulus and G_{1c} is the critical strain energy release rate, commonly called the fracture energy with units of energy per unit area (J/m^2). The LEFM theory works well when the crack length a is much greater than the deformation zone length r_p at the crack tip, as discussed in the last section, and when viscous energy dissipation in the body of the material is absent. When both viscous and plastic energy dissipation are absent in the deformation zone, a simple thermodynamic balance suggests that the fracture energy is equivalent to the new surface energy γ, created via

$$G_{1c} = 2\gamma. \tag{6.63}$$

This energy can be considered an absolute lower bound on the brittle fracture of materials and is rarely observed, even with polymers where $M \ll M_c$. Typically, γ is of order $\gamma \approx 0.1$ J/m^2 such that $G_{1c} \approx 0.2$ J/m^2, which is up to 4 orders of magnitude less than that for commodity plastics such as PP and PS with $G_{1c} \approx$ 100–1000 J/m^2. The additional energy observed during fracture comes from the plastic deformation near the crack tip, such as crazing and shear yielding, and the volumetric dissipation of viscous energy in the bulk of the plastic away from the crack tip. Because E and G_{1c} are constant in Eq. (6.62), the constant terms can be separated out on the right-hand side as

$$\sigma[\pi a]^{1/2} = [EG_{1c}]^{1/2}, \tag{6.64}$$

where the term $\sigma[\pi a]^{1/2}$ is called the critical stress intensity factor K_{1c}, and is related to G_{1c} via the LEFM relation:

$$G_{1c} = K_{1c}{}^2/E. \tag{6.65}$$

Note that K_{1c} is sometimes referred as the *toughness* since the stress concentration on the plane at a distance r in front of the crack tip behaves as $\sigma(r) \sim \sigma/r^{1/2}$, which goes to infinity at $r = 0$, such that K_{1c} gives a measure of the maximum stress that can be sustained near the crack tip. This is similar to the maximum stress σ^*, which can be sustained in the deformation zone at the crack tip using the J-integral approach discussed earlier.

Equating Eq. (6.13) for σ^* and the Griffith theory of Eq. (6.62), we obtain a useful relation for G_{1c} in terms of the cross-link density v and percolation parameters p and p_c at constant E:

$$G_{1c}/G_{1c}{}^* = [(p - p_c)/(1 - p_c)][v/v^*]. \tag{6.66}$$

For linear polymers with varying molecular weight M, p varies but v is constant, as evidenced by the independence of the M_e on M, such that we obtain

$$G_{1c}/G_{1c}{}^{*} = [(p - p_c)/(1 - p_c)], \quad (6.67)$$

which compares G_{1c}^{*} of a perfect network with that containing defects, for example, induced by chain ends for $M < M^{*}$, as discussed in the next section.

6.4 POLYMER–POLYMER INTERFACES

Polymer interfaces are ubiquitous. They play a critical role in determining the properties, reliability, and function of a broad range of materials. For example, polymer melt processing via injection molding (for plastic auto parts) and extrusion (for plastic pipe) creates many interfaces in the form of internal weld lines where the fluid fronts coalesce and weld. Compression molding and sintering (of artificial hip joints) require the coalescence of pellets or powder when their surfaces contact in the mold. Drying of latex paints and coatings (Chapter 8) entails a very large number of interfaces per unit volume, because the relatively tiny (~1000-Å) latex particles interdiffuse to form a continuous film. Construction of composites with thermoplastic matrices (aircraft bodies) requires the fiber-filled laminates to weld by an interdiffusion process at the interface. Welding of two pieces of polymer by thermal or solvent bonding is a commonly encountered example of strength development at a polymer–polymer interface; tack between uncured rubber sheets during auto tire manufacture is an important example of this kind of welding.

Numerous examples of *symmetric interfaces* exist where the same polymer occurs on both sides of the interface. In contrast, rubber toughening of glassy polymers (bulletproof glass) requires that the interface formed by the rubber particle with the glass (Plexiglas™) matrix be sufficiently strong to promote stable energy dissipation. The rubber/Plexiglas is an example of an *asymmetric interface* where two different polymers are joined. Asymmetric or dissimilar interfaces come about when polymers are blended, when dissimilar materials are laminated (in electronic materials applications), when coupling agents are used in composites, when plastic mixtures from municipal waste are recycled, when different materials are coextruded, and when photographs are developed. As our knowledge of material interfaces expands, so does our ability to fabricate more sophisticated material systems, be they artificial organs, biomedical implants, rocket motors, high-speed integrated circuits, the "Super Auto" of 2020, advanced composite space vehicles, more effective drug delivery molecules for cells, new chewing gum, plastic bridges, tractors from soybeans, or, perhaps, a better paint [1].

The subject of interfaces is very rich if we consider the wide variety of possible pairs involving polymers and materials with different chemical composition, crystalline and amorphous content, compatibility, incompatibility, molecular weight distributions, additives, surface chemistry, and so on. In this section, we examine how interfaces form, describe their structure, and provide an understanding of the relationship between structure and strength. We first examine the structure of these interfaces and then relate the structure to fracture strength through microscopic deformation mechanisms involving chain disentanglement and bond rupture. The evolution of structure at diffuse interfaces is controlled by the dynamics of the chains and the thermodynamics of chemically interacting species. Once the structure is known, a *connectivity relation* is required to relate the structure to mechanical properties. The connectivity relation for amorphous polymers is developed in terms of an entanglement percolation model [1–3]. We can then understand what is required to break the connectivity at the interface by disentanglement and/or bond rupture. The latter microscopic deformation mechanisms consume a known amount of energy from which the strength of the interface can be determined.

6.4.1 WELDING OF SYMMETRIC A/A INTERFACES

Figure 6.8 shows an interface formed by random walk chains diffusing by reptation across a polymer–polymer weld line [50]. The molecular aspects of interdiffusion of linear entangled polymers ($M > M_c$) during welding of

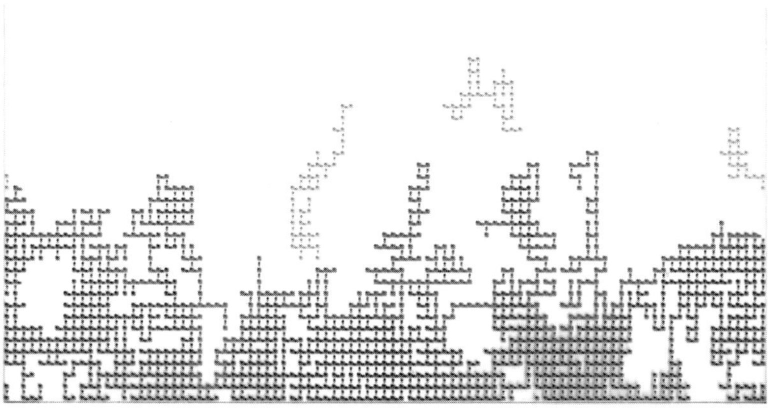

FIGURE 6.8 Polymer interface (one side) formed by random walk chains interdiffusing across the weld line at the bottom. The green chains are the connected chains, which contribute to weld strength by connecting both sides of the interface; the yellow chains are those that have interdiffused but do not contribute to strength because they are not connected to both sides. The red line is the fractal diffusion front, which divides the connected from the nonconnected chains in the diffusion gradient.

TABLE 6.3 Molecular aspects of interdiffusion at a polymer–polymer interface.

Molecular Aspect	Symbol	Dynamic Relation, $t < T_r$	Static Relation $Ht = (t)$	r, s
General property	$H(t)$	$t^{r/4}M^{-s/4}$	$M^{(3r-s)/4}$	r,s
Average contour length	$l(t)$	$t^{1/2}M^{-1/2}$	M	2,2
Number of chains	$\Sigma(t)$	$t^{1/4}M^{-5/4}$	$M^{-1/2}$	1,5
Number of bridges	$P(t)$	$t^{1/2}M^{-3/2}$	M^0	2,6
Average monomer diffusion depth	$X(t)$	$t^{1/4}M^{-1/4}$	$M^{1/2}$	1,1
Total number of monomers diffused	$N(t)$	$t^{3/4}M^{-7/4}$	$M^{1/2}$	3,7
Center of mass diffusion	X_{cm}	$t^{1/2}M^{-1}$	$M^{1/2}$	2,4
Fractal diffusion front length	N_f	$t^{1/2}M^{-3/2}$	M^0	2,6

polymer interfaces are summarized in Table 6.3 [1]. The Reptation dynamics and the interface structure relations in Table 6.3 have been demonstrated experimentally by a series of interdiffusion experiments with selectively deuterated polymers using dynamic secondary ion mass spectroscopy (DSIMS) and neutron reflectivity [51–55]. The scaling laws and the complete concentration profiles have been calculated by Kim and Wool [56] and Zhang and Wool [57]. The important result for the contour length $L \sim (t/M)^{1/2}$ (which is the basis for the early Wool theory) was also supported by the welding computer simulations of Anderson *et al.* [58]. Initially, as the symmetric ($A = B$) interface wets by local Rouse segmental dynamics, we find that rapid interdiffusion occurs to distances of the order of the radius of gyration of the entanglement molecular weight, ~30 Å. This can also occur below T_g when the top surface layer becomes more mobile than the bulk and can be explained by finite size rigidity percolation theory [59]. However, at this point, the interface is very weak and fracture can be described by the Nail solution [26]. At the wetting stage, the frictional pullout of intermeshed chain segments, which have "elbowed" their way across the interface, determines the fracture energy. As welding proceeds, Σ minor chains of length L diffuse into an interface of width X and considerable strength develops. The diffusing chains are fractal random walks and interpenetrate with chains, which are fully entangled (ignoring surface reflection configuration effects on entanglement density).

The structure of the diffuse weld interface in Figure 6.8 resembles a box of width X, with fractal edges containing a gradient of interdiffused chains, as shown by Wool and Long [50]. Gradient percolation theory [60] requires that chains, which contribute to the interface strength, straddle the interface plane during welding, such that chains in the concentration gradient which have diffused further than their radius of gyration cease to be involved in the load-bearing process at the interface. We have shown that this amounts to a very small number and, for narrow molecular weight distributions, it can be ignored [1, 56, 57]. However, for broad molecular weight distributions, the fraction of unconnected chains expressed through gradient percolation can be significant. When the local stress exceeds the yield stress, the deformation zone forms and the oriented craze fibrils consist of mixtures of fully entangled matrix chains and partially interpenetrated minor chains. Fracture of the weld occurs by disentanglement of the minor chains, or bond rupture. It is interesting to note that if the stress rises to the point where random bond rupture in the network begins to dominate the deformation mechanism, instead of disentanglement, then the weld will appear to be fully healed, regardless of the extent of interdiffusion. This can occur at high rates of testing when the minor chains cannot disentangle and bond rupture pervades the interface, breaking both the minor chains and the matrix chains.

The percolation term $[p - p_c]$ determines the number of bonds to be broken, or disentangled such that when Σ chains, each with L/L_c entanglements per chain, interdiffuse in an interface of width X, we obtain

$$[p - p_c] \sim \{\Sigma L/X - [\Sigma L/X]_c\}, \tag{6.68}$$

where $p_c \sim [\Sigma L/X]_c$. Since $\Sigma/X \sim 1/M$ (Table 6.3), it follows that $p_c \sim L_c/M$. Thus, when the interdiffused minor chain length $L \approx M_c$, we have no strength (above that of the Nail solution), and when $M \gg M_c$, $p_c \approx 0$, which we will assume henceforth for the welding analysis. In terms of a time argument, the time at which p_c is reached is controlled largely by Rouse segmental dynamics, which is much shorter than the long interdiffusion time determined by Reptation dynamics [61]. Thus, the slower interdiffusion process will dominate strength evolution versus time.

The interface of width X is composed of a fraction L/M of diffusing chains and the matrix chain fraction $(1 - L/M)$, into which the chains are diffusing. The total stored strain energy in the interface $U \sim X$ is consumed in disentangling only the Σ minor chains of length L, from the matrix chains, the latter being stretched also but they cannot disentangle at the same rate as the minor chains. We obtain $G_{1c} \sim p$ as

$$G_{1c} \sim \Sigma L/X. \tag{6.69}$$

When the matrix chains disentangle or break along with the interdiffused chains, then $p = 1$ and the virgin strength is reached. The number of diffusing chains Σ contributing to the strength at the interface is governed by gradient percolation, such that chains, which do not straddle the interface, are not counted. Also, the length L implies the number of entanglements per minor chain L/L_c, which can decrease significantly, for example, if brush-like ordering occurs at the interface, or the entanglement topology changes such that L_c becomes very large as in a solvent where $M_c \sim$ polymer concentration.

Since $\Sigma \sim X/M$ (Table 6.3) and $L \sim t/\tau^{1/2}$, we obtain the time dependence of welding as

$$G_{1c}(t) = G_{1c}{}^{*}(t/\tau^{*})^{1/2}, \tag{6.70}$$

where G_{1c}^{*} is the maximum strength obtained at $M^{*} \approx 8M_c$ and is independent of molecular weight. The average contour length $\langle L \rangle$ controls the disentanglement process via $\lambda_c \approx [\langle L \rangle/L_c]^{1/2}$. The welding time τ, to achieve complete strength, according to the Reptation model [61, 62], behaves as

$$\tau \sim M^3, \tag{6.71}$$

in which $M < M^{*}$. Recent studies [63] have indicated that while chains diffuse in a Reptation-like mode, the monomer friction coefficient (assumed constant for Reptation) may have a weak molecular weight dependence, of order $M^{0.3}$, resulting in an exponent of 3.3, instead of 3.0 in Eq. (6.71). This correction would cause a small change in the exponents for the molecular weight dependence of welding, but would not affect the time exponents. For example, the minor chain length $\langle L \rangle$, which from Table 6.3 behaves as $\langle L \rangle \sim t^{1/2}M^{-0.5}$ with $\tau \sim M^3$, would become $\langle L \rangle \sim t^{1/2}M^{-0.65}$ when $\tau \sim M^{3.3}$.

When $M > M^{*}$, the welding time is determined by the time required to diffuse a distance of order of the radius of gyration of M^{*}, such that $\tau^{*} \sim M^{*2}M$. Even though the welding time $\tau^{*} \sim M$ is shorter than $T_r \sim M^3$, the molecular weight dependence of the welding rate remains unaffected and we have for all molecular weights,

$$G_{1c}(t) \sim t^{1/2}M^{-1/2}, \tag{6.72}$$

which is consistent with the early Wool–O'Connor welding theory $G \sim L$. As the interdiffusion distance approaches R_g, the welding state becomes indistinguishable from the virgin state and Eq. (6.72) converges to Eq. (6.31), when $M < M^{*}$, or Eq. (6.36), when $M > M^{*}$. Experimental support for Eq. (6.72) was reported by Wool and O'Connor [64, 65] and Wool *et al.* [66] and reviewed in [1].

In the case of chain-end segregation to the surfaces, as can occur in crack healing and some latex particle coalescence during film formation, the number of chains Σ is constant and the percolation term becomes $p \sim L/X$, or $p \sim X$, since $X \sim L^{1/2}$. Thus, from Table 6.3, the strength development would be $G_{1c} \sim (t/M)^{1/4}$, rather than the usual $t^{1/2}$ dependence. This $t^{1/4}$ result was also predicted by Prager and Tirrell, using a crossing density analysis [67], but with a different molecular weight dependence for both the welding and virgin state.

The full interpenetration of chains (X approaches R_g) is not necessary to achieve complete strength, when $M > M^{*}$ and $\tau^{*} < T_r$. However, a cautionary note is appropriate here: Although complete strength may be obtained in terms of critical fracture measures, such as G_{1c} and K_{1c}, the durability, measured in subcritical fracture terms, such as the fatigue crack propagation rate da/dN, may be very far from its fully healed state at τ^*. We have shown that while the weld toughness K_{1c} increases linearly with interdiffusion depth X as $K_{1c} \sim X$, the fatigue crack propagation behavior of partially healed welds behaves as [1, 66]:

$$da/dN \sim X^{-5}. \tag{6.73}$$

This fatigue behavior is a very strong function of interdiffusion and underscores the penalty to pay for partial welding. Thus, the weld strength may be close to the virgin strength but the fatigue strength may be dramatically reduced below its maximum value. Thus, one should always design a welding time with respect to T_r to achieve maximum durability of welds and interfaces. The time to achieve complete strength is related to the Reptation time by

$$\tau^* = 64(M_c/M)^2 T_r, \tag{6.74}$$

such that when $M = 8M_c$, $\tau^* = T_r$. The Reptation time T_r is determined from the self-diffusion coefficient D and the end-to-end vector R by

$$T_r = R^2/(3\pi^2 D). \tag{6.75}$$

For example, when welding polystyrene at 125°C, $D \approx 4 \times 10^{-6}/M^2$ (cm^2/s) [55, 68], $R^2 = 0.45 \times 10^{-16}\,M(\text{cm}^2)$ such that $T_r = 4 \times 10^{-13}\,M^3(\text{s})$, and $\tau^* = 0.0234\,M(\text{s})$. For the case where $M = 400{,}000$ and $M_c = 30{,}000$, we have $\tau^*/T_r = 0.36$, where $T_r = 435$ min and $\tau^* = 156$ min. In this example, if the maximum weld strength were obtained at an allowed welding time of 156 min, the durability as measured by da/dN would only be about 1/5 of its virgin value compared to complete welding at $T_r = 435$ min. When plastic parts are being injection molded, laminated, sintered, or coextruded, many internal weld lines are encountered and this aspect of welding needs to be considered in designing materials with optimal durability [1].

Comment on Fatigue of Welded Interfaces

Total interpenetration of chains (X approaches R_g) is not necessary to achieve complete strength when $M > M^*$ and $\tau^* < T_r$. However, note this word of caution: Although complete strength may be obtained in terms of critical fracture measures such as G_{1c} and K_{1c}, the durability, measured in subcritical fracture terms, such as the fatigue crack propagation rate da/dN, may be very far from its fully healed state at τ^*. We have shown that while the weld toughness K_{1c} increases linearly with interdiffusion depth X, as $K_{1c} \sim X$, the fatigue crack propagation behavior of partially healed welds behaves as [1]

$$da/dN \sim X^{-5}, \tag{6.76}$$

which is a very strong function of interdiffusion and underscores the penalty to pay for partial welding. Thus, the weld strength may be near or at the virgin strength, but the fatigue strength may be dramatically reduced below its maximum value. Thus, one should always design a welding time with respect to T_r to achieve maximum durability of welds and interfaces. The following example illustrates this point.

Example 6.2 Consider welding PLA with $M = 140{,}000$ Da at 125°C, where the self-diffusion coefficient $D \sim M^{-2}$, is approximately

$$D \approx 10^{-6}/M^2(\text{cm}^2/\text{s}).$$

What is the Reptation time T_r and the welding time τ^* to achieve optimal strength? What would be the fatigue resistance *da*/*dN*, at τ^* compared to that at T_r?

Solution 6.2. The square of the end-to-end vector, $R^2 = [C_\infty Mj/M_0]b_0{}^2$. Using $C_\infty = 7$, $M_0 = 72$, $j = 3$, and $b_0 = 1.54$ A, the relation for the molecular weight dependence of R^2 is

$$R^2 = 0.69 \times 10^{-16}\, M(\text{cm}^2).$$

The Reptation time is determined by $T_r = R^2/(3\pi^2 D)$, such that the molecular weight dependence of T_r is given by

$$T_r = 2.3 \times 10^{-12} M^3(\text{s})$$

when $M = 140{,}000$, $T_r = 6311$ s, or 105 min.

The optimal welding time τ^* is given by $\tau^* = 64(M_c/M)^2 T_r$, such that with $M_c = 10{,}200$ and $M = 140{,}000$, then $\tau^* = 35.7$ min and $\tau^*/T_r = 0.34$.

The fatigue propagation rate FPR = da/dN, which is the incremental increase in crack length *da*, per cycle *dN* of fatigue, at constant driving force behaves as $da/dN \sim X^{-5}$ or $da/dN \sim (t/M)^{-5/4}$, such that we obtain

$$\text{FPR}(\tau^*)/\text{FPR}(T_r) = [T_r/\tau^*]^{5/4} = (1/64)^{5/4}[M/M_c]^{10/4}$$

Thus, $\text{FPR}(\tau^*) = \text{FPR}(T_r)[105/35.7]^{5/4} = 3.85$. This is a large penalty to pay for a weld that has reached its optimal weld strength, as measured by G_{1c} or σ, but has not attained its complete welding time, T_r. The weld strength G_{1c} is the same at τ^* and T_r, but the *da/dN* value can be significantly different. As M increases, the fatigue factor *da/dN* increases by $[M/M_c]^{10/4}$. The message here is that very high molecular weight polymers give excellent fatigue resistance, provided that their interfaces are completely welded to T_r; failure to allow full welding or sintering time carries a huge fatigue penalty. This is a rather subtle processing point, which is often not appreciated by the manufacturing industry.

Welding below T_g, as demonstrated by Boiko *et al.* [69, 70], can occur due to softening of the surface layer. We have treated the surface layer softening as a gradient rigidity percolation issue [59]. The surface rubbery layer concept controversy in thick films is interesting and this percolation theory suggests that for free surfaces there is a gradient of $p(x)$ near the surface, where $x < \xi$ and hence a gradient in both T_g and modulus E. If the gradient of p is given by $p(x) = (1 - x/\xi)$, then the value of X_c for which the gradient percolation threshold p_c occurs, and which defines the thickness of the surface mobile layer, is given by the percolation theory as

$$X_c = b(1 - p_c)/\{p_c{}^{\nu}[1 - T/T_g]^{\nu}\}, \tag{6.77}$$

in which b is the bond length and ν is the critical exponent for the cluster correlation length $\xi \sim [p - p_c]^{-\nu}$. For example, when $T = T_g - 10$ K, $T_g = 373$ K, and using $b = 0.154$ nm, $p_c = 0.4$, $\nu = 0.82$, then the thickness of the mobile layer $X^* = 3.8$ nm. This could allow for healing to occur below T_g, assuming that the dynamics are fast enough, since the mobile layers on both surfaces effectively disappear when the interface is formed. If $G_{1c} \sim X^2$ for entangled polymers, then we could deduce that for sub-T_g healing at $\Delta T = T_g - T$,

$$G_{1c} \sim [1/\Delta T]^{2\nu}. \tag{6.78}$$

This appears to be in qualitative agreement with Boiko's data [70] who examined the fracture energy of polystyrene interfaces during welding at temperatures up to 40 K below T_g.

In summary, the strength development during welding of polymers is well described by the relation

$$G_{1c} = G_{1c}{}^*(t/\tau)^{1/2}, \tag{6.79}$$

where $G_{1c}{}^*$ is the virgin strength determined by the percolation theory, and τ is the welding time, such that $\tau \sim M$ when $M > M^*$, and $\tau \sim M^3$ when $M < M^*$. Equation (6.79) reflects the scaling law for welding processes that are dominated by the diffusion stage of healing. However, as discussed in detail elsewhere [1, 64], the other stages of welding, such as surface approach, surface rearrangement, wetting, and randomization, can play a major role in the time dependence of the overall strength development. It is also important to note that G_{1c} is not a simple function of interdiffusion depth X, for all depths, since the transition from the Nail (weak-simple pullout) to the Net (strong-entangled) solution occurs at a particular value of X_c, of order R_{ge}. This transition will be important in incompatible amorphous interfaces, as discussed in the next section.

6.4.2 FRACTURE OF INCOMPATIBLE INTERFACES

Most polymers are immiscible with other polymers and form weak interfaces. An example would be starch and PLA, as discussed in Chapter 11. Consider the incompatible A/B polymer interface shown in Figure 6.9. The

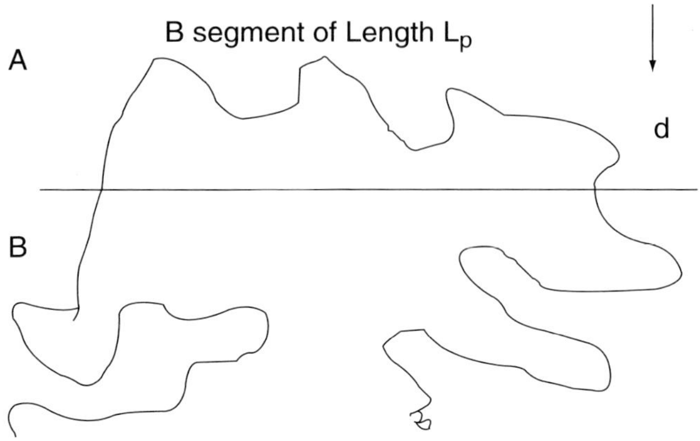

FIGURE 6.9 Chain segment of length L_p of a B-chain as it forms a bridge across an A/B incompatible interface of width $d \sim L_p{}^{1/2}$.

equilibrium interface width d, which is typically much less than R_g of either the A or B chains, can be described by the Helfand relation [71],

$$d = 2b/(6\chi)^{1/2}, \tag{6.80}$$

in which χ is the Flory-Huggins interaction parameter and b is the random walk bond length. The interface thickness d derives from a minimum in the free energy of mixing F, associated with the positive relief of entropy S, of surface confined chain segments of length $L(S \sim k \ln L)$ as they blossom forth across the interface, counterbalanced by the negative enthalpy of mixing H, of incompatible ***A/B*** segments ($H \sim \chi L$). Letting the free energy $F = H - TS$ and evaluating the free-energy minimum, $dF/dL = 0$, the equilibrium mixing length is

$$L \sim kT/\chi. \tag{6.81}$$

Since the interface width $d \sim L^{1/2}$, the equilibrium incompatible interface thickness is derived as $d \sim 1/\chi^{1/2}$, as expressed by Helfand in Eq. (6.80). With increasing compatibility, or as χ approaches zero, d approaches the normal interface width $X \sim R_g$, and the intermeshing segments become highly entangled, thereby producing a much higher fracture energy comparable to the virgin state [1, 72].

To understand the strength G of incompatible interfaces as a function of their width d, we first consider the random walk of length L, shown earlier in Figure 6.5. This length L is part of a much larger random walk chain, and is a segment that begins on the B side and traverses into the A side, and returns to the B side. In this respect, it is a bridge segment (of a larger chain) of length L_p, rather than a free chain of length L, such that the interface width is properly described by

$$d \sim {L_p}^{1/2}. \tag{6.82}$$

The number of bridges per unit area crossing the *A/B* interface is Σ_p, which is independent of molecular weight. As L_p increases, entanglements develop, crazes form, and the percolation relation $G \sim [p - p_c]$ applies. Here, the percolation parameter p is

$$p \sim \Sigma_p(L_p/L_e)/d, \tag{6.83}$$

where L_p/L_e is the number of entanglements per bridge. Since $d \sim {L_p}^{1/2}$, we obtain $p \sim d$, $p_c \sim d_c$, and hence

$$G \sim [d - d_c]. \tag{6.84}$$

Here d_c is the critical interface width corresponding to p_c, which will be of order R_{ge}, and below which no strength exists, other than that of simple pullout and surface energy terms, as described by the Nail solution. Letting the normalized width $w = d/d_c$, this equation becomes

$$G \sim [w - 1]. \tag{6.85}$$

The maximum strength G^* is determined by

$$G^* \sim [w^* - 1], \tag{6.86}$$

where $w^* \approx (M^*/M_e)^{1/4} \approx 2$. Thus, the ratio G/G^* becomes

$$G/G^* = (w - 1)/(w^* - 1). \tag{6.87}$$

To investigate the latter relation, a plot of G/G^* versus w should have a slope of $1/(w^* - 1) \approx 1$, an intercept on the w axis at $w_c = 1$, and maximum strength attained ($G/G^* = 1$) at $w^* \approx 2$, or the value of w^* corresponding to $w^* \approx 2w_c$.

Figure 6.10 shows data obtained by several investigators and analyzed by Benkoski *et al.* [73] for several asymmetric interface pairs. Here, G/G^* is plotted versus the normalized interface width $w = d/d_t$, where d_t is the Reptation tube diameter, calculated as $d_t = b(4/5N_e)^{1/2}$. Significantly, no strength develops below some critical value w_c. The magnitude of w_c is of order unity, but varies for each polymer pair due to the slight differences in their normalization procedure ($w = d/d_t$) compared to the above analysis ($w = d/d_c$). However, the slopes are of order unity, as predicted herein, and the maximum strength occurs at $w^* \approx 2$, when $w_c \approx 1$, or at $w^* = 2w_c$. The data in Figure 6.10 could be readily normalized to $w_c = 1$ to form a master curve consistent with the very simple relation

$$G/G^* = w - 1 \tag{6.88}$$

with slope of unity, intercept $w = 1$, and $w^* = 2$. This analysis differs from that provided by Benkoski *et al.* [73], who developed an interface strength theory based on the added contributions of monomer friction effects and an

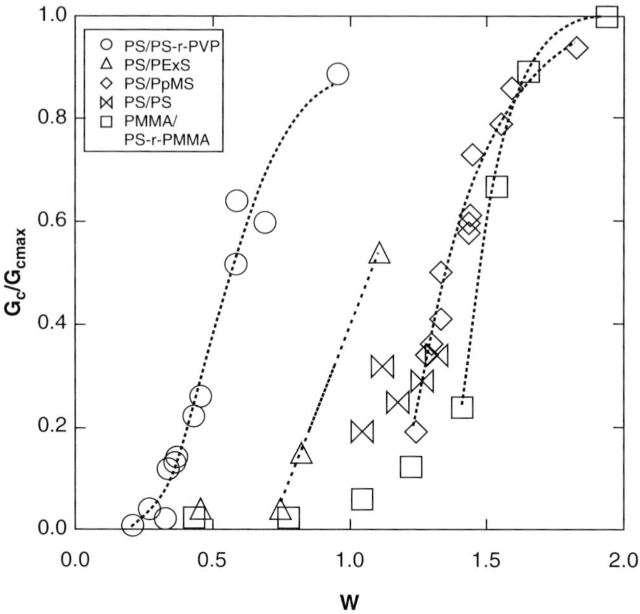

FIGURE 6.10 Data compiled by Benkoski *et al.* [73] show the interfacial fracture energy versus normalized interfacial width w for several A/B pairs. Circles represent PS/PS-rPVP; boxes, PMMA/PS-r-PMMA; diamonds, PS/PpMS; triangles, PS/PBrxS; and bowties, PS/PS.

entanglement segment length distribution. Coupled with the Brown theory of fracture [74, 75], this approach produced a more complex expression for G, which gave reasonable agreement with their data in Figure 6.10. While being significantly different in their derivation, a major fundamental difference between the theories is that the Benkoski theory requires both friction and entanglements to explain all the data in Figure 6.10, while the percolation theory requires only the entanglement effects to explain all the data, since the friction terms are effectively zero on the G/G^* scale.

When $w < w_c$, or $p < p_c$, the Nail solution applies as the Σ nonentangled chain segments of length L pull out in simple friction. However, the chain segments do not pull out as linear strings of length L, but rather as intermeshed random walks of length $L^{1/2}$; the chain segment is attached to a very long chain, which is itself entangled and, hence, will not allow the segment L to pull out as a string. Thus, the critical stress behaves as $\sigma \sim \Sigma\mu L^{1/2}$, where μ is the friction coefficient. The critical crack opening displacement behaves as $\delta \sim L^{1/2}$, such that the fracture energy for pullout is

$$G \sim \mu\Sigma L. \tag{6.89}$$

Since Σ is constant and $L \sim d^2$, it follows that in simple pullout at $w < w_c$,

$$G \sim d^2. \tag{6.90}$$

However, this fracture energy is very low and orders of magnitude lower than that obtained at $w > w_c$. Both theories based on the friction contribution agree with the quadratic dependence $G \sim d^2$, as first proposed by Willett and Wool [72].

The adhesion between immiscible polymers as a function of interfacial width was also analyzed by Cole *et al.* [76] in terms of the number of entanglements N_{ent} in the interface. They define N_{ent} in the incompatible interface of width d as

$$N_{\text{ent}} = d/L_e, \tag{6.91}$$

where L_e is the entanglement length, defined by $L_e = b[M_e/6\,M_0]^{1/2}$, in which M_0 is the monomer molecular weight and b is the bond length. They propose that the resistance to fracture is determined by

$$G \sim {N_{\text{ent}}}^2. \tag{6.92}$$

Their data are shown in Figure 6.11 (Figure 6.11 in Cole *et al.* [76]), where the slope of 2 from a plot of log G versus log N_{ent} suggests support for the quadratic dependence in Eq. (6.92). The circles in Figure 6.11 represent data

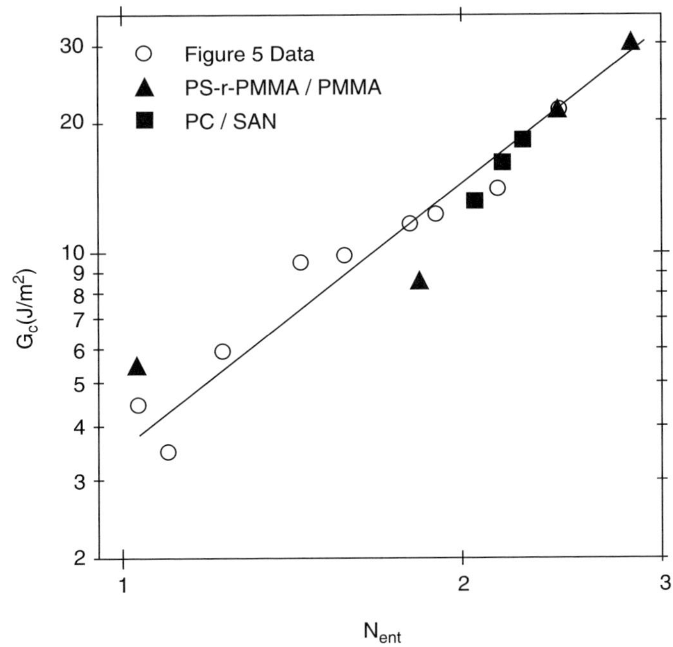

FIGURE 6.11 Fracture energy of A/B incompatible interfaces versus N_{ent}, as compiled by Cole *et al.* [76]. The power law agreement with a slope of 2 (solid line) suggests that a relationship of the form $G_c \sim {N_{\text{ent}}}^2$ adequately describes the adhesion.

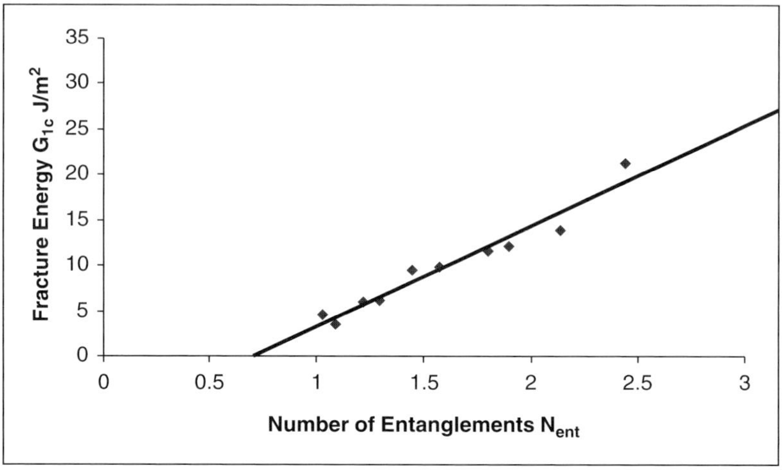

FIGURE 6.12 Plot of fracture energy versus N_{ent} using data of Cole *et al.* [76] from Figure 6.11. The line is a best fit of the data to the percolation relation $G_{1c} \sim [N_{\mathrm{ent}} - N_c]$.

obtained from interface pairs consisting of the following; PP/aPA, PS/aPA, PS/PP, PS/PEO, PS/PC, PS/PVC, PS/PE, PS/PMMA, PET/PC, using both melt and solvent lamination. The triangles in Figure 6.11 represent literature values for PS-r-PMMA from Brown [74], and the squares represent PC/SAN data obtained by Janarthanan *et al.* [75b]. Alternatively, using the percolation model, from Eqs. (6.84) and (6.91), we obtain

$$G \sim [N_{\mathrm{ent}} - N_c], \tag{6.93}$$

where N_c is the critical number of entanglements, corresponding to p_c. Normalizing this relation by the maximum strength G^* at N^*, $G^* \sim [N^* - N_c]$, we obtain

$$G/G^* = [N_{\mathrm{ent}} - N_c]/[N_c^* - N_c]. \tag{6.94}$$

Accordingly, a plot of G versus N_{ent} should give a linear plot with intercept N_c as shown in Figure 6.12, using data from Cole *et al.* (Table 3 in [76]). The linear fit correlation coefficient was $R^2 = 0.95$ (neglecting $G = 0$ points) with intercept $N_c = 0.7$ and slope of $11\,\mathrm{J/m^2}$. Cole *et al.* observed at least three $G = 0$ values in the vicinity of N_c, supporting the concept that little or no strength exists below the percolation threshold.

Thus, the data in Figure 6.12 are linear with a nonzero intercept, which meaningfully divides the data into two regions, $N_{\mathrm{ent}} < N_c$ for which $G \approx 0$, consistent with very weak interfaces, and $N_{\mathrm{ent}} > N_c$, which describes the strong interfaces. However, a power law fit with zero intercept, as required by the homogeneous function $G \sim N_{\mathrm{ent}}{}^{\beta}$, will suggest an exponent of $\beta \approx 2$, and also describes both weak and strong regions with the same function.

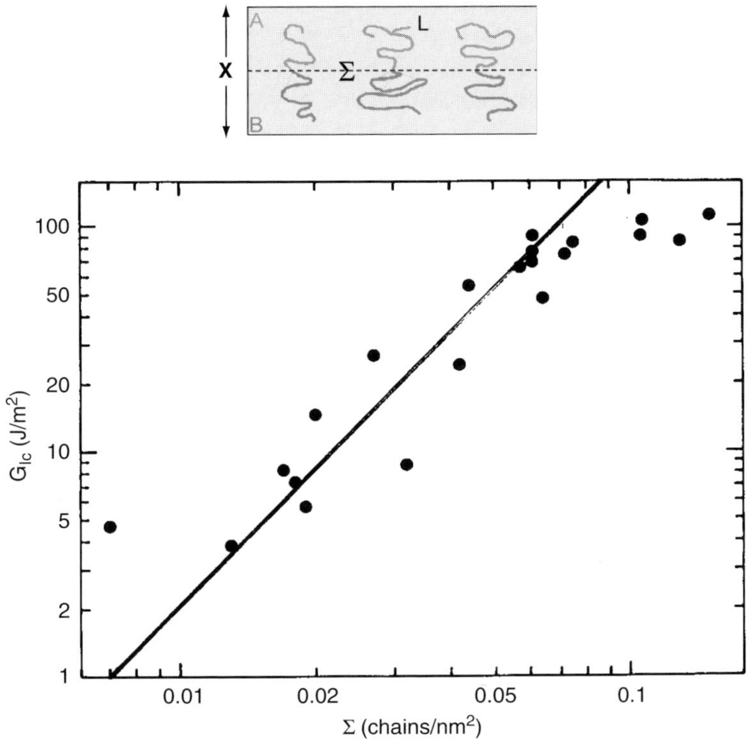

FIGURE 6.13 Fracture energy G_{1c} versus areal chain density Σ from data reported by Creton *et al.* [77] for the PS(800)-PVP(870) diblock reinforced PS/PMMA incompatible interface. The solid line was drawn with a slope of 2, suggestive of the scaling law $G_{1c} \sim \Sigma^2$.

Clearly, a plot of log G vs. log$[N_{\text{ent}} - N_c]$ would give an exponent of $\beta \approx 1$, consistent with the percolation theory. During welding, N_{ent} behaves as $N_{\text{ent}} \sim t^{3/4}M^{-7/4}$ (Table 6.3), and if one were to use the strength relation $G_{1c} \sim N_{\text{ent}}{}^2$, one would predict that $G_{1c} \sim t^{3/2}M^{-7/2}$, and $G^* \sim M^{-2}$, which is universally inconsistent with all welding and virgin state data.

6.4.3 FRACTURE OF REINFORCED INCOMPATIBLE INTERFACES

The role of A–B diblock compatibilizers or random A–B copolymers of aerial density Σ at incompatible A/B interfaces has been investigated [74–82]. Figure 6.13 shows results of G_{1c} versus Σ for PS/PMMA interfaces reinforced by PS(800)–PVP(870) diblocks. Most of the data are reasonably well described by a line with a slope of 2 on this log–log plot, suggestive of $G \sim \Sigma^2$. Brown [74, 75] analyzed this and other similar data and derived a theory of fracture, which is referred as the Σ^2 law:

$$G_{1c} \sim \Sigma^2/\sigma_{cr}, \tag{6.95}$$

in which σ_{cr} is the yield stress in the craze zone at the crack tip. If the $G_{1c} \sim \Sigma^2$ law is applied to welding, $\Sigma(t) \sim t^{1/4}M^{-5/4}$ and $\Sigma_\infty \sim M^{-1/2}$, then one obtains $G_{1c} \sim t^{1/2}M^{-5/2}$ and $G^* \sim 1/M^2$, when $\tau^* \sim M$. Despite the correct time dependence ($t^{1/2}$) of welding, the predictions are not in accord with the molecular weight dependence of welding, and particularly that of the virgin state, where contrary to all data, it is predicted that the strength decreases with increasing molecular weight. Alternatively, using Eq. (6.95), we can let $\sigma_{cr} \sim X$, such that $G_{1c} \sim t^{1/4}M^{-9/4}$. While the $t^{1/4}$ dependence is not observed in the usual case, it could occur if the chain ends were segregated to the weld surfaces, but this was not observed to occur experimentally in the HDH/DHD experiments of Welp and others [51–54] and essentially all welding data support the $t^{1/2}$ dependence. So the Brown model is incompatible with the Wool welding model, and vice versa. We can reconcile these differences within the framework of the percolation model, which predicts that $G \sim [p - p_c]$ as

$$G_{1c} \sim [(\Sigma L/X) - (\Sigma L/X)_c]. \tag{6.96}$$

Because L and X are constant, then $p_c \sim \Sigma_c$, which represents a critical number of chains required to build up the network above the percolation level. Letting $L/X \sim R_{gA}$ of the diblock ends, the percolation model predicts the following linear relation:

$$G_{1c} \sim R_{gA}[\Sigma - \Sigma_c]. \tag{6.97}$$

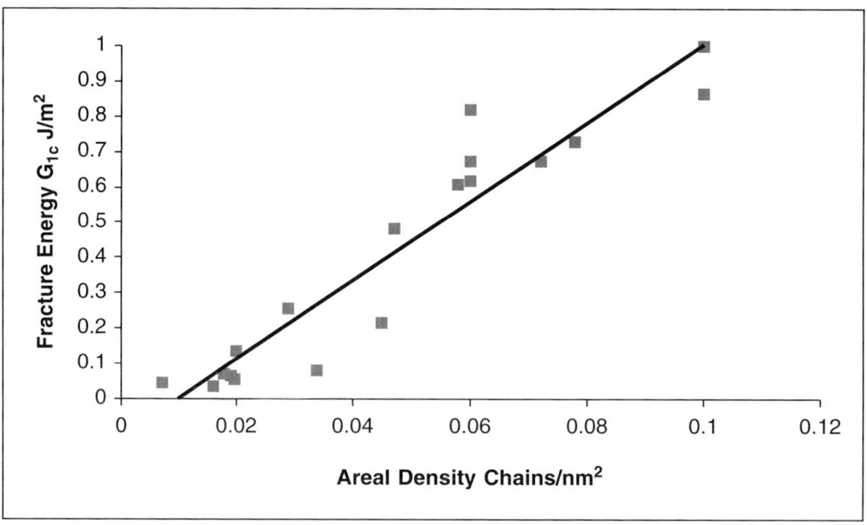

FIGURE 6.14 Fracture energy (normalized) G/G^* versus areal density Σ of A–B diblock chains in an A/B incompatible interface, using data of Creton *et al.* [77] from Figure 6.13. The line is a least squares fit to the percolation formula $G/G^* \sim [\Sigma - \Sigma_c]$.

Normalizing this relation by the maximum strength G^* at Σ^*, we have

$$G/G^* = [\Sigma - \Sigma_c]/[\Sigma^* - \Sigma_c]. \tag{6.98}$$

Figure 6.14 shows a plot of G/G^* versus Σ, using Creton's data from Figure 6.13. The fracture data were normalized by $G^* \approx 110\,\text{J/m}^2$, which is the upper range of the data presented. The linear relation for G/G^* versus Σ had a correlation coefficient of 0.9 and produced an intercept on the Σ-axis of $\Sigma_c = 0.1/\text{nm}^2$. The slope of this line is $11.1/\text{nm}^2$. The transition from Nails to Nets, or weak to strong interfaces, is demarcated by the threshold value Σ_c, which, as discussed by Creton *et al.* [77], should occur near the overlap of the diblock random coils in the interface, such that

$$\Sigma_c \approx 1/R_{gA}{}^2. \tag{6.99}$$

The radius of gyration of the PS ends with $M_n = 83,200\,\text{g/mol}$ is $R_{gA}{}^2 = 63.2\,\text{nm}^2$, such that $\Sigma_c \approx 0.016\,\text{nm}^2$, which is in reasonable accord with the experimental value $\Sigma_c = 0.1/\text{nm}^2$ in Figure 6.14. The maximum value of Σ^* at G^* can be determined from the entanglement bridge theory [1] by

$$\Sigma^* = [M_c/M^*]/2a, \tag{6.100}$$

where $a \approx 1\,\text{nm}^2$ is the cross-sectional area of a bridge segment of a diblock chain as it crosses the interface. For polystyrene, with $M_c = 30,000$ and a molecular weight of $M^* = 250{,}000$ g/mol, then $\Sigma^* \approx 0.17\,\text{nm}^2$. When brush-like ordering occurs at the interface, $L \approx 0$ as M_e increases, and G_{1c} decreases considerably.

Examining both theories, $G_{1c} \sim \Sigma^2$ and $G_{1c} \sim [\Sigma - \Sigma_c]$, as plotted in Figures 6.13 and 6.14, respectively, there is sufficient data scatter in both plots such that one could not judge, based on these data alone, which theory was more valid. However, the percolation model, in addition to describing the *A/B* reinforced interface above, is universally consistent with welding data, virgin state strength, and the transition from weak to strong interfaces. It can be deduced that the exponent of 2, reported in several instances, is an accidental consequence of inhomogeneous functions for G_{1c} versus N_{ent} with incompatible *A/B* interfaces, G_{1c} versus Σ data for reinforced *A/B* interfaces, and G_{1c} versus M for virgin strength data.

6.5 POLYMER–SOLID INTERFACES

When using sticker groups X to adhere linear chains of length M to solid substrates such as natural fiber, glass, or metal and which may contain receptor groups Y, as shown in Figure 6.15, some very interesting effects occur [83–85]. There exists an optimal number of both receptor X-groups and acceptor Y-groups to obtain maximum adhesion at the polymer–solid interface. The optimal number of X-groups is only about 3 %, whereas the optimal

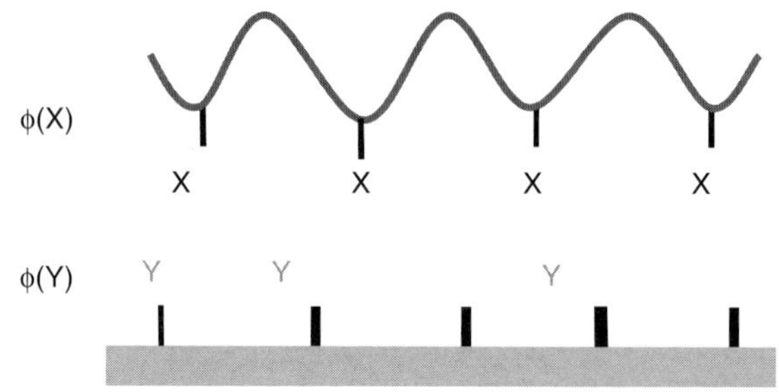

FIGURE 6.15 Schematic representation of the X–Y problem at a polymer–solid interface, where X represents specific polymer sticker groups, and Y represents specific substrate receptor groups; ϕ is the mole percent or the mole fraction of the groups.

number of Y-groups on the solid is closer to 50%. We believe these results are a balance between the connectivity of the first layer of chains to the solid substrate and the ability of this first layer to be connected to the polymer bulk above the substrate. If the polymer is too well adhered to the substrate and adopts a flat conformation, it may not be well connected to the polymer layer above, with a resultant loss of interface strength. This leads to many new concepts of adhesion design, which can be understood in terms of molecular connectivity and percolation concepts.

Lee and Wool [83, 84] and Gong and Wool [85–87] conducted experiments to investigate the influence of a small number of sticker groups ($-X$) distributed randomly along the polymer chains, adhering with receptor groups ($-Y$) distributed on the solid surface, and the X–Y interaction parameter χ. The problem was defined as the *X–Y problem* at the polymer–solid interface (see Figure 6.15), where one queried how the fracture energy G_{1c} depended on the number of sticker $\phi(X)$ and receptor groups $\phi(Y)$. In this polymer–solid interface study, G_{1c} was explored as a function of (1) interface structure, (2) surface restructuring time and temperature, (3) deformation rate in terms of polymer viscosity and non-Newtonian rheology, (4) Saffman-Taylor meniscus instability phenomena and cavitation effects in the deformation zone, (5) microscopic deformation mechanisms involving disentanglement and bond rupture, and (6) interrelationships between structure and strength of polymer–solid interfaces.

6.5.1 ROLE OF STICKER GROUPS $\phi(X)$ ON ADHESION

In the first experiment by Gong and Wool [85], the influence of $\phi(X)$ on G_{1c} at constant $\phi(Y) \sim 1$ was investigated. Carboxyl sticker groups were placed randomly on linear polybutadiene (PBD) chains and the polymer melt was adhered to aluminum (Al) foil surfaces. Pure PBD chains adhere very weakly to Al surfaces. The X–Y interaction was determined by the hydrogen bonding acid–base interaction between –COOH and aluminum oxide. It was found for this cPBD-Al interface that with increasing sticker group concentration $\phi(X)$, the fracture energy G_{1c} increased and then decreased as shown in Figure 6.16. An optimal sticker group concentration $\phi^*(X) = 3\,\text{mol}\%$ gave a maximum fracture energy of about $300\,\text{J/m}^2$. The trend shown in Figure 6.16 seemed counterintuitive in terms of current wisdom of adhesion, but was reproduced several times using different synthetic techniques and testing methods at several locations.

There were significant effects of surface rearrangement times that varied with $\phi(X)$ such that the maximum fracture energy was obtained in the shortest time at the optimal sticker group concentration $\phi(X) \sim 3\,\text{mol}\%$. Surface restructuring occurred at times several orders of magnitude longer than characteristic single-chain relaxation times (e.g., the Reptation time). Furthermore, it was found that non-Newtonian viscoelastic effects

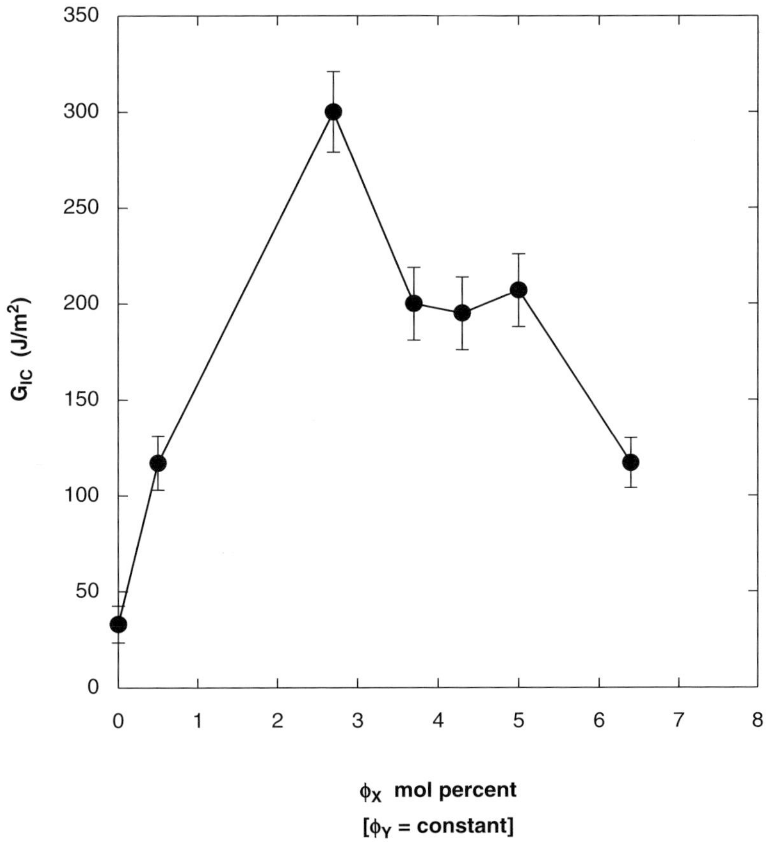

FIGURE 6.16 Sticker group (X) effect on the fracture energy of a cPBD-Al interface by Gong and Wool [85]. The M_w and M_n of the polymer were 180 kD and 98 kD, respectively. The samples were annealed at room temperature for 1000 min. The peeling rate of the samples was 30 mm/min. Here ϕ_Y(–OH) is constant and X is –COOH.

dominated the deformation zone evolution and breakdown. A microscopic analysis of the mode of failure (AFM, XPS, SEM) indicated that cohesive failure was occurring predominantly in a layer immediately adjacent to the metal surface. At low G_{1c} and $\phi(X)$ values, simple adhesive or mixed adhesive–cohesive failure occurred.

In the second experiment by Gong [87], the influence of χ was examined using aluminum oxide surfaces treated with amine terminated silanes, creating –NH2 substrate receptor groups with a much stronger X–Y interaction. Then, the sticker concentration effect on the cPBD-AlS interface was investigated at constant $\phi(Y) \sim 1$. It was found that similar trends to that shown in Figure 6.16 occurred but the maximum G_{1c} doubled to 600 J/m^2 and the optimal sticker concentration decreased to $\phi^*(X) \sim 0.5$ mol%. Again, the failure mode was cohesive at high G_{1c} G_{1c} values and adhesive near $\phi(X) \approx 0$.

6.5.2 ROLE OF RECEPTOR GROUPS $\phi(Y)$ ON ADHESION

Finally, the third experiment by Lee and Wool [84] varied the coverage of active amine receptor groups $\phi(Y)$ in the range 0–100% on the aluminum surface using mixed silanes (–CH3 and –NH2 terminated). A cPBD polymer with a constant sticker group concentration $\phi(X) \sim 3$ mol%, corresponding to $\phi^*(X)$ in Figure 6.16, was used in the peel experiments. The results of G_{1c} versus $\phi(Y)$ are shown in Figure 6.17 for surface restructuring times of 10, 100, and 1000 min. Significantly, $G_{1c}G_{1c}$ reaches its maximum value of about 600 J/m^2 at an optimal partial coverage of $\phi^*(Y) \sim 30$%. In the above experiments, the bonding dynamics were several orders of magnitude longer (up to 1000 min) than the characteristic relaxation time of the PBD bulk ($\sim$ 1 min). Additionally, the adhesion dynamics were strongly dependent on the concentrations of the sticker and receptor groups. The trend with $\phi(Y)$ was similar to that of $\phi(X)$ and suggests that design rules exist for optimizing polymer–solid interface strength.

An entanglement sink probability (ESP) model motivated by vector percolation explains the nonmonotonic influences of sticker concentration (ϕ_X), receptor concentration (ϕ_Y), and their interaction strength (χ) on the adhesion strength G_{1c} of the polymer–solid interface. The ESP model quantifies the degree of interdigitation between adsorbed and neighboring chains based on the adsorbed chain domain using an extension of the scaling treatment of de Gennes. Here, the adsorbed chain domain changes thermodynamically with respect to the energy of interaction parameter, $r = \chi\phi_X\phi_Y$. Basically, this model considers the situation of a blend consisting of a small volume fraction of adhesive molecules as a compatibilizer at the interface, where these molecules promote adhesion by adsorbing to the surface via sticker–receptor interactions. The ESP model scales solely with $r = \chi\phi_X\phi_Y$, and this parameter can be related to both the adhesive potential (G_A) and the cohesive potential (G_C). Adhesive potential G_A describes adhesive failure between adsorbed chains and the solid surface and linearly behaves as $G_A \sim r = \chi\phi_X\phi_Y$. The cohesive strength between adsorbed and neighboring chains corresponds to $G_C \sim r^{-1} = (\chi\phi_X\phi_Y)^{-1}$. When the fracture stresses for cohesive σ_c and adhesive failure σ_A are equal, as shown in Figure 6.18, the model predicts maximum adhesion strength at an optimal value of $r^* = (\chi\phi_X\phi_Y)^*$. Thus, for a given χ value, there exist optimal values $\phi_X{}^*$ and $\phi^*{}_Y$ for the sticker and receptor groups, above or below which the fracture

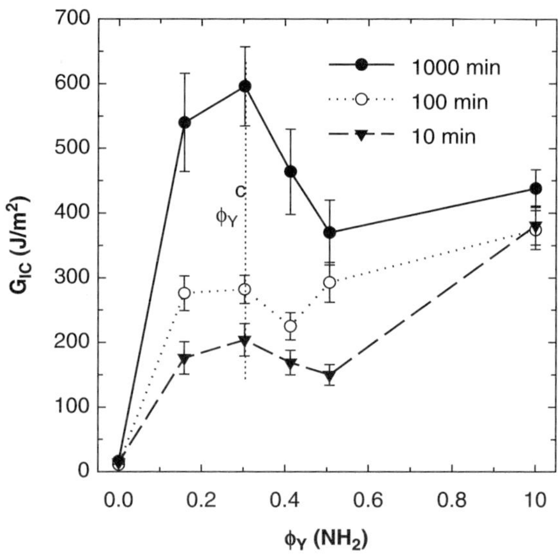

FIGURE 6.17 Receptor group (Y) effect on the fracture energy of a cPBD-AlS interface by Lee *et al.* [84]. The M_w and M_n of the polymer were 180 kD and 98 kD, respectively. The peeling rate of the samples was 30 mm/min. The samples were annealed at room temperature for various times. On the AlS, $\phi_Y(-NH_2) + \phi_Y(-CH_3) = 1$. Here $\phi_X(-COOH)$ was about 3 mol%. The data point at $\phi_Y(-NH_2) = 0$ was based on the pure dispersive forces of PBD.

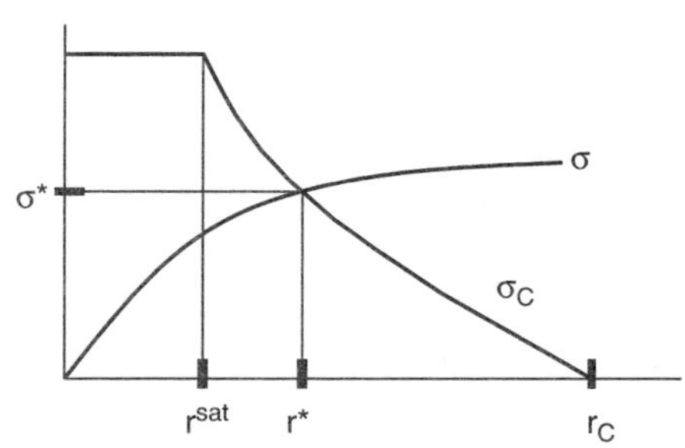

FIGURE 6.18 The crossover point of adhesive versus cohesive failure stress at the critical value of equilibrium energy of interaction at interface $r^* = (\chi\phi_X\phi_Y)^*$.

energy will not be optimized. Alternatively, if the X–Y interaction strength χ increases, then the number of sticker groups required to achieve the optimum strength decreases. Significantly, the optimum strength is not obtained when the surface is completely covered with receptor groups ($\phi_Y = 1$) and is closer to 30%. For polybutadiene, the optimum value of r^* was determined experimentally, and typically $\phi^*{}_X \approx 1-3\%$ and $\phi_Y \approx 25-30\%$.

6.5.3 DISCUSSION, ANALYSIS, AND CONCLUSION

To apply the preceding analysis to real polymer–solid interfaces, we consider first the most common polymer–solid interface in which $\phi_Y = 1$ (receptor groups totally cover the solid surface) and the polymer contains a variable number of sticker groups ϕ_X, as depicted earlier in Figure 6.15. Thus, for this interface, we have $r^* = (\chi\phi_X)^*$. Because χ is constant, then $r \sim \phi_X$ and when $\phi_X < \phi^*{}_X$, the fracture energy $G_{1c} \sim r$ where adhesive failure dominates, and we obtain

$$G_{1c} = G_{1c}{}^*\phi_X/\phi_X{}^* \qquad (r < r^*), \tag{6.101}$$

in which $G_{1c}{}^*$ is the maximum fracture energy. In Figure 6.16, $G_{1c}{}^* \approx 300\,\mathrm{J/m^2}$ at $\phi_X{}^* = 3\%$ and when $\phi_X \approx 1\%$, we predict a fracture of $G_{1c} \approx 100\,\mathrm{J/m^2}$, as noted. There exists a magic number of sticker groups $\phi_X{}^*$, which can be well approximated by the following relation:

$$\phi_X{}^* = 2M_0/M_c, \tag{6.102}$$

This is based on the premise that when an entanglement polymer network comes in contact with the solid surface, it only requires about two sticker groups per critical entanglement length to tie it down to the surface. For example, if $M_0 = 28$ and $M_c = 4000$, then $\phi_X^* \approx 1.4\%$.

However, when $r > r^*$ or when $\phi_X > \phi_X^*$, cohesive failure dominates and $G_{1c} \sim 1/r$ such that

$$G_{1c} = G_{1c}{}^*\phi_X{}^*/\phi_X \quad (r > r^*). \tag{6.103}$$

For example, in Figure 6.16, at $\phi_X = 6\%$, one expects that $G_{1c} \sim 1/2G_{1c}{}^*$, or about $150\,\mathrm{J/m^2}$, as observed. If the surface became contaminated such that ϕ_Y decreased from 100% to 80% coverage, then $\phi^*{}_X$ needs to be increased from 3% to 3.75% to obtain the same fracture energy. Therefore, with regard to sticker group variation, we see that the optimization "hill" shape is quite asymmetric due to the nature of adhesive ($G_{1c} \sim r$) versus cohesive failure ($G_{1c} \sim 1/r$). Thus, when designing interfaces for optimal adhesion in which χ and ϕ_Y are constant, values of ϕ_X on the low side of $\phi_X{}^*$ can weaken the interface more than on the high side of $\phi_X{}^*$.

In another important polymer–solid interface example, consider the case depicted in Figure 6.17, in which the sticker group concentration ϕ_X and bond strength χ are both constant but the receptor group concentration ϕ_Y

can be varied from 0% to 100% coverage. In Figure 6.17, the bonding χ-factor was increased considerably by using $-NH_2$ Y-groups to bond with the $-COOH$ X-groups, compared to the native oxides used on the aluminum surfaces (Figure 6.16). Thus, when $\phi_Y = 1$, the optimal sticker group concentration ϕ_X for this interface is determined by

$$\phi_{X1}{}^* = \phi_{X2}{}^*[\chi_2/\chi_1]. \tag{6.104}$$

For example, when $\phi_{X2}{}^* = 3\%$ for the oxide, and χ_2 for the NH_2 Y-groups is much greater than the oxide bond strength, we expect $\phi_{X2}{}^*(NH_2)$ to be much less, as noted by Gong [86], who obtained values of $\phi_X{}^* = 0.5-1.0\%$ at maximum strength. In Figure 6.17, we used $\phi_X = 3\%$, which was the same carboxylated polymer used in Figure 6.16 at the optimal strength. Therefore, to optimize the interface shown in Figure 6.17 with $\phi_X = 3.0\%$, since $r^* = (\chi\phi_X\phi_Y)^*$, it follows that the optimal receptor group concentration $\phi_{Y1}{}^*$, at constant χ is given by

$$\phi_{Y1}{}^* = \phi_{Y2}{}^*\phi_{X2}/\phi_{X1}{}^*. \tag{6.105}$$

Since $\phi_{Y2}{}^* = 1$ at $\phi_{X2}{}^* \sim 1.0\%$, when $\phi_{X1} = 3.0\%$, we expect the optimal receptor group concentration to be about 30%, as observed. It is quite remarkable that surfaces with partial coverage of receptor groups can be stronger than those that are completely covered. This suggests that surfaces that contain impurities (in the above case, up to 66%) can form stronger interfaces than the pristine clean surfaces, depending on the polymer that is adhering. For plasma modification of surfaces to promote adhesion by forming functional groups that act as receptor groups, it is clear that an optimal amount of modification is required, which typically corresponds to an optimal plasma exposure time.

In the above example, we again note that the optimization effect is asymmetric with respect to receptor groups ϕ_Y. By the same arguments, it follows that, when adhesive failure dominates,

$$G_{1c} = G_{1c}{}^*\phi_Y/\phi_Y{}^* \qquad (\phi_Y < \phi_Y{}^*). \tag{6.106}$$

When cohesive failure dominates,

$$G_{1c} = G_{1c}{}^*\phi_Y{}^*/\phi_Y \qquad (\phi_Y > \phi_Y{}^*). \tag{6.107}$$

From Figure 6.17, we can deduce from Eq. (6.106) that when $\phi_Y \approx 10\%$, $\phi_Y{}^* = 30\%$, and $G_{1c}{}^* \approx 600\,J/m^2$, one should obtain $G_{1c} \approx 200\,J/m^2$. When $\phi_Y \approx 50\%$, we predict by Eq. (6.106) that $G_{1c} \approx 360\,J/m^2$, which is close to that observed.

Given the complexity of these interfaces and the simplicity of our assumptions in this analysis, the strength predictions using values far from the optimal values are considered dubious. Overall, the qualitative and semi-quantitative predictions, using the parameter $r^* = (\chi\phi_X\phi_Y)^*$, are in accord with a range of experimental observations and suggest that we are garnering the essential features of the interface strength development, namely, the competition and interplay between adhesive and cohesive failure mechanisms as a function of ϕ_X, ϕ_Y, and χ.

6.6 SUMMARY OF FRACTURES IN BIO-BASED POLYMERS

A theory of fracture of entangled polymers was developed that was based on the vector percolation model of Kantor and Webman [6], in which the modulus E is related to the lattice bond fraction p, via $E \sim [p - p_c]^{\tau}$. The Hamiltonian for the lattice was replaced by the engineering strain energy density function of the bulk polymer, $U = \sigma^2/2E$, and p was expressed in terms of the normalized entanglement density, using the entanglement molecular weight, M_e. The polymer fractured critically when p approached the percolation threshold p_c, which was accomplished by utilizing the stored strain energy in the network to randomly fracture $[p - p_c]$ bonds. The fracture energy was found to be $G_{1c} \sim [p - p_c]$. When applied to interfaces of width X, containing an areal density Σ of chains, each contributing L entanglements, the percolation term $p \sim \Sigma L/X$, and the percolation threshold was related to Σ_c, L_c, or X_c. This gave a unified theory of fracture for the virgin state of polymers in the bulk and a variety of polymer interfaces. The percolation theory has also been applied successfully to fracture of thermosets (Chapter 7), carbon nanotubes [88], and fracture of polymer–solid interfaces [83–87]. Several important results are summarized in Table 6.4 and in the list that follows.

TABLE 6.4 Summary relations for strength of interfaces.

Property	Relation	Comment/Application
Percolation	$G_{1c} \sim P - P_c$	$P \sim$ entangle density
	$P_c = 1 - M_e/M_c$	$P \sim 1/M_e$
Interface structure $H(t)$	$H(t) = H_\infty(t/T)^{r/4}$	Dynamics of welding
Virgin state	$H_\infty = M^{(3rs)/4}$	$r, s = 1, 2,$ or 3
Fracture energy	$G_{1c} = G_\infty(t/\tau)^{1/2}$	$\tau \sim M^3$ if $M < M^*$
Symmetric welds	$G_{1c} \sim \Sigma L/X$	$(M^* = 8M_c)$
		$\tau \sim M$ if $M > M^*$
Bond rupture	$G_{1c} \sim G^*[1 - (M_c/M)]$	$M > M^*$
Disentanglement	$G_\infty \sim G^* M[1 - (M_c/M)^{1/2}]^2$	$M_c < M < M^*$
Virgin toughness	$K_{1c} \sim M^{1/2} - M_c^{1/2}$	$M_c < M < M^*$
Nail solution	$G_{1c} = 2S\Gamma + 1/2\mu_o L^2 \Sigma V^a$	Weak interfaces
Net solution	$\sigma = [ED_o(p - p_c)\rho/8M_c]^{1/2}$	Strong interfaces
Fatigue of welds	$da/dN \sim M^{-5/2}(t/(t))^{-5/4}$	Penalty for poor weld
Weld time	$\tau = 64(M_c/M)^2(t)$	$M \geq M^*$
Autohesion tack	$\sigma \sim (t/M)^{1/4}$	Linear elastomers
Compatibilizers	$G_{1c} \sim \Sigma - \Sigma_c$	Incompatible strong interface
	$P \sim \Sigma L/R$	
Fractal roughness	$N_f = H^{d/4}/M$	Gradient percolation

1. The fracture strength σ of amorphous and semicrystalline polymers in the bulk can be well described by the net solution $\sigma = [ED_{0\rho}/16M_e]^{1/2}$, and was found to be in excellent agreement with a large body of data. This requires no fitting parameters.
2. Fracture by disentanglement was found to occur in a finite molecular weight range, $M_c < M < M^*$, where $M^*/M_c \approx 8$, such that the critical draw ratio, $\lambda_c = (M/M_c)^{1/2}$, gave the molecular weight dependence of fracture as $G_{1c} \sim [(M/M_c)^{1/2} - 1]^2$. The critical entanglement molecular weight is related to the percolation threshold p_c via $M_c = M_e/(1 - p_c)$.
3. Fracture by bond rupture was in accord with Flory's suggestion $G/G^* = [1 - M_c/M]$.
4. For welding of A/A symmetric interfaces, $p = \Sigma L/X$, and $p_c \approx L_c/M \approx 0$, such that when $\Sigma/X \sim 1/M$ for randomly distributed chain ends, $G/G^* = (t/\tau^*)^{1/2}$, where $\tau^* \sim M$, when $M > M^*$, and $\tau \sim M^3$, when $M < M^*$. When the chain ends are segregated to the surface, Σ is constant with time and $G/G^* = [t/\tau^*]^{1/4}$.
5. For incompatible A/B interfaces of width d, normalized width w, and entanglement density $N_{\text{ent}} \sim d/L_e$, $p \sim d$ such that $G \sim [d - d_c]$, $G \sim [w - 1]$, and $G \sim [N_{\text{ent}} - N_c]$.
6. For incompatible A/B interfaces reinforced by an areal density Σ of compatibilizer chains, L and X are constant, $p \sim \Sigma$, $p_c \sim \Sigma_c$, such that $G \sim [\Sigma - \Sigma_c]$.
7. For polymer–solid interfaces containing ϕ_X sticker groups and ϕ_Y receptor groups of interaction (X–Y) strength χ, there exists an optimal parameter $r^* = (\chi\phi_X\phi_Y)^*$, such that when $r < r^*$, adhesive failure occurs via $G \sim r$, and when $r > r^*$, cohesive failure occurs and $G \sim 1/r$.

REFERENCES

1. Wool, R. P. *Polymer Interfaces: Structure and Strength*, Hanser Press, New York; 1995.
2. Wool, R. P. In *Adhesion Science and Engineering*, Vol. 2, Chap. 8, Chaudhury, M.; Pocius, A.V., Eds.; Elsevier, New York; 2002.
3. Hutchinson, J. W. "Linking Scales in Fracture Mechanics." in Advances in Fracture Research, *Proceedings of the Ninth International Conference on Fracture*, Sydney, Australia; Pergamon Press, Vol. 1, pages 1–14 (1997).
4. Tvergaard, V.; Hutchinson, J. W. *J. Mechan. Phys. Solids* 1993, **41**(6), 1119.
5. Tvergaard, V.; Hutchinson, J. W. *J. Phys. IV* 1996, **6**(C6), 165.
6. Kantor, Y.; Webman, I. *Phys. Rev Lett.* 1984, **52,** 1891.
7. Feng, S.; Halperin, B. I.; Sen, P. N. *Phys. Rev. B* 1987, **35**(9), 197.
8. Feng, S.; Thorpe, M. F.; Garboczi, E. *Phys. Rev. B* 1985, **31**(1), 276.
9. He, H.; Thorpe, M. F. *Phys. Rev. Lett.* 1985, **54**(19), 2107.
10. Garboczi, E. J.; Thorpe, M. F. *Phys. Rev. B* 1985, **31**(11), 7276.
11. Thorpe, M. F.; Garboczi, E. J. *Phys. Rev. B* 1987, **35**(16), 8579.
12. DeGennes, P. G., *J. Phys. (Paris) Lett.* 1976, **37,** L1.
13. Born, M.; Huang, K. *Dynamical Theory of Crystal Lattices*, Oxford University Press, New York; 1954.

14. Stauffer, D. *Phys. Rep.* 1980, **53,** 3759.
15. Wool, R. P.; Bretzlaff, R. S.; Li, B. Y.; *et al. J. Polym. Sci. Polym. Phys. Ed.* 1986, **24,** 1039.
16. Wool, R. P. *Polymer Eng. Sci.* 1980, **20,** 805.
17. Mooney, M. *J. Appl. Phys.* 1940, **11,** 582; 1948, **19,** 434.
18. Flory, P. J. *Principles of Polymer Chemistry*, Cornell University Press, Ithaca, NY; 1953.
19. Wool, R. P.; *et al.* Bio-Based Composites. In *Proceedings of the European Congress of Composites Materials*, ECCM-11, Rhodes, Greece; June 2, 2004.
20. Wool, R. P. *Macromolecules* 1993, **26,** 1564.
21. Uhlherr, A.; Doxastakis, M.; Mavrantzas, V. G.; *et al. Europhys. Lett.* 2002, **57**(4), 506.
22. Kausch, H. H. *Polymer Fracture*, 2nd ed., Springer-Verlag, Berlin; 1984.
22(a). Dorgan, J. R; *et al. J. Rheol.* 2001, **43**, 1141.
22(b). Carriere, C. J. *Cereal Chemistry*, 1998, **75**, 360.
23. McCormick, H. W.; Brower, F. M.; Kin, L. *J. Polym. Sci.* 1959, **39,** 87.
24. Pitman, G. L.; Ward, I. M. *Polymer* 1970, **20,** 897.
25. Vincent, P. I. *Polymer* 1972, **13,** 557.
26. Wool, R. P.; Bailey, D.; Friend, A. *J. Adhesion Sci. Technol.* 1996, **10,** 305.
27. Lauterwasser, B. D.; Kramer, E. J. *Philos. Mag.* 1979, **A39,** 469.
28. Kramer, E. J.; Berger, L. L. *Adv. Polym. Sci.* 1990, **91/92,** 1.
29. Argon, A. S.; Salama, M. M. *Philos. Mag.* 1977, **36,** 1217.
30. Donald, A. M.; Kramer, E. J. *J. Mater. Sci.* 1981, 16, 2977.
31. Bueche, A. M.; Berry, J. P. In *Fracture*, Averback, B. L.; *et al.*, Eds.; Wiley, New York; 1959, p. 265.
32. Chang, R.-J.; Gent, A. N.; Lai, S.-M. Effect of Interfacial Bonds on the Strength of Adhesion. PMSE Preprints, *Amer. Chem. Soc.* 1992, **67,** 41.
33. Chang, R.-J.; Gent, A. N. *J. Polym. Sci., Polym. Phys. Ed.* 1981, **19,** 1619.
34. LaScala, J. J.; and Wool, R. P. *Polymer* 2005, **46**, 61–69.
35. La Scala, J. J., *Ph.D. Thesis*, University of Delaware; 2002.
36. Lu, J.; Wool R. P. *J. Polym. Sci., Part B: Polymer Physics Ed.,* in press (2005).
37. Lu, J., *Ph.D. Thesis*, University of Delaware; 2004.
38. Lorenz, C. D.; Stevens, M. J.; Wool, R. P. Fracture Behavior of Triglyceride-based Adhesives. *J. Polym. Sci. B: Polym. Phys.* 2004, **42,** 3333.
39. Lemay, J. D.; Swetlin, B. J.; Kelley, F. N. *ACS Symposium Series* 1984, **243,** 165–183.
40. Levita, G.; Depretris, S.; Marchetti, A.; *et al. J. Mater. Sci.* 1991, **26**(9), 2348.
41. Pearson, R. A.; Yee, A. F. *J. Mater. Sci.* 1989, **24**(7), 2572.
42. Baughman, R. H., Zakhidov, A. A; de Heer, W. A. *Science* 2002, **297**(2), 787.
43. Thostenson, E. T.; Ren, Z.; Chou, T.-W. *Composites Sci. Technol.* 2001, **61,** 1899.
44. Lauterwasser, B. D.; Kramer, E. *J. Philos. Mag.* 1979, **A39,** 469.
45. Kramer, E. J.; Berger, L. L. *Adv. Polym. Sci.* 1990, **91/92,** 1.
46. Kambour, R. P. *Macromol. Rev.* 1973, **7,** 1.
47. Bjerke, T. W.; Lambros, J. *J. Mechan. Phys. Solids* 2003, **51,** 1147.
48. Dugdale, D. S. *J. Mechan. Phys. Solids* 1960, **8,** 100–104.
49. Rice, J. R. In *Fracture* 2 (3), H. Liebowitz, Ed.; Academic Press, Boston; 1968.
50. Wool, R. P.; Long, J. M. *Macromolecules* 1993, **26,** 5227.
51. Welp, K. A.; Wool, R. P.; Mays, J.; *et al. Macromolecules* 1998, **31**(15), 49.
52. Agrawal, G.; Wool, R. P.; Dozier, W.D.; *et al. J. Polym. Sci. B* 1996, **34,** 2919.
53. Russell, T. P.; Deline, V. R.; Dozier, W. D.; *et al. Nature* 1993, **365,** 235.
54. Agrawal, G.; Wool, R. P.; Dozier, W. D.; *et al. Macromolecules* 1994, **27,** 4407.
55. Whitlow, S. J.; Wool, R. P. *Macromolecules* 1989, 22, 2648; 1991, 24, 5926.
56. Kim, Y. H.; Wool, R. P. *Macromolecules* 1983, **16,** 11.
57. Zhang, H.; Wool, R. P. *Macromolecules* 1989, **22,** 3018.
58. Anderson, K. L.; Wescott, J. T.; Carver, T. J.; *et al.* Mesoscale Modeling of Polymer Welding. *Mater. Sci. Eng. A* 2004, **365**(1–2), 14.

59. Wool, R. P. Rigidity Percolation Theory of Thin Films and the Glass Transition. PMSE Preprints, Amer. Chem. Soc., Philadelphia; August 2004.
60. Sapoval, B.; Rosso, M.; Gouyet, J. F. *J. Phys. Lett.* 1985, **46,** L149.
61. de Gennes, P.-G. *J. Chem. Phys.* 1971, **55,** 572.
62. Edwards, S. F., *J. Chem. Soc. London* 1967, **92,** 9.
63. Cohen Addad, J.-P.; Guillermo, A. *Macromolecules* 2003, **36,** 1609.
64. Wool, R. P.; O'Connor, K. M. *J. Appl. Phys.* 1981, **52,** 5953.
65. Wool, R. P.; O'Connor, K. M. *J. Polym. Sci., Polym. Lett.* 1982, **20,** 7.
66. Wool, R. P.; Yuan, B.-L.; McGarel, O. *J. Polym. Eng. Sci.* 1989, **29,** 1340.
67. Prager, S.; Tirrell, M. *J. Chem. Phys.* 1981, **75,** 5194.
68. Donald, A. M.; Kramer, E. J. *J. Mater. Sci.* 1981, **16,** 2977.
69. Boiko, Y. M. *Mechan. Compos. Mater.* 2003, **31**(1), 89.
70. Boiko, Y. M.; Bach, A.; Lynaae-Jorgensen, J. *J. Polym Sci. B: Polym. Phys.* 2004, **42,** 1861.
71. Helfand, E. *Macromolecules* 1992, **25,** 1676.
72. Willett, J. L.; Wool, R. P. *Macromolecules* 1993, **26,** 5336.
73. Benkoski, J. J.; Fredrickson, G. H.; Kramer, E. J. *J. Polym Sci. B: Polym. Phys.* 2002, **40,** 2377.
74. Brown, H. R. *Macromolecules* 1991, **24,** 2752.
75. Brown, H. R. *J. Mater. Sci.* 1990, **25,** 2791.
75(b). Janarthanan, V., Stein, R.S., Garrett, P.D., *Macromolecules* 1994, **27**, 4855.
76. Cole, P. J.; Cook, R. F.; Macosko, C. W. *Macromolecules* 2003, **36,** 2808.
77. Creton, C.; Kramer, E. J.; Brown, H. R.; *et al. Adv. Polym. Sci.* 2002, **156,** 53.
78. Creton, C.; Kramer, E. J.; Hui, C.-Y.; *et al. Macromolecules* 1992, **25,** 3075.
79. Creton, C.; Kramer, E. J. *Macromolecules* 1991, **24,** 1846.
80. Cho, K.; Brown, H. R.; Miller, D. C. *J. Polym. Sci. Polym. Phys.* 1990, **28,** 1699.
81. Brown, H. R.; Char, K.; Deline, V. R.; *et al. Macromolecules* 1993, **26,** 4155.
82. Char, K.; Brown, H. R.; Deline, V. R. *Macromolecules* 1993, **26,** 4164.
83. Lee, I.; Wool, R. P. *J. Polym. Sci. Phys Ed.* 2002, **40**(20), 2343–2353.
84. Lee, I.; Wool, R. P. *J. Adhesion*, 2001, **75,** 299.
85. Gong, L.; Wool, R. P.; Friend, A.D.; Goranov, K., *J. Polym. Sci., Part A: Polym. Chem.*, 1999, **37,** 3129.
86. Gong, G.; Friend, A.D.; Wool, R.P., *Macromolecules* 1998, **31**(11) 3706.
87. Gong, L.; Wool, R. P.; *J. Adhesion*, 1999, **71,** 189.
88. Wool, R. P.; *J. Polym. Sci., Part B: Polym. Phys.* 2005, **43**, 168.

7

Properties of Triglyceride-Based Thermosets

Richard P. Wool

Triglycerides are the main component of plant oils, such as soybean oil, corn oil, and so on. They are found in plants, fish, and animals, and are commonly the subject of dietary discussions (e.g., trans fatty acids), rather than discussions about the formation of high-performance composite resins. Triglycerides are composed of three fatty acids connected by a glycerol center (Figure 7.1). The unsaturation sites on the fatty acids [1, 2] can be chemically modified in any of numerous ways, many of which can be used to make polymers, as discussed in Chapter 4. Triglyceride-based polymers have been used in inks and coatings [3, 4], photoresist processes [5], toughening agents in PVC and epoxy resins [6], and as the major component of a number of natural resins [1, 7, 8], composites [9] (Chapter 5), and pressure-sensitive adhesives [10] (Chapter 8).

The triglyceride-based polymers studied in this work are analogous to vinyl esters (VEs) and unsaturated polyesters (UPEs) [11, 12], such that chemically modified triglycerides have multiple functional sites per molecule, which allow the resins to cross-link. In addition, VE, UPE, and triglyceride (TGD) resins are typically copolymerized with a low-molecular-weight species, such as styrene, to modify the properties of the polymer [7, 11, 12]. The low viscosities of these resins make them ideal for inexpensive polymer composite fabrication processes, such as vacuum-assisted resin transfer molding.

Triglyceride-based resins are an attractive alternative to petroleum-based resins because they are inexpensive, have good properties, and are derived

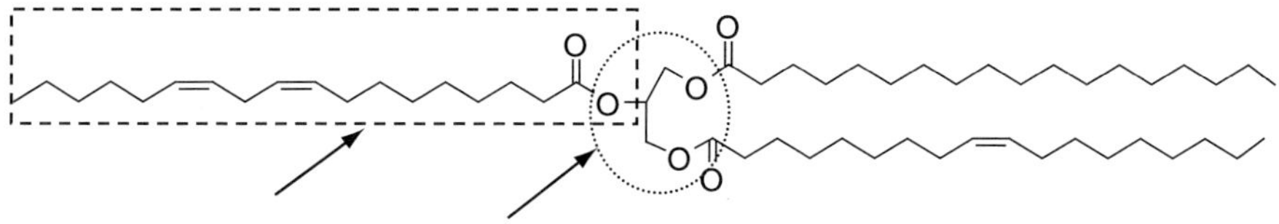

FIGURE 7.1 The molecular structure of a typical triglyceride molecule. Three fatty acids are connected to a glycerol center.

from renewable resources. Furthermore, the advent of genetic engineering technology, which can drastically change the fatty acid composition of triglycerides in plant oils, offers a large potential for inexpensively improving the properties of these polymers. This chapter effectively answers the question most often asked by genetic engineers: What is the effect of the fatty acid distribution function, triglyceride structure, and degree of chemical functionalization on the properties of polymers made with such bio-based monomers?

Typically triglyceride-based polymers form gels, which can be hard or soft depending on the level of functionalization of the triglycerides, the extent of polymerization, comonomer type, and comonomer content. In an effort to improve the properties of plant oil-based thermosetting resins, this work examines the effect of triglyceride molecular structure on the resulting polymer properties. The use of simple new models, such as vector percolation (described in Chapter 6), to predict the thermal, mechanical, and fracture properties of triglyceride-based polymers, is presented and assessed. The theoretical work is compared to model experimental studies and computer simulations of reacting triglyceride systems.

7.1 INTRODUCTION

What should be the fatty acid distribution (FAD) function of the triglycerides in a plant oil to achieve optimal polymer properties? The FAD can be controlled by genetic engineering of the plants, by blending oils from different plants, by oil refining, or by chemical modification. Control of the FAD is essential for the synthesis of polymers with known molecular architectures, such as linear chains for coatings (PSA) and thermoplastics, lightly branched linear chains for pressure-sensitive adhesives (PSAs), lightly cross-linked structures with low entanglement densities for rubbery materials, and highly cross-linked structures with high entanglement densities for rigid thermosets and foams.

LaScala and Wool [13] examined the tensile and mechanical properties of a model thermoset system consisting of acrylated epoxidized plant oil (AEPO) triglycerides in a comonomer styrene, as shown in Figure 7.2. The acrylic acid (A) groups react by free-radical polymerization with the styrene (S), which acts as a chain extender between A cross-links. Properties such as modulus E, cross-link density ν_e, and fracture stress σ were examined experimentally in terms of the level of acrylation, A, and the extent of reaction, X, of the acrylate groups and the styrene during free-radical polymerization.

The distribution of functional groups on the triglyceride molecules must be determined before percolation theory can be used to calculate the mechanical properties and before the cross-link density can be determined. Triglycerides are typically functionalized in a reaction, or series of reactions, to enable them to free polymerize radically. In this work, triglycerides are first epoxidized and then acrylated (Figure 7.3). The acrylate groups are capable of free-radical polymerization with themselves or with a comonomer such as styrene. To calculate the cross-link density of triglyceride-based polymers, the distribution of reacted acrylate groups in the polymer is required. This distribution can be calculated knowing the distribution of the unsaturation sites, epoxy groups, and acrylate groups on the triglycerides.

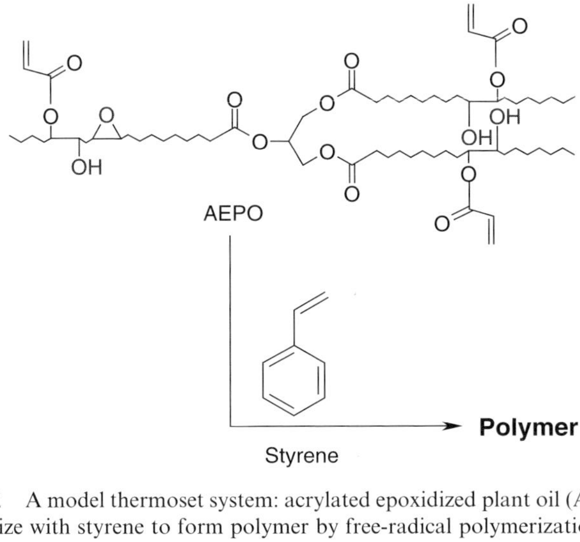

FIGURE 7.2 A model thermoset system: acrylated epoxidized plant oil (AEPO) triglycerides copolymerize with styrene to form polymer by free-radical polymerization.

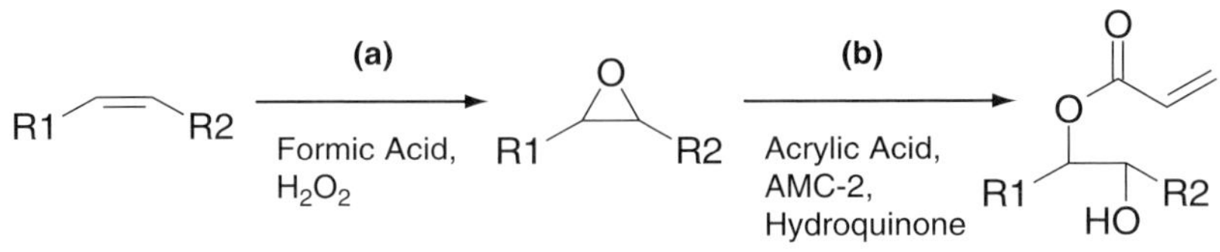

FIGURE 7.3 The (a) epoxidation and (b) acrylation of triglycerides.

7.2 DISTRIBUTION OF FATTY ACIDS AND UNSATURATION SITES IN TRIGLYCERIDES

Enzymes in the cells of plant and animals govern the production and distribution of fatty acids. The distribution of fatty acids is fairly easy to determine with techniques such as gel permeation chromatography [14–16]. Enzymes can also determine the positional distribution of fatty acids on the glycerol center [14–16]. As a result, the distribution of the resulting triglycerides is not random. Because it is experimentally difficult to determine the distribution of triglycerides, theories that predict the FAD have been developed [14]. These theories can be used to calculate the likelihood of having a given fatty acid on the 1, 2, and 3 positions of the glycerol center (Figure 7.4). The most accurate method for predicting the distribution of common fatty acids is the 1,3-Random, 2-Random hypothesis, which was proposed by Vander Wal [17] and Coleman and Fulton [18] in 1960–1961. This hypothesis assumes that two different pools of fatty acids are separately and randomly distributed to the 1,3- and 2-positions of the glycerol molecules. Thus, the sn-1 and sn-3 positions should have equivalent fatty acid distributions. The amount of each component can be calculated using this equation:

$$\%\,\text{sn-XYZ} = [\text{X at 1,3-position}][\text{Y at 2-position}][\text{Z at 1,3-position}] \times 10^{-4}, \tag{7.1}$$

where X, Y, and Z represent the constituent mol% fatty acids and sn-XYZ is a triglyceride with X in the sn-1 position, Y in the sn-2 position, and Z in the sn-3 position. However, to do this calculation, separation of the fatty acids into two different pools must be done. A means for determining separate pools of fatty acids was proposed by Evans *et al.* [19]. The rules for estimating the positional distribution of fatty acids in plant oils are known as the Evans hypothesis:

1. Saturated acids and those with chain lengths greater than 18 carbon atoms are first distributed equally at the 1- and 3-positions.
2. Oleic and linolenic acids are then distributed randomly on the unfilled 1-, 2-, and 3-positions. When the 1- and 3-positions are filled, the excess is added to the 2-position.
3. All remaining positions are filled by linoleic acid.

Analysis of plant oils has revealed a number of regular patterns in the distribution of saturated fatty acids between the sn-1, sn-2, and sn-3 positions

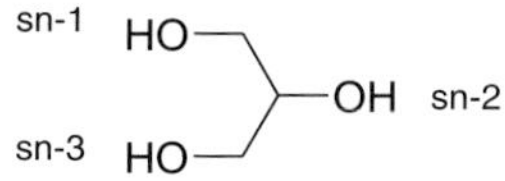

FIGURE 7.4 The positions of the three acyl attachment sites on glycerol.

that show close agreement with the Evans hypothesis. Saturated fatty acids and fatty acids with chain lengths greater than 18 carbon atoms (SA) are almost exclusively found at the sn-1 and sn-3 positions according to Eq. (7.2):

$$\begin{aligned}\%\,\text{SA fatty acid in 1-,3-position} &= 1.47\text{x},\\ \%\,\text{SA fatty acid in 2-position} &= 0.03\text{x},\end{aligned} \tag{7.2}$$

where x is the percentage of the fatty acid in the oil [14]. The error for Eq. (7.2) is less than 2% for most oils.

The level of unsaturation of a given triglyceride was simply the sum of the unsaturation sites on the constituent fatty acids. The percentage of all triglycerides with a given functionality, $P(u)$, is just the summation of the percentage of all triglycerides with functionality u:

$$P(u) = \sum_{u=i+j+k} \%\,\text{sn-}X(i)\,Y(j)Z(k), \tag{7.3}$$

where i, j, and k are the number of unsaturation sites on fatty acids X, Y, and Z. This method for calculating the unsaturation distribution was found to be very accurate [20] and is used herein. For example, soybean oil contains 15.5% saturated acids, 23.5% monounsaturated acids, 53.2% di-unsaturated acids, and 7.8% tri-unsaturated acids [21]. Using the Evans hypothesis in conjunction with the 1,3-Random, 2-Random hypothesis of Eqs. (7.1) through (7.3), the distribution of triglycerides with 0–9 unsaturation sites was calculated. Table 7.1 compares this calculated theoretical distribution with the experimentally determined fatty acid distribution for soybean oil. The results show that both the Evans hypothesis and 1,3-Random, 2-Random hypothesis accurately predict the unsaturation distribution for soybean oil, and do so for most plant oils [14, 16].

7.3 DISTRIBUTION OF FUNCTIONAL GROUPS ON TRIGLYCERIDES

In this work, triglycerides were chemically functionalized using their unsaturation sites. As a result, the distribution of these functional groups on triglycerides was governed by the distribution of unsaturation sites. Therefore, when calculating the distribution of functional groups on triglycerides, the unsaturation distribution was the starting point. The equations for calculating the distribution of functional groups can be generalized for any series of reactions. Using the distribution of reactive sites before reaction, $F(N)$, the distribution of functional groups after reaction can be determined using a binomial distribution [22]. The probability of having n functional groups on a triglyceride with N reactive sites was calculated using Eq. (7.4):

TABLE 7.1 The actual distribution of unsaturation sites in soybean oil compared to the predictions of the Evans hypothesis used in conjunction with the 1,3-Random, 2-Random hypothesis.

Unsaturation Sites per Triglyceride	Experimentally Determined Distribution	Theoretical Distribution (Evans plus the 1,3-Random, 2-Random Hypotheses)
0	0.05	0.00
1	1.03	1.27
2	5.93	6.28
3	13.92	14.19
4	24.91	25.04
5	25.42	25.10
6	20.08	20.19
7	6.85	6.91
8	0.97	0.97
9	0.05	0.05

$$P(N,\ n,\ \xi) = C(N, n)\xi^n(1 - \xi)^{N-n}F(N), \tag{7.4}$$

where ξ is the extent of reaction, $C(N, n)$ is the number of different ways the n functional groups can be arranged on the triglyceride with N reactive sites [i.e., $C(N, n)$ is the combinatorial function or binomial coefficient], and $F(N)$ is the unsaturation distribution [22]. These probabilities were calculated for all 55 possible triglyceride combinations (i.e., the number of possible combinations of having n functional groups on a triglyceride with N reactive sites). The percentage of n-functional triglycerides, $p(n,\ \xi)$, was

$$p(n,\ \xi) = \sum_N P(N, n, \xi). \tag{7.5}$$

Equations (7.4) and (7.5) were used for the epoxidation step, where $F(N)$ is the unsaturation distribution, and the acrylation step, where $F(N)$ is the calculated epoxide distribution.

This method assumed that the addition of functional groups to triglycerides was completely random. Previous work has shown that functional groups have a preference for the fatty acid to which they attach [23, 24]. On the other hand, the relative preferences of functional groups for different fatty acids were of the same order of magnitude. Furthermore, the studies that found preferential attachment of functional groups only examined the initial rates of reaction. The effect of preexisting functional groups on a fatty acid on the addition of another functional group has not been studied. The effects of preferential addition should be dampened at high conversions. Finally, we are only concerned with the level of functionality of a given triglyceride. The location of the functional groups on the fatty acid and the acid to which they are attached are not important. Therefore, the distributions of functional groups should be fairly accurate; however, they are only approximate. Furthermore, the exact percentage of triglycerides with a certain number of functional groups is not as important as the general trends observed. This method of calculating the epoxide and acrylate functional group distributions was found to be accurate using nuclear magnetic resonance (NMR) spectra [20], as shown in Table 7.2.

7.4 CROSS-LINK DENSITY

The cross-link density of model acrylated triglycerides was investigated by LaScala and Wool and was found to increase with the acrylation level for samples with and without styrene (Figure 7.5). The cross-link density was higher for samples without styrene because styrene is a linear chain extender and therefore does not contribute to cross-linking. For samples with styrene, the cross-link density increased slowly at low levels of acrylation (i.e., up to about 2.5 acrylates per triglyceride) [Figure 7.5(b)]. At higher levels of acrylation, the cross-link density increased linearly. This was a result of the fact that at low levels of acrylation, the functional groups mainly linearly extended the polymer chains rather than cross-linking them. However, at higher levels of

TABLE 7.2 The number of mono-epoxy, di-epoxy, and tri-epoxy fatty acids per molecule in oils. These numbers were calculated experimentally from the ^{1}H-NMR spectra and compared to the theoretical number.

	Number of Epoxy Fatty Acids Per Molecule					
	Mono-Epoxy		Di-Epoxy		Tri-Epoxy	
Oil	Exp.	Theory	Exp.	Theory	Exp.	Theory
Methyl oleate	0.93	0.93	0.00	0.00	0.00	0.00
Methyl linoleate	0.06	0.13	0.90	0.86	0.00	0.00
Olive oil	2.26	2.06	0.19	0.26	0.00	0.01
HOSO	2.41	2.35	0.16	0.08	0.02	0.09
Methyl linoleate	0.05	0.05	0.04	0.18	0.90	0.80
Triolein	2.79	2.79	0.00	0.00	0.00	0.00
Cottonseed oil	0.63	0.67	1.43	1.45	0.00	0.01
Canola oil	2.00	1.94	0.54	0.55	0.22	0.25
Corn oil	0.86	0.88	1.69	1.64	0.03	0.03
Soybean oil	0.72	0.71	1.51	1.60	0.22	0.23
Safflower seed oil	0.79	0.63	1.87	2.06	0.00	0.00
Trilinolein	0.18	0.17	2.81	2.82	0.00	0.00
Linseed oil	0.65	0.59	0.59	0.58	1.45	1.55

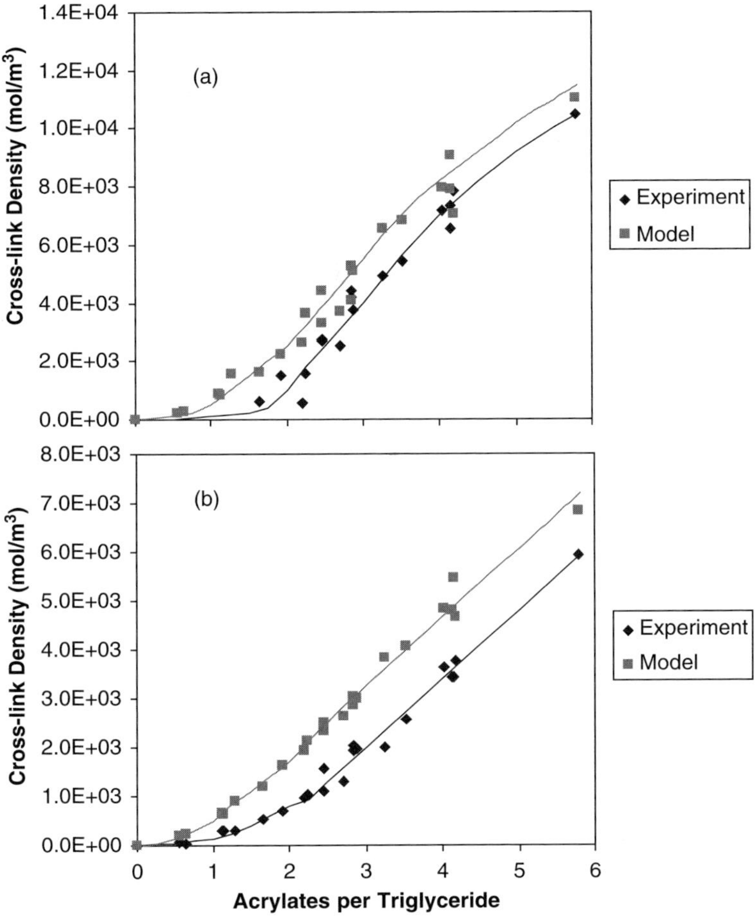

FIGURE 7.5 The cross-link density determined experimentally and predicted from the Miller and Macosko model for triglyceride-based polymers (a) without styrene and (b) with styrene (87.5 mol%).

acrylation, each additional functional group increased the cross-link density. For samples without styrene, the cross-link density increased with the level of acrylation [Figure 7.5(a)]. Samples with very low levels of acrylation (<2 acrylates per triglyceride) could not be tested because they were too brittle or did not cure. The molecular weight between cross-links ranged from 80 to 20,000 g/mol and indicates that some of these networks were very highly cross-linked, whereas some were only lightly cross-linked. The plant oils used in this work had highly varied fatty acid compositions, which results in significantly different acrylate distributions. However, the distribution of acrylate groups did not have a significant effect on the cross-link density, as can be seen by the fact that cross-link density was a single function of acrylation (Figure 7.5).

The cross-link density is defined by the density of chains or segments that connect two infinite parts of the polymer network, rather than the density of cross-link junctures. The cross-link density is affected by the functionality of the cross-linker molecule. Vinyl polymers are typically represented as having two functional groups, f, per vinyl group, n:

$$f = 2n. \tag{7.6}$$

The number of functional groups on a cross-linker that polymerizes and leads to another cross-linked molecule is m, where $m \leq f$. For example; triglycerides with $m = 1$ functional group are dangling chain ends. The functionality, m, and the number of cross-linked chains, N_x, per triglyceride are related:

$$N_x = 3m/2 - 3 = 3n - 3. \tag{7.7}$$

Therefore, with m in the range 0–18, the number of cross-link chains varies from 0 to 24 per triglyceride, depending on the functionality. The positional distribution of the functional groups does not affect N_x. Therefore, the structure of the triglyceride (i.e., three chemically linked fatty acids) has no effect on the cross-link density except that it allows each triglyceride molecule to have a different number of functional groups and cross-links.

Flory and Stockmayer, as discussed in reference [25], used the principles of branching and cross-linking to derive the cross-link density for additional polymers. The number of cross-links per molecule in divinyl polymers is

$$c = \theta {c_0}^{\sim} \tag{7.8}$$

where θ is the extent of polymerization (as determined by Fourier transform infrared spectroscopy (FTIR) and c_0 is the fraction of polymerizable groups on the cross-linking agent. Equation (7.8) is defined for cross-link sites rather than cross-link segments. Equation (7.9) accounts for this and generalizes Eq. (7.8) for a mixture of cross-linking agents with n polymerizable groups (i.e., vinyl groups):

$$c = \theta \sum_{n=2}^{n=9} \phi_n (3n - 3), \tag{7.9}$$

where ϕ_n is the fraction of triglycerides with n polymerizable groups. The molecular weight between cross-links is simply the average molecular weight of the monomers, M_{av}, divided by the number of cross-links per molecule:

$$M_c = M_{av}/c. \tag{7.10}$$

The cross-link density resulting from chemical cross-link chains can then be calculated by dividing the polymer density by M_c.

Intramolecular cyclization in finite species causes the delayed gel point relative to the predictions of Flory-Stockmayer [25]. However, intramolecular cyclization only affects the cross-link density when a growing polymer chain polymerizes with functionality on the same triglyceride before it polymerizes with functionality on another cross-linking agent. Other forms of intramolecular cyclization reactions must occur for gelation and stiffening of the thermosetting polymer to occur. Intramolecular cyclization of functional groups on the same triglyceride reduces the number of effective attachment points of this triglyceride to the rest of the network. When a triglyceride reacts with itself, it forms a loop that is not attached to the rest of the polymer network. Consequently, this reduces the cross-link density of the resulting polymer. Every time two functional groups react in this way, the effective number of functional groups per triglyceride is reduced by 2. Therefore, the Flory-Stockmayer model should overestimate the cross-link density because of intramolecular cyclization. Similar theories, such as the Miller-Macosko [26] theory, predict similar results.

Figure 7.5 shows the predictions of the Miller-Macosko model along with the experimentally determined cross-link densities. The model overpredicted the cross-link density at all levels of acrylation for polymers with and without styrene. This was expected because the model did not account for intramolecular cyclization of triglycerides. In addition, rubber elasticity theory is not expected to completely apply to these highly cross-linked polymers, but at least should show general trends in how cross-link density is affected by the acylation level. In fact, the model's predictions matched the trends in the experimentally determined cross-link densities. Both plots show that the predicted and experimentally determined cross-link densities increased linearly as the acrylation level increased above a certain value. Therefore, for the most part, the cross-link density is proportional to the acrylation level. Both plots show that the model accurately predicted that the cross-link density increased at a lower rate at low levels of acrylation. However, the model and experiment disagree on the point where the cross-link density should increase linearly because of the overprediction of the cross-link density.

The model predicted that there was some level of cross-linking regardless of the level of acrylation. However, monomers with very low levels of acrylation (<0.5 acrylates per triglyceride with styrene and <1.5 acrylates per triglyceride without styrene) did not cure. Again, this was due to the effect of intramolecular cyclization. The extent of intramolecular cyclization can be quantified by fitting the model results to the experimental results. The shift required to equate the model cross-link density, $\nu_{\text{model}}(A)$, to the experimentally determined cross-link density, $\nu_{\text{exp}}(A)$, was A_{lost}:

$$\nu_{\text{model}}(A - A_{\text{lost}}) = \nu_{\text{exp}}(A). \tag{7.11}$$

The number of acrylate groups lost toward intramolecular cyclization, A_{lost}, is plotted as a function of the level of acrylation in Figure 7.6 for acrylated triglycerides copolymerized with styrene. The amount of intramolecular cyclization increased with the acrylation level at low levels of acrylation. After ~2 acrylates per triglyceride, the amount of intramolecular cyclization was constant. Overall, the model predicted that the amount of intramolecular cyclization reduced the effective functionality by approximately 0.7–0.9 acrylates per triglyceride for polymers copolymerized with styrene. The presence of styrene in samples with low levels of acrylation actually decreased the extent of intramolecular cyclization, as expected. The concentration of triglycerides with two or more acrylates per triglyceride was very low in samples with low levels of acrylation. As a result, the probability of intramolecular cyclization should be high relative to the probability for cross-linking. However, the presence of styrene reduced the rate of cyclization at low levels of acrylation by propagating the growing radical away from the triglyceride molecule. In addition, samples with low levels of acrylation were more likely to terminate by chain transfer, which lowers the cross-link density [27].

The amount of acrylate groups lost to intramolecular cyclization was approximately 0.5 for homopolymerized triglycerides at all levels of acrylation. Any trends in the amount of intramolecular cyclization were obscured because of the scatter in the cross-link density as a function of acrylation level.

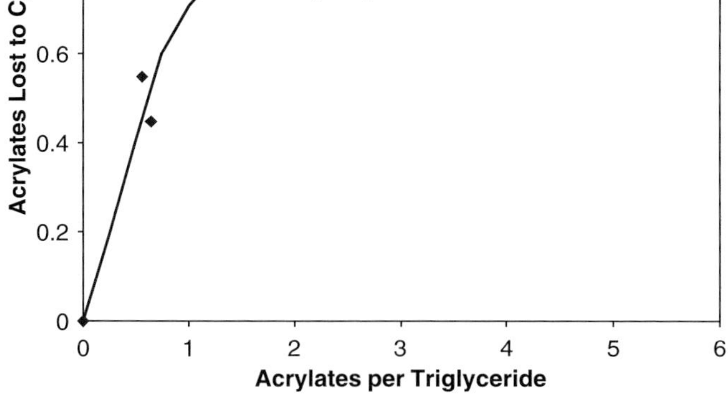

FIGURE 7.6 The number of acrylate groups lost due to intramolecular cyclization of the triglycerides for polymers copolymerized with styrene (87.5 mol%).

In all cases, the amount of intramolecular cyclization was less than 1 acrylate per triglyceride. This could indicate that one cyclization event per triglyceride was a maximum. After this cyclization event occurs, further intramolecular cyclization of acrylate groups on the same triglycerides may be sterically hindered. However, more work needs to be done to verify this.

The rubber modulus determined from dynamic mechanical analysis is a measure of the effective cross-link density of polymers. The effective cross-link density takes into account chemical and physical cross-links [28, 29]. Chain entanglements can increase effective cross-linking. However, for chain entanglements to occur in branched and linear polymer systems, the molecular weight of the polymer chains must be greater than the entanglement molecular weight, which is on the order of at least 10^4 g/mol [30]. Only the resins with 0.56 and 0.65 acrylates per triglyceride copolymerized with styrene had molecular weights between cross-links that were on that order of magnitude. Therefore, it was possible that chain entanglements increased the effective cross-link density and decreased the calculated extent of intramolecular cyclization for polymers with low levels of acrylation. However, it was unlikely that chain entanglements had much of an effect on the other samples.

7.5 TENSILE PROPERTIES

The mechanical properties of triglyceride-based polymers were strong functions of the acrylation level. The tensile properties of these polymers increased exponentially with the level of acrylation up to 3 acrylates per triglyceride (Figures 7.7 and 7.8). The properties then leveled off as the level of acrylation was increased beyond 3 acrylates per triglyceride. To an excellent first approximation, the oil type did not have a significant effect on the modulus or strength of these polymers, as linseed-, soybean-, and HOSO-based polymers with similar acrylation levels had similar properties. The mechanical properties were very low at low acrylation levels because the cross-link density was very small (Figure 7.5). As the cross-link density increased, the polymer chains became more tightly bound to each other, which increased the properties. This was indicative of a percolation phenomenon [28, 31–33]. Obviously, the fact that T_g increased as the acrylation level increased was also a reason that the modulus increased.

Percolation theory is a simple way to predict the strength of polymers as a function of cross-link density. Percolation theory relates the polymer properties to the connectivity of the polymer network. The critical stress, σ_c, required to break a thermoplastic network is [28]

$$\sigma_c = [D_0 E \nu_c (p - p_c)]^{1/2} \tag{7.12}$$

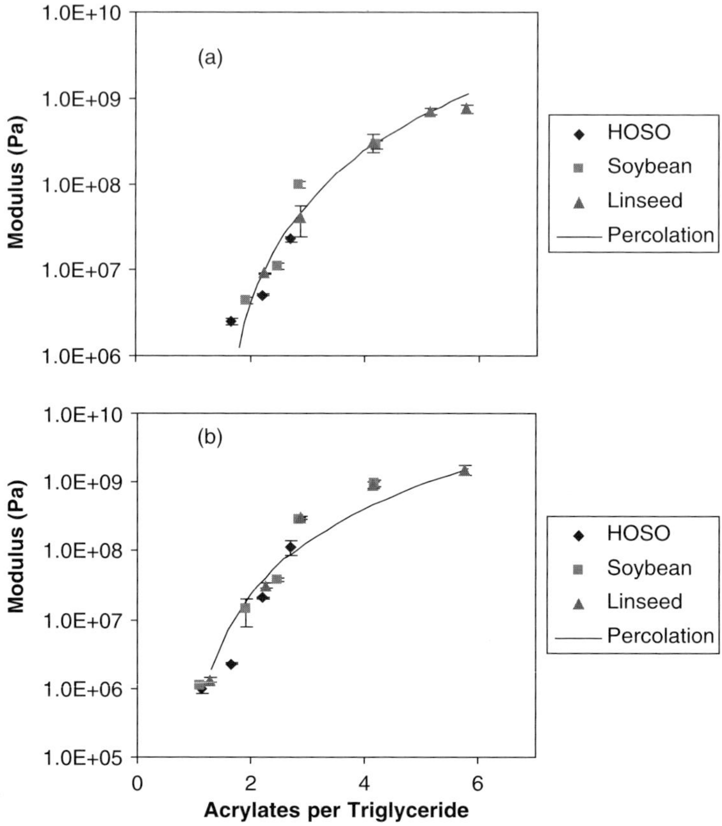

FIGURE 7.7 The tensile modulus as a function of the level of acrylation for polymers (a) without styrene and (b) with styrene (87.5 mol%) using oils acrylated to different extents.

where E is the tensile modulus, ν_c is the critical entanglement density, D_0 is a constant of proportionality representing the energy to disentangle or break bonds, p is the level of perfection of the entangled network (i.e., $p = 1$ for a perfect network), and p_c is the percolation threshold. For thermosetting systems, there is no critical cross-link density. The cross-link density can theoretically increase to as high as that found in graphite, but will more likely level off at a value of $\sim 10^4$ mol/m^3 for most cross-linked polymer systems. The extent of cross-linking exceeded the percolation threshold for all of the polymers tested, considering that the polymers had gelled and had enough mechanical integrity for mechanical testing. For examining the effect of different extents of cure, the term $p - p_c$ would be important. However, for fully cured thermosetting systems, this value is likely to be constant. The level of perfection is affected by the morphology of the system, but DMA traces indicated no differences in phase behavior and cure studies indicated that the

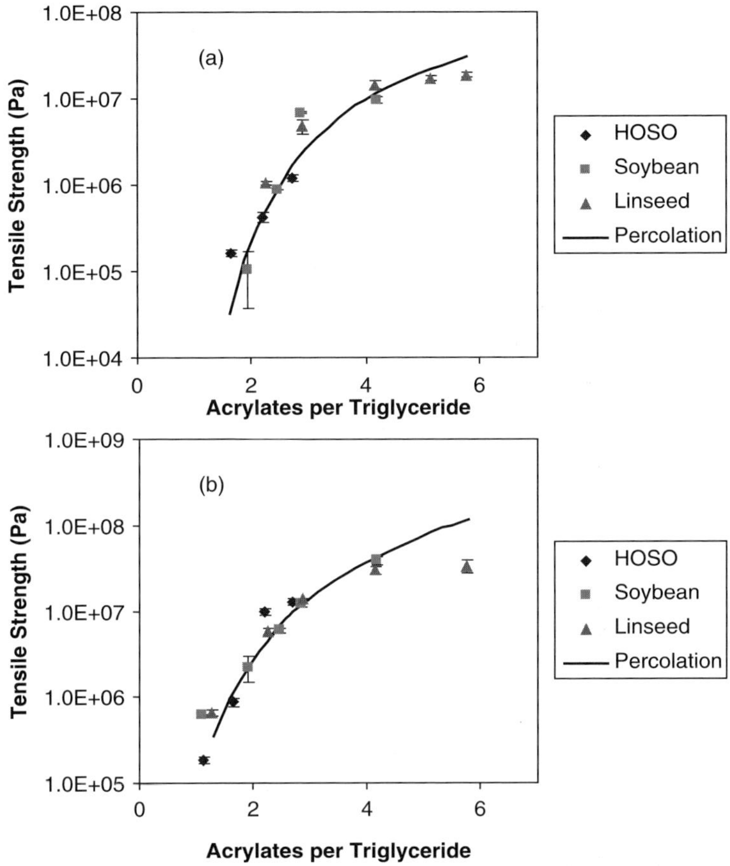

FIGURE 7.8 The tensile strength as a function of the level of acrylation for polymers (a) without styrene and (b) with styrene (87.5 mol%) using oils acrylated to different extents. The predictions of the vector percolation model are also shown.

level of triglyceride functionality has no significant effect on reactivity ratios. Therefore, the strength of these triglyceride polymers can be estimated as follows:

$$\sigma_c = [\sigma_0 E \nu]^{1/2}. \tag{7.13}$$

Percolation theory has also been used to predict the modulus [28, 31]. However, these theories do not account for T_g effects. As seen in Figure 7.9, T_g increased from below to above room temperature as the acrylation level increased, which caused the steep increase in the modulus as the acrylation level increased from 2 to 3 acrylates per triglyceride (Figure 7.7). Therefore, percolation theory was not used to predict the modulus. On the other hand,

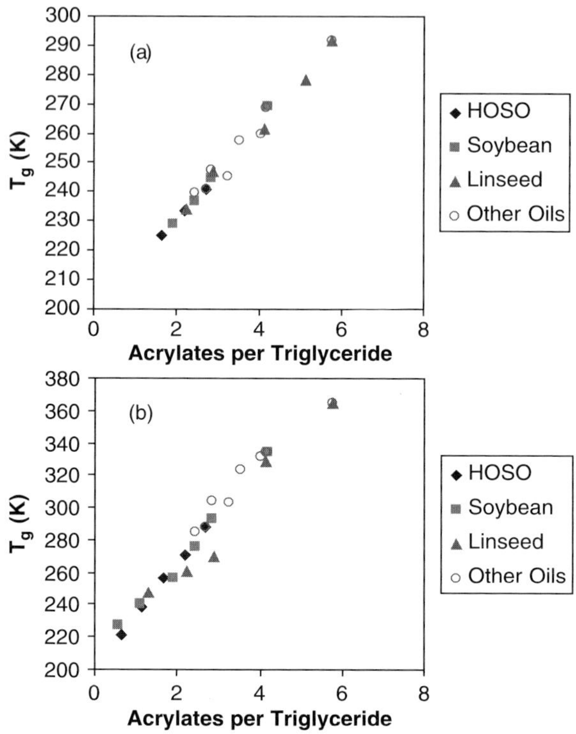

FIGURE 7.9 The effect of acrylation level on the glass transition temperature of triglyceride-based polymers (a) without styrene and (b) with styrene (85 mol%).

Eq. (7.12) is expected to be valid regardless of varying T_g because the equation separates out stiffness effects and bond energy effects.

Figure 7.8 shows that vector percolation theory adequately describes the tensile strength as a function of acrylation level. Thus, the strength improvements with acrylation level are indeed percolation phenomena. The vector percolation predictions of tensile strength were calculated using the experimentally determined cross-link densities. For both samples with and without styrene, vector percolation predictions at high levels of acrylation overestimated the strength. This may have occurred because other failure mechanisms were beginning to dominate at high cross-link densities for this triglyceride-based polymer system.

The constant of proportionality, σ_0, was 5×10^3 J/mol for samples with 87.5 mol% styrene and 1.4×10^2 J/mol for samples without styrene. This indicates that styrene improves the mechanical integrity of the network through means other than simply increasing the modulus or affecting the cross-link density. Similar thermosetting systems, such as unsaturated polyesters and vinyl esters, typically form microgels of high cross-link density at the initial stages of cure, which link up at higher extents of cure [32]. The cracks that form during failure need not travel in a straight line through the material, but instead could travel through the path of least resistance. In other words, the crack can travel through regions of much lower cross-link density and stiffness than the average across the whole polymer. Therefore, a possible explanation for the values of σ_0 is that homopolymerized triglycerides have greater microscopic variance in the stiffness and cross-link density than triglycerides polymerized with high styrene contents.

7.6 COMPUTER SIMULATIONS OF TRIGLYCERIDE STRUCTURE AND STRENGTH

Considering the complexity of triglyceride reactions, and the many degrees of freedom in selecting chemical pathways for oils with different fatty acid distribution functions, a computer simulation was designed to optimize materials properties of bio-based composites. This simulation would allow (1) the selection of the oil FAD; (2) the extent of chemical functionalization of the starting oil; (3) the amount and type of comonomer; (4) the type of initiator; (5) reactions to occur near a solid surface, such as a fiber; (6) extent of reaction to be followed with reaction kinetics; (7) examination of the resulting structure at the nanoscale; (8) determination of the cross-link density, in the bulk and at interfaces; and (9) determination of tensile properties such as modulus, fracture stress, fracture strain, and toughness. In addition to optimizing triglyceride reactions, the simulations would also give us a method to explore the validity of theoretical network connectivity models and interpret experimental data on model oil reactions. We have found this three-way analysis of complex problems to be most useful in addressing such intricate problems as polymer dynamics and strength of interfaces, as discussed in Chapter 6. The simulation work, which was initiated at the University of Delaware, was done in collaboration with C. D. Lorentz and M. J. Stevens at Sandia National Laboratory (SNL) [35, 36].

The simulation procedure used to generate the triglyceride networks and to apply the tensile strain to the networks consists of different simulation methods and algorithms. The triglyceride networks are generated by an off-lattice serial Monte Carlo code that uses the general algorithm developed by Khot [37]. First, a mixture of the triglyceride chains (23 mol%), styrene monomers (66 mol%), and the initiator molecules (11%) is randomly placed, allowing overlap. A single bead is used to represent the glycerol center and the fatty acid chains range from 5 to 7 beads in size, with each coarse-grained bead representing about 3 carbon atoms. The styrene monomers are modeled as 1 bead and the initiator as 2 beads connected by a spring, which decomposes into 2 reactive beads during free-radical polymerization. In this bead spring system, the beads interact through the standard Lennard–Jones (LJ) potential, $U(r)$, as a function of the distance r and bead diameter d, as

$$U(r) = 4u_o[(d/r)^{12} - (d/r)^6], \tag{7.14}$$

where u_o represents the LJ energy. A cutoff potential of $r = 2.5d$ was used. The initial system is a liquid mixture of different molecules at a density $\rho = 0.9d - 3$ in a rectangular box with dimensions $L_x \times L_y \times L_z$, where L_z is the separation distance between two solid walls and there are periodic boundary conditions in the L_y and L_z directions. The solid wall particles do not interact with each other but interact with the polymer beads via the same LJ potential during the reaction. A cosine potential is used to eliminate overlap of the starting monomers in this box during a span of 10,000 time steps. Then, molecular dynamics is used to equilibrate the liquid system at the simulation temperature $T_L = 1.1u_o/k$, using a Langevin thermostat, for an additional 10,000 time steps to eliminate artifacts of the initial state.

The second stage in the simulation uses a Monte Carlo (MC) simulation to form the polymer networks with three stages of initiation, propagation, and termination of the free-radical reaction. Each bead move in the system is evaluated using a Metropolis algorithm and accepted with a probability of $\exp(-\delta E/kT)$, where δE is the energy change associated with the move. The polymerization MC algorithm was run for 10,000 cycles, which is long enough for all of the initiator molecules to decay and propagate their free radical onto a monomer to cause the initiation event. The simulations took 2–3 days for a system of 25,000 beads in a $25.8d \times 30.0d \times 32.0d$ box. The molecular dynamics simulations were done on the CPlant computer at SNL using a massively parallel MD code, LAMMPS [38]. The reaction was run to extents of reaction of at least 85% of the possible bonds that could form and the resulting network was then evaluated for its tensile properties. The network was then cured at $T = 1.0u_o$ to complete the reactions and then cooled to $T = 0.2u_o$, which is well below the glass transition temperature. The walls were then moved apart in the z-direction at a constant rate $v_z = 0.01d/\tau$, corresponding to a strain rate of $3.6 \times 10^{-4}/\tau$, which is consistent with rates used in previous simulations of epoxy and glassy material deformation by Stevens [39]. The MD simulations took approximately 2 days on 16 processors for the 25,000-bead system. Eight triglyceride systems were studies with fatty acid distribution functions and unsaturation levels n, representative of soybean oil ($n = 4.6$), olive oil ($n = 2.8$), linseed oil ($n = 6.6$), and five others with n values ranging from 2 to 6.6.

In these systems, it was found that approximately 63% of the styrene reacted, independent of the unsaturation level n. The extent of reaction X of the triglycerides increased with n as $X = 0.52$ for $n = 2$, $X = 0.62$ for $n = 3$, $X = 0.68$ for $n = 4$, and $X = 0.72$ for $n = 5$. In all cases, approximately 95% of the reactive sites and 90% of styrene monomers that reacted did so to their full extent possible. It was also found that the cluster size formed during the reaction was very dependent on n. This is important for nanocomposites, for example, with nanoclays, where the intercalation reactions occur in similar dimensions (Chapter 14) as these simulations. For $n = 2$ systems, there were no spanning clusters, only small clusters with the largest consisting of about 2% of the beads. As n increased to 3, a system-spanning cluster was present about half the time, with the remainder of the systems forming about a dozen clusters, each consisting of about 7% of the beads. For $n = 4$ and $n = 5$, the resulting network was basically one large spanning cluster with the remaining beads forming very small clusters, or not reacting at all. This behavior is the same as seen in percolation, where as you increase the number of bonds in the system, the largest cluster will grow in size until it percolates the system and one cluster will dominate, with only small clusters making up the remainder of the system.

The cross-link density v was determined as a function of the number of reactive sites n per triglycerides and the results are shown in Figure 7.10. The data are well represented by the simple proportionality, $v \sim n$, as follows:

$$v = 0.119 + 0.038n, \tag{7.15}$$

which is a result similar to that obtained by LaScala (see Section 7.5), who showed that for acrylated epoxidized plant oils, $v \sim A$, where A is the level of acrylation, similar to n in the simulations. This is in very nice agreement with experiment and theory.

The general trend of the stress–strain curves for the series of triglycerides was the same. At low strains, we observe a linear elastic response, which is then followed by a peak in the stress, which represents the yield stress σ_y. The elastic region was independent of the bonding nature of the system and the yielding marked the first slippage of beads. After the yielding at a strain of 0.1, we observed a plateau in the stress and then a range of strains where the stress is increasing again as the bonds are being stretched. Void formation often accompanies the stress increase as bonds break randomly and finally the system fails at the fracture stress σ_f, and then decreases rapidly. The effect of n and bonding on the plateau region becomes apparent, where the bonds are being pulled taut, such that the length of the plateau region increases with n. The tensile stress as a function of n is shown in Figure 7.11. The behavior is represented by

$$\log \sigma_f = 0.55 + 0.45 \log n. \tag{7.16}$$

The slope of 0.45 in this log – log plot supports the percolation prediction, $\sigma \sim v^{1/2}$, and $\sigma \sim n^{1/2}$, at constant E and $[p - p_c]$ values. From Eq. (7.15), when $n^* = 9$, $v^* = 0.461$ and $n = (v - 0.119)/0.038$. Substituting for n and n^* in Eq. (7.16) and letting $p = v/v^*$, we obtain

$$\sigma/\sigma^* = 1.16\,[p - p_c]^{1/2}, \tag{7.17}$$

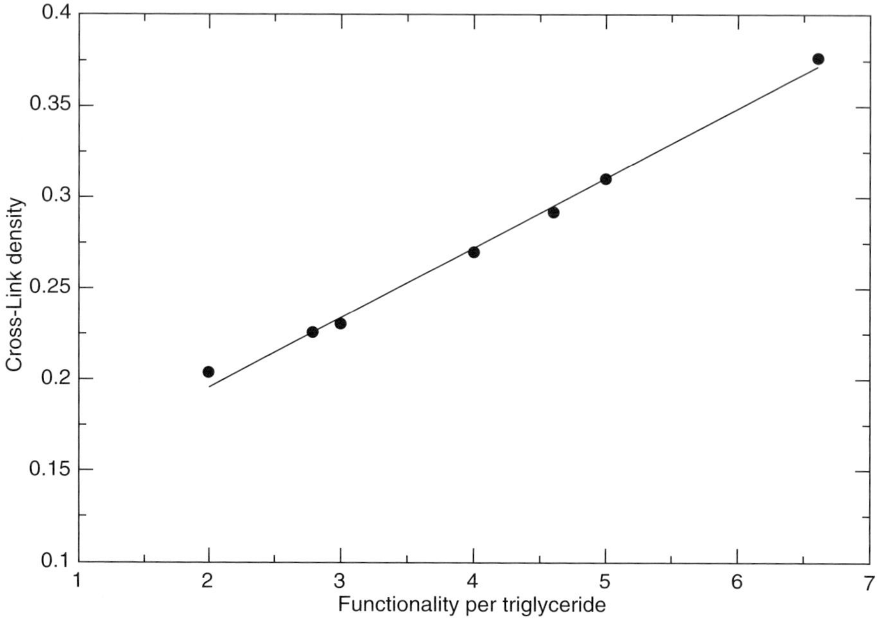

FIGURE 7.10 Simulations of cross-link density versus level of functionalization of triglycerides [36].

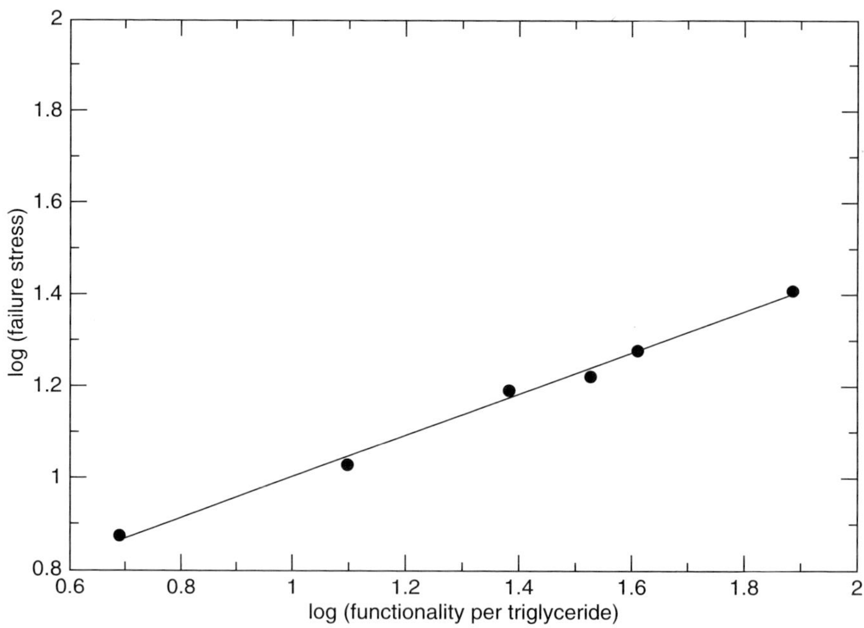

FIGURE 7.11 Simulation of fracture stress versus functionality of the triglycerides [36].

where $p_c = 0.119/0.461 = 0.258$, and the factor of 1.16 gives $\sigma/\sigma^* = 1$ when $p = 1$.

Further work is in progress to separate the interface effects from bulk fracture effects in this simulation, where it was observed that the extent of reaction of the triglycerides decreased substantially near the wall, which is a critical issue for all composites made by free-radical (unsaturated polyester, vinyl ester) or condensation (epoxy, urethane, imides) curing reactions.

Thus, to obtain high-performance properties for thermosets with triglycerides, the FAD should have a high level of unsaturation, approximately $n = 5-6$. For low levels of unsaturation, mixed chemical pathways can be used. For example, if a soybean oil has 3 acrylic acid groups per triglyceride ($A = 3$), the properties are moderately good. However, the properties can be significantly improved by maleinizing the remaining hydroxyl groups such that the functionality increases to $n = 5-6$. Although thermoset properties are dominated by the level of functionalization, they do not appear to be too sensitive to the details of the FAD. For linear thermoplastic polymers, such as those needed for PSAs, elastomers, and coatings before final cure, monofunctionality is the key to attain high molecular weight and good properties. Thus, the details of the FAD are crucial and one would like to have a monodisperse FAD of high oleic oils with $n = 3$. Simply having an average of $n = 3$ is not sufficient, as we discovered (Chapter 8), to obtain linear polymers from the fatty acids with $n = 1$. We also investigated monosubstituted triglycerides, by taking epoxidized soybean oil and controlling the acrylation reaction to give an average of $n = 1$ acrylate per triglyceride. This monosubstituted triglyceride gelled during the free-radical reaction because a sufficient number of fatty acids were present with both $n = 2$ and $n = 3$. Therefore, the best method of obtaining linear polymers is to genetically engineer a plant that has a very high oleic fraction (greater than 90%, $n = 3$), break off the fatty acids (each with $n = 1$) by a methanolysis reaction (identical to making biodiesel), and separate the oleic methyl ester from the glycerol. Fatty acid separations can also be done with broad FAD oils but are expensive. The glycerol can be reused in the SOMGMA reaction.

7.6.1 CONCLUSIONS ON STRENGTH AND STIFFNESS

To predict the properties of triglyceride-based polymers, the cross-link density of these polymers must be determined. Acrylated triglycerides were prepared from various oils and model compounds. The distributions of unsaturation sites on unmodified triglycerides were calculated using the 1, 3-Random, 2-Random hypothesis, Evans hypothesis, and experimental correlations. Using the unsaturation distribution, the acrylate distribution was calculated using a binomial distribution of acrylate groups. The cross-link densities of the resulting polymers were calculated using the extent of cure as measured with FTIR and the recursive method of Miller and Macosko from a knowledge of the acrylate distribution. The cross-link density was found to increase gradually at low acrylation levels and linearly at higher acrylation levels, and the trends in the cross-link density predictions matched the experimental results. The deviation in the experimental results and model predictions were the result of intramolecular cross-linking. Approximately 0.5 and 0.8 acrylates per triglyceride were lost to intramolecular cyclization for homopolymerized acrylated triglycerides and triglycerides copolymerized with styrene, respectively.

Equations for the level of perfection of the triglyceride networks and the percolation threshold were developed using the calculated number of acrylates lost to cyclization. Polymers with levels of perfection less than 0.1 without styrene and 0.39 with styrene did not have mechanical integrity, validating the definition of the level of perfection and percolation threshold.

The properties of triglyceride-based polymers obey vector percolation behavior. The cross-link density, tensile modulus, and tensile strength increased exponentially with the level of acrylation at low levels of acrylation and increased linearly at high levels of acrylation. The vector percolation behavior of these polymers was a result of the polymer chains becoming more tightly bound to each other as the cross-link density increased with increasing network perfection. In addition, the properties of triglyceride-based polymers can be accurately predicted using solely the number of acrylates per triglyceride (Table 7.3). In general, for triglycerides containing f functional groups per molecule, the relations between tensile strength σ, cross-link density ν, modulus E, and extent of reaction $X \leq 1$ are summarized as follows:

$$\begin{aligned} \sigma &\sim [E\nu]^{1/2}, \\ \sigma &\sim [fX - f_c]^{1/2}, \\ \sigma &\sim [X - X_c]^{1/2}, \\ \nu &\sim [f - f_c]^{1/2}, \end{aligned} \tag{7.18}$$

in which X_c is the gelation threshold extent of reaction and f_c is the critical number of functional groups required to cause cross-linking.

TABLE 7.3 The scaling relations between the level of acrylation and other polymer properties.

Polymer Property	Scaling Relation
Level of perfection of polymer network	$P - P_c \propto A$
Cross-link density	$v \propto A$
Modulus	$E \propto A^3$
Tensile strength	$\sigma \propto [Ev]^{1/2}$

7.7 GLASS TRANSITION TEMPERATURE VERSUS STRUCTURE

The glass transition temperature increased linearly as the level of acrylation increased for the triglyceride-based polymers with and without styrene (Figure 7.9). The T_g value was not dependent on the oil type, in that different oils acrylated to the same level had similar T_g values (Figure 7.9). Furthermore, the T_g values of maximally acrylated oils and partially acrylated oils with the same level of functionalization were similar. Therefore, T_g only depended on the acrylation level and comonomer content; and T_g was related to the acrylation level, A, for polymers without styrene,

$$T_g = 16.1\,A - 74.7, \quad [^\circ\mathrm{C}] \tag{7.19}$$

and for polymers with styrene,

$$T_g = 27.5 \cdot A - 65.0, \quad [^\circ\mathrm{C}] \tag{7.20}$$

Note that T_g increased with the level of acrylation because the cross-link density of the oils increased with the level of acrylation. The glass transition temperature was significantly higher for the samples with styrene because the aromatic nature of styrene imparted rigidity to the network. The T_g values of these polymers with styrene ranged from $-50\,^\circ$to $92\,^\circ$C over 0.6–5.8 acrylates per triglyceride. Without styrene, T_g ranged from $-50\,^\circ$ to $18\,^\circ$C over 1.7–5.8 acrylates per triglyceride. Therefore, the level of acrylation can be varied to make very soft through very rigid materials. It was found that T_g increased linearly through 5.8 acrylates per triglyceride. However, it would be expected that T_g would not increase higher than the T_g of the copolymer. We would expect T_g to level off at $\sim 100\,^\circ$C after some level of functionalization, but we have not seen the point where this occurs. Acrylated soybean oil and linseed oil that have been maleinized, bringing their total functionality to greater than 6, had $T_g > 100\,^\circ$C, showing that further increases in the functionality continue to increase T_g [40]. The T_g was higher than the T_g of styrene because of unreacted maleic anhydride and because the attached maleate groups affected the T_g of these systems [40].

Rather than using the level of acrylation to predict T_g, the cross-link density can be used. The value of T_g increased linearly as a function of the cross-link density for polymers with no comonomer [Figure 7.12(a)]:

$$T_g[\mathrm{K}] = 0.0055\left[\frac{\mathrm{K} \times \mathrm{m}^3}{\mathrm{mol}}\right] \cdot v + 225[\mathrm{K}]. \tag{7.21}$$

The y-intercept, 225 K, was the theoretical glass transition temperature of the triglyceride-based polymers with no cross-linking, $T_g{}^u$. This value closely matched the T_g of linear polymers made from fatty acids [10]. The value of $T_g{}^s$ increased at a decreasing rate as the cross-link density increased for

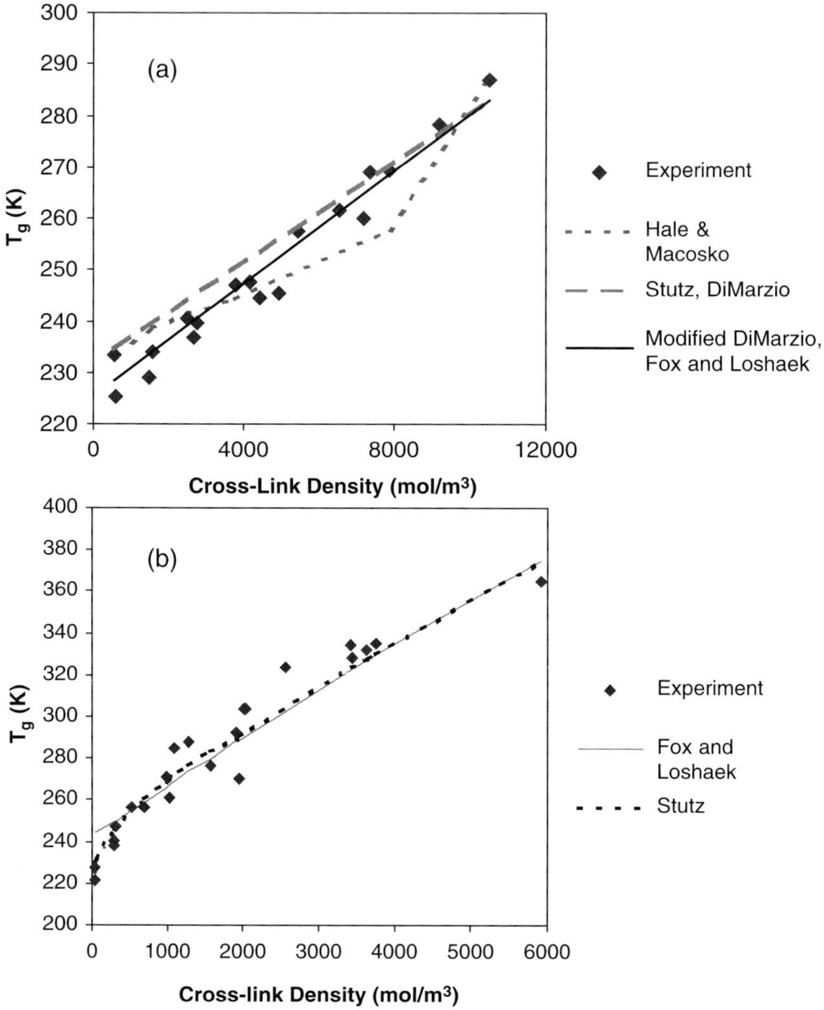

FIGURE 7.12 Glass transition temperature T_g of acrylated triglyceride-based polymers of the cross-link density as determined by experiment and various models for polymers (a) with no comonomer and (b) with comonomer (85 mol%).

samples with 85 mol% styrene [Figure 7.12(b)]. The superscript *s* indicates that these properties were for triglyceride-based polymers with styrene. The T_g^s values deviated from linearity because triglycerides with low functionality were able to polymerize with styrene, whereas they could not polymerize without styrene. As a result, $T_g^{u,s}$ (actual) = 225 K was lower than expected for these polymers. However, a linear relationship did an adequate job of fitting T_g at all cross-link densities and did a very good job of fitting at mid- to high cross-link densities:

$$T_g{}^s[\mathrm{K}] = 0.024\left[\frac{\mathrm{K} \times \mathrm{m}^3}{\mathrm{mol}}\right] \cdot v + 243[\mathrm{K}]. \tag{7.22}$$

Using this relationship, $T_g{}^{u,s}(\text{linear}) = 243\,\mathrm{K}; T_g{}^{u,s}$ (linear) was used to model T_g^s, except when noted.

7.7.1 MODELS FOR T_g VERSUS STRUCTURE

Various models predict T_g based on the cross-link density. Fox and Loshaek [41] derived an expression relating the T_g to the cross-link density by assuming the specific volume of the polymer at the glass transition to be a linear function of T_g:

$$T_g = T_g{}^u + \frac{K_n}{M_c} \quad [\mathrm{K}], \tag{7.23}$$

where K_n is a material constant. DiMarzio [42] derived an expression to describe the change in T_g caused by the formation of cross-links:

$$T_g = \frac{T_g{}^u}{(1 - K_2X)} \quad [\mathrm{K}], \tag{7.24}$$

where K_2 is a material constant and X is a measure of the cross-link density, defined as the moles of chains between cross-links divided by the total number of moles of segments. A modified version of DiMarzio's model was used, where the cross-link density, v, was used in place of X:

$$T_g = \frac{T_g{}^u}{(1 - K_2v)} \quad [\mathrm{K}]. \tag{7.25}$$

Stutz *et al.* [43] derived a model similar to that of DiMarzio, but with a slightly different form:

$$T_g = T_g{}^u\left[1 + K_2\frac{X}{1 + X}\right] \quad [\mathrm{K}]. \tag{7.26}$$

Hale and Macosko [44] described T_g of cross-linked networks taking into account the effects of non-Gaussian cross-link behavior and steric effects on the chain configurations:

$$T_g = \frac{T_g{}^u}{\left(1 - \frac{K_2X}{1-\psi X^2}\right)} \quad [\mathrm{K}], \tag{7.27}$$

where ψ is a constant that lumps together the effects of these nonidealities.

Figure 7.12(a) shows that a number of these models do a good job of predicting T_g as a function of the cross-link density for homopolymerized triglycerides. The DiMarzio and Stutz models [Eqs. (7.24) and (7.26)] underestimated the effects of cross-link density on T_g. The Hale and Macosko model [Eq. (7.27)] did not capture the linear dependency of T_g on the cross-link density. The concave curvature induced by the parameter ψ caused the deviation from linearity, and made the model inapplicable for triglyceride-based polymers with and without styrene. On the other hand, the Fox and Loshaek model [Eq. (7.23)] and the modified DiMarzio model [Eq. (7.25)] accurately predicted T_g for all cross-link densities and predicted a linear dependency that matched that of Eq. (7.21). For the Fox and Loshaek model, the parameter $K_n = 5.2 \times 10^3$ K · g/mol was used to fit the model to the experimental results. The value of $K_2 = 2.0 \times 10^{-5}$ m^3/mol was used to fit the modified DiMarzio model to the experimental results.

The DiMarzio and modified DiMarzio models predicted that $T_g{}^s$ versus cross-link density would be concave for triglycerides copolymerized with styrene, which did not follow the trends of the experimental data. The Fox and Loshaek model predicted a linear fit [Figure 7.12(b)] that matched Eq. (7.23) when $K_n = 2.5 \times 10^4$ K · g/mol. Figure 7.12(b) shows that the Stutz model did an excellent job of predicting the $T_g{}^s$ as a function of cross-link density. This model captured the nonlinear behavior at low cross-link densities. Note that $T_g{}^{u,s} = 225$ K was used because this was the $T_g{}^s$ of the uncross-linked polymers, while the relationship in Eq. (7.23) overestimated $T_g{}^{u,s}$ because it was a linear relationship. The parameter $K_2 = 18.7$ yielded the best fit of this model.

7.7.2 PERCOLATION THEORY OF T_g IN POLYMER THIN FILMS AND BULK

The glass transition temperature of thin films is an important issue for nanomaterials and thin-film coating processes, such as in many electronic materials. A significant number of papers have been published in this field dealing with the dynamics of heterogeneous media near T_g, confinement effects, surface effects, measurement methodology, and thermal and mechanical properties. We have treated this problem as a finite size vector percolation problem [34, 45]. This analysis is similar to that of the scalar percolation problem of microbial invasion of starch-filled polymers as a function of film thickness, discussed by R. P. Wool *et al.* [46–48]. The percolation threshold is reduced by the thickness of the film due to finite size clusters spanning the film.

The percolation threshold p_c for a finite-sized object is defined as the minimum concentration p (of the percolating medium) at which the contact of the bottom surface to the top surface is established. The percolation threshold p_c is different for lattices of different geometry. For 2*d* site percolation, the threshold p_c is 59.27%, whereas p_c is 31.17% for a cubic lattice. Peanasky *et al.* [48] applied the percolation theory to analyze the degradation of starch in polymer composites, where only the starch was degradable. The fraction of starch accessed as a function of volume fraction $A(p)$ is determined by the usual percolation relation:

$$A(p) \sim [p - p_c]^{\nu} \quad [p > p_c], \tag{7.28}$$

where p_c is the critical percolation concentration or percolation threshold, and ν is the critical exponent. The exponents for three-dimensional systems were computed by Peanasky *et al.* [48] using $\nu = 0.41$ and $p_c = 0.31$. Equation (7.28) describes $A(p)$ in the vicinity of p_c for an infinitely large sample, such that when $p < p_c$, $A(p) = 0$. However, for real polymer materials, especially films, $A(p)$ is not zero when $p < p_c$ since clusters of starch can be accessed from the surface. In this case, we have shown that the accessed fraction f can be described by the following relation [46]:

$$f = S(b/h)[1 - p/p_c]^{-\alpha} \quad (p < p_c), \tag{7.29}$$

where b is the particle diameter, h is the film thickness, S is the number of free surfaces ($S = 3$, 2, 1, or 0), and α is determined by

$$\alpha = \nu(D - d + 1), \tag{7.30}$$

in which D is the fractal dimension of the clusters, d is the dimension of the sample (typically $d = 2$ or 3), and ν is the cluster correlation exponent, which gives the average size of the cluster as $\xi = b|p - p_c|^{-\nu}$.

In $3d$, $\nu = 0.8$, $D \approx 2.5$, and $\alpha \approx 0.4$; in $2d$, $\nu = 4/3$, $D \sim 7/4$, and $\alpha \approx 1$. The S-factor of 2 in Eq. (7.29) refers to the two exposed surfaces of the thin material and this becomes unity if only one surface is exposed to degradation or the film is adhered to a substrate. In Figure 7.13, the surface fraction accessed is determined using $d = 2$, $p = 0.58$, $p_c = 0.5927$, $b = 1$, $h = 512$ (lattice size), $D = 7/4$, and $\alpha = 1$, such that Eq. (7.29) predicts that $f \approx 18\%$, which is in close agreement with the computer simulations (17%) at $p < p_c$. The cluster size correlation length in this case was $\xi = 79$ and complete removal of the starch would occur ($f = 100\%$) if the thickness h were reduced from 512 to 93.

To apply this theory to T_g, we replace the particles by a fraction p of Lindemann atoms (LA), which mechanically behave as a fraction p of holes in the lattice. The intermolecular bonding between atoms is anharmonic and

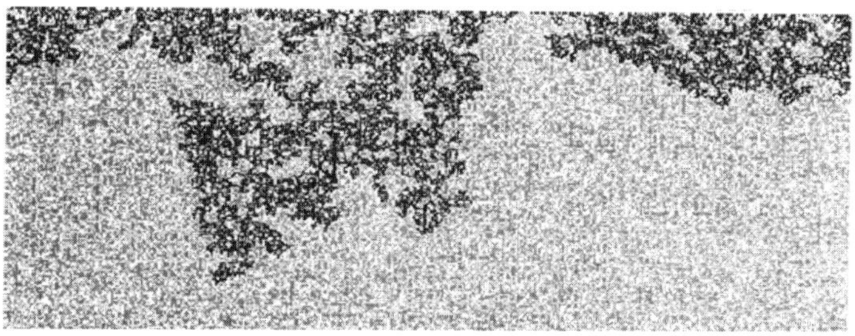

FIGURE 7.13 Surface invasion of microbes in a thin film with a starch fraction of $p = 0.58$. The accessed (dark) fraction $f = 17\%$.

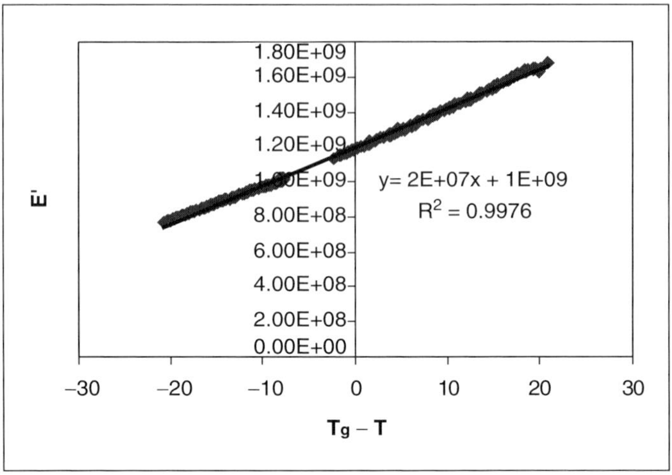

(a)

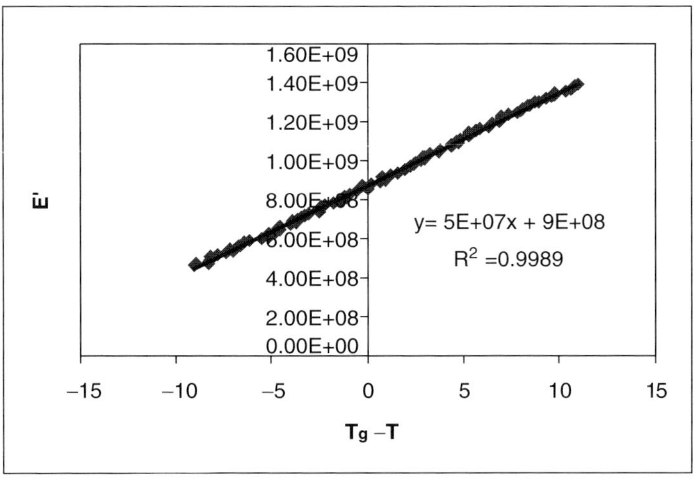

(b)

FIGURE 7.14 (a) Modulus E' versus $[T_g - T]^v$ for AELO ($A = 5.77$). The straight line through the data with an exponent $v = 1$ is fitted by the equation $E' = 0.02[T_g - T] + 1\,\text{GPa}$. (b) Modulus E' versus $[T_g - T]^v$ for olive oil ($A = 2.6$). The slope through the data is a best fit with $v = 1$ and is given by $E' = 0.05[T_g - T] + 0.9\,\text{GPa}$.

an atom no longer transmits rigidity when it has thermally expanded beyond a critical distance (~0.22), which is related to the position of the first derivative (force) maximum in the intermolecular potential energy function. Around 1910, Lindemann proposed this as a mechanism for melting due to the onset of vibrational instability in the lattice with sufficient LA atoms.

This concept was later expanded on by Born in 1939 as the *shear rigidity catastrophe* theory. We have elaborated further on the Born criterion using finite size vector percolation theory. During thermal expansion, we assume that the number of LAs is proportional to temperature and that they are in dynamic equilibrium such that their fraction $p \sim T$, and $p_c \sim T_g{}^{\infty}$, where the latter is the T_g of the bulk glass at infinite thickness. Since the modulus $E \sim [p - p_c]^{\nu}$, where the exponent $\nu \approx 1$, the glass-to-rubber transition occurs when there are a sufficient connected clusters of LA atoms at p_c and the high glass modulus decreases toward zero: E does not actually go to zero experimentally since the rubbery modulus is finite.

Figure 7.14 shows the storage modulus E' as a function of $[T_g - T]$ for (a) linseed oil, containing 5.77 acrylic acid groups per triglyceride (no styrene), and (b) olive oil with 2.6 A-groups per triglyceride, no styrene.

The exponent ν in the relation $E \sim [T_g - T]^{\nu}$ was $\nu = 1$ in both plots with a very strong correlation coefficient, and also in all other triglyceride resin systems explored by LaScala, with and without styrene. This result was valid in a range $\pm 20\,^{\circ}\mathrm{C}$ of T_g; beyond that range, the glass and rubbery plateau set in and E' became independent of T.

When heat is applied to the thin film, as implied in Figure 7.13, the free surfaces effectively have a monolayer of liquid atoms, which enhance the connectivity of the clusters at the surface. Thermal energy invades from the surface as vibrational waves with random amplitude causing intermolecular dissociation events on the amorphous "lattice" of anharmonically bonded atoms on the polymer chains. Using Eq. (7.29), and substituting for $p/p_c = T/T_g{}^{\infty}$, the finite size percolation threshold $f(h) = p^*$, and we obtain the thickness dependence of $T_g(h)$ as follows:

$$T_g(h) = T_g{}^{\infty}[1 - (B/h)^{\gamma}], \tag{7.31}$$

in which

$$B = Sb/p^*, \tag{7.32}$$

$$\gamma = 1/[\nu(D - d + 1)]. \tag{7.33}$$

The parameter B can have values of 0, 1, 2, or 3. For two free surfaces, $S = 2$ and the value of $B \approx 0.77$ is determined using $b = 0.154$ nm for a C–C bond, and a percolation threshold $p^* = 0.4$. For one free surface, for example, a thin film deposited on a neutral substrate, $S = 1$ and $B = 0.4$; for a thin film in contact with two neutral surfaces, $S = 0$ and $B = 0$, such that the thin film properties are the same as the bulk; for $S = 3$, for example, with 3d nanoparticles of volume $V \sim h^3$, then $B \approx 1.16$, which shows the greatest effect of T_g reduction with h. For strongly adsorbing thin films, the mobility of the surface layer is suppressed and T_g will actually increase relative to the bulk value. Thin films with one side free and the other side strongly adsorbed could provide some interesting local mobility battles. The value of γ is determined by the vector percolation values of ν and D and is of order unity. To explore the utility of this relation, we examine $T_g(h)$ data obtained by several investigators.

Studies by Tsui and Zhang [49] of the effects of molecular weight M, above and below the critical entanglement molecular weight M_c, on the glass transition temperature T_g of thin polystyrene films cast doubt on the surface mobile layer theory [49]. Jones and coworkers [50] proposed that a highly mobile surface layer of thickness B exists at temperatures much less than T_g and grows with increasing temperature until it spans the film of thickness h giving $T_g(h)$ as follows:

$$T_g(h) = T_g{}^{\infty}[1 - (B/h)^{\gamma}], \tag{7.34}$$

where T_g^{∞} is the T_g of the infinitely thick layer and γ is an exponent. Tsui and Zhang [49] determined T_g for thin PS films of monodisperse molecular weights 13.7 and 550 kDa. The critical entanglement molecular weight for polystyrene is about 30 kDA. They find that there is relatively no effect of entanglements on T_g versus h behavior. However, Eq. (7.34) gave an excellent fit to their $T_g(h)$ data and they obtained the values for B and γ that are shown in Table 7.4.

While their parameters differed with Jones *et al.*, both groups of investigators observed a considerable effect of film thickness on T_g such that $T_g(10\,\text{nm})/T_g{}^{\infty} = 0.96$, or about a 12 K decrease in T_g. Their values of $B \approx 0.8$ agree with our prediction of $B = 2b/p^*$. We obtain the value of $\gamma \approx 1.4$ using $\gamma = 1/[\nu(D-2)]$, when $D = 2.5$ and $\nu = 1.4$, or using $\nu = 0.82$ such that $D = 2.9$. The scalar percolation values of $D = 2.5$ and $\nu = 0.82$ predict $\gamma = 2.43$, which is not expected to agree with experiment. We expect that the ν-value for vector percolation will be greater than that for scalar percolation. If we retain the fractal dimension $D = 2.5$, then a value of $\nu = 1.42$ gives exact agreement with their value of $\gamma = 1.4$.

Tsui and Zhang obtained results for T_g versus M in terms of the Flory–Fox equation [51]:

TABLE 7.4 Parameters obtained from fitting experimental data, $T_g(h) = T_g{}^{\infty}[1 - (B/h)^{\gamma}]$.

Tsui and Zhang [49]	B (nm)	γ
$M_w = 13{,}7000$	0.87 ± 0.06	1.44 ± 0.04
$M_w = 550{,}000$	0.78 ± 0.13	1.35 ± 0.11
Jones *et al.* [50]	3.2 ± 0.6	1.8 ± 0.2
R. P. Wool [34]	0.8	2.43 – 1.4

$$T_g(M) = T_\infty[1 - m_o/M]. \quad (7.35)$$

They found that the constant m_o was not independent of film thickness. In this relation, $T_\infty = 373\,\text{K}$ is the asymptotic T_g value at high molecular weight and $m_o = 455.8$ for polystyrene. The constant m_o was associated with the mass density difference between a chain end and that of the chain segment. Tsui *et al.* [49] find that m_o for the thin films is about 40% less than the bulk value, and that results with thin film blends of high and low molecular weight suggest chain-end segregation at the surfaces. They conclude that if the proposed independence between surface chain ends and thin-film T_g were correct, the observed reduction in the T_g of the polymer films could not be due to a surface rubbery layer. The percolation theory of T_g suggests that if chain-end segregation occurs at the surface, it is a redundant substitution of the liquid monolayer already existing at the free surface. Thus, the T_g behavior in thin films with chain-end segregation will be essentially that of a higher-molecular-weight polymer, departing substantially from the Flory–Fox theory.

The surface rubbery layer concept controversy in thick films is interesting, and this percolation theory suggests that for free surfaces with $S = 1$, it exists, but there is a gradient of $p(x)$ near the surface, where $x < \xi$ as implied in Figure 7.13 and, hence, a gradient in both T_g and modulus E. If the gradient of p is given by $p(x) = (1 - x/\xi)$, then the value of X_c for which the gradient percolation threshold p_c occurs, and which defines the thickness of the surface mobile layer, is given by the percolation theory as

$$X_c = b(1 - p_c)/\{p_c^{\nu}[1 - T/T_g]^{\nu}\}. \quad (7.36)$$

For example, if $T = (T_g - 10)$, $b = 0.154\,\text{nm}$, $p_c = 0.4$, and $\nu = 0.82$, then the thickness of the mobile layer $X^* = 3.8\,\text{nm}$. During welding of polystyrene interfaces at 10–15 °C below T_g (summarized in [46]), we always noted from neutron reflection and DSIMS studies that there was an immediate interdiffusion distance of about 3 nm, which can be due to the surface mobile layer. After this initial rapid interdiffusion, diffusion ceased, consistent with the presence of an initial mobile layer that had lost its surface mobility by forming the polymer–polymer interface and, hence, the bulk polymer at $T < T_g$.

The bulk T_g effects involving chain ends, degree of cross-linking, and so on using the percolation theory essentially mirror the free volume theories of T_g, since p is proportional to free volume. However, the percolation theory offers some interesting new insights into the fractal nature of T_g, physical aging effects, the role of plasticizers and molecular weight, etc. It also provides some of the empirical constants. For example, the effect of molecular weight is given by

$$T_g(M) = T_g^{\infty}[1 - 2M_o/p_c M], \quad (7.37)$$

where M_o is the monomer molecular weight. This is identical to the Flory–Fox equation, where the constant $m_o = 2M_o/p_c$ for randomly distributed chains ends. For polystyrene, if we let $p_c = 0.4$, $M_o = 104$ Da, then $m_o \approx 500$ Da, as noted by Flory and Fox. If a fraction f of chain ends is segregated to the surface from the body of the thin film, then the constant will change from m_o to fm_o and will be less than the bulk, as noted by Tsui and Zhang [49].

The prediction for the dependence of cross-link density ν on T_g is given by the percolation theory as the linear relation,

$$T_g(\nu) = T_g^{\,o} + (T_g^{\,o} M_{ox}/p_c\rho)\nu, \tag{7.38}$$

where ρ is the density, M_{ox} is the molecular weight per backbone atom of the cross-linking chain structure, and $T_g^{\,o}$ is the T_g of the linear polymer extrapolated to $\nu = 0$. This relation is in excellent qualitative and quantitative accord with the T_g versus ν data in this chapter and the Fox–Loshaek (FL) theory. The slope in the FL theory is given by $(T_g^{\,o} M_{ox}/p_c\rho)$; see Eqs. (7.21) and (7.22).

7.7.3 CONCLUSIONS ON THE GLASS TRANSITION TEMPERATURE

Overall, the FL and the rigidity percolation models did the best job of predicting T_g for polymers with and without comonomer, especially at mid- to high cross-link densities. However, the Stutz model did a better job of predicting T_g at low cross-link densities because it predicted a more rapid drop in T_g with decreasing cross-link density. These results show that T_g values for triglyceride-based polymers were predictable based on their cross-link densities. The cross-link density was primarily a function of the acrylation level. The distribution of functional groups did have an effect on the cross-link density, which subsequently affected T_g. The percentage of fatty acid arms not attached into the polymer network did not correlate with the T_g. In fact, samples with high contents of unfunctionalized fatty acids, such as partially acrylated linseed oil, had similar T_g relative to samples with low contents of unfunctionalized fatty acids (Figure 7.9), such as maximally acrylated oils (Chapter 6). The vector percolation theory of T_g was found to be a useful theory to explain many phenomena of thin films and bulk T_g effects.

7.8 RHEOLOGY OF TRIGLYCERIDE RESINS

The processing of chemically modified triglycerides is important to the production of numerous products, including liquid molding resins [1]. Unmodified plant oils are Newtonian fluids with low viscosities (~50 cP), but when they are chemically modified (e.g., epoxidized or acrylated), the viscosities of these oils can increase significantly. Therefore, knowledge of the rheology of chemically modified triglycerides would aid in their processing and reaction to form high-performance polymeric materials.

The composites industry has established some guidelines for liquid molding resins [52]. Possibly the most stringent requirement is the resin's viscosity, which must range between 200 and 1000 cP. At viscosities lower than 200 cP, air pockets can remain in the mold after injection. At viscosities greater than 1000 cP, there can be problems with fiber wetting, voids may occur in the part, and the time required for injection increases. Increased injection time is an issue for two reasons. First, the longer it takes to mold a part, the lower the production rate of the part. The cure cycle should be shorter than 1 h to have a sufficient production rate of the composite part. In addition, if a promoter is used, the resin could cure before it is fully injected in the mold and has time to infuse between the fibers.

Triglycerides can be modified to contain various functional groups at different levels of functionality (Chapter 3). As a result of chemical modification, the polarities of triglyceride molecules change [53]. Differences in molecular polarity can have a very large effect on the rheology [53–55]. Furthermore, during chemical modification, unwanted side reactions, such as the etherification reaction, increase the molecular weight of the modified triglycerides [56–58]. Etherification can lead to gelation and thus can have a very large impact on the rheology [54, 56–58]. Changes in both the polarity and molecular weight can affect the zero-shear viscosity, the viscosity's temperature dependence, and can potentially induce shear-thinning behavior. Because we are concerned with making triglyceride-based polymers using liquid molding techniques, the effect of comonomer content on the rheology of these systems is also very important. In this section, we examine the rheological effect of epoxidizing and acrylating a series of different oils and model triglycerides with well-defined fatty acid distributions in the hope of gaining a fundamental understanding of the rheology of triglyceride-based systems.

7.8.1 CHEMICALLY MODIFIED PLANT OILS

Epoxidized, acrylated, hydroxylated, and maleinized oils and model triglycerides were prepared using the procedures described in Chapter 4. The level of functionality of the oils was measured using ^{1}H NMR. Vikoflex 7190 (Elf Atochem) (epoxidized linseed oil) and Drapex 6.8 (Witco) (epoxidized soybean oil) were used in addition to our synthetically prepared oils, and their rheology was compared. There were no significant differences between our chemically prepared epoxidized oils and the commercially available epoxidized oils. The level of functionality was varied by using different starting oils with different levels of unsaturation: olive oil, HOSO, triolein, cottonseed oil, canola oil, corn oil, soybean oil, safflower seed oil, and linseed oil. In addition, the levels of acrylation, epoxidation, and maleinization were varied by functionalizing given oils to different extents. Samples of epoxidized soybean oil that had 1.1, 2.4, 3.3, and 4.4 epoxides per triglyceride were prepared using different formic acid and peroxide concentrations. Samples of maleinized soybean oil with 2.1, 2.5, 2.9, 3.4, and 3.7 maleate groups per triglyceride were used.

7.8.2 SIZE EXCLUSION CHROMATOGRAPHY OF EPOXIDIZED AND ACRYLATED TRIGLYCERIDES

Oligomerization of triglycerides can occur during the epoxidation and acrylation reaction because of epoxy homopolymerization. Size exclusion chromatography (SEC) was used to quantify the amount of triglyceride monomers and oligomers for epoxidized and acrylated oils [37, 59]. Figure 7.15 shows that monomers, high-molecular-weight oligomers, and polymers can be detected using SEC.

7.8.3 RESIN PREPARATION

Low-molecular-weight comonomers, such as styrene, can be used to modify the viscosity of triglyceride-based resins. The rheological effects of

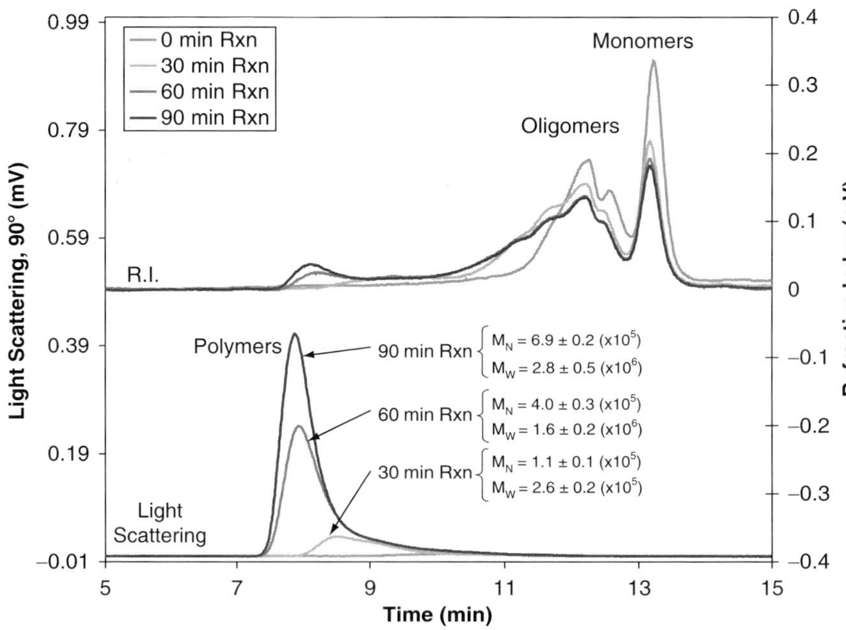

FIGURE 7.15 The SEC spectra of a highly oligomerized oil as a function of reaction time.

four different comonomers were studied: styrene, methyl methacrylate, ethyl acrylate, and butyl methacrylate. These comonomers were added to acrylated soybean oil (4.2 acrylates per triglyceride) and acrylated linseed oil (5.8 acrylates per triglyceride) in different amounts ranging from 0% to 100% comonomer. Their rheological properties are explored in the next subsection.

7.8.4 RHEOLOGY OF UNMODIFIED AND CHEMICALLY MODIFIED PLANT OILS

Steady-state shear experiments were used to determine the zero-shear viscosity, time-dependent viscosity, and shear-thinning behavior of chemically modified and unmodified oils, model triglycerides, and fatty acid methyl esters. The rheological effect of comonomers was measured by testing mixtures of the acrylated oils and one of the comonomers: styrene, methyl methacrylate, ethyl methacrylate, and butyl methacrylate. These mixtures were tested at $10\,s^{-1}$ and 25 °C. Steady-state shear experiments were performed on selected samples. The effect of temperature on the viscosity of oils was measured using a temperature step experiment on three maximally acrylated samples: acrylated high oleic soybean oil (HOSO), acrylated soybean oil, and acrylated linseed oil, and two partially acrylated linseed samples with 2.2 and 3.5 acrylates per triglyceride. The temperature was increased from 20° to 80°C and then decreased to 20 °C in increments of 2.5 °C. The literature values for the comonomer viscosities were used [60, 61]. The viscosities of styrene, methyl methacrylate, ethyl methacrylate, and butyl methacrylate are 7.0×10^{-4}, 6.0×10^{-4}, 6.5×10^{-4}, and 7.3×10^{-4} Pa · s, respectively [60, 61]. The viscosities of maximally hydroxylated oils were measured using a Brookfield II+ viscometer. The viscosities of the hydroxylated oils were measured at 50 °C because their viscosities were too high to be measured at room temperature with this viscometer. The viscosities were measured at shear rates ranging from 0.5 to $17\,s^{-1}$. The viscosities of bulk maleinized resins were measured with 33 wt% styrene.

7.9 RESULTS AND DISCUSSION

7.9.1 SEC CHARACTERIZATION

The light scattering detector did not record any significant peaks during the SEC characterization of the modified and unmodified oil samples, but the refractive index detector did (Figure 7.16). Therefore, there are no polymeric molecules within these samples. The refractive index detector observed a single peak for the unmodified samples (Figure 7.17). This peak eluted at 13.30 mL and was due to individual triglyceride molecules. No oligomers were observed in the unmodified samples. The refractive index detector observed three peaks for the epoxidized samples, 12.3, 12.65, and 13.28 mL, and two peaks for the acrylated

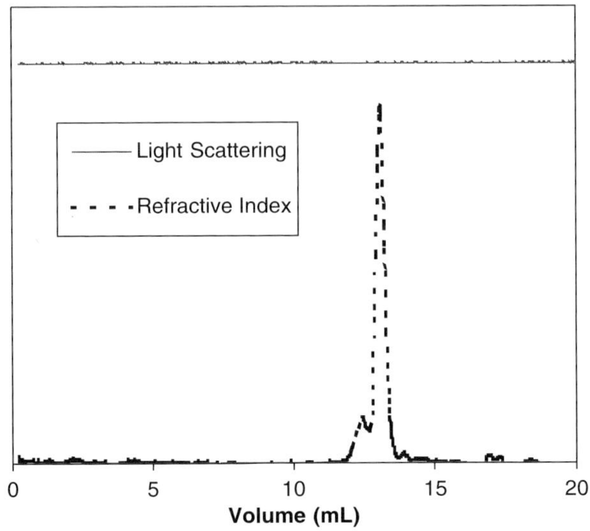

FIGURE 7.16 The SEC light scattering and refractive index spectra for acrylated safflower oil. The light scattering detector did not record any peaks while the refractive index detector did.

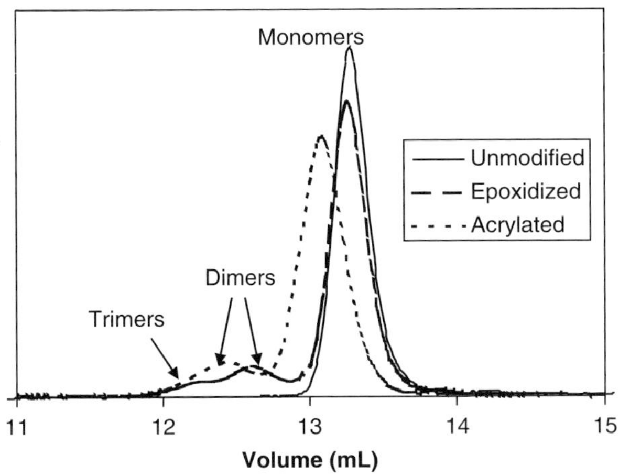

FIGURE 7.17 The SEC refractive index spectra of unmodified, epoxidized, and acrylated safflower seed oil. The SEC spectra of other oils were very similar.

samples, 12.45 and 13.10 mL (Figure 7.17). The peak eluted at 12.3 mL represents trimers. This peak was only observed in the epoxidized high oleic soybean oil and epoxidized safflower seed oil samples.

The 12.45- to 12.65-mL peak represents dimers, and the last peak eluted (Figure 7.17) represents monomers. The ether extraction process probably removed the trimers from the acrylated product. The elution time of the peaks decreased as the samples went from being unmodified to epoxidized to acrylated because the molecular weight of the triglycerides increased as a result of these reactions.

The ratios of the peak areas (Figure 7.17) were used to determine the fraction of oligomerized triglycerides, O:

$$O = \frac{A_{\text{oligomers}}}{A_{\text{oligomers}} + A_{\text{monomers}}}, \tag{7.39}$$

where A_{monomers} and $A_{\text{oligomers}}$ are the areas of the monomer and oligomer peaks, respectively. The extent of oligomerization was low, except for acrylated linseed oil, and was not a function of the level of functionalization (Figure 7.18). The percentages of oligomerized triglycerides in the epoxidized and acrylated oils, not counting acrylated linseed samples, were 7.4 $\pm$ 4.3% and 9.6 $\pm$ 3.7%, respectively. Forty-five percent of the triglycerides in acrylated linseed oil were oligomerized, but lower levels of oligomerization ($\sim$15%) were found for other maximally acrylated linseed samples, but their viscosities were not measured. The extent of oligomerization was similar for the epoxidized and acrylated samples, with the exception of the linseed oil samples. This was expected because the AMC-2 catalyst was used during the acrylation step to prevent oligomerization [62]. Therefore, most of the oligomerization occurred during the epoxidation step (Figure 7.18). A significant amount of oligomerization took place during the acrylation step for linseed oil because of the high epoxide concentration and likelihood for epoxy homopolymerization.

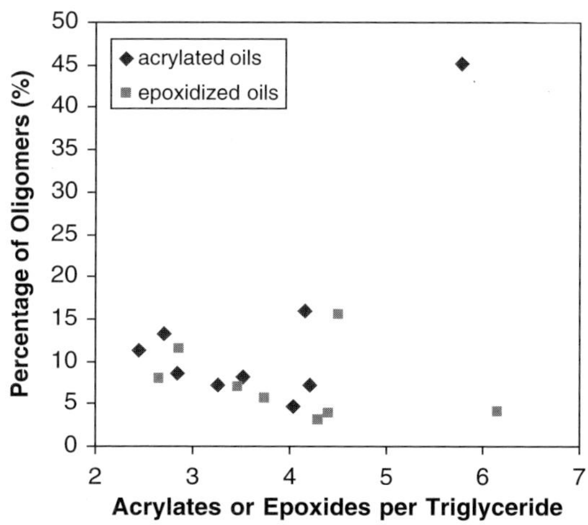

FIGURE 7.18 The percentage of oligomerized triglycerides as a function of the level of epoxidation and acrylation for maximally epoxidized and acrylated oils.

7.9.2 EFFECTS OF UNSATURATION LEVEL ON THE VISCOSITY OF UNMODIFIED TRIGLYCERIDES

The viscosity of unmodified oils decreased slightly as the level of unsaturation increased (Figure 7.19). The standard error was no more than 5% for any given sample. Equation 7.40 shows the dependence of the viscosity on the level of unsaturation U at 25 °C:

$$\eta_0 = 0.115U^{-0.60}[\text{Pa} \cdot \text{s}]. \tag{7.40}$$

The cause for this observed trend has to do with the conformation of the fatty acid chains. The *cis* character of the carbon–carbon double bonds in these oils puts kinks in the fatty acid chains that increase the average distance between fatty acid chains. Therefore, intermolecular interactions decrease as the level of unsaturation increases. Evidence of this effect is seen in the inverse relationship between melting point and level of unsaturation for triglycerides [63]. No shear thinning was observed in the range of tested shear rates.

7.9.3 EFFECTS OF EXTENT OF EPOXIDATION

Soybean oil was epoxidized to different extents and the viscosity was measured. The viscosity of epoxidized oils increased with the level of epoxidation (Figure 7.20). The error in the zero-shear viscosity was no more than 5% for any level of epoxidation. Increases in molecular weight [54, 56–58] and polarity [53–55] can cause such effects. Each epoxide group added to a triglyceride molecule increased its molecular weight by about 1.8%. This

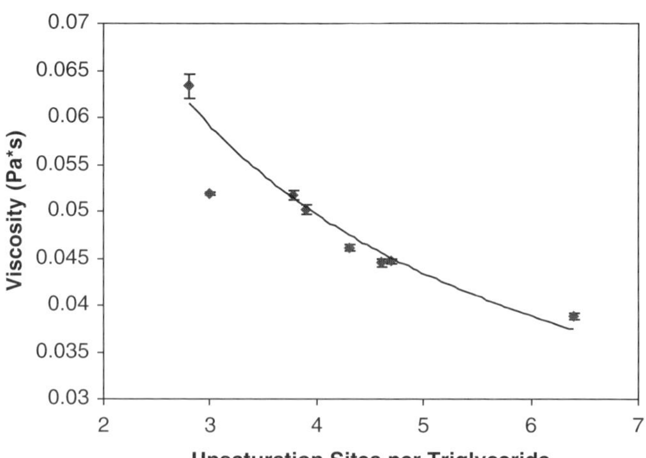

FIGURE 7.19 The viscosity of unmodified triglycerides as a function of their level of unsaturation.

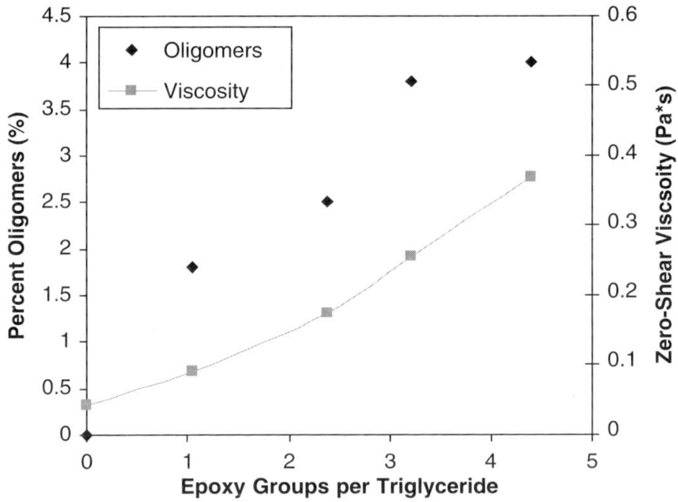

FIGURE 7.20 The percentage of oligomerized triglycerides and the zero-shear viscosity of epoxidized soybean oil as a function of the extent of epoxidation.

molecular weight change was observed in the SEC results in that the peak representing the triglyceride monomers shifted to lower elution times when the sample was epoxidized (Figure 7.17). In addition, SEC results confirmed that there was a small but significant amount of oligomerization during the epoxidation reaction (Figure 7.20).

The Rouse theory predicts that the viscosity scales with the molecular weight ($\eta_0 \sim M$), and it applies for molecules below their critical entanglement molecular weight M_c [54, 64]. Reptation theory predicts that the viscosity is a stronger function of molecular weight ($\Sigma_0 \sim M^{3.4}$), but applies to entangled polymers with a molecular weight above M_c [64]. The critical entanglement molecular weight is on the order of 10^4 g/mol for most polymers and is much higher than the molecular weight of these chemically modified triglycerides [54]; thus, the Rouse theory should apply. However, the viscosity increases predicted from the Rouse and Reptation theories were only 10% and 40%, respectively, at maximum epoxidation. These increases were considerably lower than the experimentally observed increase in viscosity (840%). Thus, the increase in molecular weight was not the main factor for the increased viscosity. In fact, failure of the Rouse and Reptation theory was expected because the chemical nature of triglycerides changes as a result of epoxidation, and these theories do not account for changes in molecular polarity. Epoxidized triglycerides are more polar than unmodified triglycerides. As a result, the polar nature of epoxide groups increased the intermolecular interactions, causing an increase in the viscosity.

7.9.4 EFFECTS OF THE LEVEL OF EPOXIDATION

The viscosity of maximally epoxidized oils increased slightly with the level of epoxidation at both 25° and 45 °C for the more highly epoxidized oils (epoxidized corn, soybean, safflower seed, and linseed oils) (Figure 7.21). The other epoxidized oils were not tested at 25 °C because they were solid at this temperature. The combination of polarity and oligomerization effects caused the viscosity to increase with the level of epoxidation. However, the viscosity of the epoxidized oils was not a function of the level of epoxidation for samples with few epoxide groups per triglyceride (Figure 7.21). The melting point of the oils decreased as the level of epoxidation increased because epoxidized fatty acids are not as able to closely pack as unmodified fatty acids. Thus, samples with fewer epoxide groups were more solid in nature. As a result, the combination of melting point effects, polarity effects, and molecular weight effects caused the viscosity to vary incoherently with the level of epoxidation. This melting point effect may have also lessened the effect of the level of epoxidation on the viscosity of the more highly epoxidized oils (i.e., epoxidized corn oil through epoxidized linseed oil). The error in the zero-shear viscosity was approximately 5% and was thus not a factor. Epoxidized oils did not exhibit any shear thinning at the measured shear rates because their molecular weight and intermolecular interactions were too small.

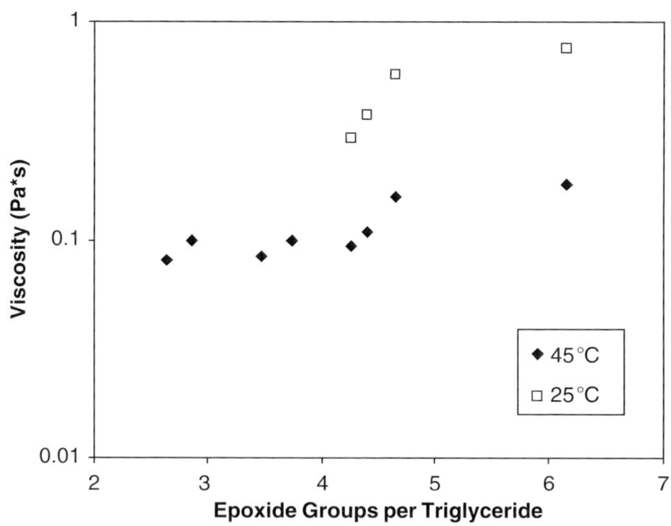

FIGURE 7.21 The viscosity of maximally epoxidized triglycerides at 45° and 25 °C as a function of the number of epoxide groups per triglyceride.

7.9.5 EFFECTS OF ACRYLATION ON OIL VISCOSITY

The extent of acrylation had a large effect on the viscosity of acrylated oils (Figure 7.22). The zero-shear viscosity of the acrylated oils increased exponentially with the level of acrylation, A, at 25 °C:

$$\eta_0 = 0.104\exp(1.34A)[\mathrm{Pa}\cdot\mathrm{s}]. \tag{7.41}$$

Acrylated triglyceride molecules contain a very polar hydroxyl group and polar ester linkage for every acrylate group added. These groups increase intermolecular interactions via hydrogen bonding and dipole–dipole interactions, and thus cause an increase in the viscosity. The dipole moment of acrylated triglycerides was calculated using the Debye equation and a group contribution method [65, 66]. The dipole moment increased linearly with the level of acrylation (Figure 7.23). Therefore, the increase in oil viscosity with the level of acrylation can be at least partially attributed to increases in molecular polarity.

Molecular weight can also have an effect on the viscosity of acrylated oils. For every acrylate that was attached to the triglyceride, its molecular weight increased by ~8%. Therefore, the molecular weight of the acrylated oils (~1300 g/mol) was up to 45% greater than the epoxidized oils (~900 g/mol) (Table 7.5).

The extent of oligomerization was fairly low (~10%) and did not change significantly with the level of acrylation. The molecular weight changes

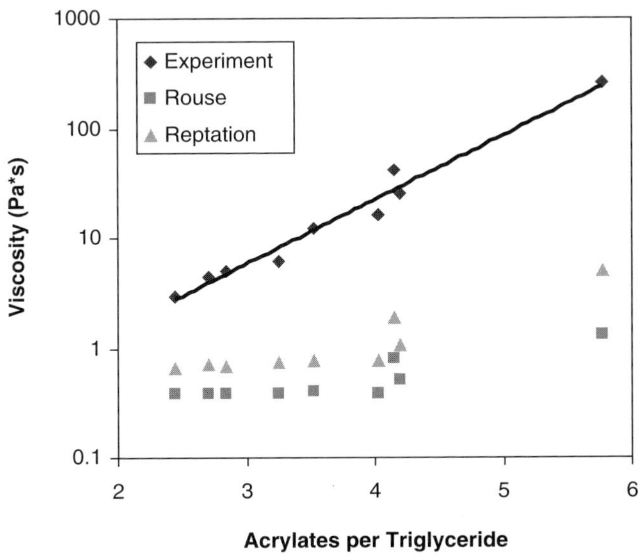

FIGURE 7.22 The viscosity as a function of the level of acrylation for maximally acrylated oils determined experimentally and as predicted by Rouse and Reptation theory.

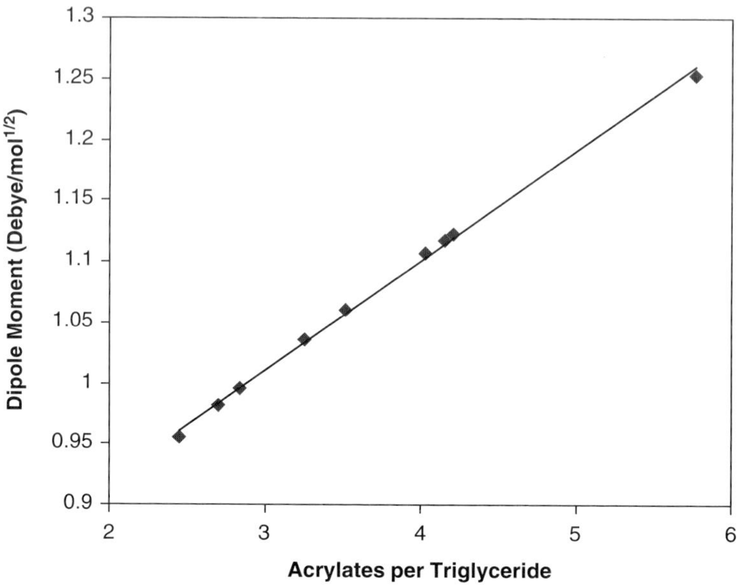

FIGURE 7.23 The dipole moment of triglycerides as a function of their level of acrylation.

TABLE 7.5 The molecular weight of maximally epoxidized and acrylated oils, factoring in oligomerization and chemical modification.

	Functional Groups per Triglyceride		Molecular Weight of Chemically Modified Oils (g/mol)	
Oil	Epoxides	Acrylates	Epoxidized	Acrylated
Olive	2.7	2.5	950	1150
HOSO	2.8	2.7	970	1190
Triolein	2.9	2.8	970	1190
Cottonseed	3.5	3.3	950	1190
Canola	3.7	3.5	970	1240
Corn	4.3	4.0	960	1260
Soybean	4.4	4.2	960	1290
Safflower seed	4.5	4.2	1020	1350
Linseed	6.2	5.8	990	1700

associated with the addition of acrylate groups and oligomerization could not cause the observed large increases in viscosity with the level of acrylation. Rouse and Reptation theories severely underpredicted the increase in viscosity (Figure 7.22). For example, Rouse theory predicted that the viscosity of maximally acrylated linseed oil (5.8 acrylate groups per triglyceride) would double relative to epoxidized linseed oil, and Reptation theory predicted the viscosity would increase by 1 order of magnitude. Yet, experiment showed that the viscosity increased by 3 orders of magnitude as a result of acrylation. The failure of the Rouse and Reptation models indicates that polarity effects caused the observed viscosity increase with acrylation level.

It is possible that the difference in polarity between the very polar functional groups and the nonpolar triglyceride backbone induced clustering of the functional groups in these oils. Neutron scattering studies would need to be performed to determine if these oils have such a microstructure. Assuming the acrylated oils did form clusters due to polar interactions, the average molecular weight of these clusters, M_{cluster}, can be calculated from the molecular weight of the epoxidized oil, M_E, the viscosity of the acrylated oil, η_A, and the viscosity of the epoxidized oil, η_E. Assuming Rouse dynamics, M_{cluster} is

$$M_{\text{cluster}} = M_E \cdot \eta_A / \eta_E. \tag{7.42}$$

Figure 7.24 shows that the molecular weight of these theoretical clusters increased exponentially with the acrylation level:

$$M_{\text{cluster}} = 10^3 \exp(A), \tag{7.43}$$

and molecular weights on the order of 10^5 g/mol were reached. The number of monomers, N_{cluster}, making up these clusters increased exponentially from as low as 2 to as many as 200 monomers per cluster:

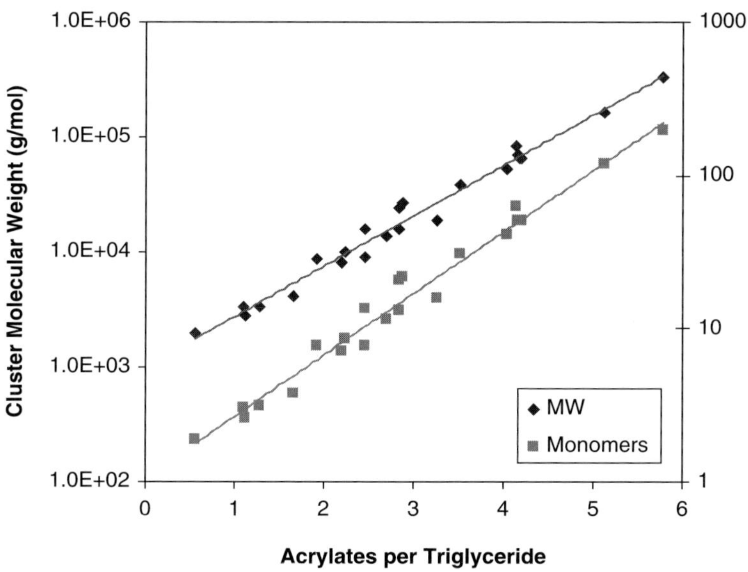

FIGURE 7.24 The molecular weight of the theoretical clusters and the number of monomers that make up the clusters for acrylated oils.

$$N_{\text{cluster}} = 1.06\exp(0.92A). \tag{7.44}$$

Cluster sizes this large should be observable using neutron scattering.

The oil type had a small effect on the viscosity of acrylated oils. The zero-shear viscosity of the oils increased exponentially with the level of acrylation (Figure 7.25), just as it did for maximally acrylated oils. Figure 7.25 also shows that the zero-shear viscosity of acrylated samples increased as the oil went from HOSO to soybean oil to linseed oil. Error bars were omitted because the error in the viscosity was only ~5% for each sample. Oligomerization did not cause this effect because the samples had similar molecular weights for the same level of acrylation (Table 7.6). Polar groups at the end of a fatty acid chain are more likely to induce intermolecular interactions, which cause an increase in viscosity, whereas polar groups nearer the glycerol center mainly increase intramolecular interactions. The distance of the polar groups (i.e., hydroxyl and acrylate groups) from the center of the glycerol center increased going from acrylated HOSO to acrylated linseed oil. HOSO contains mostly oleic acid, and the unsaturation sites are located at the 9,10-carbon atoms relative to the glycerol center [63, 67]. Linseed oil contains mostly linolenic acid where the unsaturation sites are located up to the 15,16-carbon atoms [63, 67]. Soybean oil is the intermediate of the two oils [63, 67]. Because these oils were functionalized at the unsaturation sites, the distance of the functional groups from the glycerol center increased from HOSO

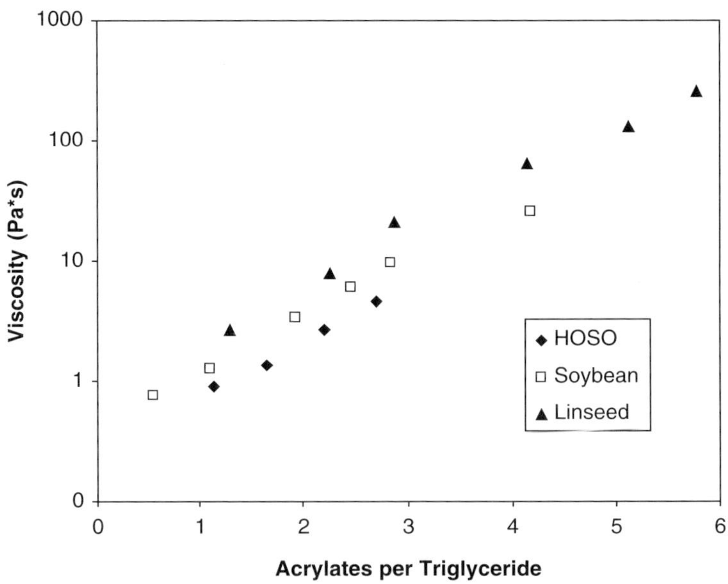

FIGURE 7.25 The zero-shear viscosity as a function of the level of acrylation for acrylated high oleic soybean oil, acrylated soybean oil, and acrylated linseed oil.

TABLE 7.6 Molecular weight of partially acrylated oils taking oligomerization into account.

Oil	Acrylate Groups per Triglyceride	Molecular Weight (g/mol)
HOSO	1.1	1070
HOSO	1.7	1100
HOSO	2.2	1140
HOSO	2.7	1190
Soybean	0.6	1020
Soybean	1.1	1060
Soybean	1.9	1110
Soybean	2.5	1150
Soybean	2.8	1180
Soybean	4.2	1290
Linseed	1.3	1090
Linseed	2.2	1170
Linseed	2.9	1210
Linseed	3.5	1260
Linseed	4.1	1310
Linseed	5.1	1390
Linseed	5.8	1700

to soybean oil to linseed oil. Therefore, linseed oil had the most functional groups at the end of its fatty acid chains, which induced intermolecular interactions and caused an increase in the viscosity, whereas HOSO had the fewest.

Shear thinning at high shear rates ($>100\ s^{-1}$) was thought to be observed for acrylated samples. The Ellis model [68] was used to fit the experimental data:

$$\eta = \frac{\eta_0}{1 + (k \cdot \gamma)^{1-n}} \tag{7.45}$$

and the fits were excellent (Figure 7.26).

However, n was equal to -0.65 for all oils, whereas the value of n is between 0 and 1 for shear thinning fluids [68]. Furthermore, decreasing the gap spacing caused the onset of shear thinning to occur at higher shear rates. As a result, it was concluded that the decrease in viscosity with increasing shear rate was due to viscous heating (Figure 7.26). Even though the temperature was controlled and monitored during the experiment, perfect uniformity and control of the temperature throughout the sample are extremely difficult to achieve at high shear rates.

The viscosity of unmodified and chemically modified triglycerides was found to be time independent. In addition, there was no evidence of shear history in the viscosity as a function of shear rate during the steady-state shear experiments. This was expected because the acrylated triglycerides were not polymeric.

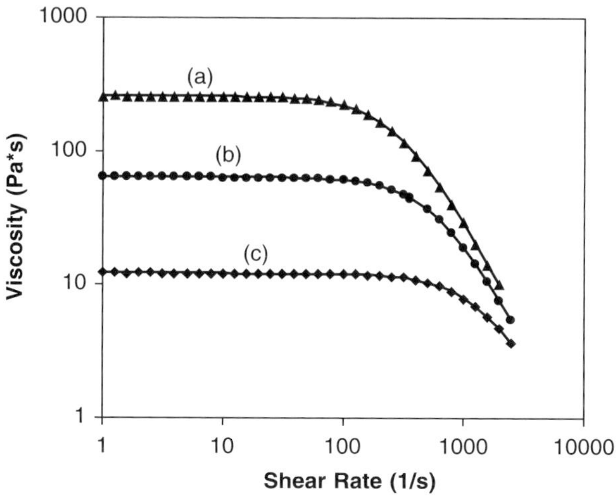

FIGURE 7.26 The viscous heating behavior determined experimentally and calculated from the Ellis model of (a) maximally acrylated linseed oil (5.8 acrylates per triglyceride), (b) partially acrylated linseed oil (4.1 acrylates per triglyceride), and (c) maximally acrylated canola oil (3.5 acrylates per triglyceride).

7.9.6 EFFECTS OF TEMPERATURE ON THE VISCOSITY OF ACRYLATED TRIGLYCERIDES

As the temperature increased, the viscosity of the resin decreased exponentially (Figure 7.27). As a result, increasing the temperature of these oils slightly can reduce their viscosities significantly, making them easier to process. The temperature dependence was accurately modeled (Figure 7.27) using an Arrhenius relationship [19]:

$$\eta = \eta_\infty \exp\left(\frac{E_\eta}{RT}\right)[\text{Pa} \cdot \text{s}], \tag{7.46}$$

where η_∞ is the pre-factor and E_η is the activation energy for viscous flow. The deviation in the viscosity was less than 5% at a given temperature. Increasing the level of acrylation caused the activation energy to increase linearly (Figure 7.28). In fact, the addition of a single acrylate group caused the activation energy to increase by 10^4 J/mol. The increase in activation energy with the level of acrylation was expected because temperature disrupts intermolecular interactions, and this has a larger effect on oils with more acrylate groups. This large change in activation energy should not be observed based on the small molecular weight increase of these oils with acrylation level. This is further evidence that polar interactions are the main cause for the increase in oil viscosity as a function of the level of acrylation. Calculating η_∞ involves extrapolating to infinite temperature. Therefore, an

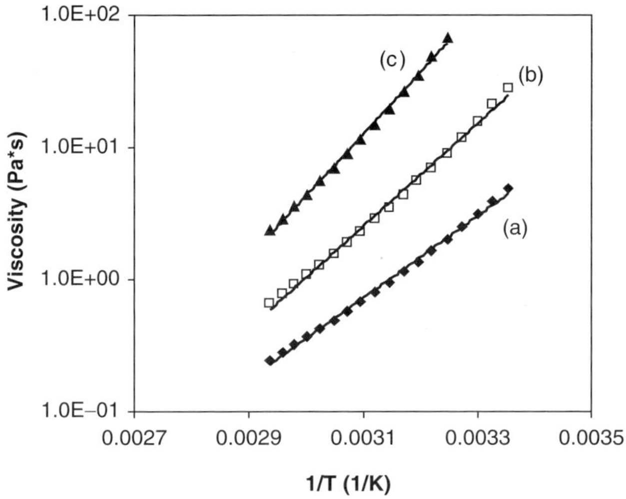

FIGURE 7.27 The experimental temperature dependence of the viscosity of (a) maximally acrylated high oleic soybean oil, (b) maximally acrylated soybean oil, and (c) maximally acrylated linseed oil compared with their Arrhenius fits.

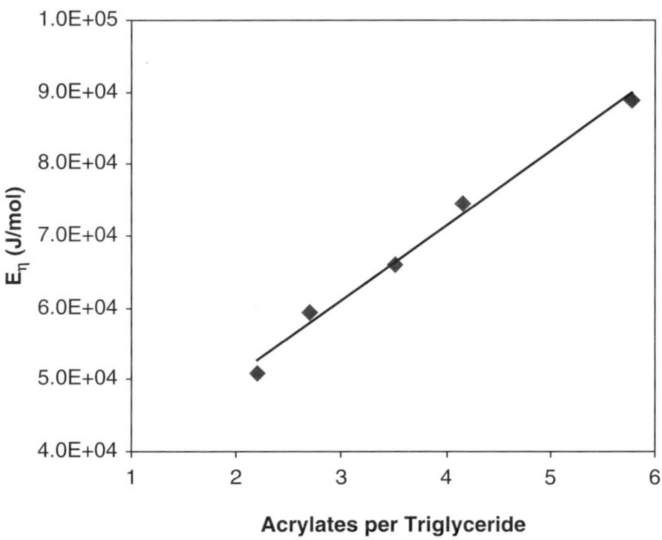

FIGURE 7.28 The activation energy for viscous flow as a function of the number of acrylate groups per triglyceride.

accurate measurement of this pre-factor is very difficult, and it requires a much more sensitive experiment than the one we performed.

7.9.7 EFFECTS OF COMONOMER ON THE VISCOSITY OF TRIGLYCERIDE-BASED RESINS

Comonomers had a very large effect on the rheological character of the acrylated triglycerides. The sample viscosity, η, was normalized based on the pure comonomer viscosity, $\eta_{comonomer}$, and pure acrylated oil viscosity, η_{oil}:

$$\eta_{normalized} = \frac{\eta}{\eta_{comonomer}\eta_{oil}}\left[\frac{1}{\mathrm{Pa\cdot s}}\right], \quad (7.47)$$

where $\eta_{normalized}$ is the normalized viscosity. As the volume fraction of comonomer, $\eta_{comonomer}$, was increased, the normalized viscosity decreased in an exponential manner (Figure 7.29). This result is very encouraging in that a small amount of comonomer can be used to make these resin systems much easier to process. Furthermore, the normalized viscosity was not a function of the oil or comonomer. Therefore, the normalized viscosity was only a function of the volume fraction of comonomer (Eq. 7.48):

$$\eta_{normalized} = 1.2 \times 10^3 \exp(-14\phi_{comonomer})\left[\frac{1}{\mathrm{Pa\cdot s}}\right]. \quad (7.48)$$

Equation (7.48) can be used to predict the viscosity of any given solution of oil and comonomer. The viscosity dependence of mixtures of vinyl esters and unsaturated polyesters with styrene [69] and the viscosity dependence of con-

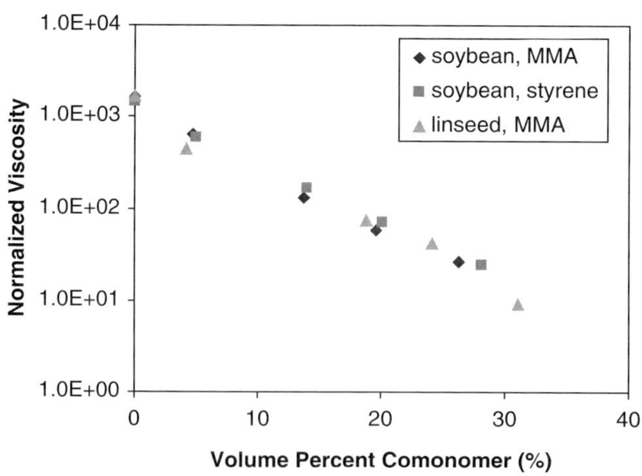

FIGURE 7.29 The normalized viscosity as a function of the volume percentage of comonomer for three samples: acrylated soybean oil with methyl methacrylate (MMA), acrylated soybean oil with styrene, and acrylated linseed oil with methyl methacrylate.

centrated polymer solutions [70] are similar to the observed results for acrylated oils. The addition of small amounts of comonomer (10 vol%) severely reduced or eliminated the viscous heating effect. The error in the viscosity was ~5% at all comonomer concentrations, as determined by repeat runs.

7.9.8 EFFECTS OF HYDROXYLATION ON OIL VISCOSITY

The viscosities of the hydroxylated oils are plotted as a function of the level of hydroxylation (Figure 7.30). The viscosity at 50 °C increased exponentially as the level of hydroxylation increased:

$$\eta_{\text{hydroxyl}} = 0.1 \exp(0.43 \cdot N_{\text{hydroxyl}}) \tag{7.49}$$

SEC was not run on hydroxylated samples, but nuclear magnetic resonance (NMR) results showed that there was only a relatively small amount of oligomerization. Therefore, the effect of hydroxylation level on the viscosity was a result of polarity effects. Figures 7.27 and 7.28 show that acrylated/hydroxylated oils had a higher dependency on the temperature as the level of functionality increased. Based on the viscosity and activation energy for acrylated oils, the viscosities of hydroxylated oils at 25 °C were estimated (Figure 7.30). The viscosity dependence on hydroxylation level was stronger at 25 °C than it was at 50 °C. Furthermore, the viscosity of hydroxylated oils was slightly higher than that of acrylated oils for a given level of total functionality, that is, the sum of acrylate and hydroxyl groups on acrylated oils (Table 7.7) and the sum of hydroxyl groups and formate esters on hydroxylated oils (Table 7.8).

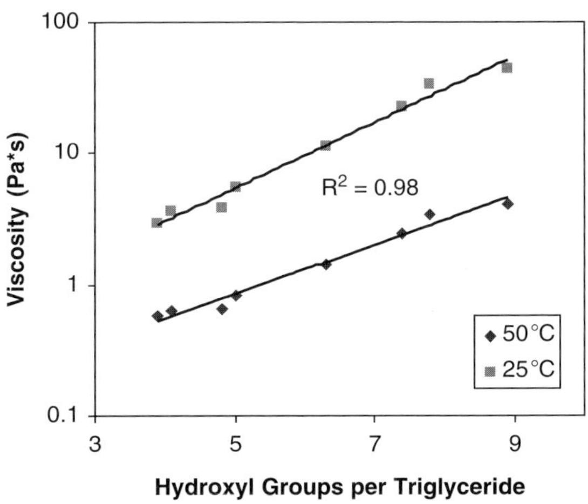

FIGURE 7.30 The viscosity of maximally hydroxylated oils as a function of the number of hydroxyl groups per triglyceride at 50 °C and as estimated at 25 °C.

TABLE 7.7 Maximum level of acrylation achievable for various oils.

Oil	Acrylates per Triglyceride	Extent of Acrylation (Based on Epoxides)	Extent of Acrylation (Based on Unsaturation Sites)
Methyl oleate	0.9	0.94	0.90
Methyl linoleate	1.8	0.94	0.90
Olive oil	2.5	0.92	0.88
HOSO	2.7	0.94	0.90
Triolein	2.8	0.97	0.95
Cottonseed oil	3.3	0.93	0.86
Canola oil	3.5	0.94	0.90
Corn oil	4.0	0.96	0.94
Soybean oil	4.2	0.95	0.91
Safflower seed oil	4.2	0.92	0.88
Trilinolein	5.7	0.94	0.95
Linseed oil	5.8	0.98	0.90

TABLE 7.8 The number of hydroxyl groups, formate esters, ether linkages, and unsaturation sites remaining on hydroxylated oils and the initial level of unsaturation.

Oil	U_I	U	N_{formate}	N_{ether}	N_{hydroxyl}	Extent of Hydroxylation
Olive	2.9	0.30	1.0	0.06	3.9	0.70
HOSO	3.0	0.30	1.2	0.06	4.1	0.69
Triolein	3.0	0.0	1.1	0.06	4.8	0.81
Cottonseed	3.8	0.36	1.7	0.11	5.0	0.66
Soybean	4.6	0.55	1.6	0.18	6.3	0.69
Safflower seed	5.1	0.45	1.6	0.20	7.4	0.73
Linseed	6.4	1.2	1.3	0.59	8.9	0.68
Conjugated linseed	6.4	1.5	1.4	0.60	7.8	0.61

7.9.9 EFFECTS OF MALEINIZATION ON OIL VISCOSITY

The effect of maleinizing soybean oil on the viscosity is shown in Figure 7.31. The viscosity increased in an exponential manner as the level of maleinization increased. The viscosities at all levels of functionalization were very high. Maleinized triglycerides have carboxylic acid and hydroxyl functionalities. Carboxylic acids are more polar than ester linkages [65, 66], and the viscosities of maleinized oils should be higher than those of acrylated oils for a given level of functionalization, as they were. The extent of oligomerization was very low, so molecular weight effects can be neglected.

The viscosities of maleinized resins with 33 wt% styrene were measured for different oils. The viscosities of these oils were not plotted as a function of the level of maleinization because these oils all had similar levels of maleinization (Table 7.9). Therefore, the polarities of these oils were similar [65, 66].

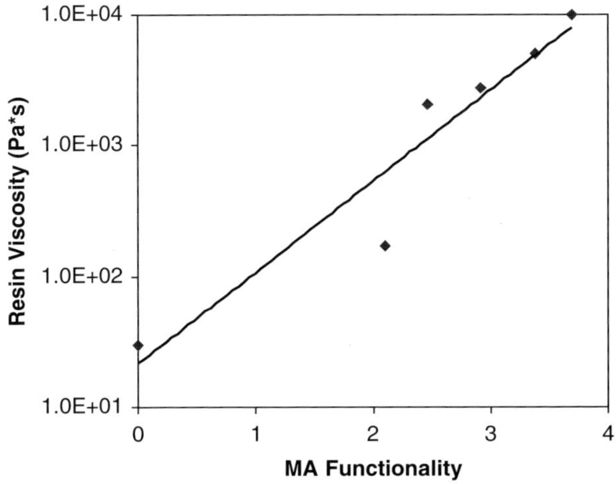

FIGURE 7.31 The viscosity of maleinized soybean oil as a function of the extent of maleinization.

TABLE 7.9 The level maleinization as a function of oil type.

Oil	Maleate Groups per Triglyceride
HOSO	2.0
Triolein	2.1
Cottonseed	2.1
Soybean	2.2
Safflower	2.3
Linseed	2.3

Figure 7.32 shows that the viscosity of the resins increased exponentially with the hydroxylation level of the oil before maleinization:

$$\eta = 0.024\exp\left[0.61 \cdot \left(N_{\mathrm{hydroxyl}}\right)_0\right][\mathrm{Pa}\cdot\mathrm{s}], \tag{7.50}$$

where $(N_{\mathrm{hydroxyl}})_0$ is the number of hydroxyl groups per triglyceride before maleinization. Some of the viscosity increase was due to increases in the hydroxylation level (Figure 7.30). Yet, the viscosity of maleinized oils alone was considerably higher than that of hydroxylated oils (Figure 7.31) because of oligomerization and polarity effects associated with the added carboxylic acid groups [65, 66]. SEC was not run on these maleinized oils, but previous work has shown that maleinized oils oligomerized to very high extents [37]. Therefore, more experiments need to be done to determine how oligomeriza-

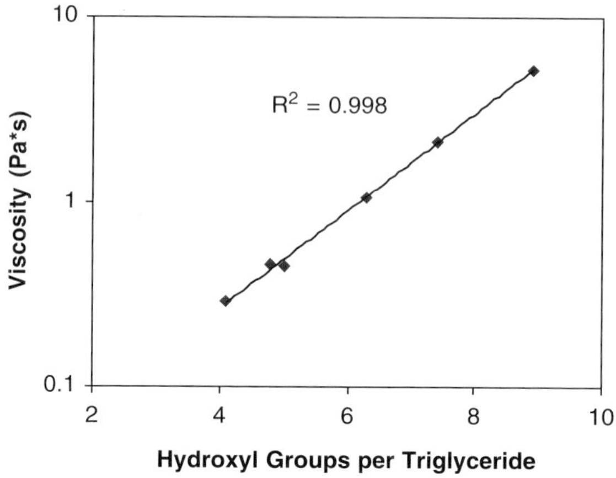

FIGURE 7.32 The viscosity of maleinized resins with 33 wt% styrene as a function of the number of hydroxyl groups per triglyceride before maleinization.

tion and polarity affect the viscosity of these maleinized resins. It is important to note that maleinized resins were so viscous that the sample with the most hydroxyl groups (linseed) had too high of a viscosity for resin transfer molding even with 33 wt% styrene.

7.10 SUMMARY OF TRIGLYCERIDE RHEOLOGY

The viscosity of triglycerides increased as a result of chemical modifications that increased the intermolecular interactions among molecules. Epoxidized triglycerides are slightly polar and had a higher viscosity than unmodified triglycerides. Acrylated triglycerides are considerably more polar and, as a result, had a significantly higher viscosity. The viscosity of acrylated oils was accurately predicted based on their level of acrylation alone. Oligomerization only occurred to a small extent in these oils and the resulting increases in molecular weight were too small to account for the observed increases in viscosity. The viscosity of hydroxylated oils also increased exponentially with the level of hydroxylation due to polarity effects. However, the viscosity of maleinized oils increased exponentially as a result of oligomerization and polarity effects. Added comonomers acted like a solvent and drastically reduced the viscosity of the triglyceride-based resin. In addition, the viscosity of a resin containing any acrylated plant oil and any comonomer was accurately predicted using only the comonomer volume fraction.

Overall, the viscosity of these oils decreased exponentially as the distance between functional groups increased, whether this occurred through the addition of a comonomer or a decrease in the functionality. As the distance between the functional groups and the glycerol center of the triglyceride increased, the viscosity of acrylated oils increased slightly because intermolecular interactions increased. Furthermore, the temperature dependence of the viscosity was accurately modeled using an Arrhenius relationship. The activation energy can be predicted based on the level of acrylation of the triglyceride. Lastly, unmodified and chemically modified triglycerides did not shear-thin and had no memory of shear history. Table 7.10 summarizes the effect of oil type and functionality on the rheological properties. Further discussion by LaScala and Wool on the rheology of chemically functionalized triglycerides can be found in reference [71].

TABLE 7.10 The effect of oil type and functionality on the rheological properties of unmodified and chemically modified oils.

Increasing Level of Functionality	Viscosity	E_η	Cause for Rheological Effect
Unsaturation	Decreased	—	Melting point
Epoxidation (same oil)	Increased	—	Polarity
Epoxidation (different oils)	Constant or increased	—	Melting point, polarity, oligomerization
Hydroxylation	Increased exponentially	Increased	Polarity
Acrylation	Increased exponentially	Increased linearly	Polarity
Maleinization	Increased exponentially	—	Polarity and oligomerization

REFERENCES

1. Wool, R. P.; Kusefoglu, S. H.; Palmese, G. R., Zhao, R.; Khot, S. N. U.S. Patent 6,121,398; 2001.
2. Nevin, C. S. U.S. Patent 3,125,592; 1964.
3. Trecker, D. J.; Borden, G. W.; Smith, O. W. U.S. Patent 3,931,075; 1976.
4. Borden, G. W.; Smith, O.W.; Trecker, D.J. U.S. Patent 3,878,077; 1975.
5. Steinberg, I. V. U.S. Patent 3,450,613; 1969.
6. Kirschenbauer, H. G. In *Fats and Oils: An Outline of Their Chemistry and Technology*, Reinhold Publishing Corp., New York; 1960, pp. 146–156.
7. Khot, S. N.; La Scala, J. J.; Can, E.; *et al. J. Appl. Polym. Sci.*, 2001, **82,** 703–723.
8. Can, E.; Kusefoglu, S.; Wool, R. P. *J. Appl. Polym. Sci.* 2002, **83,** 972–980.
9. Williams, G. I.; Wool, R. P. *Appl. Comp. Mater.* 2000, **7,** 421–432.
10. Bunker, S. P.; Wool, R. P. *J. Polym. Sci. A: Polym. Chem.* 2002, **40,** 451–458.
11. Johnson, C. F. In *Composite Materials Technology*, Mallick, P. K.; Newman, S., Eds.; Hanser Publishers, New York; 1990, pp. 149–178.

12. Monk, J. F. In *Thermosetting Plastics*, J. F. Monk, Ed.; George Goodwin Limited, Great Britain; 1981, pp. 1–25.
13. La Scala, J. J.; Wool, R. P.. *Polymer*, 2005, **46**, 61.
14. Litchfield, C. In *Analysis of Triglycerides*, Academic Press, New York; 1972, pp. 233–264.
15. Gunstone, F. D. In *Fatty Acid and Lipid Chemistry*, Blackie Academic and Professional, New York; 1996, pp. 3–39.
16. Verma, D. P. S.; Shoemaker, R. C., Eds. *Soybean: Genetics, Molecular Biology, and Biotechnology*, Cab International, Wallingford, UK; 1996, pp. 165–180.
17. Vander Wal, R. J. *J. Amer. Oil Chem. Soc.* 1960, **37,** 18.
18. Coleman, M. H.; Fulton, W. C. In *Enzymes of Lipid Metabolism*, Desnuelle, P., Ed.; Pergamon, Oxford; 1961, p. 127.
19. Evans, C. D.; McConnell, D. G.; List, G. R.; *et al. J. Am. Oil Chem. Soc.* 1969, **46,** 421.
20. La Scala, J. J. The Effects of Triglyceride Structure on the Properties of Plant Oil–Based Resins, Ph.D. *Dissertation,* University of Delaware, Newark; 2002.
21. Liu, K. In *Soybeans: Chemistry, Technology, and Utilization*, Chapman and Hall, New York, 1997, pp. 27–30.
22. Choi, S. C. In *Introductory Applied Statistics in Science*, Prentice-Hall, Englewood Cliffs, NJ, 1978, pp. 32–35.
23. La Scala, J. J.; Wool, R. P. *J. Am. Oil Chem. Soc.*, 2002, **79,** 59–63.
24. La Scala, J. J.; Wool, R. P. *J. Am. Oil Chem. Soc.*, 2002, **79,** 373–378.
25. Flory, P. J. In *Principles of Polymer Chemistry*, Cornell University Press, Ithaca, 1953, pp. 347–398.
26. Macosko, C. W.; Miller, D. R. *Macromolecules*, 1976, **9,** 199–205.
27. Miller, D. R.; Macosko, C. W. *Macromolecules*, 1976, **9,** 206–211.
28. Wool, R. P. In *Polymer Interfaces, Structure, and Strength*, Hanser Publishers, New York; 1995, pp. 102–116.
29. Miller, D. R.; Macosko, C. W. *Macromolecules*, 1978, **11,** 656–662.
30. Barrett, L. W.; Sperling, L. H.; Murphy, C. J. *J. Am. Oil Chem. Soc.* 1993, **70,** 523–534.
31. Kantor, Y.; Webman, I. *Phys. Rev. Lett.* 1984, **52,** 1891–1894.
32. Feng, S.; Sen, P. N. *Phys. Rev. Lett.* 1984, **52,** 216–219.
33. Feng, S.; Thorpe, M. F.; Garboczi, E. *Phys. Rev. B* 1985, **31,** 276–280.
34. Wool, R.P., "Rigidity Percolation Theory of Thin Film Melting and the Glass Transition," *Amer. Chem. Soc PMSE Preprints*, Philadelphia, August 2004.
35. Wool, R. P., *J. Polym. Sci., Part B: Polym. Phys.* 2005, **43**, 168.
36. Lorenz, C. D.; Stevens, M. J.; Wool, R. P. *J. Polym. Sci. A: Polym. Chem.* 2004, **42,** 3333.
37. Khot, S. N. *Ph.D. Dissertation,* University of Delaware, Newark; 2001.
38. Plimpton, S. J. *J. Comput. Phys.* 1995, **117,** 1.
39. Stevens, M. J. *Macromolecules* 2001, **34,** 2710.
40. Lu, J. *PhD. Dissertation,* University of Delaware, Newark; 2004.
41. Fox, T. G.; Loshaek, S. J. *J. Polym. Sci.* 1952, **15,** 371.
42. DiMarzio, E. A. *J. Res. Nat. Bur. Stand. A: Phys. Chem.* 1964, **68A,** 611.
43. Stutz, H.; Illers, K.-H.; Mertes, J. *J. Polym. Sci. B: Polym. Phys.* 1990, **28,** 1483.
44. Hale, A.; Macosko, C. W. *Macromolecules*, 1991, **24,** 2610.
45. Wool, R. P. Percolation Theory of Melting and the Glass Transition Temperature of Thin Polymer Films, American Physical Society, Division of High Polymer Physics; March 2002.
46. Wool, R. P.; Raghavan, D.; Wagner, G. C.; *et al. J. Appl. Polym Sci.* 2000, **77,** 1643.
47. Goheen, S. M.; Wool, R. P. *J. Appl. Polym. Sci.* 1991, **42,** 2691.
48. Peanasky, J. S.; Long, J. M.; Wool R. P. *J. Polym. Sci. B: Polym. Phys.* 1991, **29,** 565.
49. Tsui, O. K. C.; Zhang, H. F. *Macromolecules*, 2001, **34,** 9139.
50. Keddie, J. A.; Jones, R. A. L.; Cory, R. A. *Faraday Discuss.* 1994, **98,** 219.
51. Fox, T. G.; Flory, P. J. *J. Appl. Phys.* 1950, **21,** 581.

52. Johnson, C. F. *Composite Materials Technology,* Mallick, P. K.; Newman, S., Eds.; Hanser Publishers, New York; 1990, pp. 149–178.
53. Guo, A.; Cho, Y.; Petrovic, S. *J. Polym. Sci. A: Polym. Chem.* 2000, **38,** 3900.
54. Berry, G. C.; Fox, T. G. *Adv. Polym. Sci.* 1968, **5,** 261.
55. Scholnick, F.; Linfield, W. M. *Fatty Acids,* Pryde, E. H., Ed.; American Oil Chemists' Society, Champaign, IL; 1979.
56. Doyle, T.; Fekete, F.; Keenan, P. Combination Catalyst-Inhibitor for Beta Hydroxy Carboxylic Ester, U.S. Patent 3,317,465; 1967.
57. Dowd, R. T. *Epoxy Resin Technology,* Bruins, P. F., Ed.; Interscience Publishers, New York; 1968.
58. Lee, H.; Neville, K. *Epoxy Resins,* McGraw-Hill, New York; 1957.
59. Matejke, L.; Dusek, K. *Polym. Bull.* 1986, **15,** 215.
60. Yaws, C. L. *Handbook of Viscosity,* Vol. 2, Gulf Publishing Company, Houston; 1995.
61. Yaws, C. L. *Handbook of Viscosity,* Vol. 3, Gulf Publishing Company, Houston; 1995.
62. Product Bulletin for Aerojet Accelerator, AMC-2, Aerojet Chemicals; 2000.
63. Gunstone, F. D. *An Introduction to the Chemistry and Biochemistry of Fatty Acids and Their Glycerides,* 2nd ed., Chapman and Hall, Bungay, Great Britain; 1967.
64. Doi, M. *Introduction to Polymer Physics,* Clarendon Press, Oxford; 1992.
65. Huang, Y. J.; Jiang, W. C. *Polymer* 1998, **39,** 6631.
66. Krevelan, D. W. *Properties of Polymers,* 3rd ed., Elsevier, London; 1990.
67. Liu, K. *Soybeans: Chemistry, Technology, and Utilization,* Chapman and Hall, New York; 1997.
68. Macosko, C. W. *Rheology Principles, Measurement, and Applications,* Wiley-VCH, New York; 1994.
69. Zaske, O. C. *Handbook of Thermoset Plastics,* Goodman, S. H., Ed.; Noyes Publications, Park Ridge, NJ; 1986, pp. 91–93.
70. Bohdanecky, M.; Kovar, *J. Viscosity of Polymer Solutions,* Elsevier Scientific Publishing Company, New York; 1982.
71. LaScala J., Wool, R.P., *J. Appl. Polym. Sci.*, 2005, **95,** 774.

8

Pressure-Sensitive Adhesives, Elastomers, and Coatings from Plant Oil

Richard P. Wool

In this chapter, we switch from highly cross-linked polymers used in composites, as discussed in the last four chapters, to highly linear polymers made with the single fatty acids derived from the triglycerides. The rather simple linear chain architecture, so readily attainable with petroleum-based monomers with C = C functionality, such as polyethylene and polypropylene, presents challenges for triglyceride oils. The key to success is to obtain oils that are capable of providing mono-functionalized monomers, which are the fatty acids containing just one unsaturated C = C bond, such as the high oleic oils. These oils can be obtained through genetic engineering, crop selection, specialty high-oleic crops, and fatty acid separations. The fatty acid separation process, while being technically feasible at the laboratory scale, can be quite expensive to mass produce, especially if one is looking to keep prices for resin near $2/kg. As with all bio-based materials, the least costly approach is to grow the monomers (or their precursor materials) in the field using free sunlight, water, oxygen, and carbon sources and, at the same time, remove global warming gases from the air.

In this chapter we explore the development of pressure-sensitive adhesives (PSAs), elastomers, and coatings from high-oleic oils. The linear polymers with molecular weights of the order of 10^6 Da are made by emulsion polymerization using water as a solvent. The resulting water-borne latex particles are placed on a substrate such as paper or polymer, and as the water evaporates, they coalesce by interdiffusion to form thin films. The resulting PSA products are the familiar Scotch® tapes, postage stamps, name labels, duct tape, masking tapes, packaging labels, and so on. These products are typically disposable and amount to about 14 billion pounds per year. The U.S. Post Office alone uses about 11% of the total U.S. market. The importance of the development of linear polymers from triglycerides is that it provides the technology platform to make other materials requiring linear architecture such as coatings, paints, elastomers, and toughening agents.

8.1 INTRODUCTION TO PRESSURE-SENSITIVE ADHESIVES

Pressure-sensitive adhesives are almost indispensable in everyday life because they are used for labels, tapes, films, postage stamps, and many adhesive applications. Currently, the majority of PSAs are made from petroleum-based acrylate monomers, such as 2-ethylhexyl acrylate, *n*-butyl acrylate, and isooctyl acrylate [1]. To alleviate this dependency on petroleum, it is desirable to investigate the synthesis of these adhesives from a renewable resource, such as plant oil. Because most of the PSA applications are of a disposable nature, it would also be desirable to make these materials biodegradable. As discussed in Chapters 1 and 4, plant oils are triglyceride esters of fatty acids, which vary in chain length and functionality. Their chemical versatility and abundance make them an ideal starting material [2]. The most common oils have a carbon–carbon double bond functionality. An example of a triglyceride molecule is shown in Figure 8.1.

The $C = C$ unsaturation of fatty acids has traditionally been used for oxidative coupling reactions leading to "air drying" of some plant oils. This is the chemistry of the well-known alkyd resins used for paint and varnish binders and the previnyl, old-fashioned floor covering known as "Lino" (derived from cross-linked linseed oil), and now once again quite stylish, but expensive. Although there are many examples of the use of drying oils for surface coating applications that date back hundreds of years to antiquity, the unsaturation on the fatty acid is not sufficiently reactive to allow homo- or copolymerizations of the molecule directly to give resins with any degree of structural strength or stiffness. However, both triglycerides as well as individual fatty acids can be chemically modified in order to participate in free-radical polymerization reactions. The fatty acid molecule offers a number of reactive sites for functionalization. These include the double bond, the allylic carbons, the ester group, and the carbon alpha to the ester group, as shown in Figure 8.1. Typical modifying reactants include maleic acid, maleic anhydride, methacrylic acid, and acrylic acid [3], as discussed in Chapter 4. Besides conventional bulk polymerization, these components can also be polymerized

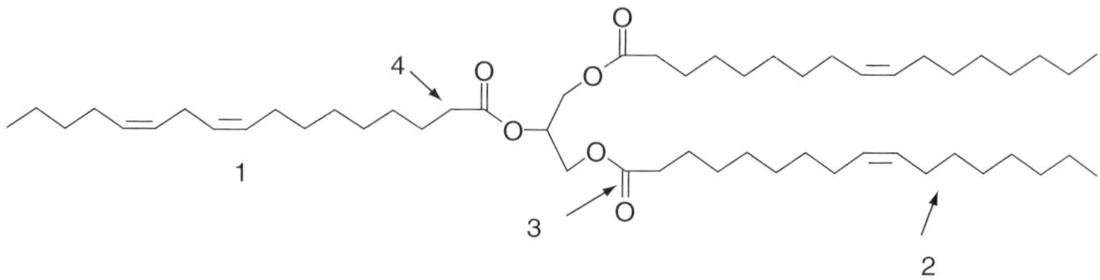

FIGURE 8.1 Diagram of a triglyceride molecule. A triglyceride is composed of three fatty acids connected at a glycerol center. The different functionalities are shown with the corresponding numbers: (1) double bond, (2) allylic carbons, (3) ester group, and (4) alpha carbon.

using emulsion polymerization, a common practice in the PSA industry. Solution polymerization should work as well. However, the PSA industry has moved toward eliminating solvent-based material for ecological and economical reasons. Therefore, this work focuses on optimizing a water-based emulsion system.

The polymer in a pressure-sensitive adhesive is a viscoelastic material that is permanently, as well as aggressively, tacky and has enough cohesive strength and elasticity to be cleanly removed from a substrate surface [1]. These polymers are typically linear polymers with a slight degree of cross-linking. The degree of cross-linking is one of the key features controlling the balance between the cohesive and adhesive strengths of the polymer, in addition to the role of sticker and receptor groups, as discussed in Chapter 6. Monomers derived from plant oils possess an inherent degree of unsaturation that varies from plant to plant. The variation of unsaturation among the various plant oils, and hence the fatty acids, can be used to advantage. Depending on the property desired in the final product, various oils, or mixtures thereof, can be used in synthesizing the monomers. Functional groups to increase adhesive strength with particular substrates can also be placed onto the unsaturation sites (see Section 8.5).

Previous work in this area by S. P. Bunker and R. P. Wool [4] focused on synthesizing a monomer from a fatty acid methyl ester that is capable of forming high-molecular-weight polymers using conventional (macro)emulsion polymerization. However, miniemulsion polymerization has several advantages over the normal emulsion technique. Miniemulsion is a good polymerization method for highly hydrophobic monomers because each droplet can be considered a minibatch reaction for the polymerization [5]. This is different from conventional emulsion polymerization, which has both monomer droplets and polymer particles. The conventional emulsion requires the transport of water-insoluble monomers from droplets to growing polymer particles, which can yield slower kinetics and, therefore, longer polymerization times [5, 6].

In the next section, the mechanical properties of the renewable resource-based dispersions are compared to two petroleum-based dispersions. Specifically, the first one is a market standard for filmic label application, Acronal® A220 (www.basf.de/dispersions/), which is known for its high transparency, excellent water resistance, and outstanding adhesion to polyolefinic substrates. The second is a model dispersion of 2-ethylhexyl acrylate (2-EHA)-co-methyl methacrylate (MMA). This system was selected because the 2-EHA has a structure similar to that of the fatty acid methyl ester-based monomer. The side-by-side comparison of the properties of the petroleum-based PSA standards with the new bio-based polymers should be reassuring to most, especially if the economics are right and the bio-based PSA has additional benefits, such as biodegradability or being less energy intensive to produce. It may also be reassuring to some readers to know that the bio-based PSA could still be made in 2084 when global oil supplies may no longer support the existing petroleum-based materials.

8.2 MACROEMULSION AND MINIEMULSION POLYMERIZATION

8.2.1 MACROEMULSION POLYMERIZATION

Acrylated methyl oleate (AMO) was synthesized using methods reported by Bunker and Wool [4] and discussed in Chapter 4. The monomer synthesis requires two steps. First, the unsaturated bond in oleic methyl ester (OME) must be epoxidized by a peroxy acid. The epoxidized fatty acid methyl ester is then acrylated using acrylic acid. The acrylate groups are able to participate in free-radical polymerization. A schematic of the monomer synthesis is shown in Figure 8.2. The OME can also be derived as a by-product from biodiesel, assuming that we have an efficient fatty acid separation process. The separation process was explored by Bunker and Wool and potentially can be done economically at large scale. This would circumvent the need for the development of specialty high-oleic oils and provide additional utilization of biodiesel plants currently being constructed in Delaware and elsewhere. From a green engineering perspective, the biodiesel is perhaps more valuable as a chemical feedstock rather than a combustible fuel feedstock and can attain this value when the current generation of internal combustion engines is replaced in the future by their fuel-cell equivalents.

The AMO monomer is polymerized using both macroemulsion (also referred to as conventional emulsion polymerization) and miniemulsion polymerization. The experimental conditions for the macroemulsion polymerization are outlined in detail by Bunker *et al.* [4]. The formulation of the macroemulsion is shown in Table 8.1. The reaction time was approximately 18 h at 70 °C.

8.2.2 MINIEMULSION POLYMERIZATION

The specific formulations for each miniemulsion polymerization are listed in Table 8.1 (samples 1 to 5). Typically, the polymerizations were conducted

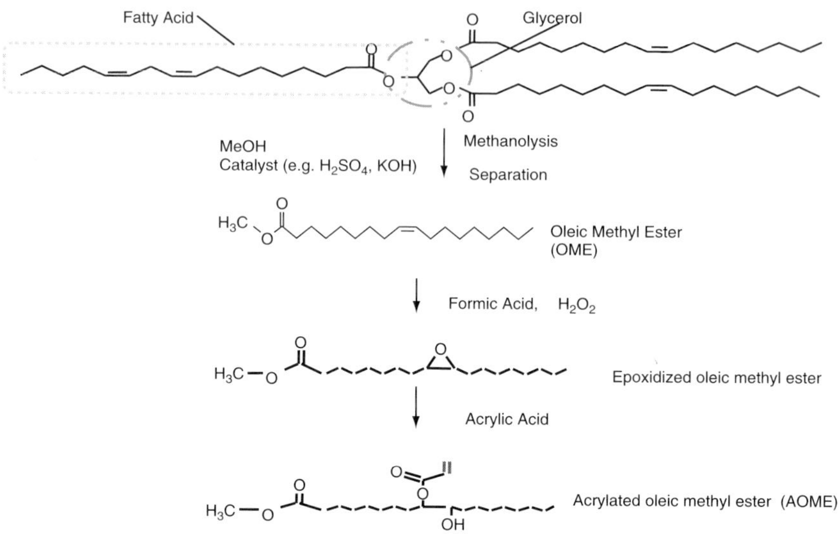

FIGURE 8.2 Schematic diagram of the monomer synthesis steps.

TABLE 8.1 Emulsion composition and the properties of the resulting polymers.

Component	Sample 1	Sample 2	Sample 3	Sample 4	Sample 5	Conventional
DDI H_2O (g)	40	40	40	40	40	30
SLS (15 wt%) (g)	1.33	1.33	1.33	1.33	1.33	2.25 g Aerosol® OT
AMO (g)	10	9	8.5	8.0	–	15
MMA (g)	–	1	1	1	1	0.5 g Acrylic Acid
BDDA (g)	–	–	0.5	1	–	—
2-EHA (g)	–	–	–	–	9	—
Vazo 67 (g)	0.1	0.05	0.05	0.05	0.05	0.05 g V-50
Emulsion droplet size (nm)	390	350	380	420	780	>1000
Particle diameter (nm)	350	350	380	400	800	>1000
K_{value}	26.8	42.7	*NA*	NA	NA	26.6
T_g (°C)	−49	−49	−50	−46	−58	−39

in a 500-mL round-bottom flask equipped with a reflux condenser, nitrogen inlet, and a Teflon stirrer. First, the initiator was combined with the monomer using a magnetic stirrer, to ensure its complete dissolution in the monomer phase. After the initiator dissolution, the surfactant and water were mixed into the system using a magnetic stirrer for approximately 10 min. The miniemulsion was then prepared by continuous ultrasonification for 5 min. During sonification, the emulsion was submerged in an ice bath to maintain a temperature below 50 °C. This ensured that the initiator did not prematurely decompose. The glass reactor containing the monomer emulsion was then placed in an oil bath and heated to 85 °C for 1 h.

8.3 POLYMER CHARACTERIZATION

The monomer conversion as a function of time is shown in Figure 8.3. This plot tracks the intensity of the peak that corresponds to the carbon–carbon double bond of the monomer as well as the carbonyl group in the developing polymer. This chart indicates that the reactive monomer groups are completely depleted after 1 h of reaction time, which corresponds to the maximum intensity of the polymer carbonyl group. This is a significant improvement over the conventional emulsion polymerization. Figure 8.4 depicts the typical conversion of monomer to polymer in a conventional emulsion reaction as a function of time, as recorded using gravitational analysis. This study indicates that 18 h of reaction is required to achieve 90% monomer conversion. Additional reaction time does not further increase this conversion.

8.4 POLYMER PROPERTIES

Table 8.1 reports the particle size distribution (PSD) of the dispersion, the *K*-value, and the glass transition temperature (T_g) of the resulting polymers. Typical dispersions prepared by miniemulsion have particle sizes between 50 and 500 nm [5]. This corresponds to the PSD of samples 1 to 4, whereas sample 5, prepared with EHA, has a larger PSD. The effect of additional

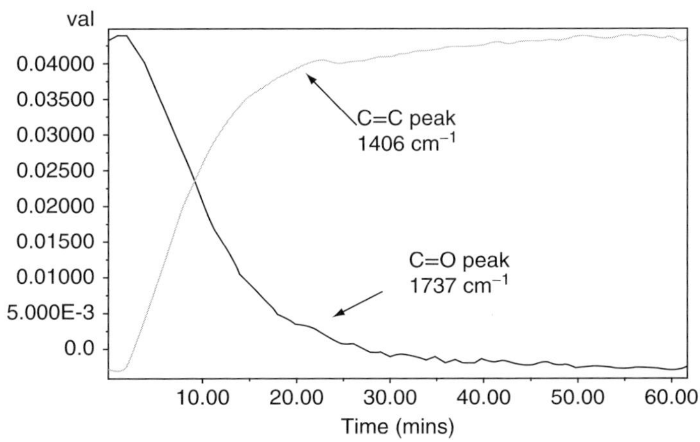

FIGURE 8.3 Conversion of monomer to polymer as a function of time for miniemulsion polymerization.

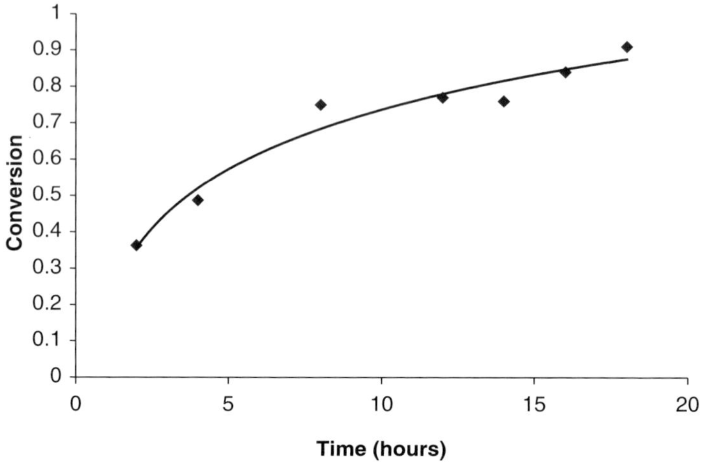

FIGURE 8.4 Conversion of monomer to polymer as a function of time for conventional emulsion polymerization.

sonification time and additional surfactant on the mean particle size was examined. The additional sonification time (up to 10 min) seemed to have little effect on decreasing the particle size. An increase in surfactant levels from 2–5 wt% also resulted in no decrease in particle size. Surfactant concentrations above this amount were not used due to the well-known detrimental effect of excess surfactant on adhesive properties [7–9]. However, in conventional emulsion polymerization, 15 wt% of surfactant was required to form a stable dispersion [4].

The *K*-value of the AMO homopolymer (sample 1 in Table 8.1) from miniemulsion polymerization is similar to that of the polymer synthesized using conventional emulsion. Examination of the effect of comonomer (sample 2) on the *K*-value indicates that the addition of comonomer greatly increases the molecular weight of the polymer. Previous research in thermosetting polymers from acrylated epoxidized soybean oil (AESO) found that a comonomer is required to increase the conversion of the AESO [10]. The comonomer behaves like a chain extender as well as a reactive diluent and reduces the mass transfer limitations associated with the reaction of the bulky AESO.

All of the glass transition temperatures presented here are significantly below room temperature. This is typical of PSA polymers because a majority of them are used at room temperature. The low T_g allows the polymer to flow and quickly form a bond to a substrate at room temperature. All of the T_g's are similar except for the polymer made using conventional emulsion polymerization, which has a higher T_g. This can be attributed to the large amount of surfactant used to stabilize the polymer particles. As previously stated, the polymers tested were not purified and therefore the surfactant remained in the polymer.

8.4.1 DYNAMIC MECHANICAL ANALYSIS

The performance of a PSA is related to the viscoelastic response of the bulk adhesive. Storage and loss moduli for each polymer are shown in Figures 8.5(a) and (b), respectively. The storage moduli of the AMO homopolymer and the AMO-co-MMA polymer are very similar. The lack of a plateau region indicates that the polymer is linear and has a molecular weight that is below or around the critical molecular weight required for physical entanglements to form. Correspondingly, the polymers will have very little cohesive strength and therefore poor shear properties, but can easily wet rough surfaces, which is important for good contact and adhesion.

To improve these properties, the molecular weight needs to be greater than the critical entanglement molecular weight of about $8M_c$. The most obvious method to accomplish this is to decrease the initiator concentration. However, on further analysis of the monomer, it was concluded that the monomer is the limiting factor of the molecular weight. Although ^{1}H NMR and ^{13}C NMR of the AMO monomer indicated that the monomer contained 95% acrylate functionality, further analysis using gas chromatography showed that only 83% of the monomer had acrylate functionality. This discrepancy can be attributed to error in the NMR analysis.

Figure 8.6 shows the gas chromatography/FTIR coupling results, which give a structure for the different side products. The monomer is composed of 83% of the acrylated methyl oleate, 13% epoxidized methyl oleate, and 3% of the starting material, methyl oleate. These results were confirmed with the GC/MS analysis. As shown in Figure 8.2, the epoxidized methyl oleate (EMO) is the intermediate product in the monomer synthesis process. Therefore, the nonreacting part, approximately 17%, will behave as a plasticizer in the polymer and reduce the mechanical and adhesive properties. Also, the nonreacting components limit the molecular weight of the polymer. Techniques to decrease the amount of EMO were explored. However, even with this limitation, the polymers synthesized with the current monomer show acceptable PSA properties, as demonstrated later.

Above the glass transition region, the storage modulus of 2-EHA-co-MMA (sample 5) exhibits a rubbery plateau with G' almost independent of temperature but at high enough temperatures viscous flow is dominating (G' falls below G''). This indicates that this is a linear but physically entangled polymer. From G' in the rubbery zone, the mean molecular weight between entanglements is calculated to be $M_e = 60\,\text{kg/mol}$. This value is in good agreement with literature data [11]. Such a high-molecular-weight physically entangled polymer is ideal for adhesive applications. Such polymers have long entangled chains that will impart cohesive strength to the system, but at the same time, the polymer chains are still mobile enough to form a good adhesive bond. These effects should show up in the application test results.

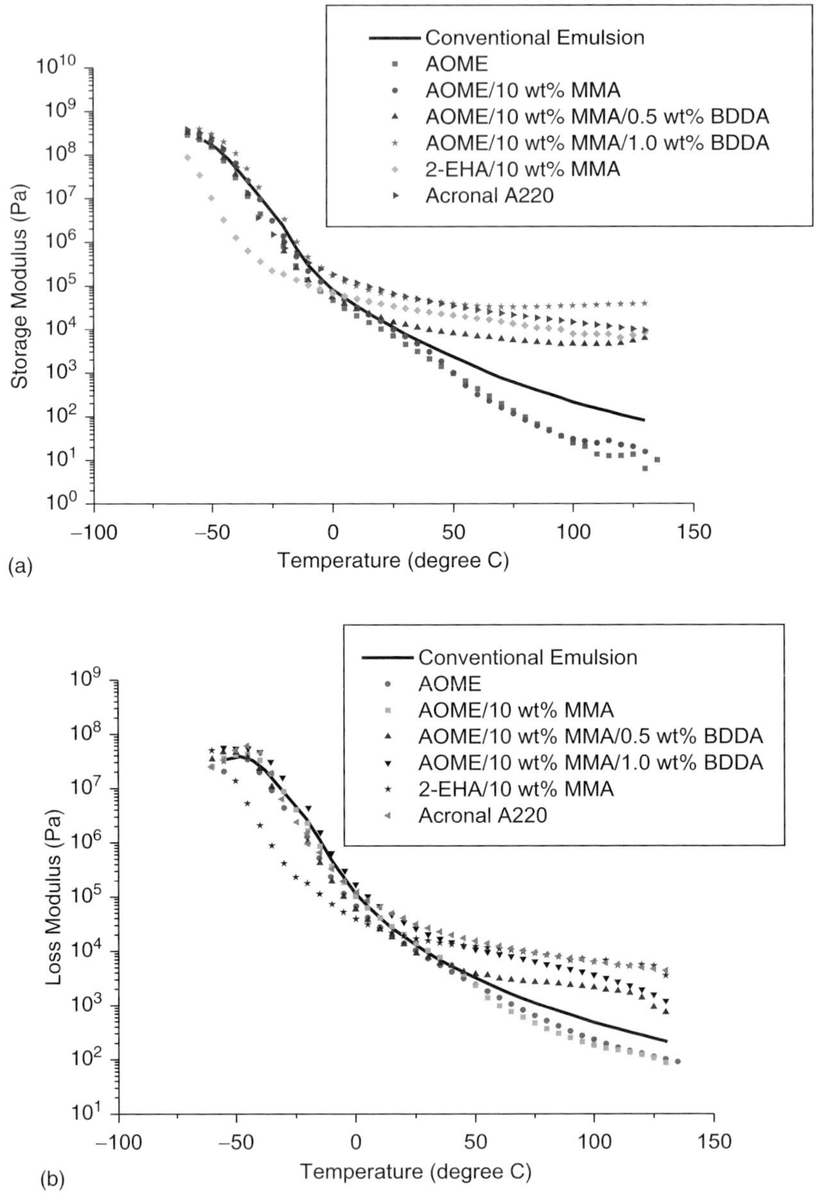

FIGURE 8.5 (a) The storage moduli of the polymers as a function of temperature. (b) The loss moduli of the polymers as a function of temperature.

The storage modulus of the polymer synthesized using conventional emulsion polymerization offers a slight improvement, indicated by the higher modulus values. Although the *K*-value of the conventional polymer and the miniemulsion AMO homopolymer are similar, the differing rheological

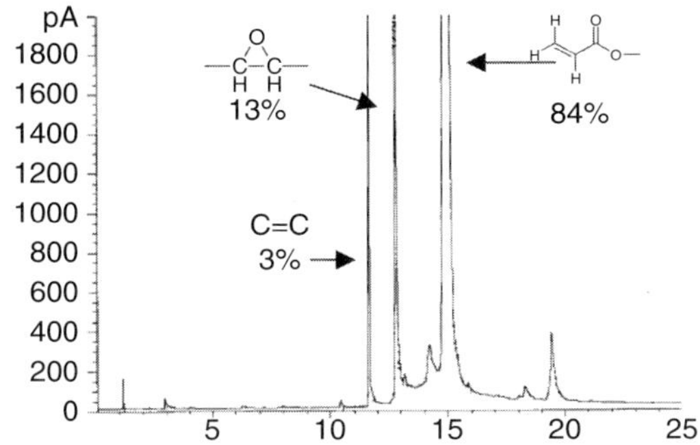

FIGURE 8.6 GC spectra of the monomer, AMO. The monomer is composed of 84% acrylate species, 16% epoxidized species, and 3% of the original OME.

properties may be explained by the different reaction times. It is well known that as polymerization reaction times increase, branching will increase due to chain transfer to polymer [12]. It was determined previously that the branching in the conventional emulsion is caused by the hydrogen abstraction of the tertiary backbone C–H bond [4]. Therefore, this chain transfer and resulting branched structure will have an effect on the dimethacrylate (DMA) properties as indicated by an increase in the storage modulus. Copolymerization of the stiff acrylic acid groups is another parameter expected to increase the storage modulus.

The addition of the cross-linking comonomer, 1,4-butanediol diacrylate (BDDA), to the miniemulsion system increased the modulus significantly. In fact, the resulting modulus of the AMO-MMA-co-BDDA (both 0.5 and 1 wt%) is comparable to the commercial Acronal® A220. Comparing the modulus profiles for these samples with the uncross-linked counterparts indicates that these polymers are indeed chemically cross-linked; G' is always higher than G'' even at high temperatures where the uncross-linked materials start to flow. Nevertheless, the plateau modulus of these slightly cross-linked polymers still fulfills the Dahlquist criterion, which states that polymers to be used for PSA applications should have a plateau modulus below 0.3 MPa [13].

8.4.2 TACK

The adhesion performance of PSAs is determined by three main properties: tack, peel strength, and shear resistance. Tack is a key property of PSAs and is defined as the ability of an adhesive to form a bond of measurable strength to another material under conditions of low contact pressure and short contact time [14]. Figures 8.7 and 8.8 are bar charts of the tack properties of the polymers tested with a polyethylene and stainless steel probe, respectively. In both cases, the conventional emulsion polymer displayed the lowest tack energy. This is attributed to the large amount of

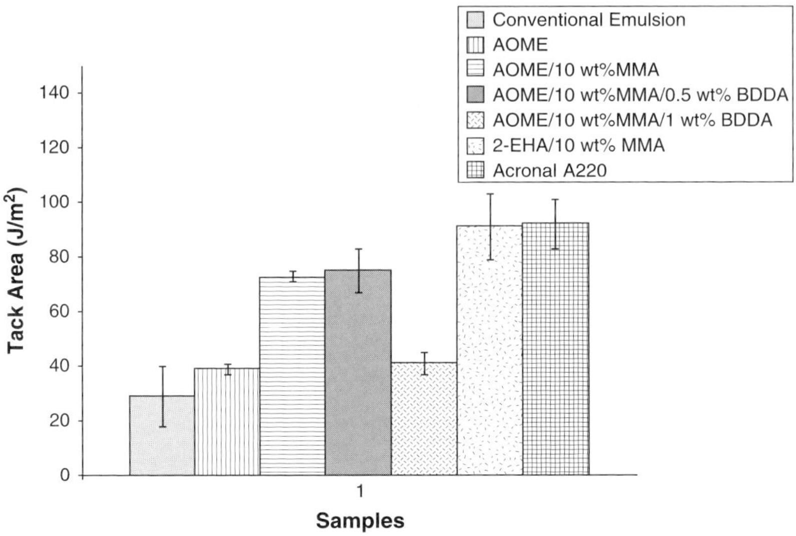

FIGURE 8.7 Tack results of polymers to a polyethylene probe.

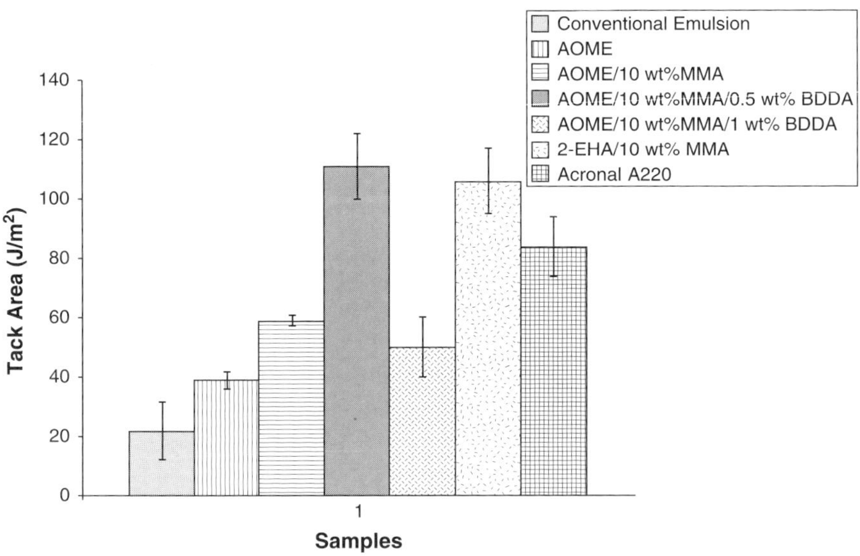

FIGURE 8.8 Tack results of polymers to a stainless steel probe.

surfactant remaining in the polymer. It is well known that residual surfactant in the polymer migrates to the polymer–air surface, which decreases the tack and adhesive properties [14].

For both testing probes, the miniemulsion polymers display increasing tack values with the addition of comonomer. Against PE, the AMO-co-MMA copolymers show tack values almost as high as the petroleum-based polymers (2-EHA-co-MMA and Acronal® A220). Against stainless steel, the highest tack is observed for the sample with 0.5% BDDA. This may be due to the two carbonyl groups on the BDDA monomer that increase its attraction to a metal surface. However, there is a limit on the increase of tack with the increase in concentration of BDDA. The polymer with 1 wt% BDDA showed a significant decrease in tack value. Again, this is attributed to the balance between a polymer that has cohesive strength while at the same time flows to make an adhesive bond quickly. The 1 wt% BDDA polymer has a tight network as indicated by the high storage modulus, providing good cohesive strength, but flow is probably so strongly restricted that it cannot form a tight contact to the probe within the short dwell time. Overall, the AMO-co-0.5 wt% BDDA polymer shows tack results comparable to both the 2-EHA-co-MMA polymer and Acronal® A220.

8.4.3 PEEL AND SHEAR TESTS

Figures 8.9(a) and (b) show the results of the peel and shear time-to-failure tests, respectively. The peel results of the conventional polymer, AMO homopolymer, and AMO-co-MMA and the 2-EHA-co-MMA are very comparable. The peel value for the conventional polymer is somewhat higher than for the other linear polymers, presumably due to the acrylic acid copolymerized in the conventional polymer. All of the linear polymers offer very little shear resistance, as observed with a time to failure on the order of minutes. These values are so low that they do not appear on Figure 8.9(b). This is even true for the 2-EHA-co-MMA polymer, which clearly had a physically entangled network based on the storage modulus, although the time to failure is significantly longer than for the AMO linear polymers.

Introducing chemical cross-links is one strategy to improve the shear strength. Consequently, the chemically cross-linked AMO-copolymers did show a large increase in shear holding time, increasing from minutes to hours. However, with the increased amount of chemical cross-linking, the peel values decreased drastically. On the other hand the 2-EHA-co-MMA and Acronal® A220 show a good balance between tack, peel, and shear properties, which is contributed to the high molecular weight of these systems. This clearly shows the future direction into which the development of the AMO polymers has to go in order to yield competitive PSA materials [15–17].

8.4.4 CONCLUSIONS ON PSA SYNTHESIS

Significant improvement of the polymerization of AMO latex was achieved by using miniemulsion polymerization. The reaction time was decreased from 18 h down to 1 h in addition to the surfactant concentration being drastically reduced from 15 to 2 wt%. The resulting polymer has shown

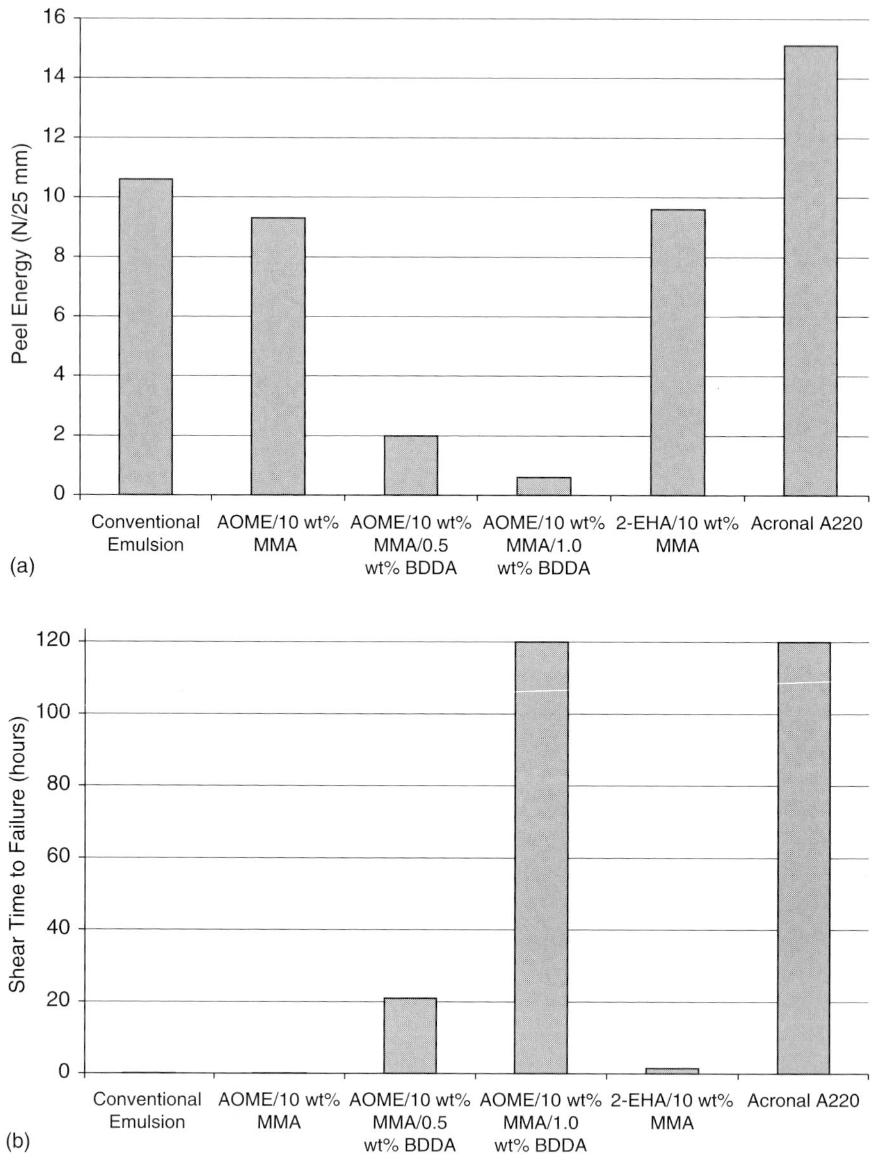

FIGURE 8.9 (a) 180° peel test results. The peel energy is in units of N/25 mm. (b) Shear time to failure, recorded in hours.

physical properties that are comparable to petroleum-based polymers. Most importantly, the polymers derived from a renewable resource display typical PSA properties. Also, these materials, when exposed to soil, completely disappeared in a few months.

8.5 POLYMER–SOLID ADHESION MODIFICATION OF PSAS

It is well known in the adhesion field that the addition of functional groups to a polymer backbone can enhance the adhesive potential of the polymer in contact with solid surfaces [1, 18]. The adhesion potential was related to the surface energetics, where the number and type of chemical groups are important in predicting the adhesion potential [19]. However, a fundamental understanding of the effect of functional groups on adhesive behavior had not previously been achieved. Recent work on model polymer–substrate systems indicates that the fracture energy of polymer–solid interfaces is not a monotonic function of the surface energetics, as previously expected [4]. Lee and Wool modeled the behavior of the peel energy as a function of the number of functional groups (i.e., sticker groups) in the polymer and receptor groups on the substrate, as well as the interaction between these two groups, as discussed in Chapter 6 [18].

Essentially, this model identifies the existence of an optimal concentration of functional sticker groups for reaching a maximum adhesive strength. The polymer–solid interface is a competition between cohesive failure of the polymer–polymer interface and adhesive failure of the adsorbed polymer–solid interfaces. The polymer–polymer interface involves the interpenetration of the adsorbed and free chains in a layer adjacent to the surface, whereas the adsorbed polymer–solid interface considers the adhesive strength of the adsorbed chains in contact with the solid surface. Thus, as the concentration of functional groups in a polymer chain increases and becomes strongly adsorbed on the surface, its conformation flattens out onto the surface, decreasing its ability to interpenetrate with the free chains in the bulk of the polymer. Cohesive failure results when the adhesive forces required to debond the adhered chains exceed the cohesive forces required to break the poorly interpenetrated polymer–polymer interface. In contrast, adhesive failure results when there are few functional–receptor group interactions and the chain is highly interpenetrated in the polymer bulk. In PSA applications, the substrate usually cannot be controlled and therefore the effect of the receptor groups cannot be taken into account. However, the PSA polymer can be designed to incorporate functional groups that will result in optimal adhesive behavior, that is, high peel energy and adhesive failure, for a given type of substrate. In this section, we examine the effect of acid functional groups on the adhesive performance on a metal substrate, that is, acid–base interactions.

Figure 8.10 shows the increase in acid number with the increase in maleic acid added in the initial monomer mixture. This indicates that the maleic acid is being incorporated into the polymer backbone. However, the magnitude of the acid number is higher than expected for a given quantity of acid initially charged to the reaction vessel. Partial hydrolysis of the methyl methacrylate and AOME monomers could account for this higher than expected value.

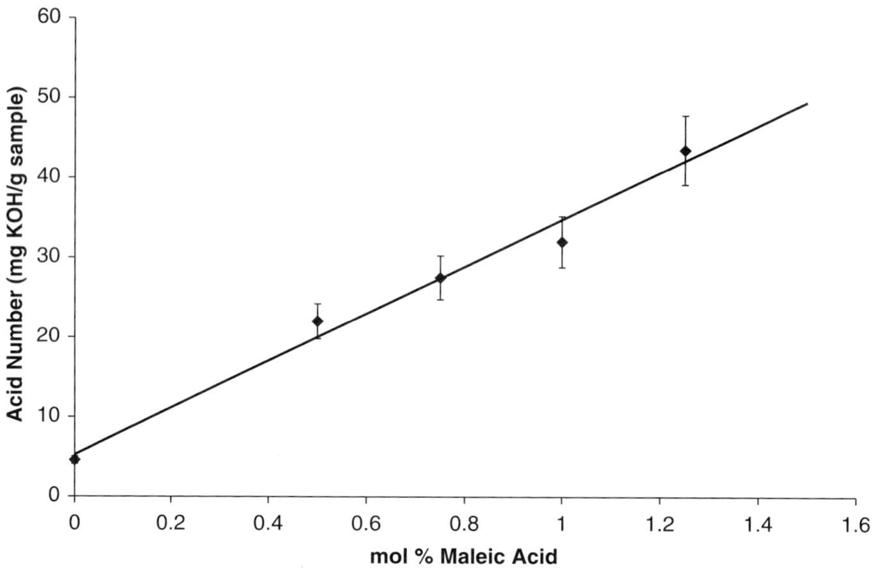

FIGURE 8.10 The experimentally recorded acid number as a function of maleic acid added to the monomer mixture.

However, the increasing trend confirms the incorporation of MA into the polymer.

8.5.1 VISCOELASTIC PROPERTIES

The storage modulus as a function of temperature at six different maleic acid concentrations is shown in Figure 8.11. These are compared to the storage modulus of a miniemulsion polymer that contains no maleic acid. The storage moduli of the AOME-co-MMA-co-MA polymers are slightly higher than that of the AOME-co-MMA polymer. This is attributed to the stiff maleic acid group that is incorporated into the polymer chain. An example of the storage and loss modulus of an AOME-co-MMA-co-MA polymer is shown in Figure 8.12. The G' is greater than the G'' at temperatures $> 0\,^{\circ}C$, which indicates that the elastic behavior of the polymer dominates the properties and that physical entanglements are present. Similar behavior is observed for all of the polymers synthesized with maleic acid.

8.5.2 ADHESION PROPERTIES

The increase in peel energy with the increase in maleic acid content is shown in Figure 8.13. The peel energy increases to an optimum concentration of maleic acid, approximately 1 mol%, with adhesive failure. An increase in maleic acid beyond this quantity results in lower peel energy as well as in

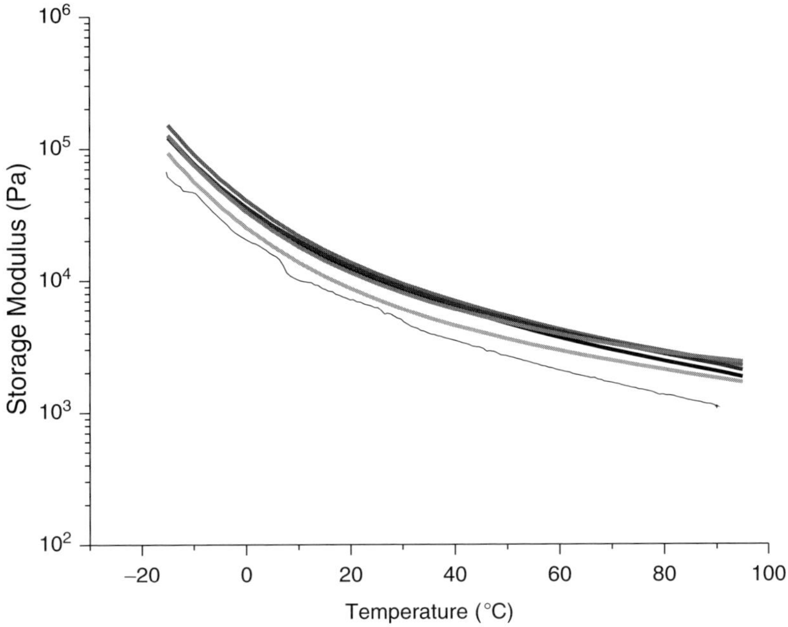

FIGURE 8.11 The storage modulus of the polymers that contain maleic acid as a function of temperature. The thin line represents a comparable polymer that contains no maleic acid.

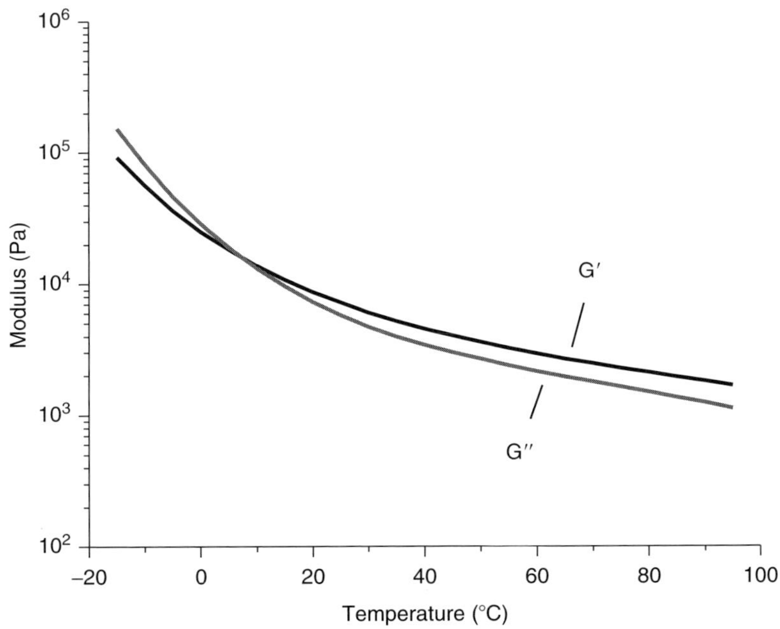

FIGURE 8.12 The storage and loss modulus of an AOME-co-MMA-co-MA polymer.

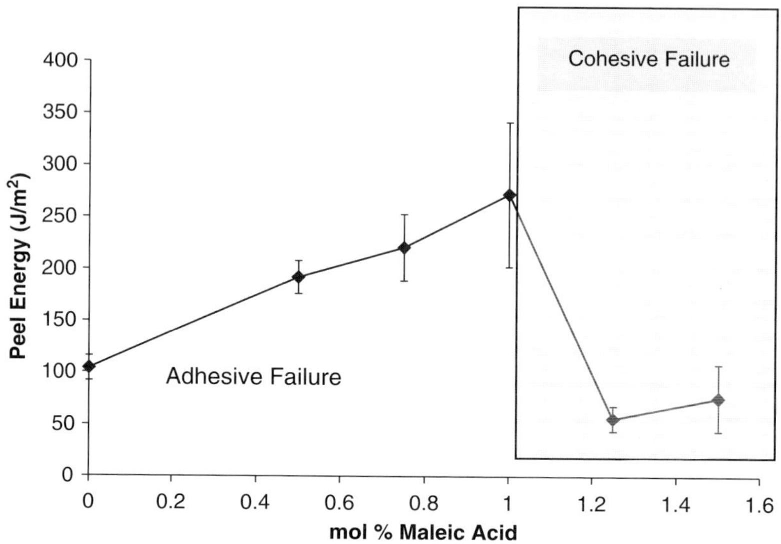

FIGURE 8.13 The peel energy as a function of maleic acid content. An optimum concentration of approximately 1 mol% maleic acid results in high peel energy and adhesive failure.

cohesive failure. This behavior is similar to previously reported results using carboxylated poly(butadiene) [18]. At 1 mol%, small amounts of cohesive failure "patches" were observed. Specifically, randomly located small quantities of adhesive were observed on the stainless steel substrate. Therefore, the optimum level of maleic acid is slightly below 1 mol%.

An entanglement sink probability (ESP) model motivated by vector percolation explains (Chapter 6 and 7) the nonmonotonic influences of functional group concentration, ϕ_x, receptor concentration, ϕ_y, and their interaction strength, χ, on the adhesion strength of the polymer–solid interface [18]. The ESP model quantifies the degree of interaction between adsorbed and neighboring chains based on the adsorbed chain domain. Specifically, the adsorbed chain domain changes thermodynamically with the energy of interaction, r, at the interface, as shown by

$$r = \chi\phi_x\phi_y. \tag{8.1}$$

This parameter can be related to both the adhesive potential (G_A) and the cohesive potential (G_C). The adhesive strength potential behaves linearly as follows

$$G_A \sim r = \chi\phi_x\phi_y. \tag{8.2}$$

The cohesive strength between the adsorbed and neighboring chains is modeled as

$$G_C \sim r^{-1.0} = (\chi\phi_x\phi_y)^{-1.0}. \tag{8.3}$$

This model predicts maximum adhesion strength when the fracture stresses for the cohesive and adhesive failure are equal at an optimal value of $r^* = (\chi\phi_x\phi_y)^*$. Therefore, for a given χ value, there exist optimal values, $\phi_x{}^*$, and $\phi_y{}^*$, above or below which the fracture energy will not be optimized. In this study, the χ and ϕ_y parameters are considered to be constant. Therefore, the adhesive strength is solely linearly dependent on the concentration of functional groups, ϕ_x, in the polymer chain, up to the optimal concentration of ~1 mol%. Beyond this, a sharp decrease in the peel energy is observed and the mode of failure becomes cohesive. In the cohesive region, the peel energy has an inverse relationship with the functional group concentration. However, the decrease in the peel energy is larger than that predicted by the preceding model.

The quantity of maleic acid can be predicted *a priori* based on single-chain scaling [20]. An amorphous linear polymer chain becomes entangled in the network when a chain intersects an arbitrary plane three times, as illustrated in Figure 8.14. The molecular weight of this single chain is defined as M_c, or the critical molecular weight of the polymer that is required for entanglements to form. Based on the schematic in Figure 8.14, from the $*_1$ starting point, the polymer chain returns to the origin of the plane two additional times. Essentially, two functional groups are required per number of monomer units, as demonstrated in Figure 8.14. Therefore, the concentration of functional groups required to anchor the polymer chain to the substrate can be estimated by the following

$$\phi_x{}^* \approx \frac{2M_0}{M_c}, \tag{8.4}$$

where M_0 is the monomer molecular weight. Because this polymer has three monomers, a weighted average M_0 is calculated as follows:

$$\overline{M}_0 = x_1 M_{01} + x_2 M_{02} + x_3 M_{03}, \tag{8.5}$$

where x_x is the weight fraction of the corresponding monomer. Based on the extreme cases of the terpolymer (0.5 and 1.5 mol% maleic acid; the quantity of AOME and MMA remained fixed), the $\overline{M}_0$ is 356 g/mol. The M_c value can be calculated using the entanglement model as follows [20]:

$$M_c \approx 30 \cdot C_\infty \cdot M_0. \tag{8.6}$$

Using the weighted average M_0, the $M_c = 107{,}000$ g/mol. Subsequently, $\phi_x \approx 0.7\%$. This estimate is slightly less than the experimentally determined value of approximately 1% but is in close agreement.

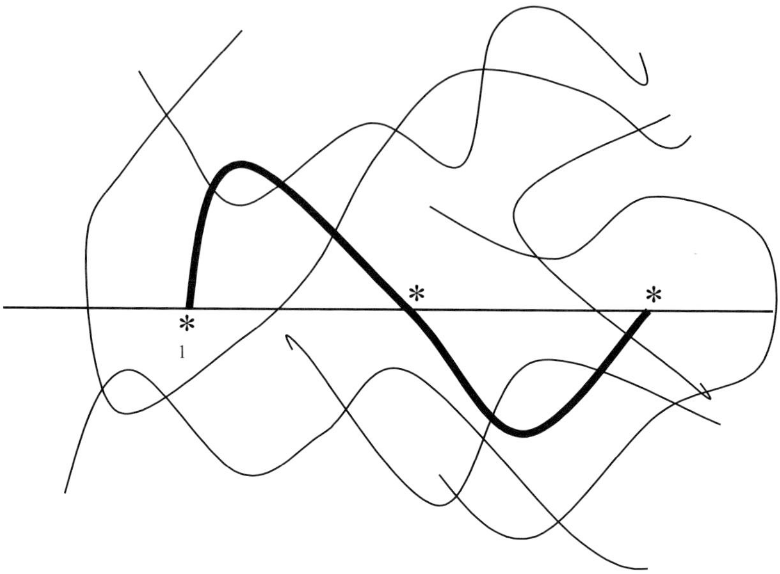

FIGURE 8.14 Entanglements in a polymer melt. The bold line represents the critical molecular weight (M_c) required to obtain a connected network. The asterisks (*) represent the points along the polymer chain that cross an arbitrary plane.

Using this fundamental understanding of the adhesion properties, the peel energy and mode of failure can be designed and controlled. This work could be extended by the incorporation of several different functional groups to enhance the adhesive properties of the polymer toward a variety of substrates.

8.6 BIO-BASED ELASTOMERS

Elastomers are widely used in automotive components, conveyor belting, transport, construction, footwear, and so on. The global market for elastomers and associated products is $40 billion [21]. A small amount of elastomers is natural, obtained from trees chiefly on plantations and small holdings in Malaysia, Indonesia, and other Asian countries [22]. Even though most of the elastomers used to be derived from the rubber tree, now most of the elastomers are synthesized from petroleum oil. Elastomers are soft (stiffness $E \sim 1$ MPa), highly extensible (~400%), and elastic. These unique properties come from the lightly cross-linked polymer network structure. The polymer chain itself must be linear so that it is flexible enough to deform in any direction. In the relaxed state, the molecules are coiled up in a random fashion. The cross-links limit the polymer chain extension but at the same time enable the network to recover to its original shape due to the entropic force generated during the deformation. The properties of an elastomer are controlled by the nature of the cross-linked network. Hardness, modulus, tensile strength, elastic recovery, and so on are all influenced by the cross-link density. The cross-link density in most lightly cross-linked rubbers is in the range of 10^{-5} to 10^{-4} mol cm^{-3} [23]. Elastomers can be further reinforced by particulate fillers, such as carbon black and nanoclays, which causes an increase in ultimate properties, such as tear and tensile strength, abrasion resistance, and modulus of elasticity.

Some interpenetrating network (IPN) systems were developed for the application of elastomers from plant oils. Athawale and Kolekar [24, 25] modified castor oil with linseed oil and tung oil and prepared urethanes and their IPNs with poly(methyl) methacrylate. In addition to this, Athawale and Raut developed new elastomers based on uralkyd resin blended with polystyrene [26], poly(methyl) methacrylate [27], or poly(butyl) methacrylate [28, 29].

Triglycerides can be easily broken down into OME, as shown earlier in Figure 8.2. After epoxidation and acrylation, we can obtain AOME. The double bond in the acrylate group is reactive and can easily undergo free-radical polymerization, as observed for the PSA in the previous section. The long fatty acid chain will give steric hindrance to the rotation of the main chain. However, at the same time, the high flexibility of the long fatty acid also allows them to work as plasticizers and makes the polymer more flexible. Therefore, it was considered by Zhu and Wool [30] to be a promising starting material for the development of new bio-based elastomers. Several molecular design strategies were developed using AOME as the starting monomer. The cross-link density was controlled by varying the cross-link ratio, cross-link agent, and reaction conditions. Various mechanical and thermal properties can be achieved by changing the network structure, and the material can be utilized in different applications.

8.6.1 ELASTOMER MOLECULAR DESIGN

In rubber elasticity theory, the stress σ, strain λ, and modulus G are all related to cross-link density ν, which we can use to design the elastic network. The tensile strength can be expressed as

$$\sigma(\lambda) = G(\lambda - \frac{1}{\lambda^2}), \tag{8.7}$$

in which the modulus $G = \nu kT$. The strain hardening draw ratio, λ_{sh}, is related to the chain length between cross-links Nb and its end-to-end vector $R = N^{1/2}b$ via [31]

$$\lambda_{sh} = \frac{Nb}{N^{1/2}b} = N^{1/2}; \tag{8.8}$$

for most elastomers, $\lambda_{sh} \approx 4$ such that the optimal value is $N_e = 16$. Using Eq. (8.6) and the monomer molecular weight $M_o = 384\,g/mol$ for AOME, then $M_e \approx M_c/2 \approx 57{,}600\,g/mol$. Because the density ρ of rubber is around $10^6\,g/m^3$, the cross-link density required to reach the strain-hardened area is around $17\,mol/m^3$. At room temperature (300 K), this would provide a rubber modulus $G = 43\,kPa$. As shown in Chapter 6, the RP model of fracture [32] also predicts that the fracture stress behaves as $\sigma \sim \nu$.

8.6.2 ELASTOMER SYNTHESIS AND PROPERTIES

To form an elastic network, cross-links were introduced into the system using ethylene glycol dimethacrylate (EGDMA) as the cross-linker [28, 29]. It has two double bonds and is miscible in AOME. By varying the amount of EGDMA, different cross-link densities can be achieved, as shown in Table 8.2. From the AOME structure, we can see that there are many polar groups in the chain that make the final polymer tacky. In addition, although the long branches increase the flexibility of the chain, they also generate a weak network structure. To optimize the mechanical properties of the network and reduce its initial tack, we introduce an MMA comonomer. The methyl group increases the rigidity of the chain and MMA has a glass transition temperature of 102.8 °C, which can reduce the tackiness of the material by increasing the glass transition temperature of the copolymer. At the same time, the polymer chain is extended due to the addition of MMA. The stress–strain relationship for different AOME elastomer samples is shown in Figure 8.15.

Compared to the results of the bulk polymer sample, the introduction of MMA increases the mechanical properties significantly. With 5% MMA, the elongation at break increases to 85% although the tensile strength decreases slightly. Increasing the MMA ratio to 10% doesn't change the elongation much but the tensile strength increases. When the MMA ratio goes up to 40%, the tensile strength and elongation at break increases 169% and 68%, respectively, compared with the sample with 5% MMA. The cross-link density is mainly controlled by the concentration of the EGDMA in this method. The tensile tests shows that samples with 0.5% EGDMA have a 163%

TABLE 8.2 Effects of cross-link density on mechanical properties.

Sample	Tensile Strength (Mpa)	Elongation at Break (%)	Gel Fraction (%)	Crosslink Density (mol/m^3)
10%MMA1%EGDMA	0.26	85.51	86.11	199.71
40%MMA1%EGDMA	0.35	143.28	77.54	130.73
5%MMA1%EGDMA	0.13	84.85	79.12	112.82
5%MMA0.5%EGDMA	0.07	223.16	52.35	15.33

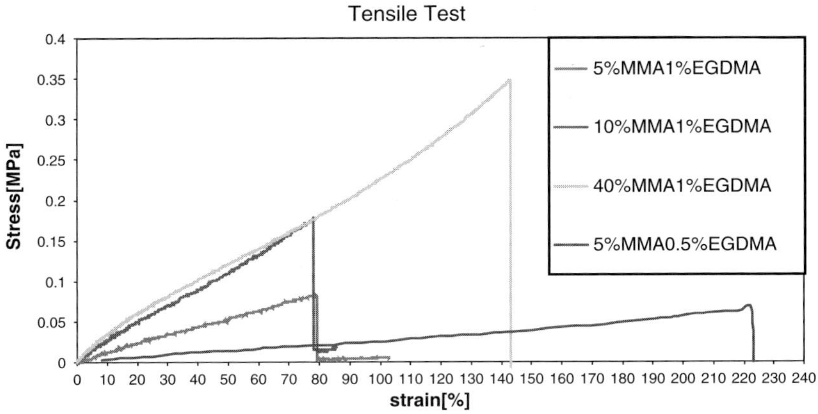

FIGURE 8.15 Stress–strain behavior of AOME elastomers with varying MMA and cross-linker.

increase in elongation but 26% decrease in tensile strength. The elongation at break reaches 223%.

8.6.3 ELASTOMERS REINFORCED WITH NANOCLAYS

Clay is widely used as a nonblackening filler in the rubber industry. It is noted for its low cost and low to moderate reinforcement [33]. Details on nanoclay structure and properties are discussed in Chapter 15. In 1987, the Toyota research group [34] replaced the inorganic exchange cations in the galleries of the native clay with alkylammonium surfactants and formed the organoclay. The surface chemistry was compatible with the hydrophobic polymer matrix and good dispersion was obtained. Nylon-6/clay nanocomposites were generated [35] and the research on organoclay-reinforced nanocomposites was extended to epoxy resins [36–44], polyamide [38, 45–48], polystyrene [39–52], polyurethane [53, 54], polypropylene [55, 56–58], and so on. These nanocomposites demonstrate an increase in tensile properties, reduced gas permeability, thermal stability, and flame retardance [59].

Kojima *et al.* [60] studied the nanoclay-reinforced nitrile rubber and found that the permeability of hydrogen and water vapor was reduced by about one third. Lopez-Manchado [61–63] prepared the organoclay nanocomposites based on natural rubber and noticed an increase in the cross-link density, degree of curing, structural order, and glass transition temperature. *cis*-1,4-Polyisoprene and epoxidized natural rubber were studied by Vu and coworkers [64]. Clays were incorporated into the elastomers by mixing the components in a standard internal blender or by mixing dispersions in toluene and methyl ethyl ketone. They found that the reinforcing effects depend on the degree of exfoliation. The morphology and mechanical properties of clay-reinforced styrene-butadiene rubber (SBR) were explored by Zhang *et al.* [65]. SBR latex was mixed with a clay/water dispersion to achieve the structure of layered bundles. The mechanical properties were increased compared with other fillers and regular rubber processing methods of mixing clay. Wang *et al.* [66] synthesized the silicone rubber/organomontmorillonite hybrid nanocomposites by a melt intercalation process. The properties of the resulting nanocomposites were quite close to the aerosilica-filled silicone rubber. Song *et al.* [67] prepared a high-performance nanocomposite consisting of a polyurethane elastomer and organoclay. An increase of 150% in tensile strength and strain was observed and the fatigue properties were improved. Pramanik *et al.* [68–70] used the solution method to obtain the thermoplastic elastomer/clay nanocomposites. The tensile strength was doubled by 4 wt% organophilic clay loading and the thermal stability was higher by about 34 °C.

Tsujimoto *et al.* [71] and Uyama *et al.* [72] developed green nanocomposites consisting of plant oils and clay. The epoxidized plant oil was cured in the presence of organophilic montmorillonite to produce the triglyceride-clay nanocomposites. A green nanocomposite coating was also developed by them. The hardness and mechanical properties were improved and good flexibility was shown as well as high biodegradability. Lu *et al.* [73] calculated the solubility parameter of the functionalized triglycerides and clay organic modifier and demonstrated their miscibility, as described in Chapter 12. Improvements were observed in the flexural modulus and thermal stability.

When polymers are associated with nanoclays, three possible structures can form, as shown in Figure 8.16. The polymer infiltrates into the clay galleries in the intercalated and exfoliated structure. In the intercalated

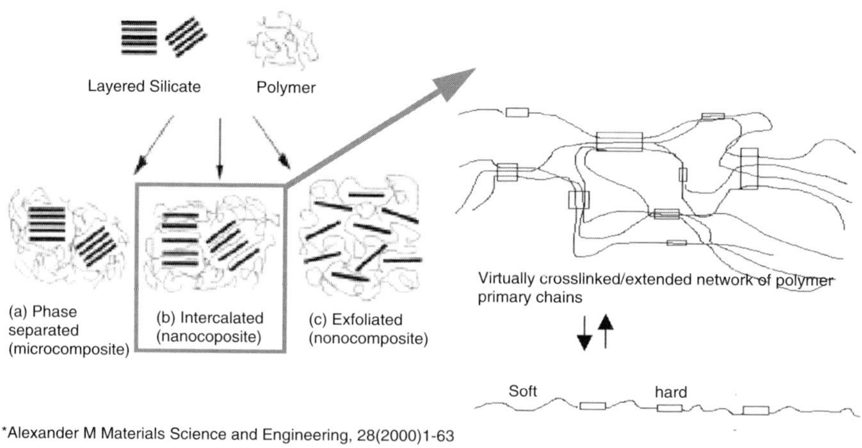

FIGURE 8.16 Structure of nanocomposites and polyurethane. (*Source:* From M. Alexander, *Materials Science and Engineering*, 2000, 28, 1–63.)

structure, the clay layers expand but still form an ordered structure. This structure is very similar to the thermoplastic polyurethane (TPU) structure in which the hard sections form physical cross-links. Therefore, in the intercalated structure, it is possible that the nanoclay generates a physical cross-linked network that has special properties.

Polymer–clay nanocomposites can be synthesized in four ways [74]: (1) exfoliation-adsorption, (2) *in situ* intercalation polymerization, (3) melt intercalation, and (4) template synthesis. Using the *in situ* polymerization method, different ratios of clay were mixed with AOME by mechanical stirring for 24 h. With the addition of 0.8 wt% cobalt naphthenate and 3 wt% Trigonox, the mixture was cured at room temperature. MMA and EGDMA were added before curing to modify the cross-link density of the final elastomer [34]. A Flory–Huggins solubility analysis (see Chapter 15) indicated that Cloisite®30B had a similar solubility parameter as poly(AOME) and was used in these experiments.

The morphology of the nanocomposite structure was measure by wide-angle X-ray diffraction (XRD) and transmission electron microscopy (TEM). Figure 8.17 shows how the degree of exfoliation depends on the clay ratio. The XRD peaks can be used to qualitatively identify the amount of the clay with certain spacing. With the increase of the clay ratios, the peak amplitude decreases at the low diffraction angles. It can be seen that Cloisite®30B remains well dispersed in AOME at loadings up to 10%. Figure 8.18(a) shows the TEM image of the distribution of clay in the polymer matrix. We can see that the clay is quite well dispersed. Some clay bundles are observed in Figure 8.18(b). The polymer infiltrated into the clay layers, but the clay still keeps its regular layered structure. The exfoliated structure is shown in Figure 8.18(c). In this state, the clay is not well aligned anymore.

Figure 8.19 shows the stress–strain curves of the 10% clay elastomers subjected to repeated loading–unloading strain cycles. It should be possible to design self-healing elastomers using the nanoclay method. The intercalated layers should be able to debond and dissipate considerable energy, and upon removing the load, the stored strain energy in the clay nanobeams would allow them to heal again, thereby restoring the original structure. Zhu and Wool [30] noted that some healing did occur, but was comparable to typical healing in filled elastomers, as discussed and measured by Wool [20]. For the nanobeam healing effect to occur, the intercalation process could be first done with a nonpolymerizing fluid and then inserted into the polymer matrix. However, the effect of the nanoclay on the mechanical properties was significant.

In Figure 8.20, we see that the elongation at break and the tensile strength increase with clay content. The nanoclay significantly improves the mechanical properties of the elastomer: With only 3% clay loading, the tensile strength increases by 180% and the strain at break by 100%. When the clay loading goes up to 10%, the tensile strength increases to 0.58 MPa and the

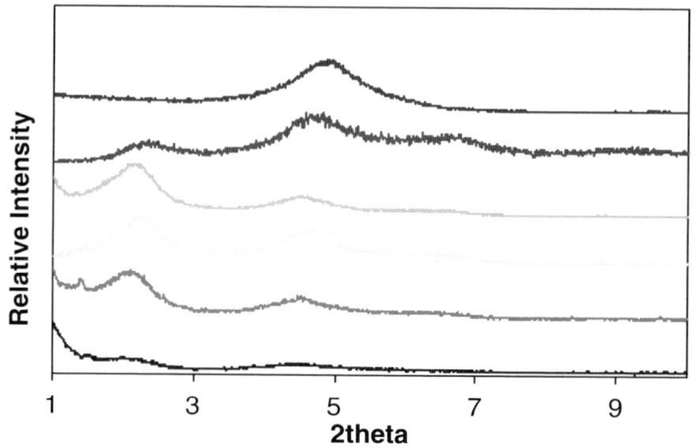

FIGURE 8.17 XRD for different clay ratios.

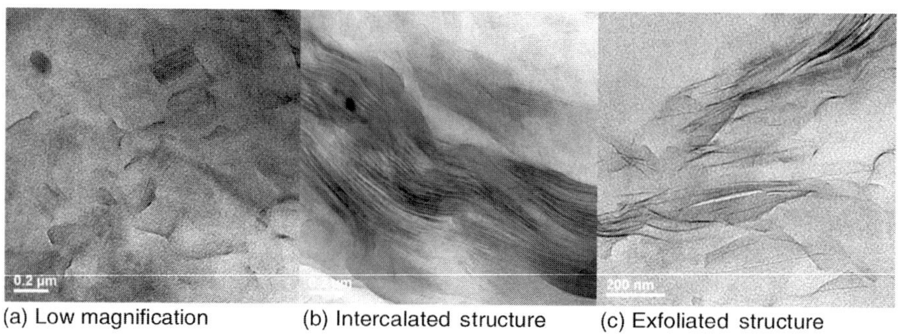

FIGURE 8.18 TEM of 5% clay-filled elastomer: (a) low magnification, (b) intercalated structure, and (c) exfoliated structure.

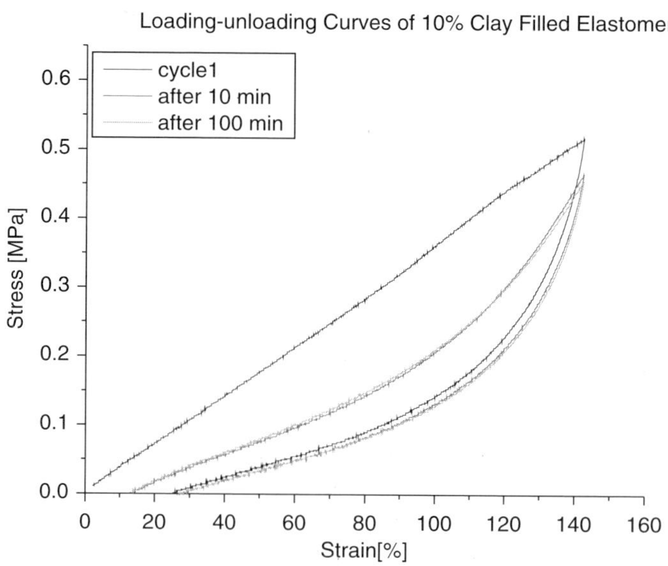

FIGURE 8.19 Loading–unloading curves of 10% clay-filled elastomers.

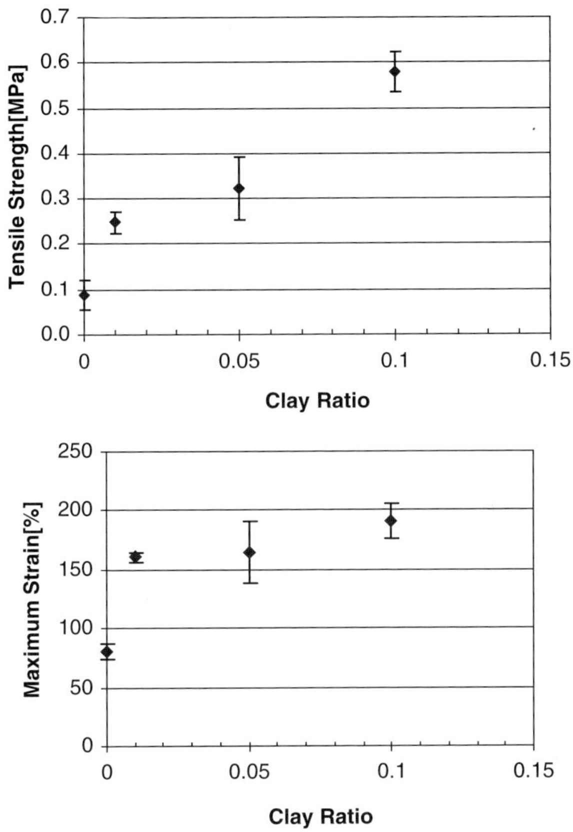

FIGURE 8.20 Fracture stress and elongation of elastomer vs. clay ratio.

maximum strain to 190%. Increasing the amount of MMA increased the tensile strength but the elongation remained about the same. The thermal decomposition initiation temperature also increased by about 12 °C from 138 ° to 144 °C with 3% nanoclay.

The fracture stress is related to the number of bonds in a percolation network, as discussed in Chapter 6. The critical stress σ required to break the network can be expressed as $\sigma = [2E\nu D_o(p - p_c)]^{1/2}$, where $[p - p_c]$ is the percolation fraction of bonds that must be broken to cause fracture in the network. Table 8.3 gives the calculated $[p - p_c]$ values. With increasing clay, $[p - p_c]$ increases linearly. Thus, the loading of the clay increases the perfection of the network by connecting the free chain end together and forming a more cross-linked structure. However, the network is still poor as indicated by the low $[p - p_c]$ value and that is the reason for the low mechanical properties. Percolation theory suggests that the optimal elastomer structure is

TABLE 8.3 Network perfection.

□	σ (MPa)	E (MPa)	ν (mol/m^3)	$p - p_c$
No clay	0.09	0.17	22.87	0.0033
3% Clay	0.25	0.19	25.56	0.0207
5% Clay	0.32	0.23	30.94	0.0239
10% Clay	0.58	0.28	37.67	0.0518

obtained with high-molecular-weight linear polymers, which are subsequently cross-linked chemically, or physically, by intercalation with nanoclays.

8.7 BIO-BASED COATINGS

With increasing interest in "green chemistry" on the part of signatory nations to the Kyoto Accord on global warming, significant efforts among the international scientific communities were directed toward the area of renewable resources [75]. Plant oils were used in varnishes and alkyd resins used in the coatings industry. Varnishes are generally physical solutions of the natural or synthetic resin in plant oils. In alkyd resins, plant oils are chemically combined with polyester resins. The chemistry of plant oils allows paints based on these resins to "air dry" through oxidative coupling reactions [76]. However, the popularity of these resins in the coatings industry is waning with the increasing demand for water-borne emulsion polymers. In 2000, the world emulsion polymer demand was $15 billion, and the coatings industry supplied more than 50% of this demand [77]. Emulsion polymers used in latex paints are the highest volume coating resins in the industry [78] and are currently produced from petroleum derivatives. Incorporating renewable plant oils in latex technology will provide a renewable and sustainable alternative for the coatings industry as well as a new market for plant oils.

Organic coatings are complex mixtures of various substances. Components include polymers or resins, volatile organic compounds (VOCs), pigments, and additives. Polymers and resins, commonly called *binders* by the coatings industry, form the continuous film that adheres to the substrate, binds other substances in the film together, and imparts film strength and durability. VOCs are used to aid in film formation. However, due to ever-increasing environmental regulations on VOC emissions, the industry is focusing on low to no VOC paint formulations. Pigments impart color, opacity, and other visual effects to the coating film. Additives enhance the properties of the final product and include dispersants, colorants, and rheology modifiers.

The polymeric binder is the main vehicle of the coating. Many types of polymeric binders are used in coating formulations; primary examples include alkyd resins, polyester resins, isocyanates (polyurethanes), drying oils, and emulsion polymers (latexes). Latexes are the primary binders used in architectural coatings, particularly in the United States [79]. They offer superior durability, lower VOC emissions, and are much easier to use than their oil-based counterparts, making them more attractive to consumers.

Latex binders are formed via emulsion polymerization. Emulsion polymerization is a free-radical polymerization in which a monomer or mixture of monomers is polymerized in an aqueous surfactant solution to form a latex [80]. Emulsion polymers used in architectural coatings are typically linear, high-molecular-weight polymers that form films under ambient conditions by the evaporation of water and solvents, and the coalescence of latex particles. Common monomers in emulsion polymers include acrylic and vinyl esters, and their selection generally depends on specific requirements of the coating and cost.

Film formation of emulsion polymers occurs by coalescence of latex particles. A schematic of this process is shown in Figure 8.21 [79, 80]. Latex binders used in coatings typically have a high solids content ranging from 20% to 50%. Solid polymer particles are dispersed in an aqueous phase. Once applied to a substrate, the water and solvents within the emulsion begin to evaporate, leading to a close-packed layer of latex particles. This is the first and longest stage of film formation, and continues until the particles make up 60–70% volume fraction. The rate of evaporation is approximately equal to the rate of evaporation of water [80].

The second stage begins when the particles are concentrated and come into irreversible contact. A clear, continuous, but still weak film is formed due to particle deformation at temperatures greater than the minimum film forming temperature (MFFT) [79, 80]. The MFFT is the lowest temperature at which coalescence occurs sufficiently to form a continuous film.

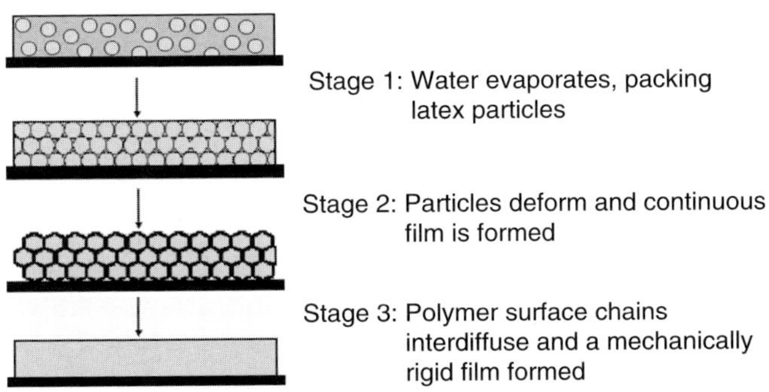

FIGURE 8.21 Film formation process of latex polymers.

The final stage of film formation occurs at temperatures above the glass transition temperature (T_g). Further coalescence transpires as polymer surface chains interdiffuse across interfaces of adjacent particles. Interdiffusion develops the mechanically coherent film, and the full strength of this film is reached when the surface chains diffuse a distance equal to their radius of gyration, R_g [20].

As mentioned earlier, plant oils are essential to oil-based coatings. These coatings form cross-linked films through autoxidation of fatty acids in the oils. The process is slow and the coatings often require high levels of solvent to aid in drying. The films also continue to oxidize and polymerize over long periods of time after application, causing eventual film degradation. Emulsion coatings are often preferable because they usually contain lower amounts of solvents and dry more rapidly. Efforts were made to extend the use of plant oils into the field of emulsion coatings in order to decrease the amount of petroleum used by the industry. Thames *et al.* [81] developed plant oil-based latex polymers for coatings that show good film-forming properties. The polymers were derived from ricinoleic acid, the primary fatty acid of castor oil. This acid comprises 90% of castor oil's triglycerides. It is a monounsaturated fatty acid with a hydroxyl functional group, as described in Chapter 4. The monomer is synthesized by converting ricinoleic acid to its methyl ester and acrylating the hydroxyl group. The monomer can then take part in free-radical polymerization of the acrylate group. The long carbon chain of the monomer also acts as a plasticizer in the films, decreasing the need for a solvent in the coating formulation. Additionally, the polymer undergoes cross-linking via oxidative polymerization of the residual double bond after surface application. This imparts additional strength to the film.

8.7.1 DESIGN OF BIO-BASED COATINGS

To obtain desired coating properties, most latex binders used in the field are actually copolymers. A critical decision in designing a latex emulsion is monomer and comonomer selection. A primary criterion for this selection is the glass transition temperature (T_g). It is crucial to select a monomer combination that produces a copolymer with the appropriate T_g; the T_g must be low enough to permit coalescence at the lowest application temperature, yet high enough to ensure coating durability [79]. Coalescence will not occur unless the temperature is at least slightly higher than the T_g. Typical T_g's of latex binders used in architectural coatings range from 0° to 25 °C. The T_g of the AOME polymer is approximately −60 °C [15]. This is too low to be used alone in formulating a latex binder for architectural coatings. A hard comonomer with a high T_g is essential to increase the T_g to an appropriate level. Alternatively, additional cross-linking reactions will also increase T_g. When using a comonomer, the amounts of different monomers necessary to produce a copolymer with the appropriate T_g can be estimated using several models discussed in Chapter 7. The comonomers considered in this work were methyl methacrylate (MMA) and styrene. The T_g of these monomers are 105 °C and 100 °C, respectively. MMA and styrene are typical comonomers used in acrylic binders. The Fox [82] equation predicts that approximately 40 wt% AOME and 60 wt% MMA or styrene is necessary to reach a common architectural coating with T_g of 15 °C. The petroleum content can be further reduced by cross-linking reactions.

The chemicals used in the miniemulsions include the AOME monomer, styrene, MMA, sodium dodecyl sulfate as the surfactant, and azodiisobutyrodinitrile as the initiator [83]. The miniemulsion was prepared by ultrasonification for 5 min and reaction at 80 °C for 1 h. The particle diameter, as determined by light scattering, ranged from approximately 90 to 170 nm, depending on the amount of comonomer used. This is consistent with the particle size for typical miniemulsions. Figure 8.22 shows the average particle diameter as a function of comonomer content. The particle size decreases with increasing styrene or MMA content.

8.7.2 COATING PROPERTIES

The storage modulus plot of the 40% styrene, 60% styrene, and 60% MMA films is shown in Figure 8.23. The glassy regions are observed for each film sample at approximately 1.5 GPa. The modulus begins to decrease for the

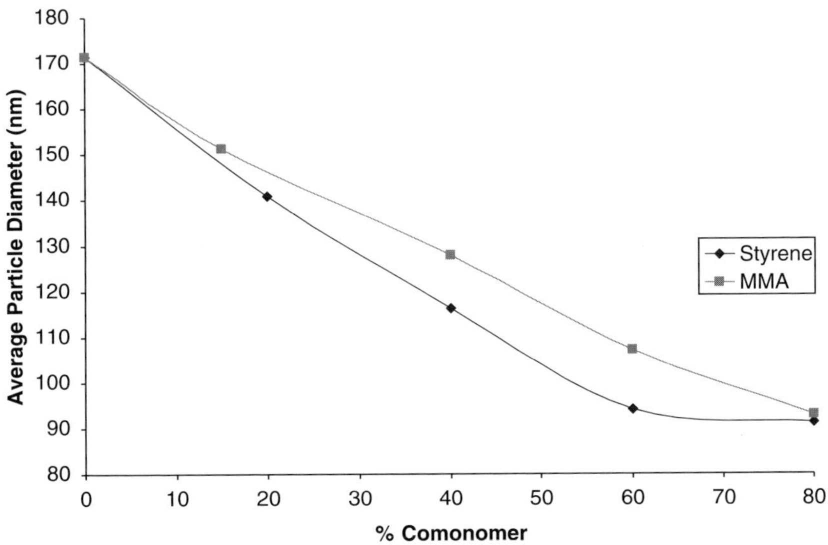

FIGURE 8.22 Average particle diameter of emulsions determined by dynamic light scattering.

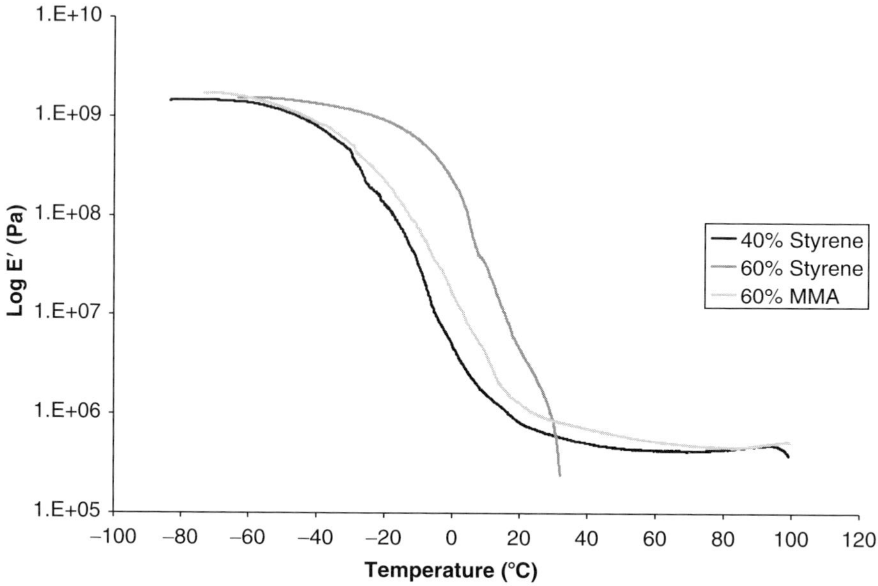

FIGURE 8.23 Storage modulus plot for films prepared from AOME copolymerized with styrene and MMA.

40% styrene film and 60% MMA film at approximately −55 °C, whereas the modulus begins to decrease for the 60% styrene film at approximately −45 °C. The storage modulus remains highest for the 60% styrene film until the temperature reaches approximately 33 °C. Here the modulus for the 60% styrene film continues to drop, whereas the other films begin to level into their rubbery plateau areas at approximately 46 kPa. This signifies that the 40% styrene and 60% MMA films are indeed cross-linked due to a higher concentration of technical-grade AOME polymer. The technical-grade OME used to produce the AOME monomer was only 70% pure. Levels of linoleic and linolenic methyl esters were present in the material. These have two and three double bonds, respectively, in the carbon chain, and these bonds have the ability to cross-link. The storage modulus of a polymer in the rubbery plateau region was used to determine the cross-link density. The cross-link density of the 40% styrene film sample at approximately 40 °C was 66.7 mol/m^3. The cross-link density of the 60% MMA film sample at approximately 50 °C was 77.1 mol/m^3.

8.7.3 NANO COATINGS

The coatings discussed in the preceding sections were quite tacky and soft in comparison to commercial latex polymers, even with rather high comonomer content. One solution may be to reinforce the coatings with mineral clays, like montmorillonite. Montmorillonite clay naturally forms stacks of platelets. These platelets are less than 10 Å thick. A solution of 10 wt% Cloisite Na+ clay to polymer in water was prepared. The films were allowed to dry under ambient conditions for approximately 16 h and then subsequently dried under vacuum to remove any excess moisture. The emulsion clay mixtures yielded hard, nontacky films. The extent of intercalation was studied through XRD. Pure Cloisite Na+ has a peak at approximately 7°. This corresponds to a *d*-spacing of approximately 12.62 Å. The 5% and 10% clay film samples showed peaks at approximately 2.4°, 4.5°, and 7° which correspond to *d*-spacing values of 36.78, 19.62, and 12.62 Å, respectively. The peak at 2.4° corresponds to clay intercalated by the polymer.

The storage moduli of the films as a function of temperature and 20%, 40%, and 60% comonomer content are shown in Figure 8.24. All specimens show a large increase in storage modulus in both glassy and rubbery plateaus. The fact that a rubbery plateau is seen on the 60% styrene film sample with 10% clay content indicates that the clay imparts properties similar to that of a cross-linked film, as discussed in the last section. The rubbery plateau is increased by three orders of magnitude to approximately 55 MPa for the MMA samples and 40 MPa for the styrene samples. The T_g for each film was obtained from the tan δ peaks and is reported in Tables 8.4 and 8.5. Overall, the addition of clay seems to either decrease or have little effect on the T_g of the films. The T_g was expected to increase due to the decrease in tackiness

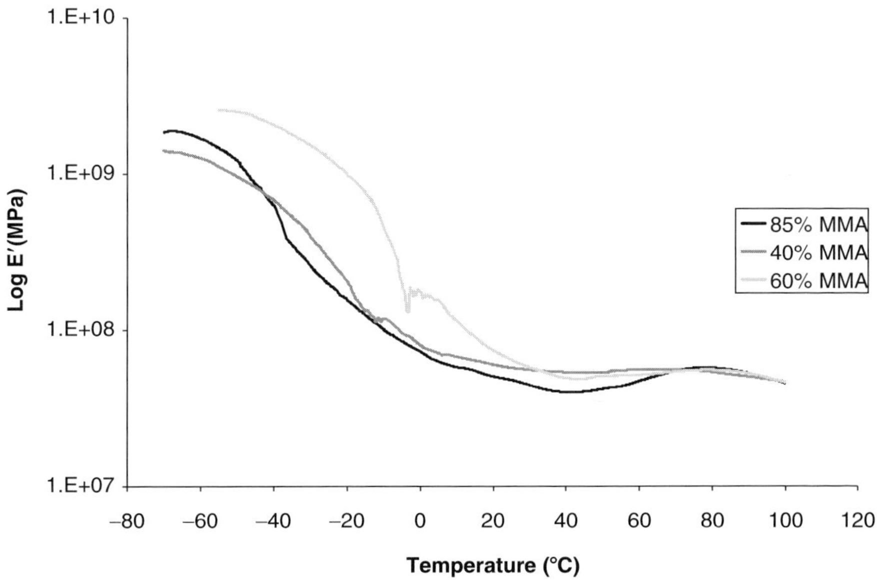

FIGURE 8.24 Storage modulus of films with 10 wt% clay to polymer, increasing MMA content.

TABLE 8.4 T_g and cross-link density of the AOME and styrene films with 10 wt% clay to polymer.

% Styrene	T_g (°C)	v_e (mol/m^3)
20	−25	10,197
40	−15	6,797
60	35	4,405

TABLE 8.5 T_g and cross-link density of the AOME and MMA films with 10 wt% clay to polymer.

% MMA	T_g (°C)	v_e (mol/m^3)
15	−36	9299
40	−23	8640

of the film upon clay addition. The measured T_g only increased for the 60% styrene film, rising from approximately 20° to 35 °C. The cross-link densities for the samples and the temperatures at which they were calculated are listed in Tables 8.4 and 8.5. There is a significant increase in the cross-link density.

The gel fraction (in benzene) of the film sample without polymer was determined to be approximately 70%. The gel fraction of the sample with clay, determined by subtracting the amount of clay in the sample from the initial and final weights, was found to be approximately 97%. The gel fraction and therefore cross-link density were greatly increased by the addition of clay, which may account for the increase in the rubbery plateau modulus.

Film hardness is often described in the coatings industry by the ASTM D-3363 Film Hardness by Pencil Test. This test employs the use of a set of pencils each with a different grade of hardness. The grade is determined by the amount of baked graphite and clay in the composition of the pencil. The grades include 9H, 8H, 7H, 6H, 5H, 4H, 3H, 2H, H, F, HB, B, 2B, 3B, 4B, 5B, 6B, 7B, 8B, and 9B. The hardest is 9H, F is the middle of the scale, and 9B is the softest. The test is performed by flattening the lead of the pencil at a 90° angle using 400-grit sandpaper. Starting with the lowest grade pencil, the pencil is held at a 45° angle to the film and pushed forward 1/4 inch using as much downward pressure as can be applied without breaking the lead. This is repeated with increasing grade pencils until the film is scratched. The hardness of the film is determined by the grade of pencil that scratches the film. Typical coatings range in hardness from 3B to 9H [84, 85]. Both emulsion films with and without nanoclay were cast on glass slides and allowed to dry overnight under ambient conditions and subsequently dried under vacuum. The pencil test was then performed as described. The nanoclay increased the film hardness of the emulsion polymers from 8B to 3B.

The addition of clay would allow films with low hard comonomer content to serve as coatings. In fact, the styrene copolymer nanocoatings showed increasing rubbery storage modulus with decreasing styrene content. This is very promising because the clay is both environmentally and economically friendly. Its use in combination with the plant oil-based AOME monomer could drastically reduce the amount of petroleum-based monomer needed in latex binders for architectural coatings.

REFERENCES

1. Satas, D., Ed. *Handbook of Pressure Sensitive Adhesive Technology*, 2nd ed.; Van Nostrand Reinhold, New York; 1989.
2. Pryde, E. H. *Fatty Acids*, American Oil Chemists' Society, Champaign, IL; 1979.
3. Wool, R. P.; Kusefoglu, S. H.; Palmese, G. R.; *et al.High Modulus Polymers and Composites from Plant Oils,* U.S. Patent 6,121,398; 2000.
4. Bunker, S. P.; Wool, R. P. *J. Polym. Sci. A: Polym. Chem.* 2001, **40,** 451–458.
5. Lovell, P. A.; El-Aasser, M. *Emulsion Polymerization and Emulsion Polymers,* John Wiley & Sons, New York; 1997.
6. Gilbert, R. G. *Emulsion Polymerization: A Mechanistic Approach,* Academic Press, Inc., San Diego; 1995.
7. Charmeau, J. Y.; Kientz, E.; Holl, Y. *Prog. Organic Coatings* 1996, **27**(1–4), 87–93.
8. Kientz, E.; Holl, Y. *Coll. Surf. A: Physiochem. Eng. Aspects* 1993, **78,** 255–270.
9. Zosel, A.; Schuler, B. *J. Adhesion* 1999, **70,** 179–195.
10. Khot, S. N. Synthesis and Application of Triglyceride-based Polymers. In *Chemical Engineering,* University of Delaware, Newark; 2000.
11. Zosel, A. *Int. J. Adhesion Adhesives* 1998, **18,** 265.
12. Lovell, P. A.; Shah, T. H. *Polym. Commun.* 1991, **32**(4), 98–103.
13. Chang, E. *J. Adhesion* 1997, **60,** 233–248.
14. Zosel, A. *J. Adhesion* 1991, **34,** 201–209.
15. Bunker, S. P. *Ph.D. Thesis,* University of Delaware; 2002.
16. Bunker, S.; Staller, C.; Willenbacher, N.; *et al. Int. J. Adhesion Adhesives* 2003, **23,** 29–38.
17. R. P. Wool; Bunker, S. P. U.S. Patent 6,646,033; 2003.
18. Lee, I.; Wool, R. P. *J. Polym. Sci. B: Polym. Phys.* 2002, **40**, 2343.
19. Fowkes, F. M.; Mostafa, M. A. *Ind. Eng. Chem. Prod. Res. Dev.* 1978 **17**, 3.
20. Wool, R. P. *Polymer Interfaces: Structure and Strength*, Hanser/Gardner Publications, Cincinnati; 1995.
21. http://www.elastomersolutions.com.
22. *Synthetic Rubber—The Story of an Industry,* International Institute of Synthetic Rubber Producers; 1973.
23. Blackley, D. C. *Synthetic Rubbers: Their Chemistry and Technology;* 1983.
24. Athawale, V.; Kolekar, S. *J. Macromolec. Sci.: Pure Appl. Chem.* 2000, **37,** 65–79.
25. Athawale, V.; Kolekar, S. *Polym. J.* 1998, **30,** 813–818.
26. Athawale, V.; Raut, S. *Polym. J.* 1998, **30,** 963–967.
27. Athawale, V.; Raut, S. *Physical Chem. Chemical Phys.* 2000, **2,** 1249–1254.
28. Athawale, V.; Kolekar, S. *J. Appl. Polym. Sci.* 2000, **75,** 825–832.
29. Raut, S.; Athawale, V. *European Polym. J.* 2000, **36,** 1379–1386.
30. Zhu, L.; Wool, R. P. *Amer. Chem. Soc Preprints*, Philadelphia National Meeting; August 2004.
31. Mark, J. E.; Erman, B.; Eirich, F. T. *Science and Technology of Rubber,* Academic Press; 1994.

32. Wool, R. P. *J. Polym. Sci. Part B: Polym. Phys.* 2005, **43,** 168.
33. Morton, M., Ed. *Rubber Technology*, Van Nostrand Reinhold Company, New York; 1987.
34. Fukushima, Y.; Inagaki, S. *J. Inclusion Phenomena* 1987, **5,** 473–482.
35. Usuki, A.; Kojima, Y.; Kawasumi, M.; *et al. J. Mater. Res.* 1993, **8,** 1179–1184.
36. Salahuddin, N. A. *Polym. Adv. Technol.* 2004, **15,** 251–259.
37. Lan, T.; Pinnavaia, T. J. *Chem. Mater.* 1994, **6,** 2216–2219.
38. Ratna, D.; Manoj, N. R.; Varley, R.; *et al. Polym. Int.* 2003, **52,** 1403–1407.
39. Pinnavaia, T. J.; Lan, T.; Wang, Z.; *et al.* In *Nanotechnology;* 1996, pp. 250–261.
40. Ratna, D.; Becker, O.; Krishnamurthy, R.; *et al. J. Polym.* 2003, **44,** 7449–7457.
41. Kang, J. H.; Lyu, S. G.; Sur, G. S. *Polymer-Korea* 2000, *24,* 571–577.
42. Daniel, I. M.; Miyagawa, H.; Gdoutos, E. E.; *et al. Exper. Mechan.* 2003, **43,** 348–354.
43. Lee, C. R.; Ihn, K. J.; Gong, M. S. *Polymer-Korea* 2003, **27,** 392–395.
44. Kornmann, X.; Lindberg, H.; Berglund, L. A. *Polymer* 2001, **42,** 1303–1310.
45. Lan, T.; Kaviratna, P. D.; Pinnavaia, T. *J. Chem. Mater.* 1994, **6,** 573–575.
46 Delozier, D. M.; Orwoll, R. A.; Cahoon, J. F.; *et al. Polymer* 2002, **43,** 813–822.
47. Yano, K.; Usuki, A.; Okada, A.; *et al. J. Polym. Sci. A: Polym. Chem.* 1993, **31,** 2493–2498.
48. Yano, K.; Usuki, A.; Okada, A. *J. Polym. Sci. A: Polym. Chem.* 1997, **35,** 2289–2294.
49. Kim, K. Y.; Lim, H. J.; Park, S. M.; *et al. Polymer-Korea* 2003, **27,** 377–384.
50. Tseng, C. R.; Wu, J. Y.; Lee, H. Y.; *et al. J. Appl. Polym. Sci.* 2002, **85,** 1370–1377.
51. Qutubuddin, S.; Fu, X. A.; Tajuddin, Y. *Polym. Bull.* 2002, **48,** 143–149.
52. Fu, X.; Qutubuddin, S. *Mater. Lett.* 2000, **42,** 12–15.
53. Chen, A. M.; Tian, Y.; Han, B.; *et al. Acta Polym. Sinica* 2003, 591–594.
54. Ma, J. S.; Qi, Z. N.; Zhang, S. F. *Acta Polym. Sinica* 2001, 325–328.
55. Wang, Z.; Pinnavaia, T. J. *Chem. Mater.* 1998, **10,** 3769.
56. Zhang, Y. Q.; Lee, J. H.; Rhee, J. M.; *et al. Compos. Sci. Technol.* 2004, **64,** 1383–1389.
57. Ma, J. S.; Qi, Z. N.; Hu, Y. L. *J. Appl. Polym. Sci.* 2001, **82,** 3611–3617.
58. Hasegawa, N.; Okamoto, H.; Kawasumi, M.; *et al. J. Appl. Polym. Sci.* 1999, **74,** 3359–3364.
59. Alexandre, M.; Dubois, P. *Mater. Sci. Eng. R: Reports* 2000, **28,** 1–63.
60. Kojima, Y.; Fukumori, K.; Usuki, A.; *et al. J. Mater. Sci. Lett.* 1993, **12,** 889–890.
61. Lopez-Manchado, M. A.; Herrero, B.; Arroyo, M. *Polym. Int.* 2003, **52,** 1070–1077.
62. Arroyo, M.; Lopez-Manchado, M. A.; Herrero, B. *Polymer* 2003, **44,** 2447–2453.
63. Lopez-Manchado, M. A.; Arroyo, M.; Herrero, B.; *et al. J. Appl. Polym. Sci.* 2003, **89,** 1–15.
64. Vu, Y. T.; Mark, J. E.; Pham, L. H.; *et al. J. Appl. Polym. Sci.* 2001, 82, 1391–1403.
65. Zhang, L. Q.; Wang, Y. Z.; Wang, Y. Q.; *et al. J. Appl. Polym. Sci.* 2000, **78,** 1873–1878.
66. Wang, S. J.; Long, C. F.; Wang, X. Y.; *et al. J. Appl. Polym. Sci.* 1998, **69,** 1557–1561.
67. Song, M.; Hourston, D. J.; Yao, K. J.; *et al. J. Appl. Polym. Sci.* 2003, **90,** 3239–3243.
68. Pramanik, M.; Srivastava, S. K.; Samantaray, B. K.; *et al. Macromolec. Res.* 2003, **11,** 260–266.
69. Pramanik, M.; Srivastava, S. K.; Samantaray, B. K.; *et al. J. Appl. Polym. Sci.* 2003, **87,** 2216–2220.
70. Pramanik, M.; Srivastava, S. K.; Samantaray, B. K.; *et al. J. Polym. Sci. B: Polym. Phys.* 2002, **40,** 2065–2072.
71. Tsujimoto, T.; Uyama, H.; Kobayashi, S. *Macromolec. Rapid Commun.* 2003, **24,** 711–714.
72. Uyama, H.; Kuwabara, M.; Tsujimoto, T.; *et al. Chem. Mater.* 2003, **15,** 2492–2494.
73. Lu, J.; Hong, C. K.; Wool, R. P. *J. Polym. Sci. B: Polym. Phys.* 2004, **42,** 1441–1450.
74. Kalgaonkar, R. A.; Jog, J. P. *J. Polym. Sci. B: Polym. Phys.* 2003, **41,** 3102–3113.
75. Anastas, P. T.; Warner, J. C. *Green Chemistry, Theory and Practice,* Oxford University Press, Oxford; 1998.
76. Solomon, D. H. *The Chemistry of Organic Film Formers,* John Wiley and Sons, Inc., New York; 1967.
77. The Freedonia Group, I. Adhesives and Sealants Industry; 2002www.**freedoniagroup**.com/coatings.html

78. Lambourne, R. *Paint and Surface Coatings: Theory and Practice,* Ellis Horwood Limited, New York; 1987.
79. Wicks, Z. W.; Jones, F. N.; Pappas, S. P. *Organic Coatings: Science and Technology,* 2nd ed., Wiley-Interscience, New York; 1999.
80. Lovell, P. A.; El-Aasser, M. S. *Emulsion Polymerization and Emulsion Polymers,* John Wiley and Sons, New York; 1997.
81. Thames, S. F.; Yu, H. B.; Subramanian, R. *J. Appl. Polym. Sci.* 2000, **77,** 8–13.
82. Fox, T. G. *Bull. Amer. Phys. Soc.* 1956, **1,** 123.
83. Kulbick, A. L., *M.S. Thesis,* University of Delaware; 2004.
84. http://www.psrc.usm.edu/macrog/mpm/composit/nano/struct2_1.htm; 2004.
85. http://www.rohmhaaspowdercoatings.com/tech/technical_briefs/techbriefs.hardpencil.jsp; 2004.

9

Thermal and Mechanical Properties of Soy Proteins

Xiuzhi Susan Sun

9.1 Structure and Thermal Behavior of Soy Protein

9.1.1 Structure of Soy Protein

Like many other plant proteins, soybean proteins are mainly storage proteins that provide amino acids during seed germination and protein synthesis. The monomer of the storage protein has the same amino acid residues as many other proteins and is linked by amide bonds into polypeptide chains as shown in Figure 9.1. The polypeptide chains are associated and entangled into a three-dimensional complicated structure by disulfide and hydrogen bonds with a molecular weight ranging from 300,000 to 600,000 Da. Soy proteins can be divided into water-soluble albumins and salt solution-soluble globulins. Most soy proteins are globulins, containing about 25% acidic amino acids, 20% basic amino acids, and 20% hydrophobic amino acids. The composition of the amino acids of a soy protein and its main components glycinin and conglycinin is given in Table 9.1.

Based on sedimentation coefficients obtained from centrifugation [1], soy proteins contain four major components, including 2, 7, 11, and 15 based on the Svedburg unit, which is calculated as the rate of sedimentation per unit field of centrifugal strength. Generally, soy protein contains about 30–50%

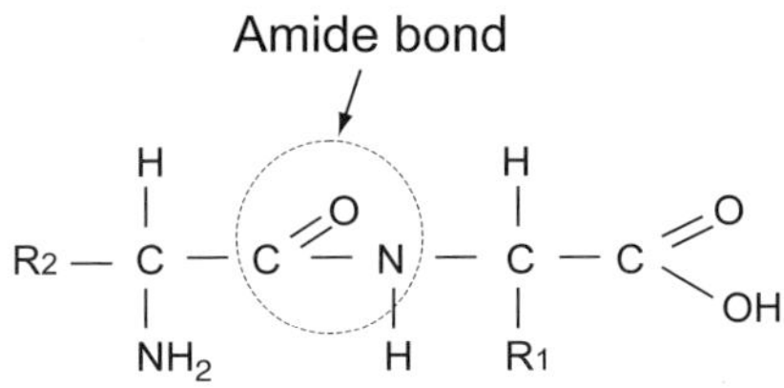

FIGURE 9.1 Structure of amide bond linking amino acid.

TABLE 9.1 Amino acid composition in percentage (%) of soy protein (determined at Bio Tech Core Lab at Kansas State University) and its components.

Amino Acids	Soy Protein	Soy Glycinin	Soy Conglycinin
Tryptophan	—	0.75	0.30
Isoleucine	2.86	4.24	6.40
Tyrosine	2.57	2.81	3.60
Phenylalanine	3.92	3.85	7.40
Proline	5.73	6.85	4.30
Leucine	8.02	7.05	10.30
Valine	3.39	4.83	5.10
Lysine	5.88	4.44	7.00
Methionine	1.23	0.98	0.30
Cysteine	0.99	1.44	0.00
Alanine	6.05	5.16	3.70
Arginine	4.68	5.81	8.80
Threonine	4.05	3.91	2.80
Glycine	9.74	7.50	2.90
Serine	7.46	6.66	6.80
Histidine	5.00	1.89	1.70
Aspartate	11.55	11.88	14.10
Glutamate	16.89	19.97	20.50

Source: Data from Nielson [72].

11S, referring to glycinin, and 20–30% 7S, referring to conglycinin. The glycinin has a molecular weight of about 200–400 kDa, and conglycinin about 100–200 kDa. Both glycinin and conglycinin contain similar secondary structures and random coils: about 6% α-helix, 38% β-sheet, and 57% random coils. Glycine, tyrosine, and tryptophan are often buried inside, whereas tryptophan in conglycinin is often exposed outside. Conglycinin contains three major subproteins: α (68 kDa), β (175 kDa), and γ (150–200 kDa) conglycinins. Glycinin has two major subproteins: basic glycinin of about 20-kDa molecular weight, and acidic glycinin of about 40-kDa molecular weight. Each subglycinin has six isoelectric points forming two-layer trimers that are connected by disulfide bonds (Figure 9.2). These subproteins are often determined using a gel electrophoresis method that is referred to as the SDS–PAGE (sodium dodecyl sulfate–polyacrylamide gel

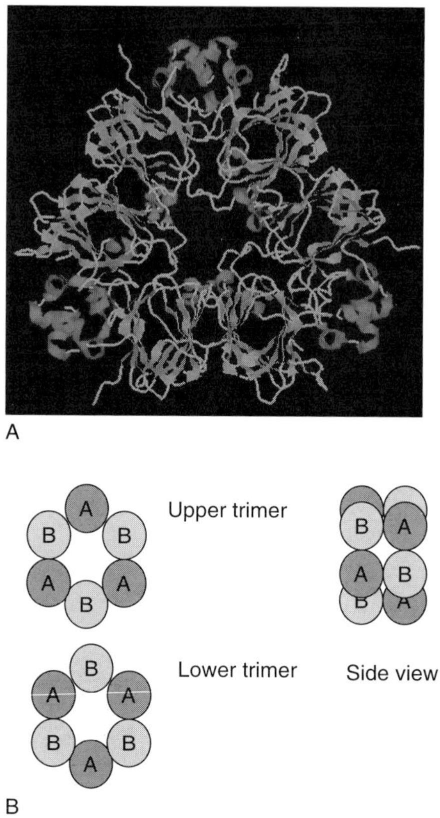

FIGURE 9.2 Structure of glycinin: (A) Stereo view of the ribbon diagram. (*Source:* Regenerated by Zheng, Mitchell, and Sun at Kansas State University in 2003; data from M. Adachi *et al.*, *J. Molec. Biol.* 2001, **305,** 291–305.) (B) Schematic diagram of glycinin dimer showing position of the acidic, A, and basic, B, subunits. (*Source:* Adapted from Bradley *et al.*, *Biochem. Biophys. Acta* 1995, **412,** 214–228.)

electrophoresis) method. A typical SDS–PAGE picture of a soy protein is illustrated in Figure 9.3. The percentage of each fraction or component can be quantified by using a scanner and computational method. The pH value of soy protein at the isoelectric point is approximately 4.5. For glycinin it is approximately 6.8, while conglycinin has a pH of approximately 4.8. The solubility of the protein is lowest at its isoelectric point.

9.1.2 THERMAL BEHAVIOR OF SOY PROTEIN

Denaturation, such as the process of boiling an egg, causes a major phase change in proteins. Protein denaturation involves structural or conformational changes from the native structure without alteration of the amino acid

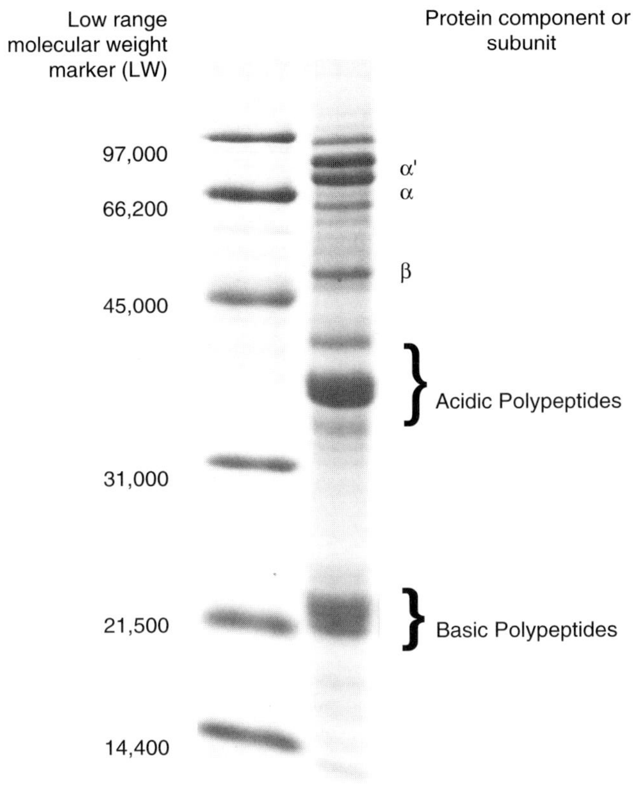

FIGURE 9.3 Soy protein fractions determined using the SDS–PAGE method. Soybean variety (K1410) was provided by Dr. W. Schapaugh (Agronomy, Kansas State University).

sequence. These changes can be induced by pH, detergents, urea, and guanidine hydrochloride, as well as by heat. Protein denaturation is apparently a highly cooperative process accompanied by considerable enthalpy changes [2, 3]. Heat-induced denaturation can be observed in the differential scanning calorimeter (DSC) scan (Figure 9.4). The first peak corresponds to conglycinin denaturation, and the second peak is caused by the denaturation of glycinin. The compact protein structure unfolds during denaturation. The unfolding is accompanied by the breaking and reforming of the intermolecular and intramolecular interactions [4]. Unfolding protein absorbs heat when it is denatured. The peak temperature is defined as the denaturation temperature (T_d) and the enthalpy is defined as the denaturation heat of fusion (H_{df}). Heat-induced denaturation disappears in the second DSC scan (Figure 9.4), which means that soy protein is an irreversible thermal set polymer.

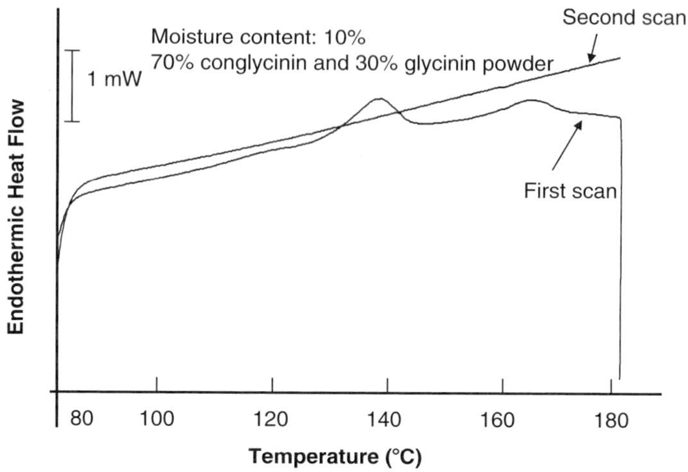

FIGURE 9.4 Thermograms of soybean protein containing 70% conglycinin and 30% glycinin proteins with 10% moisture content scanned using a differential scanning calorimeter from 30° to 200°C at a heating rate of 10°C/min.

Soy protein decomposes at 230°C (Figure 9.5). The decomposition temperatures of conglycinin and glycinin subproteins are all similar to soy protein, and the weight loss curve patterns are almost identical. The first weight loss occurs between room temperature and approximately 100°C, mainly from water evaporation. Protein weight remains constant from 100° to 200°C and then the protein starts to lose weight, primarily from the decomposition of the components in the protein. At more than 500°C, the sample weight loss slows, and remains almost constant again, and the solids left are mainly ashes.

9.1.3 INTERACTIONS OF PROTEIN WITH WATER

Water is important in maintaining the three-dimensional structure of proteins, and it affects their physicochemical and functional properties and behavior during processing, storage, preparation, and consumption. Many studies on protein–water interactions have been undertaken and they have provided significant advances in understanding the involvement of water in protein structure, stability, dynamics, and function [5–9]. With increasing water content, more water molecules interact with protein to form a hydration shell, and the molecular mobility in the system is improved. The denaturation temperature T_d and the thermal stability decrease significantly with an increase in water concentration. Water is considered a major plasticizer in the processing of soybean proteins. Generally, the denaturation temperature of soybean proteins decreases with water content [10–14].

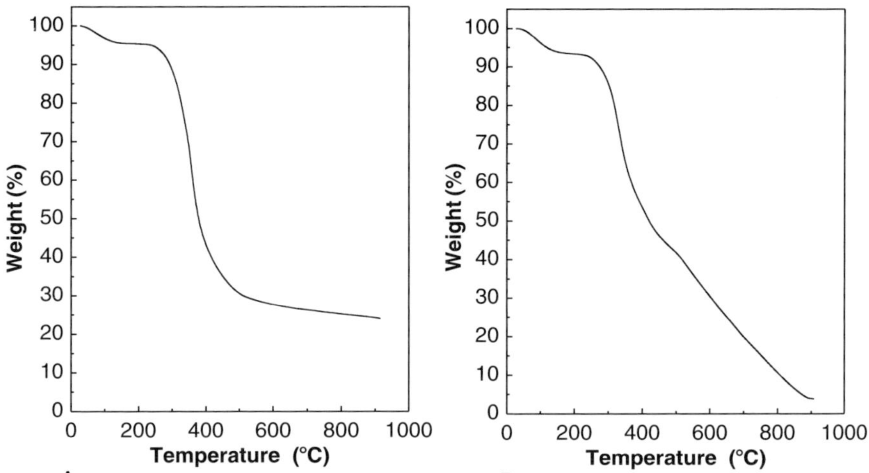

FIGURE 9.5 Thermogravimetry graphs of soy proteins: (A) Commercial soybean protein isolates with 90% protein content. (B) Soybean protein isolated from commercial soy flour with 92% protein content using laboratory procedure. Soy protein samples were thermally scanned from 30° to 900°C at a heating rate of 20°C/min in nitrogen environment.

The denaturation heat of fusion H_{df} displays some complicated relationships with water content [10–13]. Kitabatake *et al.* [12] stated that the H_{df} of soy proteins was independent of water content. Other studies showed that H_{df} for the glycinin protein decreased as water content increased, whereas H_{df} for the conglycinin protein increased first, and then decreased as water content increased [10, 13]. Zhong and Sun [15] found that the H_{df} of the soy glycinin protein increased as water content increased to 30 wt% and then leveled off at higher water content.

The increase in H_{df}, along with the increase in water content, is a common phenomenon for protein denaturation [6]. The unfolded proteins have a more extensive interface with water than the folded ones. The average conformation in the unfolded state changes as the hydration degree is reduced. The unfolded state should be more compact and have more internal bonding at lower water content, resulting in a decrease in H_{df}.

The manner of sample preparation also affects the thermal properties of a protein [15]. The denaturation process is an intramolecular change involving the destruction of internal order. Soybean protein is a complex protein having a quaternary structure, and the process of denaturation involves dissociation into subunits [16], meaning that the H_{df} includes the energy used to break down the quaternary structure and to dissociate the proteins into subunits. Mechanical shearing might cause a quaternary or tertiary structure change in soy protein [17], and protein molecules could be unfolded or uncoiled in shearing [18]. However, as studied by Zhong and Sun [15], structural changes caused by mechanical shearing did not have a significant influence on T_d (Figure 9.6). The denaturation stability of glycinin protein may be due mainly to the stability of the tertiary and/or secondary structure of the protein. Similar results were obtained for conglycinin proteins [15].

Water molecules interact strongly with protein molecules when limited water is available. The water cannot freeze under experimental conditions; such water is referred to as nonfreezable water or bond water. As water content increases, some water molecules become relatively free, not bound tightly by the protein, with properties similar to those of the bulk water. These water molecules can freeze under certain conditions as illustrated in Figure 9.7. The fusion transition of the water in the soy protein becomes detectable at around 0 °C when water content is 24 wt% or higher, and the fusion peak area increases with water content.

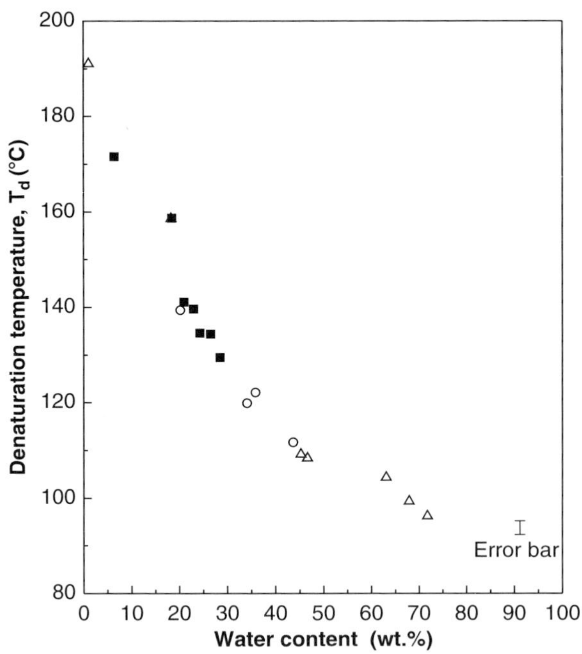

FIGURE 9.6 Relationship of denaturation temperature (T_d) of glycinin protein to water content. Samples were prepared by exposing in a desiccator with 95% moisture atmosphere (■); being dissolved first in distilled water to form 20% solution and then freeze-dried (△); and being kneaded in a glass mortar at a given moisture content and equilibrium at 25 °C for 48 h (○). The T_d values were determined with DSC by first quenching the sample to –40 °C and holding at that temperature for 1 min, scanned at 10 °C/min from 40° to 220 °C. (*Source:* Reprinted with permission from Zhong and Sun, 2000, Thermal Behavior and Nonfreezing Water of Soybean Protein Components, *Cereal Chemistry* 77:495–500. © 2000 American Association of Cereal Chemists.)

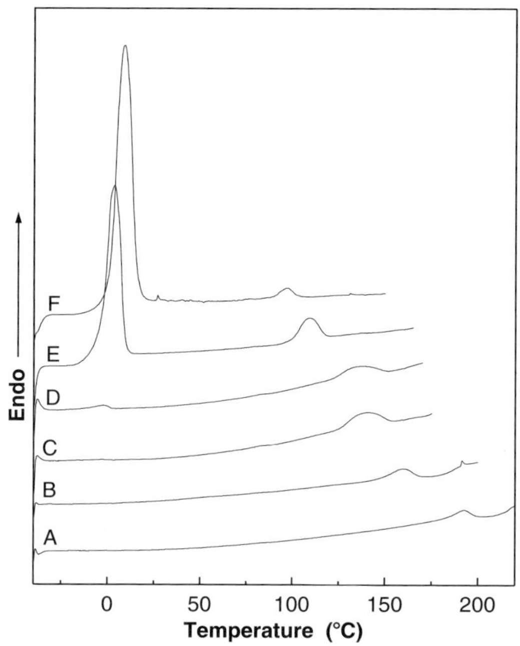

FIGURE 9.7 Thermograms of glycinin protein determined using DSC. Water content (%) of the protein sample is 1, 18, 23 , 24 , 45, and 72 (curves A–F, respectively). Sample preparation and DSC methods followed procedures similar to those described in Figure 9.6. (*Source:* Reprinted with permission from Zhong and Sun, 2000, Thermal Behavior and Nonfreezing Water of Soybean Protein Components, *Cereal Chemistry* 77:495–500. © 2000 American Association of Cereal Chemists.)

When the water content is 23 wt% or lower, all water in a protein is nonfreezing water (Figure 9.8). The arrow indicates the minimal water content below which there is no ice formation in the protein, which corresponds to 0.30–0.32 h (gram of water per gram of protein, mass ratio) [15, 19, 20]. The nonfreezing water increases rapidly as the water content increases (Figure 9.8), up to 40 wt%, mainly due to insufficient water in the protein. The amount of nonfreezing water before protein denaturation is lower than that after denaturation. As mentioned before, the compacted globular conformation of a protein becomes random coils upon denaturation. Such a process exposes some previously buried peptide bonds and amino acid side chains that can interact with water [21], resulting in an increase of 2–10% of bound water (nonfreezing water) [8].

Protein aggregation after denaturation may actually decrease water binding by replacing water–protein interactions with protein–protein interactions, which may lead to underestimating the amount of absorbed water during denaturation. At the interface between the protein surface and the bulk water, nonfreezing water is the result of competition between the protein

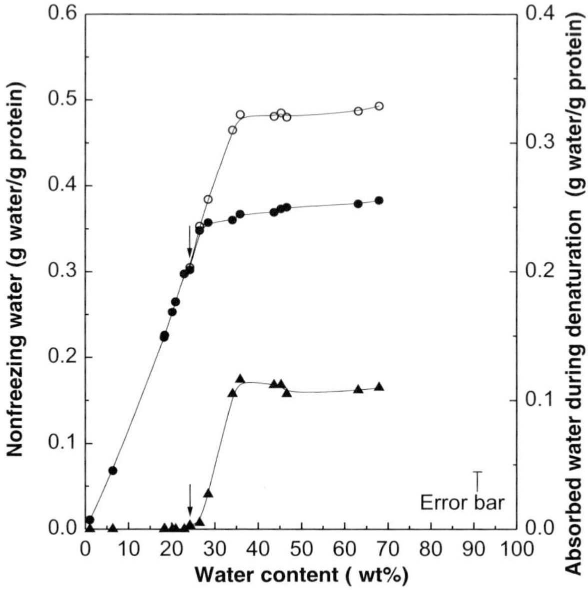

FIGURE 9.8 Variations of nonfreezing water of glycinin protein before (●) and after (○) denaturation and after absorbing water during denaturation (▲) with water content were determined from DSC measurements. Sample preparation and DSC methods followed procedures similar to those described in Figure 9.6. (*Source:* Reprinted with permission from Zhong and Sun, 2000, Thermal Behavior and Nonfreezing Water of Soybean Protein Components, *Cereal Chemistry* 77:495–500. © 2000 American Association of Cereal Chemists.)

surface and growing ice crystals. Ice crystals are expected to incorporate interface water over regions of the surface where water does not interact strongly with protein. Therefore, the measurement of nonfreezing water could be an indicator of a protein–water interaction.

9.2 CURING STRENGTH OF SOY PROTEINS

9.2.1 IMPORTANT CURING PARAMETERS

As a thermoset polymer, proteins become harder when they are cross-linked, which is called the *curing process*. Different protein structures have different thermal and flow-curing behaviors and result in different rheological properties at each thermal state. The curing tensile strength of soy protein is significantly affected by molding temperatures and initial moisture content [22–24]. For example, soy protein plastics prepared from soy protein powder with 11.7% moisture content have a tensile strength of 40 MPa at a 140°C molding temperature, whereas the tensile strength is 35 MPa at a

125 °C molding temperature [22]. Using the percolation theory in Chapter 7, $\sigma \sim \nu^{1/2}$, this corresponds to a 30% increase in cross-link density at 140 °C.

Like many polymers, soy proteins change their phases when they are exposed to large temperature changes. Phase transition temperatures are defined as the temperature ranges over which the polymers change their phases. The phase transition temperature is often affected by such factors as molecular structure, composition, and chemical or enzymatic treatment. Therefore, phase transition is temperature, time, and composition dependent, and also represented a materials-specific change in the physical state. Many properties of the polymer, especially mechanical and rheological properties, are related in particular to its phase transition temperature as well as its processing conditions. Soy proteins with different chemical treatments or different levels of plasticizers have different phase transition behavior and, consequently, different optimum processing conditions, which must be used to produce better quality products.

9.2.2 EFFECTS OF TEMPERATURE AND TIME ON CURING STRENGTH

Two main external factors, heat and pressure, influence the protein curing behavior and, consequently, the mechanical properties of soybean protein plastics [24]. Protein molecules unfold and become aggregated upon heating [25, 26], revealing association/dissociation behaviors [27, 28]. Analyzed with a dynamic mechanical analyzer (DMA), the cured modulus of a soy protein also confirms that soy protein is a thermoset polymer [29]. The modulus of a soy protein with 40% moisture reaches its lowest level between 70° and 100 °C (Figure 9.9). The curing moduli increase as temperature increases, reaching a constant level before decomposing at approximately 200 °C. The curing moduli are also affected by glycerol content, a plasticizer that also reduces phase transition temperatures. Such curing processes are temperature dependent (Figure 9.10), and the curing speed is much faster at 140 °C than at 100 °C, which is also approved in a real press molding experiment [24] as presented in Figure 9.11. It takes longer for soy protein to reach its maximum curing strength and elongation at a lower temperature than at a higher temperature. For example, at a molding temperature of 120 °C, the soy protein plastic takes about 10 min to reach maximum curing strength, whereas at 150 °C, the curing process takes about 3 min.

The color of the soy protein specimen changes from whitish yellow at low temperature, such as 100 °C, to brown and transparent at 150 °C and then to dark brown at 160 °C. Soy protein plastic prepared at 150 °C has a smooth surface (Figure 9.12B), indicating that the protein molecules have melted/unfolded, distributed homogeneously, and then interacted with each other, becoming well entangled upon curing. The plastic prepared at 100 °C has granular proteins and voids (Figure 9.12A). Plastics with a smooth,

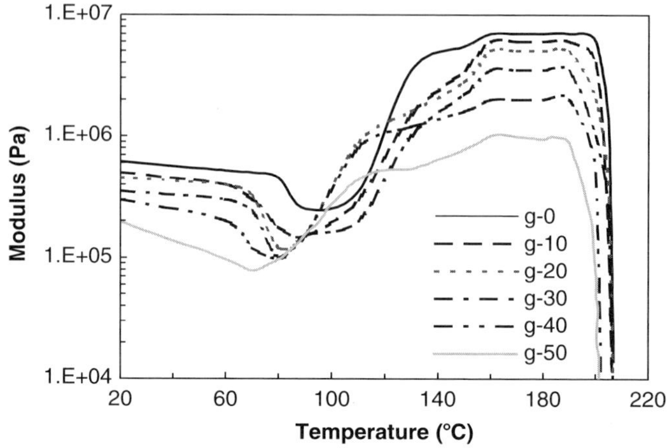

FIGURE 9.9 Storage moduli of soy protein samples with various glycerol contents (semilog coordinates). Curves are denoted as g-0, g-10, g-20, g-30, g-40, and g-50, where "g" stands for glycerin and the numbers represent concentration percentage. Samples were scanned from 20° to 200°C at the rate of 2°C/min using a dynamic mechanical analyzer. (*Source:* Copyright Feng *et al., J. Polym. Eng.* 1999, **19,** 383–393.)

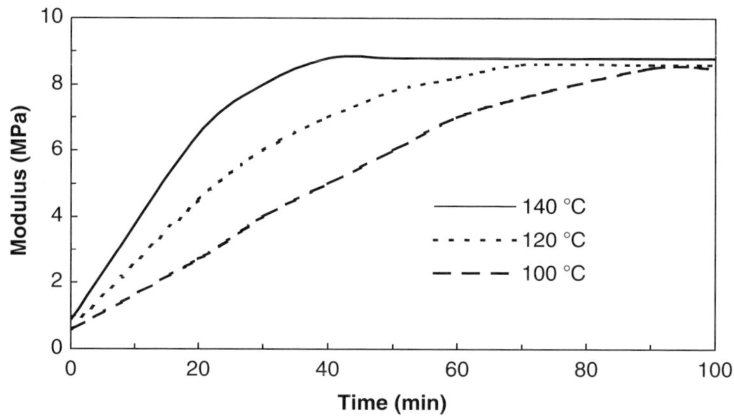

FIGURE 9.10 Isothermal curve of pure protein polymer. Samples were kept at constant temperature, and the modulus was measured every 5 sec using a dynamic mechanical analyzer. (*Source:* Copyright Feng *et al., J. Polym. Eng.* 1999, **19,** 383–393.)

continuous structure are significantly stronger in both tensile strength and elongation than plastics with a loose, void structure.

The interactive effects of molding temperature and time on tensile strength and elongation of soy proteins in the presence of 25% glycerol are similar to the results for soy protein alone. However, the highest tensile strength and elongation occur at a molding temperature of 140°C, instead of 150°C, due to the plastication effects of glycerol.

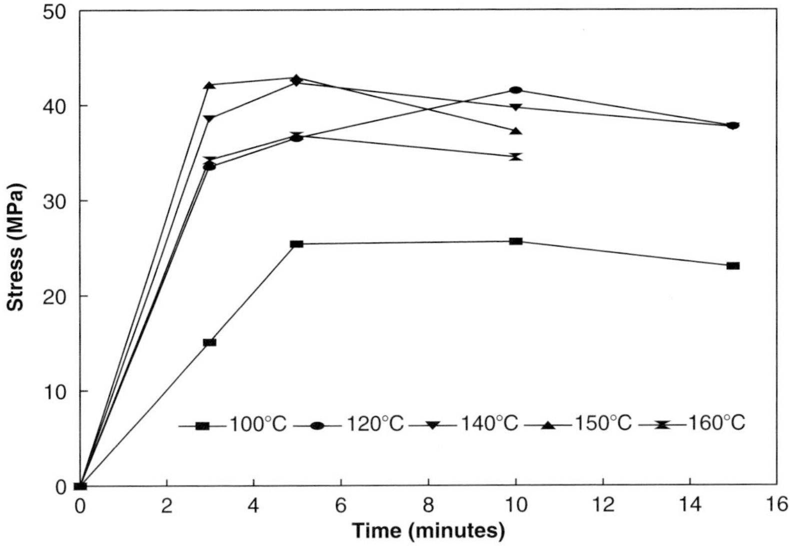

FIGURE 9.11 Effects of molding temperatures and times on tensile strength of soy protein plastics compression-molded at 20 MPa. (*Source:* From Mo *et al.* [24]. *Journal of Applied Polymer Science* © 1999. Reprinted with permission of John Wiley & Sons, Inc.)

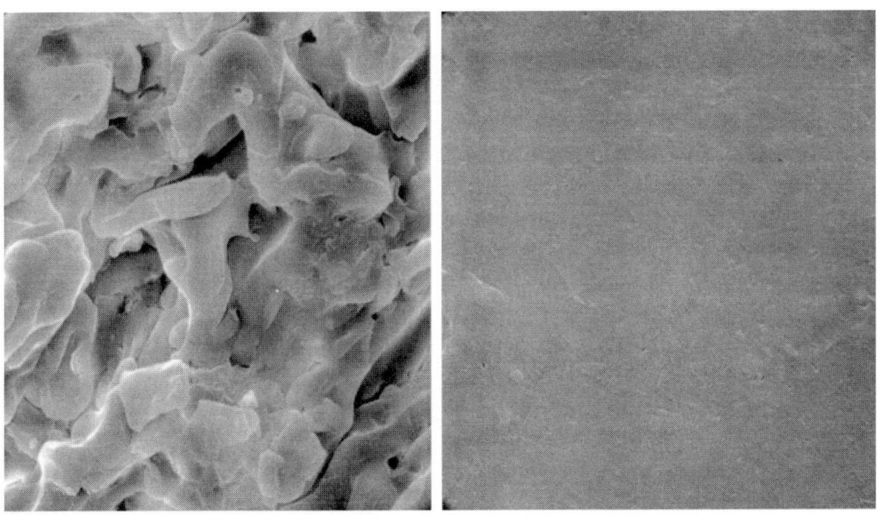

FIGURE 9.12 Microstructure of soy protein plastics compression-molded at 20 MPa for 5 min at (A) 100 °C and (B) 150 °C. The microstructure was observed using a scanning electron microscope at an accelerated voltage of 20 kV.

9.2.3 EFFECT OF PRESSURE ON CURING STRENGTH

The curing strength of protein is also greatly affected by curing pressure [24]. The curing strength increases rapidly at lower pressure, and maximum curing strength was obtained at a pressure of 20 MPa in an experiment with soy protein (Figure 9.13). Young's modulus shows a similar trend, reaching a maximum value (1156 MPa) at a pressure of 20 MPa. Elongation at maximum stress increases sharply from 0.9% to 5.4% as the pressure increases from 5 to 10 MPa.

As mentioned before, the soy protein presents two distinctive endothermal peaks (Figure 9.3). The value of T_d shifts to a much higher temperature as the moisture of the soy protein reduces. At a moisture content of approximately 10%, T_d of conglycinin protein is about 137.6°C, and T_d of glycinin protein rises to approximately 163.4°C. The curing strength and elongation of the soy glycinin protein increase as molding temperature increases up to its DSC thermal transition temperature (about 163°C). However, the curing strength and elongation of the conglycinin protein reaches its maximum value at 145°C, which is about 7°C higher than its DSC thermal transition (about 138°C), suggesting that an interaction between conglycinin and glycinin proteins could take place during curing because the conglycinin contains a certain amount of glycinin [27, 30, 31]. The decrease in curing strength and elongation at higher molding temperatures may be caused by the thermal

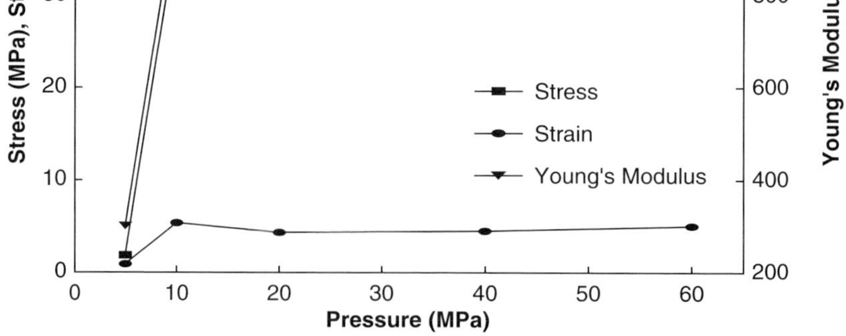

FIGURE 9.13 Effects of molding pressures on mechanical properties of soy protein plastics compression-molded at 150°C for 3 min. (*Source:* From Mo *et al.* [24]. *Journal of Applied Polymer Science* © 1999. Reprinted with permission of John Wiley & Sons, Inc.)

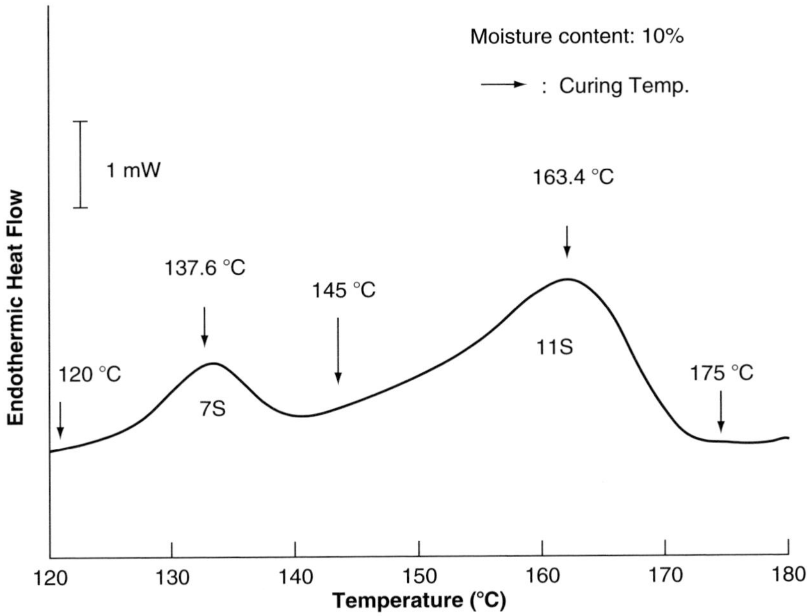

FIGURE 9.14 Thermogram of glycinin and conglycinin mixture (1:1) with 10% moisture content, determined using DSC scanned from 30° to 180°C at a heating rate of 10°C/min. (*Source:* From Sun *et al.* [29]. © 1999 The American Oil Chemists' Society. Reprinted with permission of AOCS Press.)

degradation of the protein by exposure to a temperature higher than its denaturation temperature.

9.3 MECHANICAL PROPERTIES OF SOY PROTEINS

9.3.1 SOY PROTEINS PLASTICATED WITH POLYOL-BASED CHEMICALS

Plastics made solely from soybean proteins are rigid and brittle. Polyols have often been used as plasticizers to improve their processibility, flexibility, and stretchability [32]. A plasticizer can interpose itself between the protein polymer chains and decrease the forces holding the chains together [33]. In doing so, it acts as a lubricant to facilitate the movements of the protein macromolecules over each other, disrupting the protein–protein interactions including hydrogen bonds and the Van de Waals and ionic forces, and depressing the glass transition temperature by increasing protein free volume [34].

Soybean proteins are composed of approximately 63% polar amino acids [35] and have a large molecular size distribution. Protein structural conformations are mainly stabilized by hydrogen bonds, electrostatic forces, and hydrophobic interactions. A plasticizer that contains polar groups should be compatible with soybean proteins. The plasticizing effect is influenced both by its ease of insertion and its position within a three-dimensional protein network [36]. Molecular weight, numbers, and the positions of the hydroxyl groups of a plasticizer are all variables as a plasticizer.

Mo and Sun [37] plasticated soy protein with a few polyol-based plasticizers with various carbons and hydroxyl groups. Glycerol, propylene glycol (PPG), 1,2-butanediol (Btd12), and 1,3-butanediol (Btd13) are all polyols with different molecular sizes and positions of hydroxyl groups. Glycerol and PPG have three carbons in common, but glycerol has three hydroxyl groups, and PPG has two hydroxyl groups; Btd12 and Btd13 have four carbons in common and two hydroxyl groups attached to 1,2 and 1,3 carbons, respectively. As expected, the T_d of plasticized proteins is lower than that of unplasticized proteins (Table 9.2). The H_{df} of the plasticized proteins is reduced by about 2 J/g. Soybean proteins have a three-dimensional structure, with most hydrophobic groups buried inside and hydrophilic groups facing outward.

To denature the proteins, chemical reagents must first disrupt the hydrophilic shell and then penetrate into the hydrophobic region. Polyols with polar hydroxyl groups can effectively disrupt only the hydrophilic shell. The penetration ability of polyols should be enhanced with their hydrophobicity [38]. With 3-carbon groups, PPG has one hydroxyl group and is more hydrophobic than glycerol. As a result, the protein plasticized with PPG has a lower T_d than protein plasticized with glycerol. With four carbons, Btd12 has two hydroxyl groups attached to the 1,2 carbons, which should be more hydrophobic than Btd13. The protein plasticized with Btd12 has a lower T_d than that with Btd13.

Because of a wide variety of peptide chains in the soybean protein [39], protein presents a large range of glass transition temperatures (T_g). The magnitude of the T_g depression is an indicator of the chemical compatibility

TABLE 9.2 Denaturation temperatures of plasticized soy proteins as determined using a differential scanning calorimeter with a heating rate of 10 °C/min.

Samples	Conglycinin T_d (°C)	Glycinin T_d (°C)
Control	139.7	169.9
Glycerol	117.0	140.9
1,3-Butanediol	95.7	113.8
1,2-Butanediol	89.5	109.5
Propylene glycol	82.6	107.5

Source: Adapted from Mo and Sun [37].

of a plasticizer with a protein [40]. Proteins with PPG and glycerol have lower T_g and appear to be more compatible with soybean protein than that with Btd12 or Btd13 because both glycerol and PPG have smaller molecular size and more easily insert into the protein chains, and they can establish hydrogen bonds with polar groups of proteins [41]. As a result, the protein–protein interactions decrease because of the increased plasticizer–protein interactions. Conversely, Btd12 and Btd13 have relatively large molecular size, preventing them from easily penetrating into the three-dimensional protein structures and inhibiting the interactions between the polar groups in the proteins. Btd12 depresses the T_g of the protein more than Btd13, indicating that Btd12 is more compatible with soybean proteins. The two adjacent hydroxyl groups in Btd12 could allow it to interact with proteins more easily than Btd13, which has two separated hydroxyl groups.

As expected, the mechanical strength and Young's modulus of the plasticized proteins decrease significantly. Btds have relatively long carbon chains and a relatively low plasticization effect. Proteins plasticized with Btds have higher tensile strength and lower elongation than proteins plasticized with PPG and glycerol. The proteins plasticized with Btd12 have a higher elongation and a lower Young's modulus than proteins with Btd13. The proteins plasticized with glycerol have the highest elongation, and the proteins with PPG have the lowest tensile strength and a low elongation. As discussed before, PPG has a smaller molecular size than glycerol. Therefore, it would interact easily with the polar groups associated with the peptide chains, largely weakening the protein–protein interactions and resulting in low cohesiveness in the protein network.

The microstructure of the native soy protein appears to be smooth with cracks (Figure 9.15A), suggesting a brittle-type failure, and the plasticized soy proteins appear to be rough, suggesting a ductile failure (Figure 9.15B). The protein plasticized with Btd12 (Figure 9.15B) has smaller cracks than the plastics with Btd13 (Figure 9.15C), which agrees with its thermal and mechanical properties. The cross section of the protein plasticized with glycerol (Figure 9.15D) appears to have a fluctuated and continuous matrix, whereas protein plasticized with PPG (Figure 9.15E) has relatively smooth and extensive cracks, indicating poor mechanical behavior, which is caused by overly weakened protein interactions.

9.3.2 DENATURATION AND UNFOLDING OF SOY PROTEINS

Besides plasticizers, the flexibility of soy protein can be improved by protein modification in a process that tailors protein structures through physical, chemical, and enzymatic methods. Protein modification, including denaturation, can improve its functional properties, such as solubility, foaming, emulsifying, gelation, and viscosity [42]. Urea is a commonly used denaturing agent for protein, which can unfold protein structure and make

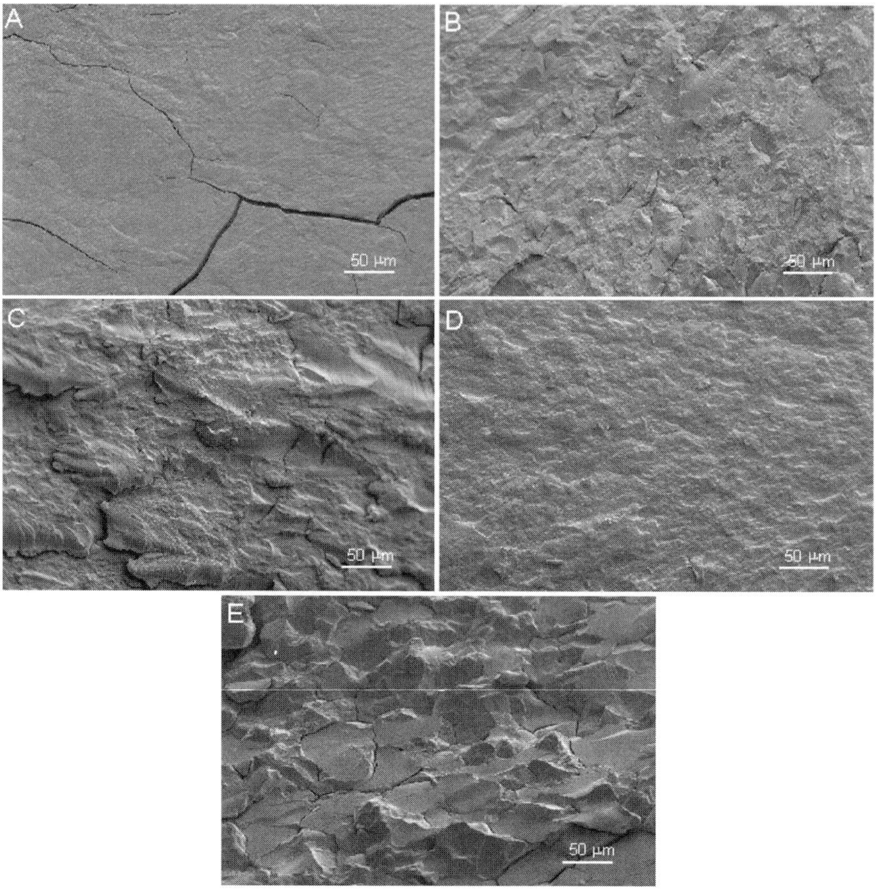

FIGURE 9.15 Surface morphology of soybean protein plastics molded at 140°C, 20 MPa for 5 min. Soy protein plastics (A) with no plasticizer, (B) with 1,2-butanediols, (C) with 1,3-butandiols, (D) with glycerol, and (E) with propylene glycol. (*Source:* From Mo and Sun [37]. © 2002 The American Oil Chemists' Society. Reprinted with permission of AOCS Press.)

protein chains more flexible by partially disrupting hydrogen bonds [43]. The urea-modified protein could have a high degree of entanglements and cross-linked structures in thermal processes to improve mechanical properties [44]. The tensile strength and Young's modulus of the urea-modified soy protein are improved (Table 9.3). Hydrogen bonds are among the major bonds maintaining the stability of protein conformation, which could be weakened at higher urea concentrations, leading to protein denaturation.

The protein modified with 2 M urea has about 71% elongation, but similar tensile strength as compared with the unmodified protein. At lower urea concentrations, urea in the protein serves mainly as a plasticizer, which

TABLE 9.3 Mechanical properties of urea-modified soy protein compression molded at 120 °C and 5.6 MPa for 3 min.

Urea Concentration (M)	Tensile Strength (MPa)	Elongation (%)	Young's Modulus (MPa)
0	13.0	1.2	1549
1	4.0	13.4	131
2	12.5	71.6	329
4	18.3	1.9	1404
8	24.1	2.1	2163

Source: Data from Mo and Sun [44].

increases the protein flexibility. Soy proteins that contain various amino residue side groups, such as amino, carboxy, and hydroxyl groups, are relatively active and may react with each other upon heating to form cross-links. A high degree of entanglements and cross-links is expected for the protein with a high degree of denaturation. At higher urea concentrations, protein can be completely denatured, resulting in high entanglement density and, consequently, an increase in stiffness and strength.

The storage moduli of urea-modified proteins exhibit a typical decrease in the glass transition zone (Figure 9.16). The two peaks observed are caused by the glass transition of conglycinin and glycinin proteins. The T_g's of the urea-modified proteins are lower than that of unmodified protein. However, the T_g of the modified proteins increases as urea concentration increases. The T_g value for a material is a function of molecular weight and the amount of plasticizer [45]. At lower concentrations, urea acts as a plasticizer by introducing free volume and increasing protein chain flexibility. However, at higher urea concentration, as protein is denatured, more extensive protein unfolding occurs and, consequently, forms high-molecular-weight aggregates, entanglements, or cross-links in the molding process, resulting in an increase in T_g.

Guanidine hydrochloride (GuHCl) and sodium dodecyl sulfate (SDS) are often used to unfold protein. The degree of protein unfolding can be controlled by chemical concentration, resulting in altered properties [46, 47]. For example, at 2.4 M of GuHCl, most ordered structures of a protein are destroyed, and no thermal denaturation transition can be observed [46, 47]. The minimum value of nonfreezing water of the modified protein occurs at about 1 M GuHCl, which is due to the interaction between the protein and GuHCl molecules. At this unfolding degree, the modified protein exhibits maximum tensile strength and elongation (Table 9.4). The increase in elongation apparently is due to the plasticization effect of GuHCl. At lower GuHCl concentration, interactions between protein and GuHCl molecules can be enhanced through hydrogen bonds. At higher GuHCl concentrations, the distance between protein molecules could be larger due to the large amount of GuHCl, reducing protein–protein interactions and, consequently, reducing

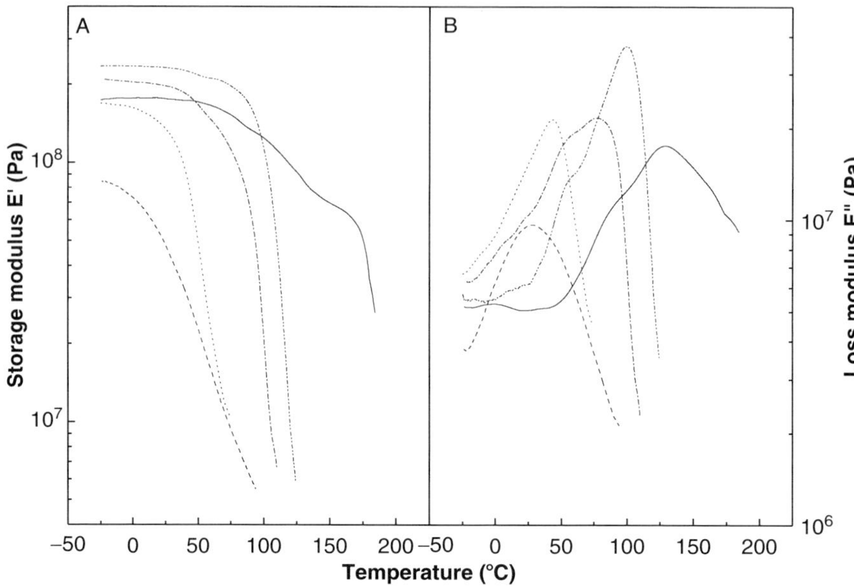

FIGURE 9.16 (A) Storage modulus and (B) loss modulus of soy protein plastics molded at 120°C, 5.6 MPa for 3 min. Soy protein modified with 0 M urea (———); 1 M urea (-------) ; 2 M urea (··········); 4 M urea (-·-·-·-·-); and 8 M urea (-··-··-··-), as determined using a dynamic mechanical analyzer scanned from –25° to 200°C at a rate of 5°C/min with 110-mN static force, 100-mN dynamic force, and 1-Hz frequency. (*Source:* From Mo and Sun [44]. © 2001 The American Oil Chemists' Society. Reprinted with permission of AOCS Press.)

TABLE 9.4 Mechanical properties of soy glycinin protein modified with GuHCl; moisture content: 2.6%; compression molded at 120°C for 4 min.

GuHCl Concentration (M)	Tensile Strength (MPa)	Elongation (%)	Young's Modulus (GPa)
0	19.3	1.40	1.43
0.5	16.4	1.26	1.38
~1	20.6	2.74	0.83
2.4	3.9	1.35	0.44

Source: Data from Zhong and Sun [46].

strength and elongation. The fracture surface of the modified protein with 1 M GuHCl is coarse and rough, a typical characteristic of tough fractures (Figure 9.17A) as compared with the microstructure of the unmodified protein shown in Figure 9.15A. The protein modified with 2.4 M GuHCl is dispersed in a GuHCl matrix (Figure 9.17B). Similar phenomena were observed when proteins were modified with SDS [47], but the mechanism, which is discussed in Chapter 10, is different. Effects of chemical concentration on adhesive strength are also discussed in detail in Chapter 10.

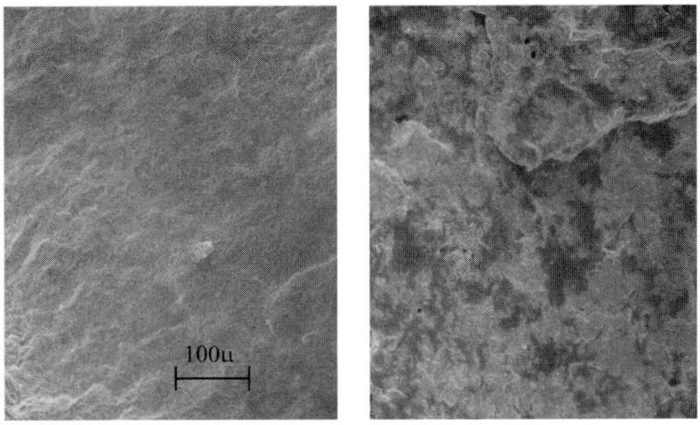

FIGURE 9.17 Microstructure of glycinin protein modified with (A) ~ 1 M GuHCl and (B) 2.4M GuHCl. The dark area represents modified protein body and the bright area represents GuHCl molecules. (*Source:* Adapted from Zhong and Sun [46].)

9.4 PHYSICAL AGING OF SOY PROTEIN PLASTICS

Physical aging is a general phenomenon that occurs over time in glassy or partial glassy polymers below their T_g and is a manifestation of the non-equilibrium nature of the glassy state [48, 49]. When a polymer is quenched from a temperature above T_g, it has larger free volume and enthalpy than when it is in an equilibrium state. The gradual rearrangement of molecular chains causes an increase in packing density and a reduction in free volume and enthalpy within the polymer, which would reduce molecular mobility, resulting in increased stiffness and decreased toughness [50–55].

Physical aging reduces both the free volume and internal energy of a polymer. The aged protein polymer absorbs the heat lost during aging and recovers the lost free volume at a temperature above T_g, presenting an endothermic peak at its glass transition (Figure 9.18) [56]. Excess relaxation enthalpy (H_{er}), a measurement of the increased number of main chain segments, becomes active at the transition. The slope of the linear regression could then be used to express the relative aging rate (Figure 9.19) [57]. The apparent linearity indicates an exponential relationship between the H_{er} and aging time, indicating that protein plastics age faster at the beginning and then aging slows thereafter. The aging behavior of soy protein plastics resembles the aging of petroleum-based polymers and is a self-retarding process [48], where the mobility of the polymer chains decreases continuously during aging. The H_{er} values are higher for the protein modified with urea because the mobility of protein peptide chains could be enhanced by urea, and the peptide chain may rearrange quickly.

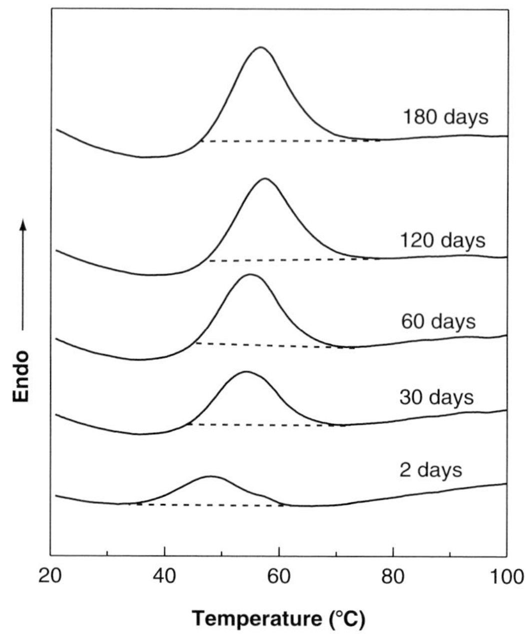

FIGURE 9.18 Thermograms of the soybean protein plastics with 25% glycerol stored at 25 °C and 50% relative humidity for various lengths of time determined using DSC scanned from 5° to 200 °C at a heating rate of 10 °C/min. (*Source:* Copyright Mo and Sun [56].)

Moisture absorption is a common phenomenon in protein storage. The modified proteins with glycerol had higher moisture (4–9%) than the unmodified proteins (about 3%) [58]. However, the moisture content of all protein plastics remains constant for the most part during 180 days of storage at 25 °C and 50% relative humidity (RH) [56].

The mechanical properties of the unmodified protein become weakened with storage time (Table 9.5). The protein plasticized with glycerol gave a maximum elongation at about 30 days of storage, but the tensile strength was about 13 MPa at 2 days and then remained relatively constant at about 8.8 MPa for the remainder of 180 days of storage. The plasticization effect of a plasticizer in the protein plastics would replace protein–protein contact with protein–plasticizer contact and decrease the rigidity of the three-dimensional protein structure [59]. The attractive force between protein chains would gradually squeeze the plasticizer out and re-form the rigid polymeric structure, such as for the protein modified with urea (Table 9.5). The elongation of the protein plasticized with glycerol remained around 170% over 180 days of storage, indicating that glycerol is a higher "solvent power" plasticizer [59] and has better interactions with soy protein than urea.

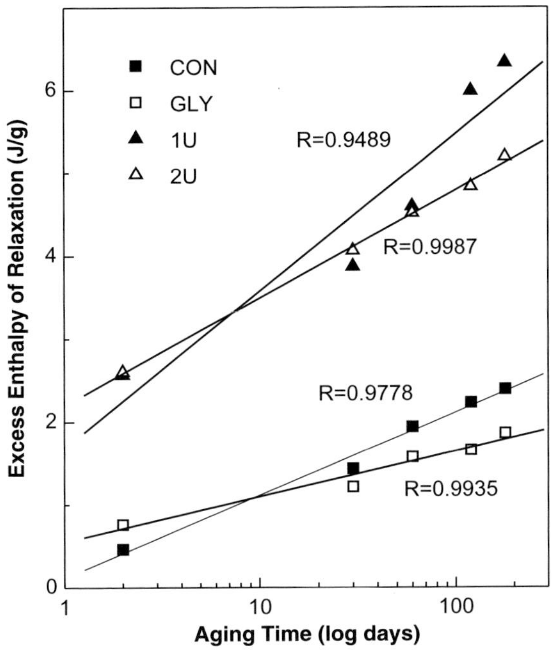

FIGURE 9.19 Excess enthalpy relaxation of soy protein plastics as affected by storage time at 25°C and 50% RH. CON, GLY, 1U, and 2U represent plastics made from soybean protein alone, with 25% glycerol, modified with 1 M urea, and modified with 2 M urea, respectively. (*Source:* Copyright Mo and Sun [56].)

9.5 COMPATIBILITY OF SOY PROTEIN WITH POLYESTER

9.5.1 BLENDING CHEMISTRY

Soy protein and polyester are both large molecules and usually do not have readily reactive functional groups [60, 61]. The blends consist of two distinctive phases whose interfaces are bound weakly with poor interaction, and this results in inferior physical properties, as discussed in Chapter 6. Coupling reagents are often introduced into the blending system to improve the compatibility of the two macromolecules. Isocyanate-based chemicals are good coupling reagents because the isocyanate groups can easily react with hydroxyl or amino groups from soy proteins and polyesters. Urethane and amide bonds could be formed among isocyanate, amino, or hydroxyl groups [62]:

$$-R_1-N=C=O + -R_2-OH \Rightarrow -R_1-\overset{H}{\overset{|}{N}}-\overset{O}{\overset{\|}{C}}-O-R_2-$$

$$-R_1-N=C=O + -R_2-NH_2 \Rightarrow -R_1-\overset{H}{\overset{|}{N}}-\overset{O}{\overset{\|}{C}}-\overset{H}{\overset{|}{N}}-R_2-$$

TABLE 9.5 Tensile strength (σ) and elongation (ε) of soybean protein-based plastics stored at 25 °C and 50% RH for various lengths of time.

	Native Protein		Plasticized with Glycerol		Modified with 1 M Urea		Modified with 2 M Urea	
Storage Time (Days)	σ (MPa)	ε (%)	σ (MPa)	ε (%)	σ (MPa)	ε (%)	σ (MPa)	ε (%)
2	34.6	2.6	13.0	172.5	27.7	5.4	4.6	120.9
30	26.6	2.7	8.8	276.2	33.0	5.5	7.0	95.2
60	24.5	2.2	8.9	175.8	31.4	3.8	8.7	78.9
120	20.9	2.1	9.0	170.3	24.5	3.0	9.6	72.8
180	19.1	1.4	8.8	168.7	13.2	1.6	10.0	72.0

Source: Data from Mo and Sun [56].

Zhong and Sun demonstrated such a reaction by blending soy protein with polycaprolactone (PCL) [63], a biodegradable polyester. The results were observed with an infrared technique (FTIR) [64]. Methylene diphenyl diisocyanate (MDI) was chosen because of its relatively environmentally friendly nature after reaction compared to other isocyanate-based chemicals. PCL shows a strong absorption band at 1727 cm^{-1} (Figure 9.20, curve A) caused by the C=O stretching vibration [64] and a weak absorption band at 3440 cm^{-1} because of the O–H stretching vibration of the hydroxyl end group of PCL. Soy protein exhibits strong absorption bands around 1550, 1660, and 3300 cm^{-1} (Figure 9.20, curve B), attributable to the N–H bending vibration (amide II band), C–O stretching vibration (amide I band), and N–H stretching vibration, respectively. The blend of protein/PCL (Figure 9.20, curve C) gave a strong sharp band at 1729 cm^{-1} caused by PCL and three strong broad bands in the vicinity of 1550, 1660, and 3300 cm^{-1} because of soy protein. The blend with 2% MDI did not give new absorption bands

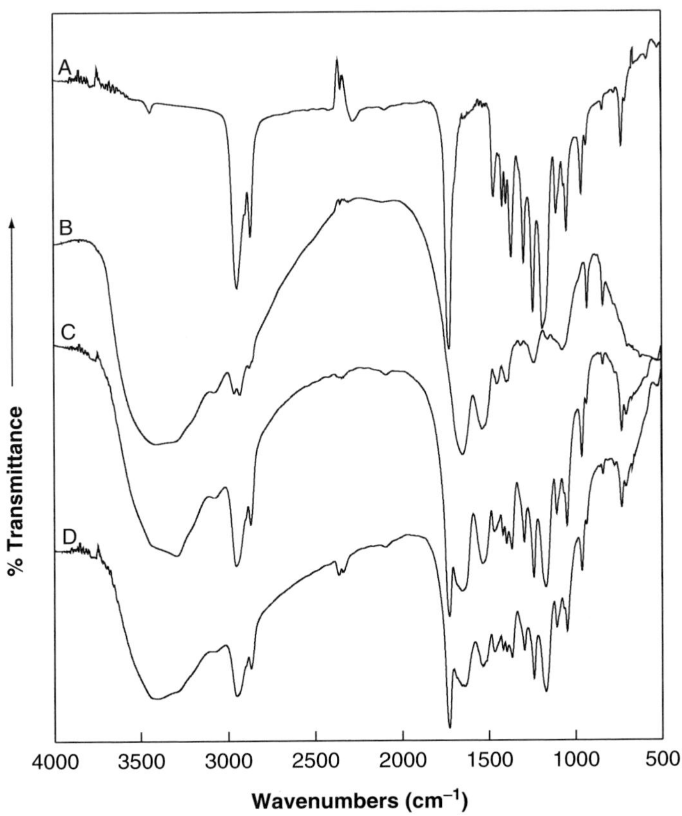

FIGURE 9.20 FTIR spectra of the (curve A) PCL, (curve B) soy protein, (curve C) soy protein and PCL blend (50/50) without MDI, and (curve D) soy protein and PCL blend (50/50) with 2% MDI. (*Source:* From Zhong *et al.* [64]. © 2003 ASAE. Reprinted with permission from the American Society of Agricultural Engineering.)

compared with the blend without MDI. However, the peak position of the N–H stretching vibration band shifted from 3293 cm^{-1} for the blend without MDI to 3409 cm^{-1} for the blend with MDI, and the shape of the absorption band also changed significantly (Figure 9.20, curves C and D). In addition, the absorption band at 1727 cm^{-1} broadened for the MDI blend.

Using a curve-fitting technique, two absorption peaks are located at 1724 and 1740 cm^{-1} for PCL alone (Figure 9.21A). These two component peaks

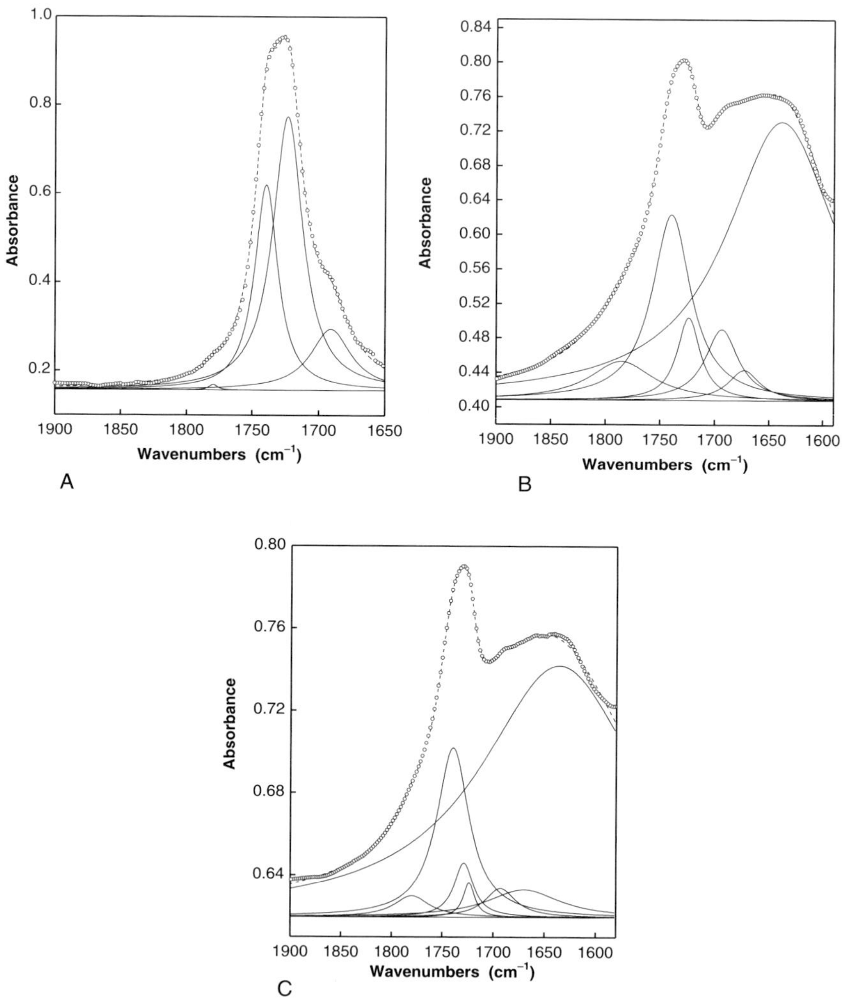

FIGURE 9.21 Curve-fitting results of FTIR spectrum presented in Figure 9.20: (A) PCL in the range of 1650–1900 cm^{-1}; (B) FTIR spectrum of the soy protein and PCL blend (50/50) without MDI in the range of 1590–1900 cm^{-1}; and (C) soy protein and PCL blend (50/50) with 2% MDI in the range of 1590–1900 cm^{-1}. Experimental data (O), calculated curve (– – – –), and components' curves (———). (*Source:* From Zhong *et al.* [64]. © 2003 ASAE. Reprinted with permission from the American Society of Agricultural Engineering.)

were due to the stretching vibration of the carbonyl group of PCL in the crystalline phase (1724 cm^{-1}) and the amorphous phase (1740 cm^{-1}) [65, 66]. The ratio of the component peak area of 1724 to 1740 cm^{-1} decreased after blending soy protein with PCL. The blend (Figure 9.21B) is further reduced for the blend with MDI (Figure 9.21C), suggesting that the crystallinity of PCL decreases because of the presence of protein and MDI. The stretching vibration of the carbonyl group in $-\phi-NH-CO-O-CH_2-$ has been reported, resulting in a peak at approximately 1730 cm^{-1} [67]. The new peak should be attributed to carbonyl vibration of the newly formed urethane linkage.

9.5.2 MORPHOLOGY

As mentioned before, the native soy protein has a smooth, homogeneous fracture surface (Figure 9.15A). The blends without MDI exhibit an obvious phase separation as well as large protein particles distributed in the PCL matrix (Figure 9.22A). The surfaces of the protein particles were smooth, with no evidence of a marked interaction between the protein and PCL phases. The PCL/protein blend with 2% MDI has a rough and heterogeneous fracture surface (Figures 9.22B–F). Flowability of the blend is obviously improved as PCL content increased (Figures 9.22B and C). The protein formed a continuous phase at low PCL content (Figure 9.22F), whereas PCL became the continuous phase at 50% PCL (Figure 9.22B). The blend with 40% PCL presents characteristics of bi-continuous phases.

9.5.3 THERMAL AND MECHANICAL PROPERTIES

The melting temperature of a protein blend with polyester is influenced primarily by the polyester (Table 9.6) because the T_d of soy protein with 1.5% moisture content is about 150°C. With 2% MDI, the T_g of soy protein in the blends decreased with PCL content, which also confirmed that there is some compatibility between soy protein and PCL. Otherwise, T_g of the protein should remain constant if the protein is completely incompatible with PCL. The decrease in melting temperature (T_m) of PCL with increasing protein content further confirms the observed phenomena.

The protein alone, with 2% MDI, typically presents a brittle fracture and no yield (Figure 9.23), whereas the blends of protein with PCL show a ductile fracture. For example, the blend with 50% PCL has a yield point and demonstrates ductile fracture compared to other blends with lower PCL content. As expected, the elongation of the blend is improved as PCL content increases at the expense of reduced tensile strength (Figure 9.24), which is also observed in the microstructure in Figure 9.22 for phase exchange. Young's modulus of the blends with 2% MDI decreased as PCL content increased because pure PCL has a low modulus of about 203 MPa. The energy of unit volume absorbed by a specimen before fracture, which is simply the area under the stress–strain curve [68], can be used as a measure

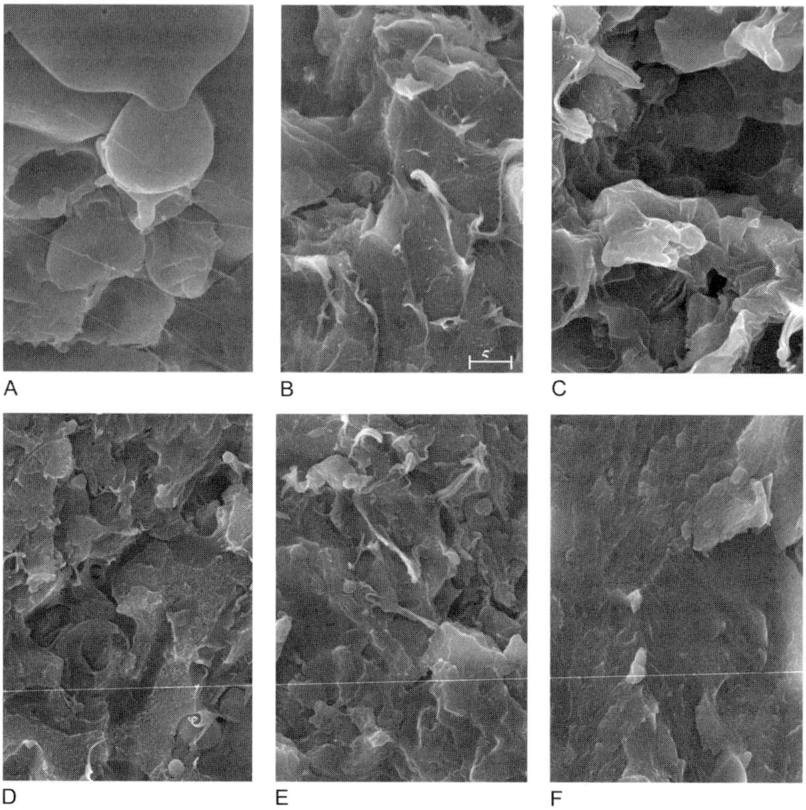

FIGURE 9.22 Microstructure of soy protein and PCL blends: (A) protein with 50% PCL without MDI; blends with 2% MDI at PCL content (B) 50%, (C) 40%, (D) 30%, (E) 20%, and (F) 10%. (*Source:* Adapted from Zhong and Sun [63].)

TABLE 9.6 Glass transition temperature (T_g), melting temperature (T_m), heat of fusion (ΔH_f), and crystallinity (X_c) of the soy protein/PCL blends with 2% MDI, measured using a differential scanning calorimeter, first quenched to –60 °C and heated to 200 °C, then quenched again to –60 °C and heated to 200 °C at 10 °C/min.

	DSC			DMA		
PCL Content	T_g (°C)	T_m (°C)	ΔH_f (J/g PCL)	X_c of PCL (%)	T_g (°C)	T_m (°C)
0%	108.4	—	—		114.2	—
10%	102.9	56.2	34.9	26	111.8	56.0
20%	97.8	57.5	39.5	29	108.0	56.2
30%	92.3	57.9	44.4	33	101.7	56.3
40%	88.6	58.5	46.1	34	96.3	57.3
50%	84.0	58.7	48.7	36	93.0	59.2

Source: Copyright Zhong and Sun, *Polymers,* 2001, **42,** 6961–6969 [63].

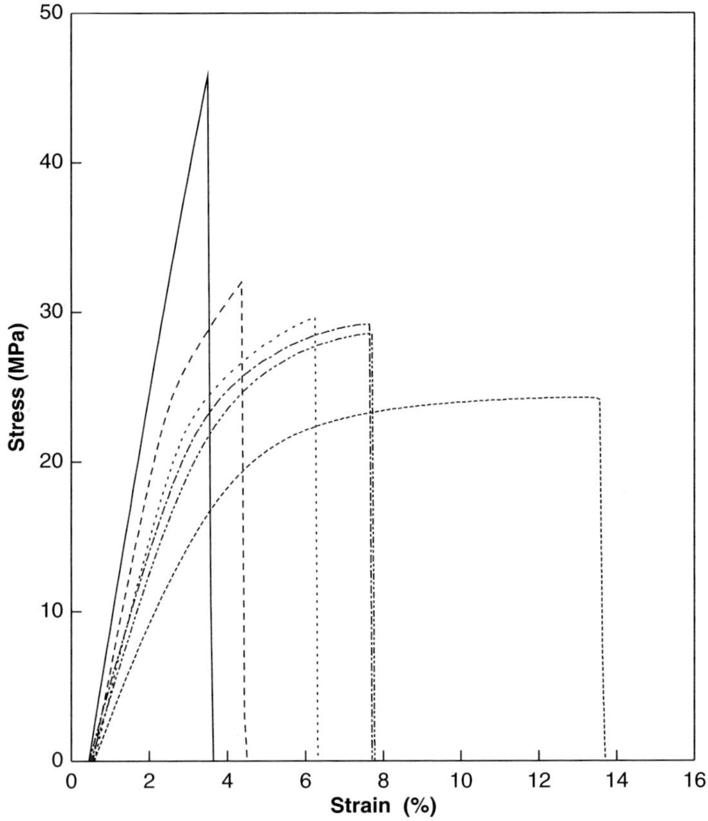

FIGURE 9.23 Stress–strain curves of the soy protein/PCL blends with 2% MDI at PCL content: (———) 0%, (- - - - - -) 10%, (··········) 20%, (-·-·-·-·) 30%, (-··-··-··-) 40%, and (- - - - - - -) 50%. Mixtures were prepared mechanically for 10 min at room temperature, followed by melt blending with an intensive mixer, and then compression-molded. (*Source:* Copyright Zhong and Sun [63].)

of toughness of a material. The toughness of the PCL/protein blends with 2% MDI increased with an increase in the PCL content, and the toughness at a 50% PCL content was four times that at a 0% PCL content.

9.6 WATER ABSORPTION OF SOY PROTEIN

Soy protein is a hydrophilic polymer, absorbing up to 468% water when soaked in water (Table 9.7). Water retention and resistance of a soy protein can be improved by protein modification or blending with hydrophobic polymers. Water absorption of a soy protein plasticized with polyols, such

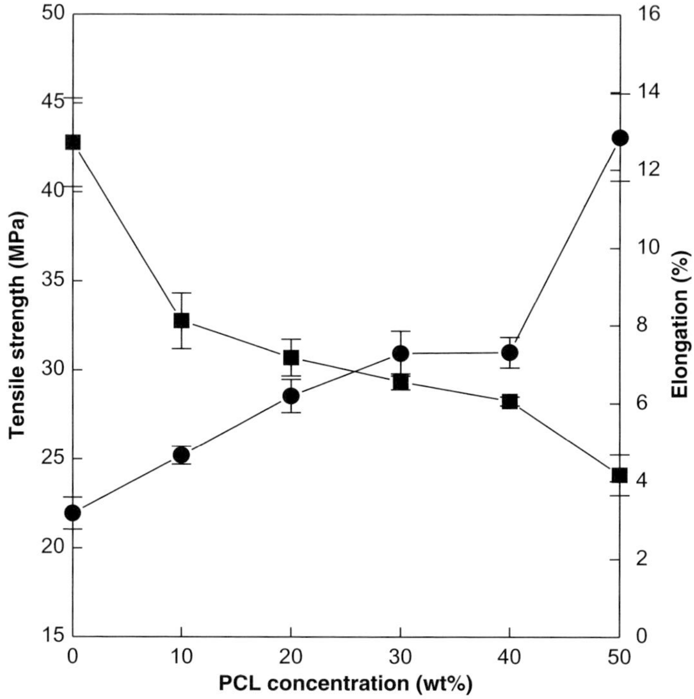

FIGURE 9.24 Tensile strength (■) and elongation (●) of the soy protein blends with PCL in the presence of 2% MDI. Mixtures were prepared mechanically for 10 min at room temperature, followed by melt blending with an intensive mixer, and then compression-molded. (*Source:* Copyright Zhong and Sun [63].)

TABLE 9.7 Water absorption of soy protein polymers and modified soy protein plastics after soaking in tap water for 24 h at room temperature.

Samples	Water Absorption (%)	Reference
Native protein	468	[37]
Plasticized with 15% glycerol	338	[37]
Plasticized with 15% propylene glycol	230	[37]
Modified with 1 M GuHCl	110	[46]
Modified with 2 M urea	40.4	[44]
Blended with 50% PCL coupled with 2% MDI	13.0	[64]

Sources: From Refs. [37], [44], [46], and [64].

as PPG, is reduced by about 50% (Table 9.7). Proteins with a large degree of aggregation or entanglement usually have low water absorption. Proteins with high denaturation temperatures often show little aggregation or entanglement after molding. The soy protein plasticized with glycerol has a higher T_d than that plasticized with PPG and would aggregate/entangle to a lower degree after molding. Thus, the plastic with glycerol has higher water absorption.

Protein unfolding also affects water absorption. Optimum unfolding exposes hydrophobic groups buried inside the protein, reducing band water and improving water resistance [46]. A modified protein with GuHCl absorbed less water than the unmodified protein (Table 9.7). The modified protein absorbed water at a minimum value at about 1 M GuHCl modification, then increased with GuHCl concentration. This variable trend was similar to that of bond water (Figure 9.25). The interaction between protein and GuHCl molecules at lower GuHCl can reduce water absorption because such interaction reduces the number of protein groups that can interact with water molecules [46]. As GuHCl concentrations increase, the distance between protein molecules increases and the structure of the molded protein is less compacted, resulting in higher water uptake.

Soy protein modified with 2 M urea can reduce its water uptake to 40% (Table 9.7) [44]. Urea can unfold the protein to a certain degree, and the protein becomes much entangled and cross-linked in the thermal molding process compared to the protein plasticized with glycerol.

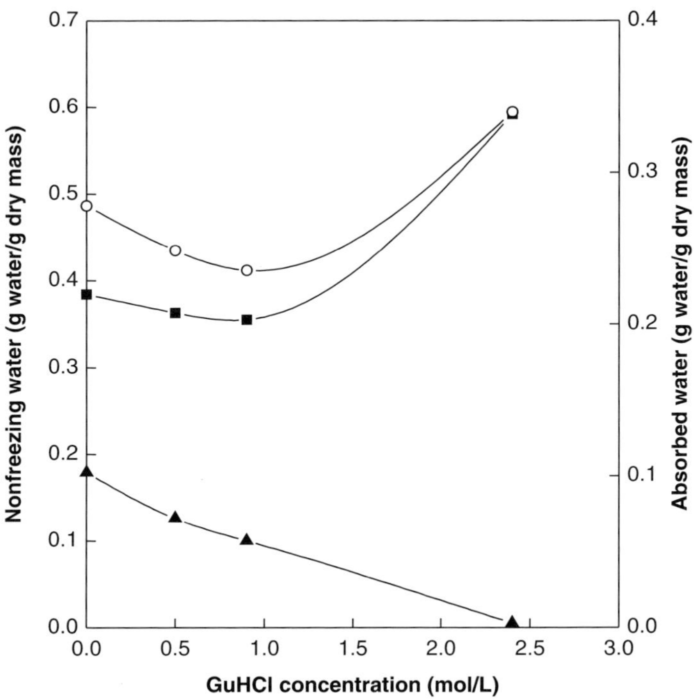

FIGURE 9.25 Nonfreezing water of soy glycinin protein solutions before (■) and after (○) denaturation and absorbed water (▲) as affected by GuHCl concentration, determined using DSC data. (*Source:* Copyright Zhong and Sun [46].)

By blending with PCL, the water resistance of soy protein is significantly improved (Table 9.7). Water resistance is further improved by introducing a coupling reagent, such as MDI (Table 9.7), and reaches equilibrium state at about 26% (Table 9.8). Soy proteins contain some polar groups, such as hydroxyl, amino, carboxyl, and ionic groups. All of these groups would interact with water. Reactions between some of these groups with the isocyanate group should reduce the number of water binding groups. In addition, the hydrophobic PCL molecules would prevent water penetrating into the protein body.

The kinetics of water absorption of soy protein and PCL blend can be described by Eq. (9.1) [69]:

$$\frac{M_t}{M_\infty} = k \cdot t^n, \tag{9.1}$$

where M_t is the amount of water absorbed at time t, M_∞ is the amount of water absorbed over a long time, k is a constant incorporating characteristics of the macromolecular network system and the diffusant, and n is the diffusion exponent, which is an indicator of the transport mechanism. For a plane sheet, Fickian behavior is defined by $n = 0.5$, and the mechanism for penetrant transport is controlled by diffusion. The mechanism is relaxation controlled for $n = 1$. For $0.5 < n < 1$, the water uptake mechanism is controlled by both diffusion and relaxation [70].

Soy protein alone has an n of 0.9, suggesting that the water sorption mechanism is mainly relaxation controlled. The n values of the blends with 50% PCL are about 0.6. With 2% MDI, n values of the blend are about 0.5. In this case, the equation derived from Fick's law can be used to calculate the approximate diffusion coefficient (D) and to evaluate water absorption properties. For a thin sheet with thickness l and constant D, the solution to Fick's second law is expressed as follows [69, 71]:

TABLE 9.8 Water absorption of soy protein with 50% PCL in the presence of 2% MDI, soaked in tap water at room condition for various times.

Soaking Time (h)	Water Absorption (%)
2	3.9
5.1	5
6.7	10
13.0	24
20.8	48
25.6	72
24.8	144
25.9	216

Source: Data from Zhong *et al.* [64].

$$\frac{M_t}{M_\infty} = 1 - \sum_{m=0}^{\infty} \frac{8}{(2m+1)^2\pi^2} \exp\left[\frac{-D(2m+1)^2\pi^2}{l^2} t\right], \tag{9.2}$$

where m is the summation number, which was from 0 to 100 in this case. Soy protein has the highest D value of 14.1×10^{-12} m^2/s. After incorporating PCL and/or MDI, D decreases. Similar to n, D has a sharp decrease as PCL content increases to 50%, which is probably caused by phase reversal because hydrophobic PCL becomes a continuous phase as its content reaches 50% or above. The experimental data obtained by Zhong *et al.* [64] fit well the predicted values calculated by Eq. (9.2).

9.7 SUMMARY

Soy protein is a globulin protein stored in soybean oilseeds. The polypeptide chains of soy protein polymers are associated and entangled into a three-dimensional complicated structure by disulfide and hydrogen bonds with a molecular weight ranging from 300,000 to 600,000 KDa. Two major protein polymers in soy protein are glycinin protein and conglycinin proteins that are about 80% of the total soy protein. Soy protein is mainly thermoset polymer in amorphous form with denaturation phase transition and decomposed at about 230°C. The curing modulus of soy protein polymers is significantly affected by curing temperature, pressure, and time. The curing strength of soy protein can reach as high as 45 MPa, but the cured soy protein resin is often brittle with about 2.5% elongation. Mechanical and thermal properties of soy protein polymers can be modified by chemical and enzymatic approaches as well as curing methods. Mechanical strength of cured soy protein polymer can be from 5 to 45 MPa with corresponding elongation of about 250 to 2.5%. Soy protein polymers have great potential for wood adhesives, coatings, paint, and glues, which are discussed in Chapter 10.

REFERENCES

1. Thanh, V. H.; Shibasaki, K. *J. Agric. Food Chem.* 1976, **24**(6), 1117–1121.
2. Privalov, P. L. Stability of Proteins: Small Globular Proteins, *Adv. Protein Chem.* 1979, **33,** 167–241.
3. Privalov, P. L.; Khechinashvili, N. N. A Thermodynamic Approach to the Problem of Stabilization of Globular Proteins: A Calorimetric Approach, *J. Mol. Biol.* 1974, **86,** 665–684.
4. Careri, G.; Giansanti, A.; Gratten, E. Lysozyme Film Hydration Events: An IR and Gravimetric Study, *Biopolymers* 1979, **18,** 1187–1203.
5. Rupley, J. A.; Careri, G. Protein Hydration and Function, *Adv. Protein Chem.* 1991, **41,** 37–172.
6. Gregory, R. B. Protein Hydration and Glass Transition Behavior. In *Protein–Solvent Interactions*, Gregory, R. B., Ed.; Marcel Dekker, Inc., New York; 1995, pp. 191–264.
7. Nielsen, N. C. Structure of Soy Proteins. In *New Protein Foods*, Vol. 5, Altschul, A. M., Ed.; Academic Press, New York; 1985, pp. 27–64.

8. Kinsella, J. E.; Damodaran, S.; German, B. Physicochemical and Functional Properties of Oilseed Proteins with Emphasis on Soy Proteins. In *New Protein Foods*, Vol. 5, Altschul, A. M., Ed.; Academic Press, New York; 1985, pp. 107–179.
9. Roos, Y. H., Ed., *Phase Transitions in Foods*. Academic Press, New York; 1995.
10. Oates, C. G.; Ledward, D. A.; Mitchell, J. R. Physical and Chemical Changes Resulting from Heat Treatment Soya and Soya Alginate Mixtures, *Carb. Polym.* 1987, **7,** 17–33.
11. Sheard, P. R.; Fellows, A.; Ledward, D. A.; *et al.* Macromolecular Changes Associated with the Heat Treatment of Soya Isolate, J. *Food Technol.* 1986, **21,** 55–60.
12. Kitabatake, N.; Tahara, M.; Doi, E. Thermal Denaturation of Soybean Protein at Low Water Contents, *Agric. Biol. Chem.* 1990, **54,** 2205–2212.
13. Sessa, D. J. Hydration Effects on the Thermal Stability of Proteins in Cracked Soybeans and Defatted Soy Flour, *Lebensm.-Wiss. u.-Technol.* 1992, **25,** 365–370.
14. Sessa, D. J. Thermal Denaturation of Glycinin as a Function of Hydration, *J. Am. Oil Chem. Soc.* 1993, **70,** 1279–1284.
15. Zhong, Z. K.; Sun, X. S. Thermal Behavior and Nonfreezing Water of Soybean Protein Components, *Cereal Chem.* 2000, 77(4), 495–500.
16. Pace, C. N. The Stability of Globular Proteins, *CRC Crit. Rev. Biochem.* 1975, **3,** 1–43.
17. Chen, R.-H.; Ker, Y.-C.; and Wu, C.-S. Temperature and Shear Rate Affecting the Viscosity and Secondary Structural Changes of Soy 11S Globulin Measured by a Cone-Plate Viscometer and Fourier Transform Infrared Spectroscopy, *Agric. Biol. Chem.* 1990, **54,** 1165–1176.
18. Pradipasena, P.; Rha, C. Pseudoplastic and Rheopectic Properties of a Globular Protein (β-Lactoglobulin) Solution, *J. Texture Stud.* 1977, **8,** 311–325.
19. Kuntz, Jr. I. D.; Kauzmann, W. Hydration of Proteins and Polypeptides, *Adv. Protein Chem.* 1974, **28,** 239–345.
20. Ruegg, M.; Moor, U.; Blanc, B. Hydration and Thermal Denaturation of β-Lactoglobulin. A Calorimetric Study, *Biochim. Biophys. Acta. Int. J. Biochem. Biophys.* 1975, **400,** 334–342.
21. Lewin, S. *Displacement of Water and Its Control of Chemical Reactions*, Academic Press, New York; 1974.
22. Paetau, I.; Chen, C.-Z.; Jane, J. *Ind. Eng. Chem. Res.* 1994, **33,** 1821.
23. Huang, H. *Thesis*, Iowa State University, Ames; 1994.
24. Mo, X.; Sun, X.; Wang, Y. *J. Appl. Polym. Sci.* 1999, **73,** 2595–2602.
25. Tolstoguzov, V. B. Some Physico-chemical Aspects of Protein Processing into Foodstuffs, *Food Hydrocolloids* 1988, **2**(5), 339–344.
26. Kitabatake, N.; Tahara, M. Thermal Denaturation of Soybean Protein at Low Water Contents, *Agr. Bio. Chem.* 1990, **54**(9), 2205–2212.
27. Babajimopoulos, M.; Damodaran, S.; Rizvi, S. S. H.; *et al.* Effect of Various Anions on the Rheological and Gelling Behavior of Soy Proteins: Thermodynamic Observations, *J. Agr. Food Chem.* 1983, **31,** 1270–1275.
28. Utsumi, S.; Kinsella, J. E. Forces Involved in Soy Protein Gelation: Effects of Various Reagents on the Formation, Hardness and Solubility of Heat-induced Gels Made from 7S, 11S, and Soy Isolate, *J. Food Sci.* 1985, **50**(5), 1278–1282.
29. Sun, X.; Kim, H.-R.; Mo, X. J. *Am. Oil Chem. Soc.* 1999, **76**(1), 117–123.
30. Utsumi, S.; Domodaran, S.; Kinsella, J. E. Heat Induced Interactions Between Soybean Proteins: Preferential Association of 11S Basic Subunits and Beta Subunits of 7S, *J. Agr. and Food Chem.* 1984, **32**(6), 1406–1412.
31. Yamagishi, T.; Miyakawa, A.; Noda, N.; *et al.* Isolation and Electrophoretic Analysis of Heat-induced Products of Mixed Soybean 7S and 11S Globulins, *Agr. Bio. Chem.* 1983, **47**(6), 1229–1237.
32. Wang, S.; Sue, H. J.; Jane J. Effects of Polyhydric Alcohols on the Mechanical Properties of Soy Protein Plastics, *J. Macromolec. Sci.: Pure Appl. Chem.* 1996, **A33,** 557–569.
33. Entwistle, C. A.; Rowe, R. C. Plasticization of Cellulose Ethers Used in the Film Coating of Tablets, *J. Pharm. Pharmacol.* 1978, **31,** 269–272.

34. Sears, J. K.; Darby, J. R. Mechanism of Plasticizer Action. In *The Technology of Plasticizers*. Sears, J. K.; Darby, J. R., Eds. Wiley-Interscience, New York; 1982, pp. 35–77.
35. Ly, Y. T. P.; Johnson, L. A.; Jane, J. Soy Protein as Biopolymer. In *Biopolymers from Renewable Resource*, Kaplan, D. L., Ed.; Springer, Berlin; 1998, pp. 144–176.
36. Cuq, B.; Gontard, N.; Cuq, J. L.; *et al.* Selected Functional Properties of Myofibrillar Protein-based Films as Affected by Hydrophilic Plasticizers, *J. Agric. Food Chem.* 1997, **45,** 622–626.
37. Mo, X.; Sun, X. S. Plasticization of Soy Protein Polymer by Polyol-Based Plasticizers, *J. Am. Oil Chem. Soc.* 2002, **79,** 197–202.
38. Fukushima, D. Denaturation of Soybean Proteins by Organic Solvents, *Cereal Chem.* 1969, **46,** 156–163.
39. Morales, A.; Kokini, J. L. Glass Transitions of Soy Globulin Using Differential Scanning Calorimetry and Mechanical Spectrometry, *Biotechnol. Pro.* 1997, **13,** 624–629.
40. Gioia, L. D.; Guilbert, S. Corn Protein-based Thermoplastics Resins: Effect of Some Polar and Amphiphilic Plasticizers, *J. Agric. Food Chem.* 1999, **47,** 1254–1261.
41. Galietta, G.; Gioia, L. D.; Guilbert, S.; *et al.* Mechanical and Thermomechanical Properties of Films Based on Whey Proteins as Affected by Plasticizer and Cross-linking Agent, *J. Dairy Sci.* 1998, **81,** 3123–3130.
42. Wu, Y. V.; Inglett, G. E. Denaturation of Plant Proteins Related to Functionality and Food Applications. A Review, *J. Food Sci.* 1974, **39,** 218–225.
43. Hayakawa, I.; Linko, Y.; Linko, P. Mechanism of High Pressure Denaturation of Proteins, *Lebensm.-Wiss. U.-Technol.* 1996, **29,** 756–762.
44. Mo, X.; Sun, X. Thermal and Mechanical Properties of Plastics Molded from Urea Modified Soy Protein Isolates, *J. Am. Oil Chem. Soc.* 2001, **78,** 867–872.
45. Fox, T. G.; Flory, P. J. Second Order Transition Temperatures and Related Properties of Polystyrene. I. Influence of Molecular Weight, *J. Appl. Phys.* 1950, **21,** 581–591.
46. Zhong, Z.; Sun, X. Thermal and Mechanical Properties and Water Absorption of Guanidine Hydrochloride Modified Soy Protein (11S), *J. Appl. Polym. Sci.* 2001, **78,** 1063–1070.
47. Zhong, Z.; Sun, X. Thermal and Mechanical Properties and Water Absorption of Sodium Dodecyl Sulfate Modified Soy Protein (11S), *J. Appl. Polym. Sci.* 2001, **81,** 166–175.
48. Struik, L. C. E. *Physical Aging in Amorphous Polymers and Other Materials*. Elsevier, Amsterdam; 1978, pp. 35–49.
49. Chartoff, R. P. *Thermal Characterization of Polymeric Materials*, Vol. 1, Turi, E. A., Ed.; Academic Press, New York; 1997, pp. 484–744.
50. Hay, J. N. *Pure Appl. Chem.* 1995, **67**(11), 1855–1858.
51. Hutchinson, J. M. *Prog. Polym. Sci.* 1995, **20**(4), 703–760.
52. Wang, S. F.; Ogale, A. A. *Polym. Eng. Sci.* 1989, **29**(18), 1273–1278.
53. Arnold, J. C. *Polym. Eng. Sci.* 1995, **35**(2), 165–169.
54. Bauwens-Crowet, C. *J. Mater. Sci.* 1999, **34**(8), 1701–1709.
55. Cunningham, P.; Ogale, A. A.; Dawson, P. L.; *et al. J. Food Sci.* 2000, **65**(4), 672–679.
56. Mo, X.; Sun, X. Effects of Storage Time on Properties of Soybean Protein-based Plastics, *J. Polym. Env.* 2003, **11,** 15–22.
57. Aref-Azar, A.; Hay, J. N. *Polymer* 1982, **23**(8), 1129–1133.
58. E. J. Pyler, *Baking Science and Technology*, Vol. 1, Sosland Publishing Company, Kansas City; 1988, pp. 46–82.
59. Paton, C. *Plasticizers, Stabilizers, and Fillers*. Ritchie, P. D.; Critchley, S. W.; Hill, A., Eds.; Butterworth, London; 1972, pp. 39–49.
60. John, J.; Tang, J.; Bhattacharya, M. *Polymer* 1998, **39**(13), 2883–2895.
61. John, J.; Bhattacharya, M. *Polymer Int.* 1999, **48,** 1165–1172.
62. Dieterich, D.; Grigat, E.; Hahn, W. In *Polyurethane Handbook*, Oertel, G., Ed.; Hanser Publishers, Munich; 1985, Chap. 2, pp. 7–41.

63. Zhong, Z.; Sun, X. Thermal and Mechanical Properties and Water Absorption of Soy Protein/Polycaprolactone Blends, *Polymer* 2001, **42,** 6961–6969.
64. Zhong, Z.; Sun, X.; Hagan, S.; *et al.* Soy Protein Isolate/Polycaprolactone Blends: Compatibilization Reactions and Water Absorption Mechanisms. *Trans. ASAE* 2005, **48**, no. 3.
65. Coleman, M. M.; Zarian, J. Fourier-transform Infrared Studies of Polymer Blends. II. Poly(ε-caprolactone)-poly(vinyl chloride) System, *J. Polym. Sci., Polym. Phys. Ed.* 1979, **17,** 837–850.
66. He, Y.; Inoue, Y. Novel FTIR Method for Determining the Crystallinity of Poly(ε-caprolactone), *Polym. Intl.* 2000, **49,** 623–626.
67. Dolphin, D.; Wick, A. E. *Tabulation of Infrared Spectra Data*, John Wiley & Sons, New York; 1977.
68. Andrews, E. H. In *Fracture in Polymers*, Aberdeen University Press, London; 1968, Chap. 2, pp. 37–73.
69. Ritger, P. L.; Peppas, N. A. A Simple Equation for Dof Solute Release. I. Fickian and Non-Fickian Release from Non-swellable Devices in the Form of Slabs, Spheres, Cylinders or Discs, *J. Controlled Release* 1987, **5,** 23–36.
70. Crank, J. *The Mathematics of Diffusion*, Clarendon Press, Oxford; 1975.
71. Rogers, C. E. In *Polymer Permeability*, Comyn, J., Ed.; Elsevier Applied Science Publishers, London; 1985, Chap. 2, pp. 11–73.

10

SOY PROTEIN ADHESIVES

XIUZHI SUSAN SUN

About 20 billion pounds of adhesives are used annually in the United States in the production of plywood, particleboard, labeling, packaging, and sizing, among other things. The various forms of wood adhesives represent an extremely large and diverse market, probably the largest in the world today [1]. Soy-based adhesives were first developed in 1923 when a patent was granted for a soy meal-based glue [2]. However, those soy protein adhesives had low gluing strength and little water resistance. Adhesives produced from petroleum-based chemicals have overcome those disadvantages, but many concerns have surfaced about air quality and environmental pollution, and even toxicity during product manufacturing, distribution, and use. Among these adhesives, about 8 billion pounds are formaldehyde-based adhesives annually used by the wood-based product industries. The greatly expanding markets for adhesives, the threat of limited world oil reserves, and increasing concerns over environmental pollution have forced industry to seek new adhesives from bio-based polymers. Protein modification is designed to improve functional properties by altering protein's molecular structure or conformation, through physical, chemical, or enzymatic agents, at the secondary, tertiary, and quaternary levels. Soy protein, as discussed in Chapter 9, has potential uses in the production of adhesives with high gluing strength and improved water resistance [3–5].

Soy proteins are good resins for binding various fibers, recycled newspapers, wood, and agricultural residue fibers. The composite made from soy proteins and waste newspapers looked like granite but handled like hard wood [6]. Adhesive foam from modified soy proteins through extrusion can be used for plywood applications [7]. Alkali is commonly used in protein adhesive preparations [1, 8]. The binding strength and water resistance of the alkali-modified soy protein adhesives can be enhanced. Alkali increases the degree of unfolding of protein molecules, resulting in an increased contact area and exposure of the hydrophobic bonds. Combination of the alkaline hydrolysis protein fractions with phenol-resorcinol-formaldehyde resin can speed up the curing process of soy protein adhesives [9], and the adhesive forms a gel within a few seconds and further polymerizes at room temperature, which could provide the strength and durability required for its successful use in structural lumber. X. Sun and coworkers at Kansas State University modified soy protein to have adhesive properties similar to those of latex adhesives for glue, packaging, sealing, labeling, wallpaper mounting, veneer plywood, foundry industry glue, and artwork.

10.1 PROTEIN ADHESION MECHANISM

Adhesion between adhesives and substrates is complicated, and no single theory accurately describes the interactions that take place at the interface. Several adhesion mechanisms have been proposed by researchers during the past century and were described by Schultz and Nardin [10]. They include mechanical interlocking, electron transfer, boundary layers and interfaces, adsorption, diffusion, and chemical bonding. For the mechanical interlocking mechanism, bonding strength is influenced by the surface morphology and physicochemical surface properties of the substrate and adhesive. For electronic theory, the substrate and adhesive may have different electronic bond structures, and an electron transfer mechanism may occur.

The theory of boundary layers is that the cohesive strength of a weak boundary layer can always be considered as the main factor in determining the level of adhesion, even when the failure appears to be interfacial. Adsorption theory has been widely used in adhesion science. This theory is based on the postulate that the adhesive will adhere to the substrate because of intramolecular and intermolecular forces established at the interface through van der Waals and Lewis acid–base interactions, provided that intimate contact is achieved [11]. This mechanism was also discussed in Chapters 6 and 8 with regard to adhesion at polymer–solid interfaces. Diffusion theory is based on the premise that a mutual diffusion of macromolecules occurs across the interface between the substrate and adhesives to form an interphase resulting in adhesion strength. For the chemical bonding theory, we know that chemical bonds formed across the substrate–adhesive interface greatly enhance the adhesion strength.

Adhesion theory between protein polymers and wood substrates is mainly attributed to a combination of three major mechanisms, including mechanical bonding, physical adsorption, and chemical bonding. The importance of each mechanism for a protein adhesive should be determined by the nature of the adhesive and the substrate.

The mechanical bonding theory describes how protein adhesives spread and wet the surface of the substrate, penetrate into the fiber cells through the capillary path, and then cure in place, acting like a mechanical anchor. Mechanical interlocking and penetration are two major contributors to mechanical bonding theory. The roughness of the wood or agricultural fiber surface and the flow behavior of the protein adhesive on the surface of the fiber are two major variables that determine the degree of mechanical interlocking effects bearing the shearing load. Penetration relies on an appropriate penetration depth into the wood surface, which is about 2–6 fiber depths [12].

Protein polymer has a certain molecular weight and distribution. The proteins with smaller molecular size can easily penetrate into the fiber cell structure and cure to form a continuous complex with fibers at depth as well as those larger protein molecules from the substrate surface. The wood surface structure, surface roughness, degree of cross-link and entanglements between protein molecules, and molecular weight and distribution contribute significantly to the bonding strength. For example, the gluing strength with pine is much lower than that for walnut, cherry, maple, and poplar samples [3]. The surface microstructure of the pine sample is smoother, with a fiber-oriented structure, than walnut wood samples (Figure 10.1). If the surface is too rough, it causes cohesive wood failure; if the surface structure is too smooth, it causes adhesion failure [13]. Protein molecules penetrate into the pores from the wood surface, forming a complex matrix upon curing. The rough surface under pressure can form random micro "finger joint" effects, enhancing gluing strength.

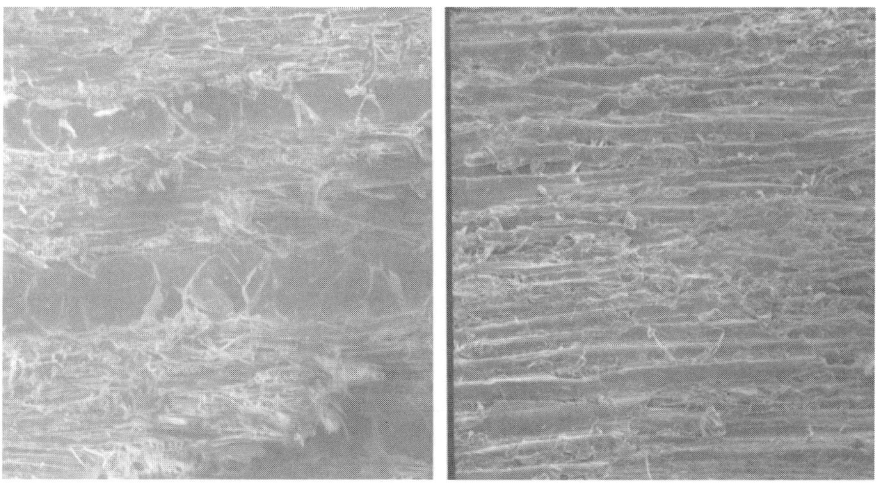

A B

FIGURE 10.1 Scanning electron micrograph of surface microstructure of (A) walnut and (B) pine. The microstructure was observed using a scanning electron microscope at an accelerated voltage of 20 kV. (*Source:* Copyright Sun and Bian [3].)

The adsorption theory describes any physical or electrostatic attraction between protein polymers and wood surfaces through hydrogen bonding and van der Waals forces. Good wetting should be established between the protein adhesive and the wood surface to enhance such physical attraction. The hydrogen bonding strength is achieved by the attraction between positive hydrogen atoms and negative oxygen or nitrogen atoms. The van der Waals forces are dipolar or electrostatic attractions between the nuclei of atoms or molecules and the electrons of other atoms or molecules, or intermolecular forces caused by the attraction between induced dipoles. Both hydrogen bonds and van der Waals forces require a certain distance. Per adsorption theory, the wood surface structure, protein structure and composition, surface wetting, contact angles, and press force should be major factors influencing adhesive strength.

Chemical bonding often occurs at the interface between adhesive and substrate, hence forming covalent chemical bonds, which are the strongest and most durable bonding. However, the chemical bonding between protein and cellulosic fibers may not be strong. Protein contains some functional groups as illustrated in Figure 10.2. These functional groups easily interact with hydroxyl and carboxyl groups from cellulosic fibers, but may not form covalent bonds unless under special reaction conditions. Protein modification is often recommended to improve protein structure conformation and surface physical properties and, consequently, improve bonding strength.

Based on the analysis and experiments conducted by Cheng [14], mechanical interlocking, penetration, and attraction are all the most important factors contributing protein adhesive strength on cellulosic materials. Chemical interaction and physical attraction would be very helpful to enhance adhesion performance. The diagram shown in Figure 10.3 can be used as a model to describe the adhesion mechanism between protein polymers and wood and fiber materials. The degree of cross-link or entanglements among

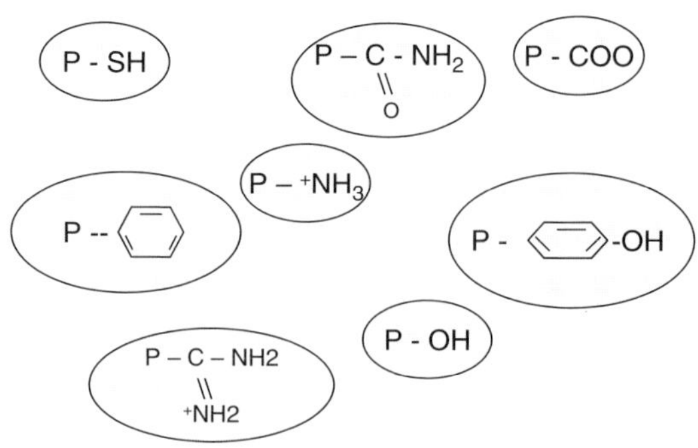

FIGURE 10.2 Most reactive chemical groups of plant protein. (*Source:* Courtesy X. Sun.)

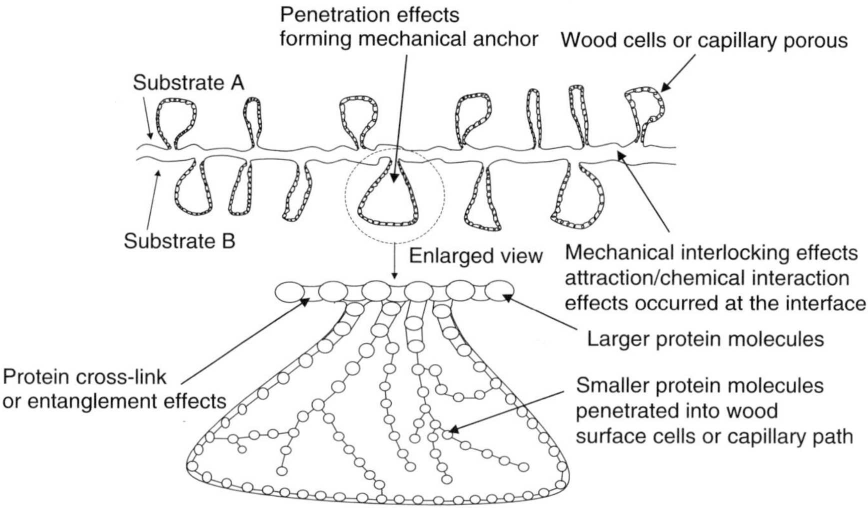

FIGURE 10.3 Diagram of proposed adhesion interface of protein adhesives and wood substrate. (*Source:* Courtesy X. Sun.)

protein molecules can enhance the strength of the protein complex formed at the interface of the conjunction of cellulosic fibers due to penetration. Such penetration can be observed using optical spectroscopy as presented in Figure 10.4. The degree of penetration should be optimum. Too much penetration can cause weaker connections between protein molecules due to the larger distance between protein polymer molecules; and less penetration can result in poor complex zone formation at the interface among protein polymers and wood cells and local wood fibers.

Hydrophobic interaction within protein molecules has an important influence on the degree of cross-link and entanglement and, consequently, on the degree of complex zone formation and the adhesion quality. Hydrophobic interactions between nonpolar or hydrophobic segments of a denatured protein occur spontaneously in polar media, such as water. Hence, proteins containing significant numbers of hydrophobic amino acids are found to have higher gluing strengths and water resistance. Recent studies show that the basic component of a glycinin protein from soy protein has much higher wet strength than the acidic component because the basic component contains a higher proportion of hydrophobic amino acids. Native corn zein protein, which also contains large amounts of hydrophobic amino acids, has strong gluing strength and high water resistance.

As mentioned in Chapters 2 and 9, protein contains about 20 amino acids, and each of the amino acids is different from another due to its various side-chain group. Amino acid composition, sequence (the order of connection), and molecular chain length are major variables determining the primary,

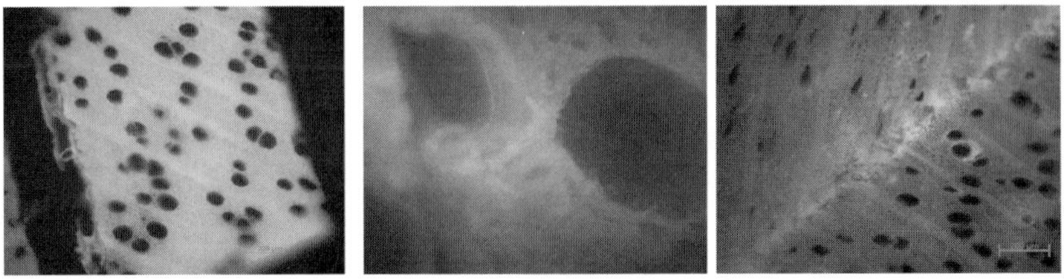

A B C

FIGURE 10.4 Protein adhesive (blue or brighter color) diffusion into pores of cherry wood surface (red or dark color): (A) Cross section of wood sample, (B) enlarged wood surface structure from part A, and (C) glued wood samples. Samples were thin-sectioned into an ~ 50-μm thickness using a Vibratome series 1000 (TPI Technical Products International, Inc., St. Louis, MO). The wood surface was examined using Axioplan 2 imaging (Axioplan 2 Imaging Universal Microscopes, Carl Zeiss MicroImaging, Inc., NY). Filter set C2 (G 365; FT395; LP420), the blue fluorescence color, was used for proteins; Filter set 15 (BP546/12; FT580; Lp590), the red color, was used for wood. (*Source:* Courtesy S. Yan and X. Sun.)

secondary, tertiary, and quaternary structures of a protein; they determine the degree of hydrophobic interaction, viscosity, surface tension, and pH-dependent behaviors and, consequently, affect all interactions and reactions shown in Figure 10.3 and eventually adhesive performance.

As discussed in Chapter 9, protein has both temperature- and pressure-sensitive polymers. Temperature and pressure are critical factors that influence adhesive performance, which is discussed in Section 10.3.

10.2 PROTEIN UNFOLDING AND ADHESIVE PROPERTIES

A modification process that changes the secondary, tertiary, or quaternary structure of a protein molecule is referred to as *denaturation* [8]. The compact protein structure unfolds or cross-links during denaturation, which is accompanied by the breaking and re-forming of the intermolecular and intramolecular interactions [15]. Protein modification can also turn some hydrophobic amino acids, which are buried inside, outward to improve water resistance. Protein denaturation can be induced by pH (alkali or acid); detergents; chemicals with reactive groups, such as amino, hydroxyl, carboxyl, isocyanate, and so on; and by heat treatment. The degree of change is influenced not only by the structure of modification chemicals but also by chemical concentration and modification procedures.

10.2.1 DETERGENT AS AN UNFOLDING AGENT

Detergent and Interaction with Protein Polymers

Detergent is defined as amphipathic molecules, containing a polar group with a long hydrophobic carbon tail. The polar group can form hydrogen bonds with water molecules, and the hydrophobic chain can interact with matter with hydrophobic groups, such as oil and protein [16]. Detergents can be categorized into cationic, anionic, nonionic, and ampholytic detergents. Detergents with ionic charge groups can denature protein at higher degrees than the nonionic or ampholytic detergents. A detergent with a negative ionic group as a head is called an *anionic detergent,* such as sodium dodecyl sulfate (SDS) ($C_{12}H_{25}NaO_4S$) and sodium dodecylbenzene sulfonate (SDBS) ($C_{18}H_{29}NaO_3S$). A detergent with a positive ionic charge at the head is called a *cationic detergent,* such as cetyl trimethyl-ammonium bromide (CTAB).

Proteins of biological origin contain sequences of hydrophobic amino acids that can determine protein folding and biological functions. The interactions of detergent can bring some hydrophobic groups buried inside to the surface of the protein, resulting in protein unfolding, and consequently change structure conformation and the properties of the protein. Such interaction is significantly affected by detergent type, concentration, pH, ionic strength, and temperature [17]. Generally, there are three major binding characters between protein and ionic detergents: specific binding, cooperative binding, and saturation. The specific binding is induced by electrostatic interactions between opposite charges from detergent and protein, which is greatly affected by the pH value of the protein because surface charges of a protein are predominantly influenced by pH. At the isoelectrostatic point of pH 4.5, soy protein has a neutral charge on the surface, meaning that all positive and negative charges are equally canceled. As pH increases, negative charge on the soy protein surface becomes predominant, and as pH falls below 4.5, the net charge on the surface turns to positive. The concentration of detergent requested for unfolding the protein is obviously dependent on the pH value.

Cooperative binding refers to the process by which the binding affinity keeps increasing by creating more binding sites between detergent and protein molecules, forming micelle-like structures of detergents on the protein surface, resulting in protein unfolding. Saturation binding between detergent and protein is generally pH independent and likely controlled by cooperative binding.

SDS is an anionic detergent, and the driving force for any degree of unfolding brought about by anion binding may arise from one or more of the following factors: (1) electrostatic repulsion between the charges of bound species, including the net charge of the protein; (2) penetration of the hydrocarbon tail into the polar regions of the protein; (3) binding-induced changes in the protein–hydrogen ion equilibrium resulting in an increase in electrostatic repulsion between charged species; and (4) a favorable ratio of the number of binding sites and protein association constants in the native form to those in the unfolded form [18]. The degree of protein unfolding increases with SDS concentration [17]. Protein modification through anion binding can move some interior hydrophobic side chains outward, where they can interact with the hydrophobic moieties of detergent molecules and form micelle-like regions [16] to increase hydrophobicity and, thus, increase water resistance.

The apparent viscosity of the SDS unfolded soy protein adhesives increases with SDS concentration [19]. Protein molecules become swollen and unfolded in SDS solution, resulting in an increase in the effective volume or hydrodynamic volume, a decrease in the distance between protein molecules, and, consequently, an increase in viscosity. In addition, swelling and unfolding can increase the axial ratio or axis of rotation of protein molecules, which also increases viscosity.

Unfolding Degree and Adhesive Properties

Hydrophobic interactions induced by soy protein modification with various SDS concentrations have been proved to have different adhesive properties [4]. With a pH of about 7.0, proteins modified with 0.5% and 1% SDS gave the highest gluing strength (Table 10.1). Soy protein modified with SDBS, another anionic detergent, in a concentration range similar to that of SDS, presented similar gluing properties, as reported by Huang and Sun [4]. The soy proteins modified with 1% SDS also had higher water resistance, showing no change in gluing strength after exposure to high relative humidity (RH) [4]. Wood specimens glued with the 1% SDS modified protein adhesive had about a 10% reduction in shear strength after three water-soaking cycles (Table 10.1).

SDS can produce a cooperative conformational change at lower concentration, and proteins can be partly unfolded, becoming partly denatured [16]. The denaturation degree of a protein can be estimated by differential scanning calorimetry (DSC) measurements. The total DSC enthalpy of the SDS modified protein decreased with SDS concentration (Table 10.2), indicating that the higher the SDS concentration, the greater the degree of protein unfolding. The lower gluing strength at higher SDS concentration shown in Table 10.1 suggests that there should be an optimum unfolding of a protein for better adhesive performance. The protein modified with 3% SDS exhibits

TABLE 10.1 Adhesive strength (MPa) of unfolded soy proteins with sodium dodecyl sulfate (SDS) on cherry wood samples. The soaked strength refers to the strength after three-cycles of 48-hr water-soaking test.

SDS Concentration (%)	Dry Strength (MPa)	Soaked Strength (MPa)	Ref
0	4.1	0	[4]
0.5	5.4	3.3	[4]
1.0	5.5	4.9	[4]
3.0	3.8	3.2	[4]

TABLE 10.2 Denaturation temperatures (T_d) and enthalpies of soy proteins estimated with differential scanning calorimetry (DSC) method. Data of denaturation temperatures and enthalpies of bulk protein are from reference [4], and of enthalpies of glycinin and conglycinin proteins are from reference [20].

			Enthalpy (J/g)		
SDS (%)	$T_{d\ glycinin}$ (°C)	$T_{d\ conglycinin}$ (°C)	Bulk Protein	Glycinin	Conglycinin
0	74.84	91.19	9.54	12.07	7.25
0.5	74.82	89.84	5.95		
1	72.81	90.17	3.39	9.03	5.43
3	—	83.52	2.57		

significant reduction in secondary structure as determined by the circular dichroism (CD) method as presented in Figure 10.5. This also suggests that the secondary structure may be favorable for protein adhesive performance. As expected, the interactions between hydrophobic groups of the protein with hydrophobic moieties of detergent molecules can form micelle-like regions to increase the water resistance of the adhesive. The results obtained on fiber cardboard with SDS-modified soy protein also indicate similar phenomena [19]. Soy protein modified with anionic detergent SDBS also exhibited similar trends among concentration and adhesive properties as observed by Huang and Sun [4].

Glycinin protein, a major component of soy protein, contains more hydrophobic groups than conglycinin, another major component of soy protein. Interactions between SDS with glycinin protein are stronger than that with conglycinin. The glycinin modified with 1% SDS gave higher gluing strength than the conglycinin protein modified at the same condition (Table 10.3) [20]. Glycinin protein requires higher energy to be denatured due to its compacted and stable globular structure compared to conglycinin (Table 10.2). The denaturation degree of glycinin at 1% SDS is lower than that of conglycinin. Although the gluing strength of glycinin is higher than that of conglycinin, the increase for conglycinin promoted by 1% SDS modification is larger than that for glycinin (Table 10.3). Conglycinin protein has a quaternary structure, which is stabilized by hydrophobic and hydrogen bonding. The extent of disulfide cross-linking is limited because there are only two to three cysteine groups per mole of protein [21]. On the other hand, the glycinin protein is tightly folded and linked via disulfide bonds. It contains approximately 48 mol cysteine, and disulfide groups account for about 37 mol per mole of proteins [22]. The high content of disulfide bonds greatly helps maintain the stability of the glycinin structure. As a result, proteins with a high glycinin content should have more ordered structure than proteins with a higher conglycinin content.

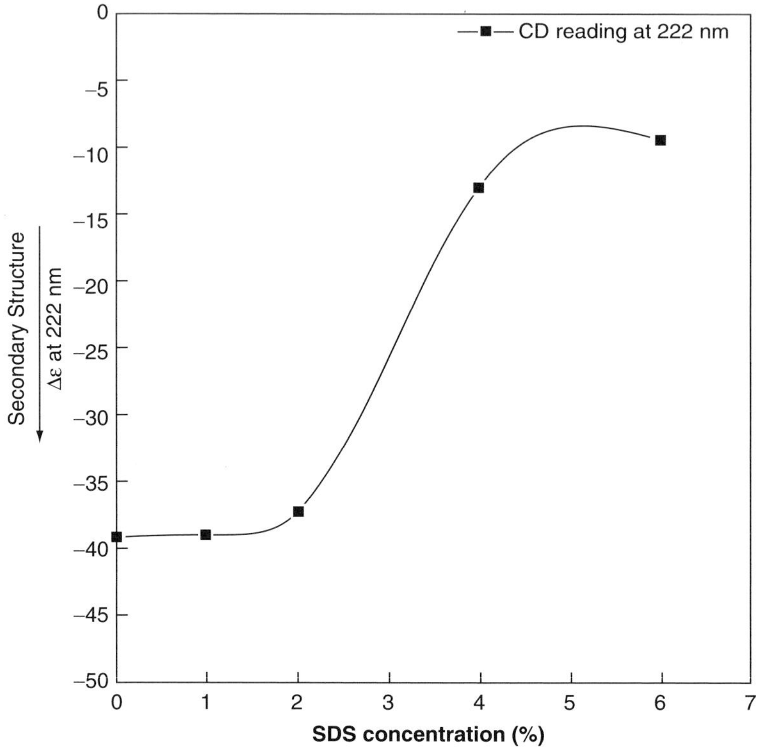

FIGURE 10.5 Secondary structure (together of α-helix and β-sheet) of soy protein unfolded with SDS estimated using circular dichroism (J-720, JASCO Corp., Tokyo, Japan). (*Source:* Courtesy Z. Zhong and X. Sun.)

TABLE 10.3 Adhesive strength of unfolded soy glycinin and conglycinin proteins with sodium dodecyl sulfate (SDS) on cherry wood samples. Wet strength was performed after soaking in tap water for 48 hours. Data source is from reference [20].

	Dry Strength (MPa)		Wet Strength (MPa)	
SDS (%)	Glycinin	Conglycinin	Glycinin	Conglycinin
0	4.7	3.8	1.5	1.1
1	5.1	4.3	1.75	1.72

10.2.2 AMINO-CONTAINING CHEMICAL AS AN UNFOLDING AGENT

Chemicals containing amino groups can interact with protein polymers. The mechanism of interaction is different from that between detergent and proteins. The oxygen, hydrogen, and nitrogen atoms of amino-containing chemicals can interact actively with hydroxyl groups of soy proteins to break down the hydrogen bonding in the protein body and, consequently, unfold the protein complex [3].

Urea is one of the amino-containing chemicals that is found to be a useful denaturation chemical to unfold the secondary helical structure of proteins [16]. Studies suggest that the complete unfolding of a protein can happen at higher urea concentrations, such as 8, 9, or 10 M [16, 23–25]. Studies also found that urea can unfold the secondary structure of a protein at higher concentrations, such as above 5 M [16, 23, 24]. Those proteins modified at relatively lower urea concentrations (1 and 3 M) can be partly unfolded and still retain a certain amount of secondary structure.

Obviously, similar to the processing of detergent modification, urea concentration also has a significant effect on adhesive strength. Soy proteins modified with urea at 1–3 M (about 6–18%) yield higher gluing strength than those modified using urea contents above 5 M (Table 10.4). The delamination percentage of the wood specimens glued using the 3 M urea-modified adhesive is 0% after three cycles of 48 h of water soaking, whereas the unmodified soy protein or protein modified with a higher urea concentration experienced 90–100% delamination under the same testing conditions (Table 10.4).

As mentioned before, the unfolding of proteins should increase their surface contact area and result in high gluing strength. Also, some of the hydrophobic amino acids that are buried inside become available on the exterior of the molecule, thus increasing water resistance. With up to 8 M of urea, the soy protein is estimated to be 100% denatured using the DSC method, which can significantly reduce the cross-linking effects during curing. The soy proteins partially denatured (about 20–50%) by 1 or 3 M urea can balance the cross-linking effects and the surface contact area as well as the water resistance on area. At lower urea concentration, urea can destabilize globular protein by forming strong hydrogen bonds with water molecules that surround the protein, perhaps protecting the protein from denaturation, while also disrupting protein hydrogen bonds, resulting in partially unfolded protein structures. The transition from the native state to

TABLE 10.4 Adhesive strength (MPa) and delamination (%) of denatured soy proteins with urea on cherry wood samples. Untreated soy protein was used as control. Apart (%) was performed after three cycles of 48-hr soaking in tap water at room temperature. Data source is from reference [34].

		Urea concentration			
Sample	Untreated	1M	3M	5M	8M
Dry strength	4.1	4.2	5.9	3.7	3.3
Apart %	100	0	0	30	100

the denatured state is often incomplete if the protein conformation is stabilized with disulfide bonds [16].

As discussed in Chapter 9, the glycinin globulin is a very heterogeneous oligomeric protein with a molecular weight of 320–360 kDa. It has a quaternary structure and consists of six subunits, as illustrated in Figure 9.2, that are joined by disulfide bonds, forming acidic–basic subunits [26]. The conglycinin globulin contains a major β-conglycinin, a trimeric glycoprotein with a molecular weight of 150–200 kDa, consisting of three types of subunits: α′ (72 kDa), α (68 kDa), and β (52 kDa) [27]. Conglycinin protein is less hydrophobic and contains more hydroxyl and amino groups based on its amino acid composition. Most subproteins of conglycinin are associated with hydrogen bonds. Therefore, conglycinin can actively interact with urea and become unfolded and easily denatured, whereas glycinin should remain or be difficult to denature. The enthalpy absorbed at denaturation is often used as an indicator of denaturation degree.

The degree of denaturation (Table 10.5) estimated by Mo *et al.* [20] shows that conglycinin is almost 100% denatured by 3 M urea and glycinin is about 75% denatured. Based on the relations between protein unfolding and adhesive properties, the adhesive strength of both glycinin and conglycinin should be improved after 3 M urea modification. However, the experimental data reported by Mo *et al.* [20] show that this is only true for conglycinin protein, but the adhesive strength of glycinin protein decreased (Table 10.6). Because of less interaction between glycinin and urea, some free urea molecules are in the glycinin protein matrix that would reduce entanglement or cross-link density during curing, and also weaken the interaction between protein and wood substrate surface and, consequently, lower adhesive strength. When the free urea molecules are removed from the urea-treated glycinin protein, the adhesive strength of glycinin is back to the level of untreated glycinin (Table 10.7). When the free urea is removed from the urea-treated conglycinin, the adhesive strength of the conglycinin protein is significantly improved compared to the untreated conglycinin. This experiment indicates that urea does not interact well with proteins associated with disulfide bonds, but strong interaction with proteins is associated with hydrogen bonds. This also indicates that many amino group-containing chemicals may not interact well with proteins containing disulfide bonds. When amino group-containing

TABLE 10.5 Denaturation enthalpies (J/g) of soy glycinin and conglycinin proteins with amino group containing chemical urea, as determined using DSC from reference [20].

Urea Concentration	Glycinin	Conglycinin
0	12.07	7.25
3 M	3.06	0.46

TABLE 10.6 Adhesive strength (MPa) of unfolded soy glycinin and conglycinin proteins with amino group containing chemical urea. Wet strength was performed after soaking in tap water at room temperature for 48 hours. Data source is from reference [20].

	Dry Strength		Wet Strength	
Urea Concentration	Glycinin	Conglycinin	Glycinin	Conglycinin
0	4.7	3.8	1.5	1.1
3 M	3.75	5.25	0.9	2.1

TABLE 10.7 Adhesive strength (MPa) of unfolded soy glycinin and conglycinin proteins with amino group containing chemical urea. Wet strength was performed after soaking in tap water at room temperature for 48 hours. (Courtesy X. Sun and X. Mo).

	Dry Strength		Wet Strength	
Urea Concentration	Glycinin	Conglycinin	Glycinin	Conglycinin
0	5.52	2.97	2.1	0.99
3 M	3.83	3.95	0.91	1.35
3 M and free urea is removed	5.38	4.30	1.83	1.20

Proteins used in this experiment are the same as those proteins used for Table 10.6, but aged for about 5 months. The variations in adhesive strengths are caused by experimental errors and also protein aging.

chemicals are used as an unfolding agent for a protein associated with both hydrogen and disulfide bonds, the amount of the chemical should be calculated based on the portion of the proteins associated with hydrogen bonds. There should be no free chemicals left in the protein system.

Guanidine hydrochloride (GuHCl) is another amino group-containing chemical. The denatured state of globular protein molecules in GuCHl solution depends on GuHCl concentrations [16, 23–25]. At lower GuHCl concentration, such as 0.3–2 M, the specific tertiary structure of a protein can be destroyed as monitored by near-ultraviolet CD spectra, whereas at 2–6 M GuCHl, the compactness of molecules in the secondary structure can be destroyed [28]. Globular proteins moved into a molten-globule state when denatured by GuCHl at lower concentrations, and it is possible that the protein could have intermediate to native and highly unordered structure [25, 28]. A globular protein in this state was nearly as compact as native proteins and had a high content of a secondary structure with a fluctuating tertiary structure [25, 29–31]. The heat capacity of the molten globule state was much higher than the heat capacity of the native state [31].

Soy proteins modified with 0.5 M and 1 M GuHCl exhibited greater gluing strength (Table 10.8) [5]. Wood specimens glued with proteins modified by

TABLE 10.8 Adhesive strength (MPa) of unfolded soy proteins with amino group containing chemical guanidine hydrochloride (GuHCl) on cherry wood samples. The soaked strength is referred to as the strength after three cycles of 48-hr water-soaking test. Apart (%) was determined as the delamination rate after three cycles of water-soaking test. Data source is from reference [5].

GuHCl Concentration (M)	Dry Strength (MPa)	Soaked Strength (MPa)	Apart (%)
0	3.0	0	100
0.5	4.9	0.9	0
1.0	6.0	1.3	0
3.0	3.6	0	100

0.5 and 1 M GuCHl also exhibited high water resistance, retaining gluing strength after humidity incubation and three water-soaking cycles (Table 10.8). Soy proteins modified with 0.5 and 1 M GuCHl had higher thermal transition temperatures and enthalpies than the unmodified proteins (Table 10.9). The soy proteins modified by 0.5 and 1 M GuCHl could be in the molten globule state, resulting in higher denaturation temperature and enthalpy (Table 10.9). Although protein in the molten-globule state is rather compact, more hydrophobic groups in the molten-globule state can be expanded in water than in the native state. Furthermore, proteins in the molten-globule state have a more labile surface than native proteins, facilitating their ability to penetrate membranes more easily [25]. This implies that the protein modified with 0.5 and 1 M GuCHl can easily penetrate the wood surface and generate more adhesion force to improve gluing strength.

10.3 EFFECTS OF CURING TEMPERATURE AND PRESSURE ON ADHESIVE STRENGTH

Proteins are temperature- and pressure-sensitive polymers. Curing temperature, pressure, and time have an important influence on the final protein

TABLE 10.9 Denaturation temperatures (T_d) and enthalpies of unfolded soy proteins with guanidine hydrochloride (GuHCl), as estimated with differential scanning calorimetry (DSC) method. Data are from reference [5].

GuHCl (M)	$T_{d\ \mathrm{glycinin}}$ (°C)	$T_{d\ \mathrm{conglycinin}}$ (°C)	Enthalpy (J/g)
0	74.94	88.79	9.97
0.5	79.49	98.52	10.28
1	79.33	96.52	10.74
3	70.81	85.85	2.55

structure reconformation, mutual diffusion at the interface, and, consequently, adhesion strength. The structure change induced by modification requires different curing conditions. We discussed in Chapter 9 the importance of temperature and time on curing strength [32].

When soy protein slurry is applied to cellulosic substrates, such as a piece of wood or fiber cardboard, it first spreads and wets the surface. Then, the polar and apolar groups of the protein molecules interact with the substrate surface by physical and/or chemical forces. The protein molecular chains may also penetrate into the substrate surface through the porous structure, which can enhance interactions between the protein adhesive and the surface. Two important factors influence gluing performance: (1) intimate contact at the interface between the substrate surface and protein adhesive, and (2) immobilization of the protein adhesive. Assembly time and press conditions, such as temperature, pressure, and time, are all major parameters affecting gluing strength [33]. As press force increases, contact at the interface between the substrate surface and protein adhesive increases, and as press time increases, immobilization of protein adhesive is enhanced, resulting in higher shear strength. In addition, the press time can promote penetration of protein molecules into the substrate surface and also enhance any chemical interaction at the interface between the adhesive and substrate. Press temperature not only enhances immobilization of the protein adhesive but also increases the possibility of chemical reactions at the interface between the protein adhesive and substrate.

One experiment reported by Sun and Bian [32] demonstrates the relations among adhesive curing strength, curing temperature, and curing time (Figure 10.6). A longer curing time is needed at a lower curing temperature. For example, with a fixed press force (20 kg/cm^2), about 5 min is needed to reach the highest gluing strength at 130 °C, but it takes about 15 min to obtain a similar curing strength at 90 °C.

In another experiment done by Zhong *et al.* [33], similar results were also obtained that indicated the gluing strength of fiber cardboard with a soy protein adhesive increased significantly as the curing temperature was increased up to 100 °C (Figure 10.7). Fiber cardboard pressed at a higher pressure and longer time had a higher gluing strength (Figure 10.8). Pressing at a high pressure for a long time also enhances the interaction between the protein adhesive and fiber cardboard and, hence, reduces the interaction between the protein adhesive and water molecules. It is difficult for water molecules to penetrate into the gap between two pieces of fiber cardboard that are glued and then pressed at high pressure for a long time, because the two pieces of fiber cardboard are in closer contact with each other. Assembly time also is a critical factor (Figure 10.9) [33]. A short assembly time prevents the protein adhesive from sufficiently wetting the fiber cardboard, which means that the protein molecules penetrate the cardboard less. A longer assembly time may lead to overevaporation of water, resulting in precuring

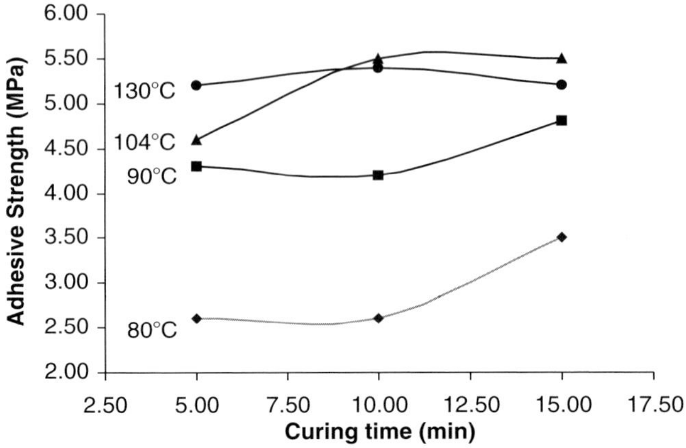

FIGURE 10.6 Effects of curing temperature and time on adhesive strength of soy protein modified with 3 mole urea on walnut wood samples. (*Source:* Data from Ref. [32].)

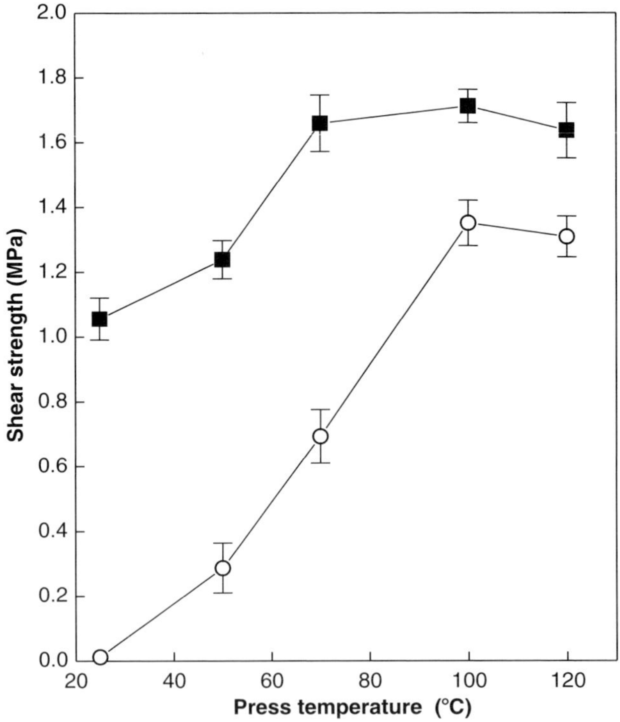

FIGURE 10.7 Effect of curing temperature on adhesive strength of fiberboards bonded with soy protein and modified with 3% SDS: dry strength (■) and wet strength (○). (*Source:* Copyright Zhong *et al.* [19].)

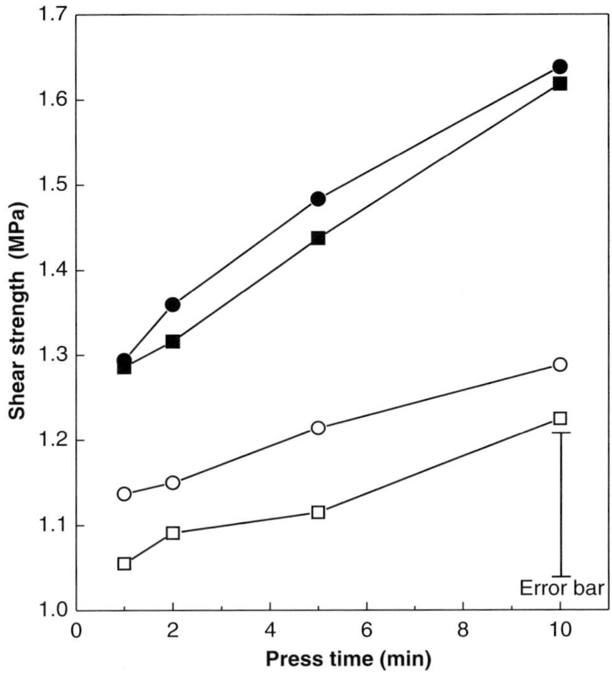

FIGURE 10.8 Effect of curing time on adhesive strength of fiber cardboard specimens glued with soy protein adhesives. Pressed at 25°C and 1 MPa (■□), 2 MPa (●○), dry strength (■●), and wet strength (□○). (*Source:* Copyright Zhong *et al.* [33].)

and/or difficulty entangling the proteins together on the two pieces of fiber cardboard after assembly, leading to low gluing strength.

10.4 VISCOSITY OF SOY PROTEIN ADHESIVES

Soy protein denatures with heat and its viscosity changes during heating [32]. The relative viscosity of native soy protein decreases at first as temperature increases from room temperature to about 65°C, and then remains constant for about 4 min (Figure 10.10). The protein molecular chain unfolds at higher temperatures, resulting in lower viscosity. As the temperature increases up to about 80°C, the protein begins thermal denaturation, and the protein molecules become unfolded, absorb water, and then swell. The viscosity increases rapidly at this point. Then, the viscosity of the protein starts to decrease immediately after reaching its maximum value at holding temperature, presenting a shear-thinning behavior. Continued spinning of the paddle of the viscometer at 95°C can destroy the entanglement structure, resulting in low viscosity again. During cooling, the viscosity of the protein

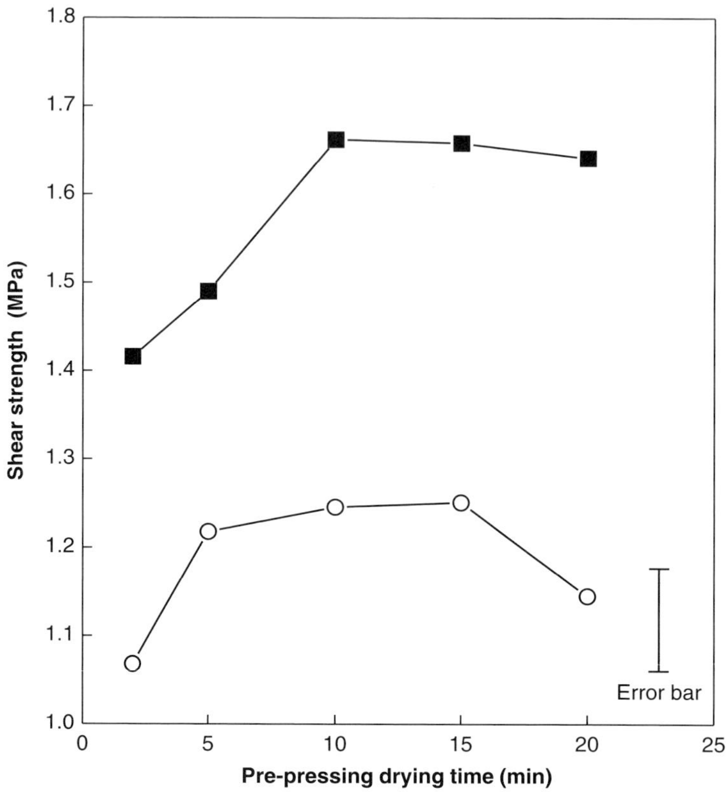

FIGURE 10.9 Effect of assembly time (prepressing time) on adhesive strength of fiber cardboard glued with soy protein adhesives: dry strength (■) and wet strength (○). (Copyright Zhong *et al.* [33].)

increases again because of gelling. Unlike the native proteins, the viscosity of the urea-modified soy proteins is low and thermally stable because of the effects of unfolding (Figure 10.10).

The apparent viscosity of soy protein unfolded with anionic detergent, such as SDS, decreases slightly as shear rate increases (Figure 10.11), exhibiting a shear-thinning behavior [34], which can be expressed by the power equation [35]

$$\tau - \tau_0 = K\dot{\gamma}^n, \tag{10.1}$$

where τ is the shear stress (N/m^2); τ_0 is the yield stress (N/m^2); $\dot{\gamma}$ is the shear rate (s^{-1}); and n and K are the flow behavior index and the consistency index, respectively. The n and K values can be obtained by plotting log $(\tau - \tau_0)$ against log $\dot{\gamma}$. The τ_0 value can be obtained using the Casson equation [36]:

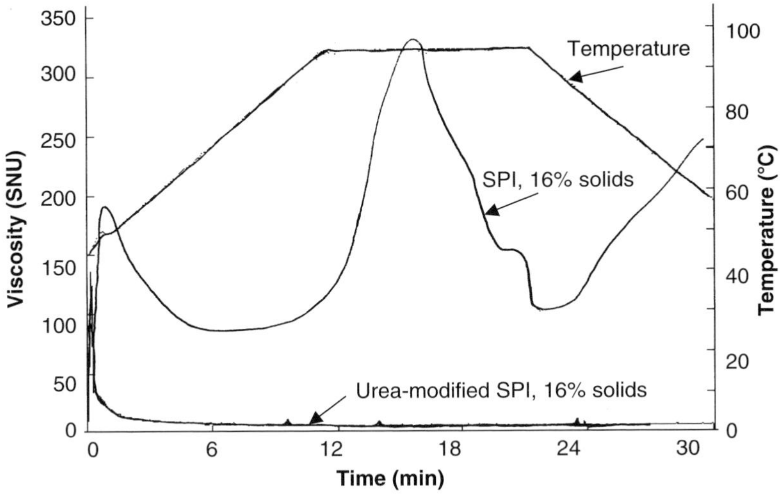

FIGURE 10.10 Relative viscosity curves of native and urea-modified soy protein isolate (SPI) slurries at 16% solid content, as determined by rapid viscosity analyzer. SNU, the stirring number, is used as the viscosity unit. (*Source:* Copyright Sun and Bian [3].)

$$\sqrt{\tau} = \sqrt{\tau_0} + \sqrt{\eta_c \cdot \dot{\gamma}}, \tag{10.2}$$

where η_c is the Casson viscosity. The value of n increases as the unfolding degree increases, but all are less than 1. At a higher unfolding degree, such as above 90%, soy protein has an almost Newtonian flow behavior [19].

10.5 NATURAL STRAW COMPOSITES WITH SOY PROTEIN ADHESIVES

Agriculture residues, such as straws, are another possible fiber source for particleboards. Large quantities of agricultural residues are available; for example, about 67 million metric tons of wheat straw are produced annually in the United States. Wheat straw often is used as fuel, cattle feed, mulch, and bedding materials for animals. Straw has a chemical composition similar to that of wood, but with more cellulose and 1–4% silica. Of all the elements contained in straw, 99% are lighter elements like carbon, hydrogen, nitrogen, and oxygen, which are feasible for particleboard [37, 38].

Urea formaldehyde (UF)-based resins are still major adhesives for the particleboard industry. Formaldehyde emission remains an environmental problem for resin processing, distribution, applications, and particleboard end users. UF resin has poor bondability with straw. The inherent nonpolar and hydrophobic characteristics of a straw surface, due to the silica and the wax components, are not compatible with the polar and hydrophilic nature of the UF resins [39, 40].

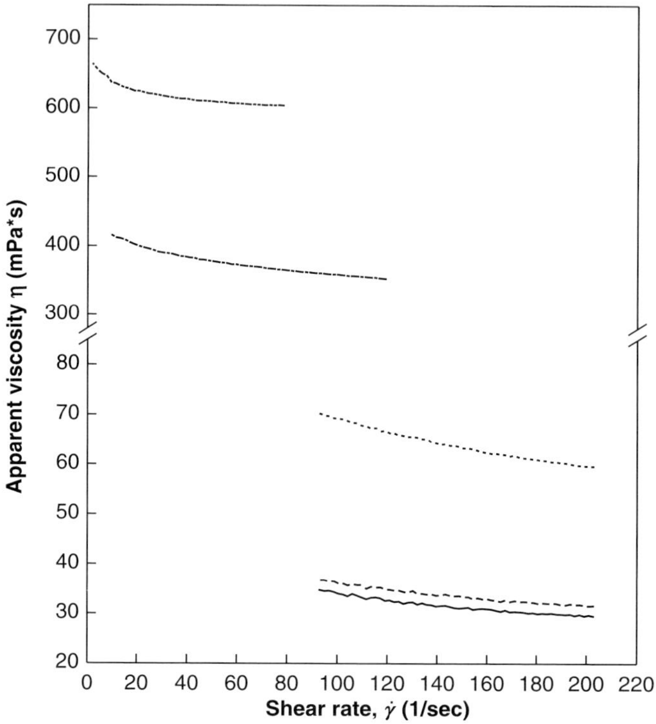

FIGURE 10.11 Shear rate dependence of apparent viscosity of soy protein unfolded with SDS at these concentrations: 0 wt% (———), 0.5 wt% (-------), 1 wt% (·········), 3 wt% (-·-·-·-·), and 5 wt% (-··-··-··-··). Rheological properties of the modified soy protein are determined using a Brookfield programmable rheometer (DV-III +) equipped with a small sample adapter (SC4-21/13R) (Brookfield Engineering Laboratories, Inc., Middleboro, MA). (*Source:* Copyright Zhong *et al.* [19].)

Soy protein-based adhesives are superior or similar to UF resin for wheat straw particleboard. Soy protein is often in slurry form, which can be easily applied to straw particles by mixing. The resin-coated straw can be hot pressed into flat panels. The factors affecting straw particleboard quality mainly include soy protein concentration, protein modification, straw particle size, mixing methods, moisture content of the resin-coated straw prior to press, and press conditions.

10.5.1 ROLE OF WATER

The moisture content of the resin-coated straw has a significant effect on board strength [41]. Resinated straw with 10% moisture content cannot form a low-density particleboard. As discussed in Chapter 9, soy protein curing requires certain pressure. In low-density board processing, the pressure is low, and is not enough for protein to form an entangled matrix. Water acts as a plasticizer and allows the protein to unfold to a greater extent and to become entangled upon curing, resulting in strong bonding strength as the water evaporates. However, at high moisture content, beyond 40%, internal cracks occur within the particleboard due to the high water vapor pressure trapped in the composite.

Particleboard bonded with isocyanate-based adhesives, such as MDI, has better mechanical properties than those made with soy protein-based adhesives [41]. MDI is more effective in wetting the surface of the straw than the soy protein-based adhesives, enhancing chemical bonding through hydrogen bonds and polyurethane covalent bonds. In addition, the isocyanate groups of MDI can react with water in the straw, producing cross-linked polyureas for better mechanical bonding [42, 43]. However, at high moisture content, water can react with most of the isocyanate groups, and fewer isocyanate groups are available for bonding with straw, often resulting in reduced bonding strength.

10.5.2 MECHANICAL PROPERTIES

The mechanical properties of a particleboard bond with soy protein adhesive vary as board density (Table 10.10). The board with a density of about 0.8 g/cm^3 or above has better or similar mechanical properties than that with UF resin. The particleboard with soy protein adhesives has lower internal bonding strength than that with MDI, but higher than that with UF. No significant differences are observable in an equilibrium moisture content test between the particleboard with modified soy protein and MDI at 30% RH and 60% RH. The particleboard with modified soy adhesive has a slightly higher equilibrium moisture content than that with MDI at 90% RH [41].

TABLE 10.10 Mechanical properties of wheat straw particleboard bond with soy protein-based adhesives. Where UF = urea formaldehyde adhesive, MOR = modulus of rupture, MOE = modulus of elasticity.

Density (g/cm^3)	Tensile Strength (MPa)	MOR (MPa)	MOE (MPa)	Reference
0.8–0.9	6.0–7.0	18–21	3000–3300	[41, 46]
0.6–0.7	3.5–4.5	7.0–9.0	1700–1900	[41, 42]
0.72 (UF resin)	3.9	6.3	1805	Courtesy E. Cheng, X. Sun
0.75 (Commercial wood particleboard)		13–17	2200–2400	[41]

For economic reasons, soy flour, containing 50% protein, is often used as a starting material for producing the adhesive for particleboard used for interior applications. Chemical modification is usually used to improve the adhesive performance of soy flour. Many modification methods for soy protein can be applied to soy flour except for urea. As discussed before, urea is an important denaturation chemical that can unfold the secondary, tertiary, and quaternary structures of a protein, resulting in enhanced adhesion bonding strength. Soy flour has an enzyme, urease, which can increase the hydrolysis rate of urea to carbon dioxide and ammonia. This reaction weakens the effect of urea modification. Urease inhibitors, such as thiophosphoric triamide (*n*BTPT, *N*-(*n*-butyl)), can be used to prevent the urease catalyst action [43], which enhances adhesive strength for particleboard application [40].

The other 50% of the soy flour is mainly cellulosic-based carbohydrate material, which can also be modified to improve adhesive performance [40]. Citric acid can improve cross-linking of cellulose. It also interacts with amino groups in soy protein. With sodium hypophosphite (NaH_2PO_2) as a catalyst, citric acid can actively interact with the cellulose in soy carbohydrates. In addition, citric acid in soy flour can create ester cross-linking within the straw complex [44]. With the adhesion provided from soy protein and curing conditions provided by a hot press, straw particles should attach to each other firmly. The mechanical strength of straw board made from soy flours treated with citric acid is significantly improved [40].

Boric acid is another chemical that can interact with carbohydrates in soy flour to create cross-links within the carbohydrate complex, which results in a significant decrease in water absorption in the case of soy plastics [40]. The tensile strength of the straw particleboard bond with soy flour treated with boric acid also is improved as demonstrated by Cheng *et al.* [40]. Weight gain, thickness swell, and linear expansion of the board are all significantly reduced. The insoluble carbohydrates in soybean include cellulose, hemicellulose, and pectin. Boric acid can interact with cellulose from both straw and soy flour, and cross-links might occur among the complexes by interaction with those long-chain molecules in the amorphous region in cellulose, thus holding the cellulose together more tightly. This inhibits the penetration of water molecules into the complex, hence reducing water absorption.

It is believed that the mechanical strength of straw board bonded with the soy flour and treated with a combination of chemicals should be enhanced, as demonstrated by Cheng *et al.* [40]. Optimal mechanical properties are obtained when soy flour is treated with 1.5 M urea, 0.4% *n*BTPT, 7% citric acid, 3% boric acid, 4% NaH_2PO_2, and 1.85% NaOH (Table 10.11). With tensile strength at 5.6 MPa, modulus of rupture at 11.89 MPa, and modulus of elasticity at 3350 MPa, its mechanical strength meets the appropriate ANSI standard (Section Industry Standard). Water resistance is also significantly improved. In conclusion, soy protein-based adhesives give results

TABLE 10.11 Mechanical strength and water resistance of straw particleboard made with soy flour adhesives modified with citric acid, boric acid, and urea with urease inhibitor. Treatment codes: control: soy flour without modification; Treat 1: soy flour modified with 1.5 M urea + 0.4% *N*-(*n*-butyl)thiophosphoric triamide + 7% citric acid + 3% boric acid + 4% sodium hypophosphite + 1.85% sodium hydroxide; Treat 2: soy flour modified with 1.5 M urea + 0.3% *N*-(*n*-butyl)thiophosphoric triamide + 5% citric acid + 2% boric acid + 4% sodium hypophosphite + 0.93% sodium hydroxide. Data source is from reference [40].

Treatment	Tensile Strength (MPa)	MOR (MPa)	MOE (MPa)	Weight Gain (%)	Thickness Swell (%)	Linear Expansion (%)
Control	3.7	9.7	2340	129.8	140.6	1.5
Treat 1	5.6	11.9	3353	95.2	83.4	1.0
Treat 2	5.6	10.3	3169	96.2	66.6	1.2

superior or similar to those of UF resin for particleboard production and are also economically viable.

10.6 PRODUCTION OF LOW-COST ADHESIVE IN POWDER FORM

Protein modification for various applications is often integrated in protein isolation processing from soybean flour or soybean meals [4–5, 8, 45–49]. The adhesive derived from modified soy proteins is in liquid form. However, industry prefers powder form, which is easier in distribution, handling, and application. The major principle of protein adhesion is to disperse the protein in water in which protein structure becomes unfolded to an optimum degree in the presence of modifiers [1]. The unfolded protein increased surface contact area; therefore, the adhesive strength can be enhanced. In addition, some hydrophobic amino acids buried inside become exposed outwards during unfolding, which improved water resistance.

Traditionally, proteins are modified in excess water content to assure complete modification. However, the excess water dilutes the protein solution, resulting in less protein–protein interaction. As mentioned before, the unfolding of proteins for adhesives does not change the primary structure, but the quarternary, tertiary, or secondary structures that make the protein molecules become unfolded and result in a new structure conformation. The unfolded protein structure partially returns back to its original conformation in drying at elevated temperature based on minimum internal energy law. This section discusses unfolding the protein molecules in a low moisture environment. The unfolded protein molecules become entangled in each other due to the shorter distance between protein molecules in a lower moisture content. Such an entangled structure remains unchanged during drying at an elevated temperature. When the entangled protein (modified protein) powder is redissolved in water, it retains the same unfolding degree. The low moisture protein unfolding can stabilize the unfolded protein structure and also reduce drying energy to remove less extra water.

10.6.1 ADHESIVE POWDER PRODUCTION

As shown in Figure 10.12, traditional procedures for protein extraction, modification, or unfolding in excess water need high water content up to 95%. For example, the defatted soy flour is added into tap water at about 14% solid content and stirring for 20 min at room temperature. Meanwhile, the pH value of the slurry was adjusted to 8.5 by adding 2 N NaOH during the 20 min stirring. The mixture is then centrifuged at 4000 rpm at room temperature for about 10 min to remove carbohydrate and some insoluble particles. Then the pH value of the supernatant is adjusted to 4.6 with 2 N HCI and centrifuged at 4000 rpm at room temperature for 10 min. The precipitate is the crude protein with about 85% protein content extracted from the defatted soy flour. The crude protein was washed three times using tap water, and then water content is adjusted to about 88%. The pH of the slurry

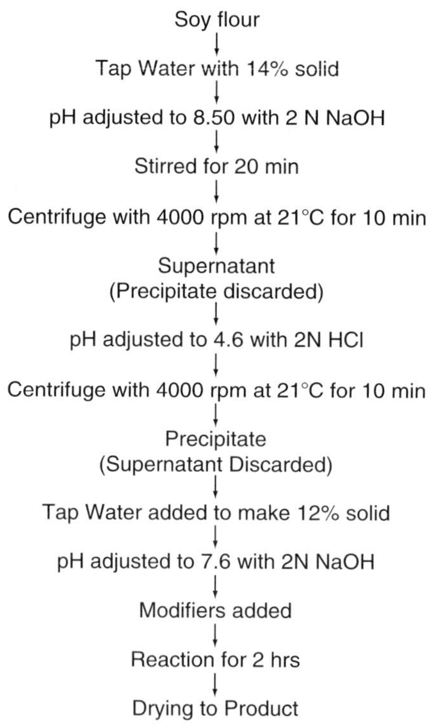

FIGURE 10.12 Procedure for soy protein extraction and modification in excess water content.

is adjusted to 7.6 with 2 N NaOH and 8.33% SDS (based on the crude protein weight) were added and stirred for 2 hours. The modified crude protein slurry was dried using a spray dryer and freeze dryers.

The protein extraction procedure is similar to that described for modification of soy protein adhesives using high moisture, but the three-time water washing procedure at the end of protein extraction is omitted. The water content of the crude protein is about 65%. The crude protein can then be unfolded and modified at this point. In one example, the crude protein is unfolded using SDS. The unfolded proteins become very viscous and elastic with dark brown color (Figure 10.13). The unfolded proteins are pelletized and dried at elevated temperature range from 55° to 120°C and milled into powder with particle size about 190 μm. The defatted soy flour with about 50% protein can be used directly as adhesive materials using the similar technology.

10.6.2 ADHESIVE PERFORMANCE UNFOLDED IN HIGH WATER CONTENT

After unfolding in high water content, the soy protein adhesive dried using freeze drying and then resolved in water as adhesive gives higher adhesive

FIGURE 10.13 Soy flour adhesives unfolded in low water content.

TABLE 10.12 Shear strength (MPa) comparison of modified soy protein adhesives prepared using spray-drying and freeze-drying methods. Press temperature was 130°C, press pressure was 1.7 MPa, and press time was 5 min.

Drying Methods	Dry Strength	Wet Strength (48 h)	Wet Strength (96 h)	Soaked Strength
Spray drying	3.04[d]	0.61[d]	0.22[d]	1.08[d]
141 inlet –75 outlet°C	5.02[ab]	1.16[b]	0.99[b]	3.89[b]
171 inlet –75 outlet°C	3.85[c]	0.92[c]	0.64[c]	2.55[c]
195 inlet –85 outlet°C	5.23[a]	1.74[a]	1.64[a]	4.64[a]
Freeze drying	5.30[a]	1.87[a]	1.81[a]	4.95[a]
LSD	0.40	0.21	0.19	0.69

[a]The least significant difference (LSD) was at probability level of $\alpha = 0.05$

strength than that dried with spray drying (Table 10.12). There is no significant difference between freeze drying and spray drying on dry strength. However, freeze-dried adhesive showed significantly higher wet strength and soaked strength than spray-dried adhesive. Freeze drying can retain the unfolded protein structure. However, in spray drying, the unfolded protein structure can be completely or partially folded back to its original state when the fine droplets go concurrently through the hot air. In addition, the unfolded protein could be further denatured in spray drying due to high drying temperature. Spray-drying temperature also has significant effect on adhesive strength. The adhesives prepared at 171–75°C spray-drying temperature have the highest shear strength among the three tested temperature profiles. The adhesives prepared at 141–75°C have the lowest shear strength. At a lower spray-drying temperature, although the heat-induced denaturation is lower, the degree of folding up of the modified protein could be higher due to slower drying rate. At a higher spray-drying temperature, such as 195–85°C, the degree of folding up of the unfolded protein can be smaller, but the heat-induced denaturation can be larger than that dried at a medium spray-drying temperature. These results may indicate that moderate denaturation may be suitable for protein drying. Particle size of the spray-dried powder also varied with drying temperature, which can be another factor influencing adhesive strength. Hollow particles are formed during spray drying indicating damage to unfolded protein structure (Figure 10.14).

10.6.3 ADHESIVE PERFORMANCE UNFOLDED IN LOW WATER CONTENT

Difference in shear strength existed between soy protein adhesives unfolded at low and high water contents (Table 10.13), especially significant for wet strength of the specimens prepared at higher press temperatures. Although the wet strength of the specimen pressed at 180°C was the highest, press temperature higher than 180°C was not recommended because the wood

FIGURE 10.14 Particle structures of soy flour adhesive particles after spray drying unfolded in excess water.

TABLE 10.13 Shear strength (MPa) of soy protein adhesives modified with SDS at low (65%) and high (85%) water content as affected by press temperature. Press pressure was 1.7 MPa, and press time was 5 min.

	Dry Strength		Wet Strength (48 h)		Wet Strength (96 h)		Soaked Strength	
Press Temperature	Low[a]	High[b]	Low	High	Low	High	Low	High
130°C	5.4	5.0	1.8	1.7	1.9	1.6	4.9	4.8
150°C	5.9	5.2	2.2	1.7	2.2	1.7	5.1	4.9
170°C	5.6	5.7	2.7	2.1	2.9	2.3	5.7	5.6
180°C	5.5	5.6	2.9	2.4	2.9	2.4	6.0	5.9
LSD[c]	0.42	0.42	0.46	0.35	0.47	0.40	0.49	0.88

[a]Low = Protein modification at low water content.
[b]High = Protein modification at high water content.
[c]The least significant difference (LSD) was at probability level of $\alpha = 0.05$.

sample became a little dark at 190°C due to the heat damage. The results indicate that the protein modified at low water content became unfolded and entangled. The entangled unfolded protein structure remained unchanged upon drying because it retained the hydrophobic amino acid groups still exposed outward, resulting in higher water resistance, and consequently gave higher wet strength. The particle size of the dried adhesive unfolded in low water content is shown in Figure 10.15, indicating that the entangled structure remained unattached. For dry strength test, 100% wood sample was broken for all cases; therefore, it is difficult to judge the real difference in

FIGURE 10.15 Particle structures of soy flour adhesive particles after oven drying unfolded in low water content.

strength between the adhesives prepared in low and high water content. Stronger wood than cherry or thicker wood samples could be used in the future to determine the real strength.

The soy flour adhesive unfolded in low water content has higher adhesive strength than unmodified and modified at high moisture content (Table 10.14). The shear strength of the modified soy flour adhesives cured at 170°C is higher than that cured at 130°C. Since the soy flour contains about 50% protein, its adhesive strength is lower than that of the modified soy protein adhesives obtained in Table 10.13. The concept of protein unfolding and entanglement described in protein modification at low water content can be applied to explain what is observed in this soy flour adhesive experiment. The modified soy flour adhesives in low moisture content have great potential to replace partially or completely urea formaldehyde resin for plywood and particle board productions, or medium density fiber board.

10.7 SOY PROTEIN LATEX-LIKE ADHESIVES

Multibillion pounds of latex-based adhesives are used annually in the United States. They include foundry adhesives, wood adhesives, school children glues, labeling adhesives, paper box packaging adhesives, and envelope adhesives. Most latex-based adhesives are synthetic chemicals and contain vinyl acetate or acetaldehyde, which are especially restricted for food and pharmaceutical-related product packaging.

TABLE 10.14 Shear strength (MPa) of soy flour adhesives modified at low and high water contents.

Type of Modification and Press Conditions	Dry Strength	Wet Strength (48 h)
Unmodified soy flour (press temperature 130 °C)	3.8 ± 0.38	apart
Modified soy flour		
At low moisture content (press temperature 130 °C)	4.7 ± 0.50	0.5 ± 0.02
At high moisture content (press temperature 130 °C)	4.2 ± 1.2	apart
At low moisture content (press temperature 170 °C)	5.9 ± 0.67	1.9 ± 0.24
At high moisture content (press temperature 170 °C)	4.7 ± 0.34	1.8 ± 0.5

Unfolded protein in low moisture content also enables protein molecules entangled in each other to form a continuous complex that has strong tack adhesion properties, which is called latex-like adhesive in this section. This adhesive is fluid-like with solid content ranging from 20 to 70% (Figure 10.16). The latex-like adhesive has a long shelf life up to 8 months at room temperature and no protein-water phase separation in storage and application. This adhesive in various formulas can be used as children's glue and color paints, wood veneer adhesive, fiber composite resin, foundry adhesives, and packaging and labeling adhesives.

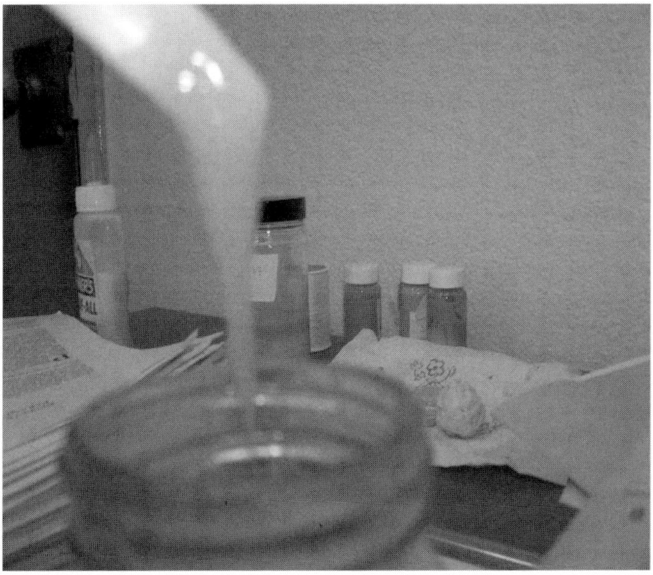

FIGURE 10.16 Soy protein-based latex-like adhesive sample with 37% moisture content.

10.7.1 ADHESION STRENGTH ON CHERRY VENEER WOOD

Adhesive strength is determined using an Instron machine (Model 4465, Canton, MA) according to ASTM D2339. For water resistance test, specimens are evaluated according to ASTM D1151 for effects of moisture and temperature on adhesive bonds, and ASTM D1183 for resistance of adhesive to cyclic lab aging. Samples are soaked in tap water at 23°C for 48 hr, and are immediately tested for wet strength. A boiling test is conducted following PS1-95 method. Place one group of specimens in a tank of boiling water, separated by wire screens in such a manner that all surfaces are freely exposed to water. The specimens are immersed at least 51 mm deep during boiling test cycle, and are kept boiling for 4 h. The specimens are then dried for 20 h at 63 ± 3°C with sufficient air circulation to reduce the moisture content (MC) of the specimens to original, within an allowable variation of ±1% MC. The 4-h boiling cycle is repeated, the specimens are removed and cooled in running tap water at 18 to 27°C for 1 h, and then the specimens are evaluated for wet strength.

The strength of the adhesives is not affected by storage temperature and time (Table 10.15). One-hundred percent cohesive wood failure (CWE) was obtained for dry strength tests. The CWF of wet and boiling test samples are not evaluated. About 90% of the glued area showed coarse fibers for wet test samples, and broken very fine fibers were observed for boiling test samples.

10.7.2 ADHESION STRENGTH ON FOUNDRY SANDS

The iron and steel casting foundry industries generate huge air pollution problems due to hazardous substance emission from adhesive during manufacturing. The used sands often need to be reconditioned for recycling, and the core has to be removed after casting. This newly discovered latex-like protein-based adhesive has great potential for the foundry industry. The adhesive becomes decomposed above 250°C and so the used sand can be removed

TABLE 10.15 Adhesive strength (MPa) of dry, wet, and boiling tests as affected by storage time and temperatures.

Samples	Dry Strength	Wet Strength	Boiling Strength
Stored at room conditions (23°C)			
0 month	6.54 ± 1.20	4.35 ± 0.34	2.78 ± 0.06
1 month	6.34 ± 1.12	4.25 ± 0.27	2.80 ± 0.14
2 month	6.64 ± 1.56	3.89 ± 0.32	2.65 ± 0.35
4 month	6.24 ± 1.65	3.69 ± 0.31	2.30 ± 0.49
Stored 4°C			
3 month	6.30 ± 1.87	3.67 ± 0.41	2.64 ± 0.21
Stored at −15°C and completed thawed at room temperature			
3 cycles	5.5–6.5 ± 2.0	3.54 ± 0.60	2.33 ± 0.48

from the core by air or a mechanical pumping system. No odor is generated, and almost no reconditioning is needed for recycling the used sands.

About 1% of the soy adhesive by solid is used to mix with sand samples provided by the foundry industry. The mixture is loaded into a cylinder mold of about 25 mm in diameter and 25 mm in length. The specimens are tested following the procedures described in the American Foundrymen's Association for core testing (Testing and Grading Foundry Sands). The soy adhesive is microwave curable. The material of the mold is polytetrafluoroethylene (PTFE) suitable for microwave curing. Multichannels are made through the cylinder wall to allow water evaporation during microwave curing. The coated sand sample is loaded with slight pressure into the mold and cured in a microwave (SAM-155, CEM Corporation, Matthews, NC) using 90% power for 2 min. The cured sample as shown in Figure 10.17 is compressively tested using Instron and has about 2.6 MPa average compressive strength without sand loading pressure, and 5.5 MPa with sand loading pressure.

10.7.3 CHILDREN'S GLUE AND COLOR PAINT

About 200 million bottles of 4–8 oz glues, such as Elmer's Glue-All, are used annually in the United States in schools, offices, universities, day care centers, and homes, which represents a large market. Most of the school glues are prepared using latex formulas containing many chemicals. Although Elmer's Glue-All is not toxic, some ingredients are defined as hazardous substances in the Federal Hazardous Substances Act. Some children' skins are sensitive or even allergic to latex chemicals. In preschools, some young children eat the glue by accident. Although the level of hazardous or toxic components in latex

FIGURE 10.17 Foundry sand testing specimen glued with soy protein latex-like adhesive.

glue is not high enough to cause any health problem, the long-term effect of latex ingredients on human health is still not fully understood. The protein-based soy latex adhesive can be designed to be edible and odorless and compatible with food color. The texture and performance of the soy latex-like adhesive are similar to those of Elmer's glue. The tacky properties and curing speed at room temperature are also similar or better than Elmer's glue (Figure 10.18). Food color can be applied to the soy latex-like adhesives to produce colored glue. The colored glue can be worked in a viscosity that makes it suitable to be used as colored paint for children's artwork (Figure 10.19).

FIGURE 10.18 Paper artwork sample glued with soy protein glue.

FIGURE 10.19 Color paint with soy protein color glue.

10.7.4 PACKAGING AND LABELING ADHESIVES

The large market of packaging containers for food, feed, and nonfood products, such as glass bottles, plastic bottles, and paper boxes, represents the demands for latex-based adhesives currently used in the United States. Adhesive with wet tack property allows a label to stick to a bottle immediately, and the glue should be strong enough to hold the label in place to survive in a fast packaging processing line. Industries usually remove labels for recycling purposes. The glued labels have to be removed by burning the bottle, which often consumes substantial energy and also releases some unpleasant vapors into the environment.

Large amounts of medium density fiberboard are used annually worldwide. The adhesives used for both making the fiberboard and sealing the fiberboard boxes are all petroleum-based adhesives, which not only cause environmental pollution during manufacturing but also limit recycling of the used fiberboard box. The soy latex-like adhesive can be designed to have desirable viscosities and tacky properties of a sealant adhesive or fiberboard-making adhesive. The glue strength of the soy latex-like adhesive is stronger than the petroleum-based adhesives currently used for fiberboard manufacture. Commercial fiberboard sealed with soy latex adhesives showed 90% cohesive failure under wet conditions and 100% under dry conditions (Figure 10.20).

10.8 ADHESIVE STRENGTH AND WATER RESISTANCE AT ISOELECTRIC pH

10.8.1 PH EFFECTS

pH value has a significant effect on the performance of soy protein adhesive, especially on the wet strength of the adhesives. One example of soy adhesives on a cherry veneer wood sample shows various adhesive strengths when the adhesive is prepared at various pH values (Table 10.16). For freeze-dried soy protein adhesives, a pH from 3.6 to 7.6 gives a greater dry strength (7.01–7.41 MPa) than that at pH 8.6 to higher (5.04–6.24 MPa). Similar results are observed for the soaked strength. For wet strength, soy protein adhesive at a pH at or close to its isoelectric point (pI = 4.6) yields the greatest tensile strength (3.29–3.45 MPa), which is significantly higher than that at pH 5.6 or higher. There was no significant difference on adhesive performance between freeze-dried and oven-dried soy proteins especially on the wet strength.

Soy proteins are complex macromolecules that contain about 18 different amino acid monomers connected through peptide bonds to form the primary structure (polypeptide chain), which dominates their properties. Positive and negative charges are distributed inside or on the surface of the protein body. Protein surface net charge becomes zero at pI pH, and negative at pH above

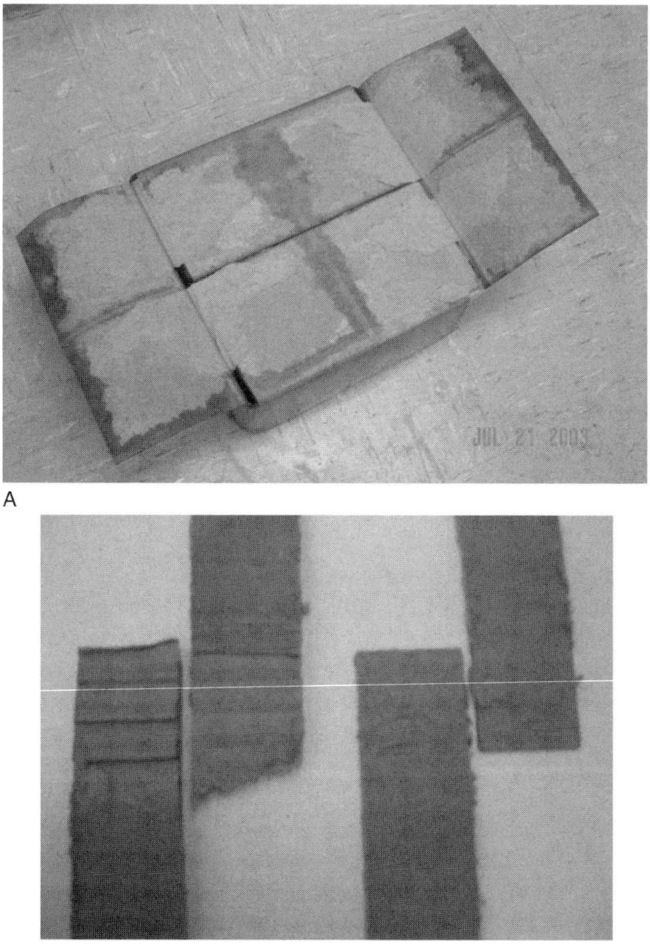

FIGURE 10.20 Fiberboard box sealed with soy latex-like adhesive: A. After raining test, the box was still sealed well and opened by force, showed 90% fiberboard cohesive failure; B. Commercial fiberboard (corrugated and solid) glued with soy latex adhesives showed 100% fiberboard cohesive failure.

pI, and positive at pH below pI point. When the soy proteins were exposed to an imposed electric field (i.e., pH at or close to pI), the cross-linked protein molecules contracted and turned into a very compact globular structure. The solubility of protein decreased severely, and supernatant appeared. Soy protein precipitated on the surface of the wood boards and formed a compact layer of soy protein. While wood boards are pressed together using a hot press at high press temperature (180°C), soy protein is further denatured and

TABLE 10.16 Effect of pH Values on the Tensile Strength (TS) of Soy Protein Adhesive

	Tensile Strength (MPa)[a]					
	Dry Strength		Wet Strength (96 h)		Soaked Strength	
pH Values	Freeze Dry	Oven Dry	Freeze Dry	Oven Dry	Freeze Dry	Oven Dry
1.6	4.92c[b]	4.71d	2.11c	1.67c	5.21bc	4.62d
2.6	5.08c	4.80d	1.62c	1.59c	5.05c	5.06d
3.6	7.20a	8.18a	3.29a	3.02a	7.11a	7.65a
4.6	7.33a	6.69c	3.45a	3.11a	7.46a	6.17c
5.6	7.41a	6.59c	3.37a	2.72b	7.55a	6.68b
6.6	7.21a	7.07bc	2.63b	2.59b	7.30a	7.30a
7.6	7.01a	7.83a	2.59b	2.36b	7.11a	7.11a
8.6	6.42a	7.34b	1.85c	2.58b	5.57b	7.24a
9.6	5.04c	6.83c	1.62c	1.69c	4.79c	7.01ab

[a]Hot pressing: 180°C, 10 min, 1.8 MPa.
[b]Means in the same column followed by different letters are significantly different at $P < 0.05$.

reacted with oxygen, and protein molecules can become unfolded again. Some hydrophobic amino acids, which are buried inside the molecule, move outward to increase water resistance.

In general, denaturation involves disruption of the non-covalent forces responsible for the organization of the native structure, although in some instances it may also include rupture of disulfide bonds. Denaturation results in an altered conformation of the protein, changes in physical and biological properties, and usually greater susceptibility to proteolysis. Denaturation may be caused by extremes of pH and heating, and the extent of denaturation depends on protein type, concentration, moisture content, ionic concentration, and composition.

Values of pH at or close to pI cause an aggregation of soy protein, and result in precipitation. Such precipitation of soy proteins at pH 4.5 results in the formation of a protein complex, some of which does not resolubilize. The acid-sensitive proteins are formed mostly from the 2-S and 7-S fractions. The quantity formed may amount to 25 to 30% of soy protein. Denaturation is classically indicated by a loss of solubility and an increase in hydrodynamic properties, which can be a reason to enhance water resistance. In addition, during hot pressing, some new chemical components can be produced and some of them contain a lot of carbonyl groups. In other words, some carbonyl compounds such as ether, ester, and ketone can appear. These compounds are difficult to dissolve in water; it could be one of the reasons why water resistance of protein adhesives is increased greatly by a high hot-press temperature.

10.8.2 pH EFFECTS IN THE PRESENCE OF CROSS-LINK REAGENTS AND CHEMICALS

Soy protein was modified with various chemicals to improve adhesive strength and water resistance. Major modifiers were chemicals containing carboxyl groups, aldehyde compounds, or epoxy groups. Examples of these chemicals include citric acid, tricarballylic acid, 1,2,7,8-diepoxyoctane, hexamethylenetetramine, polyamide-epichlorohydrin (PAE), etc. These chemicals can be used alonc or in combination. The pH of modified or unmodified adhesives should be near a soy protein isoelectric point ranging from 3.6 to 5.5. The adhesives developed from this invention have similar gluing strength and water resistance to formaldehyde-based adhesives, and can be used alone or blended with formaldehyde-based adhesives for cellulosic materials, especially for plywood and particleboards, but not limited.

Combined with isoelectric pH technology as described in Section 10.8.1, adhesive strength and water resistance of the modified soy proteins are significantly improved. For example, the soy protein adhesive modified with PAE gives the highest adhesive strength at pH around 5.5, which is the isoelectric pH of the modified protein (Figure 10.21). The soy adhesive at pH 5.5 has 3.7 MPa wet strength with 72% wood failure and 6.5 MPa dry strength with 100% wood failure. Compared to the soy adhesive at isoelectric pH 4.6 as described in Section 10.8.1, the soy protein modified with 5% PAE at isoelectric pH 5.5 has higher adhesion strength and water resistance (Table 10.17). The boiling adhesive strength of the PAE-modified soy adhesive at isoelectric pH 5.5 is about 2.6 MPa with 64% wood failure. These results are comparable to formaldehyde-based adhesives: urea formaldehyde adhesive has an average 3.5-MPa wet strength with 70% wood failure and 0-MPa boiling strength; phenol formaldehyde adhesive has an average 3.5-MPa wet strength with 81% wood failure and 2.7-MPa boiling strength with 72% wood failure.

The isoelectric pH of the PAE-modified soy protein is shifted from 4.5 to 5.5. The high dependence on pH of the ionic complex interactions can be observed from the IR scan results (Figure 10.22). The soy protein has a maximum absorbance at around pH = 4.6, corresponding to its isoelectric pH. With addition of PAE, two peaks appeared: one has almost fixed location at about pH = 4.6, and the other one varied with PAE concentration from pH = 5.4 for 3 wt% PAE, to pH = 6.1 for 10 wt% PAE. Also with addition of PAE, the peaks broadened and precipitate appeared in wider pH range. For the 15 wt% PAE-soy protein slurry, the peak is too broad to determine peak location accurately.

PAE is a cationic polymer. After adding PAE to a soy protein solution (pH ~ 7), the cationic part of PAE interacts with the anionic carboxyl group of the soy protein to form PAE-SPI complexes (Figure 10.23). At pH ~ 7, soy

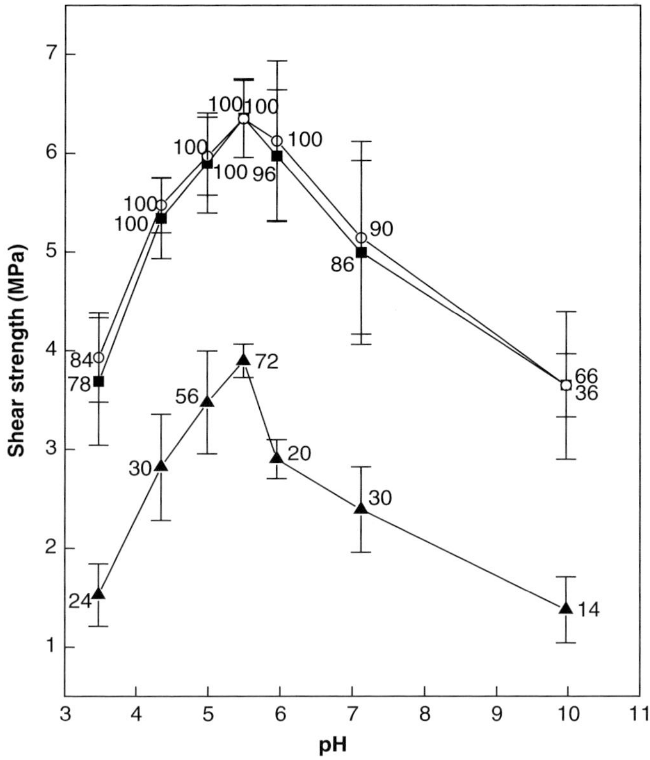

FIGURE 10.21 Effects of pH on shear strengths of cherry plywood bonded with 5 wt% PAE-modified SPI adhesives at 170°C, 1.4 MPa for 5 min: unsoaked (■), soaked (○), and wet (▲) shear strength. The data beside the symbol indicated the average percentage of wood failure. (Courtesy Z. Zhong and X. Sun.)

protein has a net negative charge. Although the PAE solution has a pH of 4.6 ~ 4.9, incorporation of a small amount of PAE into soy protein solution would not change the pH significantly. The driving force for the PAE-protein complex formation is ionic interaction, and the complex formation is reversible upon pH change. When the pH is adjusted to 9 or above, the abundant anionic OH^- group in the adhesive slurry can bind to the cationic group of PAE to release PAE from the PAE-protein complex, and hence dissolve both components. When the pH is adjusted to below 4, the carboxyl group (COO^-) of SPI would bind the proton ion (H^+) to form the COOH group, resulting in release of soy protein from the PAE-protein complex, and then the complex would disappear. At low PAE concentrations, there were not enough PAE molecules to react with SPI, resulting in the first peak at around 4.6 due to free soy protein molecules. Other peaks in the IR scan (Figure 10.22) resulted from the PAE-SPI complex formation and pH-dependence.

TABLE 10.17 Comparison of the adhesion properties between the unmodified and the 5% PAE-modified soy protein adhesives at neutral pH and isoelectric pH values. (Courtesy Z. Zhong and X. Sun.)

	Adhesion Strength (MPa)			
	Unsoaked	Soaked	Wet	Boiling
Unmodified soy protein adhesive				
pH = 7.1	3.71 ± 0.40 CFW 26%	3.18 ± 0.33 CFW 24%	0.73 ± 0.23 CFW 5%	0.31 80% Del
pH = 4.5	5.36 ± 0.21 CFW 100%	4.83 ± 0.38 CFW 100%	2.84 ± 0.22 CFW 50%	1.79 ± 0.32 CFW 50%
5% PAE-modified soy protein adhesive				
pH = 7.1	4.99 ± 0.93 CFW 86%	5.14 ± 0.98 CFW 90%	2.39 ± 0.43 CFW 30%	0.64 ± 0.46 60% Del
pH = 5.5	6.36 ± 0.40 CFW 100%	6.35 ± 0.39 CFW 100%	3.90 ± 0.17 CFW 72%	2.60 ± 0.37 CFW 64%

CFW: cohesive failure within wood; Del: delaminated.

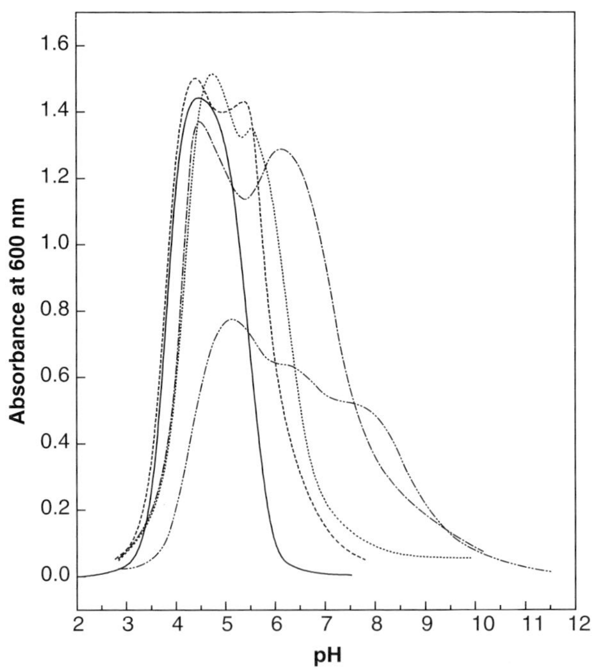

FIGURE 10.22 Absorbance profiles of the PAE-modified SPI in 10 mM Tris. The PAE concentrations were: 0 wt% (———); 3 wt% (– – – –); 5 wt% (............); 10 wt% (– · – · – · –); and 15 wt% (– · · · – · · · – · ·). (Courtesy Z. Zhong and X. Sun.)

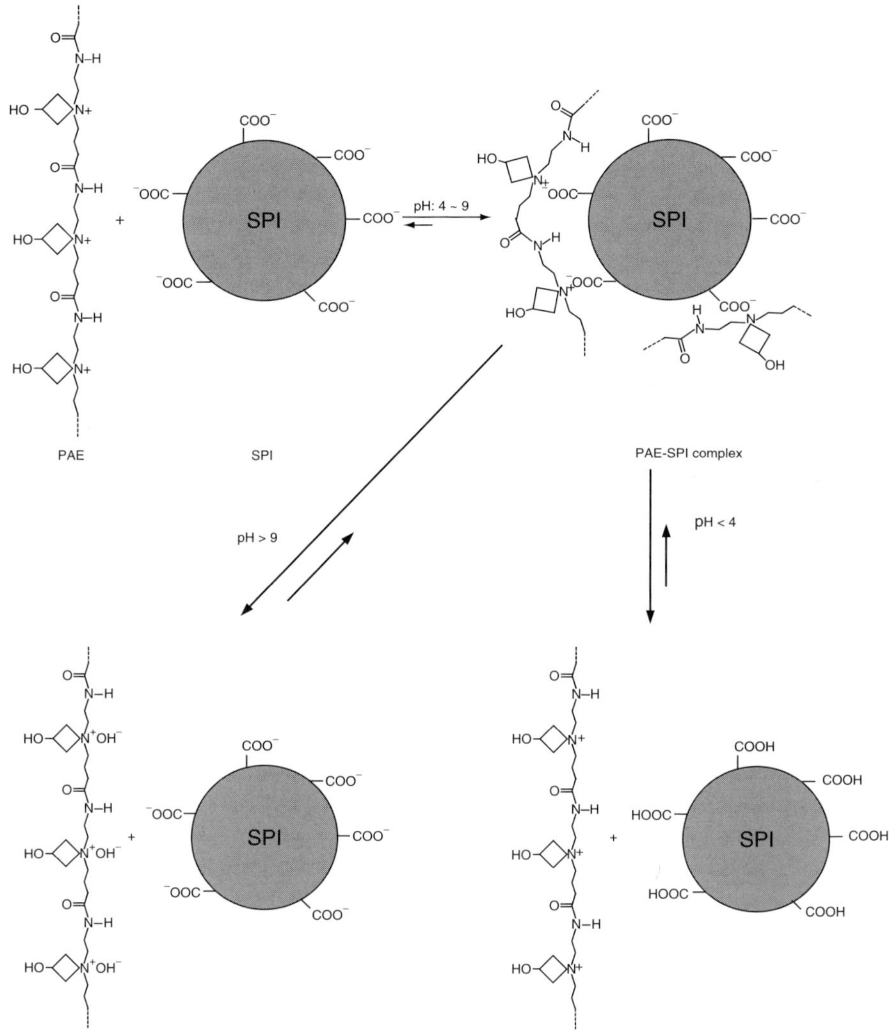

FIGURE 10.23 The schematic diagram of the formation of PAE-soy protein interpolymer complex and its disassociation (where SPI = soy protein isolate). (Courtesy Z. Zhong and X. Sun.)

The resulting special structure/conformation of the complexes should play a key role in this shift. At pH 5.5, the complexation interactions were mostly enhanced as shown by the peak in Figure 10.22. Also, the PAE-modified soy protein has the lowest net charge at pH 5.5, which greatly reduces the water resistance. Although pH also affects the chemical reactions of PAE with the amino and carboxyl groups (Figure 10.24), this dependence should not contribute much at low pH values. In the paper industry, the

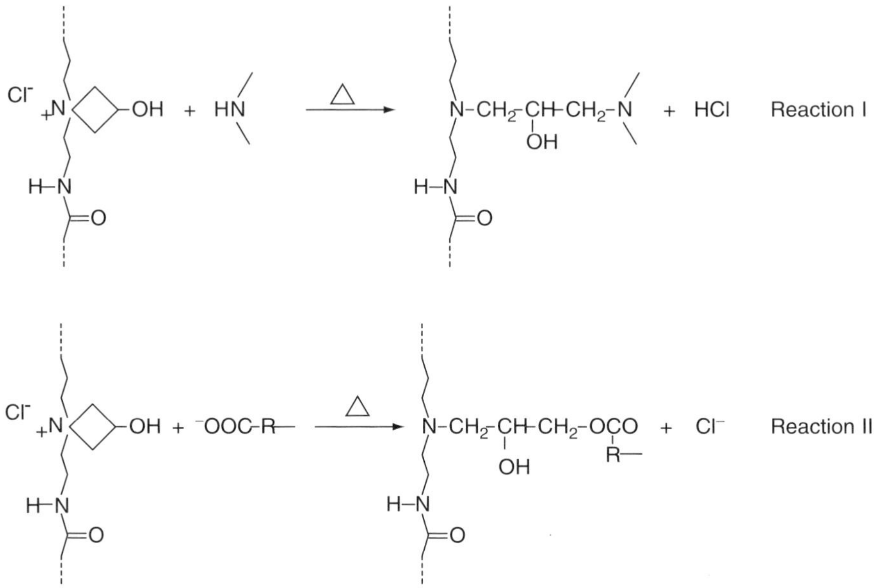

FIGURE 10.24 The schematic diagram of the chemical reactions between the azetidinium group of PAE and the primary and secondary amines and carboxyl group. (Courtesy Z. Zhong and X. Sun.)

PAE is most often used in a pH range of 6 ~ 8 [50]. It is also reported [50] that PAE could not react with the neutral carboxyl group (RCOOH), the protonated amino group (RNH_3^+), and the amino group ($R_2NH_2^+$). As pH decreases, amino groups of soy protein are protonated, the histidine has the lowest isoelectric pH of 6.5 for the positively charged amino acids, the amount of the anionic carboxylate groups ($RCOO^-$) decreases, and hence the efficiency of the chemical reactions (Figure 10.24) decreases. Therefore, the ionic complexation interaction is the main reason for the pH-dependence of the adhesion properties.

REFERENCES

1. Lambuth, A. L. Protein Adhesives for Wood. In *Handbook of Adhesive Technology*, Pizzi, A.; Mittal, K. L., Eds.; Marcel Dekker, New York; 1994.
2. Johnson, L. A.; Myers, D. J.; Burden, D. J., Early Uses of Soy Protein in Far East, *U.S. Inform.* 1984, **3**, 282–284.
3. Sun, X.; Bian, K. Shear Strength and Water Resistance of Modified Soy Protein Adhesives, *J. Am. Oil Chem. Soc.* 1999, **76**(8), 977–980.
4. Huang, W.; Sun, X. Adhesive Properties of Soy Proteins Modified by Sodium Dodecyl Sulfate and Sodium Dodecylbenzene Sulfonate, *J. Am. Oil Chem. Soc.* 2000, **77**(7), 705–708.
5. Huang, W.; Sun, X. Adhesive Properties of Soy Proteins Modified by Urea and Guanidine Hydrochloride, *J. Am. Oil Chem. Soc.* 2000, **77**(1), 101–104.

6. Haumann, B. F. Researchers Finding New Ways to Use Soy, *Inform.* 1993, **4**(12), 1324–1333.
7. Dunn, L. B.; Hojilla, M. P. Foaming Properties of Soy Proteins and Their Use in Plywood Adhesives. Abstract, in *New Industrial Products Based on Soy Proteins*, United Soybean Board, Kansas City, MO; 1998.
8. Hettiarachchy, N. S.; Kalapathy, U.; Myers, D. J. *J. Amer. Oil Chem. Soc.* 1995, **72**(12), 1461–1464.
9. Kreibich, R. E.; Hemingway, R. W.; Steynberg, J. P.; *et al.* Soy Derived Protein as Raw Material for Structural Wood Adhesives. Abstract, in *New Industrial Products Based on Soy Proteins*, United Soybean Board, Kansas City, MO; 1998.
10. Schultz, J.; Nardin, M. Theories and Mechanisms of Adhesion. In *Handbook of Adhesive Technology*, Pizzi, A.; Mittal, K. L., Eds., Marcel Dekker, New York; 1994.
11. Petrie, M. E. *Handbook of Adhesives and Sealants,* McGraw-Hill Company, New York; 2000.
12. Gollob, L.; Wellons, J. D. Wood Adhesion, In *Handbook of Adhesives,* 3rd ed., Skeist, I., Ed., Van Nostrand Reinhold, New York; 1990, pp. 598–610.
13. Machay, C. D. Good Adhesive Bonding Starts with Surface Preparation, *Adhesive Age* 1998, **41,** 30–32.
14. Cheng, E. Adhesion Mechanism of Soybean Protein Adhesives with Cellulosic Materials, *Ph.D. Thesis,* Kansas State University, Manhattan, KS; 2004.
15. Careri, G.; Giansanti, A.; Gratten, E. Lysozyme Film Hydration Events: An IR and Gravimetric Study, *Biopolymers*, 1979, **18,** 1187–1203.
16. Tanford, C. Protein Denaturation. In *Advanced in Protein Chemistry*, Anfinsen, C. B.; Anson, M. L.; Edsall, J. T., *et al.*, Eds.; Academic Press, New York; 1968, pp. 121–283.
17. Narhi, L. O.; Zukowski, M.; Arakawa, T. Stability aprA-subtilisin in Sodium Dodecyl Sulfate, *Arch. Biochem. Biophys.* 1988, **261,** 161–169.
18. Reynolds, J. A.; Herbert, S.; Polet, H.; *et al.* The Binding of Divers Detergent Anions to Bovine Serum Albumin, *Biochemistry* 1967, **6,** 937–947.
19. Zhong, Z.; Sun, X.; Feng, X.; *et al.* Adhesion Strength of Sodium Dodecyl Sulfate-Modified Soy Protein on Fiberboard, *J. Adhesion Sci. Technol.* 2001, **15**(12), 1417–1427.
20. Mo, X.; Sun, X.; Wang, D. Thermal Properties and Adhesion Strength of Modified Soybean Storage Protein, *J. Am. Oil Chem. Soc.* 2004, **81**, 395–400.
21. Peng, I. C.; Quass, D. W.; Dayton, W. R.; *et al.* The Physicochemical and Functional Properties of Soybean 11S Globulin: A Review, *Cereal Chem.* 1984, **61,** 480–490.
22. Wolf, W. J. Soybean Proteins: Their Functional, Chemical, and Physical Properties, *J. Agric. Food Chem.* 1970, **18,** 969–979.
23. Chandra, B. R. S.; Rao, A. G. A.; Rao, M. S. N. Effect of Temperature on the Conformation of Soybean Glycinin in 8M Urea and 6M Guanidine Hydrochloride, *J. Agric. Food Chem.* 1984, **32,** 1402–1405.
24. Pace, C. N. The Stability of Globular Proteins, *CRC Crit. Rev. Biochem.* 1975, **3,** 1–43.
25. Ptitsyn, O. B. Protein Folding: Hypotheses and Experiments, *J. Protein Chem.* 1987, **6,** 273–293.
26. Badley, R. A.; Atkinson, D.; Hauser, H.; *et al.* The Structure, Physical and Chemical Properties of the Soybean Protein, *Glycinin, Biochim. Biophys. Acta.* 1975, **412,** 214–228.
27. Thanh, V. H., K. Shibasaki. Beta-Conglycinin from Soybean Proteins, *Biochim. Biophys. Acta.* 1977, **490,** 370–384.
28. Nazaka, M.; Kuwajima, K.; Nitta, K.; *et al.* Detection and Characterization of the Intermediate on the Folding Pathway of Human Alpha-Lactalbumin, *Biochem.* 1978, **17,** 3752–3758.
29. Dolgikh, D. A.; Gilmanshin, R. I.; Brazhnikov, E. V.; *et al.* Alpha-Lactalbumin: Compact State with Fluctuating Tertiary Structure? *FEBS Lett.* 1981, **136,** 311–315.
30. Dolgikh, D. A.; Kolomietz, A. R.; Bolotina, I. A.; *et al.* Molten-Globule State Accumulates in Carbonic Anhydrase Folding, *FEBS Lett.* 1984, **165,** 88–92.

31. Pfeil, W.; Bychkova, V. E.; Ptitsyn, O. B. Physical Nature of the Phase Transition in Globular Proteins, *FEBS Lett.* 1986, **198,** 287–290.
32. Sun, X.; Bian, K. Adhesives from Modified Soy Protein Polymers. In *Polymers from Renewable Resources: Carbohydrates and Agroproteins*, Gross, R., Ed.; American Chemistry Society; 2001.
33. Zhong, Z.; Sun, X.; Feng, X.; *et al.* Adhesive Properties of Soy Protein with Fiber Cardboard, *J. Am. Oil Chem. Soc.* 2000, **78**(1), 37–41.
34. Wang, D.; Sun, X. Low Density Straw Particleboard, *Indust. Crops Products* 2002, **15**(1), 43–50.
35. Holdsworth, S. D. *J. Texture Studies*, 1971, **2,** 393–418.
36. Hermansson, A. M. *J. Texture Studies*, 1975, **5,** 425–439.
37. Han, G.; Zhang, C.; Zhang, D.; *et al.* Upgrading of Urea Formaldehyde-bonded Reed and Wheat Straw Particleboards Using Silane Coupling Agents, *J. Wood Sci.* 1998, **44**(4), 282–286.
38. Wang, D.; Sun, X. Composites from Agricultural-Derived Resins and Fibers, *Agro Food Indust. Hi-Tech* 2001, July–August issue.
39. Mo, X.; Cheng, E.; Wang, D.; *et al.* Physical Properties of Medium-Density Straw Particleboard, *Indust. Crops Products* 2002, **18**(1), 47–53.
40. Cheng, E.; Sun, X.; Karr, G. Adhesive Properties of Modified Soybean Flour in Wheat Straw Particleboard, *Composite A: Appl. Sci. Manufact.* 2004, **35**, 297–302.
41. Mo, X.; Hu, J.; Sun, X.; *et al.* Compression and Tensile Strength of Low Density Wheat-Protein Particleboard, *Indust. Crops Products* 2001, **14,** 1–9.
42. Chelak, W.; Newman, W. H. MDI High Moisture Content Bonding Mechanism, Parameters, and Benefits Using MDI in Composite Wood Products. In *Proc. 32nd Wash. State Univ. Int. Particleboard/Composite Materials Symp.*, 1991, pp. 205–229.
43. Watson, C. J.; Miller, H.; Poland, P.; *et al.* Soil Properties and the Ability of the Urease Inhibitor N-(n-butyl) Thiophosphoric Triamide to Ammonia Volatilization from Surface-Applied Urea, *Soil Biol. Biochem.* 1994, **26,** 1165–1171.
44. Salmoral, E.; Gonzalez, E.; Mariscal, M.; *et al.* Comparison of Chickpea and Soy Protein Isolate and Whole Flour as Biodegradable Plastics, *Indust. Crops Products* 2000, **11,** 227–236.
45. Sun, X.; Karr, G.; Seib, P. Edible and Biodegradable Feed Packaging Materials, U.S; Patent No. 6,337,097; 2001.
46. Rowe, S.; C. T. Nugent U.S. Patent No. 2,887,395; 1959.
47. Kalapathy, U., Hettiarachchy, N. S., Myers, D. J. and Hanna, M. A. 1995 *J. Am. Oil. Chem. Soc.* **72**, 507–510.
48. Kalapathy, U., Hettiarachchy, N. S., Myers, D. J., and Rhee, K. C. 1996 *J. Am. Oil Chem. Soc.* **73**, 1063–1066.
49. Espy, H. H., in: *Wet-Strength Resins and Their Application*, L. L. Chan (Ed), Ch. 2, pp. 13–61. TAPPI Press, Atlanta (1994).

11

PLASTICS DERIVED FROM STARCH AND POLY (LACTIC ACIDS)

XIUZHI SUSAN SUN

11.1 STARCH STRUCTURE

Starch is a highly hydrophilic polymer that consists of anhydroglucose units linked by α-D-1,4-glycosidic bonds [1]. There are two distinct structural molecular classes, namely, linear amylose and highly branched amylopectin. Linear amylose is linked by α-1,4-bonds and branched amylopectin is linked by α-1,6-bonds. The molecular structure of an amylopectin is illustrated in Figure 11.1; the molecular structure of amylose is similar to the linear portion of the amylopectin structure. The structure of monosaccharide D-glucose can be either in an open-chain or a ring form [2]. The ring form is highly thermodynamically stable and has a structure similar to that of sugar in solution. The aldehyde group at carbon number 1 is highly reactive, making it a reducing sugar.

Natural starch exists in a granular form. The granular shape and size are different in different plants (Figure 11.2). Corn starch granules are mainly spherical (Figure 11.2A), wheat starch has both spherical and disk-shaped granules (Figure 11.2B), and potato starch has a smooth granular surface and is mainly oval (Figure 11.2C). The disk-shaped wheat starch granule has an average diameter of 25 μm, and the spherical-shaped granule has a smaller diameter of less than 10 μm (Figure 11.3) [3]. A cornstarch granule has an average diameter of 10.3–11.5 μm [4], whereas potato starch has a larger granule size with an average diameter of about 40 μm [3].

Starch consists of amorphous and crystalline phases (Figure 11.4). As illustrated in Figure 11.4, following the arrow direction, the crystalline hard

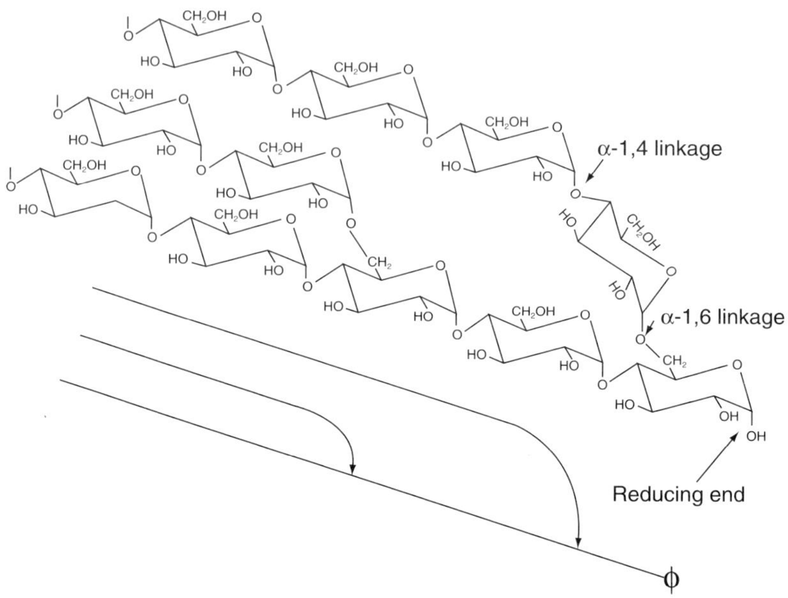

FIGURE 11.1 General molecular structure of starch. (*Source:* Courtesy Paul A. Seib.)

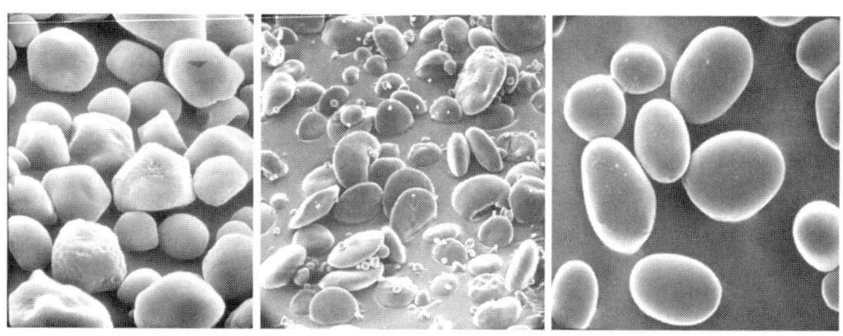

FIGURE 11.2 Granular shape of various starches: (A) cornstarch, (B) wheat starch, and (C) potato starch. (*Source:* Courtesy Paul A. Seib.)

shell layers alternate with a semicrystalline soft shell. The amorphous channels are braided in the crystalline hard shell. The crystalline hard shell consists of many blocklets with a superhelix amylopectin (AP) structure that alternately spirals between crystalline and amorphous lamellae. The amylopectin, amylose, lipids, and amylose–lipid are entangled into a complex structure in a form dictated by biological information, starch postmodification, and processing conditions. On heating, starch undergoes various phase changes from glassy to rubbery, to gelatinized, and to a melted crystal. A typical fully gelatinized starch melt presents a smooth, entangled, and mushy structure (Figure 11.5).

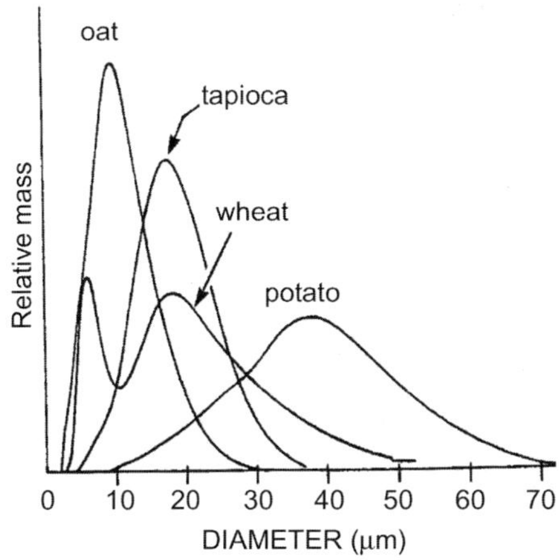

FIGURE 11.3 Starch granule size distributions. (*Source:* Copyright M. H. Moon and J. C. Gidings, *J. Food Sci.* 1993, **58,** 1166.)

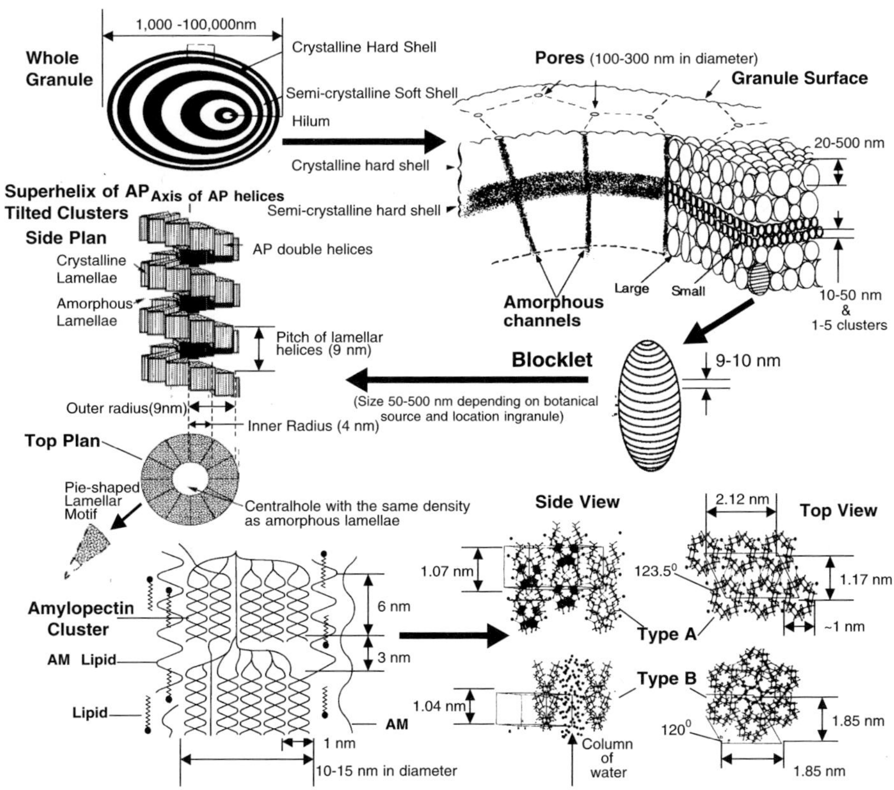

FIGURE 11.4 Overview of starch granule structure. (*Source:* Courtesy Paul A. Seib.)

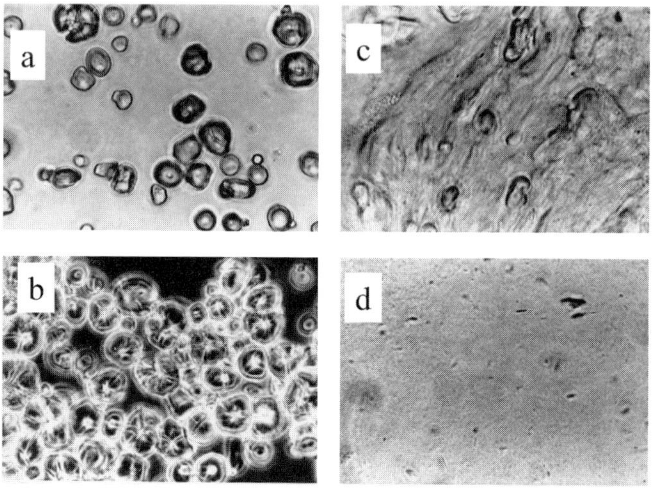

FIGURE 11.5 Optical microscopy of (a) raw cornstarch and of cornstarch extracted from starch/PLA blend with (b) 10% moisture content, (c) 30% moisture content, and (d) 50% moisture content. (*Source:* Adapted from Ke and Sun [64].)

11.2 THERMAL PROPERTIES OF STARCH

Thermal properties are important in starch when processing as thermoplastics. Water is a primary plasticizer of starch and can affect starch properties in many ways. In a starch–water system, water interacts directly or indirectly with starch. The free water becomes frozen, but the bond water remains unfrozen even at subfreezing temperatures. This water is called *nonfreezing water* and it can be used to estimate starch hydration. The nonfreezing water of a native cornstarch is almost the same as that of a gelatinized corn starch at a water content below 20% (Figure 11.6) [5]. However, at a water content of more than 30%, the nonfreezing water of the gelatinized cornstarch increases because granule swelling binds more water molecules. Nonfreezing water can also provide important information on the structure and functional properties of starch [6], and it can be a reference determining parameter for processing and storage of dehydrated and frozen products containing starch [7].

Starch, in the presence of water, undergoes a series of thermal transitions upon heating or cooling, including water crystallization, ice melting, glass transition, gelatinization transition, crystal melting, and amylose–lipid complex melting. Differential scanning calorimetry (DSC) is a useful tool to determine the thermal properties and phase transitions of starch [7]. Endothermal and exothermal changes in a DSC thermograph reveal transitions or reactions that occur during DSC testing. On gelatinization, amylopectin swells and becomes loose, and amylose leaches out of the amylopectin

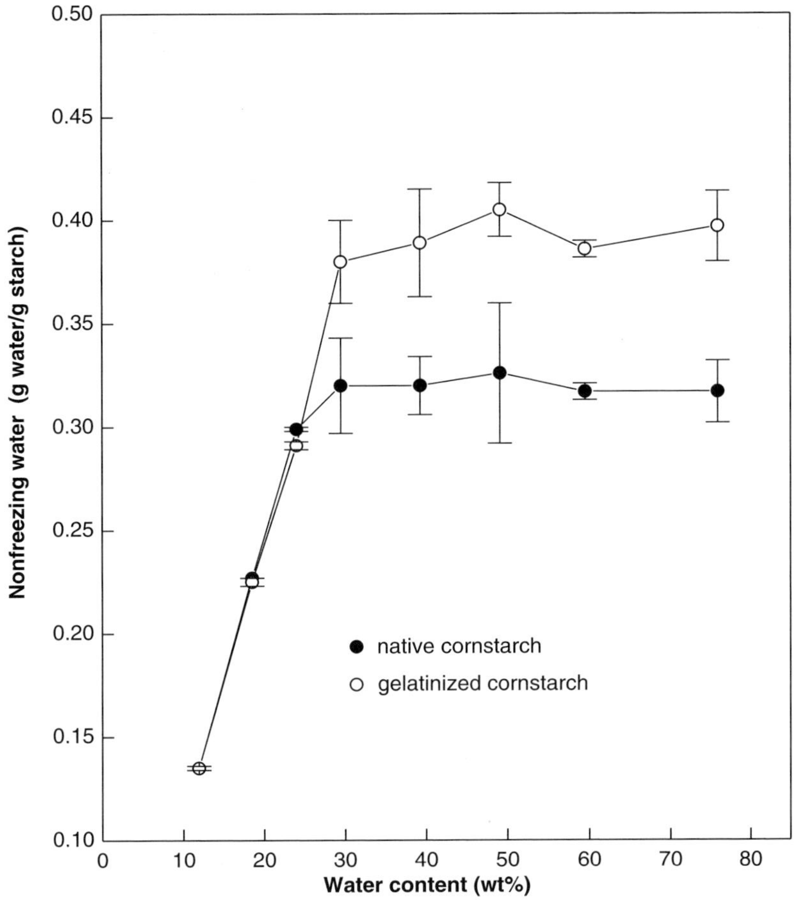

FIGURE 11.6 Nonfreezing water of the native and the gelatinized cornstarch affected by moisture content. (*Source:* Copyright Zhong and Sun [5], *J. Food Eng.* 2005, **69**, 453–459.)

shell and also swells. As a result, the starch granule becomes hydrated, swollen, and eventually starch crystal starts to melt, resulting in an increase in viscosity. To characterize starch gelatinization behavior, starch is usually mixed with 3 times its amount of water and then subjected to a DSC test. In this case, the DSC thermograph exhibits only one gelatinization peak in the temperature range of 50–80°C. Some DSC studies on wheat, potato, and rice starches with reduced water content showed two transition peaks relating to starch gelatinization and crystal melting [8–12]. For some starches that contain lipids, a peak at higher temperature can be observed that is attributable to the melting transition of the amylose–lipid complex [13].

At low moisture content, all water molecules are absorbed and bonded with starch molecules in the amorphous phase. There is not enough water available to form ice, nor to gelatinize the starch [5], which is demonstrated via DSC techniques, in which no gelatinization peak is observed (Figure 11.7, curve A). The crystal melting often occurs at an elevated temperature. At intermediate moisture content, sufficient water is available for starch granules to swell, leading to a partial loss of crystallinity, as proposed by Donovan [10] and Jenkins and Donald [14]; this is the physical point of initiation of gelatinization, and the remaining starch crystals start melting at higher

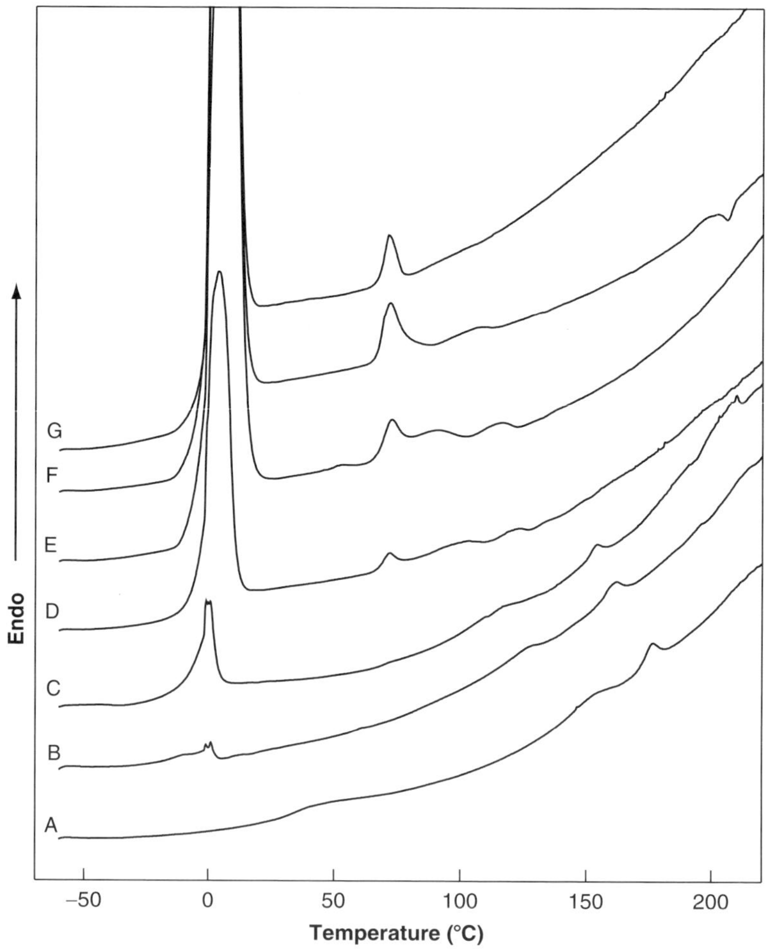

FIGURE 11.7 Thermograms of cornstarch with moisture content: (curve A) 18.5%, (curve B) 24.0%, (curve C) 29.5%, (curve D) 39.3%, (curve E) 49.1%, (curve F) 59.5%, and (curve G) 75.9%. Commercial cornstarch was exposed to 95% moisture for various times to obtain moisture content ranging from 10% to 30%. The cornstarch was mixed with an appropriate amount of distilled water to obtain samples with a moisture content greater than 30%. (*Source:* Copyright Zhong and Sun [5], *J. Food Eng.* 2005, **69**, 453–459.)

temperatures (Figure 11.7, curve C). With adequate water, enough free water is available so starch granules can fully gelatinize and swell, and more starch crystals start melting (Figure 11.7, curves D and E). When water is beyond 60%, the starch crystal melting overlaps with gelatinization, and only one peak appears (Figure 11.7, curves F and G). The glass transition temperature (T_g) of starch decreases as moisture content increases [5]. The glass transition of starch with a moisture content of more than 30% is not detectable because the T_g is around the freezing point and, hence, overwhelmed by the ice-melting transition [5]. The plasticization effect of water also depresses the melting temperatures of starch crystals and amylose–lipid complexes.

Zhong and Sun [5] reported on a state diagram (Figure 11.8) for a native corn starch–water system that can be used to predict the physical state of the system at different temperatures and compositions. Such diagrams of starches from other plants, such as wheat and sorghum, can also be developed. Starch is often processed with a lower water content as thermoplastic. Starch becomes soft above its glass transition temperature and melts completely at a temperature above point A of Figure 11.8 for amylose–lipid complex melting. No gelatinization should be expected at moisture below 30%. The diagram can be used to estimate phases at a given moisture content and temperature. For example, at room temperature (23 °C), for a native cornstarch with 15% moisture content, there exists a glassy starch–water mixture, starch crystal, and amylose–lipid complex (Figure 11.8, point A). At 80 °C and 50% moisture, there exists free water, a rubbery, partially gelatinized cornstarch–water mixture, starch crystal, and amylose–lipid complex (Figure 11.8, point B).

11.3 STARCH AS A FILLER

Fillers were first introduced to the plastics industry for economic reasons. Subsequently it was documented that, in some cases, filler could also strengthen the mechanical modulus of a composite. Fibers and particulate fillers are used more extensively in thermoplastics, such as polyethylene, polypropylene, polyvinyl chloride, and nylon, to improve mechanical properties and reduce costs of raw materials. Natural starch has no effect on the atmospheric carbon dioxide level. Starch is inexpensive and available and is often used as a filler for thermoplastic production [17–24].

Starch, by itself, has severe limitations because of its water solubility. Poly(lactic acid) (PLA) is attractive for disposable and biodegradable plastic substitutes because of its mechanical properties. PLA is a polyester polymerized from L- or D-lactic acid, either by ring opening or by direct condensation. Lactic acid can be produced by converting sugar-based carbohydrates using fermentation technology, which is an environmental friendly technology. PLA derived from L-lactic acid has a high melting point and high

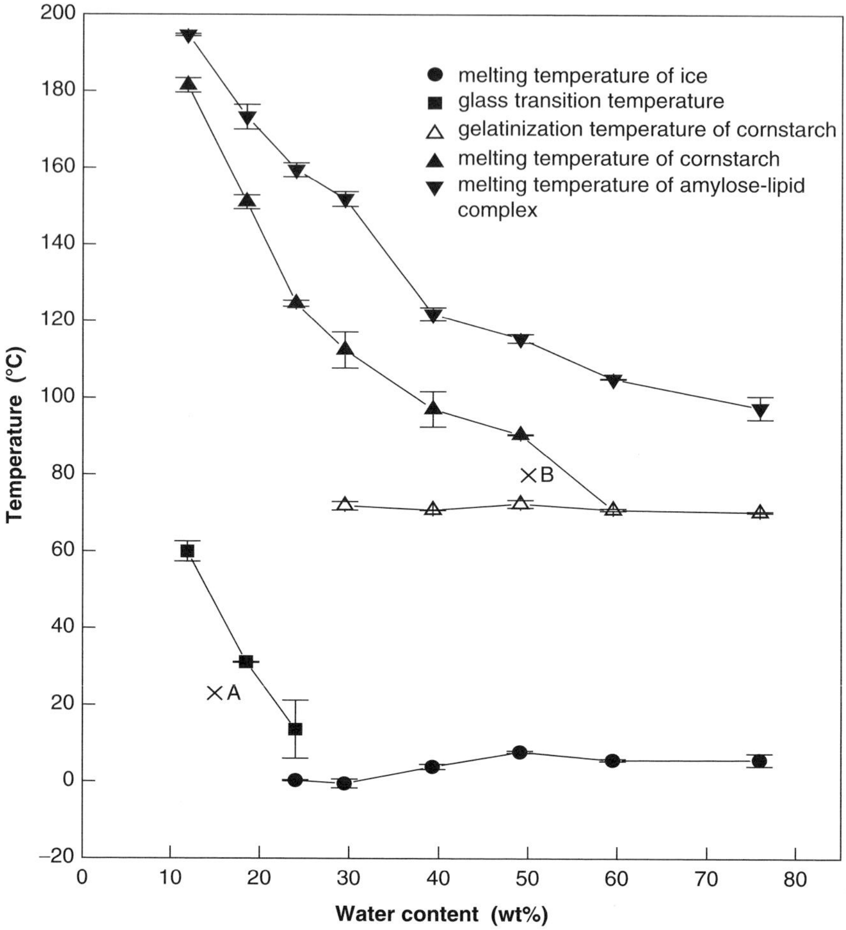

FIGURE 11.8 Phase transition diagram of native cornstarch. Point A: 23 °C and 15% moisture content; point B: 80 °C and 50% moisture content. The method of obtaining the varied moisture contents is the same as that described in Figure 11.7. (*Source:* Copyright Zhong and Sun [5], *J. Food Eng*. 2005, **69**, 453–459.)

crystallinity, but poor flowability. When the percentage of L-lactic acid is lower than 80%, PLA is almost a fully amorphous polymer [25]. Like many other petroleum-based resins, the properties of PLA can be modified depending on the application. A major drawback of PLA is that it loses stiffness and becomes soft when above its T_g, much like rubber.

Raw, pure PLA, made of mainly L-lactic acid with a molecular weight of about 120 kDa, has a T_g of approximately 60 °C, a crystallization temperature (T_c) of approximately 125 °C, and a melting temperature (T_m) of 172 °C. Figure 11.9 presents the thermal phase transitions of PLA alone and

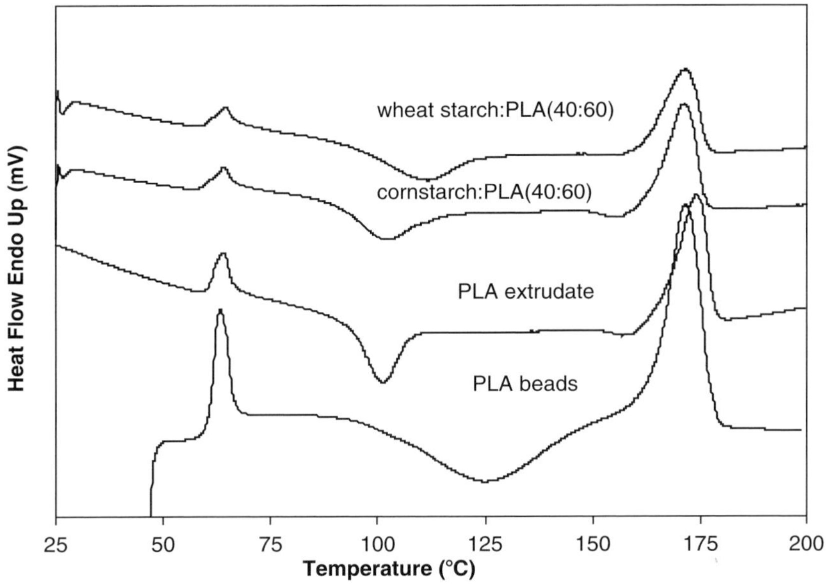

FIGURE 11.9 Thermograms of raw PLA, extruded PLA, cornstarch and PLA extrudate, and wheat starch and PLA extrudate. (*Source:* Reprinted with permission from Ke and Sun, 2000, Physical Properties of PLA and Starch Composites with Various Blending Ratios, *Cereal Chemistry* 77:761–768. © 2000 American Association of Cereal Chemists.)

PLA/starch blends [26]. The peak at the glass transition zone of around 63 °C is caused by the physical aging of PLA. When PLA is thermally treated, such as by extrusion, its T_g and T_m remain unchanged, but T_c shifts to a lower temperature of approximately 102.6 °C. A minor crystallization peak at 158.3 °C is also observed, caused by PLA recrystallization during postextrusion cooling.

When blended with dried starch, PLA serves as the matrix, and starch is the filler. The thermal transition behaviors of the starch/PLA blends are similar to that of extruded pure PLA (Figure 11.9), indicating that the thermal transition of the blends results mainly from the PLA, not the starch. PLA and starch are not compatible, with observable gaps at the interface between PLA and starch, and the PLA matrix becomes discontinuous when the starch content is more than 60% (Figure 11.10) [26]. The dried starch retains its granular shape in the PLA matrix. Therefore, the crystallinity calculation is based on DSC information that comes mainly from PLA crystallization. The crystallinity of the blends (X_c) can then be estimated using the following equation [27]:

$$X_c(\%) = (\Delta H_m + \Delta H_{ci}) \times 100\%/(93 \times X_{PLA}), \qquad (11.1)$$

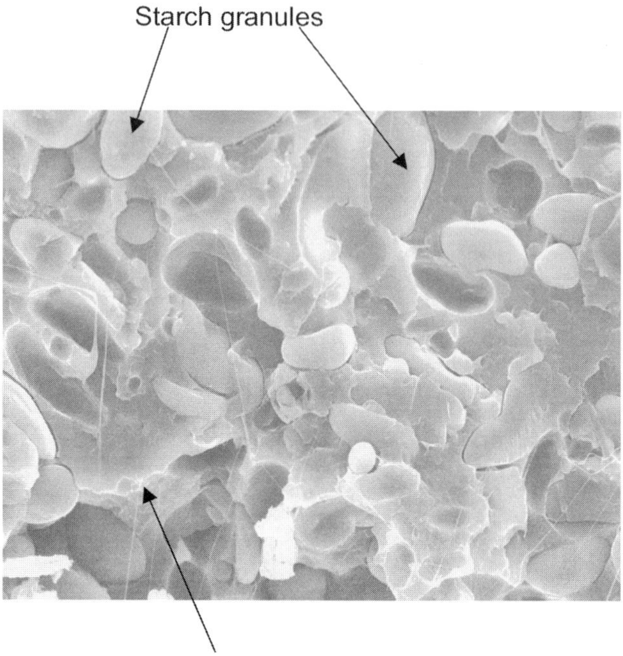

FIGURE 11.10 Morphology of extruded PLA with 45% cornstarch. (*Source:* Courtesy H. Wang and X. Sun.)

where ΔH_m and ΔH_{ci} (J/g) are the enthalpies of fusion and crystallization of PLA, respectively, and 93 J/g is the enthalpy of fusion of a PLA crystal of infinite size. The crystallinity of PLA is improved at a starch content of less than 40%, but decreases as starch content increases [28]. As the starch content increases, the mobility of the PLA molecule may be restricted by starch, and the crystal growth rate could be reduced, which is discussed in Section 11.4.

The storage modulus of the blends increases as the starch content increases [35], especially in the temperature range of T_g and T_c. The blends become soft at temperatures above T_g, the modulus increases as crystallization occurs, and then the blend melts at about 170°C. As expected, the bulk tensile strength and elongation of the blends decrease linearly as starch content increases (Table 11.1). As the blend is subjected to an external tensile load, the PLA matrix becomes the load-bearing phase. As the starch content increases, the PLA matrix becomes discontinuous, and the effective cross-sectional area of the continuous phase reduces, resulting in reduced strength and elongation. The modulus of the blend increases slightly with starch content because of the filling effects.

Fourier transform infrared (FTIR) measurements show no peak shift of the blends at the carbonyl stretching vibrating absorption [26]. To determine

TABLE 11.1 Mechanical properties of cornstarch and PLA blends as affected by starch content.

Blend of Starch/PLA	Tensile Strength (MPa)	Elongation (%)	Modulus (GPa)
0/100	61.0	6.33	1.20
20/80	42.4	2.9	1.61
50/50	32.4	2.15	1.63
70/30	18.9	1.10	1.83

Source: Data from Ke and Sun [26].

if any interactions (adhesion forces) occur between the PLA and starch phases, the geometric model for tensile strength as a function of filler concentration for a particulate-filled two-phase polymer composite can be used [29]:

$$\sigma_c = \sigma_0(1 - 1.21\phi^{2/3}), \tag{11.2}$$

where σ_c is the tensile strength of the two-phase composite, σ_0 is the tensile strength of the continuous-phase polymer (PLA, in this case), and ϕ is the volume fraction of the filler. This model is based on the assumption that the filler is distributed uniformly in the continuous phase and that the filler particles are spherical. The decrease in tensile strength, predicted by this model, is assumed to be caused by the reduction in effective cross-sectional area of the continuous phase by the spherical filler particles instead of adhesion between the two phases. The ϕ in Eq. (11.2) can be calculated as follows [30]:

$$\phi_i = \frac{\omega_i/\rho_i}{\sum \omega_i/\rho_i}, \tag{11.3}$$

where ω_i and ρ_i are the weight fraction and density, respectively, of the component in the composite. The densities are $1.4\,\mathrm{g/cm^3}$ for starch and $1.28\,\mathrm{g/cm^3}$ for PLA.

The relative tensile strength (σ_c/σ_0), from Eq. (11.2) and from the data obtained by Ke and Sun [26], can be plotted in Figure 11.11, and the slope of the predicted line is 1.21, which is in the range of 1.6–1.21 for spherical filler [31]. The relative tensile strength of the blend decreases linearly as the starch content increases. The slope of the regression line is −0.80 for cornstarch and −0.7 for wheat starch. Both are greater than −1.21, indicating that some adhesion forces probably exist between the two phases. The adhesion force between starch and PLA may be caused by polar interaction between the two phases, or by hydrogen bonding forces between the carbonyl group in PLA and the hydroxyl group in starch. A similar trend in elongation also is obtained, further indicating that some adhesion exists between starch and PLA. The blends with wheat starch have slightly higher tensile strength and

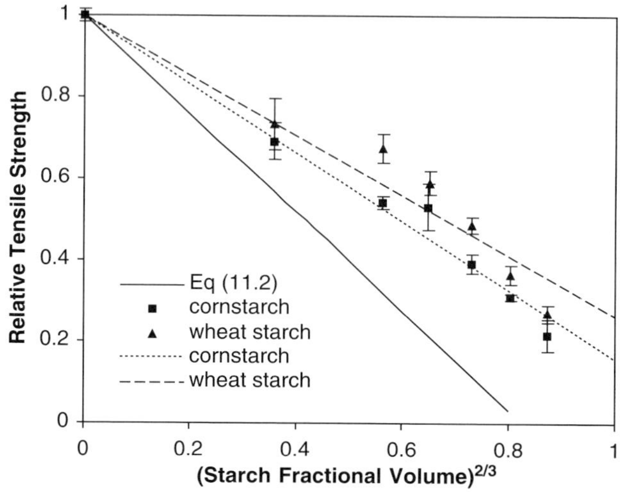

FIGURE 11.11 Relative tensile strength of predicted and experimental data of starch and PLA blends. (*Source:* Adapted from Ke and Sun, Physical Properties of PLA and Starch Composites with various blending ratios, *Cereal Chemistry* 77:761–768. © 2000 American Association of Cereal Chemists.)

elongation than the blends with corn starch [26]. This might be due to variations in starch granular size and shape.

As mentioned before, starch is a hydrophilic polymer and sensitive to water, and PLA is a hydrophobic polymer with high water resistance. The water absorption of the blends increases when the starch content is increased (Figure 11.12). Water absorption of the starch/PLA blends increases greatly on the first day and then levels off after 3 days. Water absorption increases slowly at a starch content of less than 60%, but increases rapidly when the starch content goes over 70%. Water absorption is mainly caused by starch. At a low starch content, PLA forms a very good continuous phase that covers the starch, as shown in Figure 11.10. As the starch content rises to 60% or more, the PLA phase became discontinuous, and the starch granules are not completely covered by the PLA matrix, resulting in a large water uptake.

As mentioned before, starch contains two major polymers: nearly linear amylose and branched amylopectin. When blending starch and PLA, the properties of amylose and amylopectin at or near the starch granule surface affect the properties of blends. However, the amylose content is found to not significantly affect mechanical properties [32], but amylose reduces water absorption because amylose does not swell as much as amylopectin when it is exposed to water [33].

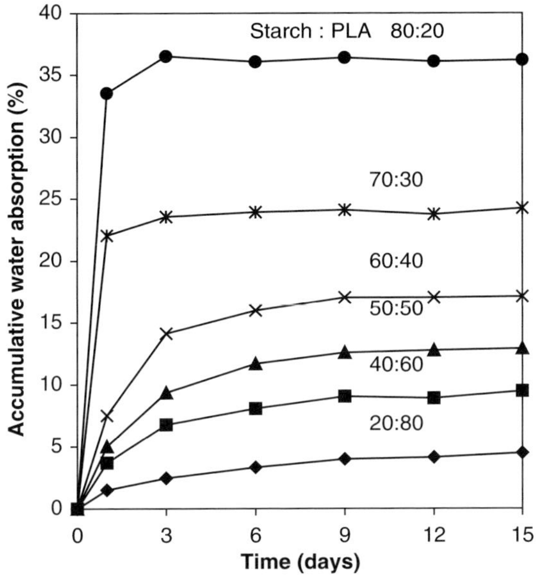

FIGURE 11.12 Water absorption of extruded PLA with various cornstarch contents soaked in tap water. (*Source:* Adapted from Ke and Sun, Physical Properties of PLA and Starch Composites with Various Blending Ratios, *Cereal Chemistry* 77:761–768. © 2000 American Association of Cereal Chemists.)

11.4 STARCH AS A NUCLEATING AGENT

PLA plays a key role in starch/PLA composites, dominating the mechanical and biodegradation properties of the blends. The microstructure of PLA, which is formed during thermal processing, has a significant effect on ultimate product quality. Based on a study by Ke and Sun [28], starch is not only a filler, but also a nucleating agent of PLA crystallization. The time required, for example, to reach 50% crystallinity for PLA at an isothermal crystallization temperature of 100°C is about 13 min, whereas it is in the range of 1.8–3.2 min for PLA blends with starch corresponding to a starch content ranging from 1% to 40%.

The Avrami equation [34] is widely used to describe the isothermal crystallization processes in polymers:

$$X(t) = X_\infty[1 - \exp(1 - Kt^n)], \tag{11.4}$$

where $X(t)$ is the volume fraction crystallinity at time t, X_∞ is the volume crystallinity after infinite time, which is estimated by using ΔH_∞, K is the overall kinetic rate constant, and n is the Avrami exponent, which depends on the nucleation and growth mechanism of the crystal. The evolution of the crystallinity with time can be estimated by using the degrees of crystallization (α) as expressed by the ratio of enthalpy determined by DSC:

$$\alpha = X(t)/X_\infty = \Delta H_t/\Delta H_\infty.$$

It can be further expressed by:

$$\alpha = \frac{\int_{t_0}^{t} \left(\frac{dH}{dt}\right) dt}{\int_{t_0}^{\infty} \left(\frac{dH}{dt}\right) dt},$$

where dH/dt is the respective heat flow; the sum of dH/dt from t_0 to t is the enthalpy at time t (ΔH_t); the sum of dH/dt from t_0 to t_∞ is the enthalpy at time $t_\infty(\Delta H_\infty)$, which can be obtained as the total area under the crystallization DSC curve; and t_0 is the time at which the sample attains isothermal conditions, as indicated by a flat baseline after the initial spike in the thermal curve. In general, the Avrami equation is often converted to the traditional linear form:

$$\ln\{-\ln[1 - X(t)]\} = \ln K + n \times \ln(t). \tag{11.5}$$

By plotting the Avrami plots, $\ln\{-\ln(1 - X(t)]\}$ versus $\ln(t)$, the values of the overall kinetic rate constant K and the Avrami exponent n can be obtained from the intercept and slope, respectively.

By plotting the degree of crystallization $X(t)$ versus t, the crystallization isotherms can be obtained. Regardless of starch loading levels, the crystallization isotherm curve shifts along the time axis (Figure 11.13). The shifting behavior along the time axis suggests that the crystallization rate of PLA is affected by isothermal temperature. The experimental data obtained by Ke and Sun [28] fit the Avrami equation [Eq. (11.5)] well for the early part of the transformation and become slightly nonlinear toward the end (Figure 11.14). The crystallization process in PLA may not meet all of the prerequisites for using the Avrami equation, and the calculation is also affected by the DSC measurement of enthalpy. Starch improves PLA's crystallization kinetics, but does not affect the crystallization behavior of PLA [28]. The Avrami method can only be used to roughly characterize the isothermal crystallization behavior of PLA and its blends with starch.

11.5 COUPLING REAGENTS FOR STARCH AND PLA BLENDS

PLA and starch are thermodynamically immiscible [35]. The interfacial fracture strength is low due to the low interdiffusion of molecules necessary for creating entanglements at the interphase between PLA and starch [35]. Interfacial adhesion plays a vital role in the mechanical properties of polymer blends. Reactive interfacial coupling agents are often used to improve

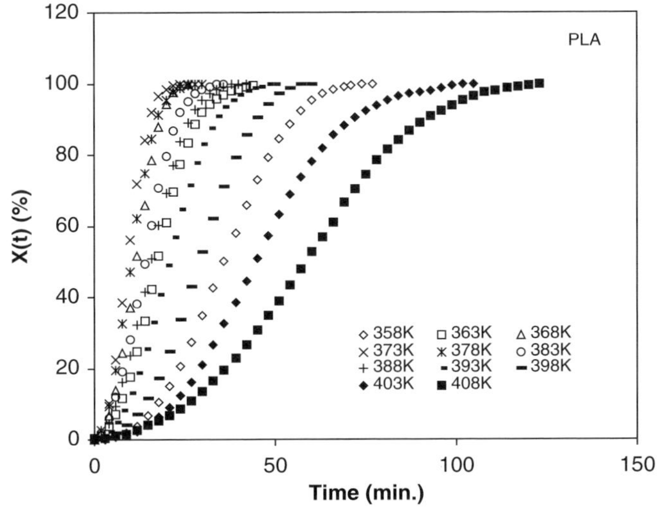

FIGURE 11.13 The relative degree of crystallization of pure PLA as a function of the isothermal crystallization time. DSC was performed by heating samples from 25° to 200°C at a heating rate of 20°C/min, holding at 200°C for 5 min, then rapidly cooling to the isothermal temperatures and holding there for 10–15 min, then reheating to 200°C at a rate of 20°C/min. (*Source:* From Ke and Sun, *Journal of Applied Polymer Science* © 2003. Reprinted with permission of John Wiley & Sons, Inc.)

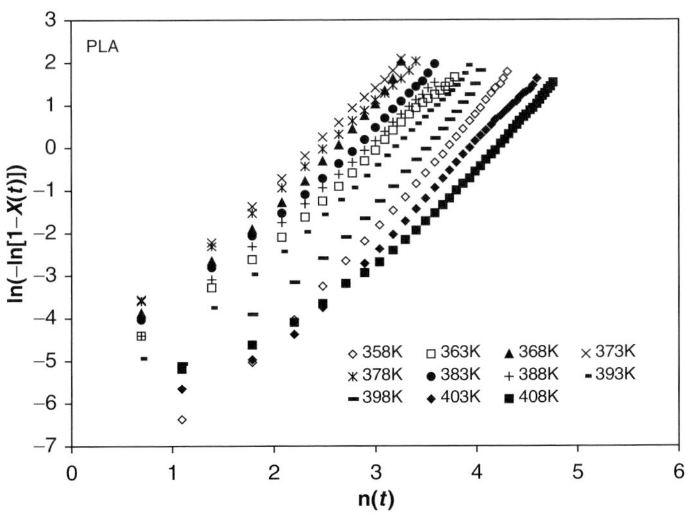

FIGURE 11.14 Plots of the degree of crystallization $\ln[-\ln(1-\alpha)]$ presented in Figure 11.13 versus the crystallization time $\ln(t)$ for isothermal crystallization of pure PLA. (*Source:* From Ke and Sun, *Journal of Applied Polymer Science* © 2003. Reprinted with permission of John Wiley & Sons, Inc.)

interfacial properties and control morphologies of polymer blends. Coupling agents containing reactive functional groups are able to generate *in situ* formation of blocks or grafted copolymers at the interface by hot melt blending. Reactive compatibilization has proved an effective method for morphology control in a variety of blend systems [36–41].

Starch is a hydrophilic polymer containing numerous hydroxy groups and has a very low level of aldehyde groups at the reducing end of the polysaccharide. PLA is a hydrophobic polymer containing both hydroxyl and carboxylic acid end groups. Chemicals with functional groups, such as isocyanate (–NCO), carboxylic acid (–COOH), and amino ($-NH_2$) groups, have the potential to be coupling reagents to improve the compatibility of a starch/ PLA blend in a hot melt system.

11.5.1 COUPLING REAGENTS WITH ISOCYANATE GROUPS

A good coupling agent should have functional groups that react with both the matrix and the filler. Methylenediphenyl diisocyanate (MDI) is highly reactive with both hydroxyl and carboxyl groups to form urethane linkages [42]. MDI is considered a toxic material but is chemically inert after reaction [43]. The reaction of MDI with water is very fast and gives solid insoluble polyurea, which is an environmentally friendly material [44]. In addition, the small amount of urethane linkage in the blend could be attacked by some fungi [45] and absorbed by soil [46]. Possible reactions between the isocyanate groups, and the hydroxyl and carboxyl groups from PLA, and the hydroxyl groups can be illustrated as is done in Figure 11.15 [35].

Amide (CONH) and urethane (COONH) groups are expected to form at the surface of starch granules during blending of PLA, dry starch, and MDI. Therefore, the compatibility between starch and PLA should be enhanced. The reaction of an isocyanate group with a carboxyl group is faster in a liquid medium than with hydroxyl. The reaction rate between an isocyanate group and a hydroxyl group would be reduced in the presence of acidic additives [48]. The following reactive mixing process can be hypothesized. At first, the liquid MDI would spread out on the surface of the starch granules and PLA particles. The MDI cannot penetrate the starch granules: Dry starch granules exclude even nitrogen molecules from their interior. The interior of the starch granules is penetrable only when the granules swell. Diffusion of MDI into the PLA particles would also be slow until they begin to melt at approximately 170°C. It seems possible that one isocyanate group on MDI would react at the surface of either starch or PLA particles during the warming/ mixing of the blend. Then the other isocyanate group may react with PLA as it melts. Diffusion of the isocyanate reagent will be slow in the viscous PLA melt, which would enhance the possibility that the second isocyanate group will react at the surface of a starch granule with which it collides. Crosslinking is likely to occur between two PLA molecules, two starch granules,

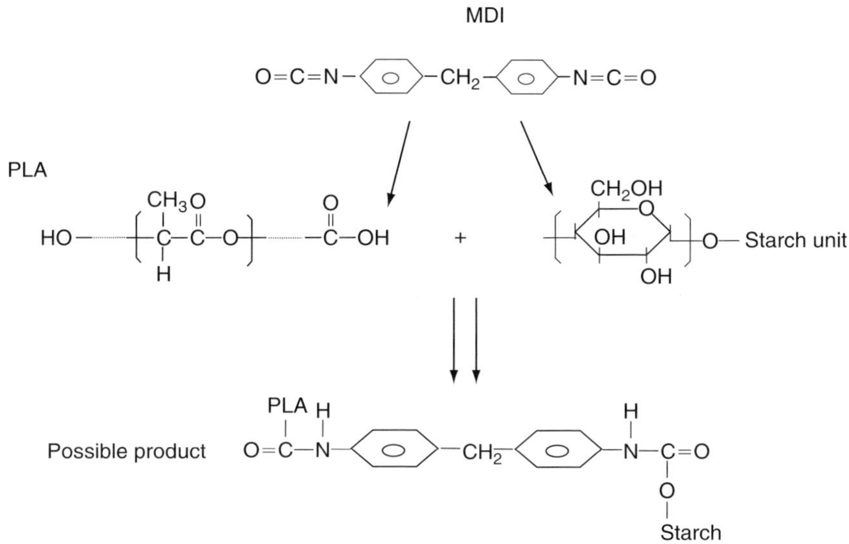

FIGURE 11.15 Scheme of possible reactions among PLA, starch, and MDI. (*Source:* Courtesy H. Wang, X. Sun, and P. Seib.)

and between PLA and starch. The PLA–starch cross-link is especially important for increasing compatibility.

For a starch/PLA blend without MDI, two phases can be clearly seen in Figure 11.10, and in the fracture process, many starch granules are pulled out of the matrix, creating large voids. Also, gaps between remaining starch granules and the PLA matrix are visible. On a blend of starch/PLA compounded with 0.5% MDI by weight, few individual starch granules can be observed, and those that are distinguishable appear to be coated with the PLA matrix (Figure 11.16). Moreover, fracturing of the blend occurs throughout the starch granules, rather than at their interfaces.

The tensile strength and elongation of the PLA/starch blend decreases significantly without MDI under ambient conditions (Table 11.2), while for the blend with MDI, the tensile strength remains almost the same as the pure PLA. The starch granules in the PLA matrix act as stress concentrators, inducing cracks and leading to low strength and elongation. The adhesion between starch and the PLA matrix in the presence of MDI is improved, as are the mechanical properties.

The spectrum of the surface structure determined via XPS (X-ray photoelectron spectroscopy, ES200B model) provides information that a chemical structure change has occurred on the surface of the PLA/starch blend coupled with MDI (Figure 11.17). The spectrum of PLA/starch without MDI (Figure 11.17, top) is almost identical to that of pure PLA, which

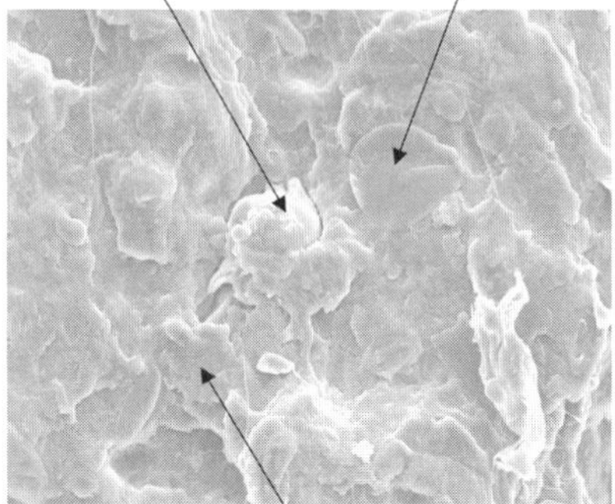

FIGURE 11.16 Morphology of the tensile fracture surface of extruded PLA with 45% cornstarch with 0.5% MDI. (*Source:* Courtesy H. Wang and X. Sun.)

TABLE 11.2 Mechanical properties of PLA, PLA/starch blends, and PLA/starch blend coupled with MDI[a].

Samples	Tensile Strength (MPa)	Elongation (%)	Young's Modulus (GPa)
100% PLA	62.1bbc	5.69a	1.41a
PLA/starch (55/45)	36.0d	2.58c	1.73c
PLA/starch/MDI (55/45/0.5)	66.7a	4.40b	1.94b

Source: Data from Wang *et al.* [35].
[a]Values in the same column followed by the same letter are not significantly different ($P < 0.05$).

means that PLA, as a continuous phase, completely covers the starch granules and no chemical change on the surface of the PLA matrix is detected. However, there is a significant reduction in C=O and OH intensities when the PLA and starch are coupled with MDI (Figure 11.17, bottom). A covalent linkage likely can be formed at the PLA/starch interface because of MDI among hydroxyl, carboxyl, and isocyanate groups, so that the interfacial adhesion would be enhanced, consequently improving tensile strength. Meanwhile, such adhesion might not cause severe restriction of elongation by forming proper entanglements, which could stretch the matrix. As expected, the Young's modulus of the PLA/starch blends in the presence of MDI is also improved, especially above its glass transition temperature.

The fracture surface of pure PLA is smooth, whereas the surface of starch/PLA blended with 0.5 wt% MDI is rough, which is typical of a compatible structure (Figure 11.16). According to the fracture theory developed by Griffin [48], fracture toughness is usually associated with fracture energy. For high toughness, the fracture energy must be at least equal to, or larger than, the surface energy at the interface of cracking, which is proportional to

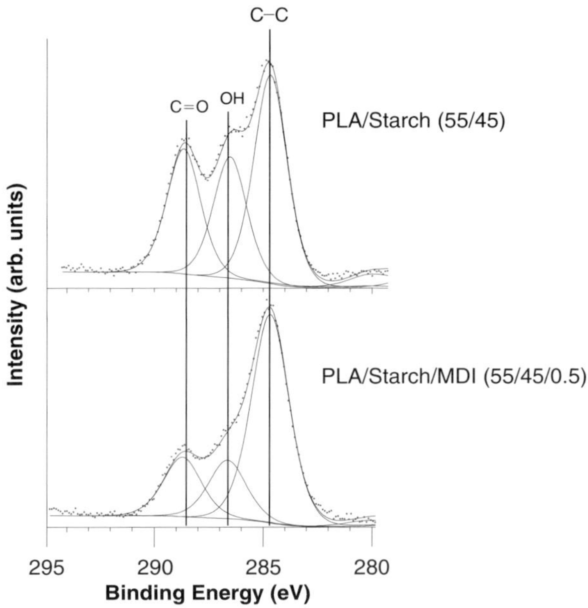

FIGURE 11.17 XPS spectrum of PLA with 45% starch (top) and PLA with 45% starch coupled with 0.5% MDI. (*Source:* Courtesy R. Moura, X. Sun, G. Claycomb, and P. Sherwood.)

the area under the stress–strain curve. The blends with MDI have higher fracture energy than those without MDI, as noted by Ke and Sun [26].

According to Wu [49], for a particulate-filled composite made with a coupling agent, the wetting and bonding at the interface significantly influenced the properties of the composite. The molecules diffuse across the interface and react chemically to create a chemical interaction and establish an adhesive bond. This diffusion is usually accelerated by low surface tension, which enhances wetting and increases the adhesive bond strength and, consequently, the fracture energy. MDI has a boiling temperature of about 208 °C [50], and the isocyanate groups are highly polar. It seems possible that the surface tension of starch could be reduced in the presence of MDI during mixing.

The storage modulus suddenly drops at approximately 65 °C because of glass transition effects. The raw PLA has the largest drop as expected, and the blend with MDI has the smallest (Figure 11.18). Both raw PLA and the PLA/starch blend without MDI show an increased storage modulus between 85° and 105 °C due to crystallization. Such crystallization behavior is not observed in the starch/PLA blend with MDI. The storage modulus of the PLA/starch blend with MDI at high temperature is high enough for the blend

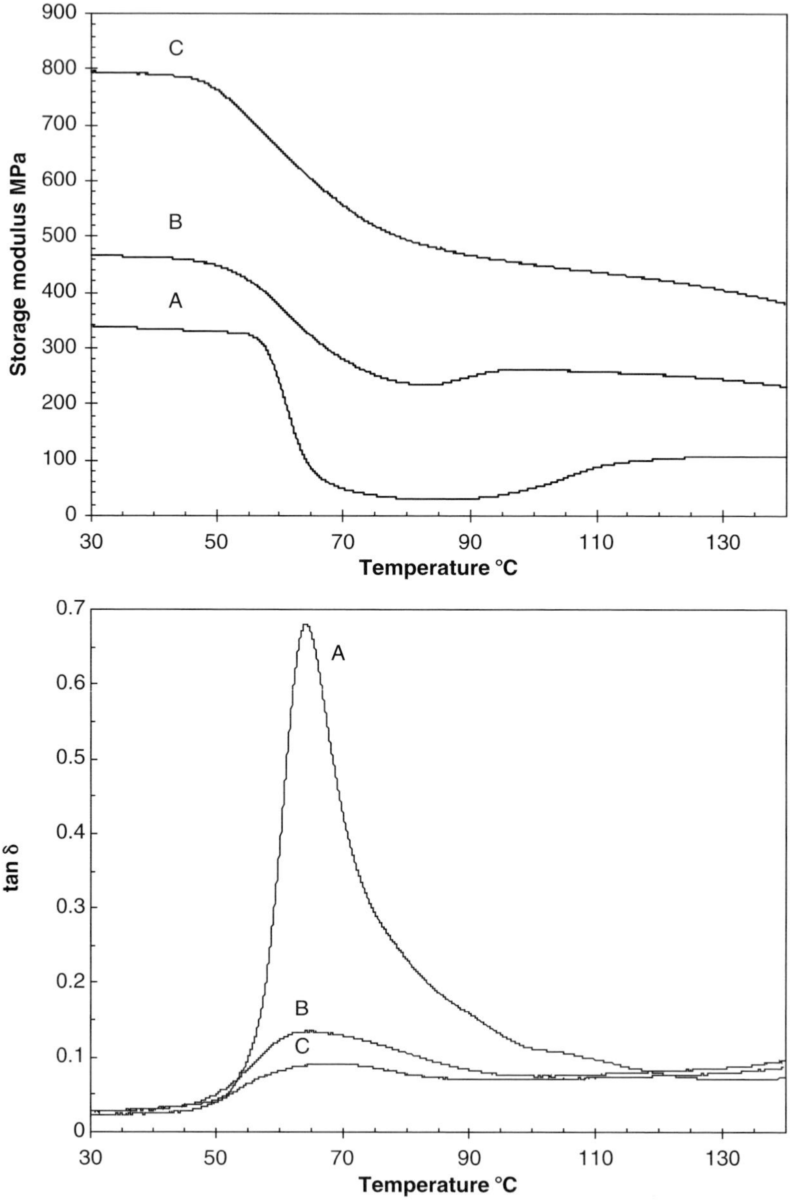

FIGURE 11.18 Storage modulus (top) and tan δ (bottom) versus temperature for PLA and PLA/starch blends (55/45 weight ratio): (curve A) raw PLA, (curve B) blend without methylenediphenyl diisocyanate (MDI), and (curve C) blend with 0.5 wt% MDI. A three-point bending rectangle method was used for DMA measurement at 1 Hz from 25° to 160 °C with heating ramp of 3 °C/min. (*Source:* Copyright Wang *et al.* [35].)

to be rigid, and almost the same as that of the blend without MDI at room temperature [35].

The mechanical damping factor tan δ, is usually associated with the inelastic manifestation in the thermal phase transition zone. Pure PLA has a sharp and high damping peak (Figure 11.18, bottom, curve A), whereas the damping of the blend with MDI is low and broad. The damping properties of a starch-reinforced blend are usually affected by starch content; the higher the starch content, the lower the value of the damping factor [51]. In a blend system with filler particles, the filler would limit the mobility of the matrix molecular chain at the interface, thereby affecting the relaxation of the matrix chains and causing a lower damping in the transition zone compared to the pure polymer matrix. In the blend with MDI, the specific interaction between filler and the polymer matrix enhances the interfacial adhesion and tends to create an absorbed layer of polymer surrounding the filler surface, which restricts the molecular motion and results in a lower damping [36, 51].

The mechanical properties of the PLA/starch blend are not significantly affected by MDI when MDI concentration is beyond 0.5%. There is a critical density C for molecules formed *in situ* by the compatibilizer at the interface [52]. The critical density is dependent on the molecular weight of the compatibilizer, $C \sim N^{-0.5}$, where N is the molecular weight of the compatibilizer. At the critical density, enough copolymers are formed at the interphase to prevent bulk polymers from penetrating the copolymer barrier for further reaction. In this case, 0.5% may be the critical density of MDI for the PLA and starch system at a ratio of 55:45.

Thermal transition temperatures of PLA mainly remain the same in the addition of either starch or MDI. Starch with almost 0% moisture will not have any crystal melting before it reaches its decomposition temperature of around 230°C. The addition of MDI improves the interface adhesion between PLA and starch, but starch still remains in its granule form in the system, as shown in Figure 11.16.

No significant difference in water absorption occurs between the blends with and without MDI (Figure 11.19). When starch is soaked in excess water at 25°C, it can take up to about 50% of its dry weight of water [53]. Therefore, the theoretical maximum water content absorbed by the blend containing 45% starch phase should be about 23% when submerged in excess water at 25°C. The water absorption for both blends with or without MDI increases greatly during the first 15 days and then levels off at about 14%.

11.5.2 MALEIC ANHYDRIDE AS A COUPLING REAGENT

Maleic anhydride (MA) is also often used as a compatibilizer in two immiscible polymer blends [36–41]. MA is highly reactive with PLA free radicals induced by an initiator, such as 2,5-bis(*tert*-butylperoxy)-2,5

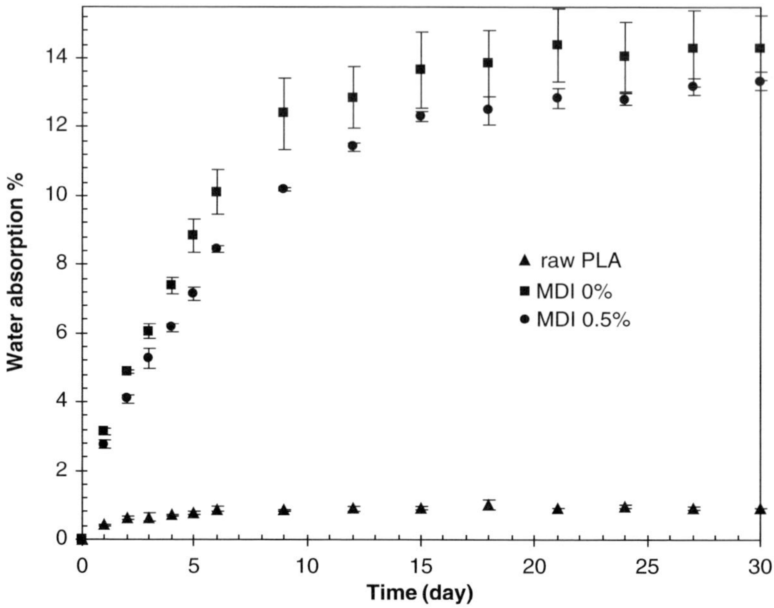

FIGURE 11.19 Water absorption of raw PLA, extruded PLA with 45% cornstarch and with 0% or 0.5% MDI. (*Source:* From Wang *et al.* [35]. *Journal of Applied Polymer Science* © 2001. Reprinted with permission of John Wiley & Sons, Inc.)

dimethylhexane (L101) [54, 55], and the anhydride group can react with hydroxyls from starch to form ester linkages, as schematically shown in Figure 11.20. The carboxylic groups arising from the hydrolyzed anhydride can also form hydrogen bonding with the hydroxyl groups [56]. The function of the initiator is to induce free radicals of PLA that can react with MA.

The mechanical properties of the PLA/starch blend are significantly improved by incorporating MA with L101 initiator [55]. The maximum tensile strength of PLA/starch is reached at 1% MA with 0.1% L101 initiator. In the blend of starch/PLA (55/45), 1% MA with 0.1 % L101 would be enough to attain the critical copolymer density as discussed in Section 11.5.1. Thus, the mechanical properties of the blend cannot be further enhanced if MA and L101 concentrations increase beyond the critical level. Similar to the blend with MDI, a uniform dispersion of starch granules in the PLA matrix is observed (Figure 11.21). The surface-to-surface intergranular distance (matrix ligament thickness) is even, and local deformation in the PLA matrix near the starch granules increases. The PLA matrix exhibits slightly plastic deformation with some extended strands, indicating a strong adhesion between PLA and starch.

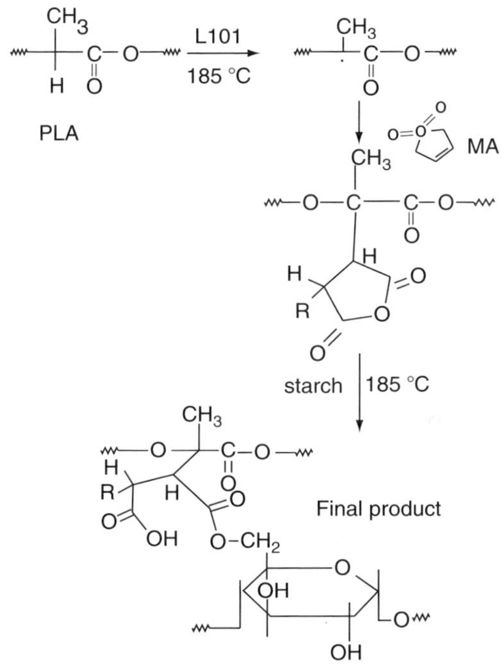

FIGURE 11.20 Scheme of possible chemical reactions among PLA, starch, maleic anhydride (MA), and initiator L101. (*Source:* Copyright J. F. Zhang and X. Sun [55], *Biomacromolecules*. 2004, **5**, 1446–1451.)

11.5.3 INTERFACIAL THICKNESS PREDICTION

The thickness of the interfacial zone between PLA and starch should be important to the mechanical properties of their blending system. The interfacial thickness between PLA and starch can be estimated using a method developed by Dedecker and Groeninckx [58] with the following assumptions: (1) The dispersed starch granules are perfectly spherical in shape with a uniform size; (2) the dispersed granules have a core that consists of starch and an interphase zone in which the composition gradually changes from pure starch, to pure coupling reagent, and then to pure PLA; (3) the concentration gradient in the interphase is linear; (4) the molecules of the coupling reagent are all completely reacted with starch and PLA; (5) the interphase has a uniform thickness; and (6) the PLA matrix surrounding the starch granule and the interphase formed between the PLA and starch are perfectly circular. Then the thickness of interaction between PLA and starch can theoretically be calculated with regard to the percentage of starch according to this equation [57]:

$$\alpha = M_i/(M_s + M_i), \qquad (11.6)$$

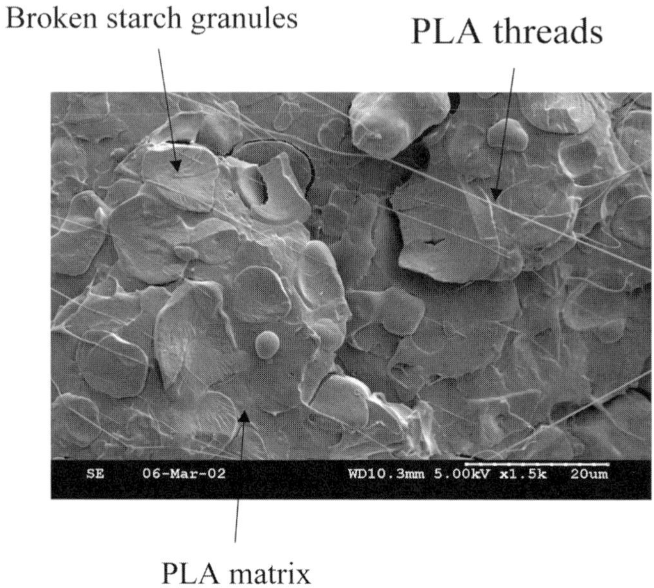

FIGURE 11.21 Morphology of PLA with 45% cornstarch coupled with 1% maleic anhydride (MA) and 0.1 initiator L101. These materials were blended simultaneously using a twin-screw extruder. (*Source:* Adapted from Zhang and Sun [5].)

where α is the percentage of starch in the interface region, M_i is the amount of starch in the interfacial region, and M_s is the amount of starch in the starch granule zone. The values of M_s and M_i can be theoretically estimated by Eqs. (11.7) and (11.8), respectively:

$$M_s = 4\pi \int_0^{r_1} r^2 dr, \tag{11.7}$$

$$M_i = 4\pi \int_{r_1}^{r_1+t_1} \frac{r_1 + t_1 - r}{t_1} r^2 dr, \tag{11.8}$$

where r_1 is the radius of the starch granule, which is assumed to be 18 μm, r is the radius at any point from the center of the starch granule, and t is the thickness of the interface. The interface thickness increases as the percentage of starch in the interface region increases, but the interface thickness is fairly thin, ranging from 0 to 72 μm, which corresponds to 0–80% starch in the region. If no reaction occurs in the PLA/starch blend, there should be no interface zone between these two components and the thickness t should be zero. If 100% starch is located in the interface zone, the thickness t of the interface becomes infinity. Between these two extremes, there is some reaction among coupling reagent, starch, and PLA, and the thickness of the interface zone should vary with the degree of reaction.

Using PLA/starch coupled with MA as an example, the amount of starch reacting with MA in the PLA/starch blend system can be determined with dimethyl sulfoxide (DMSO) using the following equation [56]:

$$m_s = W_b \times 45\% - (W_b - W_r), \tag{11.9}$$

where m_s is assumed to be the amount of starch reacting with MA, W_b is the PLA/starch sample weight, W_r is the residue weight of the blend containing PLA and starch reacting with MA, and 45% is the percentage of starch in the blend, which is based on a uniform dispersion of starch and PLA. The amount of PLA reacted with MA can be determined by extracting PLA with chloroform using the following equation [56]:

$$m_{\mathrm{PLA}} = W_b \times 55\% - (W_r - W_{rr}), \tag{11.10}$$

where, m_{PLA} is the amount of PLA reacting with MA; W_b and W_r are the same as for Eq. (11.9), W_{rr} is the residual weight of the blend containing residual PLA and starch reacted with MA extracted by DMSO and chloroform, and 55% is the percentage of PLA in the blend.

Taking the measured residual amount of starch as the percentage of starch in the interface region, the total interface thickness of the blend can be estimated using the curve given in Figure 11.22 [56]. For example, for the blend with 1% MA and 0.1% L101, the residual starch amount is 35.4%, and the estimated interphase thickness, including physical entanglement, is about 10.2 μm.

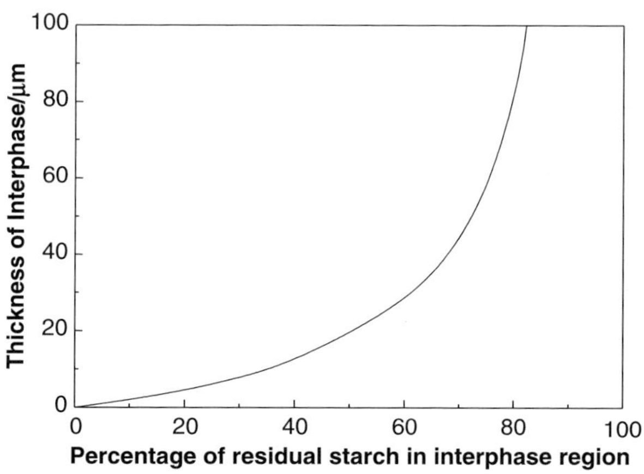

FIGURE 11.22 Theoretical calculation of interface thickness t_1 containing starch and PLA. (*Source:* Copyright J. F. Zhang and X. Sun [55], *Biomacromolecules*. 2004, **5**, 1446–1451.)

The average thickness of each starch granule, at the interphase, should be divided by the number of starch granules in the blend. The average thickness is approximately $1.01 \times 10^{-7}\,\mu m$, with the assumption of a circular starch granule and $1.4\,g/cm^3$ density. Though only $1.01 \times 10^{-7}\,\mu m$, the thickness is enough to withstand and transfer a certain load between the PLA matrix and the starch granules, which is confirmed by the improved mechanical properties with increasing interphase thickness.

11.6 ROLE OF WATER IN STARCH AND PLA BLENDS

Starch is a highly hydrophilic polymer and sensitive to water, as discussed earlier in this chapter. Starch granules become swollen and dispersed in water and then gelatinize when heated. During the blending process, the starch is partially or fully gelatinized, or destructurized in the presence of moisture, heat, and shear force [58]. Water is often used as a plasticizer to destructure starch in blends with various polymers, as well as to achieve a fine dispersion and, consequently, to obtain desirable product properties [59–61]. PLA is a hydrophobic synthetic biopolymer that is depolymerized in the presence of water at elevated temperatures, which results in poor mechanical properties [62]. Therefore, starch is often dried before blending with PLA in order to avoid PLA degradation, which means the starch remains mostly in its granular state in the blends, as shown earlier in Figure 11.10. Water is a convenient, economical, and effective plasticizer of starch, and the moisture content in starch affects the morphology and mechanical properties of the blend [63].

The surface appearances of the extrudates of starch and PLA blends are significantly affected by moisture content. The surface of the extrudates becomes smooth as the moisture content increases up to 20%, and then becomes rough as the moisture increases up to 50%. At high moisture contents, such as 40% or 50%, foaming can be observed at the die because of the high water-vapor pressure formed right before the die. The final moisture content of the extrudates, after natural cooling to ambient temperature, is less than 2%, regardless of initial moisture content. This indicates that all of the extra water vapor produced in the extrusion barrel is removed immediately at the die because of the high extrusion temperature.

Moisture has little effect on the melting behavior and crystallinity of PLA and its blends with starch, but the glass transition T_g shifts a few degrees to a lower temperature [63]. The thermogravimetric analysis (TGA) measures the change in weight of a sample caused by decomposition, reaction, and volatilization, among other things. A decrease in the decomposition temperature results in a more thermally unstable product. The thermal decomposition of PLA begins at 351.3°C and ends at 389.8°C (Figure 11.23). For starch, the corresponding temperatures are 299.8° and 331.7°C. As expected, the

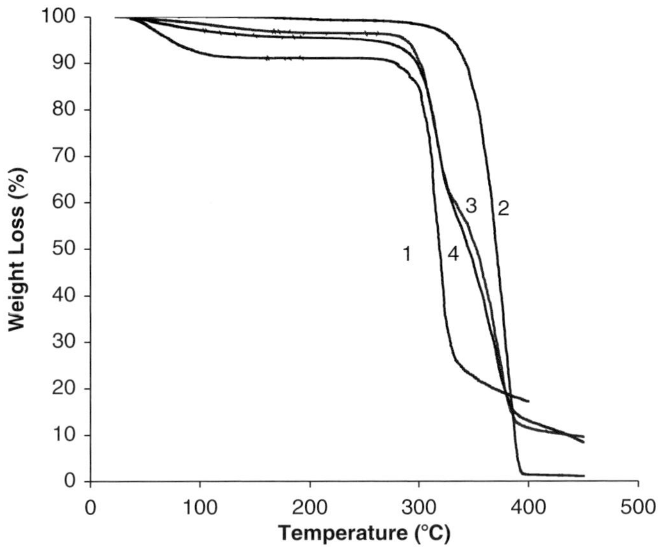

FIGURE 11.23 TGA thermograms: (curve 1) native cornstarch, (curve 2) pure PLA and blends of cornstarch and PLA at a 40:60 ratio with (curve 3) 11.9%, (curve 4) 30%, and (curve 5) 50% moisture content. (*Source:* From Ke and Sun [64]. *Journal of Applied Polymer Science* © 2001. Reprinted with permission of John Wiley & Sons, Inc.)

decomposition curves of the starch and PLA blends are in between those of native starch and pure PLA. For the PLA extracted from the blends with 10% moisture, the decomposition temperature is about 3–5 °C lower than for pure PLA, indicating that PLA is slightly degraded in the presence of water during processing. The PLA thermal degradation remains the same as the moisture increase up to 30%, but increases slightly as the moisture increases up to 50% [63].

At 10% moisture, most of the extruded starch still remains in granular form (Figure 11.5b), although some granules are ruptured by heat and shear force during extrusion, unlike native starch granules (Figure 11.5a). As the moisture increases, starch become more gelatinized, and intact starch granules are reduced and become swollen (Figure 11.5c). As the moisture content reaches 40% or 50%, the starch becomes fully gelatinized and no granular structure can be observed (Figure 11.5d).

The blend with 20% moisture content has slightly higher tensile strength and elongation (Table 11.3). As the moisture increases up to 50%, a significant decrease in mechanical properties is observed. The morphology of the gelatinized starch in the blends would affect the effective cross-sectional area of the continuous PLA phase, and the voids in the blends would concentrate stress and act as crack initiators. As moisture increases, the degree of starch gelatinization increases, and the starch becomes more dispersible, resulting in

TABLE 11.3 Mechanical properties of starch and PLA blends (40/60) at varied moisture contents (MC.)[1]

Sample	Tensile Strength (MPa)	Elongation at Break (%)	Modulus (GPa)
Pure PLA	64.0a	7.40a	1.12a
Blends with MC			
0	35.8b	4.39b	1.06b
11.9	36.6bc	4.38b	1.03b
20	38.3cd	4.76b	1.00b
30	37.3cd	4.40b	1.00b
40	37.9cd	4.30b	0.99b
50	32.9e	4.87b	0.81c

Source: From Ke and Sun [63]. *Journal of Applied Polymer Science* © 2001. Reprinted with permission of John Wiley & Sons, Inc.
[1]Values in the same column followed by the same letter are not significantly different ($P < 0.05$).

a continuous phase. PLA becomes more degraded because of such a large amount of water at an elevated temperature.

The tensile strength and elongation of a PLA/starch blend with 0.5% MDI are reduced as starch moisture increases [35]. At moisture levels between 10% and 15%, the blends show almost no appreciable difference in mechanical properties. However, at 20% moisture, the mechanical strength properties drop dramatically and are almost the same as those of the PLA/starch blend without MDI. Water competes with PLA and starch in reactions with MDI. At 20% moisture, apart from the reaction of water with MDI, the starch granules swell, hence greatly reducing the mechanical strength and elongation.

11.7 PLASTICIZATION OF STARCH AND PLA BLENDS

11.7.1 MECHANICAL PROPERTIES

Blends of starch and PLA are usually brittle, having an elongation range of 2–5%; this range can be improved by using plasticizers [64, 65]. A plasticizer is commonly used to improve the processibility, flexibility, and stretchability of a polymer because it can reduce the intermolecular force and increase the intermolecular mobility. A good plasticizer should be chemically compatible with the polymer to yield a stable, homogeneous mixture. Starch alone in a dry state is not considered a true thermoplastic in the sense that it degrades on heating rather than melting on its own. In the presence of a plasticizer, however, starch behaves as a thermoplastic [66]. Triethyl citrate (TC) is a small molecular citrate ester; it is synthesized from naturally occurring citric acid and is commonly used as a plasticizer for many polymers. TCs are often recommended for materials that come into contact with

TABLE 11.4 Tensile strength and elongation of starch and PLA blends as affected by triethyl citrate concentration.

(%) Triethyl Citrate (%)	Tensile Strength (MPa)				Elongation			
	5	10	15	20	5	10	15	20
PLA/starch (55/45)	27.3	24.8	16.5	14.7	5.8	8.5	12.7	16.8
PLA/starch/MDI (55/45/0.5)	51.5	21.8	12.7			5.4	50.8	89.8
PLA/starch/MA.L101 (55/45/1/0.1)	31.5	20.3	10.1			5.1	40.3	131.7

Sources: Data from Ke and Sun [65, 69] and Zhang and Sun [71].

food because of their favorable physiologic properties (e.g., nontoxicity and nonirritation) [67]. Citrate esters are miscible with PLA and effectively improve PLA's flexibility [68, 69]. The elongation of the extruded and molded blends increases with TC concentrations (Table 11.4). TC effectively increases the mobility of PLA, yielding a large increase in elongation. As expected, the strength of the PLA/starch blend is reduced (Table 11.4). When the starch and PLA is coupled with MDI or MA, the plasticization effects of TC are also enhanced [64, 70] (Table 11.4). Coupling reagents improve adhesion between starch and PLA, but as the TC content increases, the coupling effect decreases, resulting in phase separation [71].

With a content of 6–8% of plasticizer, elongation of the blend increases sharply and the strength drops suddenly [64, 70] (Figure 11.24a). In polymer blends in the presence of a plasticizer, the tensile strength, elongation at break, and elastic modulus often change sharply at a critical volume fraction of plasticizer. The concentration of plasticizer at the critical volume fraction is then often referred to as the *percolation threshold*. This sudden drop (or jump) is due to the formation of a critical continuous network or phase [72, 73]. Elongation can be expressed as follows [73]:

$$\begin{cases} \varepsilon_b \approx \varepsilon_0 & \text{when} \quad p < p_c, \\ \varepsilon_b \propto (p - p_c)^{\mu} & \text{when} \quad p \geq p_c \end{cases} \tag{11.11}$$

where ε_b is the elongation at plasticizer concentration p, ε_0 is initial elongation, p_c is the percolation threshold, and μ is the exponent for cubic dimensions.

Starch granules are coated with PLA; however, it is the TC content that determines how much PLA adheres to starch granules. With a TC content of less than 6.5%, more PLA adheres to the granules than with a TC content of more than 8%. The critical threshold level of TC for the PLA/starch (55/45) blend could be somewhere close to 7%. The percentage of starch also significantly influences the critical threshold of plasticizer. For instance, at a constant TC content of 15%, the blend with 45% starch has approximately 130% elongation; with 40% starch, 94% elongation; and with 0% starch (pure

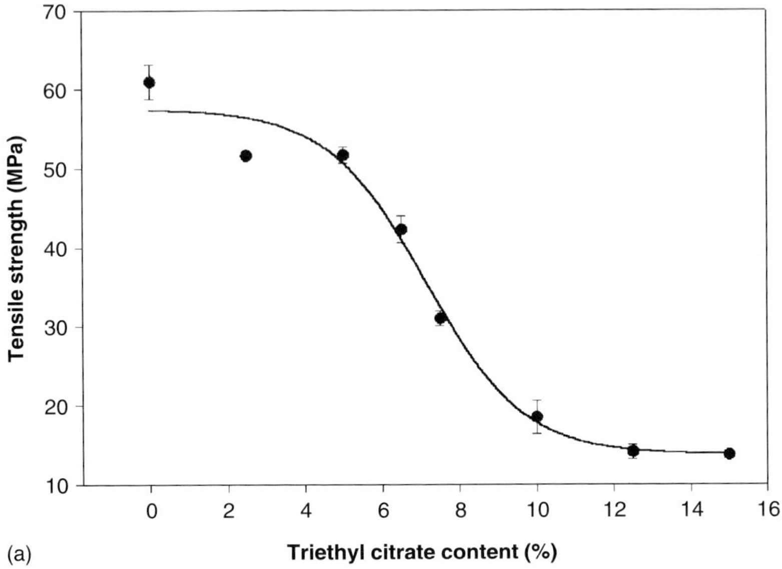

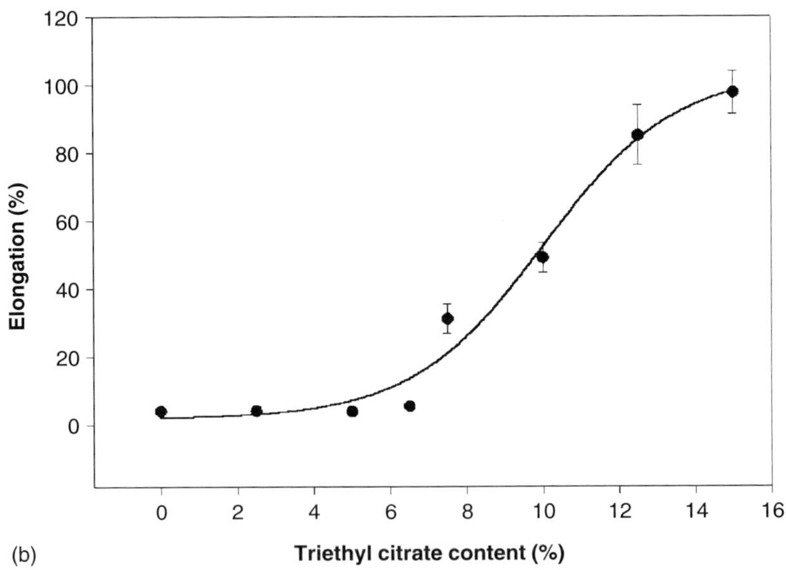

FIGURE 11.24 Tensile mechanical properties of extruded PLA with 45% cornstarch coupled with 0.5% MDI with various plasticizer contents: (a) tensile strength, (b) elongation, and (c) modulus. (*Source:* Adapted from Ke and Sun [64].) *Continued*

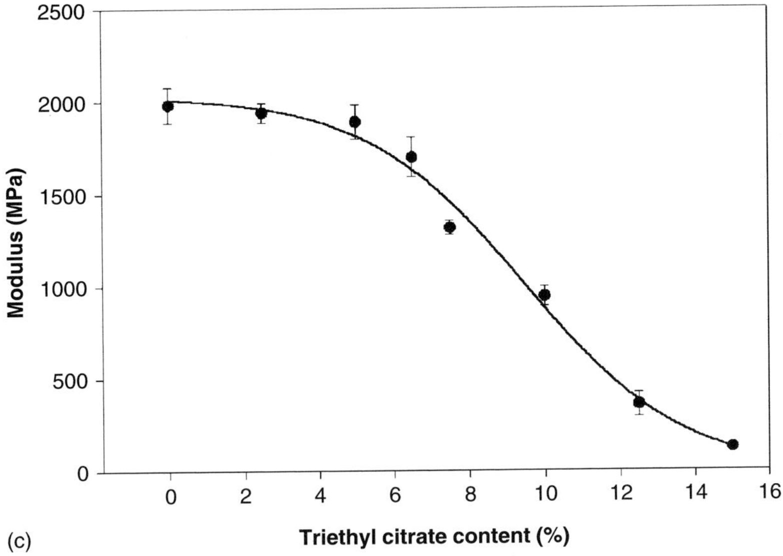

FIGURE 11.24, cont'd

PLA), 81% elongation. In addition, TC is a good plasticizer for PLA; therefore, TC is selectively dispersed in PLA rather than in starch, so starch granules remain in a TC plasticized PLA/starch blend. Increasing the amount of starch in the blend squeezes some TC from the starch phase, dispersing it into the PLA and raising the relative TC content in PLA. Further, as a plasticizer, TC molecules penetrate into the PLA matrix and starch granules, and then destroy the binding force between starch and PLA macromolecules, which makes it easy for molecular chains to slide and move, consequently increasing the elongation.

Glycerol and sorbitol are polyhydric alcohols that are miscible with starch and are good plasticizers of starch [71–73]. Some hydroxyl groups of glycerol and sorbitol interact with the ester group from PLA in some degree, such as hydrogen bonding. Poly(ethylene glycol) (PEG) and poly(propylene glycol) (PPG) are polyethers with low molecular weights that are also widely used as plasticizers [64, 74]. Starch granules can be observed in the PLA/starch plasticized with blends with TC, PEG, and PPG, and some voids exist [68] (Figure 11.25). The blends containing glycerol and sorbitol have a uniform and smooth morphology, indicating that the starch is destructurized (or plasticized).

Glycerol significantly reduces the tensile strength of the blends although the plasticized starch of the blend with glycerol above 15% is in continuous phase (Table 11.5). However, the tensile strength of the blends containing sorbitol increases slightly as sorbitol content increases. Sorbitol has less effect on the mobility of starch than glycerol because sorbitol is in a solid form at room temperature. The morphology of the blend containing sorbitol is more

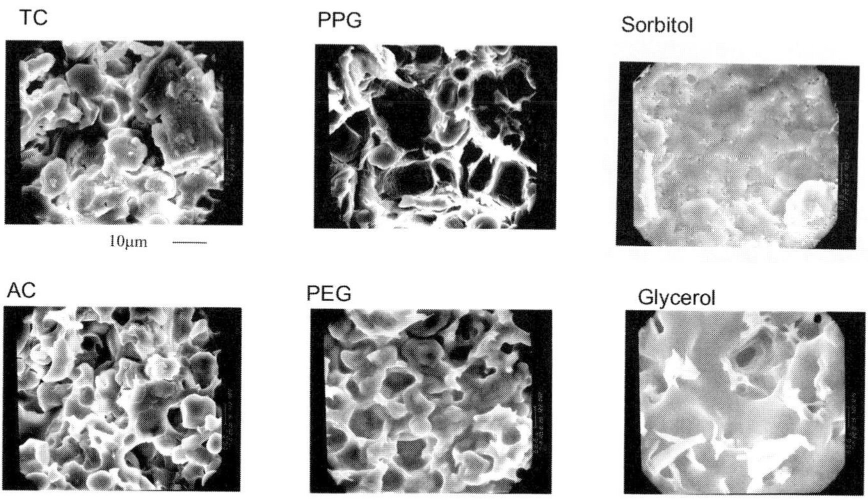

FIGURE 11.25 Morphology of tensile fractured extruded PLA with 40% starch containing 25% of various plasticizers (1000×). PLA/starch/plasticizer blends were extruded by a twin-screw extruder and injected molded. TC, triethyl citrate; AC, acetyl triethyl citrate; PEG, poly(ethylene glycol); and PPG, poly(propylene glycol). (*Source:* From Ke and Sun [68]. © 2001 ASAE. Reprinted with permission from the American Society of Agricultural Engineering.)

TABLE 11.5 Tensile strength of PLA/starch (55/45) blend as affected by plasticizer concentration.

Platicizers	PEG	PPG	Glycerol	Sorbitol
Concentration (%)				
5	25.1	20.3	25.6	27.8
10	12.4	15.3	23.5	32.3
15	7.8	8.2	20.3	33.1
20	7.7	7.9	16.7	35.4
25	6.4	6.3	16.2	35.5

Source: Data from Ke and Sun [68]. PEG, poly(ethylene glycol); PPG, poly(propylene glycol).

solid than those of the blends containing other plasticizers (Figure 11.25). PEG and PPG slightly improve the flexibility of the blends (Table 11.6), and glycerol does not improve the elongation of the blend either. However, at 25% glycerol, the elongation of the blend is improved to some degree. Sorbitol has no effect on elongation, although elongation decreases slightly because of the increase of filler volume as sorbitol content increases.

11.7.2 GLASS TRANSITION TEMPERATURES

The glass transition temperature (T_g) of PLA is depressed by adding TC because TC and PLA are compatible [69]. The T_gs of the blends decrease rapidly

TABLE 11.6 Elongation of PLA/starch (55/45) blend as affected by plasticizer concentration.

Platicizers	PEG	PPG	Glycerol	Sorbitol
Concentration (%)				
5	3.2	2.8	2.9	4.3
11	3.6	2.9	3.0	4.2
16	7.4	7.5	3.2	3.1
20	7.8	7.6	3.1	3.3
25	7.7	8.7	6.8	3.4

Source: Data from Ke and Sun [68]. PEG, poly(ethylene glycol); PPG, poly(propylene glycol).

as PEG content increases. PEG contains a repeating unit of an ether group, which is compatible in some degree with the repeating unit of the ester group in PLA, suggesting that PLA and PEG have good miscibility [68, 74]. Glycerol, sorbitol, and PPG have almost no effect on the T_gs of the blends. Glycerol and sorbitol, which are polyol compounds and are immiscible with PLA, do not influence its mobility. In structure, PPG contains a repeating unit of ether like PEG, but also contains a side methyl group. This group may hide the "ether bond" in the PPG main chain, decreasing PPG's miscibility with PLA.

11.7.3 PLASTICIZER MIGRATION

Thermally induced migration of plasticizers from the blend to the environment can cause changes in the mechanical properties of the blend. Plasticizer migration can be described using the following kinetic equation of first order [70, 75]:

$$\ln\frac{C}{C_0} = -K \cdot t\,, \tag{11.12}$$

where C is the residual concentration of the plasticizer, C_0 is the initial concentration of plasticizer in the blend, K is the general constant of the migration process, and t is the time of migration. Migration is a concentration-driven process that should be fast at the beginning and then slow down. As shown in Figure 11.26 (redrawn at 135°C 5% and 11% TC only) [70], experimental data fit Eq. (11.12) well; the TC weight loss is rapid in the first 100 min and then slows down dramatically. The TC migration rate is much faster at the 11% concentration than at 5%.

The temperature dependence of the migration rate can be well described by the Arrhenius-like Eq. (11.13):

$$K = K_0 \exp\left(\frac{-\Delta C_a}{RT}\right), \tag{11.13}$$

where K_0 is a constant, ΔC_a is the activation energy related to concentration of the plasticizer, R is the gas constant, and T is the absolute temperature.

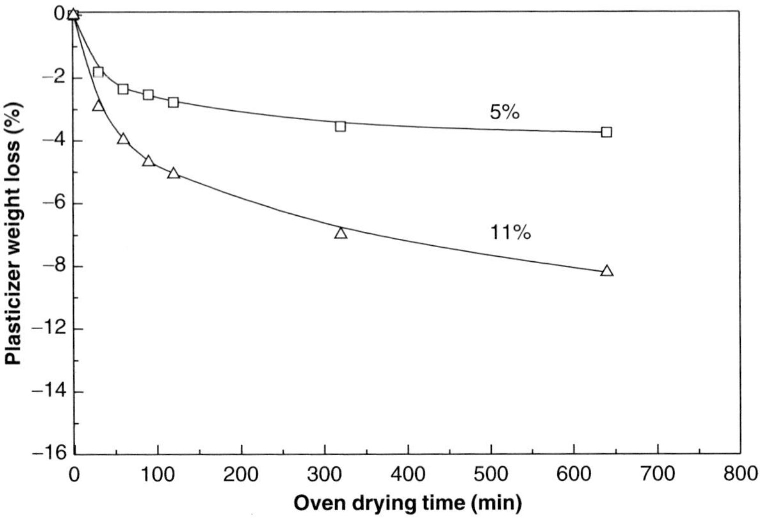

FIGURE 11.26 Thermally induced weight loss of TC from the extruded PLA with 45% cornstarch coupled with 1% MA and 0.1% initiator L101 with 5% and 11% of the TC plasticizer at temperature 135 °C, respectively. (*Source:* Courtesy J. F. Zhang and X. Sun.)

High temperatures ease the migration of a plasticizer, whereas low temperatures inhibit the migration. The value of K can be determined using Eq. (11.12) by plotting $\ln C/C_0$ against time. Therefore, ΔC_a can be determined by the Eq. (11.14):

$$\Delta C_a = \frac{R \ln (K_1/K_2)}{\frac{1}{T_2} - \frac{1}{T_1}}, \tag{11.14}$$

where K_1 and K_2 represent the migration rate at temperature T_1 and T_2. The activation energy is obviously reduced with the increasing TC content as demonstrated by Zhang and Sun [70], which further confirms that a high plasticizer content makes migration much easier than low content.

Migration of plasticizer is also affected by the molecular size of the plasticizer. The polymeric plasticizer, dioctyl maleate (DOM), reduces the migration rate (Figure 11.27) [76]. As a high molecular weight, DOM overcomes the resin hardening and minimizes the leaching kinetics. For thermally induced low-molecular-weight plasticizers, for instance, TC, migration and loss from the bulk specimen are extensive and the mechanical properties are, therefore, dramatically influenced.

DOM also improves the flexibility of the PLA/starch blends. With 10% DOM, the elongation of the blend with 45% starch is about 24%, and tensile strength is about 20 MPa [76]. The elongation then increases to 36% and tensile strength reduces to 16.2 MPa at 15% DOM. DOM has a structure

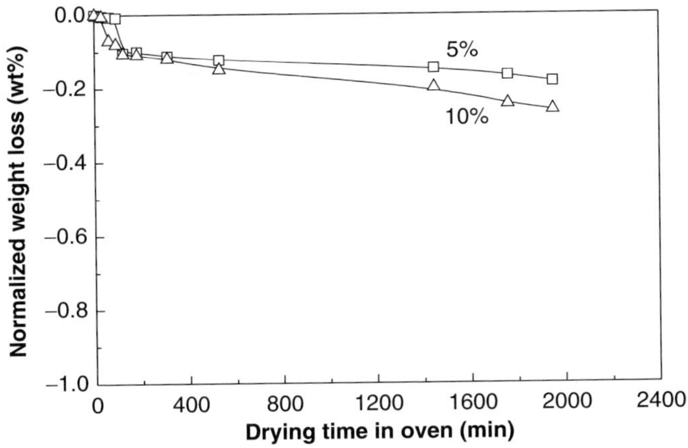

FIGURE 11.27 Thermally induced weight loss of extruded PLA with 45% cornstarch in the presence of 5% and 10% DOM, isothermally treated in a conventional oven at 135°C. (*Source:* Courtesy J. F. Zhang and X. Sun.)

similar to that of MA and acts as a compatibilizer at lower concentrations, whereas at higher concentrations, DOM forms liquid droplets, suppressing its compatibilizer ability by enhancing its chain mobility and operating as a plasticizer [76].

11.8 PHYSICAL AGING OF STARCH AND PLA BLENDS

Physical aging is an inherent characteristic of the amorphous phase in glassy or partially glassy polymers and usually occurs around its glass transition temperature (T_g) [77, 78]. Aging involves spontaneous changes in the thermodynamic state of a material, changes that are completely reversible. When a polymer is cooled from a melt state, its molecular chain becomes frozen if the temperature falls below its T_g. The polymer is in a nonequilibrium state, having large volume, enthalpy, and entropy [77]. At temperatures near, or slightly below, T_g, the free volume will reduce spontaneously toward an equilibrium thermodynamic state, resulting in a reduction in enthalpy. This process is referred to as physical aging, and it is time dependent.

The concept of free volume is usually used to describe the physical aging process. Free volume controls the molecular mobility of large segments of the polymer chains, which in turn influence the physical and mechanical properties, such as shrinking, stiffness, and brittleness, and damping decreases [78]. Physical aging can be monitored by DSC [79]. The endothermic enthalpic recovery peaks around T_g in DSC scans are often used to indicate a reduction in free volume.

The interfacial interaction between PLA and starch becomes weakened due to physical aging (Figure 11.28). Many threads and gaps can be observed in Figure 11.28 compared with the photographs presented in Figures 11.10 and 11.16. Crystalline regions are more rigid, restrained, and less extensible than amorphous regions. As aging proceeds, the PLA matrix shrinks, resulting in a gap around the interface, where some molecular chains are less restricted by the interfacial interaction. Hence, during tensile stretching, the less crystallized PLA molecular chains around the interface are more stretchable than those away from the interface.

The excess enthalpy of relaxation of the blend at its T_g increases as aging proceeds (Figure 11.29). The aging is initially fast and then slows as storage time increases [79] as illustrated in Figure 11.30. The aging rate of the blends is independent of the starch source but is influenced by the addition of MDI. The blends with MDI show a much slower aging rate than those without MDI. The excess enthalpy of relaxation correlates to the free volume, which reduces slowly for the blend with MDI because of the stronger interfacial adhesion.

The mechanical properties of the blends decrease as the aging time increases in the first 90 days (Table 11.7). As discussed, the free volume reduces

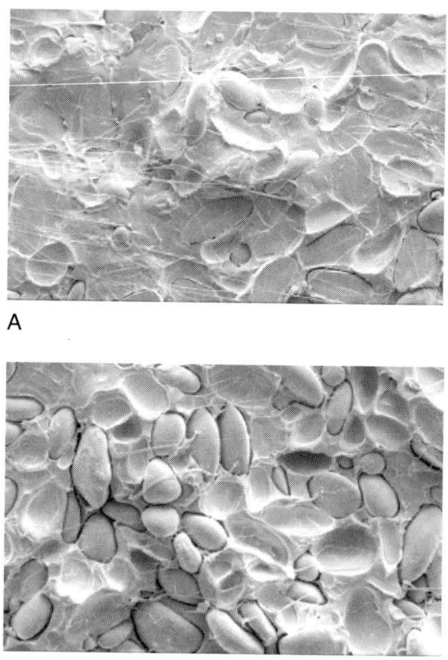

A

B

FIGURE 11.28 Morphology of the tensile fracture surface of extruded PLA with 45% wheat starch with (A) 0.5% MDI and (B) without MDI aged at 23°C and 55% relative humidity for 180 days. (*Source:* Courtesy H. Wang and X. Sun.)

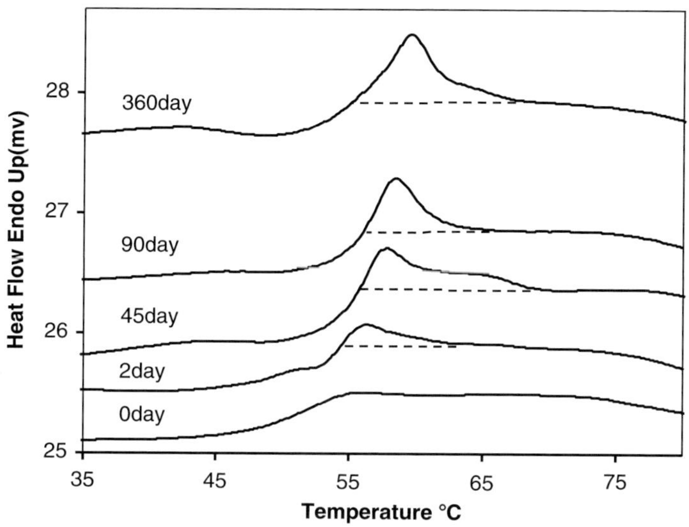

FIGURE 11.29 DSC thermograms of the PLA/wheat starch blend (w/w, 1/1) with 0.5% MDI aged at 25 °C and 50% relative humidity for various lengths of time (0, 2, 45, 90, and 360 days). (*Source:* From Wang *et al.* [80]. *Journal of Applied Polymer Science* © 2003. Reprinted with permission of John Wiley & Sons, Inc.)

as the aging proceeds, resulting in a reduced molecular mobility, and hence a decrease in mechanical properties; damping factor tan δ is reduced as well [79].

11.9 SUMMARY

Starch is a highly hydrophilic polymer that consists of linear amylose and highly branched amylopectin. Starch can be easily isolated using environmentally friendly processing technologies as described in Chapter 3. Besides food and paper uses, starch has many application potentials as thermoplastics. However, starch is brittle and hydrophilic. Another drawback of starch is its low flowability in extrusion and molding process. Blending with hydrophobic polymers, such as poly(lactic acid) (PLA), starch becomes a useful material for disposable uses. PLA is a biobased synthetic polymer derived from starch or sugar-based materials and has about 60 MPa tensile strength and 5% elongation. Simply blending of starch and PLA by mechanical method, mechanical properties of PLA are reduced to 50%. Small amounts of coupling reagents are helpful to improve mechanical properties. MDI and MA are two identified coupling reagents for the PLA and starch system. At 0.5% MDI, for example, a blend with 45% starch and 54.5% PLA had about 65 MPa tensile strength and 4.6% elongation. Triethyl citrate (TC) and

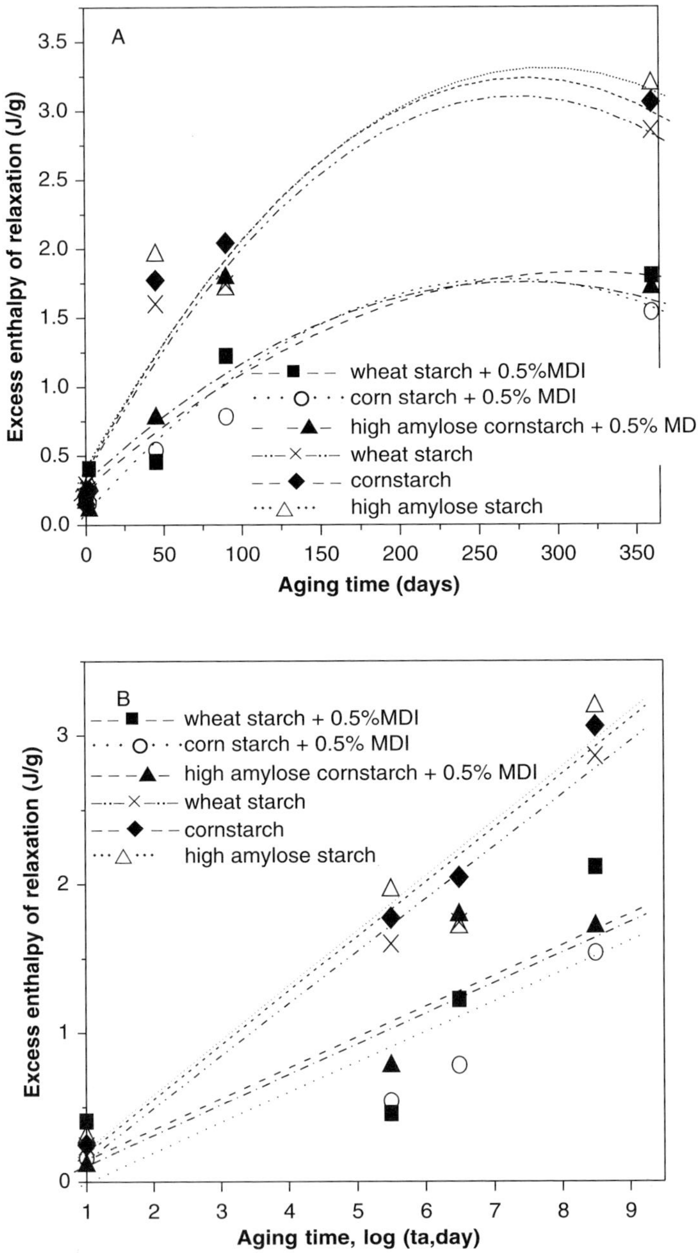

FIGURE 11.30 (A) Excess enthalpy relaxation and (B) long aging time as affected by aging time for a blend of PLA containing 45% starch without or with 0.5 wt% MDI aged at 25 °C and 50% relative humidity for various days. (*Source:* From Wang *et al.* [80]. *Journal of Applied Polymer Science* © 2003. Reprinted with permission of John Wiley & Sons, Inc.)

TABLE 11.7 Mechanical properties of cornstarch and PLA blends with 0.5% MDI stored at 25°C and 50% relative humidity.

	Native Cornstarch		
Storage Time (days)	Tensile Strength (MPa)	Elongation (%)	Modulus (GPa)
2	58.8	5.3	1.72
45	60.6	4.8	1.71
90	57.8	5.0	1.64
360	53.1	4.4	1.50

Source: Data from Wang *et al.* [80].

dioctyl maleate (DOM) are compatible plasticizers of the starch and PLA system, so that the mechanical properties of the starch and PLA blends can be adjusted to meet application specifications. Application potential of the starch/PLA plastics includes packaging rigid or flexible containers, agriculture mulch film, packaging foam, service utensils, and other disposable items for single or short uses.

REFERENCES

1. Whistler, R. L.; BeMiller, J. N. *Carbohydrate Chemistry for Food Scientists*, American Association of Cereal Chemists, St. Paul, MN; 1997.
2. Thomas, D. J.; Atwell, W. A. *Starches*, American Association of Cereal Chemists, St. Paul, MN; 1999.
3. Shi, Y.; Seib, P. A. *Is Wheat Unique*, Pomeranz, Y., Ed.; American Association of Cereal Chemists, St. Paul, MN; 1988.
4. Eliasson, A. C.; Gudmundsson, M. In *Carbohydrates in Food*, Eliasson, A. C., Ed.; Marcel Dekker, New York; 1996, pp. 431–503.
5. Zhong, Z.; Sun, X. Thermal Behavior and Phase Diagram of Cornstarch, *J. Food Eng.*, 2005, **69**, 453–459.
6. French, D. Organization of Starch Granules. In *Starch: Chemistry and Technology*, Whistler, R. L.; BeMiller, J. N.; Paschall, E. F., Eds.; Academic Press, London; 1984, pp. 183–247.
7. Roos, Y. H. *Phase Transitions in Foods*, Academic Press, New York; 1995.
8. Burt, D. J.; Russell, P. L. Gelatinization of Low Water Content Starch–Water Mixtures, *Starch*, 1983, **35**, 354–360.
9. Chungcharoen, A.; Lund, D. B. Influence of Solutes and Water on Rice Starch Gelatinization, *Cereal Chem.* 1987, **64**, 240–243.
10. Donovan, J. W. Phase Transitions of the Starch–Water System, *Biopolymer* 1979, **18**, 263–275.
11. Donovan, J. W.; Mapes, C. J. Multiple Phase Transitions of Starches and Nägeli Amylodextrins, *Starch*, 1980, **32**, 190–193.
12. Eliasson, A. C. Effect of Water Content on the Gelatinization of Wheat Starch, *Starch* 1980, **32**, 270–272.
13. Kugimiya, M.; Donovan, J. W.; Wong, R. Y. Phase Transitions of Amylose–lipid Complexes in Starches: A Calorimetric Study, *Starch* 1980, **32**, 265–270.

14. Jenkins, P. J.; Donald, A. M. Gelatinization of Starch: A Combined SAXS/WAXS/DSC and SANS Study, *Carbohydr. Res.* 1998, **308,** 133–147.
15. Maurice, T. J.; Slade, L.; Sirett, R. R.; *et al.* Polysaccharide–Water Interaction: Thermal Behavior of Rice Starch. In *Influence of Water on Food Quality and Stability,* Simatos, D.; Multon, S. L., Eds.; Nijhoff M. Publ., Dordrecht, Netherlands; 1985, p. 211.
16. Zeleznak, K. J.; Hoseney, R. C. The Glass Transition in Starch, *Cereal Chem.* 1987, **64,** 121–124.
17. Arvanitoyannis, I.; Biliaderis, C. G.; Ogawa, H.; *et al.* Biodegradable Films Made from Low-Polyethylene (LDPE), Rice Starch and Potato Starch for Food Packing Applications: Part 1, *Carbohydr. Polym.* 1998, **36,** 89–104.
18. Bagley, E.B.; Fanta, G.R.; Burr, R.C. *et al.* 1977. Graft Copolymer of Polysaccharides with Thermoplastic Polymers, *Polym. Eng. Sci.* **17,** 311–316.
19. Griffin, G. J. L. *Particular Starch Based Products in Chemistry and Technology of Biodegradable Polymers,* Griffin, G. J. L., Ed.; Chapman & Hall, London; 1994, pp. 18–47.
20. Griffin, G. J. L. Starch Polymer Blends, *Polym. Degrad. Stab.* 1994, **45,** 241–247.
21. Otey, F. H.; Mark, A. M.; Mehltretter, C. L.; *et al.* Starch Based Film for Degradable Agricultural Mulch, *Ind. Eng. Chem. Prod. Res. Develop.* 1974, **13,** 90.
22. Patil, D. R.; Fanta, G. R. Synthesis and Processing of Graft Copolymers from Cornstarch and Methyl Acrylate: Physical and Mechanical Properties, *Starch,* 1994, **46,** 142–146
23. St-Pierre, N.; Favis, B. D.; Ramsay, B. A.; *et al.* Processing and Characterization of Thermoplastic Starch/Polyethylene Blends, *Polymer* 1997, **38,** 647–655.
24. Vaidya, D. R.; Bhattacharya, M. J. Properties of Blends of Starch and Synthetic Polymers Containing Anhydride Groups, *J. Appl. Polym. Sci.* 1994, **52,** 617–628.
25. Sinclair, R. G. The Case for Polylactic Acid as a Commodity Packaging Plastic, *J. Macromolec. Sci.: Pure Appl. Chem.* 1996, **A33,** 585–597.
26. Ke, T. Y.; Sun, X. S. Physical Properties of PLA and Starch Composites with Various Blending Ratios, *Cereal Chem.* 2000, **77**(6), 761–768.
27. Fisher, E. W.; Sterzel, H. J.; Wegner, G. *Kolloid Z.Z. Polym.* 1972, **251,** 980–990.
28. Ke, T.; Sun, X. *J. Appl. Polym. Sci.* 2003, **89,** 1203–1210.
29. Nicolais, L.; Narkis, M. Stress–Strain Behavior of Styrene-Acrylonitrile/Glass Bead Composites in the Glass Region, *Polym. Eng. Sci.* 1971, **11,** 194–199.
30. Willett, J. L. Mechanical Properties of LDPE/Granular Starch Composites, *J. Appl. Polym. Sci.* 1994, **54,** 1685–1695.
31. Nielsen, L. E. *Mechanical Properties of Polymers and Composites,* Marcel Dekker, New York; 1974, pp. 386–414.
32. Ke, T. Y.; Sun, X. S. Blending of PLA and Starches Containing Varying Amylose Content, *J. Appl. Polym. Sci.* 2003, 89(13):3639–3646.
33. Tester, R. F.; Morrison, W. R. *Cereal Chem.* 1990, **67**(6), 551.
34. M. Avrami, *J. Chem. Phys.* 1939, **7,** 1103.
35. Wang, H.; Sun, X.; Seib, P. *J. Appl. Polym. Sci.* 2001, **82,** 1761–1767.
36. Rocco, A. M.; Pereira, R. P.; Felisberti, M. I. *Polymer* 2001, **42**(12), 5199.
37. Kim, S. J.; Shin, B. S.; Hong, J. L.; *et al. Polymer* 2001, 42(9), 4073.
38. Pagnoulle, C.; Jerome, R. *Polymer* 2001, **42**(5), 1893.
39. Son, Y.; Ahn, K. H.; Char, K. *Polym. Eng. Sci.* 2000, **40**(6), 1385.
40. Dedecker, K.; Groeninckx, G. *Pure Appl. Chem.* 1998, **70**(6), 1289.
41. Cho, K.; Li, F. *Macromolecules* 1998, **31,** 7495.
42. Dieteroch, D.; Grigat, E.; Hahn, W. In *Polyurethane Handbook*, Oertel, G., Ed.; Hanser Publishers, New York; 1985, p 7.
43. Yakabe, Y. Fate of Methylenediphenyl Diisocyanate and Toluene Diisocyanate in the Aquatic Environment, *Env. Sci. Technol.* 1999, **33**(15), 2579–2583.
44. Gilbert, D. S. Fate of TDI and MDI in Air, Soil, and Water; International Isocyanate Institute, Inc., Parsippany, NJ; 1987.

45. Owen, S.; Masaoka, M.; Kawamura, R.; *et al. J. Macromolec. Sci.: Pure Appl. Chem.* 1995, **A32,** 843.
46. Cowen, W. F.; Gastinger, A.M.; Spanier, C. E.; *et al. Environ. Sci. Technol.* 1998, **32,** 598.
47. Zhitinkina, A. K.; Shoshtaeva, M. V. Reaction of Hydroxyl-Containing Compounds of Varying Acidity with Diisocyanate, *Sin. Fiz. Khim. Polim.* 1968, abstract No. 5117.
48. Griffin, G. J. L. In *Fillers and Reinforcements for Plastics,* Deanin, R. D.; Schott, N. D., Eds.; American Chemical Society, Washington, DC; 1974, Advances in Chemistry Series Vol. 134, p 159.
49. Wu, S. *Polymer Interface and Adhesion,* Marcel Dekker, New York; 1982.
50. Schauerte, K. In *Polyurethane Handbook,* Oertel, G., Ed.; Hanser Publishers, New York; 1985, p. 62.
51. Dufresne, A.; Cavaille, *J. J. Polym. Sci. B: Polym. Phys.* 1998, **36,** 2211.
52. O'Shaughnessy, B.; Sawhney, U. *Phys. Rev. Lett.* 1996, **76,** 3444.
53. French, D. In *Starch: Chemistry and Technology,* 2nd ed., Whistler, R. L.; Bemiller, J. N.; Paschall, E. F., Eds.; Academic Press, New York; 1984, p. 183.
54. Carlson, D.; Nie, L.; Narayan, R.; *et al. J. Appl. Polym. Sci.,* 1999, **72,** 477.
55. Zhang, J. F.; Sun, X. S. Mechanical Properties of PLA/Starch Composites Compatibilized by Maleic Anhydride, *Biomacromolecules,* 2004, **5**, 1446–1451
56. Vaidya, U. R.; Bhattacharya, M. *J. Appl. Polym. Sci.* 1994, **52,** 617.
57. Dedecker, K.; Groeninckx, G. *Macromolecules* 1999, **32,** 2472.
58. Colonna, P.; Tayeb, J.; Mercier, C. In *Extrusion Cooking,* Mercier, C.; Linko, P.; Harper, J. M., Eds.; American Association of Cereal Chemists, St. Paul, MN; 1989, p. 247.
59. Lay, G.; Rehm, J.; Stepto, R.F.; *et al.* Polymer Compositions Containing Destructurized Starch, U.S. Patent 5,095,054; 1992.
60. Tomka, I. A Thermoplastically Proccessible Starch and a Process for Making It, PCT Patent WO 90/05161; 1989.
61. Wittwer, F.; Tomka, I. Polymer Composition for Injection Molding, U.S. Patent 4,673,438; 1987.
62. Gruber, P. R.; Kolstad, J. J.; Hall, E. S.; *et al.* U.S. Patent 5,539,081; 1996.
63. Ke, T. Y.; Sun, X. S. Effects of Moisture Content and Heat Treatment on Physical Properties of Starch and Poly(Lactic Acid) Blends, *J. Appl. Polym. Sci.* 2001, **81,** 3069–3082.
64. Ke, T.; Sun, X. Thermal and Mechanical Properties of Poly(Lactic Acids)/Starch/MDI Blending with Triethyl Citrate, *J. Appl. Polym. Sci.* 2002, **88**(13), 2947–2955.
65. Simmons, S.; Thomas, E. L. Structural Characteristics of Biodegradable Thermoplastic Starch/Poly(Ethylene-Vinyl Alcohol) Blends, *J. Appl. Polym. Sci.* 1995, **58,** 2259–2285.
66. Gachter, R.; Muller, H. *Plastics Additives Handbook,* 3rd ed., Hanser Publishers, New York; 1990, p. 398.
67. Labrecque, L.V.; Kumar, R. A.; Dave, V.; *et al.* Citrate Esters as Plasticizers for Poly(Lactic Acid), *J. Appl. Polym. Sci.* 1997, **66,** 1507–1513.
68. Ke, T.; Sun, X. Thermal and Mechanical Properties of Poly (Lactic Acid) and Starch Blends with Various Plasticizers, *Trans. ASAE* 2001, **44**(4), 945–953.
69. Lawton, J. W.; Fanta, G. F. Glycerol-Plasticized Films Prepared from Starch-Poly(Vinyl Alcohol) Mixtures: Effect of Poly(Ethylene-co-Acrylic Acid), *Carbohydr. Polym.* 1994, **23,** 275–280.
70. Zhang, J. F; Sun, X. Physical characterization of coupled poly(lactic acitd)/starch/maleic anhydride blends plasticized by triethyl citrate, *Macromolecular Bioscience,* 2004, **4**: 1053–1060.
71. Yu, J.; Chen, S.; Gao, J.; *et al.* A Study on the Properties of Starch/Glycerin Blend, *Starch/Stärke* 1998, **50,** 246–250.
72. Arvanitoyannis, I.; Nakayama, A.; Aiba, S. Edible Films Made from Hydroxypropyl Starch and Gelatin and Plasticized by Polyols and Water, *Carbohydr. Polym.* 1998, **36,** 105–119.

73. Younes, H.; Daniel, C. Phase Separation in Poly(Ethylene Glycol)/Poly(Lactic Acid) Blends, *Eur. Polym.* J. 1998, **24**(8), 765–773.
74. Sheth, M.; Kumar, R. A.; Dave, V.; *et al.* Biodegradable Polymer Blends of Poly(Lactic Acid) and Poly(Ethylene Glycol), *J. Appl. Polym. Sci.* 1997, **66,** 1495–1505.
75. Zhang, J. F.; Zheng, Q.; Yang, Y. Q.; *et al. J. Appl. Polym. Sci.* 2002, **83,** 3112–3116.
76. Zhang, J. F.; Sun, X. Mechanical and thermal properties of poly(lactic acid)/starch blends with dioctyl maleate, *J. Appl. Polym. Sci.*, 2004, **94**, 1697–1704.
77. Parker, S. *Biomaterials* 1998, **19,** 1695.
78. Chartoff, R. P. In *Thermal Characterization of Polymeric Materials,* Vol. 1, *Thermoplastic Polymers,* Turi, E. A., Ed.; Academic Press, New York; 1997, p. 549.
79. Celli, A.; Scandola, M. *Polymer* 1992, **33,** 2699.
80. Wang, H.; Sun, X.; Seib, P. Mechanical Properties of Starch/Poly(Lactic Acid) Blends as Affected by Physical Aging, *J. Appl. Polym. Sci.* 2003, **90**(13), 3683–3689.

12

Bio-Based Composites from Soybean Oil and Chicken Feathers

Richard P. Wool

Novel bio-based composite material that is suitable for electronic, automotive, and aeronautical applications can be developed from soybean oils and avian feather fibers. The avian feathers can be derived from poultry, such as chicken and turkey, and birds in general. The feather fiber, when removed from the quill, can be used as the reinforcement in composites in its natural state, or carbonized by pyrolysis to give higher performing fiber. This environmentally friendly, low-cost composite can be a substitute for petroleum-based composite materials. Keratin fibers (KF) are a hollow, light, and tough material and are compatible with several soybean (S) resins, such as acrylated epoxidized soybean oil (AESO). In this chapter, we show that the new KFS composite is lightweight and its density can be less than 1 g/cm^3. The dielectric constant of the composite material is unusually low, in the range of 1.7–2.7, which makes it suited to printed circuit boards for electronic materials. The incorporation of keratin fibers in the soybean oil polymer enhanced the mechanical properties such as storage modulus, fracture toughness, and flexural properties. However, it will be shown that when the feather fiber is carbonized by specific process pathways, the resulting low-cost carbon fiber can have much improved mechanical properties.

12.1 Introduction

Composite materials from plant oil and keratin feather fibers offer both economic and environmental advantages [1–7]. The use of renewable materials contributes to global sustainability and the diminution of global warming gases. As the number of applications of composite materials continues to increase, an alternative source of petroleum-based composites becomes important. Chapter 4 demonstrated that the resins from soybean oil can be a substitute for liquid molding resins, such as unsaturated polyester resins, vinyl esters, and epoxy resins [1, 3, 5]. The keratin fiber is also a possible replacement for synthetic reinforcing fibers [8, 9].

Soybean resins are based on triglycerides, which are the major component of plant and animal oils. Triglycerides are composed of three fatty acids joined at a glycerol juncture [10], as discussed in Chapter 4. Although unmodified triglycerides do not readily polymerize, the chemical functionality necessary to cause polymerization can be easily added to the triglycerides. The active sites on the modified triglycerides can be used to introduce polymerizable groups, using the same synthetic techniques that are applied in the synthesis of petrochemical-based polymers [1, 3, 5–7]. With suitable chemical functionalization and viscosity, the molding process of soybean resins is similar to that of conventional thermosetting liquid molding resins, using resin transfer molding (RTM), vacuum-assisted resin transfer molding (VARTM), sheet molding compound (SMC), and so forth. The triglyceride-based materials display the necessary rigidity and strength required for structural applications. Also, some, but not all, soybean resins can be made to be biodegradable [11].

The U.S. poultry industry generates more than 1 billion kilograms of feathers annually as a by-product of poultry production [12]. Disposal of the feather waste is expensive and difficult. For example, poultry waste is burned, buried, or recycled into animal feed. These methods are environmentally unsound and restricted. A more expensive disposal method is to produce a low-quality protein animal feed; demand for this product is low. Some efforts have been made to develop processes for making fiber materials from feather wastes [8, 9]. Feathers are made from the protein keratin and there are two forms of microcrystalline keratin in feathers: the fiber and the quill. The thermal energy required to perturb the molecular order of the quill is lower than that required for the fiber [13]. Thus, the feather fibers, with an α-helical structure at the molecular level, are light and tough enough to withstand both mechanical and thermal stress. Due to the hollow structure of the keratin fibers, a given volume of the fiber innately contains a significant volume of air, resulting in a low density (SG $\sim$ 0.80) and low dielectric constant ($k \sim 1.7$) suitable for composite and electronic materials [14]. The use of avian feathers in composites as reinforcing fibers offers an environmentally benign solution for feather disposal, and also presents to poultry producers the option of reducing waste disposal costs and gaining a profit from feather waste.

The overall objective in this chapter is to discuss the development of composites from soybean (S) oil (Figure 12.1) and keratin feather fibers

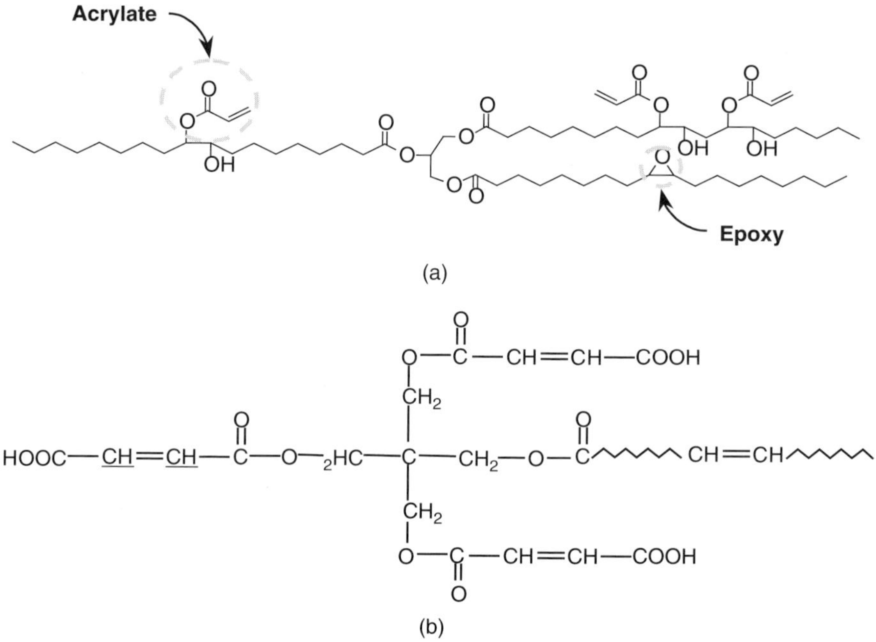

FIGURE 12.1 The molecular structure of (a) a typical acrylated epoxidized soybean oil (AESO) [1] and (b) soybean oil pentaerythritol glyceride maleates (SOPERMA) [16].

(KF) and their properties. The development of low-*k* dielectric materials is considered to be one of the main issues in modern high-speed microelectronics. The strength, stiffness, vibration damping, and density are important properties in automotive, trucking, farming equipment, civil infrastructure, defense, aerospace, housing construction, and electronic materials applications.

12.2 PROCESSING OF CHICKEN FEATHER FIBER COMPOSITES

12.2.1 WETTING AND COMPATIBILITY TEST OF KFS COMPOSITES

A critical issue for composite fabrication and good property development is the ability of the resin to wet, or be compatible with, the fibers. Keratin fibers have a typical diameter of 6 μm and length of 8 mm, with an aspect ratio of about 1000, as shown in Figure 12.2. The nodes and hooks on the hollow keratin fibers can improve the structural properties and increase the surface area in the composite. Chemically, the keratin fibers are composed of

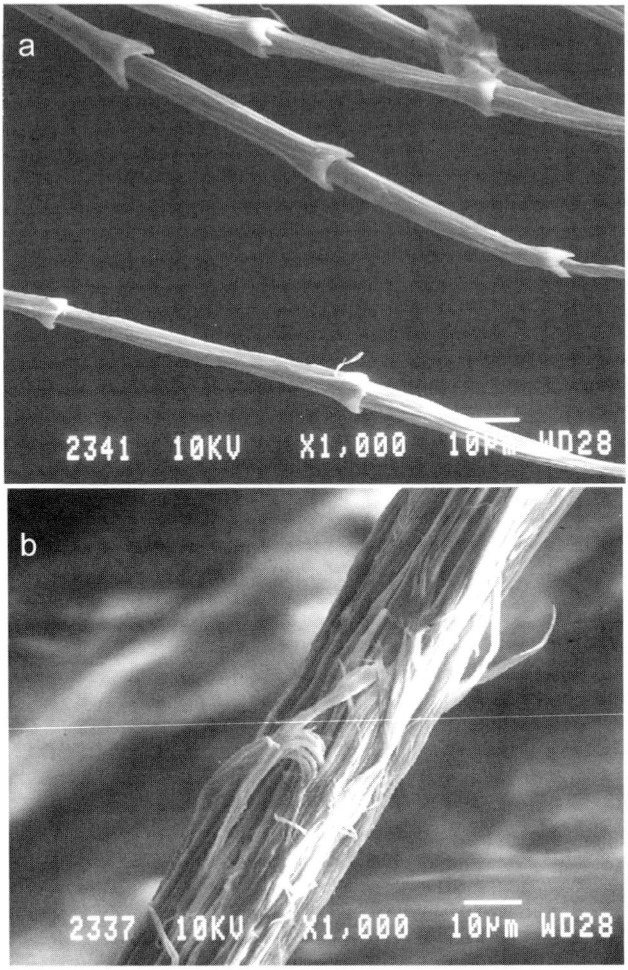

FIGURE 12.2 Scanning electron microscope (SEM) micrographs of feathers: (a) keratin fibers and (b) a small quill.

amino acids, the building blocks of protein. The amino acid sequences that cause the primary α-helical structure of feather keratin are similar among several avian species [13]. The preparation of the fiber mats from raw feathers includes washing, sanitizing, and mechanical shredding/shearing. However, the original molecular properties of the fiber are preserved during these mat-forming procedures [8, 13].

Figure 12.3 shows the wetting test of both AESO resin and water on the KF mat. The KF mat is made by a process similar to wet paper matting and was supplied by Tyson Food. The AESO resin droplet spread out easily on

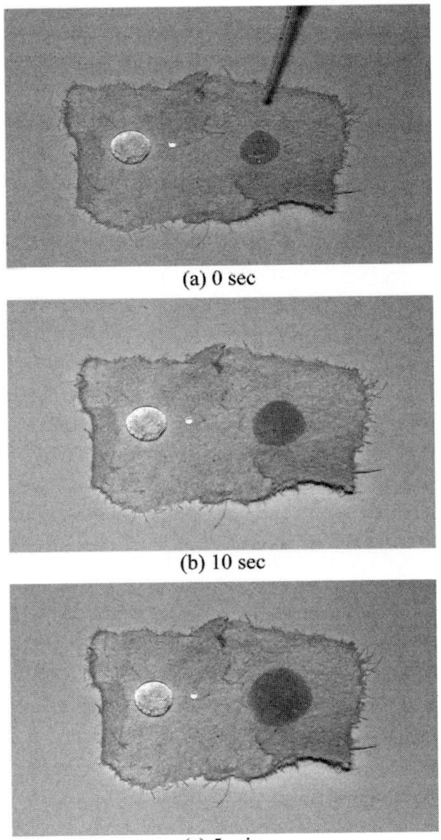

(a) 0 sec

(b) 10 sec

(c) 5 min.

FIGURE 12.3 The wetting test of water (left) and AESO resin (right) on the KF mat: (a) 0 sec, (b) 10 sec, and (c) 5 min.

the mat, whereas the water droplet did not during the 5-min time interval. The water droplet continued to remain for several hours until it evaporated. The AESO drop spread very rapidly, essentially doubling its wetted area in 10 sec (Figure 12.3b), before attaining its final wetted area, which is a competition between spreading and imbibition of the drop in the KF mats, as discussed in Ref. [15]. The degree of resin spreading and imbibition in the mat depends on a number of parameters, including the surface tension and rheology of the resin, the surface chemistry, and the porosity of the mat. It is further known that for the long-time spreading in quasi-static regimes, adsorption kinetics and dynamic interfacial tensions are the factors that control wetting [17]. Note in Figure 12.3(c), the very small drop of water to the right of the larger water drop, which also has not wetted the keratin fibers.

Therefore, keratin fibers are compatible with AESO resin and are hydrophobic. However, the SOPERMA resin has high viscosity and a hydrophilic structure due to a high acid number (237 mg KOH/g) [15], as shown in Figure 12.1(b). A small drop of SOPERMA resin (described in Chapter 4) spread in the KF mat but the imbibition of the resin during a VARTM process did not take place due to less compatibility. Also, the keratin fibers contain a significant volume of air in the hollow microcrystalline structure. The wetting inside of the fiber is a critical issue for the low dielectric constant and the density of the composites.

12.2.2 EFFECTS OF INITIATORS AND CURE CONDITIONS

The objective of this experiment was to investigate the effects of chemical initiators and cure temperature on the properties of soybean oil-based KF composites. The free-radical, chain growth copolymerization reactions (or curing) in soybean oil-based resin can be carried out either by thermal decomposition of the free-radical initiators at elevated temperatures, or by redox decomposition of the free-radical initiators using a metal promoter (accelerator) at low temperatures. For room temperature (RT) curing, the concentration of the initiators and an accelerator for the resins was varied to check the curability in 24 h. The results are shown in Table 12.1 for AESO and Table 12.2 for SOPERMA. AESO was cured with a cumyl hydroperoxide initiator with various amounts of accelerator, but was not cured by Methyl Ethyl Ketone (MEK) peroxides. The SOPERMA resin, however, was cured by MEK peroxides, but was not cured by cumyl hydroperoxides. With a low concentration of initiator and accelerator, a longer time is needed to complete room temperature curing of the resins. As the extent of cure decreases, the glass transition temperature and modulus of the polymer decrease [18].

The glass transition temperature and storage modulus for RT and high-temperature (HT) cured AESO and SOPERMA resins at various concen-

TABLE 12.1 Curability test of AESO resin at room temperature.

Cumyl Peroxides 2.5%	CoNap 0.3%	Partially Cured
3.0%	0.8%	Cured
3.5%	0.8%	Cured
MEK peroxides 2.0%	CoNap 0.5%	Not cured
2.0%	0.6%	Not cured
2.5%	0.6%	Not cured
3.0%	0.8%	Not cured
3.0%	1.0%	Not cured

trations of initiator and accelerator are given in Tables 12.3 and 12.4, respectively. Among the RT-cured polymers, the AESO properties were optimal at 3.0 wt% of cumyl hydroperoxide and 0.8 wt% of CoNap (after curing), and those for SOPERMA were optimal at 3.0 wt% of MEK peroxides and 0.8 wt% of CoNap (after curing).

The glass transition temperature and storage modulus of the HT-cured polymer are higher than those of the RT-cured sample for both resins. This trend was also observed on the KFS composites. At high temperature, higher rates of reaction and higher conversions of double bonds in the soybean resin are expected. However, a high-temperature reaction can present some process difficulties. It was hard to keep a vacuum during a high-temperature

TABLE 12.2 Curability test of SOPERMA resin at room temperature.

Cumyl Peroxides 2.5%	CoNap 0.3%	Not Cured
3.0%	0.8%	Not cured
3.5%	0.8%	Not cured
MEK peroxides 2.0%	CoNap 0.5%	Not cured
2.0%	0.6%	Partially cured
2.5%	0.6%	Partially cured
3.0%	0.8%	Cured
3.0%	1.0%	Cured

TABLE 12.3 T_g and storage modulus of RT- and HT-cured AESO.

Cure Condition (Trigonox, %: CoNap, %)	T_g(°C)	E' at 40°C (GPa)
3.0 : 0.8, RT cured	66	1.247
3.0 : 0.8 (RT cured), after curing	70	1.313
3.5 : 0.8, RT cured	64	1.095
3.5 : 0.8 (RT cured), after curing	69	1.215
HT cured	79	1.642

TABLE 12.4 T_g and storage modulus of RT- and HT-cured SOPERMA.

Cure Condition (MEKP, %: CoNap, %)	T_g(°C)	E' at 40°C (GPa)
2.0 : 0.6 (RT cured), after curing	121	0.966
2.5 : 0.6 (RT cured), after curing	122	1.019
3.0 : 0.8, RT cured	107	0.650
3.0 : 0.8 (RT cured), after curing	128	1.039
3.0 : 1.0, RT cured	113	0.693
3.0 : 1.0 (RT cured), after curing	129	1.027
HT cured	141	1.158

VARTM process because of the evaporation of the styrene monomer. Also, the styrene vapor made voids or bubbles inside the composites during high-temperature curing, leading to deterioration of mechanical properties. Generally, the selection of cure condition is motivated by processing and economic considerations. The low-temperature process can reduce the manufacturing cost. The focus of this study will be on the RT cure reaction with 3.0 wt% of initiator and 0.8 wt% of accelerator for both resins.

12.2.3 EFFECTS OF THE COMONOMER CONCENTRATION

Soybean oil-based resins can be blended with a comonomer such as styrene to improve their processability and to control the composite properties until they reach a range acceptable for structural applications. Styrene is used as a comonomer in commercial thermosetting resins and it cures well with the soybean resins. By varying the amount of comonomer, it is possible to produce composite materials with a different modulus, strength, and glass transition temperature, as shown in Chapter 4 [19]. Figure 12.4 shows the storage modulus of RT-cured AESO polymer at various styrene contents, as a function of temperature. The storage modulus at low temperature increased with an increasing amount of styrene. In this rubbery region, the modulus is dominated by the copolymer effect of the higher modulus of polystyrene. However, at higher temperature (the plateau region), thermoset polymers behave according to the rubber elasticity theory, and the cross-link density, ν, is related to the modulus of elasticity, E [20]:

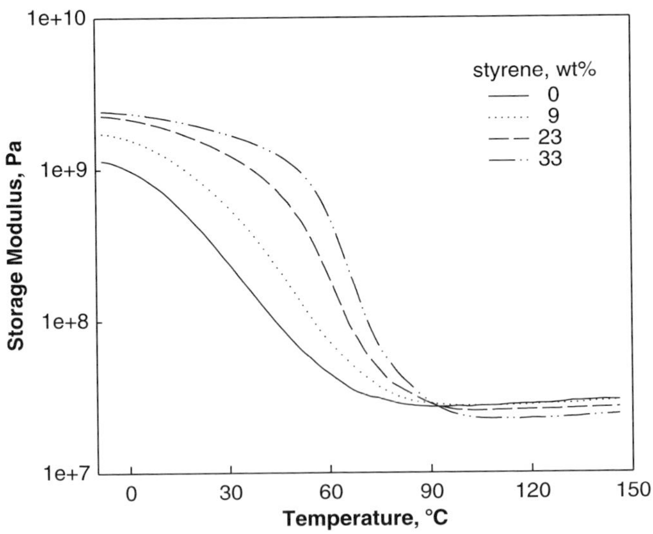

FIGURE 12.4 Storage modulus of RT-cured AESO at various styrene contents.

$$E = 3\nu RT, \tag{12.1}$$

where R is the ideal gas constant and T is the absolute temperature. The modulus determined from dynamic mechanical analysis is a measure of the effective cross-link density of polymers. Based on the fatty acid distribution of soybean oil, it is also possible to estimate the cross-linking ability of the triglycerides. The number of cross-links in the network is a function of the composition of the resin and the functionality of the triglycerides. It is already known that AESO (Ebecryl 860) contains about 3.4 acrylates per triglyceride and the cross-linking functionality is 6.8 in the network for every triglyceride [21]. Therefore, if all of the acrylates on the triglycerides react, the cross-link density in the system is calculated [22] by

$$\nu = \frac{\text{Mass fraction(AESO)} \times 6.8 \times \rho}{\text{MW}_{\text{AESO}}}, \tag{12.2}$$

where ρ is the density of the polymer and MW_{AESO} is the molecular weight of the AESO (1186 g/mol [22]). The calculated values are compared with experimentally determined cross-link densities in Figure 12.5. The cross-link density decreased linearly as the styrene mass fraction increased. As more styrene is introduced into the system, the cross-link density is lowered because styrene is a linear chain extender and reduces the amount of cross-linking; thus, the modulus is lower in the plateau region of Figure 12.4. The theoretical cross-link densities are higher than the experimental values, which can be explained by intramolecular cyclization of triglycerides [21]. The

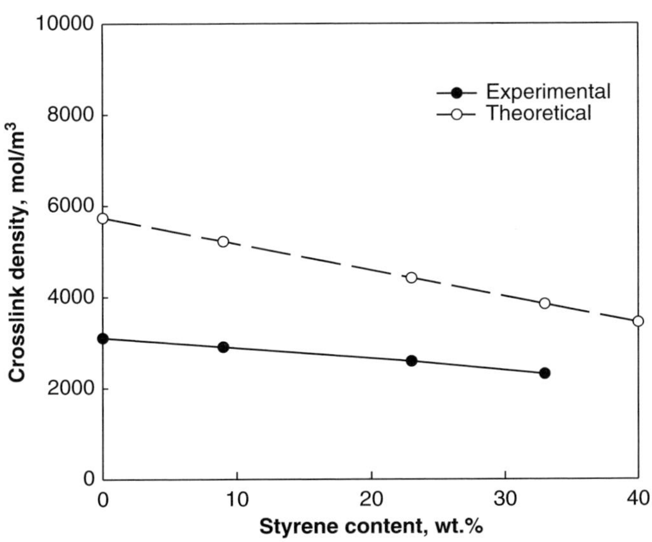

FIGURE 12.5 Comparison of experimental cross-link density of AESO resin with the theoretical values as a function of styrene content.

intramolecular cyclization is the reaction of a functional group on a triglyceride with another functional group on the same triglycerides. The plasticization effect [22] by saturated fatty acids during the thermomechanical measurement can also be a reason for lower experimental cross-link densities.

In the case of random copolymers, or miscible blends of polymers, the glass transition temperature (T_g) of the copolymer or blend can be expressed by the Fox–Flory equation [23]:

$$\frac{1}{T_g} = \frac{w_1}{T_{g1}} + \frac{w_2}{T_{g2}}, \tag{12.3}$$

where w_1 and w_2 are the weight fraction of components 1 and 2 with T_{g1} and T_{g2}, respectively. The T_g of polystyrene is 373 K [24] and that of the pure triglyceride polymer is measured. Figure 12.6 shows the experimental T_g values measured by Dynamic Mechanical Analysis (DMA) and calculated values using the Fox–Flory equation. The T_g was increased for the samples with an increase in styrene because the aromatic nature of styrene imparted rigidity to the network. Also, the samples had slightly higher T_g values than predicted. As comonomer was added, the efficiency of the polymerization increased, causing an increase in T_g. The cross-linking contribution can also be considered a reason behind the T_g increase [19]. The properties of these polymers can be controlled by different comonomers, by the degree of

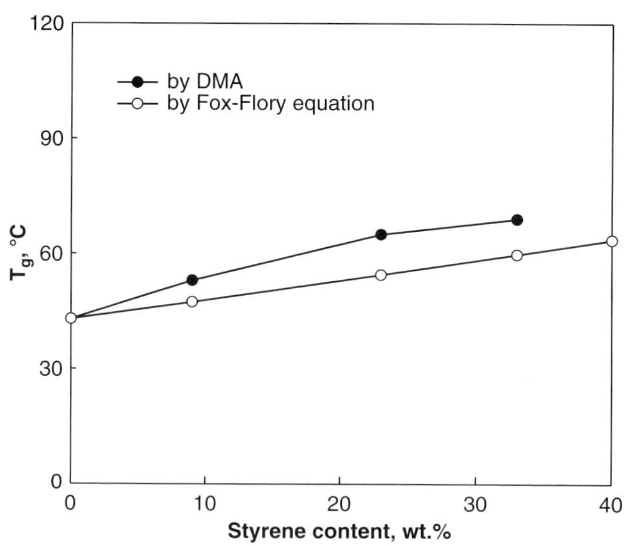

FIGURE 12.6 Glass transition temperature of RT-cured AESO as a function of styrene content.

chemical functionality of the triglycerides, and by the extent of reaction during cure [23].

12.2.4 VIBRATION-DAMPING PROPERTIES

Figure 12.7 represents the vibration-damping properties of RT-cured AESO resins at various concentrations of styrene comonomer and AESO-30 wt% KFS composite, as a function of frequency. The passive damping materials and acoustic complex techniques are frequently used in various applications in automotive and aerospace structures to reduce vibrations and noise. As shown in Figure 12.7, the material damping, or its ability to dissipate vibrational energy, is reduced with increasing styrene concentration in the resin. The triglycerides consist mostly of long aliphatic fatty acids that differ from the aromatic styrene monomer. The aromatic styrene monomer is much more restrictive to movement than the flexible aliphatic chains of the triglycerides. A larger number of relaxation modes are also available in triglycerides due to the broad fatty acid distribution [22]. As the amount of styrene in the copolymer increases, the chain mobility in the networks decreases, the modulus increases, and, hence, the vibration damping decreases. The major damping mechanisms in vibration are due to the extensional and shear deformations of the structure's viscoelastic material [25].

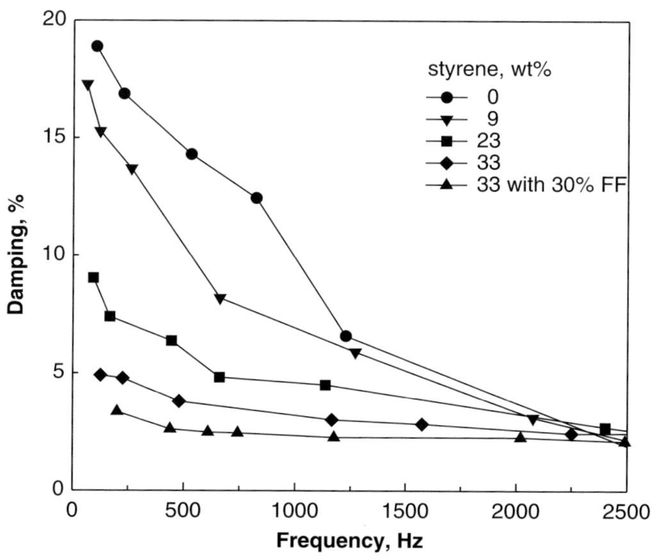

FIGURE 12.7 Vibration damping properties of RT-cured AESO at various amounts of styrene and AESO-30 wt% KF composite.

The damping property of the AESO-30 wt% KFS composite is also shown in Figure 12.7. The material damping is lowered with the addition of keratin fibers. The increase in modulus (or stiffness) caused a corresponding reduction of damping properties of the composite material. The increase in modulus of the KFS composites is discussed in the thermomechanical property section.

12.2.5 BULK DENSITY OF THE COMPOSITES

Figure 12.8 shows the experimental and theoretical density of RT-cured AESO and SOPERMA composites as a function of the keratin fiber content. The theoretical density values are calculated using the mixing rule,

$$\rho = w_1\rho_1 + w_2\rho_2, \tag{12.4}$$

where w_1 and w_2 are the weight fraction of components 1 and 2 with ρ_1 and ρ_2, respectively. The density of the hollow keratin fiber is 0.80 g/cm^3, and those of AESO and SOPERMA resins are 1.08 and 1.13 g/cm^3, respectively. The density of the composites decreased (Figure 12.8) with an increase in the keratin fiber concentration for both resins. This contrasts with the density of typical composites from synthetic reinforcing fibers. The keratin fibers are hollow and light materials, and contain air in the hollow structure. Lightweight materials make a significant impact on fuel consumption for automotive and trucking applications. Also, the density of the composite can be

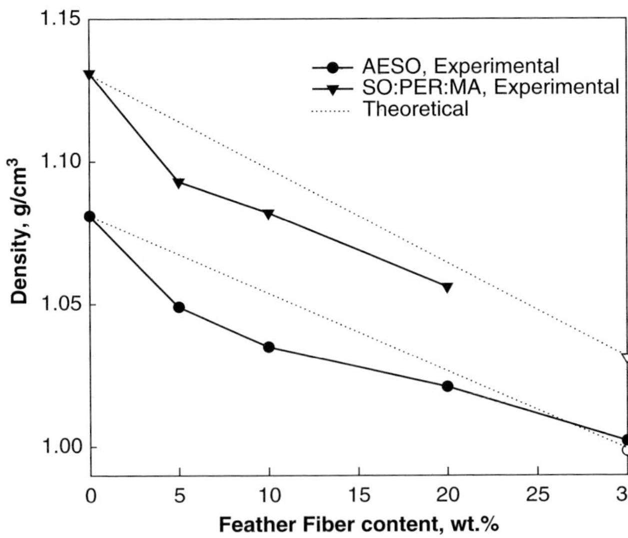

FIGURE 12.8 The density of RT-cured AESO and SOPERMA composites.

made to be less than 1 g/cm^3 at about 30 wt% of keratin fibers for AESO resin and at about 40% for SOPERMA. As shown in Figure 12.8, the experimentally measured values are lower than the theoretical density for physically mixed samples (5, 10, and 20 wt%), but the measured density is slightly higher for a VARTM-processed sample. During physical mixing, trapped and entrained air between the keratin fibers remains in the composite, resulting in a lower density than expected. However, in the case of the VARTM sample, a vacuum process sucks the air between the fibers out and some of the hollow fibers are filled by resin infusion. The experimental density for the AESO-30 wt% KFS composite (prepared by the VARTM process) was 1.001 g/cm^3, which is the theoretical value for the 28.4 wt% KF composite. This means that about 5% filling of the hollow keratin fibers occurred during the VARTM process, but the rest of the fibers (95%) in the composite still contained a significant volume of air.

The AESO-30 wt% KFS composite was heated in water to determine if there was air inside the composite and to explore the unique possibility of convective cooling, as shown in Figure 12.9. One can see lots of air bubbles on the surface of the composite at elevated temperatures, and the number of air bubbles from the composite increased with increasing temperature. The blank (AESO, no KF) sample did not show similar behavior, which could have been explained by simple bubble nucleation phenomena on the surface of the resin. The convective cooling due to the flow of air from the hollow feathers could be determined from the mass flow rate of air, the heat capacity

(a) at 20°C (b) at 50°C

(c) at 60°C (d) at 70°C

FIGURE 12.9 AESO-30 wt% KF composites in water at a heating rate of 1°C/min.

of air (42 J/mol · °C), and the temperature change. The low-density, high-air content of the KFS composites makes them suited to electronic applications, as well as automotive and aeronautical applications.

12.3 ELECTRONIC MATERIALS FROM FEATHER COMPOSITES

12.3.1 DIELECTRIC PROPERTIES OF THE COMPOSITES

In a typical microchip, performance gain is mostly limited by the intra- and interlayer capacitance, dictated primarily by the dielectric constant (k) of the insulators, known as *dielectrics* [26]. A decrease in the k value of the insulator containing the printed circuits increases the operating speed, minimizes the "cross-talk" effects between metal interconnects, and diminishes the power consumption [27]. The delay time of the electronic signal is proportional to the square root of k and values close to $k = 1$ are most desirable. Figure 12.10 shows the k values of the KFS composites developed from hollow keratin fibers and AESO resin, at a temperature of 25 °C. The k values decrease linearly from 2.7 to 1.7, with an approximate increase of the keratin fiber content as shown here:

$$k_{\mathrm{KFS}} = k_{\mathrm{AESO}}(1 - w_{\mathrm{KF}}) + k_{\mathrm{KF}} w_{\mathrm{KF}}, \tag{12.5}$$

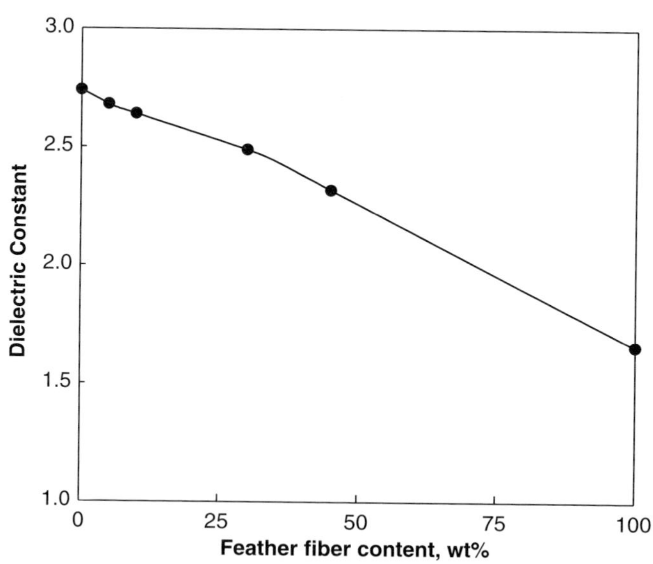

FIGURE 12.10 The dielectric constants of AESO composites at a temperature of 25 °C.

where w_{KF} is the weight fraction of hollow keratin fibers. The KFS composite has a lower dielectric constant than conventional semiconductor insulators such as silicon dioxide ($k = 3.8–4.2$ [27]), epoxies, polyimides, and other dielectric materials. The measured k_{KF} value of the KF mat itself was 1.7 because hollow keratin fibers contain a significant volume of air. The ideal minimum k value is 1.0, as represented by air and, therefore, a porous or high-air content material may have dielectric constants in the ultra-low-k (<2.2) region [26]. The low-cost KFS composite has the potential to replace the dielectrics in microchips and circuits boards in the ever-growing electronic materials field, in addition to many applications as a new lightweight composite material. Thermosetting epoxy resin has a k value of 4.1 and is used for electromagnetic component (EMC), printed circuit board (PCB), and resin-encapsulating-type semiconductor devices [28]. The k value of AESO resin itself is 2.7 and the molding process of soybean resin can follow the lead of conventional unsaturated polyester or vinyl ester resins. The low-viscosity resins are also capable of being spun into nano-sized thin films. Thus, a low-polarity soybean resin alone can also be a substitute for petroleum-based resins in many electronic material applications.

The temperature dependence of the dielectric constant of the AESO-KF composites is shown in Figure 12.11. The dielectric constant of the KFS

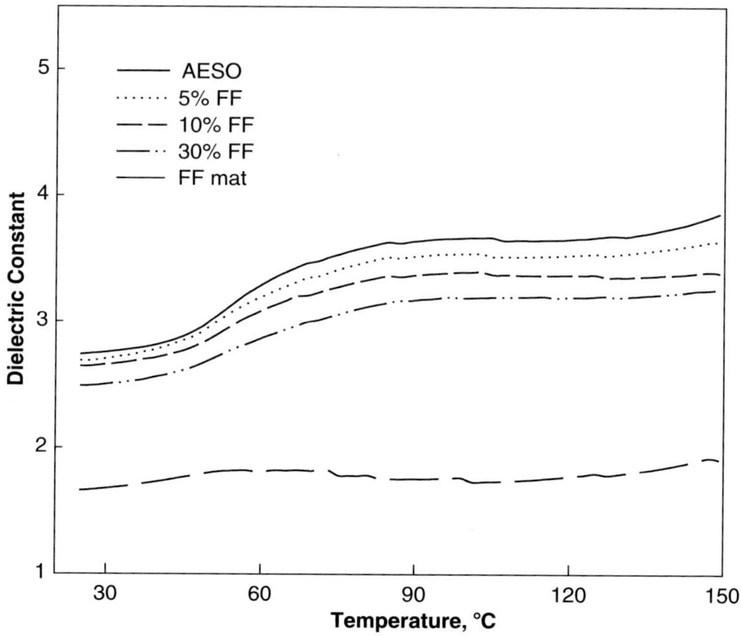

FIGURE 12.11 The temperature dependence of the dielectric constants of AESO composites.

composite slightly increased with increasing temperature, resulting from the alignment of the dipoles when the composite softened with temperature. However, because the effect of temperature on the k value of air is minimal, we expect relatively little change as noted for the pure KF mat (lower curve) in Figure 12.11.

12.3.2 THERMAL EXPANSION OF THE COMPOSITES

The coefficients of thermal expansion (CTE) of the AESO-KF composites are shown in Table 12.5. The CTE is defined as the fractional change in length divided by the change in temperature:

$$\alpha = \frac{K}{L_0}\left(\frac{\Delta L}{\Delta T}\right), \tag{12.6}$$

where α is the CTE ($\mu m/m \cdot {}^\circ C$), L_0 is the initial sample length (m), ΔL is the change in sample length (μm), ΔT is the change in temperature, and K is the cell constant (normally 1.000). From an atomic perspective, the CTE reflects an increase in the average distance between atoms with increasing temperature [29]. The change in bond length is due to the anharmonicity of the bonding and, generally, weaker bonds have a higher CTE value. The CTE of the KFS composites decreases with increasing keratin fiber content and these values are quite low, especially for the 30 wt% sample. The 30 wt% KF sample is a VARTM sample using the KF mats and the fibers are highly oriented in the in-plane direction. The CTE below the glass transition temperature of the 30 wt% composite is 67.4 $\mu m/m \cdot {}^\circ C$ (or ppm/°C). The value is low enough for electronic applications and similar to the value of silicon material (66 ppm/°C) [30] or polyimides (~59 ppm/°C) [31]. However, the CTE of the AESO is higher and changes abruptly above its glass transition temperature (T_g). The thermal expansion of AESO resin is compensated with the thermal contraction of keratin fibers in the longitudinal (oriented) direction. Interactions between resin and keratin fibers also act to resist

TABLE 12.5 CTE values of KF-AESO composites.

KF Content (wt%)	α below T_g($\mu m/m \cdot {}^\circ C$)	α above T_g($\mu m/m \cdot {}^\circ C$)
0	127.2	205.8
5	106.1	200.6
10	100.4	196.3
20	93.1	193.2
30	67.4	69.6

changes in dimension with increasing temperature. The CTE can be further decreased by increasing the glass transition temperature and cross-linking density of the network structure, resulting in increasing chain stiffness. The AESO resin molecular structure used in these examples was not optimized in accord with the rules for property optimization suggested by LaScala and Wool [21], and much room for improvement exists.

12.3.3 WATER ABSORPTION OF THE KFS COMPOSITES

The water absorption of RT-cured AESO KFS composites is shown in Figure 12.12, as a function of time. At equilibrium (after 6 days), the AESO polymer (0 wt% of KF) increased in weight by 0.5%, whereas the 30 wt% KF composite increased by 6%. The AESO resin and keratin fibers have hydrophobic properties. However, the exposed hollow keratin fibers soaked up water by capillary action and the feathers have both hydrophobic and hydrophilic properties at the molecular level [13]. The measured water uptake is dependent on the keratin fiber content, and the dependence of the initial slope of the sorption curves indicates a significant dependence of the diffusion coefficient, D, on the fiber concentration. The kinetic study of the sorption in polymers as a means of determining the diffusion coefficient has been

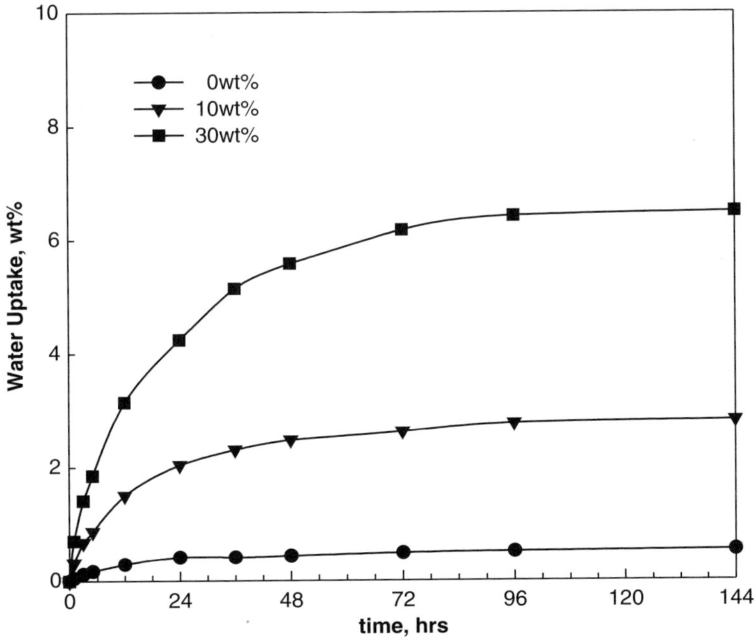

FIGURE 12.12 The water absorption of RT-cured AESO-KF composites.

widely used. In general, the partial differential equation for mass transfer (diffusion) is expressed as follows:

$$\frac{\partial C}{\partial t} = D\left(\frac{\partial^2 C}{\partial x^2}\right), \tag{12.7}$$

where C is the concentration at time t and distance x from the polymer surface. For an infinite slab with a constant D and at short times [32],

$$\frac{M_t}{M_\infty} = \frac{4}{L}\left[\frac{Dt}{\pi}\right]^{1/2}, \tag{12.8}$$

where M_t is the amount of water sorption at time t, M_∞ is the equilibrium sorption, and L is the thickness of the sample. The initial slope method from Fick's second law is commonly used to determine the diffusion coefficient (D) from the initial gradient of a graph of M_t/M_∞ as a function of $\sqrt{t}$. The diffusion coefficient for the pure AESO resin is $D = 4.43 \times 10^{-9}$ cm^2/sec; for 10% KFS composite, $D = 5.28 \times 10^{-9}$ cm^2/sec; and for 30% KFS composite, $D = 5.48 \times 10^{-9}$ cm^2/sec. The AESO resin reached equilibrium in 36 h, whereas the 30% KF-composite required 96 h. The measured water diffusion coefficient in the composites is dependent on the keratin fiber content, and sorption through the fiber is dominant. In comparison with other natural fibers, the water uptake of the AESO-30 wt% KFS composite (6%) is less than that of the AESO-30 wt% flax composites, which has a 12.4% increase at equilibrium [2].

12.4 MECHANICAL AND FRACTURE PROPERTIES

12.4.1 THERMOMECHANICAL PROPERTIES OF THE COMPOSITES

Dynamic mechanical measurements over a wide temperature range are useful in the understanding of the viscoelastic behavior and provide valuable insights into the relationship between structure and properties of composite materials. Figure 12.13 shows the storage modulus E' of RT-cured AESO KFS composites as a function of temperature at various concentrations of keratin fibers. The storage modulus was significantly improved with the addition of keratin fibers over the whole range of the testing temperature. A change in the modulus indicates a change in rigidity and, hence, strength of the composite.

Tan δ, the ratio of viscous to elastic properties, of RT-cured AESO KFS composites versus temperature is shown in Figure 12.14. The maximum value of tan δ decreases with an increase in the fiber content, indicating the increasing trend of composite rigidity. The lowering of tan δ values, the damping energy ratio, suggests the restraint effect of the fibers on matrix mobility [33], and this restriction is enhanced with increasing fiber content.

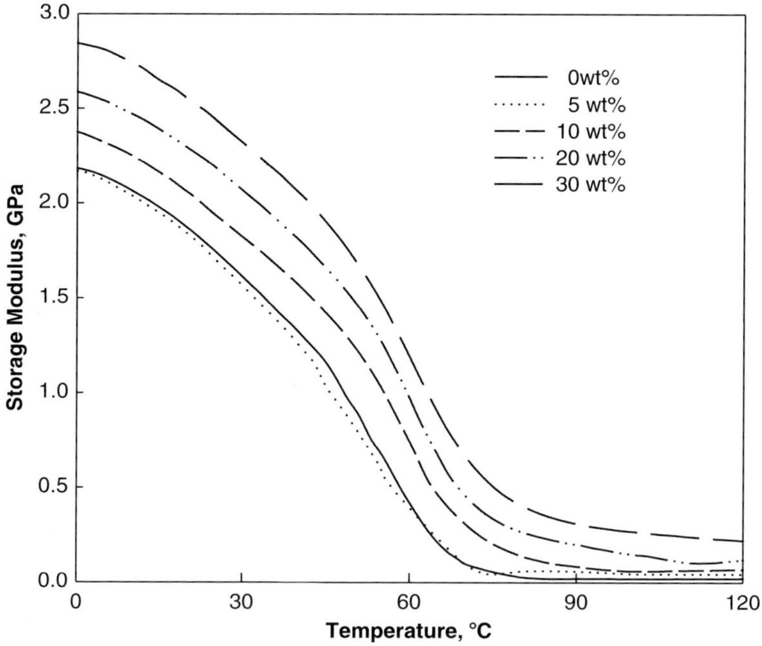

FIGURE 12.13 Storage modulus of RT-cured AESO-KF composites.

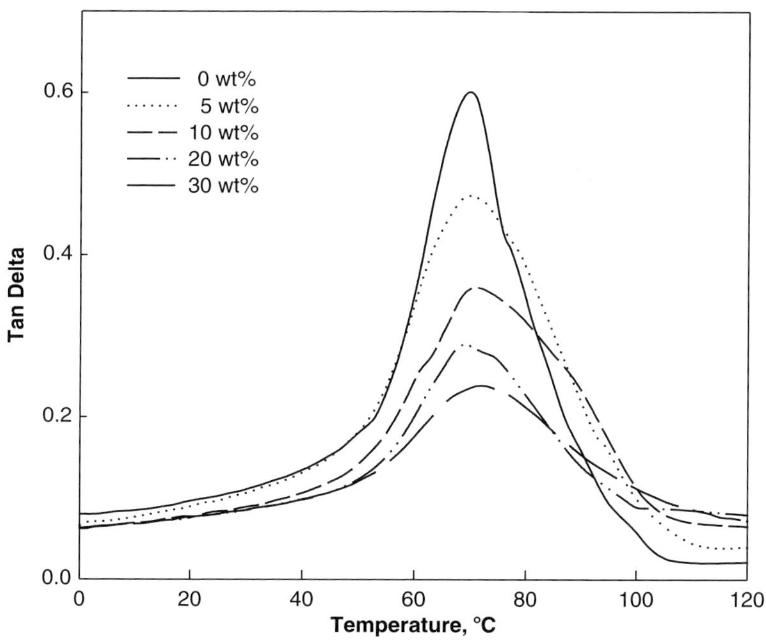

FIGURE 12.14 Tan δ of RT-cured AESO-KF composites.

This is the reason why vibration damping was lowered with the addition of keratin fibers, as shown in Figure 12.7. Also, the damping peak becomes broader with increasing fiber content due to many kinds of relaxation modes of polymer chains due to the reinforcing fibers.

The storage modulus and glass transition temperature of the RT-cured AESO composites are tabulated in Table 12.6. The storage modulus of the 30 wt% KFS composite improved 58.8% in comparison with pure AESO polymer. This is important for electronic materials, since the desired addition of air to obtain low-k materials typically results in reduced strength and stiffness. The glass transition temperature was not significantly influenced by the addition of keratin fibers. Table 12.7 shows the storage modulus and glass transition temperature of RT-cured SOPERMA composites. The properties showed the same trend as the AESO composites. However, the enhancement of the modulus is much lower than that of AESO composites.

12.4.2 MECHANICAL AND FRACTURE PROPERTIES OF THE COMPOSITES

The addition of keratin fibers improves the mechanical properties of the KFS composites. The fracture properties of RT-cured AESO composites are shown in Table 12.8. The fracture toughness (K_{ic}) and the fracture energy (G_{ic}) increased with an increase in the keratin fiber content. The K_{ic} of AESO resin increased 21.3% and the G_{ic} improved 35.7% by adding 30 wt% keratin fibers. This result is gratifying because the introduction of such natural fibers with high-air content and potential defects could have resulted in a considerably

TABLE 12.6 Storage modulus (E') and T_g of RT-cured KFS composites.

KF Content (wt%)	E' at 40°C (GPa)	T_g(°C)
0	1.313	70
5	1.291	70
10	1.598 (+21.7%)	71
20	1.836 (+39.8%)	70
30 (KF mat)	2.085 (+58.8%)	71

TABLE 12.7 Storage modulus (E') and T_g of RT-cured SOPERMA composites.

KF Content (wt%)	E' at 40°C, GPa	T_g(°C)
0	1.039	128
5	1.082 (+4.1%)	129
10	1.139 (+9.6%)	128
20	1.072 (+3.2%)	127

TABLE 12.8 Fracture properties of RT-cured AESO composites.

KF Content (wt%)	Maximum Load N	Fracture Toughness K_{ic}(MPa m$^{1/2}$)	Fracture Energy G_{ic}(KJ/m^2)
0	97.6	1.458	1.420
5	97.1	1.455 (−0.2%)	1.610 (+13.4%)
10	103.0	1.524 (+4.5%)	1.759 (+23.9%)
20	113.0	1.672 (+14.7%)	1.820 (+28.2%)
30 (KF mat)	120.1	1.768 (+21.3%)	1.927 (+35.7%)
30 (hybrid mat)	130.7	1.931 (+32.4%)	1.945 (+37.0%)

weaker material. The improvement in properties also attests to the evolution of high-performance fibers required for long-duration flights by the avian species. The composites prepared using hybrid mats also showed better properties than the pure KF mat composite. The hybrid mat is a mixture of 15 wt% of glass fibers with keratin fibers. Glass fibers are the most widely used synthetic fibers for general reinforcement of polymers. The high stiffness and strength (1.5 GPa) of glass fibers gave the composite higher toughness properties.

Scanning electron microscopy (SEM) micrographs of the fracture surfaces of the RT-cured AESO composite with 5 wt% keratin fibers are shown in Figure 12.15. The keratin fibers were broken without complete pullout during the fracture process, which indicates that adhesion between AESO resin and keratin fibers is quite good for reinforcing. The nodes and hooks on the feather fibers increase the wetted surface area and improve the structural properties of the composite. The mechanical properties of composites depend on the properties of the matrix and the fiber and on the bond strength, among other factors.

The fracture energy G_{ic} of a keratin fiber can be evaluated by using the Nail solution [34]. The composites are nailed together by Σ fibers per unit area, of length L, as shown schematically in Figure 12.16. The fracture strength of the fibers is obtained as follows [34]:

$$G_{ic}(f) = \frac{1}{2}\mu_o \Sigma L^2 V^{\alpha}, \tag{12.9}$$

where μ_o is the unit length friction coefficient and V is the pullout velocity. For static friction control of the pullout process ($\alpha = 0$),

$$G_{ic}(f) = \left(\frac{1}{2}\mu_o L\right) \cdot L \cdot \Sigma. \tag{12.10}$$

Here, $(1/2\ \mu_o L)$ is the force required to pull out a single fiber and is the distance. Therefore, the total energy for fibers in the composite is represented as

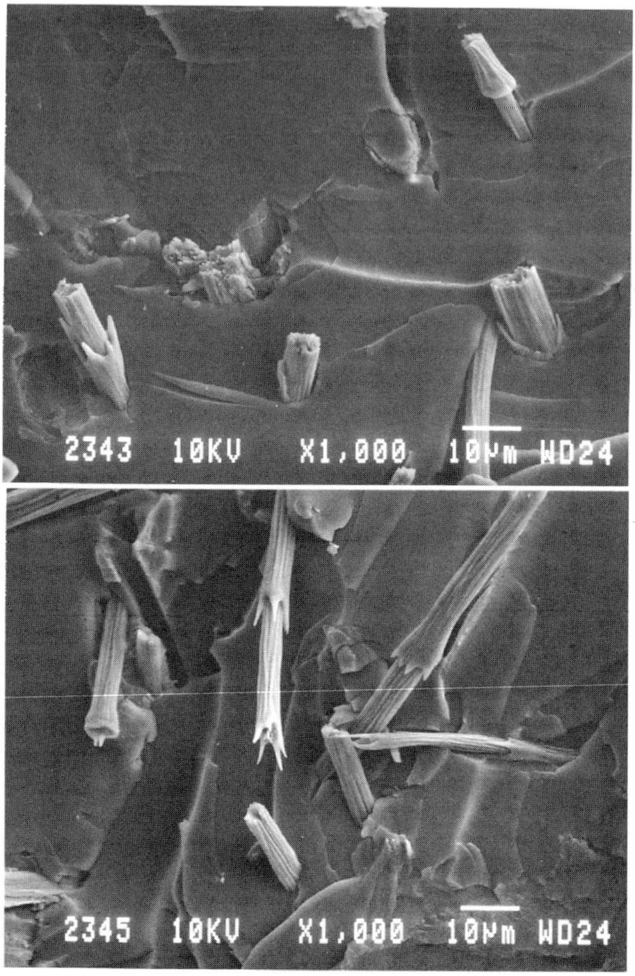

FIGURE 12.15 SEM micrographs of the fracture surface of 5 wt% AESO-KF composites.

$$G_{ic}(f) = G_1\Sigma. \quad (12.11)$$

It was assumed that complete pullout of fibers is not achieved since they fracture at some critical length. Here, G_1 is the energy required to break one fiber:

$$G_1 = G^o{}_f \cdot a_f, \quad (12.12)$$

where $G_f{}^o$ is the fracture energy of the composite when the fiber volume fraction is 1 and a_f is the area of a fiber with a diameter, d, represented as

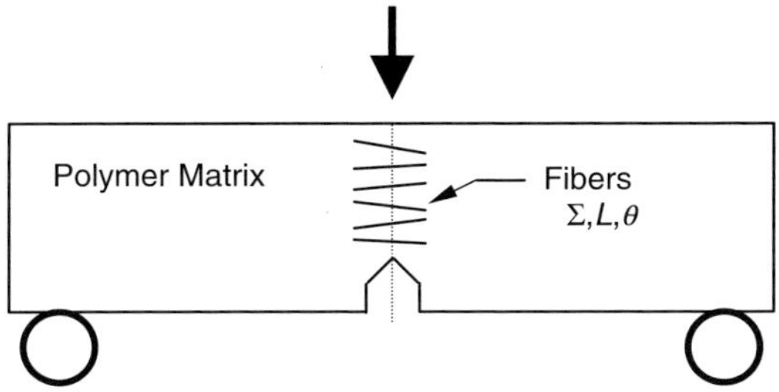

FIGURE 12.16 Schematic of the composite nailed by Σ number of the keratin fibers.

$$a_f = \frac{\pi d^2}{4}\beta, \tag{12.13}$$

where β is a geometric factor ($= 1/\cos^2\theta$) and θ is the angle of fiber orientation. The term Σ can be represented as the volume fraction of the fibers (ψ):

$$\Sigma = \phi h = \frac{\psi}{v_1}h = \frac{4\psi\cos^2\theta}{\pi d^2} \tag{12.14}$$

where ϕ is the number of the fibers per unit volume, $h = L\cos^2\theta$ and v_1 is the volume of a fiber. Therefore, the total energy for fibers is

$$G_{ic}(f) = G_1\Sigma = \psi G_f{}^o. \tag{12.15}$$

The total energy for composites (polymer matrix and fibers) is

$$G_{\text{total}} = G_{ic}(M) + G_{ic}(f) \tag{12.16}$$

and

$$G_{ic} = (1-\psi)G_M{}^o + \psi G_f{}^o = G_M{}^o + (G_f{}^o - G_M{}^o)\psi. \tag{12.17}$$

The fracture energy of the AESO composites is plotted in Figure 12.17 as a function of volume fraction of keratin fibers. The intercept is the fracture energy of the polymer matrix ($G_M{}^o$) and the slope represents ($G_f{}^o - G_M{}^o$). From Figure 12.17, $G_f{}^o = 2.805\,\text{kJ/m}^2$. The fracture stress ($\sigma_f = G_f{}^o/\delta$) of keratin fiber can be calculated as $\sigma_f = 93.5 - 187.0$ MPa ($\delta \approx L_{\max} = 30 - 15\,\mu\text{m}$). The experimentally measured tensile fracture strength of keratin fiber was in the range of 41.4–129.7 MPa due to the heterogeneity of the fibers.

Table 12.9 shows the toughness properties of HT-cured AESO composites. The properties were improved by adding keratin fibers and the HT-cured AESO polymer (0 wt%) has a higher toughness property than RT-cured resin. However, the voids and cracks, due to the evaporation of styrene at

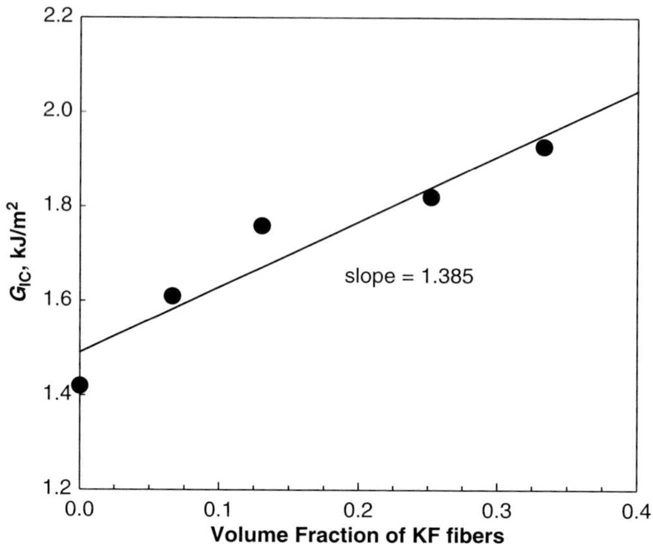

FIGURE 12.17 The fracture energy of RT-cured AESO composites as a function of volume fraction of keratin fibers.

TABLE 12.9 Fracture properties of HT-cured AESO composites.

KF Content (wt%)	Maximum Load N	Fracture Toughness $K_{ic}(\text{MPa} \cdot \text{m}^{1/2})$	Fracture Energy $G_{ic}(\text{kJ/m}^2)$
0	114.3	1.725	1.510
5	121.0	1.800 (+ 4.3%)	1.965 (+ 30.1%)
10	126.4	1.872 (+ 8.5%)	1.979 (+ 31.1%)
20	126.6	1.873 (+ 8.6%)	1.677 (+ 11.1%)

high temperature, lead to deterioration and fluctuation of the composite properties. Table 12.10 gives fracture properties of the RT-cured SOPERMA composites with various concentrations of keratin fibers. The fracture toughness and the energy are greatly enhanced with an increase in the fiber content, in spite of lower compatibility. The toughness properties of the SOPERMA resin itself are much lower than those of AESO resin. Therefore, the reinforcing effect of keratin fibers is greater on SOPERMA composites.

Table 12.11 shows the flexural yield strength and flexural modulus of the RT-cured AESO composites. The flexural properties were significantly enhanced with the addition of keratin fibers. The sample prepared using hybrid mats (15% glass fibers) shows much higher flexural properties, due to the high stiffness of glass fibers. Therefore, the keratin fiber itself or a hybrid is a

TABLE 12.10 Fracture properties of RT-cured SOPERMA composites.

KF Content (wt%)	Maximum Load N	Fracture Toughness $K_{ic}(\text{MPa} \cdot \text{m}^{1/2})$	Fracture Energy $G_{ic}(\text{J/m}^2)$
0	26.9	0.402	63
5	32.9	0.507 (+18.6%)	148 (+134.9%)
10	45.4	0.688 (+66.7%)	340 (+439.7%)
20	51.3	0.778 (+88.2%)	515 (+717.5%)

TABLE 12.11 Flexural properties of RT-cured AESO composites.

KF Content (wt%)	Flexural Yield Strength (MPa)	Flexural Modulus of Elasticity (GPa)
0	34.807	0.896
5	36.178 (+3.9%)	0.971 (+8.4%)
10	36.762 (+5.6%)	1.288 (+43.8%)
20	38.782 (+11.4%)	1.489 (+66.2%)
30 (KF mat)	45.225 (+29.9%)	1.588 (+77.2%)
30 (hybrid mat)	57.791 (+66.0%)	1.938 (+116.3%)

possible substitute for synthetic fibers used for reinforcement of composite materials [35].

12.5 CARBON FIBERS FROM CHICKEN FEATHERS

12.5.1 GRAPHITIZED FEATHERS BY PYROLYSIS

In this section, we utilize chicken feather fiber (CFF) in a carbonized form as a fibrous component in a bio-based composite material [36]. This study investigates the use of CKFin, a carbonized form in a predeveloped AESO-based resin. The first section focuses on the development of carbonized chicken feather fiber (CCFF), including experimental techniques necessary to design appropriate carbonization cycles. The second section investigates the properties of the resulting composite material with an analysis of CCFF properties from the composite characterization. These could be unique hollow graphitic fibers with interesting properties.

12.5.2 INTRODUCTION TO CARBONIZATION

Carbonization is the process of harvesting carbon-rich residues from an organic material. Generally, the material is exposed to high temperatures in an oxygen-depleted atmosphere for long periods of time. These conditions provide the activation energy necessary to break the non–carbon–carbon bonds without the reagents needed to oxidize the carbon structure. First and foremost, carbonization is a relatively cost-efficient method for deriving a new compound from a bulk process. For fibers, if the overall structure and integrity can be maintained while the mass is reduced, a significant advantage could be gained in the strength-to-weight ratio. This strategic advantage could allow for lighter and stronger composite materials [37]. Also, completely new properties have evolved out of a carbonized material. For example, extracted fir tree fibers that were submitted to carbonization show an increased affinity and selectivity in the sorption of various types of oils [38]. These two possibilities earmark the specific goals of the CFF and justify investigating carbonization as a treatment method. If the strength-to-weight ratio can be enhanced while creating a favorable interface between the carbonized fiber and plant oil resins, the potential applications for the resulting composite could be far-reaching.

In the case of chicken feather fiber, the organic structure is composed of approximately 92% keratin protein. Table 12.12 shows a complete listing of the amino acids and the relative frequency (by weight fraction) of those acids in the feather fiber protein [39]. Additionally, Table 12.12 shows results of the calculation detailing the expected residue of carbon after carbonization. From these weight fractions and the formula for each residue, it was calculated that 37% of the protein structure is composed of carbon, including the side chains present in the amino acid structure. This calculation is shown in the "% Carbon" column of Table 12.12. However, because the backbone of this structure

TABLE 12.12 Frequency (by weight fraction) of amino acids in chicken feather fiber.

Amino Acid	w/w% in Fiber	% Carbon	% Net Carbon
Cysteine	17.5	5.2	3.5
Serine	11.7	4.0	1.3
Glutamic acid	11.1	4.5	1.8
Threonine	6.9	2.8	1.4
Glycine	6.5	2.1	1.0
Leucine	6.1	3.4	2.8
Caline	5.9	3.0	2.4
Arginine	5.6	2.3	1.9
Aspartic acid	5.0	1.8	0.5
Alanine	4.8	1.9	1.3
Proline	3.6	1.9	1.5
Isoleucine	2.7	1.5	1.2
Tyrosine	1.9	1.1	0.9
Phenylalanine	1.4	0.9	0.8
Histidine	0.8	0.4	0.3
Methionine	0.5	0.2	0.2
Totals	92.0%	37.0%	22.8%

contains polypeptide bonds created by carbon–carbon and carbon–nitrogen linkages, the process of carbonization becomes complicated because the integrity of the carbon–nitrogen bonds must also be maintained to create a fibrous product. Additionally, each peptide bond contains one carboxyl carbon and many of the amino acid side chains contain oxygen in either alcohol or carboxylic acid forms. Further, if it is assumed that these oxygen atoms detach from the keratin molecule as carbon monoxide, the remaining carbon structure constitutes approximately 22% of the original feather weight. This calculation is shown in the "% Net Carbon" column in Table 12.12.

Unfortunately, the exact residue fraction required for optimal carbonization becomes unclear because of the desire to keep the peptide bonds intact and because of the uncertainty in the interactions of the amino acid functional groups during the carbonization process. Therefore, it was necessary to experimentally determine the various parameters for the carbonization cycles. The following sections highlight the importance of thermogravimetric analysis and differential scanning calorimetry in the design of the cycles. Both of these techniques were used in the analysis of the CFF system.

Thermogravimetric analysis (TGA) measures the fractional mass remaining in the residue as the temperature is ramped over a given interval. During a TGA trace of an organic such as CFF, the little mass lost as the temperature approaches 100 °C is primarily due to the evaporation of free water present. As the temperature continues to ramp upward, the trace will show the temperature at which the first bonds begin to degrade—the carbonization initiation temperature. As the trace continues, the residue fraction will continue to decrease until a minimum is approached. An ideal cycle temperature and goal for the residue fraction can be established for the bulk CFF carbonization cycle by balancing the rate of degradation with the increasing temperatures needed to reach a certain residue. In this way, a carbonization cycle for the production of CCFF can be designed.

Differential scanning calorimetry (DSC) measures the exothermic or endothermic heat duty necessary to maintain a linear heat ramp profile through a range of temperatures. For our purposes, this technique will prove vital in the determination of a solid–liquid phase change for the CFF. If the phase of the system is unchanging, the heat duty required to maintain a linear heating profile versus time should also be linear. However, during the course of a phase change, an additional heat duty is necessary to facilitate the change in macromolecular structure. Therefore, an increase in the endothermic direction for an increasing temperature ramp signals a solid–liquid phase change for the system.

12.5.3 TGA RESULTS

Figure 12.18 shows the result of the TGA trace completed for chicken feather fiber. The CFF loses approximately 2% of its mass up to 100 °C,

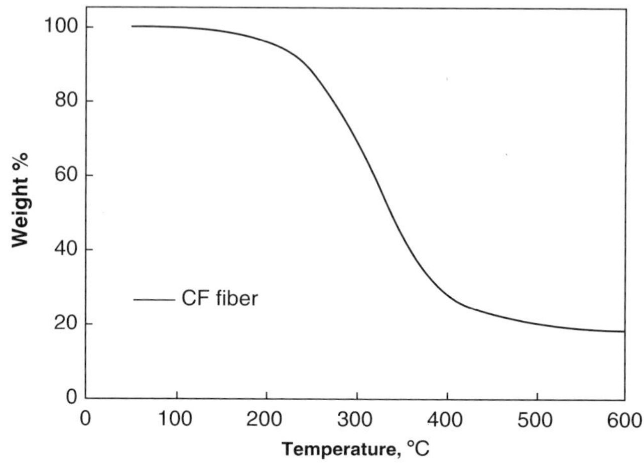

FIGURE 12.18 TGA trace (residue w/w% vs. temperature) for chicken feather fiber.

telling us that the fiber is composed of about that fraction of water. Initiation of the degradation of the fibers appears to begin at approximately 150°C. This is the point that represents a 5% further reduction of the 100°C residue. The trace begins to level at 550°C at a weight fraction of 21% of the original mass.

This trace is useful because it shows that the minimum residue that can be left is in the range of 20–25%. Additionally, the inflection point of the sigmoidal region of the trace occurs at a residue of approximately 60%, roughly indicating the midpoint of the carbonization potential for this temperature region. Therefore, an effort will be made to design and compare the properties of two carbonization efforts. The first target will be a residue of 60% and the second will be a residue of 25%. A comparison of the two will be useful in determining the optimal scenario for different properties, but also in determining the increase or decrease in interfacial interaction with the resin due to the change in the chemical composition of the fiber.

12.5.4 DSC RESULTS

The actual quantity of CFF used in the DSC analysis was 1.89 mg. Figure 12.19 shows the values of the heat flow and the derivative of the heat flow with respect to temperature. The first domain present in the graph shows a generally linear first derivative heat duty with respect to temperature. This represents the heat capacity of the system composed of the CFF in the naturally occurring phase, and the components of the DSC located within the vacuum bubble. This domain occurs between 50° and 225°C. The second domain begins around 240°C and represents the heat capacity for the DSC

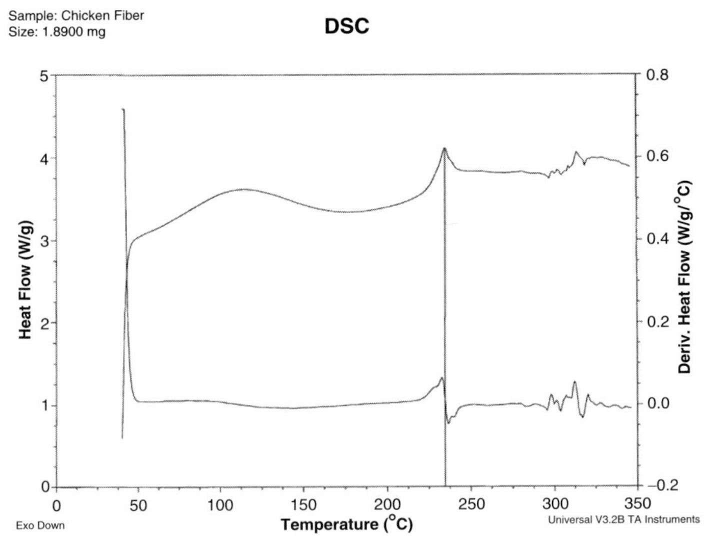

FIGURE 12.19 Heat flow and the derivative of heat flow versus temperature from the DSC analysis of chicken feather fiber.

components and a new phase of CFF. The vertical tie line present in the figure indicates the change in domains and, therefore, the temperature at which the phase change occurs.

This information will be very useful in the design of the carbonization cycles. Because we want to maintain the initial structure of the CFF, that is, the fiber aspect ratio and the semitubular form, a well-designed cycle will avoid the problems that approaching the phase change causes. Therefore, an important point for the carbonization cycles will be to notice the difference in fiber appearance between those that were gradually carbonized using a set of incremental temperature increases and those that were exposed to a temperature shock that will likely cause a phase change.

12.5.5 CARBONIZATION CYCLE RESULTS

Several carbonization cycles were investigated using the data obtained from the TGA and DSC experiments. All cycles were completed under a flowing N_2 atmosphere to limit the oxidation of the CFF material. Table 12.13 shows the temperature set points and soak times used for the different cycles, as well as the weight fractions of the resulting residues.

Each subsequent cycle was designed with the residue fraction and observation of the CCFF appearance in mind. Cycle 1 was designed to compare the equilibrium carbonization of the fiber to the temperature ramp carbonization experienced in the TGA. A temperature of 225°C was specifically

TABLE 12.13 Carbonization cycles and resulting fiber residues.

Cycle #	Heating Cycle	Residue Fraction
1	1 hr Ramp to 225°C 4 hr Soak at 225°C	0.714
2	1 hr ramp to 225°C 5 hr Soak at 225°C	0.696
3	1 hr Ramp to 190°C 1 hr Soak at 190°C 30 min Ramp to 210°C 3.5 hr Soak at 210°C	0.833
4	1 hr Ramp to 220°C 24 hr Soak at 220°C	0.646
5	1 hr Ramp to 220°C 26 hr Soak at 220°C	0.571
6	3 hr Ramp to 500°C 1 hr Soak at 500°C	0.004
7	1 hr Ramp to 220°C 26 hr Soak at 220°C 2 hr Ramp to 450°C 1 hr Soak at 450°C	0.206

chosen because it represents the hottest temperature before the phase change indicated by the DSC analysis. With a residue fraction of 0.714 at 225°C, it can be determined that there exists an equilibrium extent of carbonization that produces a lower residue fraction than the values expressed in the TGA trace. Therefore, caution should be used in the interpretation of the TGA plot, because the values represent a temperature- *and* time-based carbonization. Furthermore, Cycle 2 was evaluated to determine the degree to which carbonization is dependent on time. By adding an extra hour at 225°C to Cycle 1, the residue fraction dropped from 0.714 to 0.696, or an extra 2.5%.

The physical appearances of the products formed by each cycle were very similar. Both products were dark brown to black in color, while the sample exposed to the longer soak time was slightly darker. Although the samples seemed to maintain the fibrous structure of the original batting, the new structure was very rigid and exhibited a noticeable degree of cracking. Fractures appeared in the bulk portions of the material and small particles broke from the main structure. Additionally, small sections of the material seemed to experience fusion of the fiber units, indicating the onset of the phase transition suggested by the DSC. Both of these last two observations indicate that Cycles 1 and 2 were too aggressive and that a lower temperature than 225°C should be used.

Cycle 3 was an attempt to develop a less aggressive strategy to carbonization. In an attempt to reproduce the residue fraction of Cycle 1, an incremental temperature scheme was utilized, using lower temperatures but slightly longer times. For this cycle, the sample was soaked at 190° and 210°C for 1 and 3.5 h, respectively, producing a residue fraction of 0.833. This operation yielded a much different material structure than the previous cycles. The material created by Cycle 3 retained the pliability of the original feather fiber batting. Additionally, the color of the material was a much lighter brown than in the previous cases, similar to the color of cardboard.

To reach a cycle in line with the goal of ~55% dictated by the inflection of the sigmoidal curve of the TGA trace, a temperature between 210° and 225°C needs to be used. Additionally, much longer times should be utilized to increase the approach to the equilibrium value for a given temperature. From these conjectures, a temperature of 220°C was chosen for a soak time of 24 h. Cycle 4 follows these parameters and produces a residue of 0.646. Because this residue is very close to the goal of ~55%, another carbonization experiment was completed where the residues of the samples were taken after every hour increment of time past 24 h. At a total time of 26 h at 220°C (Cycle 5), the residue fraction was 0.571. On inspection, the resulting structure from these two cycles appeared very similar. There was less cracking and fracture than in the samples treated by Cycles 1 and 2. The fiber form of the material remained intact because there was no evidence of either the fusion of fibers or the disintegration of feather portions vital to the integrity of the fibers. Additionally, the material was very rigid and brittle. From these observations and the residue obtained from Cycle 5, the decision was made to use this material in the analysis of the composite properties.

In addition to the fiber-free and ~55% fiber composites, a third composite structure was desired that was composed of CCFF at a residue of ~25%. Therefore, a second carbonization cycle had to be designed. To test (and support) the theory that exposing the CFF to a shock increase in temperature above the phase change temperature indicated by the DSC analysis will induce an unfavorable phase change in the fiber, Cycle 6 was designed by submitting the material to a temperature of 500°C for 1 h after a 3-h ramp to that temperature. The resulting material consisted of small particles of black CCFF at a residue of approximately 0.004%. Obviously, this method does not produce a fibrous material that resembles anything similar to the original structure. The formation of distinctly separate beads of particles represents a complete phase change followed by the disintegration of the peptide bonds linking the amino acid structure of the protein-like fiber.

Cycle 7 from Table 12.13 shows the second fiber residue that will be analyzed in a composite material. By using an incremental temperature scheme, the phase change temperature was shifted far enough so that the material did not undergo such an operation during the heating to the hotter treatment temperatures. Therefore, a combination of 220°C for 24 h (Cycle 4) followed by a 2-h ramp to 450°C and soaking at that temperature for 1 h produces a residue of 0.206. The material produced by such an operation retains the fiber characteristics present in the material from Cycle 5.

Additionally, it meets the specification of approaching the minimum carbonization found by the convergence of the TGA plot. This material will be used to complete the study of the effects of CCFF fiber in composite materials.

12.5.6 SEM MICROGRAPHS OF CCCF

SEM micrographs were taken at 1000× magnification for a comparison between the structure of the carbonized and noncarbonized forms of chicken feather fiber. Figures 12.2(a) and (b) earlier in the chapter show the untreated CFF. Figures 12.20(a) and (b) show micrographs of the high-temperature treatment (0.21 residue) of carbonized chicken feather fiber. Additionally, Figure 12.21 is another micrograph of CCFF at 2500× magnification.

The micrographs of the untreated feather fiber in Figures 12.2(a) and (b) reveal several significant structural properties of chicken feather fiber. In Figure 12.2(a), the repeating, segmented nature of the feather fiber can be readily observed. Further, Figures 12.2(a) and (b) show the texture of the feather fiber surface. Generally speaking, the surface of the feather fiber appears grooved along the axial direction. Finally, Figure 12.2(b) shows the "hook-and-loop barbs" that represent the method of physical binding that occurs in feathers to maintain integrity under high stress, such as flight. From a fiber analysis aspect, the recovery of these properties is very important. The segmented structure represents the strength and integrity of the fiber. If the segmentation is broken, the strength of the fiber will decrease and will not serve its purpose as fiber reinforcement in the composite material. The grooved surface and the hook-and-loop structure provide additional surface areas for potential chemical interaction with the resin in the composite matrix.

Figures 12.20(a) and (b) and 12.21 indicate the retention of many of the features expressed in Figures 12.2(a) and (b). The segmented structure is apparent under 1000× magnification. A major difference, however, is that there seems to be a significant amount of shrinkage in both the radial and axial dimensions. In the untreated case, the diameter and length of each segment are approximately 5 and 50 μm, respectively. In the carbonized case, the approximate diameter is 3 μm and the axial length is 20 μm. In Figure 12.21, the grooves along the surface of the fiber are very distinct, preserving the high surface area for potential interfacial interactions. Unfortunately, the high magnification shows that the hook-and-loop structures of the barbs have virtually disappeared, with only a small fraction present near the joints of the segmented structure.

12.5.7 DISCUSSION ON CARBONIZED CHICKEN FEATHERS

The carbonization cycles show that the best alternative for the "low-temperature treatment" is 220 °C for 24 h. This yields a residue fraction of

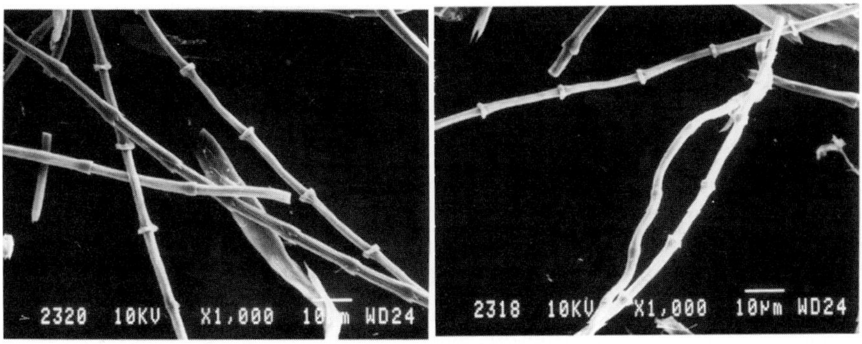

(a) (b)

FIGURE 12.20 SEM micrographs of carbonized chicken feather fiber at 1000× magnification.

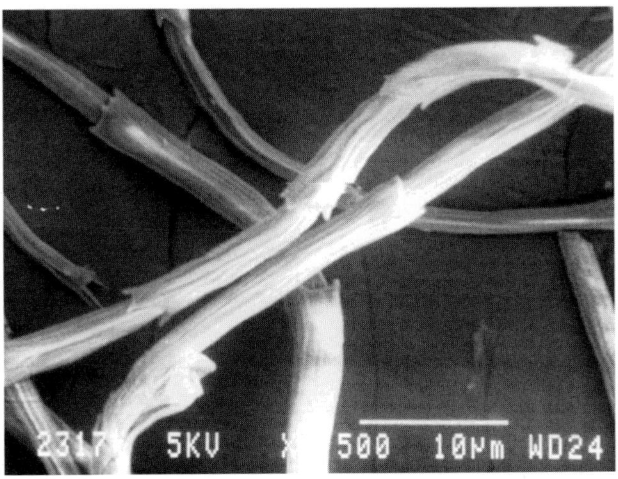

FIGURE 12.21 SEM micrograph of carbonized chicken feather fiber at 2500× magnification.

approximately 57% and retains the integrity of the CFF structure, as shown in the SEM micrographs. Because exposing the fibers to a temperature above the phase transition completely changes the structure of the CCFF, it is necessary to choose a "high-temperature treatment" that avoids this phenomenon. Therefore, the high-temperature treatment will utilize the fibers created by the low-temperature treatment with an added soak time of 2 h at a set point of 450°C. The resulting residue of this process is 21% by weight of the original fiber. Furthermore, incrementing the temperature with multiple soak times allows for the gradual carbonization of the fiber. This shifts the phase change temperature past the temperature required to reach the goal of ~25% residue fraction for the second set of fibers. Under SEM analysis, the segmented structure of the feather fiber seems to remain intact. However, a noticeable difference in the CFF and CCFF micrographs occurs in the apparent shrinkage in both the radial and axial dimensions of the fibers. The mechanical properties of these fibers are analyzed and discussed in the next section.

12.5.8 MECHANICAL PROPERTIES OF CCFF COMPOSITES

Table 12.14 contains the averages for storage modulus, loss modulus, peak height of tan δ, and the T_g for each composite system. Table 12.15 shows the increase in the storage modulus with respect to the zero fiber reference material for each composite.

This trend is very supportive of the theory that the CCFF acts as a reinforcing agent in the polymer matrix. Further, it is clear that the high-temperature carbonization has a distinct advantage in terms of the stiffening effect. Intuitively, this phenomenon is expected because the added carbonization serves to further strip down the amino acid chain to leave behind a tighter carbon–nitrogen backbone.

In all of the systems, the height of the viscous to elastic peak remained constant throughout the systems at a value of about 0.6. Further, there appeared to be no appreciable difference in the loss modulus. Although the material containing the high-temperature–carbonized fibers appears to have a large increase in the loss modulus, there does not seem to be a trend that explains why this occurs. Another interesting point is that the glass transition temperature appears to drop significantly for the final material. Again, however, there is no trend to support a conclusion other than experimental

TABLE 12.14 DMA results of composite systems.

	Storage Modulus 35°C (G Pa)	Loss Modulus 35°C (G Pa)	Tan δ (Max Ratio)	Tg (°C)
No Fiber	0.730	0.133	0.64	71.7
1.06 w/w% Low Temp	0.773	0.121	0.59	69.0
3.05 w/w% Low Temp	0.913	0.126	0.62	72.7
2.98 w/w% High Temp	1.106	0.162	0.60	64.5

TABLE 12.15 Percent increase in storage modulus.

	Percent Increase in Storage Modulus
1.06 w/w% Low Temp	5.9%
3.05 w/w% Low Temp	25.1%
2.98 w/w% High Temp	51.5%

error. The carbonized fiber surfaces may also be interfering with the free-radical reactions, and preferential absorption of the AESO triglycerides may be occurring on the carbon surfaces, similar to the carbon nanotubes that will be discussed in Chapter 14.

Maybe the most significant value contained in either Table 12.14 or Table 12.15 is the 50+% increase in storage modulus when using the high-temperature treatment fibers instead of the fiber-free material. Further, by using only these data, it is possible to estimate a range of values for the modulus of the individual fiber. Because the storage modulus is defined as the elastic portion of the tan δ signal, the storage modulus behavior of the composite can be modeled as a system of perfectly elastic springs. Specifically, there are two spring models, springs in parallel and springs in series, that could serve as a bound for the calculation. Using these basic models, the range of the high-temperature treatment CCFF is bounded by 13.5 GPa for the parallel model and 66.1 GPa for the series model. For the low-temperature treatment, the range of values were 4.8 and 10.4 GPa, respectively. Whereas these models in no way accurately describe the behavior of the CCFF in the polymer matrix, they provide an analysis that describes the order of magnitude of the fiber modulus. The carbonized CFFs have the potential to provide a source of low-cost carbon fibers that could be used in many composite applications such as SMC.

12.6 SUMMARY OF KFS COMPOSITES

In this chapter, affordable, bio-based, and environmentally friendly composite materials from soybean resins and hollow keratin fibers from chicken feathers were discussed and their fundamental properties investigated. Keratin fibers are hollow, light, hydrophobic, and compatible with AESO and SOPERMA resins. The moduli of the AESO resin in the rubbery region and glass transition temperature were increased with increasing styrene comonomer content. However, the cross-link density decreased linearly with an increase in the styrene content. The density of the composites decreased with increasing keratin fiber content and can be less than 1 g/cm^3. About 5% of hollow keratin fibers were filled by the resin infusion process, but the composite still contained a significant volume of air in the hollow structure of the fibers. The dielectric constant k value of the AESO-KFS composite was found to be in the range of 1.7 to 2.7, depending on the hollow fiber fraction. The k values were lower than that of a conventional semiconductor insulator material such as silicon dioxide, epoxies, polyimides, and other dielectric materials.

The coefficient of thermal expansion of the new composite material (67.4 ppm/°C) was low enough for electronic applications and similar to the value of silicon material or polyimides. The measured water absorption of

AESO resin was 0.5%, and the diffusion coefficients of AESO-KFS composites were dependent on the keratin fiber content such that sorption through the fiber was dominant. The storage modulus of AESO composites was significantly improved by the addition of keratin fibers. The damping peak of the composite was lowered and the peak became broader with an increase of the fiber content. The fracture toughness and fracture energy of the composites were increased with increasing fiber content. The fracture energy of a keratin fiber in the composites was evaluated using the Nail solution. The mechanical properties of the new composite materials are in the acceptable range for composite applications. Carbonized chicken feathers showed improved properties above the normal feathers and present some exciting future possibilities for low-cost carbon fibers.

REFERENCES

1. Khot, S.; LaScala, J.; Can, E.; *et al. J. Appl. Polym. Sci.* 2001, **82,** 703.
2. Williams, G. I.; Wool, R. P. *Appl. Compos. Mater.* 2000, **7,** 421.
3. Can, E.; Kusefoglu, S.; Wool, R. P. *J. Appl. Polym. Sci.* 2001, **81,** 69.
4. Thielemans, W.; Can, E.; Morye, S.; *et al. J. Appl. Polym. Sci.* 2002, **83,** 323.
5. Can, E.; S.; Kusefoglu, Wool, R. P. *J. Appl. Polym. Sci.* 2002, **83,** 972.
6. LaScala, J.; Wool, R. P. *J. Am. Oil Chem. Soc.* 2002, **79**(1), 59.
7. LaScala, J.; Wool, R. P. *J. Am. Oil Chem. Soc.* 2002, **79**(4), 373.
8. Gassner, G., III; Schmidt, W.; Line, M.; *et al.* U.S. Patent No. 5,705,030; 1998.
9. Gassner, G. U.S. Patent No. 6,027,608; 2000.
10. Liu, K. *Soybeans: Chemistry, Technology, and Utilization*, Chapman & Hall, New York; 1997, pp. 25–95.
11. Mohanty, A. K.; Misra, M.; Hinrichsen, G. *Macromol. Mater. Eng.* 1999, **276/277,** 1.
12. McGovern, V. *Environ. Health Persp.* 2000, **108**(8), A366.
13. Schmidt, W. F. In *Advanced Fibers, Plastics, Laminates and Composites*, Wallenberger, F. T.; Weston, N. E.; Ford, R.; *et al.*, Eds.; Materials Research Society; 2002, pp. 25–32.
14. Wool, R. P.; Hong, C. K. "Low Dielectric Constant Materials from Keratin Fibers and Soyoil, U.S. Patent pending.
15. von Bahr, M.; Kizling, J.; Zhmud, B.; *et al.* Surfaces for Digital Printing (S2P2), Report #S2P2 SS4.
16. Wool, R. P.; Can, E.; U.S. Patent pending.
17. von Bahr, M.; Tiberg, F.; Yaminsky, V. *Colloids Surf. A* 2001, **193,** 85.
18. Goodman, S. H. In *Handbook of Thermoset Plastics*, Goodman, S. H., Ed.; Noyes Publications; 1986, pp. 1–17.
19. Auad, M. L.; Aranguren, M.; Borrajo, J. *J. Appl. Polym. Sci.* 1997, **66,** 1059.
20. Flory, P. J. *Principles of Polymer Chemistry*, Cornell University, Ithaca; 1975.
21. LaScala, J.; Wool, R. P. *Polymer 2005,* **46,** 61.
22. Khot, S. N. Synthesis and Application of Triglyceride Based Polymers, *Dissertation,* University of Delaware, Newark; 2001.
23. Fox, T. G. *Bull. Am. Phys. Soc.* 1956, **1,** 123.
24. Brandrup, J.; Immergut, E.; Grulke, E., Eds. *Polymer Handbook*, 4th ed., John Wiley & Sons, New York; 1999 p. VI 211.
25. Heylen, W.; Lammens, S.; Sas, P. *Modal Analysis Theory and Testing*, Katholieke Universiteit Leuven, Belgium; 1997.
26. Miller, R. D. *Science* 1999, **286,** 421.

27. Treichel, H.; Withers, B.; Ruhl, G. *et al.* In *Handbook of Low and High Dielectric Constant Materials and Their Applications*, Vol. 1, Nalwa, H. S., Ed.; Academic Press, San Diego; 1999, Chap. 1.
28. Ogura, I. In *Handbook of Low and High Dielectric Constant Materials and Their Applications*, Vol. 1, Nalwa, H. S., Ed.; Academic Press, San Diego; 1999, Chap. 5.
29. Callister, W. D. *Material Science and Engineering: An Introduction*, 3rd ed., Wiley, New York; 1994.
30. Martin, S. J.; Godschalx, J. P.; Mills, M. E. *et al. Adv. Mater.* 2000, **12,** 1769.
31. Eichstadt, A. E.; *et al. J. Polym. Sci. B: Polym. Phys.* 2002, **40,** 1503.
32. Crank, J. *The Mathematics of Diffusion*, 2nd ed., Oxford University Press, Oxford; 1975.
33. Yu, S.; Hing, P. *J. Appl. Polym. Sci.* 2000, **78,** 1348.
34. Wool, R. P. *Polymer Interfaces: Structure and Strength*, Hanser Publishers, New York; 1995.
35. Hong, C. K.; Wool, R. P. *J Appl. Polym. Sci.* 2005, **95,** 1524.
36. McChalicher, C.; Hong, C.K., Wool, R. P. Paper presented of the American Physical Society, Philadelphia, August 2004.
37. Tzeng, S.-S.; Chr, Y.-G. Evolution of Microstructure and Properties of Phenolic Resin–Based Carbon/Carbon Composites During Pyrolysis, *Mater. Chem. Phys.* 2002, **73,** 162–169.
38. Inagaki, M.; Kawahara, A.; Konno, K. Sorption and Recovery of Heavy Oils Using Carbonized Fir Fibers and Recycling, *Carbon* 2002, **40,** 105–111.
39. Farner, D. S.; *et al. Avian Biology,* Vol. 6, Academic Press, New York; 1982.

13

Hurricane-Resistant Houses from Soybean Oil and Natural Fibers

Richard P. Wool

The development of hurricane-resistant housing by the ACRES group was prompted by a visit from a South Carolina congressional delegation to the laboratories at the University of Delaware; two back-to-back hurricanes had just created considerable damage in that state. A problem was posed: Could we use the new low-cost green materials to develop high-performance housing structures that would be more hurricane resistant? Given the level of devastation and the multibillion dollar costs, the solution appeared daunting, especially if the new housing materials were to be made with paper, straw, chicken feathers, and soybeans. However, on examination of the nature of the damage, it was clear that the engineering solution was not that complex but would require a radical new approach to both the roof design and method of construction. Inspection of the damage in South Carolina and Florida in the aftermath of Hurricane Andrew revealed that most of the damage to roofing was not of a catastrophic nature, but rather a gradual process of removing particle boards from the roof by the wind-induced pressure drop: As the hurricane winds sweep over the roof at 150 mph (Category 5), the roof acts like an imperfect airfoil and creates a vacuum on the sheltered or lee side of the roof. This vacuum is only of the order of 1 psi (max vacuum is 14.7 psi), but this is sufficient to peel off the 8- × 3-foot particleboards, which are nailed to the joist boards of the A-frame substructure, as shown in Figure 13.1. By comparison, a Boeing 747 jumbo jet takes off and lands at 200 mph with a pressure drop of only about 1 psi on the top side of the wings. The force to peel off the particleboards is proportional to

FIGURE 13.1 The most common roof damage when exposed to hurricane forces.

the square root of the number of nails holding them in place. Thus, the solution involves the following: Either use an enormous number of nails or make the whole roof as a monolithic piece, like the airfoils used in aircraft wings or like the sturdy molded hull of a boat. A multidisciplinary ACRES group addressed this engineering problem with bio-based materials. Their basic concept for the roof design in shown in Figure 13.2. It consists of a foam core composite structure with an integrated webbing of I-beams in the foam core. In this chapter, we explore the materials and the design of the roof as a molded-in-place monolithic piece, using low-cost composites derived from soybean oil and natural fibers. This particular application of bio-based composites to housing is potentially the world's largest utilization of fibers and resins from renewable resources.* The same bio-based low-cost materials can also be used with a different design to make emergency shelters, or small houses, which can be rapidly deployed to survivors of tsunamis, floods, earthquakes, hurricanes and such major disasters.

13.1 INTRODUCTION AND BACKGROUND

New efforts in the application of natural composite materials in the building products sector show significant potential. An analysis completed

*As this book was being finished, the first prototype hurricane-resistant roof was planned for construction in Delaware.

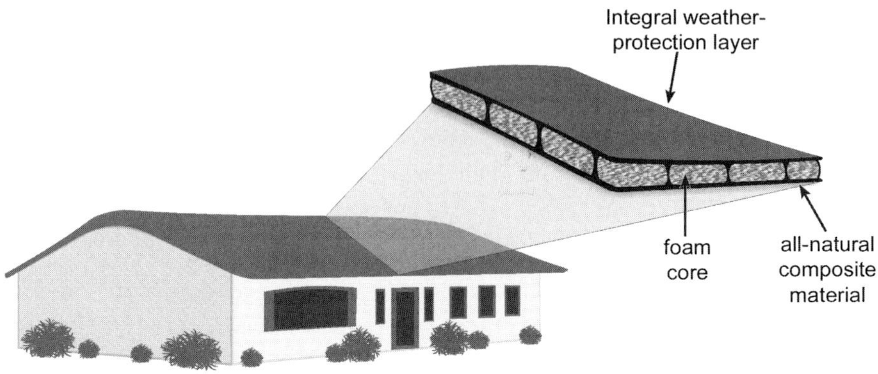

FIGURE 13.2 A schematic showing a monolithic house roof made of natural composites as a hurricane-resistant roof. (*Sources: Newsweek,* October 27, 2003; *Architectural Record,* November 2003.)

by Kline & Company, Inc., in 2001, forecasts the building market for natural fibers for the years 2000–2005 to grow 60% per year. More specifically, by 2005, the market in building products is projected to require 3 billion pounds of natural fibers for applications, led by decking and followed by new applications in siding and shingles [1]. One area of specific interest to this project is the use of these natural fiber-reinforced composites for structural applications, particularly roofing applications.

In recent years, the United States has been hit by several major hurricanes—Hugo (1989), Andrew (1992), and Iniki (1992)—that caused more than $27.5 billion dollars cumulatively in damage to insured property [2]. In 2004, the damage from the four hurricanes that impacted Florida (Charley, Frances, Ivan, and Jeanne) was expected to reach at least $30 billion dollars, and according to the NOAA hurricane forecasters, there exists an ominous potential for continued hurricane damage in the next several decades. Typically, when houses are exposed to hurricane forces, roofs are most susceptible to damage, followed by walls and openings, and then foundations. The most common roof damage is the loss of cladding or sheathing (tiles, shingles, etc.) resulting from very high suction pressures that develop at the roof/wall interface, as shown in Figure 13.1. Once the sheathing is lost, the roof no longer acts as a diaphragm, and the lateral load carrying of the structure is compromised. In addition, the loss of sheathing can result in costly water damage. Researchers at the Insurance Institute for Property Loss Reduction and the Insurance Research Council write that:

> A 40-year period of relatively benign weather left southern Florida with a false sense of security regarding its ability to withstand hurricanes. This led to complacency about hurricane risk, leading to "helter-skelter" development, lackluster code enforcement, building code amendments, shortcuts

> in building practices, and violations that seriously undermined the integrity of the [building] code and the quality of the building stock. Conservative estimates from claim studies reveal that approximately 25 percent of Andrew-caused insurance losses (about $4 billion) were attributable to construction that failed to meet the code due to poor enforcement, as well as shoddy workmanship. At the same time, concentrations of population and property exposed to hurricane winds in southern Florida grew manyfold [3].

The inability of the traditional roof structure to maintain its structural integrity during a hurricane creates an opportunity to completely overhaul the structure by developing a monolithic, all-natural composite roof system. The so-called "hurricane resistance" of this roof will be obtained by making the structure monolithic (one-unit) without cladding (see Figure 13.2). In doing so, the composite will not rely on numerous small sheathing panels to form the diaphragm, therefore becoming "hurricane resistant."

By using vacuum-assisted resin transfer molding (VARTM) processing to make the composite structures, the complicated three-dimensional architecture of a roof can be created. Furthermore, the use of natural fibers and soybean oil resins results in a very cost-efficient composite material [4–8]. These advantages of the all-natural alternative roof will create significant consumer demand. Moreover, mass production of this novel roof structure will require significant quantities of natural fiber composite raw materials, so these composites can compete with the conventional synthetic composite materials.

Recent advances in the use of natural fibers (e.g., flax, cellulose, jute, hemp, straw, switch grass, kenaf, coir, bamboo) in composites were reviewed by several authors [9–33]. Bledzki and Gassan [14] reported that natural fibers were used as early as 1908 in the fabrication of large quantities of sheets, where paper or cotton was used to reinforce sheets made of phenol- or melamine-formaldehyde resins. In ancient times, the Egyptians, Turks, and Asians found that straw embedded in mortar and mud resulted in considerable reinforcement of building structures, many of which still exist today after thousands of years' exposure to the elements. Hemp fibers were once the world's largest agricultural crop in the early 19th century, but the demand for the material has declined with advances in the field of synthetic fibers and the manipulation of conservative governments by right-wing lobbyists. Global environmental issues have led to a renewed interest in such bio-based materials with the focus on renewable raw materials that can be made recyclable or biodegradable at a reasonable cost.

Natural fibers were used to reinforce traditional thermoplastic polymers in automotive applications [14]. Polypropylene has often been used as the matrix material. The influence of surface treatments of natural fibers on the interfacial characteristics was also studied by several authors [27–31] and reviewed by Mohanty *et al.* [30]. In this chapter, new low-cost natural fiber mats were used to reinforce soybean oil resin for large-volume applications such as housing roofs, building construction material, and furniture.

13.2 BIO-BASED MATERIALS

13.2.1 SOYBEAN OIL RESIN

Extensive research was devoted to the development of polymers from triglyceride oils as a natural alternative to petroleum-based polymers [4–7]. Each triglyceride contains three fatty acid chains joined by a glycerol center, as discussed in Chapter 4. To produce a rigid cross-linked thermoset, the triglycerides must be chemically functionalized [4–7] and have a sufficiently low viscosity (less than 1000 cp) to be infused by vacuum into the fiber mats. The preferred degree of functionalization for soybean oil resins is from four to six groups per triglyceride. The selection of plant oil (soy, corn, linseed, sunflower, genetically engineered high oleic, etc.), the optimization of the chemical functionalization, and its effect on the mechanical and thermal properties of the resin were analyzed by LaScala and Wool [34] and discussed in Chapter 7.

13.2.2 NATURAL FIBER MATS

Natural fibers serve as the low-cost reinforcement of the resin in biocomposites, improving both the strength and stiffness of the resulting composites. Natural fibers are typically grouped into four different types depending on their source: leaf, bast, fruit, and seed. The leaf and bast fibers are generally used in composite processing. Examples of leaf fiber include sisal, henequen, and pineapple leaf fiber (PALF). Bast fiber examples are flax, hemp, ramie, cellulose, and jute. One of the major difficulties of natural fibers is that their properties are intrinsically dependent on where they are grown (locality), what part of the plant they are harvested from (leaf or stem), the maturity of the plant (age), and how the fibers are harvested and preconditioned in a form of mats or chopped fibers, woven or unwoven. These factors result in significant variation in properties compared to their synthetic fiber counterparts (glass, aramid, and carbon). Figure 13.3 shows images of some natural fibers, and Table 13.1 presents all the different natural fibers used in this study including descriptions of the fiber mats in terms of contents and processing conditions. Other fibers such as rice straw from California, switch grass from the Midwest, and cotton fibers from the southern United States can also be used. The advantage of the hurricane-resistant roof design is that essentially any indigenous fiber, including newspapers, and plant oil can be utilized to make the low-cost composites.

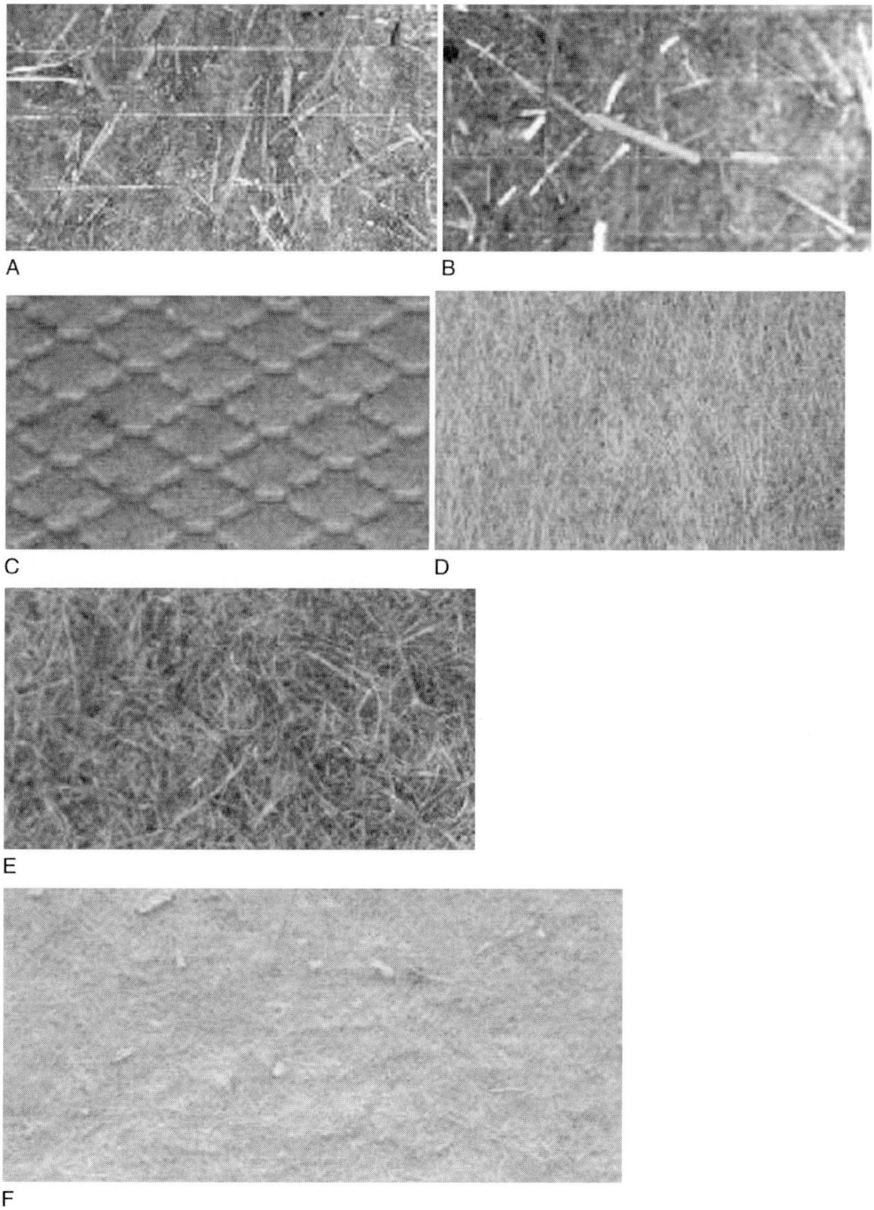

FIGURE 13.3 (A and B) Flax mat, (C) cellulose mat, (D) jute, (E) hemp, and (F) chicken feathers mat.

13.2.3 FOAMS

Elfoam, T300, manufactured by Elliott Company in Indiana, was used as a core in manufacturing the structural composites in this study. The foam is a closed-cell polyisocyanurate that is chemically similar to but higher performing than polyurethane foam. Outstanding thermal and moisture resistance make this foam a highly effective insulating material. The foam is widely used in composite sandwich construction due to its light weight. T300 foam has a density of $44.85\,kg/m^3$ ($2.8\,lb/ft^3$) and operating temperature range from -297 to $+300\,°F$. Experiments at the University of Delaware on the production of bio-based foams via supercritical CO_2 using acrylated epoxidized soybean oil (AESO) and other soy-based monomers, as described in Chapter 5, are intended to eventually replace the petroleum-based polyurethane foams in this high-volume application.

TABLE 13.1 Natural fiber reinforcements used in AESO composites.

Chart Reference	Description
Flax/PET 40/40	40% flax, 40% PET, 20% starch binder, supplied by Cargill Ltd.
Flax mat 60/40	60% flax, 40% binder, supplied by Cargill Ltd.
Flax mat 85/15 20 oz.	85% flax, 15% binder 20 oz., supplied by Cargill Ltd.
Cellulose $200\,g/m^2$	Air-laid $200\,g/m^2$ 84% cellulose, 16% binder, supplied by Concert Fabrication, Canada
CTMP pulp	Chemical thermal mechanical pulp, supplied by M&J Fibretech a/s, Denmark
Fluff pulp	Wet-laid fluff pulp low-density mat, 100% cellulose $640\,g/m^2$, supplied by Rayonier
Chemically treated pulp	Chemically treated pulp containing a hydrophobic debonder, supplied by Rayonier
Caustic treated pulp	Caustic treated pulp, 100% cellulose high-porosity mat used for filtration products, supplied by Rayonier
Cellulose $150\,g/m^2$	Air-laid $150\,g/m^2$ 82% cellulose, 18% binder, supplied by Concert Fabrication, Canada
Flaxcraft hemp	$550\,g/m^2$ nonwoven hemp, supplied by Flaxcraft
FlaxTech flax	Flax distribution of flax fibers with varying lengths and low binder content, supplied by FlaxTech
Newspaper	Newspaper
Recycled paper	$110\,g/m^2$ recycled paper from cardboard boxes, supplied by Interstate Resources, PA, USA
Chicken feathers mat	97% chicken feathers ground into about 6-μm-diameter, 8-mm-long fibers with 3% low-molecular-weight polymeric binder, solid feather density is $0.8\,g/cm^3$ [42]

13.3 COMPOSITE PROCESSING AND MANUFACTURING

13.3.1 NATURAL COMPOSITE PANEL MANUFACTURING USING VARTM

The VARTM process is a variant of vacuum-infusion RTM process in which one of the solid tool faces is replaced by a flexible polymeric film (Figure 13.4). This process or a modified version has also been called SCRIMP [35]. The VARTM process is a very clean and economical manufacturing method. The process draws resin into a dry reinforcement on a vacuum bagged tool, using only the partial vacuum to drive the resin (Figure 13.4). The process increases the component mechanical properties and fiber content by reducing void percentage, when compared to other large-part manufacturing processes, such as hand layup [36]. Very large structures can be made in this manner.

For higher performance composites, VARTM offers the potential for reduced tooling costs where matched tooling is being used, such as that used in RTM or compression molding. Because one of the tool faces is flexible, the molded laminate thickness depends in part on the compressibility of the fiber–resin composite before curing and the vacuum's negative pressure is applied. Recent advances on room-temperature VARTM processing, large-scale process automation, and flow modeling behavior have been reported [37–39].

Composite panels were manufactured from AESO and various natural fiber reinforcements using the VARTM process. The composite panel

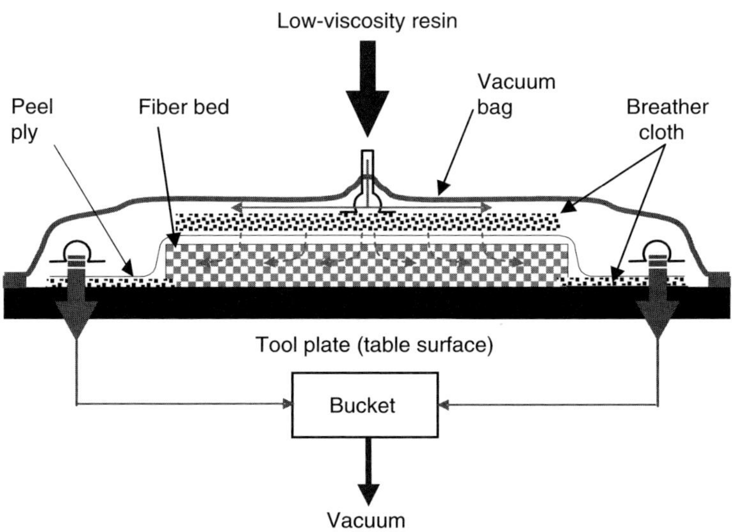

FIGURE 13.4 Schematic showing VARTM process and resin flow into the fiber bed.

specimens were manufactured with the dimensions $30.5 \times 30.5 \times 0.635$ cm ($12 \times 12 \times 0.25$ in.). The preform is vacuum bagged on a one-sided mold as shown in Figure 13.4 and the resin is drawn into the preform under the negative pressure created by the vacuum. The resin is cured at room temperature, which is important for housing construction, and gels after approximately 3–5 h. However, the panel specimens were left under vacuum overnight to improve consolidation and then demolded. Full vacuum, if reached, is equivalent to atmospheric pressure (14.7 psi or 101.3 kPa). This level of pressure may seem to be low, but it is very significant when parts with a large surface area (such as a house roof) are molded. A simple calculation shows that exerting this pressure over a 7- × 10-m roof panel would require a body force of about 709,000 kg.

13.3.2 PERMEABILITY MEASUREMENTS

To manufacture large composite structures using VARTM, permeability becomes a very important parameter that dictates the processing and quality of the final parts. Permeability is defined as the volume of a fluid of unit viscosity passing through a unit cross section of the medium in unit time under the action of a unit pressure gradient. It is a constant determined by the structure of the medium. The natural fiber mats used in this work can have random or oriented fibers, with or without binders; they can be processed using an air laid or wet laid process and the fiber length can be varied. The mat permeability plays a key role in determining the fiber content of the resulting composite.

Permeability is used in characterizing the flow behavior of a resin as it impregnates a preform or fiber bed. Darcy's law is commonly used to model the fluid flow through a porous media and to predict permeability parameters [40]. Darcy's law in one dimension describes the average velocity, $\bar{u}$ (m/s), and can be written as follows:

$$\bar{u} = \frac{K}{\mu} * \frac{dP}{dx}, \tag{13.1}$$

where K is the permeability parameter (m^2), μ is the viscosity of the injected fluid (resin) (Pa · s), P is the pressure (Pa), and x is the flow direction (m).

At constant flow rate, Q (m^3/s), the unsaturated permeability in a fiber bed with known dimensions and fiber content can be written as follows:

$$K_{\text{uns}} = \left(\frac{Q}{A}\right)^2 * \frac{\mu}{(1 - V_f)} * \frac{1}{\frac{dP}{dt}}, \tag{13.2}$$

where K_{uns} is the unsaturated permeability parameter (m^2), A is the cross-sectional area of the mold cavity (m^2), V_f is the fiber volume fraction in the cavity, μ is the viscosity of the injected fluid (resin) (Pa · s), and dP/dt is the pressure change rate (Pa/s).

This law describes the macroscopic relationship between the flow rate Q and the pressure change rate dP/dt rather than using the momentum conservation equations, which require more detailed information concerning the geometry of every channel present in a fiber bed. Therefore, Darcy's law accounts for the average ease of flow through these channels via the permeability parameter K, which characterizes the flow through the fiber-porous medium. Table 13.2 shows permeability readings for some of the natural fibers used in this study.

13.3.3 DYNAMIC MECHANICAL ANALYSIS (DMA)

Composite panels were made out of the 14 different fiber mats listed in Table 13.1. The storage modulus, E', the loss modulus, E'', and the glass transition temperature, T_g, were measured at a temperature range of 35.0–150.0 °C for the various room-temperature–cured AESO natural fiber composites. The glass transition temperature was obtained from the maximum point of the tan δ curve. The storage modulus E' of the neat resin was 1.1 GPa, and with natural fiber reinforcements, E' increased up to more than 5 GPa at approximately 50 wt% fiber. The highest E' values were obtained for cellulose derived from newspaper or recycled paper. Significantly, the recycled paper was the cheapest of all the natural fibers examined in this work and is, therefore, an excellent candidate for use in high-volume large structures, such as houses.

Values of the loss modulus were observed at two different temperatures corresponding to a temperature of 37 °C and also at the temperature at which the loss modulus achieves its maximum value. Similar behavior was observed for the loss modulus; the neat resin had a loss modulus of ~190 MPa, and with fiber reinforcement it increased to about 430 MPa. In the future, the carbonized chicken feathers discussed in Chapter 12 could add significantly higher reinforcement at low cost.

The structural or material damping of a composite material may also be analyzed using DMA testing. Tan δ is the ratio of the loss modulus to storage modulus or the ratio of the energy lost to the energy retained during a loading

TABLE 13.2 Permeability measurements.

Fiber Mat	Fiber Volume Fraction (%)	Unsaturated Permeability (m^2)
Recycled paper	42.3	3.60×10^{-13}
Flax mat 85/15 20 oz.	31.0	1.02×10^{-10}
Cellulose 200	18.3	6.00×10^{-10}
Chemically treated pulp	29.2	3.12×10^{-10}

cycle. Tan δ values were measured at 37 °C and also at its maximum value (the glass transition temperature). The most significant result was obtained from the cellulose composites with a maximum tan δ of approximately 0.3. This result indicates that natural fiber-reinforced (cellulose-based) composites have good structural damping properties that could be useful in the automotive industry.

Because the main objective of this study is utilizing natural and low-cost products to make composite materials, recycled paper from corrugated cardboard boxes was considered a cheap source for cellulose fiber. Old newspaper was initially considered and tested and despite the flow problem associated with making large parts using newspaper, the resulting composites exhibited very good mechanical properties at the lab scale. The positive experience with recycled newspaper directed our attention to a more porous recycled (cardboard) paper that showed no flow problems, and the resin perfectly infiltrated and bonded into the paper bed. The resin infusion can also be enhanced with more permeable layers of chicken feathers. Mechanical testing showed that these natural composite materials are suitable for housing applications such as roofs, walls, and construction lumber. Due to their reduced weight, environmental survivability, and noise suppression, several automotive applications can also be considered.

Last, the glass transition temperature of room-temperature–cured AESO is 66 °C and was relatively unaffected by fiber reinforcement. The T_g of AESO resin is directly related to the level of functionalization [*A*] and increases linearly with [*A*], as shown in Chapter 7. This value can also be improved by changing the curing conditions of the resin (i.e., high-temperature curing or postcuring), by altering the chemical structure of the resin, or by improving the adhesion between the fiber and the matrix with mixtures of other more polar soybean oil resins. Table 13.3 summarizes the DMA results at 37 °C for all the natural fibers in the AESO resin.

13.3.4 EFFECT OF FIBER CONTENT ON COMPOSITE STIFFNESS

Table 13.3 shows that the storage modulus increased with increasing fiber weight fraction in the different composite panels. Taking into account the complexity and irregularity of the natural fiber mats, the fiber modulus was back calculated using the rule of mixtures for every composite panel, depending on the experimental composite modulus, the pure resin modulus, and the fiber volume fraction. The fiber volume fraction is also a complex parameter because natural fibers are very porous materials and the resin may partially fill the voids in the natural fibers when the composites are made. This makes the actual density of the fiber in the composite a very difficult parameter to define. However, in this work the solid density of cellulose was used (1.5 g/cm^3) for the purpose of calculating fiber volume fraction. Natural fiber mats usually contain binder and sometimes other polymeric products to

TABLE 13.3 Summary of dynamic mechanical analysis data according to fiber mat wt% for the various room-temperature–cured AESO composites.

Composite Reinforcement/AESO	Description	Fiber Mat (wt%)	T_g (°C)	E′ (MPa)	E″ (MPa)
Pure AESO resin	Neat resin	0.0	66.0	1108	68
Cellulose 200	Cellulose 200 g, by Concert, Canada	19.1	73.5	2176	201
Cellulose 150	Cellulose 150 g, by Concert, Canada	20.0	57.0	1989	246
Flax mat 85/15 20 oz.	Flax mat 85/15 20 oz., by Cargill, Durafiber	23.0	57.5	2090	260
Flax mat CCF200C	Flax/PET 40/40, by Cargill, Durafiber	23.5	68.0	1404	171
Flaxtech flax	Flax supplied FlaxTech	29.5	67.0	1493	199
Flax mat 60/40	Flax mat 60/40, by Cargill, Durafiber	30.0	70.0	2072	200
Rayfloc XJE	Chemically treated pulp, by Rayonier	45.9	71.3	3673	341
Rayfloc JLDE	Fluff pulp, by Rayonier	47.8	73.0	4550	320
CTMP pulp	Chemical thermal mechanical pulp, by M&J Fibretech a/s, Denmark	53.4	73.5	3776	285
Recycled paper	Recycled corrugated cardboard, by Interstate Resources, PA, USA	55.2	62.5	5242	361
Porosanier JHP	Caustic treated pulp, by Rayonier	57.7	73.0	4556	287
Flaxcraft hemp	Hemp mat, by Flaxcraft Inc.	—	65.5	2160	271
NewsJournal	Newspaper	—	68.0	4874	375
E-glass fiber	Woven E-glass fiber	75.2	61.5	17,947	916

keep the mat together. In this case, the binder and other tackifiers in the mats were considered part of the matrix. The fiber volume fraction was based on the amount of pure cellulose content in the mats. Calculation of the fiber modulus for every natural fiber composite panel yielded an average value of $E_f = 8.9\,\text{GPa}$. This value was then used to model the storage modulus of the composite E, following the rule of mixtures [Eqs. (13.3) and (13.4)] and the Halpin-Tsai [Eq. (13.5)] micromechanical models [41]. The pure resin modulus was experimentally measured as $E_m = 1.1\,\text{GPa}$.

Rule of mixtures:

$$E = E_f V_f + E_m V_m \qquad \text{(upper bound)} \tag{13.3}$$

$$1/E = V_f/E_f + V_m/E_m \qquad \text{(lower bound).} \tag{13.4}$$

Halpin-Tsai model:

$$E = E_m \left(\frac{1 + \xi \chi V_f}{1 - \chi V_f} \right), \tag{13.5}$$

where

$$\chi = \frac{E_f - E_m}{E_f + \xi E_m}, \tag{13.6}$$

in which E is the modulus of the composite (MPa); E_f and E_m are the modulus of the fiber and the matrix, respectively (MPa); V_f and V_m are the volume fraction of the fiber and the matrix, respectively; and ξ is the Halpin-Tsai parameter. The ξ parameter serves to adjust the modulus value between the upper and lower bound solutions. When $\xi = \infty$, the Halpin-Tsai model gives the upper bound rule of mixtures [Eq. (13.3)] in which the fiber stiffer modulus dominates, and when $\xi = 0$, the Halpin-Tsai model gives the lower bound rule of mixtures [Eq. (13.4)], where the less stiff resin modulus dominates. The ξ-value has often been related to the aspect ratio of the fibers such that high aspect ratios approach the upper bound, while lower aspect ratios approach the lower bound, as expected. Figure 13.5 shows that the upper bound rule of mixtures and the Halpin-Tsai analysis, with $\xi = 100$ or higher, provides the best fit for this experimental data. Discrepancies in the experimental data are due to the nature of the reinforcement fiber mats as discussed earlier and due to the fact that each point came from different experiments and different natural fiber mats. Further work with rice straw and cotton/flax mixtures is planned.

13.4 APPLICATIONS: HOUSING CONSTRUCTION MATERIAL

13.4.1 BEAM DESIGN

To study the possibility of manufacturing the roof presented in Figure 13.2, beams of the proposed material were first manufactured and tested for strength requirements. Figure 13.6 presents a schematic of the prototype beam studied in the investigation. The beam is of a sandwich construction with a top horizontal face sheet, bottom face sheet, and two vertical webs, essentially resembling two I-beams. The overall dimensions of the beam are 1067 × 89 × 203 mm (42 × 3.5 × 8 in.); the face sheets as well as the webs have a nominal thickness of 6.4 mm (0.25 in.). The foam core is required for

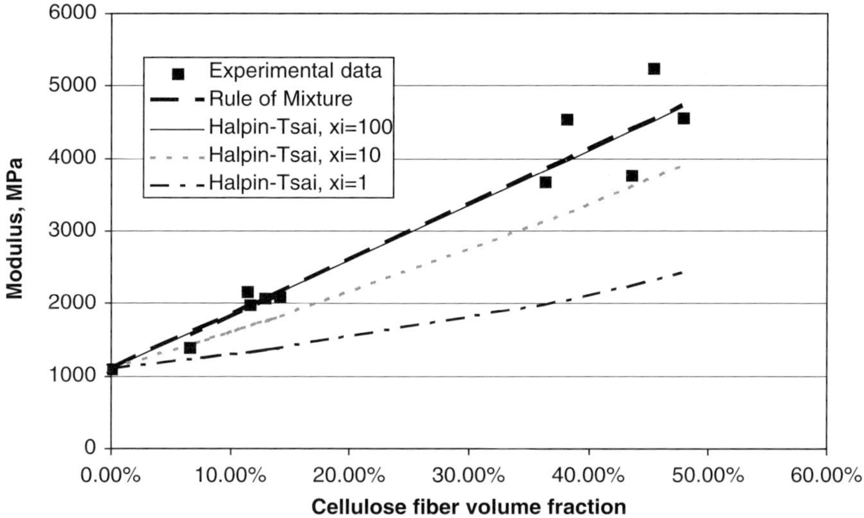

FIGURE 13.5 Modeling the storage modulus as a function of fiber volume fraction using the composites rule of mixtures and Halpin-Tsai model changing ξ between 1 and 100.

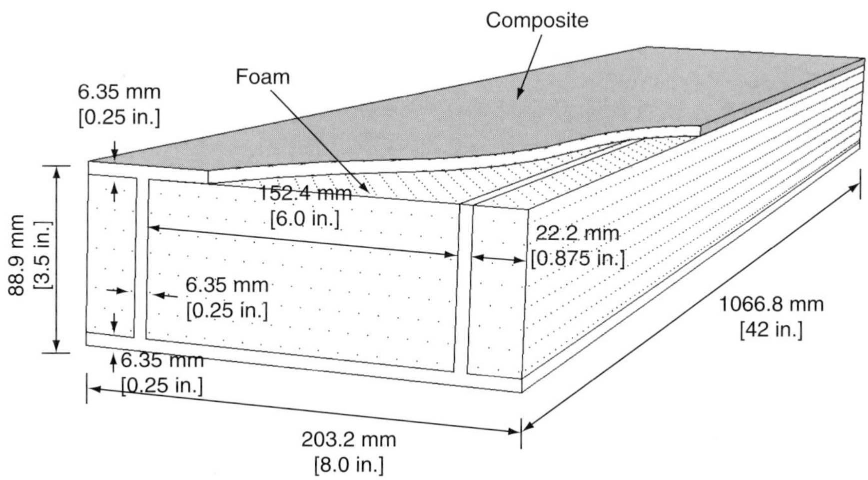

FIGURE 13.6 A schematic diagram showing the dimensions of the test beams.

the manufacture of the beam and is integral to it, but while it contributes significantly to thermal and sound insulation, it is not expected to contribute significantly to the strength and stiffness of the member. The beam was designed as a flexural member to carry loads transverse to the longitudinal axis of the beam. A detailed structural engineering analysis was conducted by the civil engineering team of Tripp Shenton and Bo Hu

[12, 13]. Note that for this preliminary investigation by M. Dweib *et al.*, the beams were not designed to satisfy any specific structural design criteria. Instead, the specimen was designed to be of a size and shape that would facilitate testing and expose any limitations or difficulties in the fabrication of the member. However, data presented in a later section show a comparison of the composite beams with some typical structural sections made from conventional construction materials.

13.4.2 STRUCTURAL COMPOSITE MANUFACTURING USING VARTM

The VARTM process as discussed earlier in Section 13.3.1 of this chapter was also used to manufacture the foam-core structural composites, and will be used for the entire roof. As stated earlier, recycled paper produced surprisingly good flat sheet composites. However, the first attempt at making a three-dimensional structure was not successful due to resin flow problems through the web, which are the vertical sections connecting the top and bottom plates of the I-beam units. To overcome the flow problem, other porous fibers were used in small quantities along with the main reinforcement (recycled paper) to provide flow channels for the resin, especially through the beam web. Figure 13.7 is a schematic showing how porous fiber mats were used in combination with the recycled paper to provide better flow. A properly infused beam made with mixed fibers, as shown in Figure 13.8, gave the best results. Three different types of porous fiber mats, or fibers with an open channel structure, were used for this purpose:

- Chicken feather mats
- A corrugated form of the same recycled paper
- Woven E-glass fiber

The materials were selected to improve processing, strength, or to reduce cost.

The concept for using chicken feathers as a polymer composite reinforcement originated in discussions between the ACRES group and Tyson, Ltd., a major producer of chicken products in the region. Chicken feathers, as described in Chapter 12, are a waste product and are generated in large quantities. Tyson Ltd. agreed to make mats of about 97% chicken feathers and a 3 wt% low-molecular-weight binder polymer, an example of which was shown earlier in Figure 13.3. The chicken feather mats were then used by the ACRES group to manufacture composites [42]. They improved the resin modulus of the neat resin by a factor of 2 and decreased the density and dielectric constant, while increasing thermal and sound resistance, as discussed in Chapter 12. In this application, however, the mats, which are very porous, were used to provide flow channels for the resin to better distribute

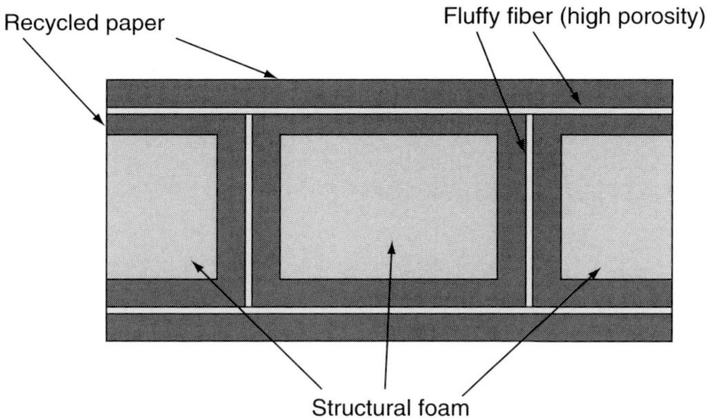

FIGURE 13.7 A schematic showing the porous fiber in the preform.

FIGURE 13.8 A view of the recycled paper/chicken feathers beam.

the resin to the dense and compacted recycled paper. Chicken feathers were used primarily because they are an inexpensive, available waste product and were proven to be good for providing the channels needed for the resin flow while also increasing the resin strength.

As a second alternative in the design, a single ply of corrugated cardboard, made from the same recycled paper, was used to provide the channels needed to facilitate flow of the resin through the preform web. The advantage of using the corrugated paper is that all of the reinforcement material for the beam is from one source, rather than two or more sources. As the final modification to the design, woven E-glass was used to provide the needed channeling for the resin. The E-glass was added to study the effect of a small amount of glass on the strength, stiffness, and ductility of the beam.

13.4.3 PROPERTIES OF ROOF BEAMS

Four-point bending tests (Figure 13.9) were done on each of the structural beams, which were loaded to failure. This testing gives load versus deflection results and strain measurements. All data were obtained using a data acquisition system. The specimen was first loaded reversibly three times in the elastic region, and then was taken to failure; Figure 13.10A shows a broken beam and the fracture surface after testing. This type of failure was noticed for flax beam, recycled paper with chicken feathers, and recycled paper with corrugated recycled paper. The different beams were tested to failure, and Table 13.4 shows results for three beams made of (1) recycled paper from old cardboard boxes and three different interlaminar or integral distribution media in a form of chicken feathers mat, (2) one ply corrugated paper, and

FIGURE 13.9 Flax/soy resin beam under four-point bending loading.

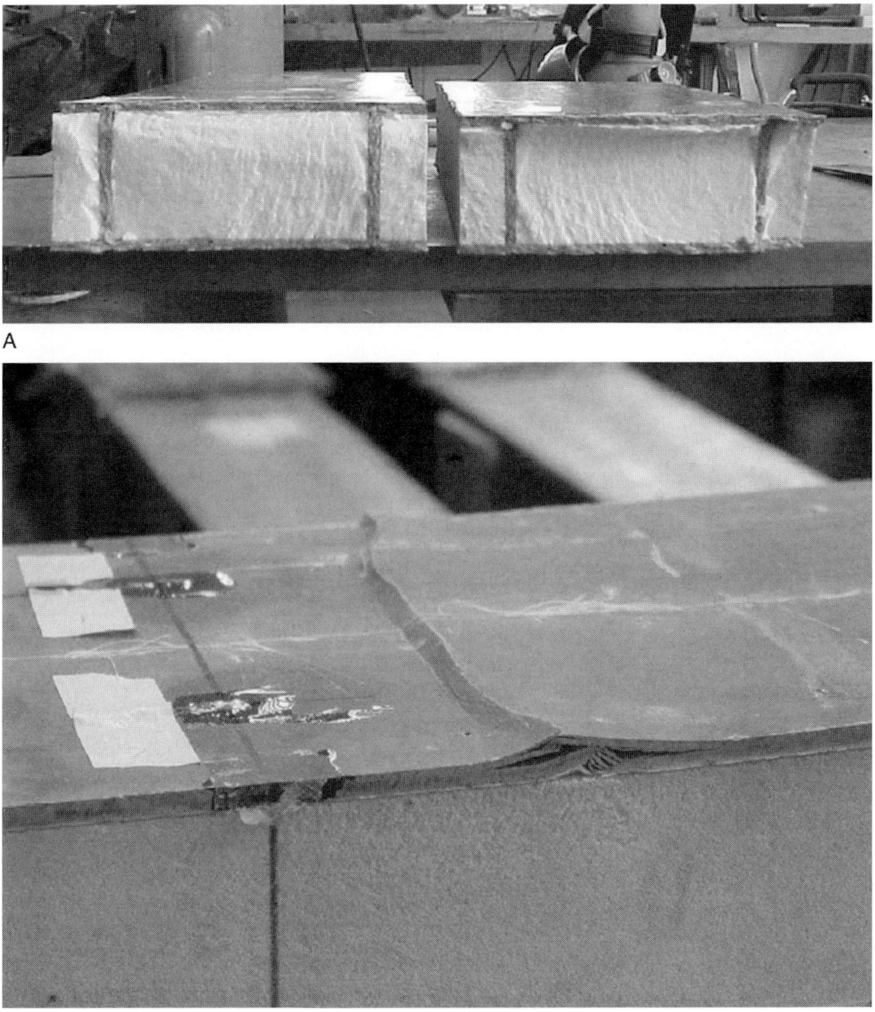

A

B

FIGURE 13.10 Fracture surface showing failure modes: (A) brittle failure of flax beam and (B) ductile failure of recycled paper with one ply of woven E-glass fiber.

(3) one ply of woven E-glass fiber. Table 13.4 also presents a comparison between the beams and the three most common wood structures used in building construction and shows that the newly developed material properties matched or superseded that of the wood structures. Using woven E-glass fiber ply as an interlaminar integral distribution media provided ductility and prevented the undesired brittle failure. Figure 13.10B shows the failure mode of the beam made of recycled paper and one ply of woven E-glass fiber. Additional improvements to the hurricane-resistant roof design, including

TABLE 13.4 Comparison of composite beam properties versus typical wood section properties.

Beam	Flexural Rigidity EI(kN-m^2)	Strength (kN)
Composite Beams		
Recycled paper/chicken feathers	12.4	24.2
Recycled paper/corrugated paper	14.8	25.8
Recycled paper/E-glass fiber	19.9	25.6
Wood Beams		
Douglas fir	18.0–30.3	15.4–29.7
Spruce	16.0–25.0	10.7–24.5
Cedar	10.0–26.4	9.5–28.8

the use of nanoclay coatings (described in Chapter 15), for the weather-resistant top surface, are being developed by the ACRES group.

13.5 DESIGN OF THE BIO-BASED COMPOSITE ROOF

The bio-based composite roof was designed for a hypothetical ranch-style house. The plan size of the house is 7.32 m (24 ft) by 15.24 m (50 ft). The elevation of the eave is 4.75 m (15 ft), and the rise of the roof is 3.05 m (10 ft). Instead of a conventional rafter/truss and plywood roof, the roof will be a monolithic composite web core sandwich panel with one-way webs running from the eave to the ridge. A schematic of the hypothetical house is shown in Figure 13.11. It should be emphasized here that the sandwich panel roof is complete; there is no need for rafters or roof trusses to support the panel.

The design loads for the roof were determined in accordance with ASCE 7–93 (*Minimum*, 1993). The building classification is Category I. The ground snow load was assumed to be 2633 N/m^2 (55 psf), which is an average level for the New England area, and a design wind speed of 177 km/h (110 mph) was assumed, which is the highest value in the ASCE 7 basic wind speed chart. For this phase of the study, the house was assumed to be in a region of low seismicity and, therefore, earthquake loads were ignored.

Per the geometry of the house and the roof panel, the roof live load was calculated to be 570 N/m^2 (12 psf), and the slope-roof snow load was 1010.5 N/m^2 (21.2 psf). For the determination of wind load, two methods were used: One was to consider the roof as a component or cladding, and the other was to consider it a part of the main wind-force-resisting system. The most unfavorable result from these two methods was adopted in the load combination. The maximum wind load is the uplift pressure on the leeward side of the roof, and has a value of -2772 N/m^2 (–57.9 psf), while the downward pressure on the windward side is 1920 N/m^2 (40.1 psf). Rain

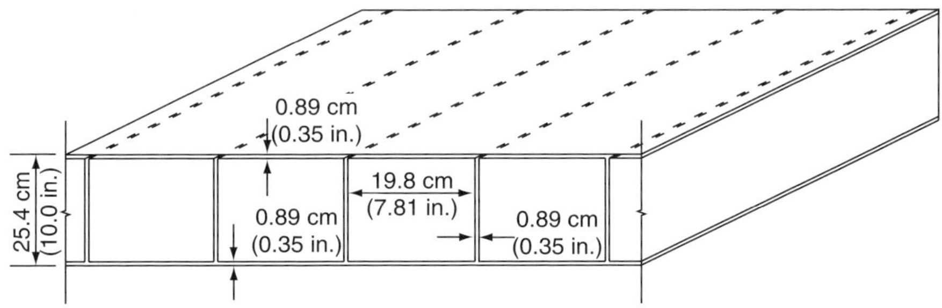

FIGURE 13.11 Cross section of monolithic roof panel and its dimensions.

load was neglected because of the unspecified drainage conditions. The load combinations considered included:

- Dead load
- Dead load + live load + (roof live load or snow load)
- Dead load + wind load
- Dead load + live load + (roof live load or snow load) + wind load

The controlling load combination was found to be the uplift wind load. The final design normal load, ignoring the dead load, which is later shown to be insignificant, for the roof panel was $2772\,N/m^2$ (-57.9 psf).

The support conditions and aspect ratio of the hypothetical roof panel are such that, for design, it can be envisioned as a one-way slab; therefore, the roof can be modeled as a beam of unit width. However, because of the orthotropic and discontinuous nature of the proposed sandwich panel, the unit beam should include the recurrence of webs. The design of the roof panel will depend on the available mechanical property data, the unit beam test results [12], and engineering judgment.

The deflection limit criterion for the roof is $L/240$ (AITC 1994), which yields an allowable deflection of 1.98 cm (0.78 in.) for the 4.76-m (15.62-ft) span. Ignoring the contribution of webs to the overall bending stiffness, imagine a beam of width 30.5 cm (1 ft) is cut from the panel. The design load for the unit width beam is 845 N/m (57.9 lb/ft). The depth of the beam and thickness of the face sheets were determined by optimizing the cross section, considering the strength and stiffness design criteria, and the following constraints: (1) The thickness of the face sheets was minimized to reduce the amount/cost of resin and reinforcement; and (2) the overall depth is in the range of 20.3 cm (8 in.) to 30.5 cm (12 in.), in order to provide good insulation and a depth that is comparable to, but not greatly in excess of, the total depth of a conventional roof. The overall depth of the roof panel was selected to be 25.4 cm (10 in.). To satisfy the deflection criterion, the thickness of the top and bottom faces was determined to be 0.89 cm (0.35 in.). Using the tensile modulus of the material reported by Anne O'Donnel *et al.* [10] [5.97 GPa (866 ksi)] the estimated flexural deflection for the panel is 1.08 cm (0.43 in.). Although the deflection due to shear deformation could not be precisely predicted at this stage of the design, the theoretical flexural deflection amounts to only 55% of the maximum allowable deflection, thus providing some allowance for uncertainties due to shear deformation and other factors that could contribute to the total deflection in the actual roof.

The web thickness was selected to be 0.89 cm (0.35 in.), the same as the top and bottom face sheets, for simplicity of manufacturing. The distance between webs will affect the shear rigidity of the section, the local bending of the top and bottom face sheets, and the stability of the webs; however, at this stage of the investigation only flexure was considered in detail. A web spacing of 19.84 cm (7.8 in.) was chosen based on engineering judgment. A schematic of the final sectional dimensions of the roof panel is shown in Figure 13.11.

The roof design must satisfy both strength and stiffness design criteria. With the above design detail, the theoretical maximum bending stress in the faces is 3.47 MPa (504 psi), which is about 10% of the ultimate tensile strength (45.7 MPa) of the material. The maximum design shear stress in the web is 0.60 MPa (87.9 psi), which is far less than the maximum shear stress of 13.34 MPa (1935 psi) developed in the test of specimen RW-3 as reported in Ref. [12]. Hence, as expected, strength is not a governing factor in the design of the roof. This is also supported by the results observed in the test of specimen RW-3 in Ref. [12], in which a very large deflection/span ratio (1/31) was reached at failure, which indicates that the strength criteria, such as maximum normal stress and maximum shear stress, will not be critical in the design of the roof panel.

The dead load of the roof panel is 45.3 kg/m^2 (9.3 lb/ft^2) based on the composite density of 1.2 g/cm^3 and the foam density of 0.045 g/cm^3. The estimated deflection due to dead load is 0.17 cm (0.067 in.). The stresses caused by the dead load in the faces and webs are 0.56 MPa (81 psi) and 0.096 MPa (13.9 psi) respectively, which are insignificant compared to the material strength.

13.6 DESIGN, TESTING, AND EVALUATION OF A MODEL BEAM

13.6.1 BEAM DESIGN

Once the roof had been designed, the next phase of the study involved fabricating and testing a unit width section of the roof. A full-scale unit width of the roof panel could not be manufactured, however, because of limitations with existing facilities, which limited the maximum span of any specimen to 2.29 m (7.7 ft). Instead, a scale model specimen of span length 2.13 m (7 ft) was designed, fabricated, and tested. To provide a direct correlation between the scale model specimen and a full-scale roof, appropriate scaling laws were followed as described next.

For a simply supported beam made of isotropic material, under transverse uniformly distributed load, the maximum deflection (Δ) and the maximum stress (σ) developed at the midspan can be expressed as follows:

$$\Delta = f(E, G, I, w, A, L), \tag{13.7}$$

$$\sigma = g(E, G, I, w, A, L), \tag{13.8}$$

where E = Young's modulus, G = shear modulus, w = the magnitude of the distributed load, I = moment of inertia, A = cross-section area, and L = span length.

According to the Buckingham pi theorem [43], these relationships can be rewritten in a dimensionless form that can be used to design the scale model specimen:

$$\frac{\Delta}{L} = f_d\left(\frac{w}{EL}, \frac{I}{L^4}, \frac{A}{L^2}, \frac{G}{E}\right), \tag{13.9}$$

$$\frac{\sigma}{E} = g_d\left(\frac{w}{EL}, \frac{I}{L^4}, \frac{A}{L^2}, \frac{G}{E}\right). \tag{13.10}$$

Note that if the same material is used for the prototype and the model beam, the last dimensionless parameter in the preceding equations will automatically be equal for both. In fact, even for anisotropic materials, because the different directional elasticity properties have the same units, using the same material will eliminate the need for extra dimensionless parameters to carry out a scale model test, and the above dimensionless form for an isotropic material will still hold. In that case one would only need to change the Young's modulus to one of the directional elasticity constants.

From Eqs. (13.9) and (13.10), we can also observe that if the same material is used and all the dimensionless parameters are equal for both the prototype and the model specimen, the deflection ratio observed (deflection/span length) in the test is the deflection ratio for the prototype, whereas the measured stress level is the expected stress level in the prototype structure. In this way the results of the model test can be directly correlated to the full-scale beam roof.

Based on the fabrication constraints and the Buckingham pi theorem, a scale model test specimen was designed. The scaling factor was $7/15.62 = 0.4481$. A schematic of the model beam section is shown in Figure 13.12. The thickness for all faces and webs was 0.40 cm (0.16 in.), the depth of the beam was 11.4 cm (4.5 in.), and the width of the beam was 18.59 cm (7.32 in.).

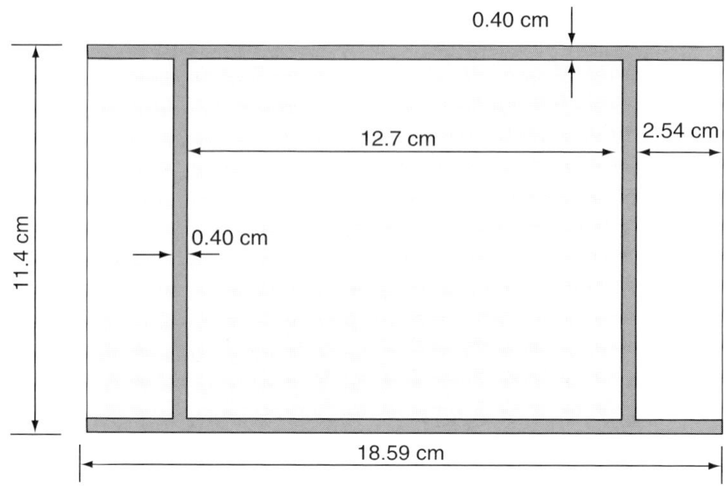

FIGURE 13.12 Section of the test beam.

The scale model beam was fabricated in the same manner as specimen RW-3 in Ref. [12]. Layers of recycled paper, with corrugated paper as the resin flow channel, were first wrapped around the foam cores. A vacuum bag was then sealed around the preforms and the AESO resin infused. A more detailed description of the material preparation and fabrication process can be found in Dweib *et al.* [9, 11] and Hu *et al.* [12]. A photograph of the cured beam after fabrication is shown in Figure 13.13. The edges of the beam shown in Figure 13.13 were trimmed to create the section as designed.

13.6.2 BEAM TESTING

Tests were conducted of the model beam for the purposes of (1) estimating the flexural and shear rigidities of the beam, (2) verifying that the design satisfies the deflection limit criteria, and (3) measuring the ultimate load capacity and mode of failure of the beam. The flexural and shear rigidities were estimated from the results of several three- and four-point bending tests, the deflection criterion was confirmed in a quasi-distributed load test, and the beam was tested to failure in three-point bending.

Stiffness Results

While the load in this test was applied manually and in increments, the strain data were continuously recorded. The time-dependent deformation of the beam can be observed; however, after a short period of unloading the full viscoelastic deformation is restored.

FIGURE 13.13 Test beam after debagging.

Key results of the three- and four-point bending tests are presented in Tables 13.5 and 13.6, respectively. Midspan deflection and average midspan strain are presented for each loading. All values are the instant elastic deformation and do not include the time-dependent viscoelastic component. In a quasi-distributed load test, the midspan elastic deflection was 0.556 cm (0.219 in.) and the average elastic midspan strain was 501 $\mu\varepsilon$.

Strength Results

Load versus midspan deflection from the test is shown in Figure 13.14. Note that since the time to restore the viscoelastic deformation was limited, the initial deflection for loading to the next level is not zero. Hence, the measured deflection is the sum of the elastic deformation and a residual deformation from the past loadings. Relaxation can also be observed during the period of constant displacement: When the displacement of the loading head is held constant at 69.4 mm for 98 s, the applied load decreases from 5.92 to 5.38 kN, a decrease of 9.1%. After two cycles of loading, the beam failed on its way to the third loading level. An ultimate load of 6583 N (1480 lb) was reached, with a simultaneous midspan deflection of 8.76 cm (3.45 in.). The corresponding maximum reading from the four strain gauges at the midspan section was 13,254 $\mu\varepsilon$ in tension. The beam failed in a brittle manner, breaking into two parts with little warning. The location of the

TABLE 13.5 Three-point bending test summary

Span Length (m) (1)	Load (N) (2)	Midspan Deflection (mm) (3)	Average Midspan Strain (4)
2.13 (7 ft)	103.2	0.81	88
	206.8	1.61	181
	310.9	2.49	274
1.83 (6 ft)	103.2	0.49	75
	206.8	1.04	150
	310.9	1.59	240
1.52 (5 ft)	103.2	0.30	60
	206.8	0.60	120
	310.9	0.91	187
1.22 (4 ft)	103.2	0.15	46
	206.8	0.33	94
	310.9	0.53	147

TABLE 13.6 Four-point bending test summary

Span Length (m) (1)	Load (N) (2)	Midspan Deflection (mm) (3)	Average Midspan Strain (4)
2.13 (7 ft)	207.7	1.22	120
	413.7	2.54	239
1.83 (6 ft)	207.7	0.79	102
	413.7	1.57	205
1.52 (5 ft)	207.7	0.43	86
	413.7	0.97	172
1.22 (4 ft)	207.7	0.30	76
	413.7	0.58	147

break was almost exactly the midspan, where the maximum strain is expected to develop.

13.6.3 MODELING AND ANALYSIS OF THE MODEL BEAM

Including the effects of shear deformation, the deflection at midspan for three-point bending and four-point bending is as follows:

$$\Delta_{3p} = \frac{PL^3}{48EI} + \frac{PL}{4kGA}, \tag{13.11}$$

$$\Delta_{4p} = \frac{Pa}{4EI}\left[\frac{L^2}{4} - \frac{a^2}{3}\right] + \frac{Pa}{2kGA}, \tag{13.12}$$

where Δ_{3p} and Δ_{4p} are the midspan deflection under three-point and four-point bending, respectively; P is the total load; a is the distance from the load

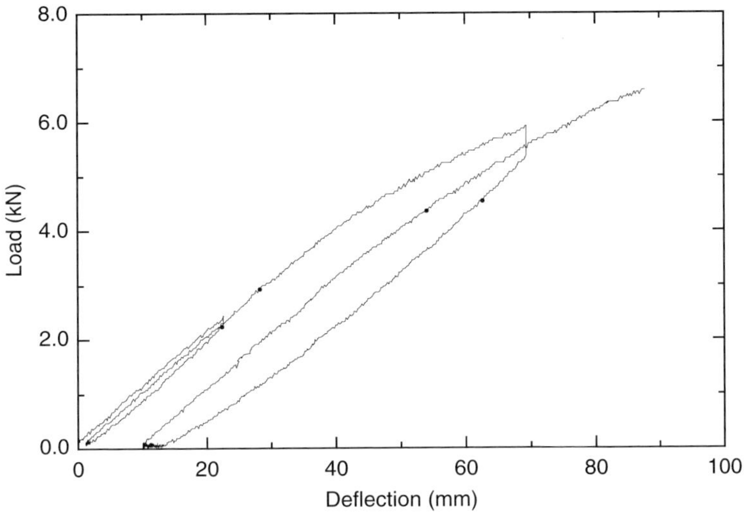

FIGURE 13.14 Load–deflection plot for strength evaluation.

point to the support point, and $a = L/3$ for all tests reported herein; EI is the flexural rigidity of the cross section; GA is the shear rigidity of the cross section; and k is the shear coefficient to reflect the nonuniform distribution of shear stress in the cross section.

Allen [44] presented a method to estimate the flexural and shear rigidities of a sandwich beam based on deflection test data of the same beam with different span lengths. The idea of using multiple-span deflection tests was adopted by Sims *et al.* [45], Bank [46], Giroux and Shao [47], and Roberts and Al-Ubaidi [48] and has been adopted in ASTM C393. Nagaraj and Gangarao [49] obtained the theoretical cross-sectional rigidities from micromechanical methods and then combined them with the test deflection and strain data to estimate the shear and flexural rigidities of pultruded GFRP beams. Zureick *et al.* [50] used the deflections at midspan and the loading points of a four-point bending test to estimate flexural and shear rigidities of FRP members.

In the following, the method proposed by Allen [44] is adopted for estimating the flexural and shear rigidities of the beam. Furthermore, a new approach is presented that makes use of both deflection and strain data from three- and four-point bending tests of different span lengths. The results from the two methods are compared.

Flexural and Shear Rigidities Based on Deflection Data

The load–deflection relationships in Eqs. (13.11) and (13.12) can be rearranged as follows:

$$\frac{\Delta_{3p}}{PL} = \frac{L^2}{48EI} + \frac{1}{4kGA}, \tag{13.13}$$

$$\frac{\Delta_{4p}}{PL} = \frac{23L^2}{1296EI} + \frac{1}{6kGA}. \tag{13.14}$$

Letting L^2 be the independent variable, Eqs. (13.13) and (13.14) are linear equations in which the slope of the line is a function of *EI* and the *y*-intercept is a function of GA. Plotting Δ/PL versus L^2 for different span lengths, the slope of the best fit line is used to calculate the flexural rigidity, while the intercept is used to calculate the shear rigidity [44].

Flexural and Shear Rigidities Based on Deflection and Strain Data

Although shear deformation has little effect on the measured longitudinal strain on the top and bottom faces of the beam, the deflection of the beam consists of both the bending and the shear components, per Timoshenko beam theory. A new approach for estimating the flexural and shear rigidities is presented here that uses deflection and strain data from the tests at different span lengths.

The average strain ($\bar{\varepsilon}$) on the top and bottom faces of a beam section is

$$\bar{\varepsilon} = \frac{|\varepsilon_{\text{top}}| + |\varepsilon_{\text{bottom}}|}{2} = \frac{1}{2}\frac{M}{EI}h_t + \frac{1}{2}\frac{M}{EI}h_b = \frac{M}{EI}\frac{h_t + h_b}{2} = \frac{M}{EI}\frac{h}{2}, \tag{13.15}$$

where h_t is the distance from the top face to the neutral axis, h_b is the distance from the bottom face to the neutral axis, h is the depth of the cross section, and M is the applied moment. Hence, regardless of the location of the neutral axis, the flexural stiffness of the cross section can be experimentally estimated:

$$EI = \frac{M}{\bar{\varepsilon}}\frac{h}{2}. \tag{13.16}$$

Once the flexural stiffness is obtained using Eq. (13.16), the shear stiffness can be calculated from the measured deflection using Eq. (13.11) for three-point bending or Eq. (13.12) for four-point bending. For three-point bending, the estimated flexural and shear stiffnesses are

$$EI = \frac{PLh}{8\bar{\varepsilon}} \quad \text{and} \quad kGA = \frac{3PLh}{12h\Delta - 2L^2\bar{\varepsilon}}, \tag{13.17}$$

and for four-point bending are

$$EI = \frac{PLh}{12\bar{\varepsilon}} \quad \text{and} \quad kGA = \frac{18PLh}{108h\Delta - 23L^2\bar{\varepsilon}}. \tag{13.18}$$

Shear Component of the Deflection

It is worthwhile to examine the actual contribution of the shear deformation to the total deflection in the composite beam. The deflection caused by shear deformation can be obtained by subtracting the flexural deflection from the measured total deflection using Eq. (13.11) or (13.12) for the three- and four-point bending test, respectively. Figure 13.15 shows the calculated result of shear deflection in a percentage form compared to the total deflection for both the three-point and four-point bending tests. In the plots, the general decrease of the shear component with the increase in span length is in accordance with theoretical prediction. Four-point bending causes less shear deformation than three-point bending because of the central constant moment region. For the longest span length of 2.13 m (7 ft), which has a span/depth ratio of about 19/1, the shear component is between 25% and 30% for three-point bending and approximately 20% for the four-point bending test.

Correlation of the Scale Model Results to Full Scale

The quasi-distributed load test is the true scale model test for the designed roof panel under the design load; therefore, the test results can be used to predict the behavior of the full-scale roof. The observed maximum deflection in the quasi-distributed load test was 0.56 cm (0.219 in.), which corresponds to a deflection of $0.56/0.4481 = 1.25$ cm (0.488 in.) in the full-scale structure. This is 63% of the maximum allowable deflection under the design load of 1.98 cm (0.78 in.) for the full-scale roof. The midspan bending moment for the quasi-distributed load test, that is, the scaled design load, was 307 N-m (2720 lb-in.). The bending moment at failure, based on the ultimate load test, was 3512 N-m (31,080 lb-in.). Therefore, the strength of the beam is about 3512/307, or about 11 times greater than the design load. These facts, along with the scaling law, reveal that for the hypothetical house, a roof panel with the proposed material and sandwich construction would satisfy the design requirements and is somewhat conservative.

13.7 BUILDING A BIO-BASED COMPOSITE ROOF

Based on the results presented, a demonstration roof was designed and two-dimensional panels were fabricated using the VARTM process. Figure 13.16 shows the roof under construction with a 2.59- × 1.52- × 0.089-*m* roof panel, made of recycled paper, soybean oil-based resin, and structural foam mounted on a demonstration house made of timber. To attach the composite roof to the house walls, wood inserts were used in the core. The 127-mm (5 in.) wood blocks were placed exactly in the area where the roof would be attached to the walls and a ridge beam. This roof was being completed at the time this book was published.

13.7.1 OYSTER LA VISTA ROOF DESIGN

Because the roof is molded in place on the house, it is possible to have many different shapes, in addition to the usual A-frame structure. For

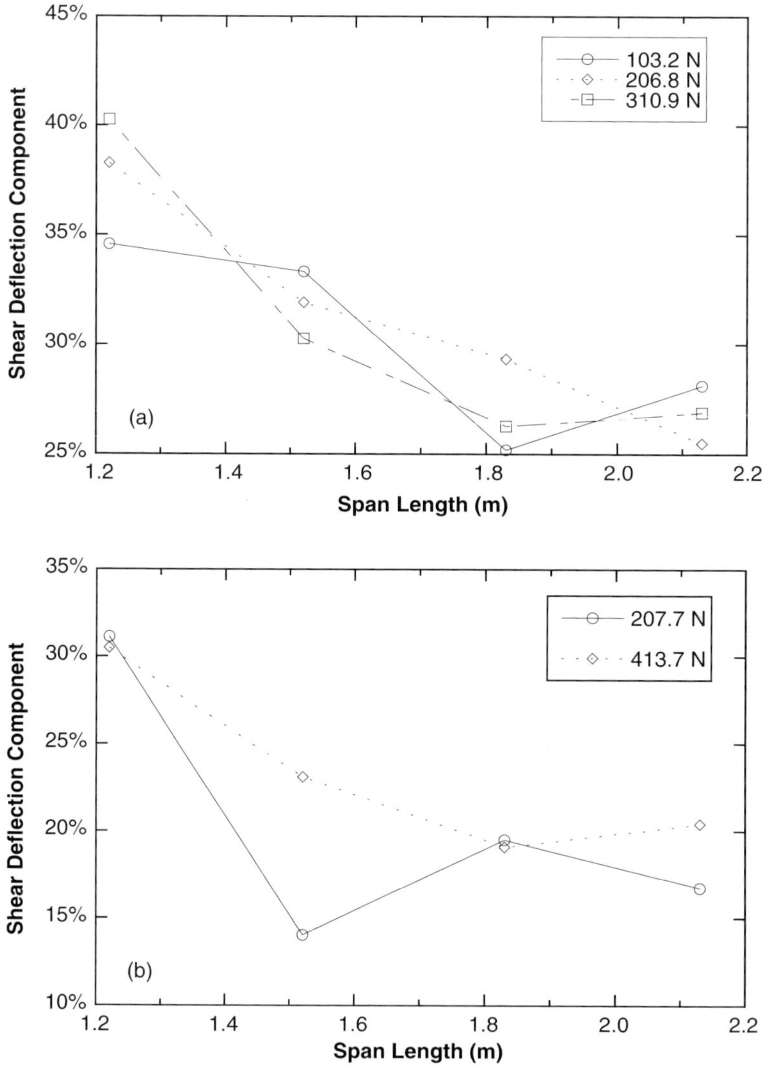

FIGURE 13.15 Shear deflection component at midspan for (a) three-point bending and (b) four-point bending.

FIGURE 13.16 The composite panel is shown as a roof top on an experimental house made of wood.

example, Figure 13.17 shows a roof design in the shape of an oyster, created by Liz Linstrom, of NY Lizzyloo Designs. The design was intended to be more amorphous than rectilinear and have a soft aerodynamic profile. Its amorphous nature is very different than the typical A-frame construction but it can have considerable aesthetic appeal and weather resistance.*

13.8 OTHER POTENTIAL APPLICATIONS

Natural composites developed in this research program are potentially cheaper than petroleum-based resin products, available in large quantities, and mostly dependent on annual crops like soybean, flax, jute, hemp, and other cellulose-rich plants. This has an advantage over using wooden structures that need fully grown trees that take tens of years to reach that stage. In addition to the economic advantage, there are also environmental and energy advantages to using natural composites. Natural composites are not only good as housing structural material, as presented earlier, but also can be used as infrastructural material such as bridge decking forms, and in furniture manufacturing to replace traditional woods and make it easier to mold furniture components, saving the intensive labor usually invested in the woodwork.

*The "Oyster la Vista" name was selected 3 years before Arnold Schwartzeneger entered politics.

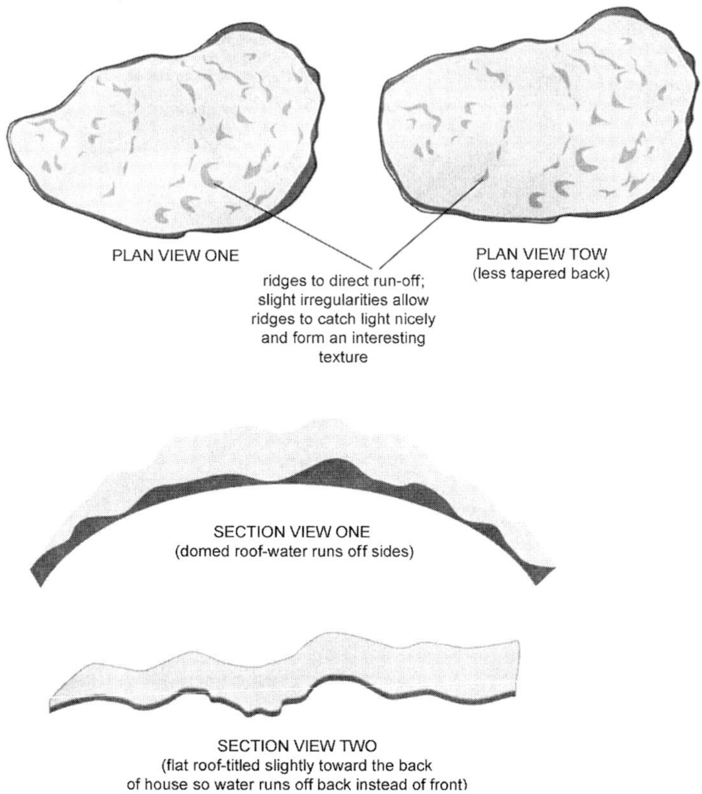

FIGURE 13.17 "Oyster La Vista" Roof, *Architectural Record*, p. 198, November 2003. (*Source:* Courtesy of Liz Linstrom.)

13.8.1 STAY-IN-PLACE BRIDGE DECKING FORM

Stay-in-place (SIP) bridge forms are corrugated sheets of material that span the distance between bridge girders. SIP forms are formwork for the concrete bridge deck and are designed to carry the dead load of the deck while the concrete cures. Before SIP forms were invented, wooden formwork was used for the same purpose, however, wooden forms are labor intensive, requiring scaffolding to be built from the ground up and then requiring removal after the concrete has cured. The first revolution in form design occurred when forms were made of corrugated steel to replace the wooden formwork. This was the birth of SIP forms. It occurred more than 16 years ago, and SIP forms are widely used today. SIP forms are made of light-gauge steel and are approximately 0.610 m (2 ft) wide by 1.22 m (4 ft) to 3.05 m (10 ft) long. They are screwed into angles that are welded onto bridge stringers.

The benefits of SIP forms include speed and ease of installation and decreased labor costs because the forms do not have to be removed after the decking cures. However, conventional SIP forms do have several disadvantages. The first of these is that the steel pans can collect moisture, keeping it trapped between the steel and the bottom of the concrete deck. This can cause corrosion of the concrete rebar, leading to possible failure of the deck. Another problem with steel SIP forms is that the form itself prevents bridge inspectors from inspecting the deck since the SIP form hides cracking or corrosion that can occur at the bottom of the deck. There are, however, several advantages of a natural composite SIP form over a conventional steel form. The first of these is that the form would be able to "breathe," or allow water to pass through and away from the concrete deck, therefore reducing the risk of corrosion. Second, the form would be biodegradable, and would break down naturally to allow bridge inspectors to examine the bottom of the deck. Finally, the forms would be lightweight compared to their steel counterparts, and would allow for faster installation and lower labor costs.

The first SIP form manufactured at the University of Delaware was made of soybean oil-based resin and woven E-glass fiber. The form was successfully manufactured using the VARTM process and then was tested in three-point bending and *in situ* by pouring concrete as shown in Figure 13.18 and Figure 13.19. Mass production of the SIPs could also be readily done using the sheet molding compound approach discussed in Chapter 5.

13.8.2 FURNITURE APPLICATION

Natural resin and flax fiber mats were used to manufacture a chair at the CCM laboratory using the VARTM process. A mold designed by graduate students at the University of the Arts in Philadelphia was made out of solid material to assemble the chair shape and the fiber was laid-up and bagged for vacuum infusion. Figure 13.20 shows a picture of the finished chair after it

FIGURE 13.18 Stay-in-place testing in three-point bending and wooden frame for pouring concrete.

FIGURE 13.19 Concrete experiment setup and the final 8-in-thick concrete slab after curing.

FIGURE 13.20 All-natural composite chairs made of soybean oil-based resin and flax fiber mats.

was attached to a metal frame. This process could also be used to replace wood fiber in MDF materials used for making desks, tables, and other furniture.

REFERENCES

1. Kline & Company, Inc., *6th International Conference on Wood Fiber–Plastic Composites*; May 14, 2001.
2. Ayscue, J. K. Natural Hazards Research Working Paper #94; November 1996; http://www.colorado.edu/hazards/wp/wp94/wp94.html.

3. Insurance Institute for Property Loss Reduction (IIPLR) and Insurance Research Council, *Coastal Exposure and Community Protection: Hurricane Andrew's Legacy,* IIPLR, Boston; 1995.
4. Wool, R. P.; Kusefoglu, S. H.; Zhao, R.; *et al.* U.S. Patent 6,121,398, Date N/A.
5. Can, E.; Kusefoglu, S.; Wool, R. P., *J. Appl. Polym. Sci.* 2001, **81,** 69.
6. LaScala, J. J.; Wool, R. P., *J. Amer. Oil Chem. Soc.* 2002, **79**(1), 59.
7. Khot, S. N.; LaScala, J. J.; Can, E.; *et al. J. Appl. Polym. Sci.* 2001, 82, 703.
8. Williams, G. I.; Wool, R. P. *Appl. Compos. Mater.* 2000, **7,** 421.
9. Dweib, M. A.; Hu, B.; O'Donnell, A.; *et al.* All-Natural Composite Sandwich Beams for Structural Applications, *Compos. Struct.,* 2004, **63**(2), 147–157.
10. O'Donnell, A.; Dweib, M. A.; Wool, R. P. Natural Fiber Composites with Plant Oil-Based Resin, *Comp. Sci. Technol.* **64**(9) 1135–1145, 2004,
11. Dweib, M. A.; O'Donnell, A.; Hu, B.; *et al.* Natural Composites Materials for Structural and Automotive Applications. In *Proc. ICCM-14,* San Diego, CA; July 14–18, 2003. Society of Manufacturing Engineers, Dearborn, MI.
12. Hu, B.; Dweib, M. A.; Wool, R. P.; *et al.* Bio-Based Composite Roof for Residential Construction: Preliminary Investigation, *J. Compos. Construct.* (submitted).
13. Hu, B.; Dweib, M. A.; Wool, R. P.; *et al.* Bio-Based Composite Roof for Residential Construction: Design, Testing and Analysis, *J. Compos. Construct.* (submitted).
14. Bledzki, A. K.; Gassan, J. Composites Reinforced with Cellulose Based Fibres, *Prog. Polym. Sci.* 1999, **24,** 221–274.
15. Mwaikambo, L. Y.; Ansell, M. P. Chemical Modification of Hemp, Sisal, Jute, and Kapok Fibers by Alkalization, *J. Appl. Polym. Sci.* 2002, **84,** 2222–2234.
16. Gassan, J. A Study of Fibre and Interface Parameters Affecting the Fatigue Behaviour of Natural Fibre Composites, *Compos. A: Appl. Sci. Manufact.* 2002, **33**(3), 369–374.
17. Ruys, D.; Crosky, A.; Evans, W. J. Natural Bast Fibre Structure, *Int. J. Mater. Prod. Technol.* 2002, **17**(1–2), 2–10.
18. Mishra, S.; Tripathy, S. S.; Misra, M.; *et al.* Novel Eco-Friendly Biocomposites: Biofiber Reinforced Biodegradable Polyester Amide Composites—Fabrication and Properties Evaluation, *J. Reinforc. Plast. Compos.* 2002, **21**(1), 55–70.
19. Kandachar, Prabhu; Brouwer, Rik. Applications of Bio-Composites in Industrial Products, *Mater. Res. Soc. Symp. Proc.* 2002, **702,** 101–112.
20. Anon, The Competitiveness of Natural Fibers Based Composites in the Automotive Sector: The Sisal Agribusiness in Brazil, *Mater. Res. Soc. Symp. Proc.* 2002, **702,** 113–139.
21. Carlo Santulli, Post-Impact Damage Characterisation on Natural Fibre Reinforced Composites Using Acoustic Emission, *NDT & E Int.* 2001, **34**(8), 531–536.
22. Van de Velde K.; Kiekens, P. Thermoplastic Pultrusion of Natural Fibre Reinforced Composites, *Compos. Struct.* 2001, **54**(2–3), 355–360.
23. Gassan, J.; Chate, A.; Bledzki, A. K. Calculation of Elastic Properties of Natural Fibers, *J. Mater. Sci.* 2001, **36**(15), 3715–3720.
24. Eichhorn, S. J.; Baillie, C. A.; Zafeiropoulos, N.; *et al.* Current International Research into Cellulosic Fibres and Composites, *J. Mater. Sci.* 2001, **36**(9), 2107–2131.
25. Iannace, S.; Ali, R.; Nicolais, L. Effect of Processing Conditions on Dimensions of Sisal Fibers in Thermoplastic Biodegradable Composites, *J. Appl. Polym. Sci.* 2001, **79**(6), 1084–1091.
26. Braun, D.; Braun, A. Natural Thermosets, *Kunststoffe Plast Europe* 2001, **91**(2), 36–38, 83–86.
27. Corbière-Nicollier, T.; Laban, B. G.; Lundquist, L.; *et al.* Life Cycle Assessment of Biofibres Replacing Glass Fibres as Reinforcement in Plastics, *Resources, Conserv. Recycl.* 2001, **33,** 267–287.
28. Hepworth, D. G.; Hobson, R. N.; Bruce, D. M.; *et al.* The Use of Unretted Hemp Fibre in Composite Manufacture, *Composites A* 2000, **31,** 1279–1283.

29. Zafeiropoulosa, N. E.; Williams, D. R.; Bailliea, C. A.; *et al.* Engineering and Characterisation of the Interface in Flax Fibre/Polypropylene Composite Materials. Part I. Development and Investigation of Surface Treatments, *Composites A* 2002, **33,** 1083–1093.
30. Mohanty, A. K.; Misra, M.; Drzal, L. T. Surface Modifications of Natural Fibers and Performance of the Resulting Biocomposites: An Overview, *Compos. Interfaces* 2001, **8**(5), 313–343.
31. Mohanty, A. K.; Drzal, L. T.; Misra, M. Engineered Natural Fiber Reinforced Polypropylene Composites: Influence of Surface Modifications and Novel Powder Impregnation Processing, *J. Adhes. Sci. Technol.* 2002, **16**(8), 999–1015.
32. Silva, J. L. G.; Al-Qureshi, H. A. Mechanics of Wetting Systems of Natural Fibres with Polymeric Resin, *J. Mater. Process. Technol.* 1999, **92–93**, 124–128.
33. Cichocki, F. R., Jr.; Thomason, J. L. Thermoelastic Anisotropy of a Natural Fiber, *Compos. Sci. Technol.* 2002, **62,** 669–678.
34. LaScala, J. J.; Wool, R. P. Polymer, 46, 61 (2005).
35. Seemann, W. H. Plastic Transfer Molding Techniques for the Production of Fiber Reinforced Plastic Structures, U.S. Patent No. 4902215, filed 30 March 1989.
36. Williams, C. D.; Grove, S. M.; Summerscales, J. The Compression Response of Fibre-Reinforced Plastic Plates during Manufacture by Resin Infusion under Flexible Tooling Methods, *Composites A* 1998, **29**A, 111–114.
37. Mathur, R.; Heider, D.; Hoffmann, C.; *et al.* Flow Front Measurements and Model Validation in the Vacuum Assisted Resin Transfer Molding Process, *Polym. Compos.* 2001, **22**(4), 477–490.
38. Heider, D.; Hofmann, C.; Gillespie, J. W., Jr. Automation and Control of Large-Scale Composite Parts by VARTM Processing. In *Proc. 45th International SAMPE Symposium/ Exhibition, Bridging the Centuries with SAMPE's Materials and Processes Technology,* Long Beach, CA, May 21–25, 2000.
39. Hsiao K.-T.; Mathur, R.; Advani, S. G.; *et al.* A Closed Form Solution for Flow During the Vacuum Assisted Resin Transfer Molding Process, *J. Manufact. Sci. Eng.* 2000, **122,** 463–475.
40. Advani, S. G.; Sozer, E. M. *Process Modeling in Composites Manufacturing,* Marcel Dekker, Inc.; 2003.
41. Whitney, J. M.; McCullough R. L. *Micromechanical Materials Modeling,* Delaware Composites Design Encyclopedia, Vol. 2, Technomic Publishing Co.; 1990.
42. Wool, R. P.; Hong, C. K. Low Dielectric Constant Materials from Plant Oils and Chicken Feathers, U.S. Patent application No. 60/396319, 2002 (pending).
43. Sabnis, G. M.; *et al. Structural Modeling and Experimental Techniques,* Prentice-Hall; 1983.
44. Allen, H. G. *Analysis and Design of Structural Sandwich Panels,* Pergamon Press; 1969.
45. Sims, G. D.; Johnson, A. F.; Hill, R. D. Mechanical and Structural Properties of a GRP Pultruded Section, *Compos. Struct.* 1987, **8**(3), 173–187.
46. Bank, L. C. Flexural and Shear Moduli of Full Section Fiber Reinforced Plastic Beams, *J. Test Eval.* 1989, **17**(1), 40–45.
47. Giroux, C.; Shao, Y. Flexural and Shear Rigidity of Composite Sheet Piles, *J. Compos. Construct.* 2003, **7**(4), 348–355.
48. Roberts, T. M.; Al-Ubaidi, H. Flexural and Torsional Properties of Pultruded Fiber Reinforced Plastic I-Profiles, *J. Compos. Constr.* 2002, **6**(1), 28–34.
49. Nagaraj, V.; GangaRao, H. V. S. Static Behaviour of Pultruded GFRP Beams, *J. Compos. Constr.* 1997, **1**(3), 120–129.
50. Zureick, A.; Kahn, L.; Bandy, B. Tests on Deep I-Shape Pultruded Beams, *J. Reinforced Plast. Compos.* 1995, **14**(4), 349–361.

14

Carbon Nanotube Composites with Soybean Oil Resins

Richard P. Wool

The field of nanomaterials has virtually exploded on the scientific community with a myriad of new materials and devices whose properties and functions are uniquely at the nanoscale of 1–100 nm, where 1 nm is 10 Å. Nanofibers, that is, fibers whose diameter is of the order of a few nanometers, such as carbon nanotubes and nanocellulose, has the potential to provide composites with unique strength and stiffness. Nanoclays, which have structures similar to a deck of cards, can be exfoliated to the individual "cards" that are about a nanometer thick but microns long and wide. Such nanostructures are typically stiff and strong and create considerable internal surface area, $\sim 100\,m^2/g$. The latter can be used to store mechanical impact energy by debonding and to promote self-healing of such impact-resistant structures.

Bio-based materials have enormous potential for the development of new nanocomposites, either through the use of nanofillers, such as nanotubes, clays, and lignin, or by the use of nanosized molecules such as proteins, triglycerides, and random coil polymers. Triglyceride molecules, when suitably functionalized, have the unique ability to self-assemble into a hierarchical structure, as noted by the cell structures of our own bodies, which are composed of self-assembled lipid layers. Thus, such molecules should behave in a very synergistic manner with nanosized objects such as carbon nanotubes and nanoclays. Their ability to self-assemble on flat surfaces should allow them to behave as surfactants for carbon nanotubes, which are very difficult to solubilize into stable dispersions, and also intercalate and exfoliate nanoclays as desired. In the next three chapters, we explore the development of new nanocomposites using several functionalized triglycerides interacting with carbon nanotubes, nanoclays, and lignin.

14.1 INTRODUCTION TO CARBON NANOTUBES

Carbon nanotubes (CNTs) have been studied extensively due to their unique mechanical [1, 2], optical [3, 4], electronic [5, 6], and gas storage [7] properties. Their high Young's modulus (~1 TPa) and aspect ratio (10–1000) makes them ideally suited as reinforcements [8, 9]. For the full exploitation of these extraordinary properties, the CNTs have to form stable dispersions [10]. Dispersions have been obtained with a variety of solvents [11, 12], surfactants [13, 14], conjugated polymers [15–17], sugars [18–20], and biological molecules [21, 22], and by chemically functionalizing the CNT surface. Chemical functionalization, involving the addition of acid functionality [23, 24], silica [25], fluorine [26, 27], and alkanes [28], has the inherent effect of altering the properties of the CNTs by altering the sp^2 hybridization of the carbon atoms, and thus the CNT properties [29]. Therefore, a benign means of dispersing nanotubes is desired. Richard *et al.* [30] showed the appearance of a self-assembled structure on the surface of carbon nanotubes by synthetic lipids. These lipids have a chemical structure similar to the fatty acids that comprise the triglycerides of natural oils. It can therefore be assumed that natural oil triglycerides will have a favorable effect on the dispersion of CNTs.

This work focused on the potential of acrylated epoxidized soybean oil (AESO, described in Chapter 4) to disperse CNTs. Both single-walled carbon nanotubes (SWNTs) and multiple-walled carbon nanotubes (MWNTs) have been used. This work was performed in cooperation with researchers at Trinity College Dublin in Ireland. SWNT work was largely directed at the ability of AESO to disperse these CNTs. MWNT work concentrated on the ability to make nanocomposites in an inexpensive fashion, using unpurified CNTs. The use of unpurified CNTs reduces their cost significantly, which is important in large-scale applications. While CNTs are currently made from petroleum-, gas-, and coal-derived products, it is conceivable that CNT precursors could be derived from naturally renewable sources. This makes CNTs a potentially renewable material and it thus fits into the concept of this discussion to increase the use and properties of thermosetting polymers with renewable materials. On the other hand, when improving the properties of the AESO-based thermosets, CNTs can help to unlock a variety of potential applications previously not achievable because of the limitations of polymer properties. At the end of this chapter, a lattice model developed by Flory and Ronca for rod-like polymers has been applied to CNT-solvent dispersions to get an understanding of their dispersion behavior.

14.2 SINGLE-WALLED CARBON NANOTUBE COMPOSITES

14.2.1 DISPERSION OF SWNTS IN TRIGLYCERIDES

SWNTs from two different production processes were used: SWNTs from catalytic chemical vapor deposition, obtained from Nanocyl, S.A. (Namur, Belgium) and SWNTs produced by high-pressure decomposition of carbon monoxide (HiPco process) [31], which were obtained from Carbon Nanotechnologies, Inc. (Houston, Texas). Nanocyl SWNTs are generally longer (up to 10 μm) and have a broader diameter distribution (1.0–2.0 nm) compared with HiPco SWNTs, which are shorter (up to 1 μm) and have a smaller diameter distribution (0.7–1.4 nm). Two different grades of SWNTs were obtained from Nanocyl: Batch AF57.1 and Batch FL22. Batch AF57.1 is a purified powder consisting of more than 70% SWNTs. The remaining 30% are specified as MWNTs, DWNTs (double-walled nanotubes), cobalt catalyst, pyrolytic carbon, and other carbon nanoparticles. Batch FL22 is unpurified and is specified to contain a minimum of 5% SWNTs. To disperse the SWNT in AESO, 20 mg of SWNT (HiPco or Nanocyl Batch 57.1) and 550 mg of AESO (UCB Chemicals) were dissolved in 20 mL toluene (Aldrich Chemicals). The suspension was sonicated for several hours in a low-power (60-W) sonic bath and allowed to settle for 2 days to allow any impurities to sediment out. The low-power sonication step resulted in a dispersion that was stable over a period of months. In addition, the dispersion may be dried and redispersed in toluene, resulting in the reformation of a stable dispersion. The result that low-power sonication was able to form stable SWNT dispersions is quite remarkable and points toward a good dispersive capacity for AESO. Note that SWNTs without AESO did not disperse in pure toluene [32].

Characterization of the dispersion with transmission electron microscopy (TEM), atomic force microscopy (AFM), and resonance Raman spectroscopy was performed. Low-resolution TEM micrographs of protruding CNTs coated with AESO can be seen in Figure 14.1. Some MWNTs and "bamboo-like" nanotubes were also present, consistent with the manufacturer's specifications that the CNT mixture contains small quantities of MWNTs and other nanostructures. The thickness of the polymer coating is approximately 10–15 nm, suggesting that polymerization occurs at the CNT surface, induced by the heat generation during sonication. This is very important since polymerization during processing should be avoided. Tapping-mode AFM identified the existence of nanotube bundles and the inclusion of catalytic particles in the dispersion. Several microscale structures could be found in the solution cast film (Figure 14.2) for HiPco/AESO samples. Height analysis showed a thickness of 20–30 nm, slightly larger than values obtained from TEM. However, surface interactions and nanotube flattening on the surface introduce some error in the AFM height analysis results.

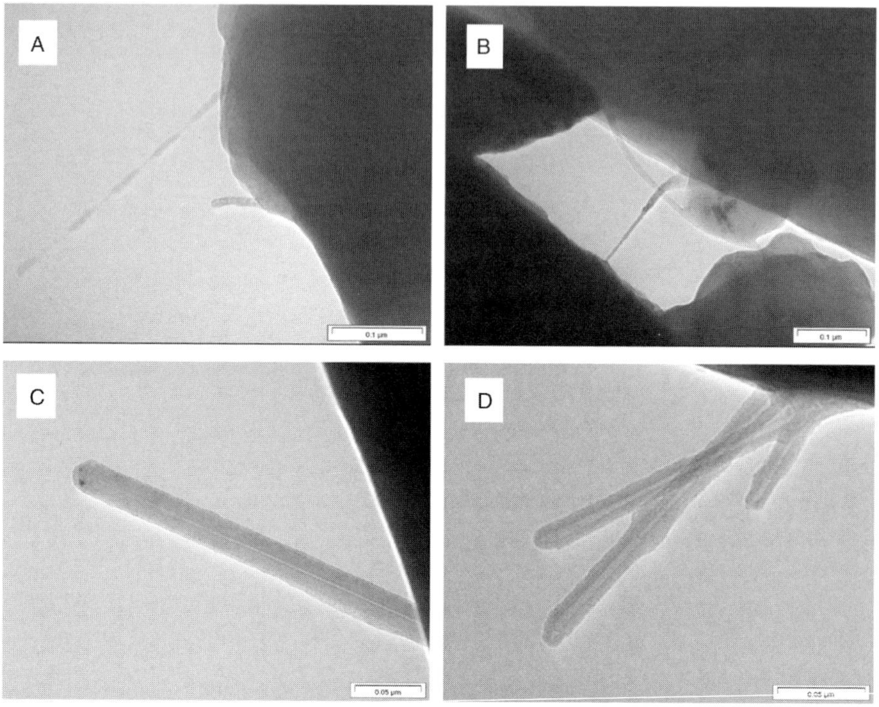

FIGURE 14.1 TEM micrograph of Nanocyl/AESO composites: (A and B) individual SWNTs, (C) individual MWNT, and (D) "bamboo-like" nanotubes.

Nanotube–AESO interactions were investigated with resonance Raman spectroscopy. The three prominent signals in the Raman spectra are the radial breathing mode (RBM) and the D-line and G-line for which values are well documented [33]. The RBM signals, existing at wavenumbers $125-300\,\mathrm{cm}^{-1}$, are extremely useful because they are related to the CNT diameter and the nanotube environment [34]. Changes in the environment are visible by upshifts in the RMBs. The RBM signals shown in Figure 14.3 display a larger frequency range for Nanocyl SWNTs, compared to HiPco SWNTs, consistent with the larger diameter range of Nanocyl SWNTs. The spectra for the SWNTs are in agreement with other published spectra for these nanotubes [21, 22]. In the composite samples, RBM upshifts are seen for both SWNTs. The HiPco/AESO sample displays smaller upshifts ($1.4-2.6\,\mathrm{cm}^{-1}$) than does the Nanocyl/AESO sample ($3.5-6.4\,\mathrm{cm}^{-1}$). Upshifts in RBM have previously been ascribed to debundling of nanotubes by conjugated polymers [17] and peptides [22].While debundling will affect the resonant absorption conditions of SWNTs, the net effect of debundling will result in an apparent upshift in RBM when measured with the same laser excitation [34]. The Raman spectroscopy data therefore suggest that

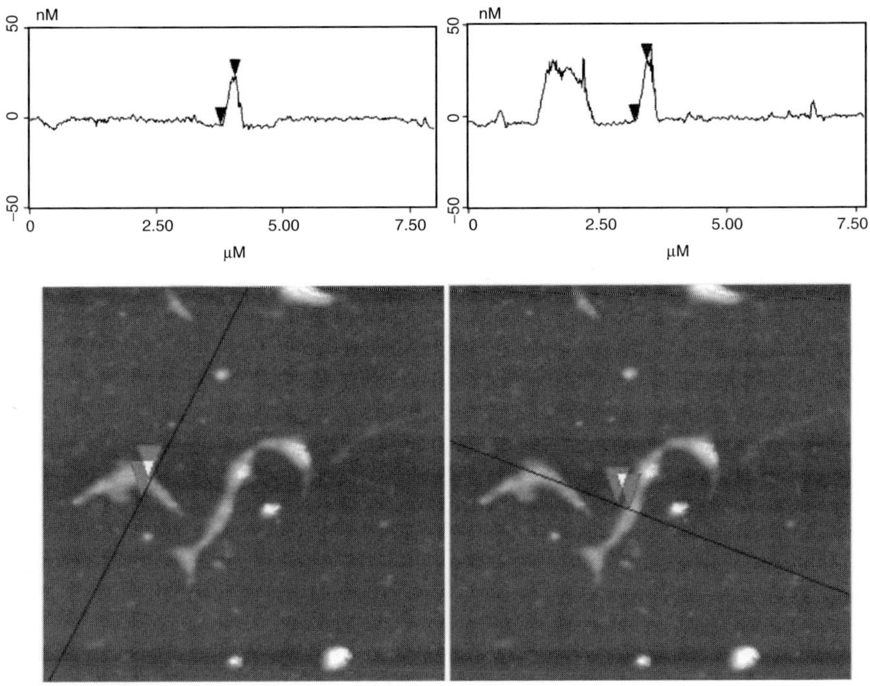

FIGURE 14.2 Tapping-mode AFM imaging on HiPco/AESO solution cast film. Height analysis along the lines shown can be seen above the AFM images.

AESO intercalates the Nanocyl/SWNT bundles to a large extent. The very small upshift seen for HiPco/AESO is then due to a mere coating of SWNT bundles by AESO without significant intercalation. This is in agreement with the TEM and AFM observations where individual and coated Nanocyl SWNTs could be found and bundles of HiPco SWNTs were more prominent. The Nanocyl SWNTs can therefore be expected to be a better reinforcement for AESO polymers.

14.2.2 COMPOSITE PROPERTIES

With the previous dispersion results in mind, a composite sample of 3 wt% Nanocyl Batch FL22 SWNTs in a 65/35 weight ratio solution of AESO/styrene was made. The suspension was sonicated with a high-power sonic tip for 2 min. Sonication was performed with the sample in an ice bath to prevent excessive temperature rise. After sonication, the sample was further cooled with cold running water. A second sonicating step was performed by placing the sample in a low-power sonic bath for 2 h. The sample was subsequently allowed to settle. The liquid dispersion was decanted to separate any

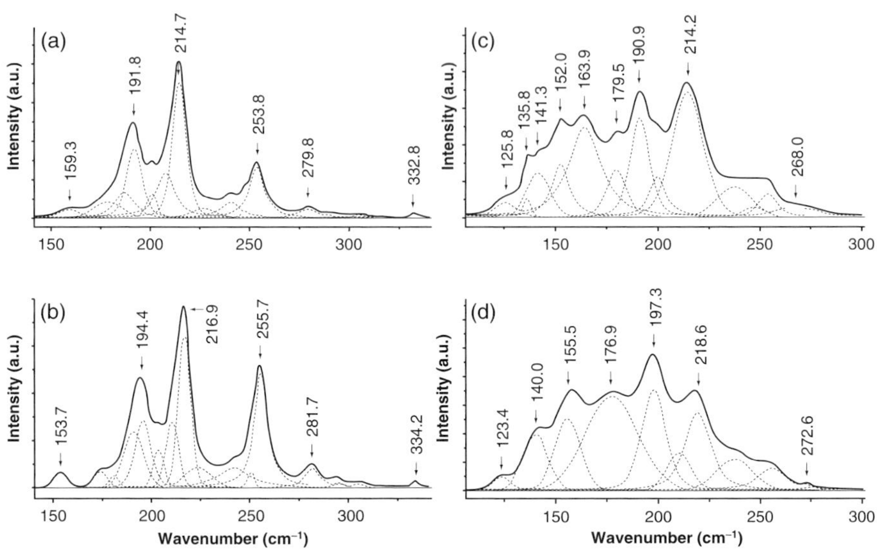

FIGURE 14.3 RBM analysis of resonance Raman spectroscopy performed at 633-nm excitation. The spectra shown are for (A) HiPco SWNT, (B) HiPco/AESO composite, (C) Nanocyl SWNT, and (D) Nanocyl/AESO composite. The dotted lines show the deconvoluted spectra, fitted with a minimum number of frequencies without fixing peak position or width.

sediment and 1.5 wt% *tert*-butyl peroxy benzoate, a free-radical polymerization initiator, was added. The dispersion was purged for 2 min with nitrogen to remove free oxygen and polymerized in a silicone mold at 110 °C for 2 h, followed by a postcure at 180 °C for an additional 2 h. A reference sample without nanotubes was also prepared and polymerized at the same time for comparison.

Figure 14.4 shows the storage modulus as a function of the temperature obtained from DMA by Thielemans [32, 35]. The SWNT sample exhibited a flexural modulus increase of 43%, a flexural strength increase of 9%, and an increase in glass transition temperature of 9%. The storage modulus at 25 °C displayed a lower increase (30%) than the flexural modulus. Using a modulus of 1 TPa for SWNTs [8, 9], some expected modulus reinforcements could be calculated. The lower and upper bound mixing rules are defined respectively as:

$$\frac{1}{E_c} = \frac{\phi_M}{E_M} + \frac{\phi_F}{E_F}, \tag{14.1}$$

$$E_c = \phi_M E_M + \phi_F E_F, \tag{14.2}$$

where E and ϕ denote modulus and volume fraction, respectively, and subscripts C, M, and F denote composite, matrix, and fiber reinforcement.

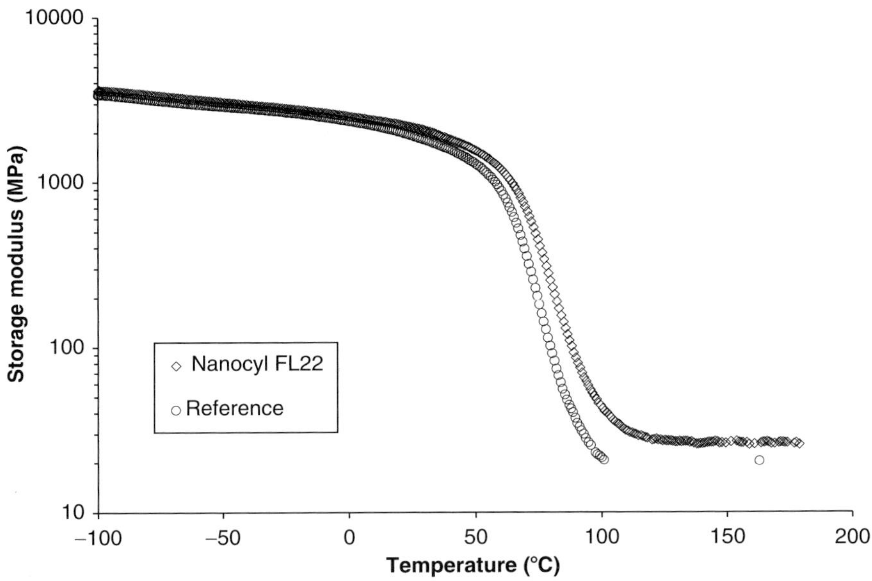

FIGURE 14.4 Storage modulus of 3 wt% Nanocyl SWNT and reference samples as a function of temperature obtained by DMA at 1 Hz in three-point bending mode.

A 3 wt% SWNT is equivalent to 1.5 vol%, assuming the SWNT density to be equal to the graphite density of 2.2 g/cm^3. With the polymer modulus of 1.33 GPa, lower and upper bound improvements are 1.35 GPa (+1.5%) and 16.1 GPa (+1112%) respectively. Using the minimum manufacturer's specifications of 5% SWNTs in the mixture, the calculated lower and upper bound improvements are 1.331 GPa (+0.07%) and 2.1 GPa (+55.6%) respectively. As expected, the improvement seen here falls within the upper and lower boundaries, and using the SWNT purity, comes close to the upper limit. The improvement capability of SWNTs for the AESO/styrene resin system thus appears quite promising. The lower improvements for the storage modulus compared to the flexural modulus may be due to the much lower deformation for DMA than for flexural testing. Deformations applied during the DMA experiments were 15 μm, whereas the flexural modulus is obtained from sample deformations on the order of 1 mm. The molecular weight between cross-links for the AESO/styrene resin was obtained from the rubbery plateau of the storage modulus to be about 570 g/mol. The assumption that we have all C–O bonds (0.141 nm) in the network results in a length between cross-links of 6.7 nm, whereas all C–C bonds (0.154 nm) result in a length between cross-links of 7.3 nm. The actual network is constituted of a combination of both bond types. This spacing is much larger than the SWNT diameter (1.0–2.0 nm). Very small deformation could thus result in a partial disconnect between the SWNT and polymer, and result in a smaller experimentally determined modulus in DMA. This could be very important when making a direct comparison of DMA and flexural data.

These results underscore the potential of using SWNTs as the reinforcement phase for the AESO/styrene resin system. However, the object of this study was only to study the potential and, clearly, more work has to be done to optimize this system with varying amounts of SWNTs and SWNTs with varying aspect ratios. The use of highly pure SWNTs should yield important information on the CNT–matrix interactions and the strength of the CNT–polymer interface. The findings presented here on the coating of SWNTs by AESO and the intercalation of Nanocyl SWNTs during sonication suggest a good interface and that stress transfer between fiber and matrix should be attainable.

14.3 MULTIPLE-WALLED CARBON NANOTUBE COMPOSITES

MWNTs are generally easier to disperse due to smaller van der Waals tube–tube interactions. After the positive results obtained with SWNTs, an initial investigation of the potential of MWNT reinforcement of AESO/styrene polymers was performed. Because it was important to keep affordability and high-volume applications in mind, unpurified MWNTs were used, combined with simple mixing. Simple mixing has already been used successfully to disperse MWNTs in epoxies [36]. The MWNTs were prepared using the arc-discharge Krätschmer-Huffman process at Trinity College Dublin, Ireland [37]. The crude powder contained between 30% and 40% nanotubes. The remaining 60–70% carbon soot contains largely pyrolytic carbon and some other carbon nanoparticles. For consistent nomenclature, carbon soot will denote the 60–70% non-nanotube carbon and graphitic particles, whereas MWNT will address only the 30–40% MWNTs in the crude powder. This powder was used as received without any purification steps. The MWNTs in the powder have an average diameter of 24 nm and length of 800 nm, resulting in an aspect ratio of 33.3. These dimensions were confirmed by high-resolution SEM and TEM.

14.3.1 MWNT-TRIGLYCERIDE COMPOSITES

AESO and styrene were premixed in a 65/35 weight ratio by Thielemans [32, 35]. The crude carbon powder and AESO/styrene mixture were added to 20-mL glass vials to obtain 1, 3, and 5 wt% crude powder in AESO/styrene. The total sample weight was approximately 7.5 g. They were mixed for various amounts of time at roughly 1150 rpm. The 1 wt% sample was stirred for 24 h, then left to settle for 24 h. Sedimentation started to occur after 24 h so it was stirred again for 24 h after which it was left to settle again for 24 h.

No sedimentation was seen afterward. The 3 and 5 wt% samples were initially stirred for 48 h and left to settle for 24 h. Only the 5 wt% sample showed signs of sedimentation and was stirred for an additional 24 h. No sedimentation was seen after 24 h of settling time. All samples were left to settle for an additional week and no visible sedimentation occurred. The dispersion thus appeared to be stable and could be polymerized to make composites, as described by Thielemans [32, 35].

14.3.2 COMPOSITE ANALYSIS

Thermogravimetric analysis (TGA) was used to determine the exact amount of crude powder in the composite samples. The TGA data obtained under air atmosphere for all samples are combined in Figure 14.5. The decomposition of the pure polymer starts around 250 °C. The maximum rate of decomposition is reached at 430 °C, while a local maximum at 375 °C can be observed as a small dip in the decomposition profile. An extra shoulder around 545 °C can also be seen in the pure polymer sample. This decomposition behavior is similar to the results found for the thermal decomposition of styrene-containing commercial polyesters [38]. The first decomposition at 375 °C is attributed to the rupture of cross-links and the formation of linear chains [39]; the second and largest weight loss region, around 430 °C, is due to random scission of the linear chains into smaller fragments [39]; and the high-temperature decomposition at 545 °C is characteristic for the complete decomposition of the polyester chain [38].

The thermal decomposition of the composite samples is largely unchanged from the pure polymer sample, except that the crude powder decomposition appears as a large shoulder between 600° and 800 °C. Little change is seen in the positions of the decomposition regions for the polymer matrix. The thermal decomposition of the polymer finishes before the decomposition of the crude powder starts. This allows us to calculate the amount of MWNTs and carbon soot (total crude powder) in the sample by the relative area under the crude powder first derivative peak and the first derivative polymer peaks [40]. Absolute differences in peak heights seen in Figure 14.5(b) are due to differences in initial sample weight and are not due to differences in decomposition behavior. TGA does not allow us to distinguish between MWNTs and carbon soot because both decompose over the same temperature range. One can thus only obtain the amount of crude powder in the composite sample.

The TGA analysis results are combined in Table 14.1. The total amount of MWNT and carbon soot found for the 1 wt% sample was deemed inaccurate due to the small amount of crude powder in the sample. The calculated crude powder fractions show a significantly lower content in the composite than in the original monomer dispersion. This can only be due to sedimentation of either nanotubes or carbon soot particles, even though no apparent sedimentation

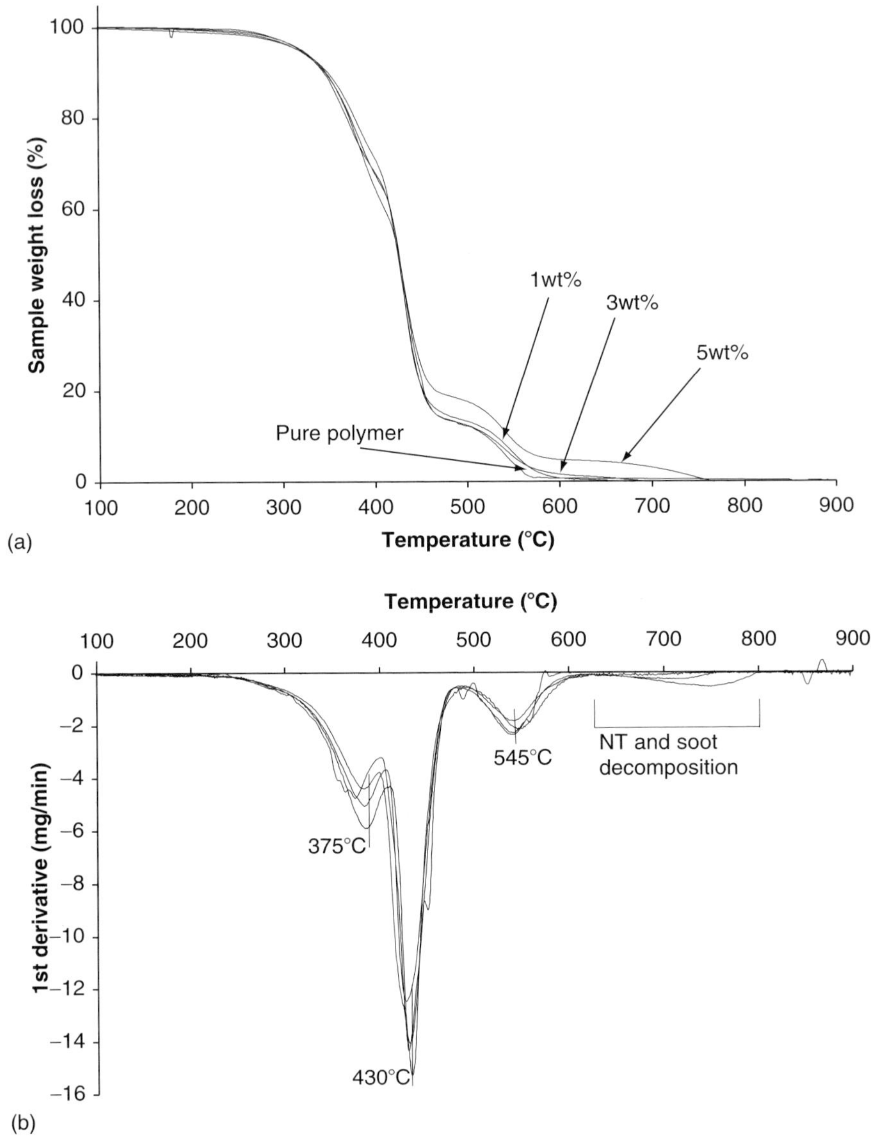

FIGURE 14.5 TGA data for pure polymer and composite samples under air atmosphere: (a) decomposition data and (b) first derivatives. Plot labels refer to amount of crude powder (MWNTs and carbon soot) in original dispersion.

TABLE 14.1 Comparison of amount of MWNT added to monomer mixture, and MWNT in composite after sedimentation and polymerization as determined by TGA under air atmosphere.

Initial Nominal Crude Powder Amount (wt%)	Crude Powder in Composite (wt%)
1	0.2[a]
3	1.8
5	4.4

[a]Deemed inaccurate due to small crude powder amount.

occurred from the liquid dispersions. The TGA data also show a significant amount of crude powder that was lost in the 3 wt% sample. That could be an indication of a dispersion that was less stable than previously assumed, such that sedimentation did occur either during the settling time or the polymerization stage (before gelation).

The anaerobic decomposition, or pyrolysis, performed under a helium atmosphere (Figure 14.6), shows very similar results as found with an active oxygen supply (Figure 14.5). The maximum decomposition rate temperatures shift toward higher temperatures and no local maximum at a low temperature can be seen. The polymer decomposed at the maximum decomposition

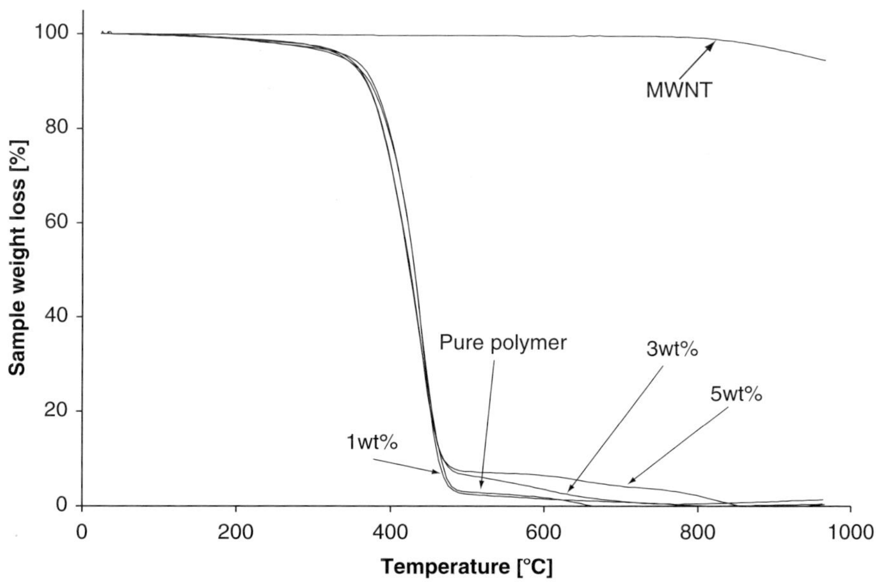

FIGURE 14.6 TGA data for the decomposition of pure polymer, MWNT, and composite samples under helium atmosphere. Plot labels refer to amount of crude powder (MWNTs and carbon soot) in original dispersion.

rate at 445°C. The shoulder at higher temperature shifts to a maximum at 645°C. There is no significant change of the polymer decomposition behavior on addition of the MWNTs and carbon soot or with varying amounts of the crude powder. The appearance of longer shoulders with crude carbon powder addition at higher temperatures, with local maxima of decomposition rates at 690° and 830°C, can be seen. The high-temperature polymer decomposition and the MWNT decomposition, however, cannot be deconvoluted to determine the crude powder content in the composite, as was the case for the experiment under air flow. Therefore, pyrolysis cannot be used to determine the total crude powder (MWNTs and carbon soot) amount in the polymer sample.

The composite sample dynamic mechanical analysis (DMA) was done on a TA Instruments DMA 2980 in dual cantilever mode. Figure 14.7 shows the storage and loss modulus for the various samples tested. The storage modulus, E', at 35°C, as well as the maxima in loss modulus E'' and tan δ are combined in Table 14.2. A significant and comparable increase in E' is seen for the 1 and 5 wt% samples, whereas the 3 wt% sample does not show any increase compared to the pure polymer samples. The most considerable increases in E''_{max} and tan δ_{max} are also seen for the 1 wt% sample. This is a clear indication that the polymer chains in this sample are the most restricted in their movement. This can only be due to dispersed MWNTs. The absence of change in modulus for 3 wt% points toward complete aggregation of carbon soot particles and MWNTs. It has been shown that the addition of carbon filler does not affect the modulus of thermosetting polymers [41]. On the other hand, carbon filler is expected to increase the glass transition temperature [42] as is seen in the increase in maxima in E' and tan δ. It thus appears that all crude powder (MWNTs and carbon soot) is aggregated in the 1.8 wt% composite sample, whereas some MWNTs are dispersed in the other composite samples. This apparent aggregation is also in agreement with the TGA data, where a loss of 1.2% of the initial crude powder amount was recorded for 3 wt% dispersion sample. Because this loss can only be due to sedimentation, which is preceded by aggregation, aggregation of most, if not all, added MWNTs and carbon soot could be expected. The small amount of loss of crude powder for the original 5 wt% solution, determined by TGA, points toward a good original dispersion. It has been shown, however, that carbon black has a tendency to aggregate in clusters during free-radical polymerization [43]. Aggregation has also been shown to occur during dynamic compounding and processing of thermoplastic polymer melts [44]. It is also well known that carbon black reduces the polymerization rate of *tert*-butyl peroxy benzoate–initiated free-radical polymerization [45]. The combination of slower polymerization and higher agglomeration rate for the 5 wt% crude powder dispersion, both due to the higher concentration of carbon soot, will result in a significantly larger amount of cluster formation, compared to the 1 wt% sample. The apparent result is the encapsulation of a

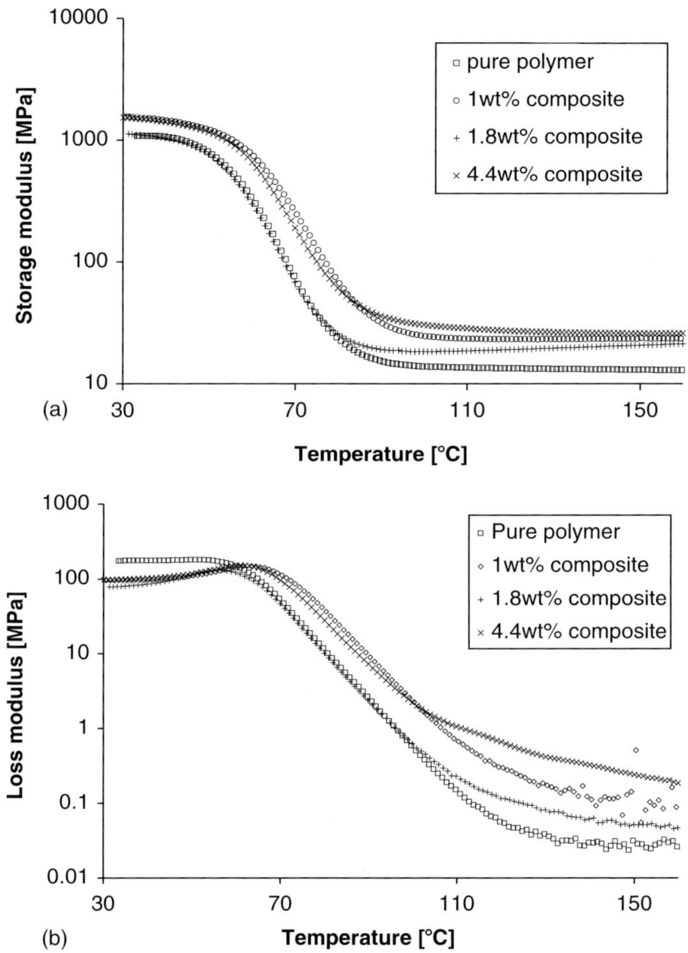

FIGURE 14.7 (a) Storage modulus and (b) loss modulus of AESO/styrene (65/35 by weight) composites for varying crude powder (MWNT + carbon soot) contents as measured by DMA in cantilever mode. Crude powder contents were obtained from TGA (except 1 wt%).

large amount of MWNTs in these aggregates, such that they do not contribute to reinforcement. This effect appears to be significant enough to result in a reduced mechanical improvement compared to the 1 wt% composite sample. So even when the polymerization is initiated from an initially well-dispersed sample, aggregation is expected and will strongly affect the mechanical properties of the obtained composite sample.

All the composite samples show an increase in the rubber plateau modulus above the glass transition temperature. The similar rubbery plateau for 1 and

TABLE 14.2 Compilation of important DMA data for the different composite samples. The variation for the pure polymer sample is the standard deviation between the two polymer samples.

Crude Powder in Solution (wt%)	Crude Powder in Composite (wt%)	E' at 35°C (MPa)	Max E'' (°C)	Max tan δ (°C)
0	0	1.113 ± 0.073	53.4 ± 3.7	70.9 ± 0.5
1	1	1.516	63.8	77.3
3	1.8	1.090	56.7	72.1
5	4.4	1.482	61.5	73.7

5 wt% MWNT composite samples points to a similar amount of dispersed nanotubes, also seen in a similar low-temperature storage modulus. The additional nanotubes in the 5 wt% sample, determined by TGA where 4.4 wt% of crude powder was in the composite sample, are thus expected to be aggregated.

The loss modulus for 1 and 5 wt% MWNT composites (Figure 14.7b) was found to behave similarly below the glass transition temperature. In the rubbery regime (above the glass transition temperature), the ratio of the loss modulus E'' and the storage modulus E', which equals tan δ, for the various composite samples is found to be related according to the following relation:

$$\left(\frac{E''}{E'}\right)_{5\,\mathrm{wt\%}} > \left(\frac{E''}{E'}\right)_{1\,\mathrm{wt\%}} > \left(\frac{E''}{E'}\right)_{3\,\mathrm{wt\%}} > \left(\frac{E''}{E'}\right)_{\text{pure resin}}. \tag{14.3}$$

Aggregates thus increase this ratio slightly, while dispersed MWNTs have a more pronounced effect on tan δ (following the relations between 5, 1, and 3 wt% samples).

Several reinforcement models can be used to describe the reinforcing behavior of carbon nanotubes. The most common models used are the upper bound rule of mixtures [Eq. (14.1)] and the Halpin-Tsai equation [46–48]. The Halpin-Tsai equation for composites reinforced with randomly oriented carbon nanotubes is given by [49–51]:

$$E_C = \left[\frac{3}{8}\frac{1 + 2a_{\mathrm{CNT}}\eta_L\phi_{\mathrm{CNT}}}{1 - \eta_L\phi_{\mathrm{CNT}}} + \frac{5}{8}\frac{1 + 2\eta_T\phi_{\mathrm{CNT}}}{1 - \eta_T\phi_{\mathrm{CNT}}}\right]E_M,$$
$$\eta_L = \frac{\left(\frac{E_{\mathrm{CNT}}}{E_M}\right) - 1}{\left(\frac{E_{\mathrm{CNT}}}{E_M}\right) + 2a_{\mathrm{NT}}}, \quad \eta_T = \frac{\left(\frac{E_{\mathrm{CNT}}}{E_M}\right) - 1}{\left(\frac{E_{\mathrm{CNT}}}{E_M}\right) + 2} \tag{14.4}$$

where $a_{\mathrm{CNT}} = (l_{\mathrm{CNT}}/d_{\mathrm{CNT}})$ is the aspect ratio with l_{CNT} being the length and d_{CNT} the diameter of the nanotubes, ϕ_{NT} is the CNT volume fraction and E_{CNT}, E_M, and E_C are the modulus of carbon nanotubes, matrix, and composite, respectively. Thostenson and Chou [48] suggested a correction to the CNT modulus due to its hollow core, and based on micromechanics:

$$E_{\mathrm{CNT,eff}} = \frac{4t}{d} E_{\mathrm{CNT}}, \tag{14.5}$$

where t is the nanotube wall thickness and d is the diameter. Equation (14.5) is only valid for $(t/d) < 0.25$, and their correlation [48] of the tube wall thickness with diameter shows that this constraint is only valid for CNTs with diameters of less than 13 nm. Because the average diameter of the MWNTs used in this work was 24 nm, this correction is not applicable. The success of the Halpin-Tsai model and rule of mixtures to describe the reinforcement of composites with randomly oriented CNTs is quite varied. Fitting of the Halpin-Tsai model to experimental data, using the aspect ratio as a variable, commonly results in large overpredictions of the aspect ratio. The fitted CNT lengths can be several orders of magnitude larger than what is determined by electron microscopy [46]. The rule of mixtures has a more varied record, with the fitted CNT modulus being both over- and underpredicted from reported values [46]. Cadek *et al.* [46] recently developed a new model based on the CNT surface area:

$$E_c = \left(\frac{4k}{d}\phi_{\mathrm{CNT}} + 1\right)E_M, \tag{14.6}$$

where k is a parameter describing the strength of the CNT–polymer interface. Other parameters are consistent with previous definitions. This model is quite interesting in that it does not include the CNT modulus. The reasoning is the relative weakness of the CNT–matrix interface, compared to the CNT modulus, such that the interface is expected to fail before the CNT does. The full reinforcement potential of CNTs, thus, cannot be attained. Equation (14.6) was successful in describing the tensile modulus for a wide variety of different CNTs and CNT weight fractions in a polyvinyl alcohol (PVA) matrix [46]. A universal k-parameter was found for these composites of 468 ± 114 nm. This would suggest that small-diameter CNTs (resulting in larger surface area per unit volume) would be the most efficient reinforcements.

Figure 14.8 shows the experimental results compared to different reinforcement models: the upper bound rule of mixtures [Eq. (14.1)], lower bound rule of mixtures [Eq. (14.2)], Halpin-Tsai model [Eq. (14.4) with $a_{\mathrm{CNT}} = 33.33$], and the Cadek *et al.* model [Eq. (14.6) with $k = 468$ nm]. The MWNT modulus was taken to be 1 TPa [47]. The polymer and MWNT density were taken to be 1.07 and 1.9 g/cm^3, respectively. The MWNT density has been reported by Thostenson and Chou [48] to be lower than the graphite density of 2.2 g/cm^3. The weight fraction of MWNTs used in the calculation was obtained by multiplying the TGA-determined crude powder amount with the MWNT purity, taken

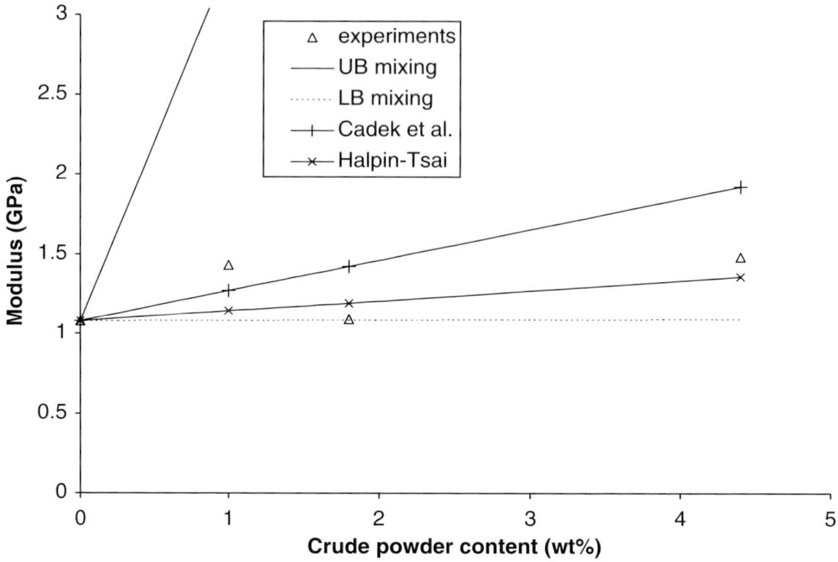

FIGURE 14.8 Comparison of experimental storage modulus at 35°C with models described in the text. UB mixing is the upper bound rule of mixtures of Eq. (14.1), LB mixing is the lower bound rule of mixtures of Eq. (14.2), Halpin-Tsai is described by Eq. (14.3), and the Cadek *et al.* model was calculated from Eq. (14.5) with $k = 468$ nm.

to be 40%. The experimental data fall within the boundaries of the two rules of mixtures as could be expected. The Halpin-Tsai underpredicts the experimental results, which is expected due to the need for higher than actual aspect ratios to fit experimental results [46]. The Cadek *et al.* model underpredicts the experimental result for 1 wt%. Since the maximum MWNT content in the crude powder has been used for the volume fraction calculations, this can only be due to a higher interfacial strength between the polymer and dispersed MWNTs. The k-value used in the model predictions was experimentally determined for thermoplastic PVA composites [46]. The interface is largely dependent on the van der Waals interactions between polymer and MWNT [52, 53]. The considerable increase for the 1 wt% sample suggests a strong polymer–MWNT interface, possibly due to some grafting during polymerization. Active radicals can react with the sp^2-hybridized structure of the CNTs and form covalent bonds. This has been shown to occur during free-radical bulk polymerization methyl methacrylate [54, 55]. These results clearly show an enormous potential for MWNT-AESO/styrene composites. To get an idea of the level of aggregation in the polymer samples, microscopy was performed, which is discussed in the next section.

14.3.3 MICROSCOPY OF CNT COMPOSITES

Figure 14.9 shows optical pictures taken of the composite samples at various magnifications in transmission mode. Optical microscopy shows an increase in the amount of black regions as the crude powder content increases. The size of these regions increases with increasing crude powder content as well. These regions are believed to be aggregates of MWNTs and carbon soot. The original size of the dark regions (at 1 wt% MWNT) was found to be between 1 and 5 μm. As the amount of MWNT increases to 4.4 wt%, the size of these regions increases to the range of 10–30 μm.

TEM images on samples of 80-nm thickness were obtained by microtoming at room temperature. The TEM micrographs of 1 wt% MWNT composite show very good dispersion for the MWNTs (Figure 14.10). Several MWNTs can be found together and aligned. Strong van der Waals forces hold them together and they cannot be easily broken apart by the shear forces of stirring. Aggregates

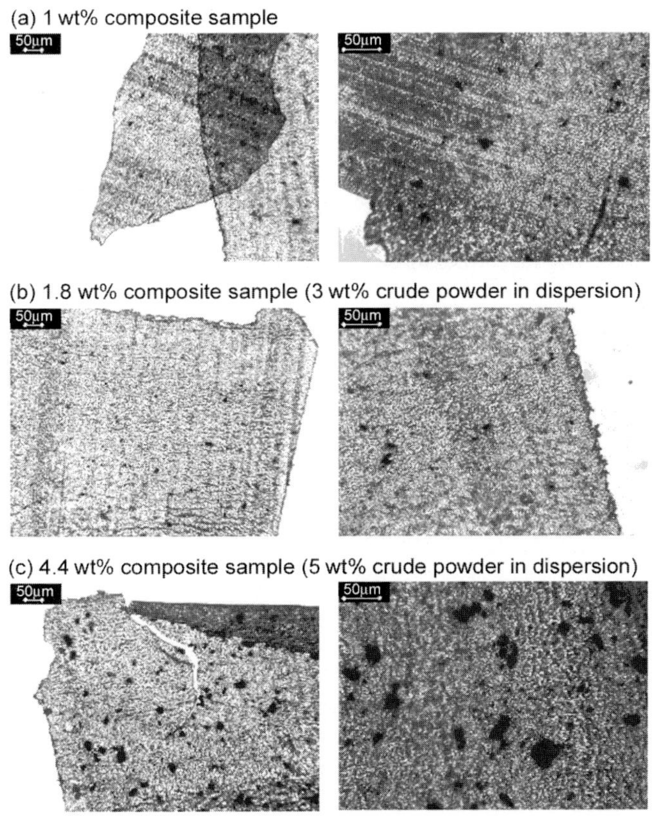

FIGURE 14.9 Optical microscopy pictures of the various composite samples.

consist largely of graphitic particles (carbon soot) and contain a very limited amount of MWNTs. Their size is around 1 μm, agreeing with optical microscopy. The significant amount of dispersed MWNTs, whether or not in sets of two aligned tubes, explains the significant improvement in storage modulus. The dispersed MWNTs will act as reinforcements while the MWNTs found in the aggregates are not expected to contribute.

The micrographs of the 4.4 wt% composite samples show large clusters of aggregates (Figure 14.11). A large fraction of the clusters was made up of MWNTs. A significant number of dispersed nanotubes were also found. These dispersed nanotubes would explain the improved properties, when compared to the pure polymer samples. Similar improvements in the 1 and 4.4 wt% composite samples point toward a similar amount of dispersed nanotubes. This is seen in the larger MWNT fraction in the aggregates. Micrographs for the 1.8 wt% composite did not show any significant amount of dispersed MWNTs.

Figure 14.12 shows TEM micrographs of ruptured sections of the composite samples. The sections were ruptured during microtoming. One can clearly see the polymer sticking to the nanotube when the polymer matrix was pulled apart, indicating a strong interface between the polymer matrix and the dispersed nanotubes. The strong interface may be partially due to grafting during the polymerization reaction. A strong interface is paramount for mechanical improvement. The cross-linking of the polymer, however, limits the deformation the matrix can attain, such that the carbon nanotube does not stay completely covered by the polymer during composite rupture. If grafting does occur, it is thus limited or the whole surface of nanotubes would be expected to stay covered. A large amount of grafting is not desired, however, since it changes the CNT sp^2 hybridization, which negatively affects the CNT properties [29].

14.3.4 WIDE-ANGLE X-RAY SCATTERING OF CNT COMPOSITES

Wide-angle X-ray scattering (WAXS) was performed on the composite samples, pure polymer samples, and crude powder. Figure 14.13(a) shows the spectrum obtained for the crude powder. Deconvolution signals were allowed to be a mix of Gaussian and Lorentzian fits as is common for these patterns [56]. This Lorentzian component is attributed to the distribution of interlayer spacing. The combination of Gaussian and Lorentzian signals resulted in the best agreement between deconvoluted and experimental profiles. The labels shown in the crude powder profile are the distances related to the scattering angle position of the maxima of the deconvoluted peaks. These distances were calculated using Bragg's law. The broad signal with a maximum at $2\theta = 23.6°$ ($d = 3.76$ Å) is the background scattering of the quartz tube and is unrelated to the scattering of the crude powder (MWNTs + graphitic particles/carbon soot). Two sharp signals are seen at $2\theta = 26.25°$ and

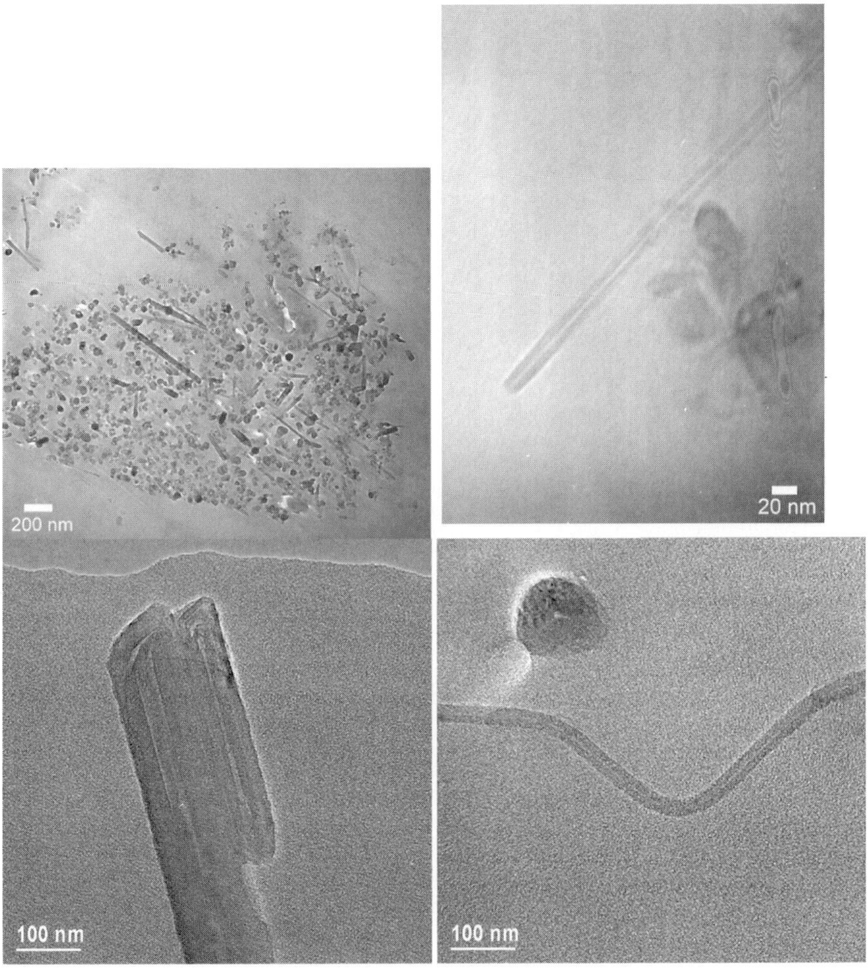

FIGURE 14.10 TEMs of the 1 wt% composite sample.

27.85° (d = 3.39 and 3.2 Å, respectively). The first signal has been reported as the diffraction signature of the distance between walls in MWNTs (d = 3.4 Å) [10, 57]. The second signal at 3.2 Å is attributed to CNT–CNT alignment. Thess *et al.* [58] obtained a lattice constant and tube diameter of SWNT bundles from which a spacing of 3.2 Å can be calculated. Charlier *et al.* [59] calculated the most stable SWNT packing distance to be 3.14 Å, found in a hexagonal packing. Girifalco *et al.* [60] calculated the equilibrium distance between infinitely long aligned SWNTs as a function of their diameter. Their calculations found a distance between nanotube walls for tubes of 24 nm of 3.13 Å. It can therefore be concluded that this signal is due to aligned

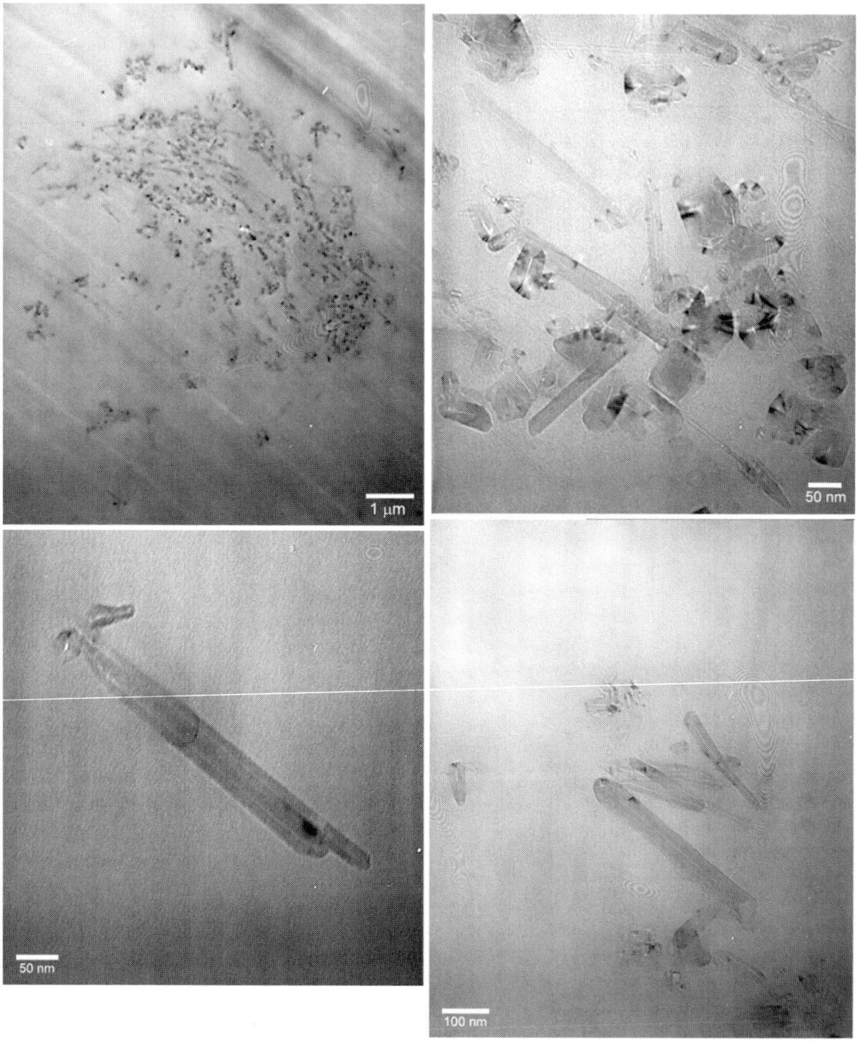

FIGURE 14.11 TEMs of the 4.4 wt% composite sample (5 wt% crude powder in original dispersion).

MWNTs in the sample. The signal $2\theta = 43.4^\circ$ ($d = 2.08$ Å) is attributed to the MWNT interwall spacing as well [61]. The two MWNT interwall signals discussed ($2\theta = 26.25^\circ$ and 43.4°) have been determined to be the reflections of the (002) and (004) graphitic planes, respectively [61]. Finally, the signals at lower angles ($2\theta = 4.23^\circ$ and 0.78°, or $d = 20.89$ and 113.8 Å) are due to low-angle scattering of amorphous carbon and graphitic particles [62, 63]. The lowest angle signal has to be viewed with caution since it is too close to the

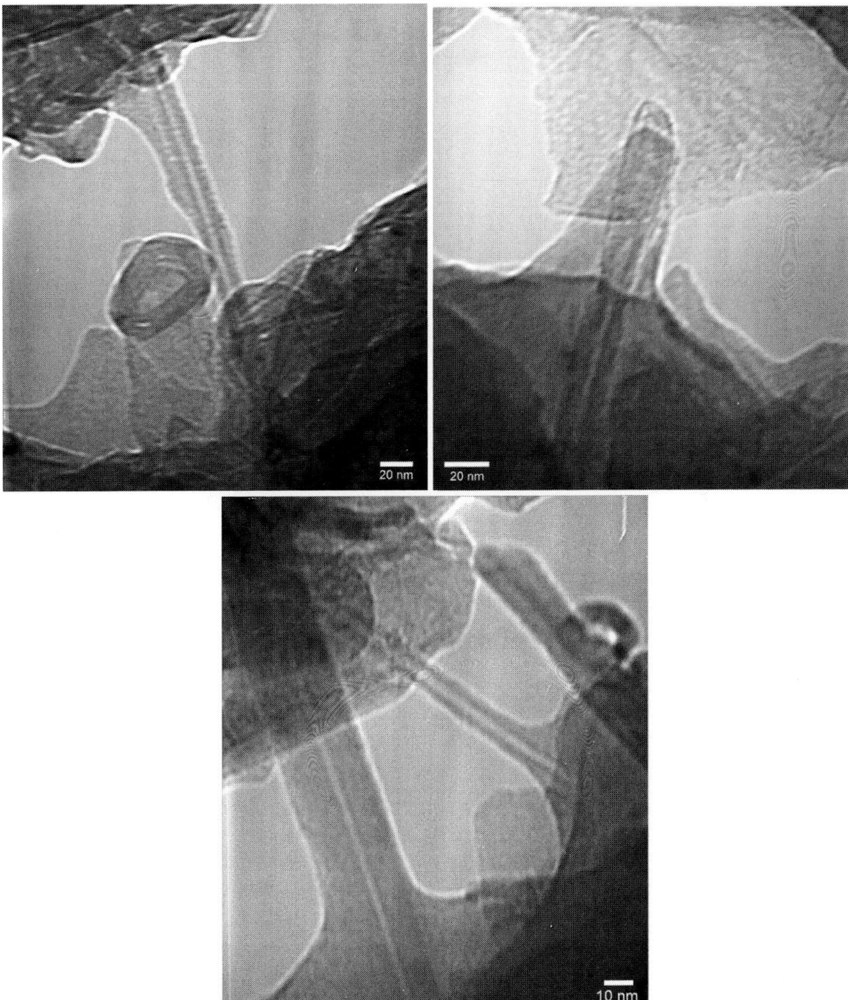

FIGURE 14.12 TEMs of MWNTs sticking out of the polymer matrix at ruptured composite sections. The nanotubes are partially covered by the polymer. Because the thermosetting polymers cannot easily deform, sticking of the polymer matrix to the dispersed nanotubes is a sign of substantial interfacial strength, potentially due to some grafting during polymerization.

limit of the detector. Duclaux *et al.* [62] performed a variety of WAXS experiments in the low-angle limit. A signal at 4.22° was assigned in this work to reflections of the (002) plane in amorphous carbon and graphitic articles. These particles have been shown to be present in the crude powder in large quantities.

Figure 14.13(b) shows the diffraction profiles of the pure polymer and composite samples. The profiles are shifted vertically for easier interpretation.

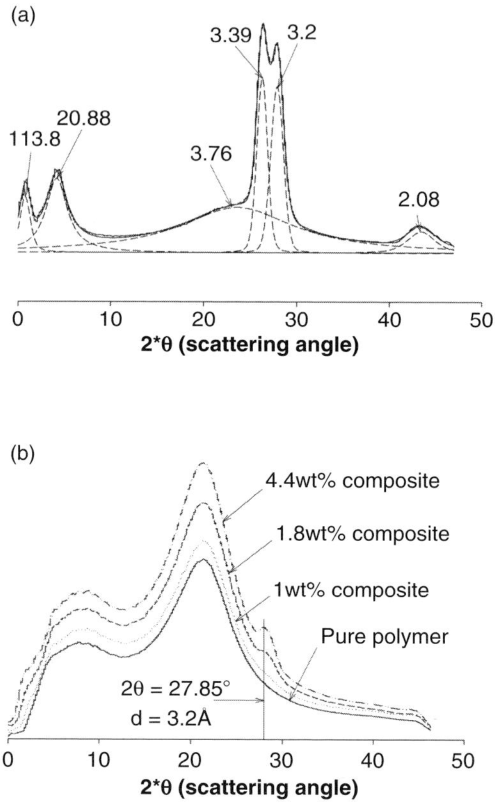

FIGURE 14.13 Wide-angle X-ray scattering profiles versus 2*θ (scattering angle): (a) Signals from the crude powder, with deconvoluted peaks. The labels refer to the distances related to the scattering angle by Bragg's law and have units of Angstroms. (b) Diffraction profile shows the signals for the polymer and composite samples. These spectra are offset vertically for easier interpretation.

The pure polymer sample shows the expected halo seen for all amorphous polymers. Two maxima in this amorphous halo can be seen at $2\theta = 21.56°$ and $8.48°$. Using Bragg's law, these correspond to distances of 4.12 and 10.42 Å, respectively. The shortest distance of 4.12 Å can be attributed to the van der Waals contacts between carbon atoms of paraffin chains in random chain configurations [64] found in the triglyceride molecule of AESO. The value found here is marginally lower than the generally reported distance range of 4.2 to 4.65 Å [64], where the lower values are signs of some molecular ordering such as crystallization. This range of values, however, was found to be valid for aliphatic thermoplastic polymers. The incorporation of styrene and cross-linking result in a decrease in free volume, apparent in the marked increase in density upon cross-linking and the higher density of polystyrene compared to aliphatic polymers [65]. This decrease results in an increase in packing density and thus the lower value of 4.12 Å.

The second distance, 10.42 Å, correlates with the length of the branches on the main chains [64]. Considering the acrylate side groups on the AESO molecule, the number of carbon atoms can be correlated to a spacing distance of 10 Å, agreeing well with the experimental value. Deconvolution of the polymer diffraction profile could not be performed since it was too diffuse. It was found that deconvolution results were too dependent on the choice of the amount of signals. This can be expected since the total profile is the sum of a large amount of small scattering signals due to the amorphous polymer matrix. The interwall MWNT peak at $2\theta = 26.25°(d = 3.4$ Å) cannot be seen for any of the composite samples due to the amorphous halo of the polymer matrix. The WAXS results of the composite samples do show an increasing signal representing the 3.20-Å distance with increasing MWNT content. It was seen in the TEM micrographs that there were still a significant amount of aligned MWNTs in the composite samples, consistent with these scattering results. It can also be expected that the amount of aligned MWNTs increases with increased aggregation. With complete aggregation of the 1.8 wt% composite sample, it is possible to determine a dispersion efficiency, ε_d. This parameter measures the relative amount of MWNT alignment. The relative signal height at 3.2 Å, corrected for the amorphous halo interference, for the 1.8 wt% composite sample is used to define the maximum amount of alignment in the 1.8 wt% sample. Together with a 3.2-Å signal height of zero for the pure polymer, it is possible to calculate the expected relative signal height for the 1 and 4.4 wt% composite samples with maximum alignment. The dispersion efficiency, ε_d, is then defined as the relative difference between this maximum alignment height and the actually measured, corrected signal height:

$$\varepsilon_d = \frac{\left(\frac{A_{\text{corr}}(3.2\,\text{Å})}{A(4.12\text{Å})}\right) - \left(\frac{A_{\text{corr}}^{1.8\,\text{wt\%}}(3.2\,\text{Å})}{A^{1.8\,\text{wt\%}}(4.12\text{Å})}\right) * \frac{\text{wt\%}}{1.8\,\text{wt\%}}}{\left(\frac{A_{\text{corr}}^{1.8\,\text{wt\%}}(3.2\,\text{Å})}{A^{1.8\,\text{wt\%}}(4.12\,\text{Å})}\right) * \frac{\text{wt\%}}{1.8\,\text{wt\%}}}, \tag{14.7}$$

where $A(i)$ is the signal height at distance i, with the subscript corr denoting that the signal height should be corrected for the polymer halo interference, and the superscript 1.8 wt% denoting the signal for the 1.8 wt% composite sample. The term corrects the signal height for maximum alignment for the 1.8 wt% composite sample to the value for the sample under investigation. The 3.2-Å signal will also include dispersed aligned nanotube signals. This is acceptable since alignment of dispersed CNTs will also reduce the reinforcement capability of CNTs. The calculated dispersion efficiencies for the 1 and 4.4 wt% composite samples are 70.7% and 10%, respectively. Taking the purity of the crude powder to be 40%, the amount of effectively dispersed MWNTs in the composites is 0.283 and 0.174 wt%, respectively, for the 1 and 4.4 wt% composite samples. These values can only be expected to be rough estimates of the amount of dispersed MWNTs. Nevertheless, they are consistent with the storage modulus measured with DMA.

14.4 ON THE DISPERSIBILITY OF CARBON NANOTUBES

As mentioned earlier and as is evident from the previous experimental results, good dispersion of CNTs is paramount to full utilization of their extraordinary properties. Ideally, we would like to be able to design and predict the dispersive ability of solvents, which would allow the tailored use of CNTs in such applications as fiber spinning [66], self-assembly of nanostructures [67, 68], and high-performance composites [69]. A useful classification of solvents has been made based on their solvatochromic parameters α (the hydrogen bond donation parameter), β (electron pair donation), and π (solvochromic parameter) [11]. This method however, does not allow for an estimation of a phase diagram and does not predict a difference in dispersibility between SWNTs and MWNTs. It is commonly known that dispersion of smaller diameter SWNTs is much more difficult than dispersion of larger diameter MWNTs.

Flory developed a thermodynamic model for solutions of rod-like polymers based on lattice theory [70]. The entropic contribution in this model was later refined by him and Ronca [71]. This model determines the free energy of mixing based on a decoupling of the enthalpic and entropic contributions:

$$\Delta G_{\text{mix}} = \Delta H_{\text{mix}} - T\Delta S_{\text{mix}}. \tag{14.8}$$

Stability of the solution/dispersion is then obtained by [72]:

$$\Delta G_{\text{mix}} = \Delta H_{\text{mix}} - T\Delta S_{\text{mix}} \leq 0, \tag{14.9}$$

$$\frac{\partial^2 \Delta G_{\text{mix}}}{\partial \phi_i^2} \geq 0, \tag{14.10}$$

with ϕ_i the volume fraction of species i. Without deriving the model, which has been explained in detail in the referenced works [70, 71], it will be introduced in the next sections. The lattice-based description has been described as one of the more accurate models to describe rod-like particle solutions. Other theories based on Onsager's theory [73] start from a virial expansion of the free energy, which is truncated at the second- or third-order term. Virial expansions are generally valid at dilute concentrations, and the theory's accuracy in describing ordered, concentrated solutions is questionable, even with the inclusion of a limited amount of higher order terms, which complicates the mathematics considerably. The disadvantage of the lattice theory used here is the presumption of equal size for the lattice sites. This makes the model calculations presented later in this chapter only valid for

SWNTs which have a diameter of the same order as solvent molecules. Rederiving the model to allow for rod-like particles with larger diameters is a work in progress. The formation of a nematic phase at higher CNT concentrations and the isotropic–anisotropic phase transition has recently been shown to occur in an acidic (pH 3–3.5) aqueous CNT dispersion [74]. The model can thus be expected to describe, at least qualitatively, the behavior of the CNT dispersions.

14.4.1 ENTROPIC CONTRIBUTION

The entropy of mixing ΔS_{mix} for rod-like particles is split up into two contributions: a combinatorial and an orientational factor. The combinatorial contribution describes the steric effects of dissolving rod-like particles in the solvent, while the orientational contribution accounts for the degree of disorientation. The entropy of mixing is then:

$$\Delta S_{\text{mix}} = \Delta S_{\text{comb}} + \Delta S_{\text{orient}}, \tag{14.11}$$

where the subscripts comb and orient naturally denote combinatorial and orientational. A statistical mechanical description of all possible configurations of a set amount of rigid rods in a lattice in the limit of a large amount of particles, such that Stirling's approximation can be used, leads to a closed form for the combinatorial entropy term:

$$\frac{-\Delta S_{\text{comb}}}{R} = n_s \ln \phi_s + n_x \ln \frac{\phi_x}{x} - (n_s + \bar{y} n_x) \ln \left[1 - \phi_x \left(1 - \frac{\bar{y}}{x} \right) \right] + n_x(\bar{y} - 1), \tag{14.12}$$

where ϕ_i is the volume fraction and n_i is the number of moles of species i with s denoting the solvent and x the rigid particle. The parameter x is the amount of solvent segments that fit along the length of the rigid particle and $\bar{y}$ is the average degree of disorientation, which will be defined later. Values of $\bar{y}$ vary between x for complete disorientation (isotropic phase) and 1 for perfectly aligned rods. The orientational entropy is derived from the same description as the combinatorial entropy term to describe the accessible situations for particles with a degree of disorientation $\bar{y}$ [71]. The obtained expression is

$$\frac{\Delta S_{\text{orient}}}{R} = n_x \left[\ln(f_1 \sigma) - \bar{y} \ln \left[1 - \phi_x \left(1 - \frac{\bar{y}}{x} \right) \right] \right]. \tag{14.13}$$

The parameters f_1 and σ will be discussed later. The combination of Eqs. (14.12) and (14.13) results in an expression for the total entropy of mixing:

$$\frac{-\Delta S_{\text{mix}}}{R} = n_s \ln \phi_s + n_x \ln \frac{\phi_x}{x} - n_s \ln \left[1 - \phi_x \left(1 - \frac{\bar{y}}{x} \right) \right] + n_x(\bar{y} - 1) - n_x \ln(f_1 \sigma). \tag{14.14}$$

The parameter σ is arbitrary, but for perfectly aligned rods, for which $\bar{y} = 1$, Eq. (14.14) reduces to the ideal mixing law, as needed, for $\sigma = x^2$. This stipulation has been adopted generally and will also be adopted here [70, 71]. It is, however, unnecessary for calculation of the phase behavior since the σ term cancels when equaling the chemical potentials in each phase. At complete disorientation, when $\bar{y} = x$ and $f_1 = 1$, Eq. (14.14) reduces to

$$\frac{-\Delta S_{\text{mix}}}{R} = n_s \ln \phi_s + n_x \ln \frac{\phi_x}{x} + n_x(x-1) - n_x \ln x^2, \tag{14.15}$$

where the relation $\sigma = x^2$ has been used. The degree of disorientation $\bar{y}$ can be described as a function of the azimuthal angle ψ between the axis of the rod-like particles and the domain axis [71]. This treatment leads to an expression for the average degree of disorientation as a function of the volume fraction of rod-like particles ϕ_x:

$$\bar{y} = \left(\frac{4}{\pi}\right) x \frac{f_2}{f_1}, \tag{14.16}$$

with

$$f_1 = \int_0^{\pi/2} \sin(\psi) \exp(-\alpha \sin \psi) d\psi, \tag{14.17}$$

$$f_2 = \int_0^{\pi/2} \sin^2(\psi) \exp(-\alpha \sin \psi) d\psi, \tag{14.18}$$

$$\alpha = -\left(\frac{4}{\pi}\right) x \ln\left[1 - \phi_x\left(1 - \frac{\bar{y}}{x}\right)\right]. \tag{14.19}$$

Equations (14.14) through (14.19) allow for the calculation of the entropic contribution of the free energy of mixing for rod-like particles in a single solvent. For solvents constituted of various components, solvent terms in Eq. (14.12) have to be adjusted to account for a multicomponent system, which can be done in a straightforward fashion [75].

14.4.2 ENTHALPIC CONTRIBUTION

The heat of mixing ΔH_{mix} in Eq. (14.8) can be related to the Flory–Huggins interaction parameter χ, and the interaction energy density B, by using a van Laar–type expression [70, 76]:

$$\Delta H_{\text{mix}} = \chi RT x n_x \phi_S = B \phi_x \phi_S V_{\text{total}}, \tag{14.20}$$

where $B = \chi RT/v_s$ and v_s as the solvent molar volume and V_{total} is the total volume of the solution. This form for the enthalpic contribution to the free energy of mixing has been shown to be valid for both coiled and rigid rod-like polymers [70, 76].

14.4.3 THE INTERACTION ENERGY DENSITY *B*

The interaction energy density B in the CNT–solvent mixture in Eq. (14.20) can be found from the binary interaction model, developed by Paul and Barlow [77]. The enthalpic interactions are expressed as interactions between the individual species of each phase. The overall interaction between each of the phases is the sum of the interactions between the species in different phases minus the interactions within each phase. The interaction energy density is expressed as:

$$\boldsymbol{B} = \sum_{\substack{i,j \\ i>j}} \boldsymbol{B}_{ij}\boldsymbol{\phi}_i^I\boldsymbol{\phi}_j^{II} - \sum_{\substack{i,j \\ i>j}}\sum_{k=I,II} \boldsymbol{B}_{ij}\boldsymbol{\phi}_i^k\boldsymbol{\phi}_j^k - \sum_i \boldsymbol{B}_{ii}(\boldsymbol{\phi}_i^I)^2 - \sum_{\mathrm{i}} \boldsymbol{B}_{ii}(\boldsymbol{\phi}_i^{II})^2, \tag{14.21}$$

where $\boldsymbol{\phi}_{ij}$ refers to the volume fraction of species i within phase j, such that $\sum_i \boldsymbol{\phi}_i^j = 1$. The superscripts I and II refer to the two phases between which the interaction is calculated. Phase I is the solvent phase S, while II is the CNT phase. If the CNT phase only consists of CNTs with a specific diameter, one obtains $\phi_{\text{CNT}}^{\text{CNT}} = 1$, and Eq. (14.21) becomes

$$B = \sum_{\substack{i,j \\ i>j}} B_{ij}\phi_i^S - \sum_{\substack{i,j \\ i>j}} B_{ij}\phi_i^S\phi_j^S - \sum_i B_{ii}(\phi_i^S)^2 - B_{\text{CNT-CNT}}. \tag{14.22}$$

Equations (14.21) and (14.22) were kept universal in order to allow for a mixture of species in the solvent phase. For single solvents, Eq. (14.22) reduces to

$$B = B_{\text{CNT-S}} - B_{\text{CNT-CNT}}. \tag{14.23}$$

where $B_{\text{CNT-S}}$ and $B_{\text{CNT-CNT}}$ are the binary interaction energies between CNTs and solvent, and between individual CNTs, respectively. The binary interaction parameters can be obtained from the solubility theory, first introduced by Hildebrand, Prausnitz, and Scott [78], with the separation of the overall solubility parameter into three different contributions [65, 79]:

$$\boldsymbol{B}_{ij} = (\boldsymbol{\delta}_i - \boldsymbol{\delta}_j)^2 = (\boldsymbol{\delta}_{di} - \boldsymbol{\delta}_{dj})^2 + (\boldsymbol{\delta}_{pi} - \boldsymbol{\delta}_{pj})^2 + (\boldsymbol{\delta}_{hi} - \boldsymbol{\delta}_{hj})^2, \tag{14.24}$$

where δ_i is the overall solubility parameter of species i, δ_{di}, δ_{pi}, and δ_{hi} are the dispersive, polar and hydrogen bonding contributions to the overall solubility parameter, respectively. These binary interaction parameters were also introduced in earlier chapters to discuss lignin solubility.

14.4.4 SOLUBILITY PARAMETER DETERMINATION

The solubility parameter δ can be predicted from the molecular structure of each species by using various models. Two of the more accurate are the Hoftyzer-van Krevelen (HvK) model [65] and the Hoy model [65, 80], both of which employ a group contribution approach. Taking the average values of the solubility parameters of both models results in improved accuracy [65]. CNTs are made up of single aromatic carbon atoms. The solubility parameters of CNTs were determined as the parameters for a single aromatic carbon atom. Because only a single carbon is used, large variations in predictions can be expected. Therefore, the average values determined from the HvK and Hoy models were used to reduce these errors. The contributions per aromatic carbon for the HvK model were calculated from the values given from a phenyl and single substituted phenyl unit, since this model does not list the individual contributions of aromatic carbons [65]. The CNT chirality does not have any effect on the solubility parameters. The obtained average values are shown in Table 14.3.

The solubility parameters for several solvents used in dispersing CNTs [11] were calculated using the HvK and Hoy models and the average values are shown in Table 14.3 as well. These solvents can be divided into three distinct groups [11]: the good dispersing media, such as ε-caprolactone, cyclopentanone, dimethylformamide (DMF), and *N*-methylpyrrolidone (NMP); the intermediate dispersants such as 4-chloroanisole (4-CA); and the poor, or nondispersing, media such as toluene, 1,2-dichlorobenzene (1,2-DCB) and 1,2-dimethylbenzene (1,2-DMB), which are good dispersants of C_{60} and C_{70} fullerenes but not of CNTs [11, 81]. The calculated average solubility parameters for AESO and styrene are added to the bottom of Table 14.3 for

TABLE 14.3 Solubility parameters for CNT and selected solvents, obtained as average values of the Hoftyzer-van Krevelen and the Hoy model [32].

	Dispersive Contribution $\delta_d(J^{1/2}/cm^{2/3})$	Polar Contribution $\delta_p(J^{1/2}/cm^{2/3})$	Hydrogen Bonding Contribution $\delta_h(J^{1/2}/cm^{2/3})$
CNT	20.87	11.61	14.45
ε-Caprolactone	16.38	8.91	9.85
Cyclopentanone	16.12	9.32	8.57
Dimethylformamide	14.61	12.44	11.09
N-Methylpyrrolidone	16.05	10.97	10.56
4-Chloroanisole	17.80	7.78	6.17
Toluene	17.99	4.79	1.37
1,2-Dichlorobenzene	18.46	8.23	3.23
1,2-Dimethylbenzene	17.11	4.08	1.84
Styrene	17.37	3.23	5.15
AESO	16.49	9.26	5.75

comparison. The solubility parameters of AESO and styrene with the other solvent values would classify styrene as a nonsolvent due to its very low polar and hydrogen bonding contributions. Styrene was experimentally seen not to disperse CNTs. AESO is on the borderline of an intermediate and good solvent with good dispersive and polar contributions but a rather low hydrogen bonding contribution. The previous findings, however, suggest that AESO helps the dispersion of CNTs. The only explanation would be the large AESO structure with localized hydrogen bonding centers around the glycerol center and the acrylate groups. This results in local polarity, which by solubility parameter predictions is averaged incorrectly over the whole molecule. From the CNT solubility parameter predictions, it can thus be hypothesized that the polar centers (glycerol center and alcohol groups) are directed toward the CNT with the aliphatic arm pointing outward.

14.4.5 CNT–CNT INTERACTION PARAMETER

Equation (14.23) shows directly that same-species interactions are always assumed to be zero (same solubility parameters), and attractive interactions are impossible due to the inherent positive value of B_{ij} (related to a positive χ_{ij} Flory–Huggins interaction parameter). Therefore, to obtain the attractive CNT–CNT interaction energy required in Eq. (14.23), we need an additional method.

To account for the attractive van der Waals interaction between individual CNTs, the cohesive energy, E_{coh}, between CNTs was used. Girifalco *et al.* [60] obtained an expression for E_{coh} per unit tube length as a function of the tube diameter d, based on a Lennard–Jones interaction potential:

$$E_{\mathrm{coh}}(d) = -0.08\sqrt{d} + 9.39 * 10^{-3}\mathrm{eV/\mathring{A}}, \tag{14.25}$$

in which the tube diameter is expressed in Angstroms (Å). The binary interaction parameter $B_{\mathrm{CNT-CNT}}$ can then be expressed as follows:

$$B_{\mathrm{CNT-CNT}} = \frac{4E_{\mathrm{coh}}(d)}{\pi d^2}\frac{\mathrm{J}}{cm^3}. \tag{14.26}$$

Equation (14.26) was obtained for SWNTs only. One could assume that only the outer wall of MWNTs is important for CNT–CNT interactions due to shielding of the inner walls by the outer wall. One can thus use the outer wall diameter to obtain a value for their van der Waals attractive forces. The value for $B_{\mathrm{CNT-CNT}}$ versus tube diameter is shown in Figure 14.14. The attractive forces (and, thus, the negative interaction parameter values) are very substantial for the smaller diameter tubes, while they become small at diameters >20 nm. The inverse dependence of $B_{\mathrm{CNT-CNT}}$ on the diameter due to stronger attractive forces for thinner tubes is consistent with experimental findings that they are harder to disperse and that their dispersions are less stable.

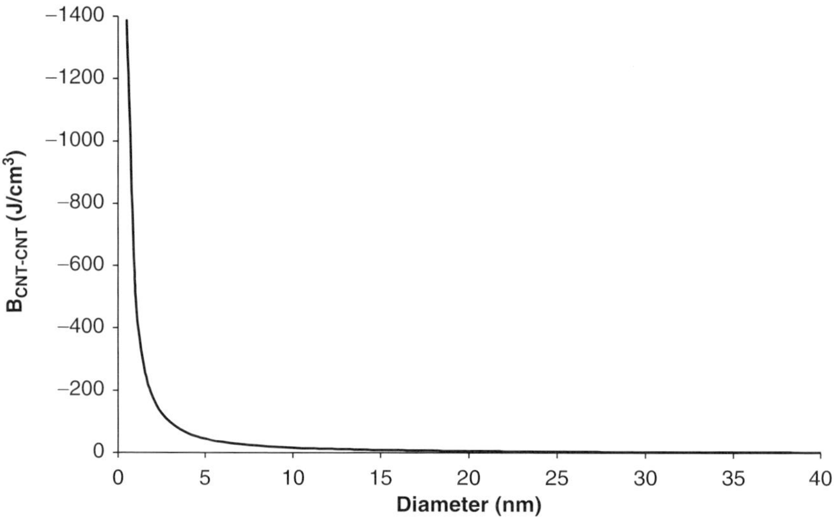

FIGURE 14.14 Dependence of the CNT–CNT binary interaction parameter on the nanotube diameter.

14.4.6 χ-PARAMETER CALCULATION

From the description of solubility in Chapter 16, the Flory–Huggins χ parameter can be calculated as a function of the solvent and the nanotube diameter by use of Equations (14.23) through (14.26) and Eq. (16.3) relating B and χ. Table 14.4 presents the solvent molar volume needed in these calculations for the solvents discussed, some of the calculated χ values without CNT–CNT interaction, and χ values with CNT–CNT interactions for CNTs with various diameters. The molar volumes were calculated from the densities (and molecular weights) given in their respective Materials Safety Data Sheets (obtained from Aldrich Chemicals). Even without the inclusion of CNT–CNT interactions (column $B_{\text{CNT-CNT}} = 0$), the obtained χ parameters are enormous for a polymer–solvent system. As a general rule, a polymer will only result in a stable solution for $\chi < 0.5$. For the rigid rods, one can expect this critical value to be even less. The values, however, lead to a good qualitative prediction. The good solvents ε-caprolactone, cyclopentanone, DMF, and NMP have the lowest χ-parameters, while the bad solvents toluene 1,2-DCB and 1,2-DMB have the highest. The intermediate solvent 4-CA has an intermediate χ-value.

The high χ values are most likely due to the predicted CNT solubility parameters. Since they were predicted from a single carbon atom, it could be understood that the solubility parameters are very dependent on the model assumptions. For general molecules, these assumptions are smoothed out by the combination of several groups. Viscometric experiments are currently

TABLE 14.4 Solvent molar volumes v_s and χ parameter values for different CNT diameters and without inclusion of CNT–CNT interactions.

	v_s (cm^3/mol)	χ			
		$B_{\text{CNT-CNT}} = 0$	$d = 1$ nm	$d = 2$ nm	$d = 10$ nm
ε-Caprolactone	110.82	2.21	24.89	10.32	2.95
Cyclopentanone	89.1	2.28	20.52	8.80	2.87
DMF	77.4	1.63	17.47	7.29	2.14
NMP	96.6	1.54	21.31	8.61	2.18
4-CA	122.5	4.66	29.73	13.62	5.47
Toluene	106.9	9.91	31.79	17.74	10.62
1,2-DCB	113.1	6.64	29.79	14.92	7.40
1,2-DMB	121.2	11.44	36.24	20.30	12.24

being prepared in our research group to determine the CNT solubility parameters from the viscosity increase on dispersion. Two different relations are widely used [82]:

$$\eta_{\text{SP}} = A\left[(\delta_{d,s} - \delta_{d,\text{CNT}})^2 + (\delta_{p,s} - \delta_{p,\text{CNT}})^2 + (\delta_{h,s} - \delta_{h,\text{CNT}})^2\right] + C \quad (14.27)$$

and

$$[\eta] = D \log|\delta_s - \delta_{\text{CNT}}| + E, \quad (14.28)$$

where η_{SP} is the specific viscosity and $[\eta]$ is the intrinsic viscosity. The parameters δ_s and δ_{CNT} are the overall solubility parameters $(\boldsymbol{\delta} = \sqrt{\boldsymbol{\delta}_d^2 + \boldsymbol{\delta}_p^2 + \boldsymbol{\delta}_h^2})$ of solvent (s) and CNT. Parameters A and C are concentration dependent, and all parameters are solvent independent. Using the four good CNT solvents from Table 14.3, an accurate determination of the CNT solubility parameters can be obtained by viscosity measurements at different CNT concentrations and the presented χ-values should be revised, as noted by Thielemans [32].

14.4.7 COMPLETE MODEL

The complete description of the Gibbs free energy of mixing for the CNT–solvent system is then the combination of Eqs. (14.8), (14.14), and (14.20) with the inclusion of a variety of other equations defining the included parameters. The Gibbs free energy of mixing is then

$$\begin{aligned}\frac{\Delta G_{\text{mix}}}{RT} =& \chi x n_x \phi_S + n_s \ln \phi_s + n_x \ln \frac{\phi_x}{x} - n_s \ln\left[1 - \phi_x\left(1 - \frac{\bar{y}}{x}\right)\right] \\ &+ n_x(\bar{y} - 1) - n_x \ln(f_1 \sigma),\end{aligned} \quad (14.29)$$

or per unit volume (by dividing by n_0, the total amount of lattice sites):

$$\Delta\tilde{G}_{\text{mix}} = \chi\phi_x\phi_s + RT\left\{\phi_s \ln\phi_s + \frac{\phi_x}{x}\ln\frac{\phi_x}{x} - \phi_s \ln\left[1 - \phi_x\left(1 - \frac{\bar{y}}{x}\right)\right]\right.$$
$$\left. + \frac{\phi_x}{x}(\bar{y} - 1) - \frac{\phi_x}{x}\ln(f_1\sigma)\right\}. \quad (14.30)$$

Direct determination of a biphasic equilibrium can be done by equating the chemical potentials between the two phases. In a solution of rod-like particles, the phases in equilibrium would be the isotropic phase and an anisotropic, or nematic, phase (in some instances two nematic phases can be in equilibrium as well but this will also follow from the following description) [71]. The chemical potentials are simply given by the partial derivatives of the Gibbs free energy of mixing with respect to n_1 and n_x for the solvent and rod-like particle chemical potentials, respectively:

$$\frac{\mu_i - \mu_0{}^i}{RT} = -\left(\frac{\partial \Delta G_{\text{mix}}}{\partial n_i}\right)_{n_j \neq n_i}. \quad (14.31)$$

Calculation of the chemical potentials for $\bar{y}$ (nematic phase) and $\bar{y} = x$ (isotropic phase) and equating the solvent chemical potentials results in the following equation:

$$\ln\left(\frac{1-\phi'_x}{1-\phi_x}\right) - \left(1 - \frac{1}{x}\right)\phi_x + \frac{(\bar{y}-1)}{x}\phi'_x - \ln\left[1 - \phi'_x\left(1 - \frac{\bar{y}}{\mathrm{x}}\right)\right] + \chi\left(\phi'^{\,2}_x - \phi_x{}^2\right) = 0. \quad (14.32)$$

Equating the isotropic and nematic chemical potentials for the rod-like particles results in

$$\ln\left(\frac{\phi'_x}{\phi_{\mathrm{x}}}\right) - (x-1)\phi_x + (\bar{y}-1)\phi'_x - \ln f_1 + \chi x\left[(1-\phi'_x)^2 - (1-\phi_x)^2\right] = 0, \quad (14.33)$$

where the prime denotes the volume fraction in the anisotropic (nematic) phase. Solution of Eqs. (14.32) and (14.33) together with Eqs. (14.16) through (14.19) allows for the determination of the volume fractions of coexisting phases.

14.4.8 MODEL CALCULATIONS

Calculations by Thielemans [32] for the previously described model were performed for SWNTs with aspect ratios of 700 and 1400. The chosen ratios fall within the range of aspect ratios generally reported for SWNTs. The calculations can easily be done for other aspect ratios. Model calculations were performed with Mathcad 2001 Professional (MathSoft, Cambridge, Massachusetts). For model consistency, the degree of disorientation $\bar{y}$, which at high CNT concentration is calculated to be less than 1 by Eq. (14.16), was set to 1 for these CNT concentrations. This is also consistent with Flory's analysis [70]. Figure 14.15 shows the behavior of $\bar{y}$ as a function of the rigid particle volume fraction ϕ_x. The degree of disorientation $\bar{y}$ starts at the value x and stays constant until anisotropic phase separation. It then decreases rapidly to values around 1. The χ parameter was varied between 0 and 0.04, at which value almost complete phase separation between solvent and a highly concentrated CNT phase was calculated. The phase diagram as a function of the χ parameter is shown in Figure 14.16, while Figure 14.17 shows the solvent chemical potentials as a function of the SWNT volume fraction for selected χ values. All states shown are stable as the values for the free energy and its second derivative obey Eqs. (14.9) and (14.10). Two main features appear in the phase diagram: a region of isotropic–anisotropic phase separation and a region of phase separation between two anisotropic phases. The isotropic–anisotropic phase separation was calculated using the previously described Eqs. (14.32) and (14.33) together with Eq. (14.16). The possibility of the existence of an anisotropic–anisotropic phase separation region can be predicted from the shape of the solvent chemical potential, shown in Figure 14.17. To obtain values for the CNT volume fractions of both phases, the chemical potentials of solvent and CNTs for two anisotropic phases were set equal, resulting in the following two expressions needing to be solved:

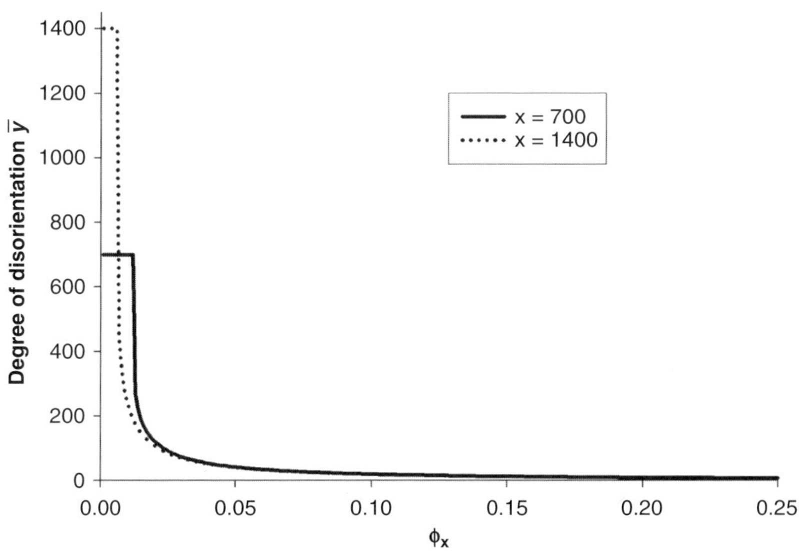

FIGURE 14.15 Degree of disorientation $\bar{y}$ as a function of the volume fraction of rod-like particles. The graph has been cut off at $\phi_x = 0.25$ to show the initial decrease more clearly.

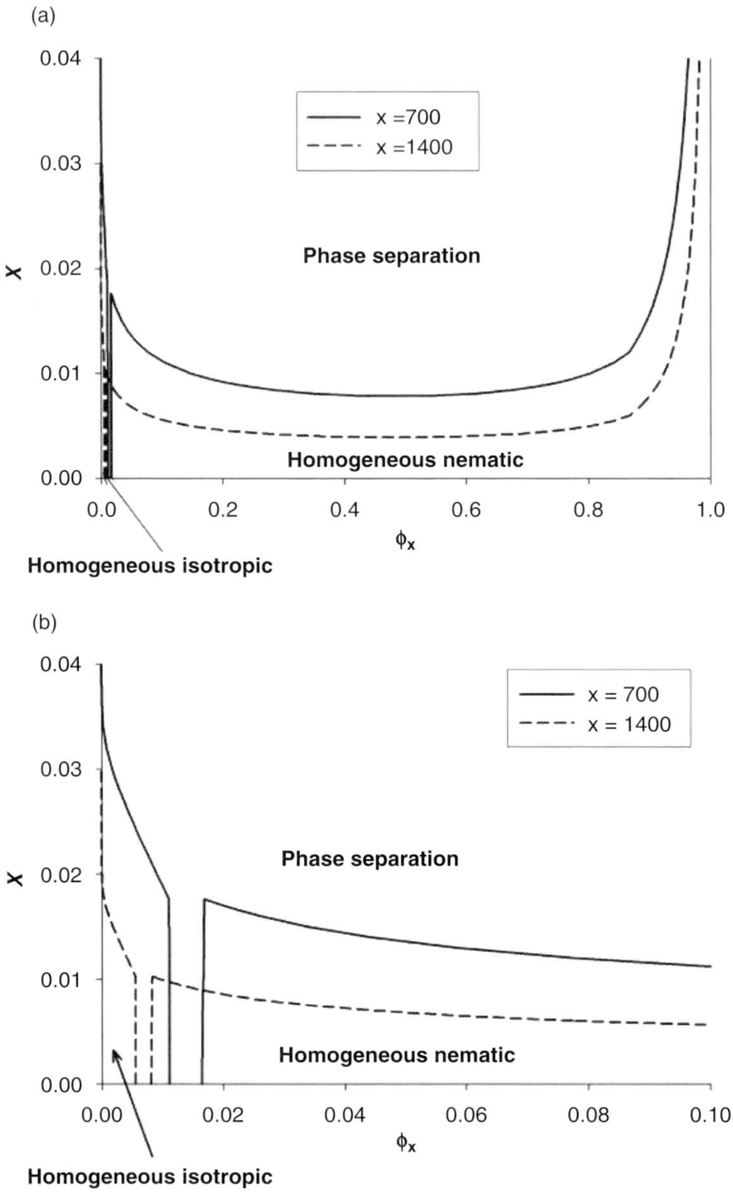

FIGURE 14.16 Phase diagram of SWNTs as a function of Flory–Huggins χ parameter for aspect ratios of 700 and 1400. Part (b) shows the isotropic–anisotropic phase separation region at low ϕ_x values in more detail.

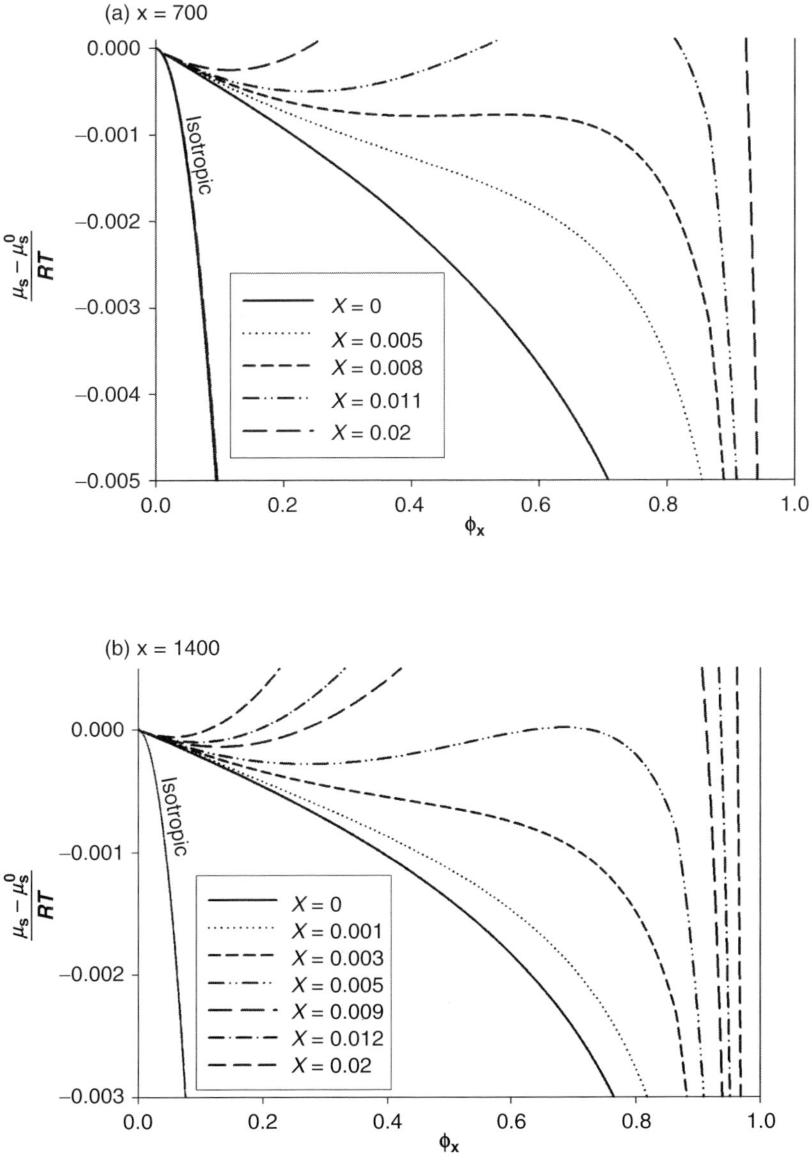

FIGURE 14.17 Solvent chemical potential as a function of the SWNT volume fraction and various values of χ for (a) $x = 700$ and (b) $x = 1400$. The solvent chemical potential in the isotropic phase varies very little with the small variation of the χ parameter. The lines not labeled *Isotropic* are the solvent chemical potentials for the anisotropic (nematic) phase.

$$\ln\left(\frac{1-\phi'_x}{1-\phi_x}\right) - \left(\frac{\bar{y}-1}{x}\right)\phi_x + \frac{(\bar{y}'-1)}{x}\phi'_x + \ln\left[1-\phi_x\left(1-\frac{\bar{y}}{x}\right)\right] - \ln\left[1-\phi'_x\left(1-\frac{\bar{y}'}{x}\right)\right] + \chi(\phi'^2_x - \phi^2_x) = 0 \tag{14.34}$$

and

$$\ln\left(\frac{\boldsymbol{\phi}'_x}{\boldsymbol{\phi}_x}\right) - (\bar{y}-1)\boldsymbol{\phi}_x + (\bar{y}'-1)\boldsymbol{\phi}'_x + \ln f_1 - \ln f'_1 + \boldsymbol{\chi}x\left[(1-\boldsymbol{\phi}'_x)^2 - (1-\boldsymbol{\phi}_x)^2\right] = \mathbf{0} \tag{14.35}$$

together with Eq. (14.16). Primes denote values for the second anisotropic phase. The χ range for stable anisotropic phase separation is very small. The χ parameter ranges between 0.007835 and 0.0176 for $x = 700$ and between 0.00392 and 0.0103 for $x = 1400$. At χ values above the point where the isotropic–anisotropic and anisotropic–anisotropic phase separation regions merge, the CNT volume concentration for stable isotropic dispersions is seen to decrease considerably with increasing χ. At χ values of 0.057 and 0.029 for $x = 700$ and $x = 1400$, respectively, the maximum CNT volume fraction for a stable isotropic dispersion becomes less than 10^{-10}. These χ values are extremely small, indicating that only very limited thermodynamic repulsion between solvent and CNT or attraction between CNTs is allowed. At zero χ values, the CNT volume fractions for the isotropic and anisotropic phases at phase separation are $\boldsymbol{\phi}_x = 0.01125$ and $\phi'_x = 0.016467$ for $x = 700$ and $\boldsymbol{\phi}_x = 0.0056315$ and $\boldsymbol{\phi}'_x = 0.008248$ for $x = 1400$. These values indicate that even without repulsion, the volume fraction of CNTs that can be isotropically dispersed is very small. If higher volume fractions are needed, surfactant may be required.

The phase diagram of CNTs can be a very serious issue for fracture of CNT composites, where the increase in fracture stress above that of the matrix can be related to $\phi_x\sigma(\text{CNT})$. McAninch and Wool [84] found essentially no change in the fracture stress of AESO–CNT composites with weight fractions up to 5%, even though individual CNTs were observed to fracture and pull out across the fracture surface. This is to be expected when phase separation occurs above the percolation threshold, and the sample, while becoming very black, shows no enhancement in strength, despite the addition of nanofibers with a modulus of 1 TPa and fracture stresses of 100 GPa (see Chapter 6) costing $250,000/lb! Thus, the future of CNTs in composite reinforcement via dispersion mixing is not too promising. However, the preceding phase diagram could be used in several fruitful applications such as CNT fiber processing, selective deposition on chemically templated surfaces, and drug delivery.

14.4.9 DISCUSSION OF DISPERSION MODEL

The χ values for stable isotropic dispersions are several orders of magnitude smaller than the values obtained for the good CNT solvents using solubility theory. The χ values for stable isotropic dispersion are also much larger than the χ values only including the CNT–CNT interactions ($\chi = B_{\text{CNT-CNT}} v_s / RT$). This could point to some assumptions made in using this model that could be too simplistic. First, the constant interaction parameter for CNT–CNT interactions and the separation of entropic and enthalpic interaction can be expected to be inconsistent with the physical behavior. Flory and Ronca adjusted their model, used here, to include orientation dependent particle–particle interactions instead of lumping them into the χ parameter as isotropic interactions [83]. This would correct the CNT–CNT interactions to physically more accurate predictions. Application of this correction is currently in progress in our research group. The predicted solvent–CNT interaction parameters are too large for isotropic phase stability. As mentioned earlier, viscosity measurements are under consideration to obtain more accurate solubility parameters for CNTs. On the other hand, it may well be possible that the reported dispersions are actually unstable. Thermodynamic calculations do not provide any information on the rate of phase separation. It could thus be that the rate of phase separation in the *good* solvents is long enough due to the low rate of diffusion of the long CNTs, in order to give the impression of phase stability. The magnitude of the enthalpic repulsion would be a measure of the driving force of phase separation, and thus the rate of phase separation, explaining the classification of the solvents. These enthalpic interactions can be measured by Raman spectroscopy, which is a useful tool in determining the dispersive capability of solvents and was used in the classification of the solvents discussed here [11]. The phase diagram behavior of CNTs does not bode well for their application in high-performance composite materials.

14.5 SUMMARY

The work presented here on CNTs allows us to draw some very important conclusions. AESO has been shown to be a good dispersing agent for CNTs. It was able to disperse CNTs in SWNTs in the nonsolvent toluene. Composite samples made with SWNTs dispersed in AESO and styrene showed some improvement in mechanical properties. The use of impure CNTs has to be monitored very carefully. Results from composite samples showed that the impurities, graphitic particles, tend to aggregate from a stable dispersion during polymerization. The impurity aggregation also facilitates unwanted

CNT aggregation. It can thus be concluded that impure CNTs can only be used in small quantities such that the aggregation is limited. The 1 wt% sample appeared to have lost only a very small amount of MWNTs due to aggregation, while for the higher weight percent samples, MWNT aggregation was much more significant. The reinforcement potential of the AESO/styrene resin system, however, is very promising. Thermodynamic calculations of dispersions set a clear limit on the maximum amount of SWNTs that can be dispersed isotropically. This limit decreases significantly with increasing CNT aspect ratio. Predicted CNT solubility parameters appear to be largely overpredicted and need to be determined using experiments. This work is being pursued by Ian McAninch and R. P. Wool [84].

REFERENCES

1. Troianni, H. E.; Miki-Yoshida, M.; Camacho-Bragado, G. A.; *et al. Nano Lett.* 2003, **3,** 751–755.
2. Thostenson, E. T.; Ren, Z.; Chou, T.-W. *Compos. Sci. Technol.* 2001, **61,** 1899–1912.
3. Saito, R.; Kataura, H. *Topics Appl. Phys.* 2001, **80,** 213–247.
4. Wan, X.; Dong, J.; Xing, D. Y. *Phys. Rev. B* 1998, **58,** 6756–6759.
5. Tang, Z. K.; Zhang, L. Y.; Wang, N.; *et al. Synthet. Met.* 2003, **133–134,** 689–693.
6. Ouyang, M.; Huang, J.-L.; Lieber, C. M. *Acc. Chem. Res.* 2002, **35,** 1018–1025.
7. Darkrim, F. L.; Malbrunot, P.; Tartaglia, G. P. *Int. J. Hydrog. Energ.* 2002, **27,** 193–202.
8. Wong, E. W.; Sheehan, P. E.; Lieber, C. M. *Science* 1997, **277,** 1971–1975.
9. Treacy, M. M.; Ebbesen, T. W.; Gibson, C. M. *Nature* 1996, **381,** 678–680.
10. Hilding, J.; Grulke, E. A.; Zhang, Z. G.; *et al. J. Disper. Sci. Technol.* 2003, **24,** 1–41.
11. Ausman, K. D.; Piner, R.; Lourie, O.; *et al. J. Phys. Chem. B* 2000, **104,** 8911–8915.
12. Dai, H. *Acc. Chem. Res.* 2002, **35,** 1035–1044.
13. Jiang, L.; Gao, L.; Sun, J. *J. Colloid Interf. Sci.* 2003, **260,** 89–94.
14. Gong, X.; Baskaran, S.; Voise, R. D.; *et al. Chem. Mater.* 2000, **12,** 1049–1052.
15. McCarthy, B.; Coleman, J. N.; Czerw, R.; *et al. J. Phys. Chem. B* 2002, **106,** 2210–2216.
16. in het Panhuis, M.; Maiti, A.; Dalton, A. B.; *et al. J. Phys. Chem. B* 2003, **127,** 478–482.
17. Dalton, A. B.; Stephan, C.; Coleman, J. N.; *et al. J. Phys. Chem. B* 2000, **104,** 10012–10016.
18. Bandyopadhyaya, R.; Nativ-Roth, E.; Regev, O.; *et al. Nano Lett.* 2002, **2,** 25–28.
19. Chen, J.; Dyer, M. J.; Yu, M.-F. *J. Am. Chem. Soc.* 2001, **123,** 6201–6202.
20. Chambers, G.; Carroll, C.; Farrell, G. F.; *et al. Nano Lett.* 2003, **3,** 843–846.
21. Panhuis, M.; Salvador-Morales, C.; Franklin, E.; *et al. J. Nanosci. Nanotech.* 2003, **3,** 209–213.
22. Dieckmann, G. R.; Dalton, A. B.; Johnson, P. A.; *et al. J. Am. Chem. Soc.* 2003, **125,** 1770–1777.
23. Georgakilas, V.; Tagmatarchis, N.; Pantarotto, D.; *et al. Chem. Commun.* 2002, **24,** 3050–3051.
24. Banerjee, S.; Wong, S. S. *J. Phys. Chem. B* 2002, **106,** 12144–12151.
25. Whitsitt, E. A.; Barron, A. R. *Nano Lett.* 2003, **3,** 775–778.
26. Lee, Y. S.; Cho, T. H.; Lee, B. K.; *et al. J. Fluorine Chem.* 2003, **120,** 99–104.
27. Khabashesku, V. N.; Billups, W. E.; Margrave, J. L. *Acc. Chem. Res.* 2002, **35,** 1087–1095.
28. Boul, P. J.; Mickelson, E. T.; Huffman, C. B.; *et al. Chem. Phys. Lett.* 1999, **310,** 367–372.
29. Garg, A.; Sinnott, S. B. *Chem. Phys. Lett.* 1998, **295,** 273–278.
30. Richard, C.; Balavoine, F.; Schultz, P.; *et al. Science* 2003, **300,** 775–778.
31. Nikolaev, P.; Bronikowski, M.; Bradley, R.; *et al. Chem. Phys. Lett.* 1999, **313,** 91–97.

32. Thielemans, W., "Lignin and Carbon-Nanotube Utilization in Bio-Based Composites, Ph.D. Thesis, university of Delaware, Newark DE 19716, (2004).
33. Dresselhaus, M. S.; Dresselhaus, G.; Jorio, A.; *et al. Carbon* 2002, **40,** 2043–2061.
34. Rao, A. M.; Chen, J.; Richter, E.; *et al. Phys. Rev. Lett.* 2001, **86,** 3895–3898.
35. Thielemans W., Wool R. P., paper submitted for publication.
36. Sandler, J.; Schaffer, M. S. P.; Prasse, T.; *et al. Polymer* 1999, **40,** 5967–5971.
37. Cadek, M.; Murphy, R.; McCarthy, B.; *et al. Carbon* 2002, **40,** 923–928.
38. Kandola, B. K.; Horrocks, A. R.; Myler, P.; *et al. ACS Symp. Ser.* 2001, **797,** 344–360.
39. Das, A. N.; Baijal, S. K. *J. Appl. Polym. Sci.* 1982, **27,** 211–223.
40. Coleman, J. N.; O'Brien, D. F.; Dalton, A. B.; *et al. Chem. Commun.* 2000, 2001–2002.
41. Perminov, V. P.; Modyanova, A. G.; Ryabkov, Y. I.; *et al. Russ. J. Appl. Chem.* 2002, **75,** 636–640.
42. Witort, I.; Tadeusz, S.; Wladyslaw, C.; *et al. Polymery* 1976, **21,** 155–157.
43. Narkis, M.; Rafail, I.; Victor, G.; *et al. J. Appl. Polym. Sci.* 2000, **76,** 1165–1170.
44. Narkis, M.; Zilberman, M.; Siegman, A. *Polym. Advan. Technol.* 1997, **8,** 525–528.
45. Sweitzer, C. W.; Lyon, F.; Grabowski, T. S. *Ind. Eng. Chem.* 1955, **47,** 2380–2385.
46. Cadek, M.; Coleman, J. N.; Ryan, K. P.; *et al. Nano Lett.* 2004, **4,** 353–356.
47. Qian, D.; Dickey, E. C.; Andrews R.; *et al. Appl. Phys Lett.* 2000, **76,** 2868–2870.
48. Thostenson, E. T.; Chou, T.-W. *J. Phys. D: Appl. Phys.* 2003, **36,** 573–582.
49. Halpin, J. C.; Kardos, J. L. *Polym. Eng. Sci.* 1976, **16,** 344–352.
50. Talreja, R.; Manson, J.-A. E. In *Comprehensive Composite Materials*, Vol. 2: *Polymer Matrix Composites*, Kelly, A.; Zweben, C. H., Eds.; Elsevier Science Ltd., New York; 2000, p. 307.
51. Cadek, M.; Coleman, J. N.; Barron, V.; *et al. Appl. Phys. Lett.* 2002, **81,** 5123–5125.
52. Li, C.; Chou, T.-W. *J. Nanosci. Nanotechnol.* 2003, **3,** 1–8.
53. Li, C.; Chou, T.-W. *Compos. Sci. Technol.* 2003, **63,** 1517–1524.
54. Jia, Z.; Wang, Z.; Xu, C.; *et al. Mat. Sci. Eng.* 1999, **A271,** 395–400.
55. Park, S. J.; Cho, M. S.; Lim, S. T.; *et al. Macromol. Rapid Commun.* 2003, **24,** 1070–1073.
56. Reznik, D.; Olk, C. H.; Neumann, D. A.; *et al. Phys. Rev. B.* 1995, **52,** 116–124.
57. Saito, Y.; Yoshikawa, T.; Bandow, S.; *et al. Phys. Rev. B* 1993, **48,** 1907–1909.
58. Thess, A.; Lee, R.; Nikolaev, P.; *et al. Science* 1996, **273,** 483–487.
59. Charlier, J.-C.; Gonze, X.; Michenaud, J.-P. *Europhys. Lett.* 1995, **29,** 43–48.
60. Girifalco, L. A.; Hodak, M.; Lee, R. S. *Phys. Rev. B* 2000, **62,** 13104–13110.
61. Dore, J.; Burian, A.; Tomita, S. *Acta Physica Polonica A* 2000, **98,** 495–504.
62. Duclaux, L.; Salvetat, J. P.; Lauginie, P.; *et al. J. Phys. Chem. Solids* 2003, **64,** 571–581.
63. Maniwa, Y.; Fujiwara, R.; Kira, H.; *et al. Phys. Rev. B* 2001, **64(7), No.** 073105, pages 1–4.
64. Jones, A. T. *Makromolekulaire Chemie* 1964, **71,** 1–32.
65. Van Krevelen, D. W. *Properties of Polymers: Their Correlation with Chemical Structure; Their Numerical Estimation and Prediction from Additive Group Contributions,* Elsevier Science, Amsterdam. 1997.
66. Dalton, A. B.; Collins, S.; Munoz, E.; *et al. Nature* 2003, **423,** 703.
67. Oh, S. J.; Cheng, Y.; Zhang, J.; *et al. Appl. Phys. Lett.* 2003, **82,** 2521–2523.
68. Gao, X.; Hu, T. J.; Liu, L. Q.; *et al. Chem. Phys. Lett.* 2003, **370,** 661–664.
69. Ajayan, P. M.; Zhou, O. Z. *Top. Appl. Phys.* 2001, **80,** 391–425.
70. Flory, P. J., *Proc. R. Soc. London Ser. A* 1956, **234,** 73–89.
71. Flory, P. J.; Ronca, G. *Mol. Cryst. Liq. Cryst.* 1979, **54,** 289–309.
72. Sandler, S. I. *Chemical Engineering Thermodynamics,* 2nd ed., John Wiley & Sons, New York; 1989, pp. 443–453.
73. Onsager, L. *Ann. N. Y. Acad. Sci.* 1949, **51,** 627–659.
74. Song, W.; Kinloch, I. A.; Windle, A. H. *Science* 2003, **302,** 1363.
75. Kamide, K.; Dobashi, T. *Physical Chemistry of Polymer Solutions: Theoretical Background,* Elsevier Science, New York; 2000.
76. Flory, P. J. *Principles of Polymer Chemistry,* Cornell University Press, Ithaca, NY; 1953.

77. Paul, D. R.; Barlow, J. W. *Polymer* 1984, **25,** 487–494.
78. Hildebrand, J. H.; Prausnitz, J. M.; Scott, R. L. *Regular and Related Solutions,* Van Nostrand-Reinhold, Princeton, NJ; 1970.
79. Hansen, C. M. *Ind. Eng. Chem. Prod. Res. Dev.* 1969, **8,** 2–10.
80. Hoy, K. L. *J. Coat. Fabr.* 1989, **19,** 53–67.
81. Zhou, X. H.; Liu, J. B.; Jin, Z. X.; *et al. Fullerene Sci. Technol.* 1997, **5,** 285–290.
82. Barton, A. F. M. *CRC Handbook of Solubility Parameters and Other Cohesion Parameters,* CRC Press, Boca Raton, FL; 1983, pp. 304–308.
83. Flory, P. J.; Ronca, G. *Mol. Cryst. Liq. Cryst.* 1979, **54,** 311–330.
84. McAninch, I.; Wool R. P. In *Proc. of the American Physical Society Annual Meeting,* Philadelphia, August 2004.

15

Nanoclay Biocomposites

Richard P. Wool

The mechanical properties of triglyceride-based polymers presented in Chapter 4 are slightly below those of the commercial polymers, such as unsaturated polyesters, which greatly restricts the commercialization and applications of these new bio-based materials. Further chemical modification of triglyceride molecules may not be wise due to the difficulty of processing; in addition, cost can be another critical issue. Reinforcement of polymers with a second phase, whether inorganic or organic, to produce a polymer composite is a common practice in industry.

New polymer nanocomposites represent a radical alternative to the conventional polymer composites. The researchers at the Toyota research center in Japan showed that the tensile modulus and strength were doubled for nylon-layered silicate nanocomposites containing as little as 2 vol% clay, and the heat distortion temperature of the nanocomposites increased 87 °C, extending the use of this polymer to under-the-hood structural parts in the engine compartment [1]. Since then, polymer-layered silicate nanocomposites have been successfully developed with most conventional polymers. These nanocomposites exhibit markedly improved mechanical, thermal, optical, and physicochemical properties when compared with the pure polymer or conventional composites [2, 3]. Bio-based nanocomposites have attracted significant interest because these materials have significant social and environmental advantages [4, 5], as discussed in Chapter 4. Most research has concentrated on epoxidized plant oils [5, 6]. Initial studies showed the formation of intercalated or exfoliated structures and a reinforcement effect through clay addition. The addition of clay to triglyceride-based polymers to form nanocomposites can broaden the applications of these new bio-based materials by improving their mechanical properties. The extremely large surface area and high aspect ratio (between 30 and 2000) of the clay make it possible for property improvements resulting from the formation of a nanocomposite.

The most commonly used layered silicate in nanocomposites is the natural clay from the smectite family: montmorillonite (MMT). The layer structure of montmorillonite was deduced by Hofmann *et al.* on the basis of its similarity to that of pyrophyllite [7], as shown in Figure 15.1. Their crystal lattice consists of two silica tetrahedral sheets fused to an edge-shared octahedral sheet of either aluminum or magnesium hydroxide. Isomorphous substitution of Al^{3+} for other cations (e.g., Mg^{2+}, Fe^{2+}, Fe^{3+}) in octahedral sites and, less frequently, of Si^{4+} for Al^{3+} in the tetrahedral lattice causes an excess of negative charges within the MMT layers, which are counterbalanced by hydrated alkali or alkaline earth cations situated between the layers. Normally, the silicate surface is hydrophilic, which hinders the homogeneous dispersion in organic matrix. Ion exchange reactions with cations render silicate surfaces organophilic, which makes them organophilic and compatible with the polymer matrix.

Depending on the nature of the components used and the preparation method, three main types of composites can be formed, as shown in Figure 15.2. When the polymer is unable to intercalate between the silicate sheets, a phase separation occurs that results in traditional microcomposites. The formation of conventional composites improves rigidity, but they often sacrifice strength, elongation, and toughness. Beyond the conventional compos-

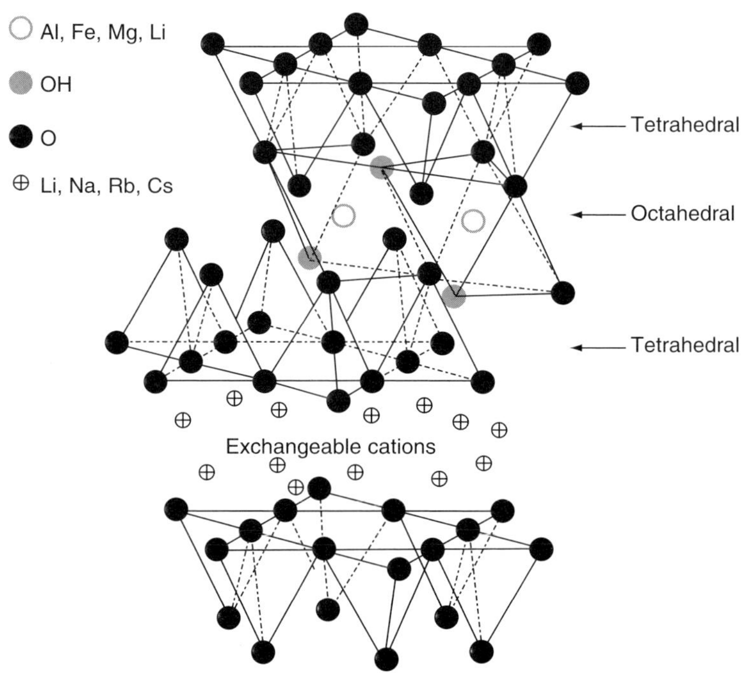

FIGURE 15.1 The layer structure of montmorillonite according to Hofmann *et al.* [7].

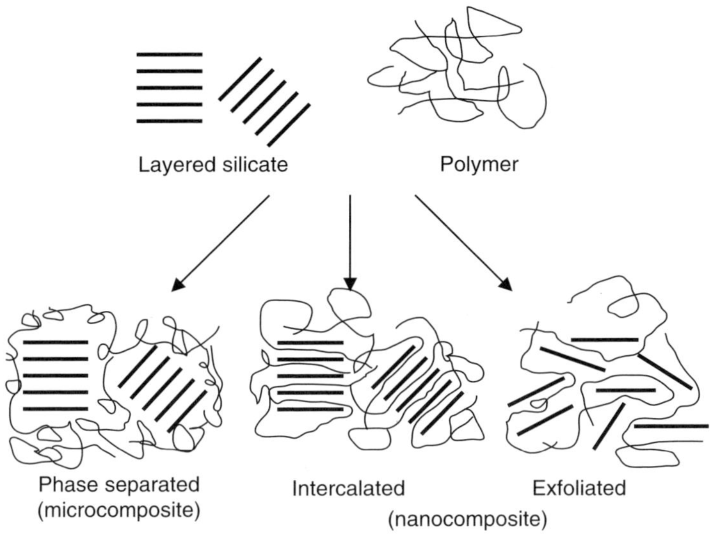

FIGURE 15.2 Schematic representations of different types of composite prepared from the layered silicates and polymers.

ites, two types of nanocomposites are possible: an intercalated structure in which a single or more extended polymer chain is intercalated between the silicate layers, resulting in a well-ordered multilayer with alternating polymeric and inorganic layers, and an exfoliated or delaminated structure, in which the silicate layers are completely and uniformly dispersed in a continuous polymer matrix. The formation of nanocomposites optimizes the number of available reinforcing elements for carrying an applied load and deflecting cracks, which results in improving stiffness, strength, and toughness. Additionally, they are lighter compared to the conventional composites because they use far less inorganic material. They also exhibit outstanding diffusional barrier properties, which enhances chemical resistance and flame retardance and reduces solvent uptake.

Essentially, three different approaches are used to synthesize polymer–clay nanocomposites: melt intercalation, solution, and *in situ* polymerization. The synthesis of nanocomposites using triglyceride-based resins is included in the last category, as are epoxy resins and unsaturated polyester resins [8, 9]. The nanocomposites are prepared by first swelling the organo-modified clay with the monomers, followed by the cross-linking reactions. The miscibility of monomer or polymer with organic modifier at the swelling stage can be assessed by their solubility parameters [10–12]. It is believed that the exfoliation of silicate layers is related to the polarity of the monomers [13]. Often an intercalated structure is formed when nonpolar styrene is polymerized with clay [14, 15], but most epoxy-based nanocomposites show an exfoliated structure [9, 16]. In addition, the self-polymerization of epoxy resin in organophilic clays due to the presence of the alkylammonium ions facilitated the formation of exfoliated structure [17]. Unsaturated polyester resins have also been well studied by several researchers [9, 18, 19]. Unlike epoxy resin, this bicomponent system offers less polarity for the molecules to intercalate to the clay layers. Although the exfoliated structure was confirmed, there is no significant impact on the dynamic mechanical properties, mechanical properties, and thermal stability [9, 18, 19]. Because triglyceride-based monomers optionally have polar groups such as hydroxyl and carboxylic acid groups, these molecules are favored in the formation of a possible exfoliated structure [4]. Triglycerides also offer the ability to control the number of polar groups through chemical modifications. For the research discussed in this chapter, we used three different functionalized triglyceride monomers: acrylated epoxidized soybean oil (AESO), malleinized acrylated epoxidized soybean oil (MAESO), and soybean oil pentaerythritol maleates (SOPERMA) [20], as shown in Figure 15.3. The effect of molecular structure on the miscibility with organic modifiers of clays and the morphology of resulting nanocomposites is examined, and the properties of these new triglyceride-based nanocomposites are reported.

15.1 PREPARATION OF NANOCLAY–SOYBEAN OIL COMPOSITES

15.1.1 SYNTHESIS OF MAESO

AESO is fully acrylated with approximately 3.4 acrylates per triglyceride and an average molecular weight of 1200 g/mol. Maleic anhydride, *N,N*-dimethylbenzylamine (BDMA), hydroquinone, and styrene were all obtained from the Aldrich Chemical Co., Milwaukee, WI, and used as received. To synthesize MAESO, 50 g of AESO and 0.05 g of hydroquinone were first heated to 70 °C, while being stirred. Maleic anhydride (8.17 g) was finely ground and added to the reaction at 70 °C. The reaction was then heated up to 80–85 °C, at which point the maleic anhydride dissolved, forming a homogeneous solution. The BDMA catalyst was then added in the amount of 1.0 g. The reaction was stopped after 6 h, before gelation occurred.

15.1.2 SYNTHESIS OF SOPERMA

In the alcoholysis reaction, 400 g of soybean oil (Aldrich) was mixed with 186.51 g pentaerythritol (Aldrich) and 5.87 g $Ca(OH)_2$ in a three-necked 1-L round-bottom flask equipped with a mechanical stirrer, a nitrogen gas inlet, and a calcium drier. The reaction mixture was heated to 230–240 °C and agitated under N_2 atmosphere for 2 h. The reaction product at room tem-

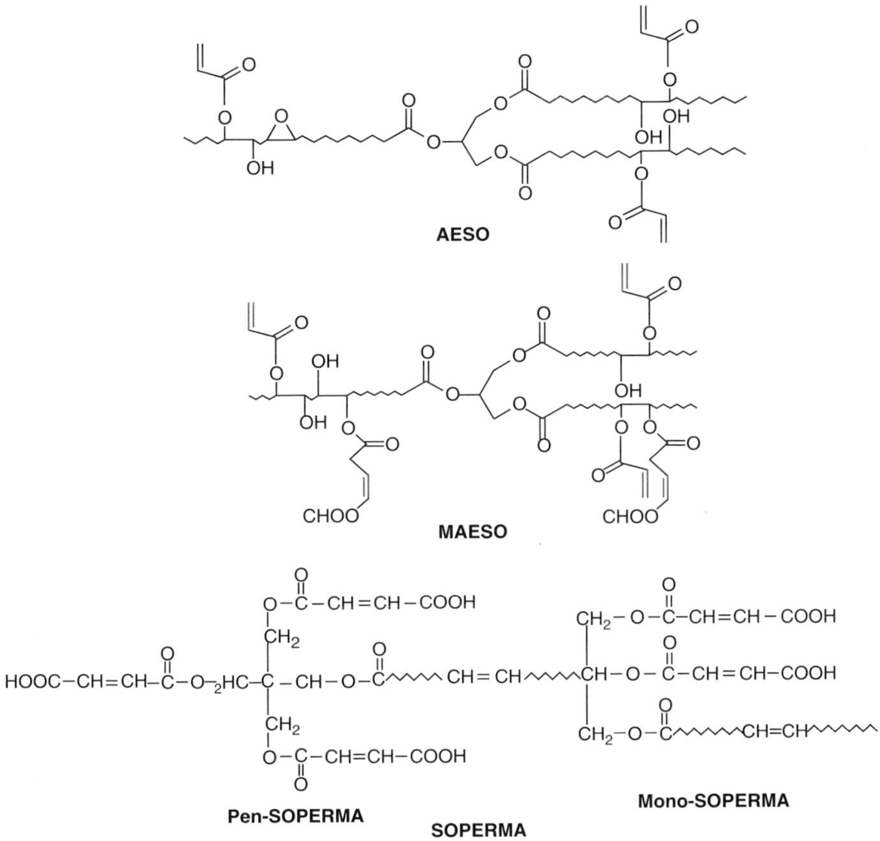

FIGURE 15.3 Molecular structure of functionalized triglycerides used for nanocomposites.

perature was a light brown, viscous liquid that separated into two layers with time. In the malleinization reaction, 540 g of the soybean oil pentaerythritol alcoholysis product and 311.86 g of maleic anhydride were placed in a 1-L flask equipped with a thermometer and a mechanical stirrer. The mixture was heated in an oil bath with stirring until the maleic anhydride melted and mixed with the soybean oil pentaerythritol alcoholysis product. Then 8.52 g of BDMA and 0.852 g of hydroquinone were added and the reaction mixture was heated to 95°C. The mixture was agitated at this temperature for 2 h. Both the IR and ^{1}H NMR spectroscopy of the product confirmed the consumption of maleic anhydride and the formation of maleate half-esters.

15.1.3 SWELLING TEST

The solubility parameters of triglyceride-based polymers were determined by swelling experiments. Fifteen grams of the functionalized triglycerides were cured with 1.5 wt% initiator at 110 °C for 2 h and postcured at 150 °C for 2 h. The weight of the cured samples was measured and they were placed in the bottles with different solvents, such as cyclohexane, methyl methacrylate, toluene, benzene, chloroform, acetone, and acrylic acid. The weight of the swollen samples was measured spontaneously until there was no weight gain.

15.1.4 PREPARATION OF CLAY NANOCOMPOSITES

The organo-treated MMTs were obtained from Southern Clay Products, Inc., Gonzales, TX, and were prepared by ion exchanging Na-MMT with alkylammonium cations. They are methyl tallow bis-2-hydroxyethyl quaternary ammonium MMT (C30B), dimethyl benzyl hydrogenated tallow quaternary ammonium MMT (C10A), dimethyl dihydrogenated tallow quaternary ammonium MMT with a modifier concentration of 125 mEq/100 g clay (C15A), and dimethyl dihydrogenated tallow quaternary ammonium MMT with a modifier concentration of 95 mEq/100 g clay (C20A). Table 15.1 shows the product information for these organo-modified clays [21]. Several different clay concentrations in the triglyceride-based resins were investigated (3, 5, 7.5, and 10 wt% based on the total weight of the resin). The composition of the resin was 66.7 wt% functionalized triglyceride monomers and 33.3 wt% styrene. The desired amount of organo-treated clay was added to the resin and mechanically stirred for 24–48 h with the flask well sealed to prevent the evaporation of styrene. Both room-temperature and high-temperature curing were used to cure the sample. For the high-temperature curing, 1.5 wt% *tert*-butyl peroxy benzoate (TBP) radical initiator was added, and the mixture was cured at 110 °C for 3 h (SOPERMA/styrene was cured at 120 °C) and postcured at 150 °C for 2 h. For room-temperature curing, 3 wt% of free-radical initiator (Trigonox 239 A for AESO, and Hi-point 90 for the MAESO and SOPERMA systems) and 0.8 wt% of cobalt naphthalate was added. Samples were cured at room temperature for 24 h and then postcured at 150 °C for 2 h. To prevent oxygen free-radical inhibition, the resin was purged with nitrogen gas prior to curing.

TABLE 15.1 General information for organo-treated clay.

Clay	Organic Modifier	Modifier Concentration	Specific Gravity (g/cc)	Particle Size
Cloisite® 30B (C30B)	MT2EtOT	90 mEq/100 g clay	1.98	90% < 13 μm
Cloisite® 10A (C10A)	2MBHT	125 mEq/100 g clay	1.90	90% < 13 μm
Cloisite® 15A (C15A)	2M2HT	125 mEq/100 g clay	1.66	90% < 13 μm
Cloisite® 20A (C20A)	2M2HT	95 mEq/100 g clay	1.77	90% < 13 μm

Source: From Ref. [21].

15.2 SOY-NANOCLAY COMPOSITES

15.2.1 MISCIBILITY OF CLAY IN SOY RESINS

The solubility parameter approach to the prediction of nanocomposite morphology was used with some degree of success in studies of polymer–clay modifier miscibility [11, 12]. Based on a simple equation of the free energy of mixing:

$$\Delta G_m = \Delta H_m - T\Delta S_m, \tag{15.1}$$

where ΔG_m is the change in Gibb's free energy, T is the absolute temperature, and ΔS_m is the entropy of mixing. A negative value for ΔG_m indicates that the solution process will occur spontaneously. The term $T\Delta S_m$ is always positive because there is an increase in the entropy on mixing. Therefore, the sign of ΔG_m depends on ΔH_m, the enthalpy of mixing. For a binary mixture, Hildebrand has related the enthalpy contribution ΔH_m to the difference of the solubility parameters of the polymers as follows:

$$\Delta H_m = (\delta_1 - \delta_2)^2 V_m \phi_1 \phi_2, \tag{15.2}$$

where V_m is the volume of the mixture, ϕ_i is the volume fraction of i in the mixture, and δ_i is the solubility parameter of the i component. In the Flory–Huggins theories, the interaction parameter χ, which is responsible for the enthalpic contribution, is related to the solubility parameters via [22]

$$\chi = 0.34 + \frac{V_1}{RT}(\delta_1 - \delta_2)^2, \tag{15.3}$$

where V_1 is the reference volume. For better miscibility, the polymers should have similar solubility parameters. This is in accordance with the general rule that chemical and structural similarity favors solubility. Experimentally, the solubility parameters of cross-linked polymers can be determined by the swelling test. The swelling coefficient, Q, is defined by [23]

$$Q = \frac{m - m_0}{\rho_s m_0}, \tag{15.4}$$

where m is the weight of the swollen sample, m_0 is the dry sample, and ρ_s is the density of the swelling agent. In Figure 15.4, the swelling coefficient Q for AESO and MAESO is plotted against the solubility parameters of various solvents. By curve fitting, the solubility parameters for AESO and MAESO are 19.5 and 19.6 $\text{MPa}^{1/2}$, respectively, as shown in Table 15.2. Using the same method, the solubility parameter of SOPERMA was 20.1 $\text{MPa}^{1/2}$ [24]. The solubility parameter can also be calculated theoretically. In this study, the solubility parameter values for functionalized triglycerides and clay modifiers were calculated by the same theoretical model for the comparison basis.

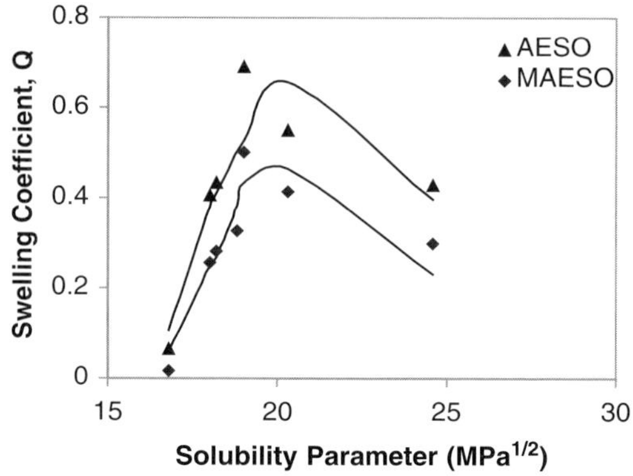

FIGURE 15.4 Solubility parameter of AESO and MAESO from the curve fitting of a swelling experiment.

TABLE 15.2 Molecular weight and solubility parameter values for monomers.

				SOPERMA		
	SO	AESO	MAESO	Mono-	Pen-	Styrene
Molecular weight (g/mol)	871	1167.9	1390.4	547.9	689.9	104
δ (Hoy model)	18.73	20.72	20.87	20.17	20.51	20.50
δ_d	17.41	16.19	15.81	16.34	15.76	17.07
δ_p	5.75	10.20	11.04	9.82	10.33	9.33
δ_h	3.82	7.94	7.97	6.57	8.11	6.45
δ (Experimental)		19.5	19.6	20.1		19.0

Note: The solubility parameter is given in units of (MPa)$^{1/2}$.

The method used was the one proposed by Hoy [25], which is very similar to the Hanser [26, 27] and H-VK models [28] in which they assume the solubility parameter directly depends on dispersive, permanent dipole–dipole interactions (polarity) and hydrogen bonding forces [28]:

$$\delta_t^2 = \delta_d^2 + \delta_p^2 + \delta_h^2, \tag{15.5}$$

where δ_d, δ_p, and δ_h are dispersion, polar, and hydrogen bonding components, respectively. Using the group contribution method, the solubility parameter is given by [29]

$$\delta = \rho \frac{\sum_i F_i}{M}, \tag{15.6}$$

where ρ is the density of the polymer, M is the molar mass of the polymer, and $F_i = (E_i V_i)^{1/2}$ where E_i and V_i are the cohesion energy and molar volume of the group considered. According to Table 15.2, the three functionalized triglycerides and styrene have very similar solubility parameters, and δ_p

values. The calculated solubility parameters are relatively higher than the experimental values. The solubility parameters for the organic modifiers of clay were calculated using the same model.

The molecular structures of organic modifiers for these clays are shown in Figure 15.5. Note that the organic modifier for C30B has polar hydroxyl groups, which presumably provide a good wetting surface with functionalized triglycerides, whereas the modifier for C10A has a benzyl unit similar to styrene, and the structure affinity may play an important role on the intercalation of monomers. Table 15.3 shows that the solubility parameters for these organic modifiers are in the range of 17.57 to 20.38 $MPa^{1/2}$, which is close to the literature range of 18 to 28 $MPa^{1/2}$. Ho *et al.* [10] found that the solubility value for C15A was 22.3 $MPa^{1/2}$ by small-angle neutron scattering technique, which is slightly higher than the value calculated by the Hoy model. By comparing Table 15.3 to Table 15.2, the solubility parameter of C30B is the closest to those of triglyceride-based monomers. Because the solubility parameters of triglyceride-based monomers and organic modifiers are very close, they would be expected to be miscible. In the mixing stage, triglyceride molecules can diffuse between the clay layers and expand the spacing between the layers.

The dispersibility of organo-treated clay in triglyceride-based monomers and styrene was examined by direct observation of optical microscopy and

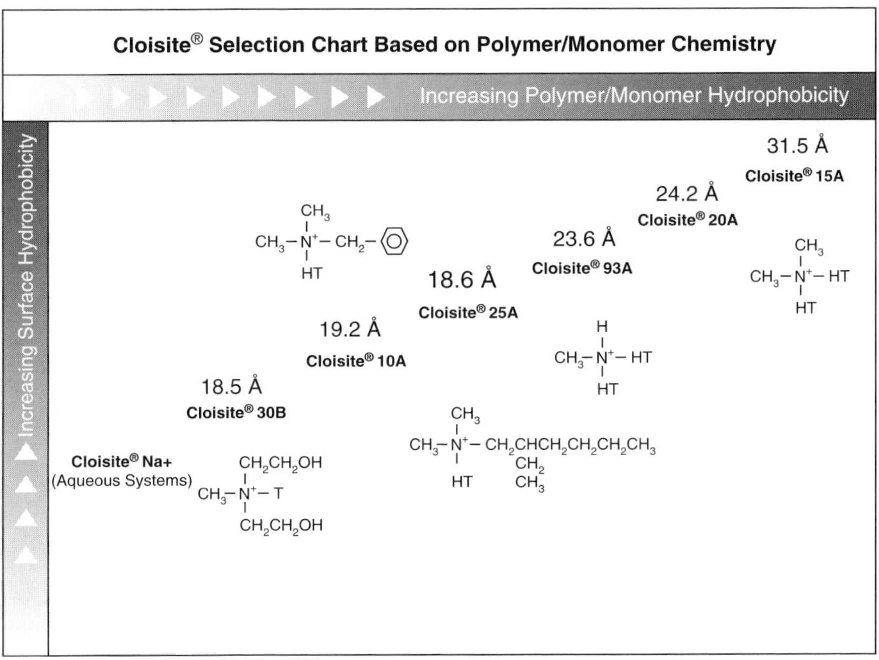

FIGURE 15.5 A series of organo-treated clay from Southern Clay Product, Inc. T represents tallow (~65% C18; ~30% C16; ~5% C14); HT represents hydrogenated tallow. (*Source:* From Ref. [21].)

TABLE 15.3 Solubility parameter values for organic modifiers.

Solubility Parameter	C30B	C10A	C15A & C20A	C25A	C93A
δ (Hoy model)	20.38	18.29	17.99	17.57	18.29
δ_d	16.68	17.12	15.73	16.86	17.11
δ_p	9.48	5.96	4.01	4.95	5.96
δ_h	6.87	2.46	7.75	0	2.46
δ (Experimental)			22.3		

Note: The solubility parameter is given in units of $(MPa)^{1/2}$.

X-ray diffraction (XRD). Figure 15.6 shows an optical micrograph for C30B dispersed in AESO and styrene mixture. The dispersion is homogeneous with the particle size in the range of 1 to 2 μm, although, obviously, the clay layers are not fully exfoliated. Thus X-ray diffraction was used to detect if monomers penetrated into the clay layers during mixing. Figure 15.7 shows a comparison of X-ray diffraction patterns for C30B and C10A when dispersed in styrene, AESO, or both. The pure C30B shows a peak at 4.78°, corresponding to d_{001} plane. Using the Bragg formula, the gallery spacing of C30B is calculated to be 18.2 Å. The gallery spacing of C10A is 19.2 Å, which is slightly higher than that of C30B. From Figure 15.7, C10A has a better dispersion in styrene monomer than C30B because of the molecular structure affinity between styrene monomer and the organic modifier of C10A, whereas C30B has a better dispersion in AESO monomer because of the similar solubility parameters and the possible interaction of the carboxylic acid group on triglycerides with the hydroxyl group of the organic modifier. In general, they show a very similar XRD pattern when mixed with the mixture of AESO and styrene. Table 15.4 summarizes the basal spacing changes when organo-treated clays are mixed with AESO and styrene monomers. The larger the basal spacing shift, the more monomers penetrated. Based on the difference of basal spacings, C10A and C30B give much better dispersion in AESO and styrene compared to C15A and C20A, which is expected from the solubility parameters analysis and from observation of molecular structures. Therefore, the following discussion concentrates on C30B and C10A.

Figure 15.8 shows XRD patterns for C30B when dispersed in different monomers. When 5 wt% C30B is dispersed in styrene, AESO, and SOPERMA, the d_{001} peak of the clay shifts to low angles corresponding to an increase in the d-spacing from 18.2 Å to 45.2, 40.6, and 50.6 Å, respectively. The d-spacing of clay dispersed in SOPERMA is higher than others; with the addition of styrene as a diluent, the peak for the pristine C30B completely disappears, and there is only one peak showing at 2.07°, which indicates a highly disordered intercalated structure. The better dispersion of clay in SOPERMA may be because the diffusion of the intercalatant molecules is not only related to the miscibility factor, but also to the flexibility of

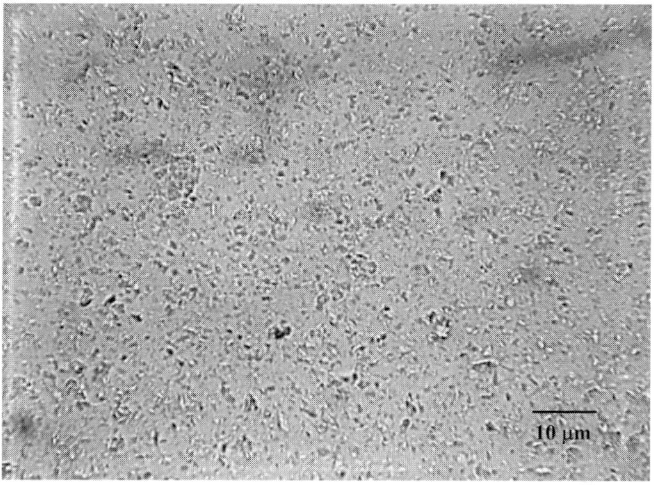

FIGURE 15.6 Optical microscopic picture of C30B dispersed in AESO and styrene mixture.

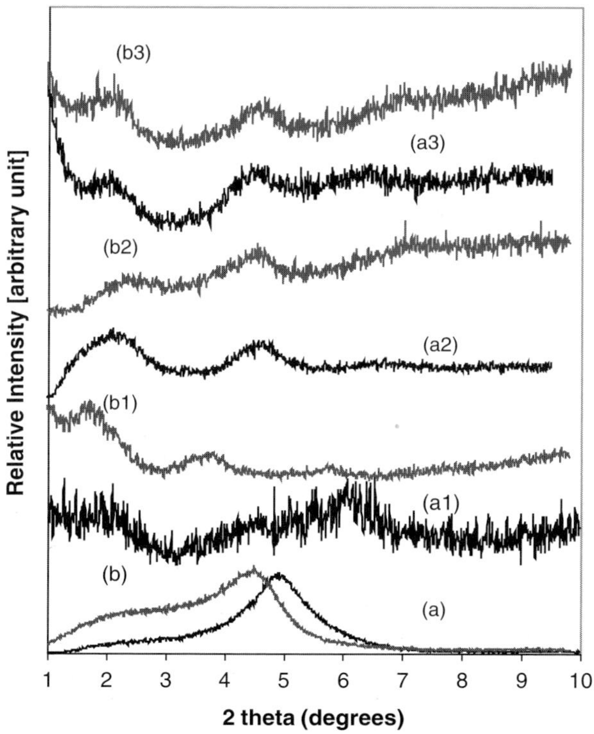

FIGURE 15.7 XRD patterns of the mixture of 5 wt% organo-treated clays in monomers: (a) pristine C30B, (b) pristine C10A, (a1) C30B/styrene, (b) C10A/styrene, (a2) C30B/AESO, (b2) C10A/AESO, (a3) C30B/AESO + styrene, and (b3) C10A/ AESO + styrene.

TABLE 15.4 The basal spacings of organo-treated clay when mixed with monomers at liquid stage.

Clay	Monomer	Original d-Spacing	d-Spacing after Mixing	Δd
C30B	Styrene	18.5	40.2/14.6	21.7
	AESO		40.6/19.0	22.1
	AESO/styrene		43.3/19.5	24.8
C10A	Styrene	19.2	47.6/22.5	28.4
	AESO		35.6/18.4	16.4
	AESO/styrene		44.0/18.7	24.8
C15A	Styrene	31.5	Not detected	—
	AESO		37.2/18.8	5.7
C25A	Styrene	24.2	Not detected	—
	AESO		36.3/18.8	12.1

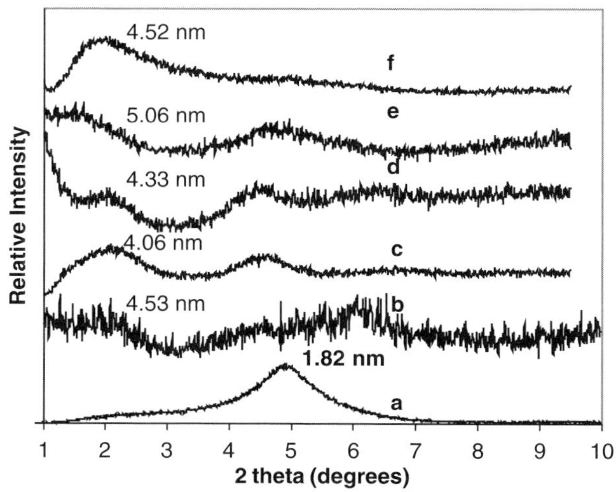

FIGURE 15.8 XRD data of the liquid samples with 5 wt% C30B: (a) pristine C30B, (b) styrene, (c) AESO, (d) AESO/styrene, (e) SOPERMA, and (f) SOPERMA/styrene.

the molecules, because the SOPERMA molecules have only one fatty acid chain, resulting in a much lower molecular weight compared to AESO and SOPERMA, as shown earlier in Table 15.2. Therefore, it is much more favored during the swelling stage. The increase in d-spacing of the silicate layers confirms monomer intercalation.

Further evidence, observed during mixing, is supported by the viscosity increase and the mixture's color change from opaque to semitransparent. It is very difficult to obtain valuable data from XRD for liquid samples because of the curved surface of liquid samples. Thus, the *in situ* measurement of monomer intercalation was not successful; in addition, the evaporation of styrene during testing may lead to error peaks as these clay layers reassemble.

Compared to unmodified soybean oil (SO), the chemical modification of triglycerides increases the solubility parameters and polarity of the molecules. As seen in Table 15.2, both the values of δ_p and δ_h of modified triglycerides are doubled, resulting in the solubility parameters being very similar to the organic modifier parameters. Therefore, the chemical modification when combined with the theoretical model can be used as an effective method to control the miscibility between these triglyceride-based molecules and the clay surface.

15.2.2 STRUCTURE AND MORPHOLOGY

The morphological state of the cured nanocomposites was investigated using XRD and transmission electron microscopy (TEM). The XRD scans of AESO–clay composites (Figure 15.9) show that the morphology of the nanocomposites does depend on the clay loading. The peak of the clay disappears in the scattering curves for the 3 wt% C30B–AESO nanocomposite. When the clay content goes up to 5 wt%, two peaks are seen. One is a weak peak at approximately 4.64° and the other is at 2.12°. These 2θ values correspond to interlayer spacings of 19.04 and 41.58 Å, respectively. The XRD data for both MAESO and SOPERMA–clay nanocomposites exhibit the same trend, as shown in Figure 15.10. From XRD data, these triglyceride-based nanocomposites show an exfoliated structure at 3 wt% clay load, and an intercalated sheet at high clay load (>5 wt%).

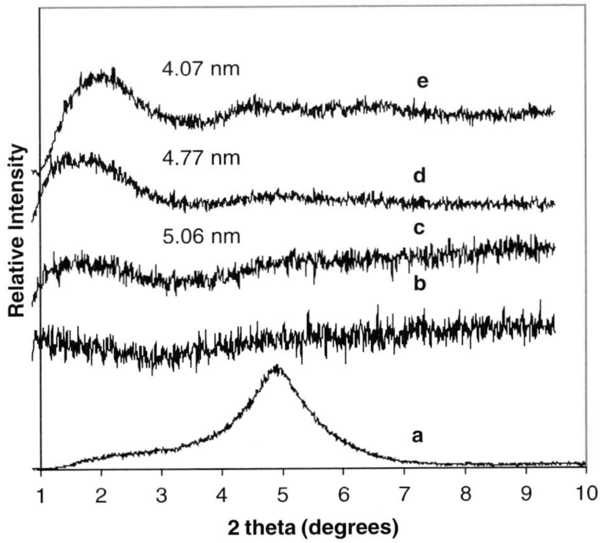

FIGURE 15.9 XRD data of AESO-based nanocomposite system with various amounts of C30B: (a) pristine C30B, (b) 3 wt%, (c) 5 wt%, (d) 7.5 wt%, and (e) 10 wt%.

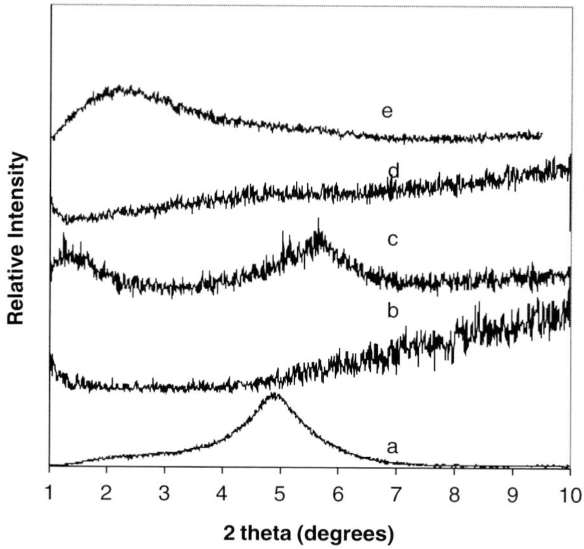

FIGURE 15.10 XRD data of triglyceride-based nanocomposite: (a) pristine C30B, (b) MAESO–3 wt% C30B, (c) MAESO–5 wt% C30B, (d) SOPERMA–3 wt% C30B, and (e) SOPERMA–5 wt% C30B.

The morphology of triglyceride-based nanocomposites was further investigated by TEM of a thin section (100 nm). The TEM images of AESO–clay nanocomposites show a completely exfoliated structure at 3 wt% clay loading and a mix of intercalated and partially exfoliated structure above 5 wt% (Figure 15.11). For the 3 wt% clay nanocomposite, well-dispersed individual silicate layers are shown. However, for 5 wt% clay in an AESO matrix, the individual sheets of clay are separated by triglyceride molecules (intercalated structure) and some of the clay layers are partially exfoliated. Also, micro-sized aggregates of clay sheets are observed at low magnification above 5 wt% clay loading. Based on the thermodynamic view, the d-spacing of nanocomposites is expected to be independent of the polymer-to-silicate composition [30–32]. A simple space-filling calculation assuming the polymer density to be unaffected by confinement of the layers suggests that for hybrids with more than 30 wt% polymer, excess polymer exists that is not intercalated. In our case, clay filler is not more than 10 wt%. However, thermodynamics can only predict the equilibrium structure; the insufficient mixing could be a possible reason because this process is very time and mixing strength dependent.

15.2.3 THERMOMECHANICAL PROPERTIES

Table 15.5 shows the storage modulus and glass transition temperature (tan δ) for AESO, SOPERMA, and MAESO nanocomposites. The storage

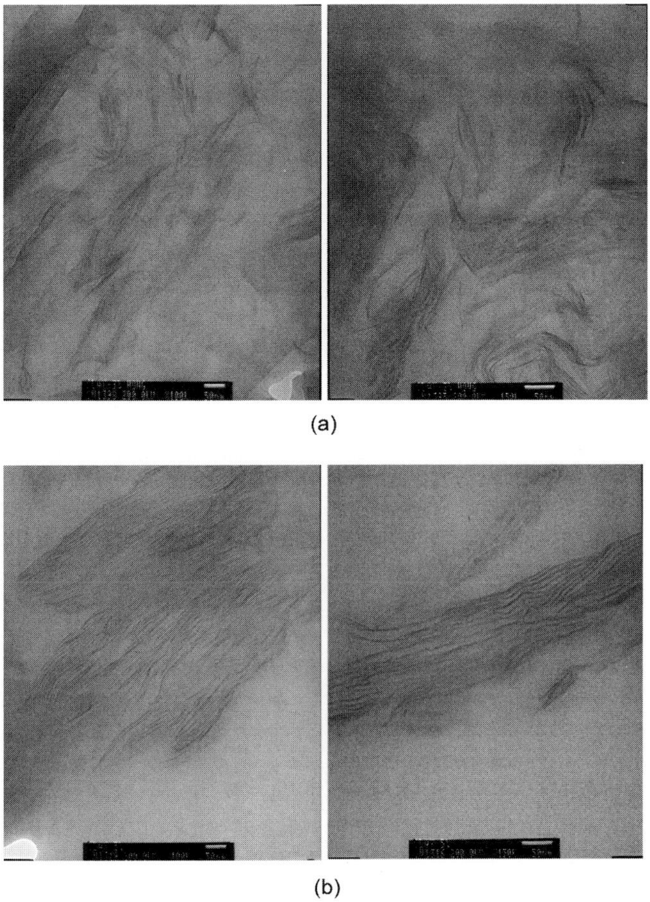

FIGURE 15.11 TEM micrographs of AESO–clay (C30B) nanocomposites, where the scale bars represent 50 nm: (a) 3 wt% AESO–clay nanocomposite and (b) 5 wt% AESO–clay nanocomposite.

modulus of all triglyceride-based nanocomposites was improved with the addition of clay. A change in the modulus indicates a change in the rigidity and, hence, strength of the nanocomposites. The SOPERMA nanocomposite has a higher improvement in storage modulus than the AESO or MAESO nanocomposites. Although SOPERMA, AESO, and MAESO have similar solubility parameters, more SOPERMA molecules are intercalated to the clay layers due to the flexibility of molecules, as shown earlier in Figure 15.8. The higher intercalated yield results in the higher storage modulus increase.

The glass transition temperature (T_g) is measured by dynamic mechanical analysis because the transition is too broad to determine accurately using

TABLE 15.5 Dynamic mechanical properties of triglyceride-based nanocomposites (percent improvement within parentheses).

Resin	Clay Content (wt%)	Storage Modulus (GPa)	$(\tan\delta)_{max}$ (°C)
AESO	0	1.258	71
AESO	3	1.321 (+5.0%)	69
AESO	5	1.368 (+8.7%)	65
SOPERMA	0	1.680	130
SOPERMA	3	2.009 (+19.6)	126
MAESO	0	2.041	136
MAESO	3	2.306 (+13.0%)	130

Differential Scanning Calorimetry (DSC) for these triglyceride-based polymers [33]. Figure 15.12 shows the temperature dependence of tan δ, the ratio of viscous to elastic properties, for MAESO nanocomposites at various clay contents. The tan δ peak is shifted to lower temperature with increasing clay content. AESO and SOPERMA nanocomposites show the same behavior as that shown in Table 15.5. The intensity of the tan δ peak also diminishes with increasing clay content. This is expected because the formation of a nanostructure restricts the molecular motions, which causes the amount of energy that could dissipate throughout the nanocomposite to decrease dramatically. This could cause the T_g to shift to higher temperatures. On the other hand, in this bicomponent system, styrene comonomer can more easily penetrate into clay layers than the triglyceride-based monomers, due to its small size. The aromatic nature of styrene imparts rigidity to the network, and the loss of styrene outside of the clay layer decreases the rigidity of the nanocomposite, resulting in the decreased T_g. Indeed, Suh and coworkers [19] demonstrated that the loss of styrene outside the clay layers causes the decrease of T_g for unsaturated polyester-based nanocomposites. They proposed a sequential mixing method, which successfully increased the T_g of the nanocomposites. Another possible reason could be the increased free volume due to the interaction of polymer matrix with clay, which disrupts packing of the molecular chains. Thus, the transition of chains participating in the relaxation is reduced in temperature.

15.2.4 FLEXURAL PROPERTIES

The flexural properties of AESO nanocomposites at different clay content are shown in Figure 15.13. The flexural modulus and strength of triglyceride polymers are significantly enhanced with the addition of clay. A 3 wt% nanocomposite shows better flexural properties than a 5 wt% nanocomposite. From TEM results, a completely exfoliated structure was observed at 3 wt% clay loading and a mix of intercalated and partially exfoliated structure was

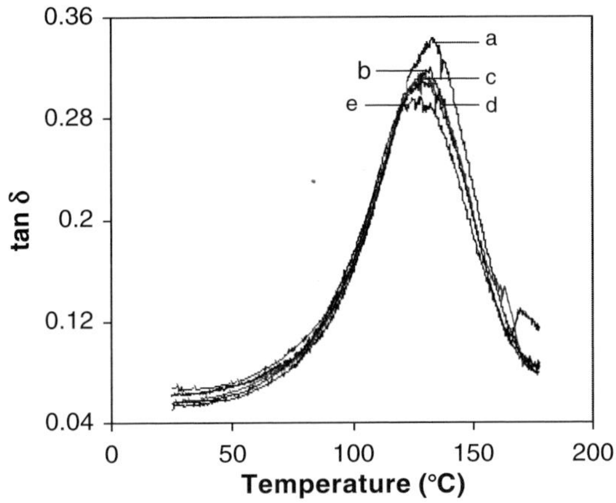

FIGURE 15.12 Tan δ of MAESO–clay nanocomposites at different clay content as a function of temperature: (a) pure MAESO, (b) 3 wt% C30B, (c) 5 wt% C30B, (d) 7.5 wt% C30B, and (e) 10 wt% C30B.

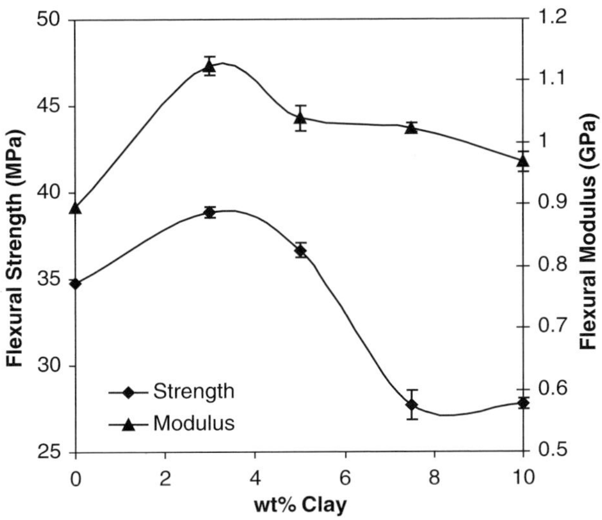

FIGURE 15.13 The effect of clay concentration on the flexural properties of an AESO-based nanocomposite.

shown at 5 wt%. Therefore, the degree of exfoliation is a control mechanism to increase the flexural properties of clay nanocomposites. Figure 15.14 shows the increase of flexural modulus for MAESO–clay nanocomposites. The flexural modulus increases significantly with increasing clay load up to 7.5 wt%, which only corresponds to 4.0 vol%; the increase in modulus is 30%. After that, the modulus decreases slightly. In practice, the modulus for the conventional composite with flake-like inclusion can be estimated using

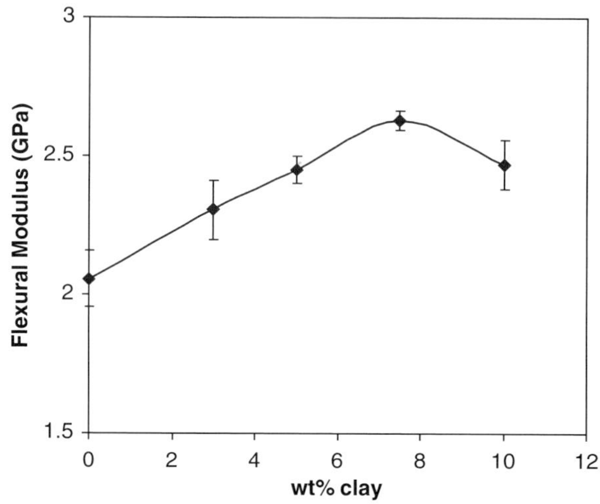

FIGURE 15.14 Flexural modulus for MAESO–C30B nanocomposites as a function of clay content.

simple empirical rules, such as the Halpin–Tsai equations, which predict a strong dependence of composite modulus on filler aspect ratio [34]:

$$\frac{E}{E_0} = \frac{1 + AB\phi}{1 - B\psi\phi}, \tag{15.7}$$

where

$$A \cong 1.33\alpha^{0.645} \quad B = \frac{\frac{E_1}{E_0} - 1}{\frac{E_1}{E_0} + A} \tag{15.8}$$

and

$$\psi \cong 1 + \frac{1 - \phi_m}{\phi_m^2}\phi, \tag{15.9}$$

where E, E_0, and E_1 are the moduli of the composite, matrix, and inclusion, respectively; ϕ is the volume fraction of the inclusion; and α is the aspect ratio of the inclusion, which is given by the ratio of the length over thickness of the inclusion. The factor takes into account the maximum packing fraction ϕ_m of the inclusion and is assumed to be 0.66. If clay and polymer matrix are connected in series (i.e., $A = 0$), the lowest possible modulus can be obtained, and Eq. (15.7) reduces to [35]

$$\frac{1}{E} = \frac{1 - \phi}{E_0} + \frac{\phi}{E_1}. \tag{15.10}$$

Figure 15.15 shows that the experimental data do not follow the prediction by the Halpin-Tsai model, which is possibly due to the extreme difference between the modulus of the silicate layer and that of the polymer matrix (silicate platelets have an in-plane modulus of 178 GPa); but the experimental data are still higher than the lowest possible modulus. Hui and Shia [36] developed new equations for composites with aligned platelet inclusions. The model is valid for the entire range of modulus and aspect ratios because the expressions were derived only by making the assumptions that the Poisson's ratios of the inclusions and the matrix are the same and equal to 1/2:

$$\frac{E}{E_0} = \frac{1}{1 - \frac{\phi}{4}\left[\frac{1}{\xi} + \frac{3}{\xi + \Lambda}\right]}, \tag{15.11}$$

$$\xi = \phi + \frac{E_0}{E_1 - E_0} + 3(1 - \phi)\left[\frac{(1 - g) - (g/2)\alpha^2}{1 - \alpha^2}\right], \tag{15.12}$$

$$\Lambda = (1 - \phi)\left[\frac{3(1 + 0.25\alpha^2) - 2}{1 - \alpha^2}\right], \text{ and} \tag{15.13}$$

$$g = \frac{\pi}{2\alpha}. \tag{15.14}$$

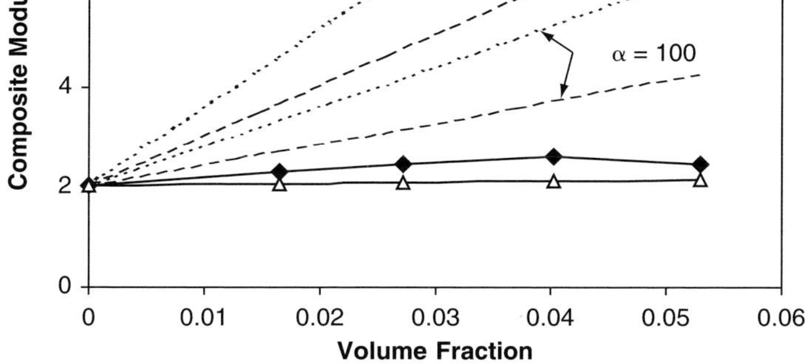

FIGURE 15.15 Experimentally measured modulus values and theoretical predictions by two models: Halpin–Tsai and Hui–Shia.

However, the experimentally measured modulus values are much lower than the values predicted by these equations for the aspect ratios in the range of 100–1000 as shown in Figure 15.15. This could be due to the imperfect bonding between the clay surface and polymer matrix, which reduces the reinforcing efficiency of the clay. Similar results were observed by other researchers [37, 38]. Solving the Halpin–Tsai and Hui–Shia equations with the experimental modulus data yields effective aspect ratios that are much lower than the aspect ratio for fully exfoliated silicate layers, as shown in Figure 15.16. The effective aspect ratio decreases with an increasing content of clay, which is consistent with the morphology shown earlier in the TEM images of Figure 15.11. The effective aspect ratio decreases significantly at 10 wt% clay loading, which may suffer from poor dispersion of clay during mixing at high clay content and aggregation of clay particles. Burnside and Giannelis [39] have attributed this effect to the deterioration of the bonding at the matrix/inclusion interface as the volume fraction of inclusion increases. The flexural strength increases slightly at low clay content and then decreases above 5 wt% clay loading as shown in Figure 15.17. Figure 15.18 shows that the flexural strain decreases with increasing clay content as the formation of the nanostructure restricts the flexibility of the cross-link network.

15.2.5 NANOCLAY AS TOUGHENING AGENT

As shown in Figure 15.19, a significant increase is seen in the fracture toughness with the addition of a small amount of C30B. The G_{1c} value is doubled at 5 wt% clay load, which corresponds to 2.6 vol%. With a further

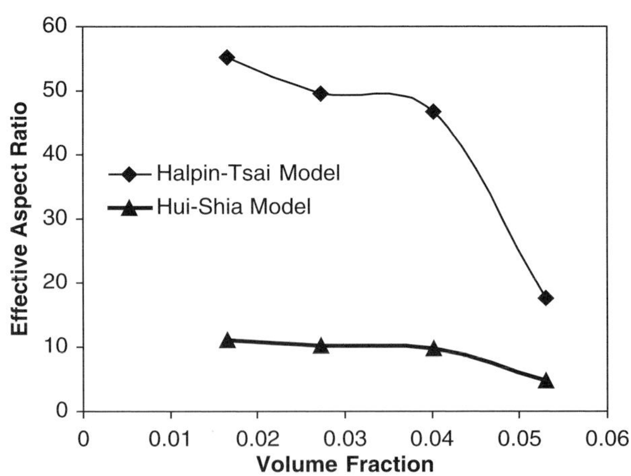

FIGURE 15.16 The effective aspect ratios calculated from the Halpin–Tsai and Hui–Shia models based on the experimental data.

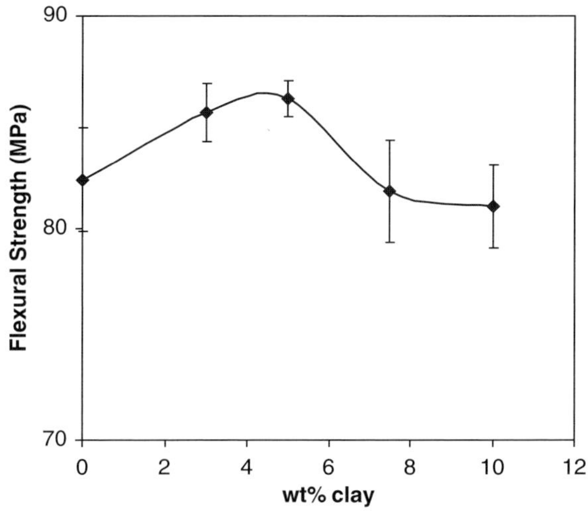

FIGURE 15.17 Flexural strength for MAESO–C30B nanocomposites as a function of clay content.

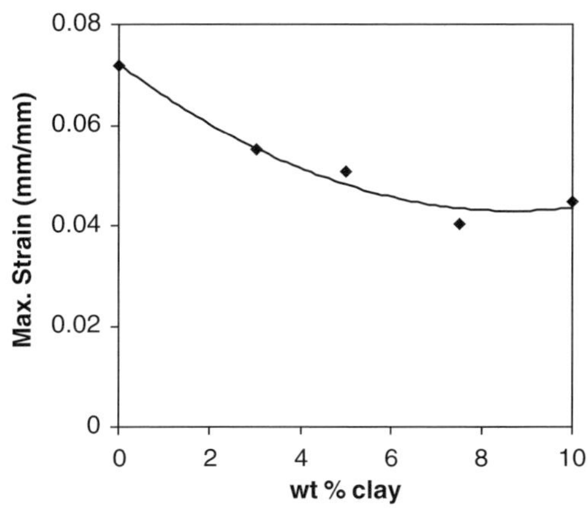

FIGURE 15.18 Flexural strain of MAESO–C30B nanocomposites as a function of clay content.

increase in clay loading, the fracture toughness decreases. Several mechanisms are available for toughening by means of an inorganic filler in highly cross-linked brittle polymers. These include crack trapping, crack-face bridging, crack-path deflection, and crack-tip shielding by microcracking [40]. According to Zilg *et al.* [41], the supramolecular assemblies, which are evidenced by the presence of anisotropic laminated nanoparticles in intercalated nanocomposites, are responsible for the increase in toughness, while a

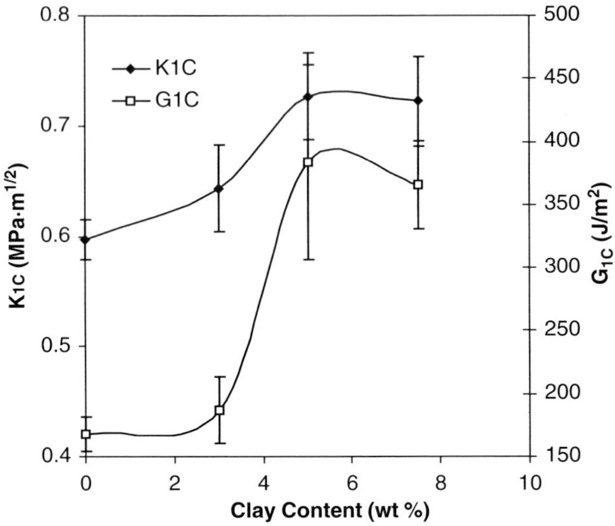

FIGURE 15.19 Fracture toughness K_{IC} and G_{1c} for MAESO nanocomposites as a function of clay content.

completely exfoliated structure accounts for poor toughness. As shown in Figure 15.20(b), the fracture surface of a 7.5 wt% nanocomposite shows a stress-whitening zone and is very rough, compared to the neat MAESO polymer, which has a glassy smooth surface as shown in Figure 15.20(a).

Figure 15.21 shows TEM micrographs of a 5 wt% clay nanocomposite. At low magnification, we clearly see the dispersion of micro-sized aggregates of clay sheets in the matrix. By increasing the magnification, we can see that the clay layers are intercalated by a layer of polymer and strongly bonded to the polymer matrix. These clay sheets are too strong to break during crack growth, thus they try to impede the growth of the crack in the MAESO polymer matrix, which leads to crack front trapping. As shown in Figure 15.22, the micrograph of nanocomposites shows a paraboloid fracture surface. Each paraboloid has a concentrator, which could be the micro-sized clay bundles, although the clay layers are not detectable. In addition, because of the strong polymer intercalation, these clay bundles are not easily debonding from the polymer matrix, thus the crack tip blunting mechanism is not the case here. Instead, these clay bundles act as obstacles, forcing the crack to bow [42] and causing deflection of the crack path, which can be one of the main toughening mechanisms for these clay-reinforced nanocomposites. Based on a simple model, the fracture toughness of composites can be proportional to the inverse of the interparticle spacing ($2c$) [43]:

$$G_{\text{comp}} = G_{\text{resin}} + \frac{T}{2c}, \tag{15.15}$$

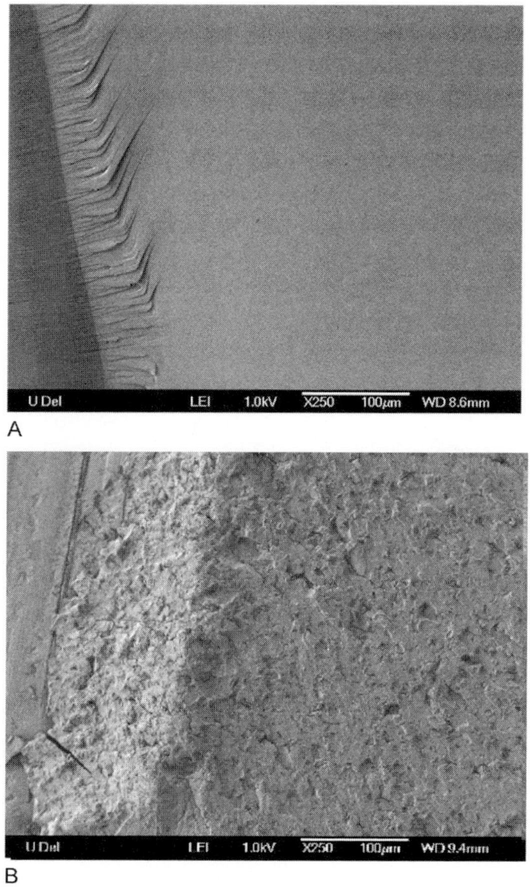

FIGURE 15.20 SEM micrographs of the fracture surfaces of (a) pure polymer and (b) 7.5 wt% clay nanocomposite at low magnification.

where T is the line tension effect, which is related to the particle size ($2r$) by

$$T = 2r/3(G_{\text{resin}}), \tag{15.16}$$

and the ratio of the particle size to the particle spacing is proportional to the volume fraction by $V_p/(1 - V_p)$. Thus the fracture toughness of a nanocomposite is proportional to $V_p/(1 - V_p)$. As shown in Figure 15.23, the experimental values deviate from the model prediction. At 3 wt% clay load, the fracture energy is much lower than expected, because the morphology is more like an exfoliated sheet, as shown in Figure 15.11, whereas the 5 wt% nanocomposite has much higher toughness, which may be due to the supramolecular assembly besides crack trapping. These nanoparticles are possibly oriented on exposure to mechanical stresses, which produces nanovoids and initiates shear yielding of the polymer interlayers [41]. Kormann *et al.* [8]

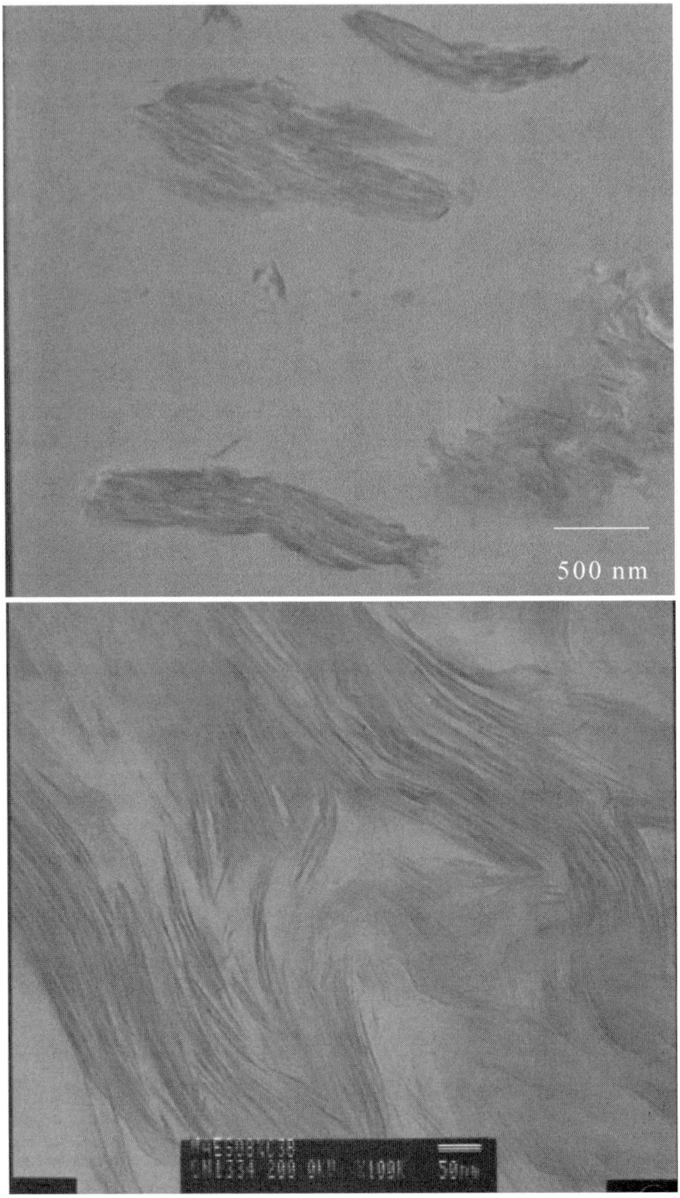

FIGURE 15.21 TEM micrographs of MAESO–C30B nanocomposites.

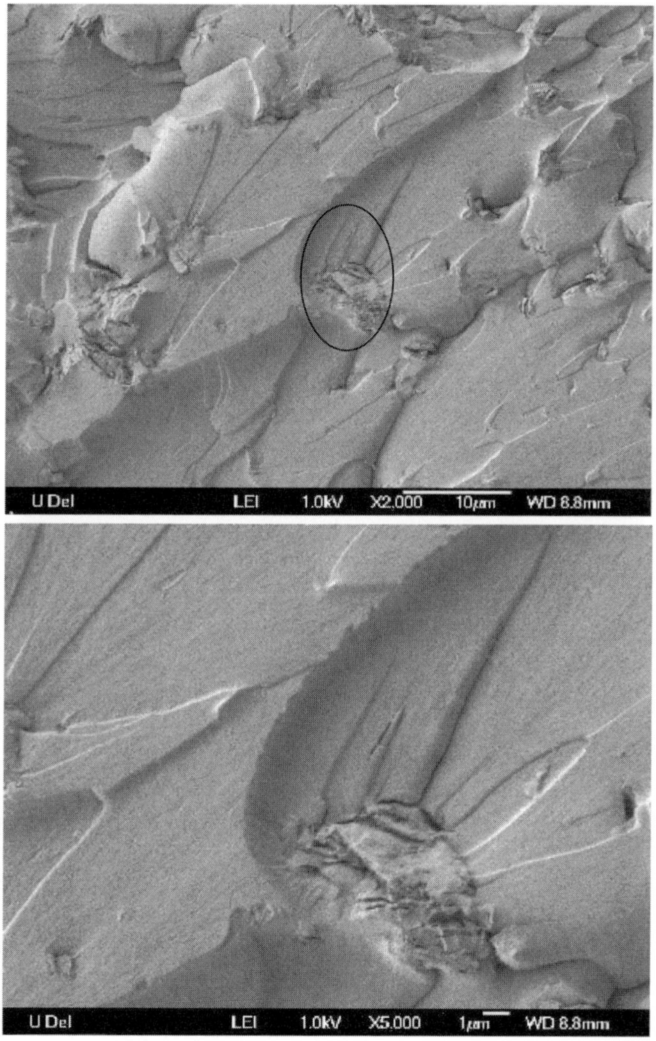

FIGURE 15.22 SEM micrographs of the fracture surface for 5 wt% of an MAESO-based clay nanocomposite at high magnification.

found that the conventional 7.3 vol% kaolin–UP shows a smaller toughness improvement than does the 1.5 vol% clay nanocomposite. Thus, these intercalated structures show better fracture toughness.

15.2.6 THERMAL STABILITY

Tyan *et al.* [44] found that the silicate layers can exert very effective retardation on the thermal degradation of polymers. Figure 15.24 shows

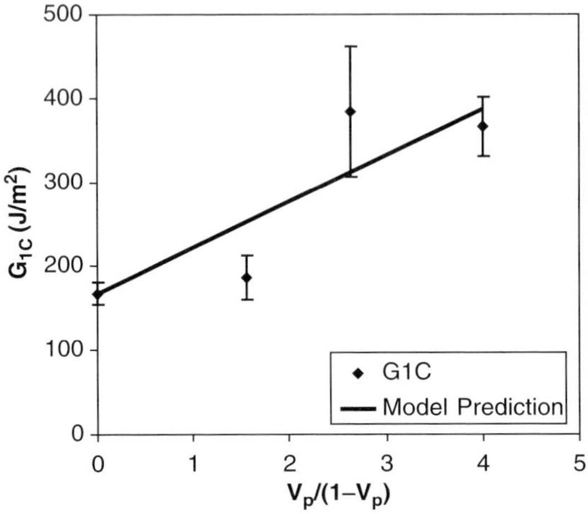

FIGURE 15.23 The variation in the fracture energy is not proportional to $V_p/(1-V_p)$, where V_p is the volume fraction of clay.

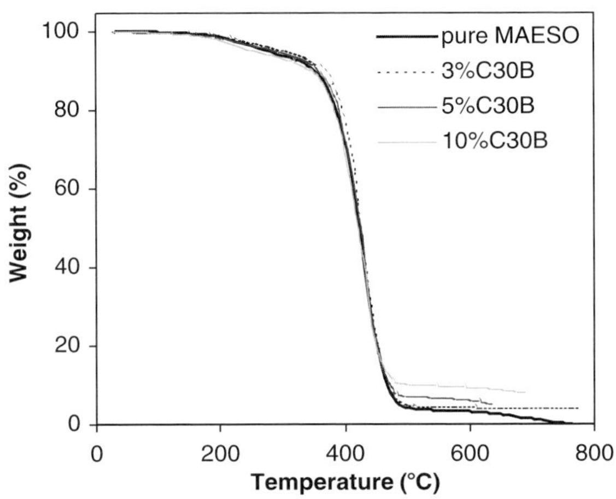

FIGURE 15.24 TGA thermal degradation profiles for MAESO–C30B nanocomposites at different concentrations of clay.

thermogravimetric analysis (TGA) traces for the cross-linked MAESO–clay nanocomposites. There is only a single degradation process under air atmosphere for all samples. The degradation happens in the temperature region from 300° to 450°C, which is due to the degradation of the intercalated agent followed by the decomposition of the cross-linked polymer network. From Table 15.6, we see that there is only a slight difference in thermal stability for the pure polymer and nanocomposites, the TGA derivative maximum occurs at 420°C for MAESO polymer, and there is an approximately 5°C increase

TABLE 15.6 TGA data of MAESO–clay nanocomposites.

Sample	TGA Derivative Maxima (°C)
Pure MAESO	420.00
MAESO–3 wt% C30B	425.61
MAESO–5 wt% C30B	426.31
MAESO–10 wt% C30B	424.32

for nanocomposites. From Figure 15.24, with the addition of 3 wt% clay, the onset of degradation is slightly slowed because of the constrained region in the nanocomposite. With more clay content, the degradation is hastened because of the presence of the unfavorable hydroxyl groups in the organic modifier. After ~500 °C, the curves all become flat and mainly the inorganic residue (i.e., Al_2O_3, MgO,SiO_2) remains.

15.3 SUMMARY

Clay nanocomposites were successfully prepared from plant oil, one of the abundant, cheap, renewable resources. Organo-modified clay was dispersed in functionalized triglyceride monomers and styrene, followed by cross-linking reactions. The specific organo-treated clay was selected based on the solubility parameters of organic modifier and molecular structure affinity. X-ray diffraction shows the d-spacing increase in the clay layers as a result of the intercalation of monomers. The attractive forces between the clay organic modifier and monomer molecules can facilitate the penetration of monomers into the clay layers. The formation of nanostructures was confirmed both by XRD and TEM images. The morphology can be described as a mix of intercalated and partially exfoliated sheets. The exfoliated nanocomposites significantly increase the flexural properties. The increase in modulus is not consistent with the theoretical models for traditional flake-like inclusion composites. The glass transition temperature is lowered with increasing clay content, which may result from the styrene loss outside of clay layers and increased free volume. The formation of an intercalated clay nanocomposite has a significant effect on the fracture toughness. The G_{1c} value is doubled at 5 wt% of clay load, which may be due to the supramolecular assembly and crack trapping. The thermal degradation is slightly hastened with increasing clay load.

REFERENCES

1. Kojima, Y.; Usuki, A.; Kawasumi, M.; *et al. J. Mater. Res.* 1993, **8**(5), 1185–1189.
2. LeBaron, P. C.; Wang, Z.; Pinnavaia, T. J. *Appl. Clay Sci.* 1999, **15**(1–2), 11–29.

3. Alexandre, M.; Dubois, P. *Mater. Sci. Eng. R: Reports* 2000, **28**(1–2), 1–63.
4. Lu, J.; Thielemans, W.; Can, E.; *et al.* In *Proc. ICCM 14*, San Diego, CA; 2003.
5. Uyama, H.; Kuwabara, M.; Tsujimoto, T.; *et al. Chem. Mater.* 2003, **15**(13), 2492–2494.
6. Miyagawa, H.; Misra, M.; Mohanty, A.; *et al.* In *Proc. AICHE*, San Francisco, CA; 2003.
7. Hofmann, U.; Bendell, K.; Wilm, D. *Z. Kristallog. Kristallgeom. Kristallphys. Kristallchem.* 1933, **86,** 340–348.
8. Kornmann, X.; Berglund, L. A.; Sterte, J. *Polym. Eng. Sci.* 1998, **38**(8), 1351–1358.
9. Kornmann, X.; Lindberg, H.; Berglund, L. A. *Polymer* 2001, **42**(4), 1303–1310.
10. Ho, D. L.; Briber, R. M.; Glinka, C. J. *Chem. Mater.* 2001, **13**(5), 1923–1931.
11. Ho, D. L.; Glinka, C. J. *Chem. Mater.* 2003, **15**(6), 1309–1312.
12. Ishida, H.; Campbell, S.; Blackwell, J. *Chem. Mater.* 2000, **12**(5), 1260–1267.
13. Kormann, X.; Lindberg, H.; Berglund, L. A. *Polymer* 2001, **42**(10), 4493–4499.
14. Doh, J. G.; Cho, I. *Polym. Bull.* 1998, **41**(5), 511–518.
15. Zeng, C. C.; Lee, L. J. *Macromolecules* 2001, **34**(12), 4098–4103.
16. Lan, T.; Pinnavaia, T. J. *Chem. Mater.* 1994, **6**(12), 2216–2219.
17. Weimer, M. W.; Chen, H.; Giannelis, E. P.; *et al. J. Amer. Chem. Soc.* 1999, **121**(7), 1615–1616.
18. Bharadwaj, R. K.; Mehrabi, A. R.; Hamilton C.; *et al. Polymer* 2002, **43**(13), 3699–3705.
19. Suh, D. J.; Lim, Y. T.; Park, O. O. *Polymer* 2000, **41**(24), 8557–8563.
20. Wool, R. P.; Can, E. Rigid Thermosetting Liquid Molding Resins from Plant Oils. U.S. Patent pending.
21. www.neveonsolutions.com.
22. Mark, J. E., Ed. *Physical Properties of Polymers Handbook*, American Institute of Physics, Woodbury, New York; 1996.
23. Sperling, L. H. *Introduction to Physical Polymer Science*, John Wiley & Sons, New York; 1986.
24. Can, E. PhD Thesis, University of Delaware, Newark, DE, 2004.
25. Hoy, K. L. *J. Coated Fabrics* 1989, **19,** 53–67.
26. Hansen, C. M. *J. Paint Technol.* 1967, **39**(505), 104–117.
27. Hansen, C. M. *J. Paint Technol.* 1967, **39**(511), 505–510.
28. Van Krevelen, D. W.; Hoftyzer, P. J. *Properties of Polymers*, Elsevier, New York; 1997.
29. Mark, J. E., Ed. *Physical Properties of Polymers Handbook*, AIP Press, New York; 1996.
30. Vaia, R. A.; Jandt, K. D.; Kramer, E. J.; *et al. Macromolecules* 1995, **28**(24), 8080–8085.
31. Vaia, R. A.; Giannelis, E. P. *Macromolecules* 1997, **30**(25), 7990–7999.
32. Vaia, R. A.; Giannelis, E. P. *Macromolecules* 1997, **30**(25), 8000–8009.
33. Khot S. N.; Lascala, J. J.; Can, E.; *et al. J. Appl. Polym. Sci.* 2001, **82**(3), 703–723.
34. Halpin, J. C.; Kardos, J. L. *Polym. Eng. Sci.* 1976, **16**(5), 344–352.
35. Nielsen, L. E.; Landel, R. F. *Mechanical Properties of Polymers and Composites*, Marcel Dekker, Inc., New York; 1994.
36. Hui, C. Y.; Shia, D. *Polym. Eng. Sci.* 1998, **38**(5), 774–782.
37. Messersmith, P. B.; Giannelis, E. P. *Chem. Mater.* 1994, **6**(10), 1719–1725.
38. Priya, L. *J. Polym. Sci. B: Polym. Phys.* 2003, **41,** 31–38.
39. Burnside, S. D.; Giannelis, E. P. *Chem. Mater.* 1995, **7**(9), 1597–1600.
40. Singh, R. P.; Zhang, M.; Chan, D. *J. Mater. Sci.* 2002, **37**(4), 781–788.
41. Zilg, C.; Mulhaupt, R.; Finter, *J. Macromolecular Chemistry and Physics* 1999, **200**(3), 661–670.
42. Roulin-Moloney, A., Ed. *Fractography and Failure Mechanisms of Polymers and Composites*, Elsevier Applied Science, New York; 1988.
43. Lange, F. F. *Philos. Mag.* 1970, **22**(179), 983.
44. Tyan, H. L.; Liu, Y. C.; Wei, K. H. *Chem. Mater.* 1999, **11**(7), 1942–1947.

16

LIGNIN POLYMERS AND COMPOSITES

RICHARD P. WOOL

The most abundantly available renewable resources are cellulose, lignin, and plant oils, in this order. Wood is made up mainly of three macromolecular species: cellulose, hemicellulose, and lignin [1]. Minor components include pectin, fats, wax, moisture, and water solubles [2]. Of these three, cellulose is the best known and most abundantly available. The estimated yearly production of cellulose by photosynthesis is 830 million tons [3]. Cellulose is a polysaccharide with an extensive amount of intra- and intermolecular hydrogen bonding, rendering it highly insoluble [4]. The hydrogen bonding and rigid structure of cellulose [Figure 16.1(a)] result in a high degree of crystallinity [4]. The cellulose in wood and natural fibers is found as fibers, consisting of helically wound crystalline microfibrils. The cellulose polysaccharide is constituted of anhydroglucopyranose units [Figure 16.1(b)] with a reducing end group (right side), where a ring opening results in the formation of an aldehyde, and a nonreducing end group (left side) in the form of a secondary alcohol.

Hemicellulose is a low-molecular-weight polysaccharide with a degree of polymerization ranging from 70 to 200, with lower degrees of polymerization found in softwoods [5]. Hansen and Bjorkman [6] showed, using the solubility theory of Hildebrand and Scott [7] and Hansen [8], that hemicellulose acts as a compatibilizer between cellulose and lignin. There is, however, no clear relation between the lignin and hemicellulose content in wood [9]. Isolated hemicellulose is found to be amorphous and either water soluble or it shows strong swelling behavior in water. Hemicelluloses are heteroglycans built up from a limited amount of sugar residues [10, 11]. Hardwood hemicellulose is predominantly partially acetylated acidic xylan, with a small percentage of

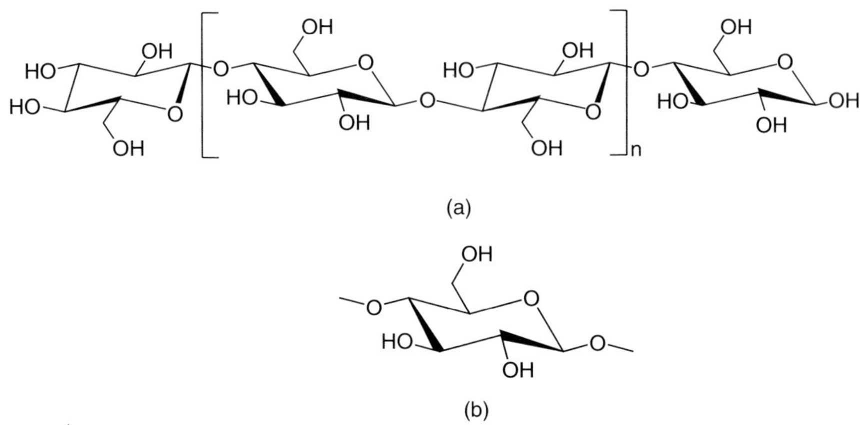

FIGURE 16.1 Chemical structures of (a) cellulose and (b) the repeat unit anhydroglucopyranose. The left end group in structure (a) is nonreducing, whereas the right end group is reducing and an aldehyde can be formed by ring opening.

mannan. Softwood hemicelluloses are predominantly partially acetylated galactoglucomannans, with a low amount of xylan similar to the hardwood xylan [10].

Lignin is found as a cell wall component in all vascular plants and in the woody stems of arborescent angiosperms (hardwoods) and gymnosperms (softwoods) [13]. The lignin content in woody stems varies between 15% and 40%. Lignin acts as water sealant in the stems and plays an important part in controlling water transport through the cell wall. It also protects plants against biological attack by hampering enzyme penetration. Finally, lignin is also a permanent glue, bonding cells together in the woody stems and thus giving the stems their well-known rigidity and impact resistance. Table 16.1 gives the cellulose-hemicellulose-lignin composition of a limited number of lignocellulosics. A rough guide for lignin content is 18–25% for hardwoods and 25–35% for softwoods. The cellulose content commonly falls between 41% and 45% [11].

16.1 INTRODUCTION TO LIGNIN

16.1.1 LIGNIN STRUCTURE

Anselme Payen [14, 15] discovered lignin in 1838 by treatment of wood with nitric acid and alkaline solutions. These treatments yielded an insoluble fraction, designated cellulose, and a soluble fraction, which he called *incrustant*. This soluble material was later named *lignin* by Schulze [16]. Lignin is generally defined as *polymeric natural products arising from an enzyme-*

TABLE 16.1 Relative composition of a limited number of lignocellulosics. These numbers do not contain minor components such as pectin, fats, and waxes.

Species	Type	Cellulose (%)	Hemicellulose (%)	Lignin (%)
White spruce	Softwood	44	29	27.1
Eastern hemlock	Softwood	42	32.5	26
Eastern white cedar	Softwood	41	26	31
American elm	Hardwood	51	23	24
White birch	Hardwood	42	38	19
Flax	Bast fiber	75.9	20.7	3.4
Jute	Bast fiber	84	15.7	0.3
Hemp	Bast fiber	79.9	19.2	0.9
Coir	Seed fiber	46.7–48.7	0.2–0.3	51–53

Sources: Data compiled from [2, 9, 11, 12]

initiated dehydrogenative polymerization of three primary precursors: trans-*coniferyl,* trans-*sinapyl, and* trans-p-*coumaryl* [13]. The chemical structures of these precursors are depicted in Figure 16.2. The only difference between the precursors is the number of methoxyl groups ($-OCH_3$) present on the aromatic ring. Lignin is generally used to address the lignin extracted from wood, whereas protolignin is used for lignin still associated with cells. The enzyme-initiated polymerization results in bonds of exceptional stability: biphenyl carbon–carbon linkages between aromatic carbons, alkyl-aryl carbon–carbon linkages between an aliphatic and aromatic carbon, and hydrolysis-resistant ether linkages (see Figure 16.3). The only linkage relatively weak and hydrolyzable is the α-aryl ether bond [Figure 16.3(c)]. These stable linkages make lignin very resistant against degradation.

Figure 16.4 represents a tentative structure for beech lignin as proposed by Nimtz [17]. The plant type determines the relative amounts of the respective precursors and thus the final structure of lignin. However, due to the similar base structure of the lignin precursors, every lignin molecule consists of sequences of phenyl-propane units. Functional groups present on the lignin molecule are therefore generally reported per phenyl-propane unit (PPU), a C_6–C_3 unit, to exclude molecular weight effects. The carbon atoms within each PPU are numbered according to a common notation, shown in Figure 16.5.

Lignins are generally divided into two major classes: guaiacyl and guaiacyl-syringyl lignins [18]. The guaiacyl lignins include the majority of gymnosperm lignins, while all angiosperm and herbaceous lignins belong to the guaiacyl-syringyl lignin class [9, 19]. This division is not absolute because different lignins can coexist, even within the same plant. As can be deduced from the class names, guaiacyl lignins only contain *p*-hydroxyphenyl propane (no $-OCH_3$ on the aromatic ring) and guaiacyl propane units (one $-OCH_3$

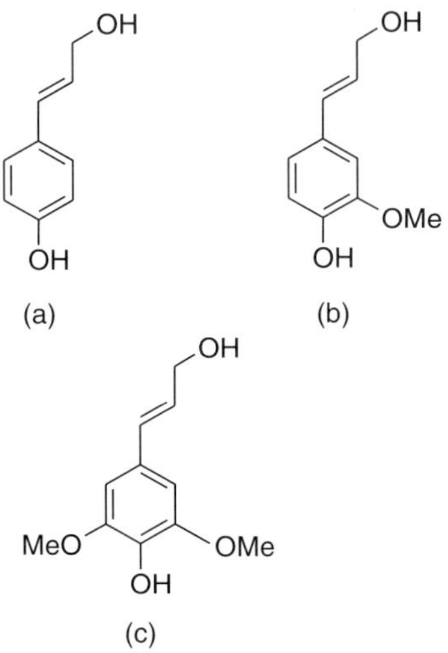

FIGURE 16.2 Primary precursors of lignin: (a) *trans-p*-coumaryl, (b) *trans*-coniferyl, and (c) *trans*-sinapyl.

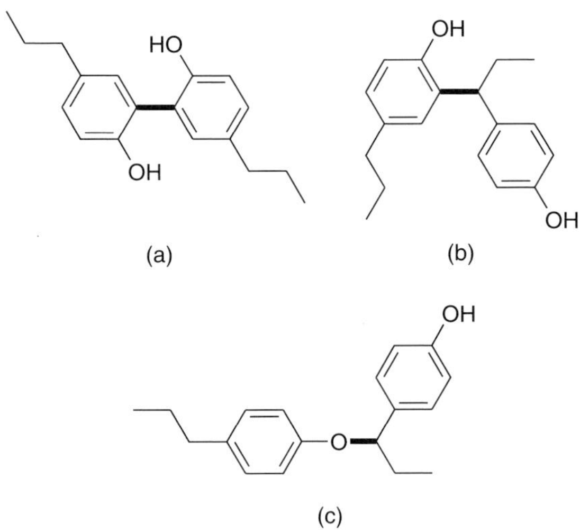

FIGURE 16.3 Representation of different linkages formed during enzymatic dehydrogenation between two *trans*-coniferyl units: (a) biphenyl carbon–carbon linkage, (b) alkyl-aryl carbon–carbon linkage, and (c) easily hydrolyzable α-aryl ether bond.

group on the aromatic ring), while the guaiacyl-syringyl lignins also contain syringyl propane units (two –OCH3 on the aromatic ring). The lignin content in plants varies greatly with species as well as the environment in which the species thrives. The largest lignin concentrations are found in the woody stems, in between the cellulosic fibers.

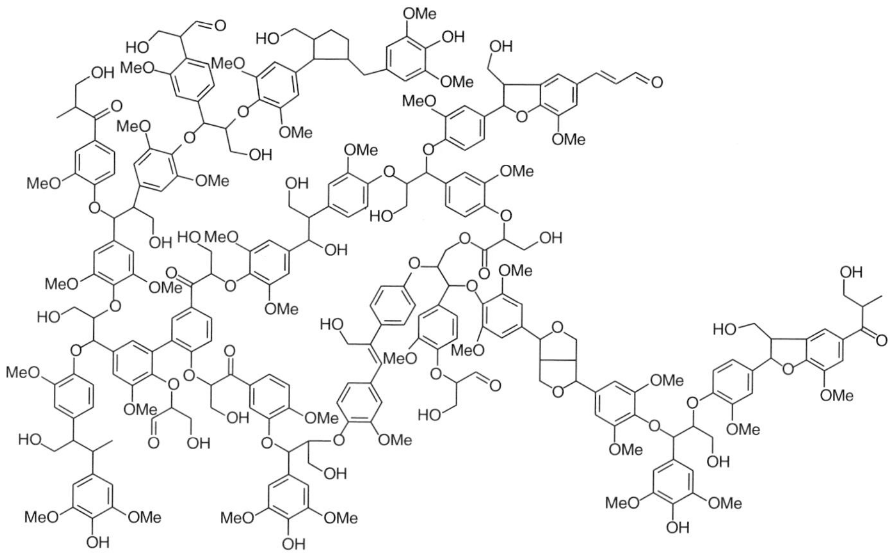

FIGURE 16.4 Proposed structure of beech lignin [17].

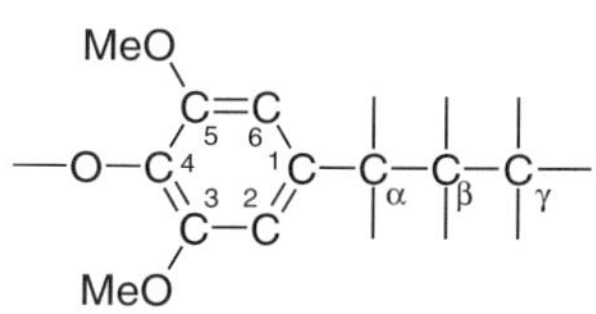

FIGURE 16.5 Common notation for addressing the different carbon atoms within each phenyl propane unit. Me denotes a methane group (CH_3).

16.1.2 DELIGNIFICATION

Delignification, the process of extracting lignin from plant sources, can be done using a variety of methods. Its aim is the disintegration of the lignocellulosic structure into its fibrous components [20]. The delignification processes can be divided into two major classes: chemical and solvent processes. The two conventional and industrially the two most widely used are sulfite and alkaline pulping, both of which are classified as chemical pulping processes [21]. They form the two extremes on the pH scale. The alkaline or kraft pulping process forms alkali- or thio-lignin, while the acidic sulfite process forms a sulphonic acid of lignin [22]. The lignin sulphonic acid is generally referred to as lignosulfonate. Details of these pulping processes and the effects on lignin under the harsh alkaline or acidic conditions have been well documented [23–26]. It is important to mention that kraft pulping has become the industrially dominant pulping process. Solvent-based or

Organosolv pulping processes are either in the developmental stage or are used in small-scale commercial production. These processes include ASAM (alkali-sulfite-anthraquinone-methanol), Organocell (sodium hydroxide, methanol, and anthraquinone) [27], Alcell (water and ethanol) [27], Formacell (acetic acid, formic acid) [28, 29], and Milox (multistage peroxyacid treatment) [29, 30]. These processes, with the exception of the ASAM process where the inorganics are primarily responsible for delignification, delignify lignocellulosics using organic solvents. They generally promise higher efficiency, fewer by-products, lower capital costs, and/or lower emissions. Organosolv delignification is in general more selective for hardwoods than softwoods due to the differing chemical structure of their respective lignins [31, 32].

A relatively new delignification method is *clean fractionation,* which uses steam explosion of biomass before solvent treatments and/or distillation [33]. While steam explosion has been around for centuries, only in the last two decades has it been considered as an economical alternative for production of fuel and chemicals from biomass [34]. Clean fractionation is an Organosolv method that has the advantages of a high organic solvent recovery (>99%), a higher energy efficiency than the current industrial processes and, very importantly, it offers the possibility to recover all constitutive wood components without destructive degradation of any of these components [35]. The high solvent recovery eliminates downstream solvent evaporation, and thus odorous emissions, and reduces the downstream effluent treatment. It also reduces the high capital investment needed for other delignification processes [33]. However, this relatively new technology has yet to be used on a large industrial scale for lignin production and is currently only limited to laboratory-scale experiments. The lignin used in this work is kraft lignin obtained from MeadWestvaco (Stamford, CT). Kraft lignin is discussed in more detail in the next section.

16.1.3 KRAFT LIGNIN

As mentioned in the previous section, kraft lignin is obtained by treatment of lignocellulosic materials with kraft pulping liquor. This kraft pulping liquor contains mainly sodium hydroxide and sodium sulfide, with other sodium salts, such as sodium carbonate and sodium thiosulfate, as minor components [20]. This liquor is used industrially for delignification of both softwoods and hardwoods. The kraft pulping process is very complex with a variety of elementary chemical and physical factors influencing the obtained lignin structure and yield [25]. This complexity ensures that a significant amount of research is being done to fully understand the effect of varying process conditions on the pulping products, yield, and efficiency [36–41]. Differences in pulping behavior between different wood species add an extra dimension of complexity to the pulping process [42]. A significant amount of alkali solutes is consumed by reaction with the dissolved lignin. A consumption of a quarter of initial alkali solute by lignin is common [20]. Other sources of alkali consumption are neutralization of degraded carbohydrates and hydrolysis of acetyl groups. The initial alkalinity has to be high enough to keep the final alkalinity above pH 9 to guarantee lignin dissolution and avoid sedimentation.

The essential step of kraft delignification is the fracture of linkages in the protolignin due to chemical attack by the alkali solutes and other nucleophiles [20]. It is largely the relatively weak ether bonds (carbon–oxygen linkages) that are ruptured under these conditions, resulting in an increase in hydroxyl groups. The simple cleavage of the relatively weak aryl-alkyl ether bond results in a significant increase in aromatic hydroxyl groups.

A variety of studies were performed by applying kraft pulping conditions to pure lignin model compounds of only two and three units. These studies confirmed the easy hydrolysis of phenolic α-aryl ether linkages [43], and nonphenolic β-*O*-4 ether linkages if free alcohols are present on the neighboring α or β-positions, or if the α-position contains a carbonyl group [44–46]. Nonphenolic α-aryl ether linkages could not be readily hydrolyzed unless free aliphatic hydroxyl groups were present in the neighboring β-position [47], while hydrolyzation of phenolic β-*O*-4 ether bonds was in competition with the formation of alkali-stable enol-aryl ether [48, 49]. These studies also demonstrated a variety of intermediates formed during base-catalyzed hydrolyzation.

The rupture of various linkages during extraction of the protolignin from lignocellulosics results in a kraft lignin with a number of average molecular weight in the range of 900–2000 g/mol and polydispersities between 2 and 3.5 [50, 51]. Figure 16.6 depicts the structure of pine kraft lignin as proposed by Marton [20]. The solubility of kraft lignin in dimethyl sulfoxide and dioxane suggests that it is not cross-linked. The cross-linked density of the native protolignin was estimated to be only 0.052, or 1 in 19 PPU [52].

16.2 LIGNIN APPLICATIONS

Cellulose is widely used in paper production, and plant oils find their largest applications in the food industry. Lignin, however, is produced as a by-product of paper pulping and is considered a waste product. The annual sales of lignin as specialty chemicals in 1998 amounted to a mere 1% of total lignin production [53]. The remaining 99% is burned in an energy recovery step or disposed of in waste streams. In 1990 global specialty chemicals from lignin amounted to 138.5 kilotons/year [54]. Extrapolating these numbers leads to an estimated 10 million tons of lignin produced worldwide every year. Lignin production is expected to rise significantly with the current shift toward biorefineries in which lignin is again a by-product. The low percentage of lignin being used in higher value applications than as a fuel

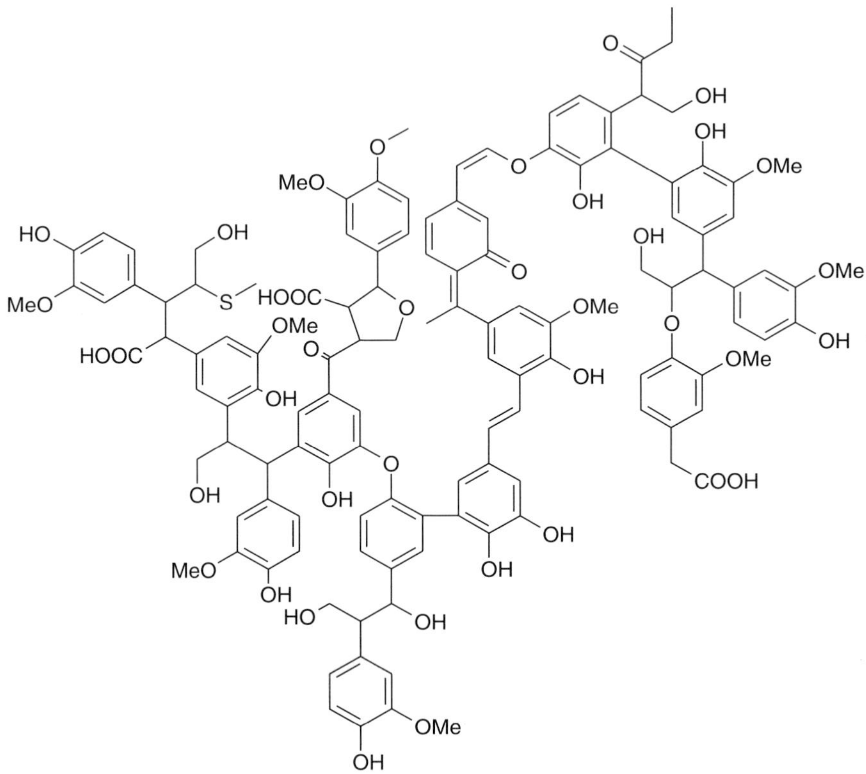

FIGURE 16.6 Structure of pine kraft lignin proposed by Marton [20].

can largely be attributed to this by-product status. The lignin properties are an effect of the treatment of the lignocellulosic plant. It is thus not suitable to change the extraction parameters in order to obtain lignin with specific chemical and physical characteristics. The utilization of lignin as a specialty chemical is thus limited by the physical and chemical properties it exhibits after the pulping process. Alterations to this structure have to be performed after lignin has been obtained and cannot be achieved by changes in the upstream process.

Most of the isolated lignins are lignosulfonates [53]. About 50% of these lignosulfonates are added to concrete for water reduction, quicker strength development, and improved workability [55]. They also reduce the amount of needed air-entraining agents [56]. Lignin can also be found as binding agents and lubricants in animal feed, using its nontoxicity [20, 55], and as dust control agents for road and mineral ore de-dusting. Lignosulfonates are also used as oil well drilling products, mud thinners, rheology modifiers, clay conditioners, and fluid loss additives. In the more specialized markets, lignin is currently used as feedstock for vanillin production; as dispersant and binder in pesticides, gypsum, inks and dyes, water treatment formulations, industrial cleaners, and various emulsions; as complexing agents with calcium, iron, copper, etc. to deliver nutrients to plants; and as an expander or surface modifier on the negative plate in lead acid batteries [53, 55]. It is being investigated for its enzyme stabilization characteristics [57] and antiviral [58, 59] and antibiotic activity [60].

16.2.1 LIGNIN IN POLYMERS

The use of kraft lignin as a copolymer or polymer additive has also received a considerable amount of attention [53, 61]. A recent review of industrial applications of lignin in polymers was published by Lora and Glasser [62]. The most straightforward application is the use of lignin as a filler material in thermoplastic [63–67] and thermosetting [68, 69] polymers and rubbers [70] with limited positive to negative effects on mechanical properties with lignin addition.

16.2.2 LIGNIN–POLYMER BLENDS

Lignin was found to increase the polymer modulus while decreasing the melt temperature of various crystallizing thermoplastics. The addition of plasticizer to lignin–thermoplastics blends decreased the degree of lignin association in the blend, previously found to negatively impact mechanical blend properties [71]. The decrease in lignin association resulted in significant mechanical improvements [72]. A very interesting application of lignin–polymer blends is their potential as a precursor for carbon fibers [73]. Carbon fibers were obtained by carbonization of dry-spun fibers of lignin dissolved in an alkaline solution and plasticized with poly(vinyl alcohol) [74]. Various chemical modifications were also used to obtain a molten viscous ligneous material that could be used for thermal spinning followed by carbonization [75, 76]. Carbon fibers were also obtained from Organosolv lignin by direct carbonization after removing the lignin infusible high molecular mass fraction. The resulting carbon fibers displayed properties in the midrange of currently available carbon fibers [77, 78]. Kadla *et al.* obtained general grade carbon fibers from kraft and Alcell lignin without any chemical modification by blending lignin with less than 5% poly(ethylene oxide) [79].

16.2.3 LIGNIN AS COPOLYMER

The lignin chemical structure (Figures 16.3 and 16.4), with its phenolic base structure, suggests its potential as a comonomer in phenolic thermosetting polymers [80, 81]. Lignin can be added to phenol-formaldehyde resins without significant deterioration of mechanical properties up to a load of about 15% [82, 83]. Co-reaction of lignin with epoxy resins [84–86], polyurethane precursors [87–96], and polyester precursors [97, 98] has also been investigated. Kraft lignin was added to an epoxy adhesive resin based on a mixture of a diglycidyl ether of bisphenol A with a polyamine [85]. Maximum strength for the polymerized sample was obtained with a 20 wt% lignin load after which the polymer was no longer homogeneous (appearance of a double glass transition temperature) and also lost strength. It was assumed that chemical bonds did form between lignin and amine groups on the polyamine hardener. The adhesive bond strength for the 20 wt% lignin sample was higher than for a similar volume fraction of silica filler.

Another epoxy resin–lignin copolymer was obtained by polymerization of a water-soluble epoxy compound [poly(ethylene glycol) diglycidyl ether] and a polyamine hardener in an alkaline solution [86]. A cross-linked gel was obtained over a lignin content range of 20–80%. The glass transition temperature increased with lignin content, ranging from $-15°$ to $85\,°C$ for 0–60% lignin content, allowing for tailoring of properties by a simple choice of the initial reactant ratios. The incorporation of lignin in various polyurethanes through the alcohol-isocyanate reaction resulted in improved mechanical properties and higher glass transition temperatures compared to the polyurethanes without lignin [89, 94, 96]. Lignin addition also resulted in improved thermal stability [90]. High lignin contents (>30%) resulted in brittle polyurethanes, regardless of lignin molecular weight and alcohol-to-isocyanate ratio [92]. The incorporation of lignin in elastomeric polyurethanes results in a loss of flexibility and a strength increase [91]. As for lignin incorporated in polyesters, copolymerization of lignin with acid chlorides resulted in rigid polyester materials [98]. Incorporation of lignin in the polyester by co-reaction of the lignin hydroxyl groups with polyethylene glycol and acid chlorides gave rise to elastomeric polyesters due to the soft segment effect of polyethylene glycol [97].

16.2.4 LIGNIN GRAFTING

Studies concentrated on the free-radical copolymerization of lignin with unsaturated polymers are very limited, concentrating primarily on grafting of lignin [99, 100]. This is largely due to the free-radical inhibiting capability of the phenolic hydroxyl groups in lignin [101, 107], capable of forming quinonic structures stabilized by resonance over the whole lignin molecule, as well as the absence of suitable double bonds reactive toward free-radical polymerization. Additionally, lignin displays incompatibility with styrene and most other monomers used as reactive diluents in unsaturated polyester resins. Lignin was grafted with methyl methacrylate and vinyl acetate [108–110], styrene [111–113], acrylonitrile [114], acrylic acid and acrylamide [115], and maleic anhydride [94, 96] among others [117, 118]. However, the inhibiting capability of the phenolic hydroxyl groups of lignin, coupled with resonance stabilization of any quinonic structures formed [102, 107], make lignin a rather inefficient grafting target. The grafting efficiency for radiation-induced radical formation, expressed by a frequency of active site formation per 100 eV of radiation, G_R, was reported to be 0.6–0.7 [108, 119]. As a comparison, this is about one-tenth the G_R value for methyl methacrylate, while styrene has a similar to lower G_R value of 0.3–0.7. Masking the phenolic hydroxyl groups results in a significant increase in graft rate, reported up to threefold [112].

The use of polar solvents such as alcohols also has a beneficial effect on the grafting efficiency by swelling the lignin molecules, improving lignin accessibility. Methanol, with a high G_R value of 13.5, can act both as initiator and chain transfer agent [120] and can increase the grafting efficiency tenfold [112]. Increasing the methanol concentration resulted in an improved grafting efficiency and rate of grafting of lignin with styrene [99]. At low grafting ratios of lignin to monomer (<100), grafting occurs predominantly on the lignin aromatic rings. At higher graft ratios, grafting will also occur on the aliphatic lignin chains [111]. The resulting product, however, is largely homopolymer and only short polymer chains are grafted onto lignin.

Chemically initiated grafting, using a decomposing initiator molecule to generate active sites, was found to be more effective than photo-induced grafting [112] for hydrochloric lignin. Whereas all polymerization reactions could be used to graft lignin molecules, only free-radical chain and step growth polymerization reactions were carried out on lignin [118]. Despite an increased efficiency, the amount of polymer attached to lignin is generally still limited and, therefore, most of the monomers are consumed by homopolymerization. Some progress was made with the graft copolymerization of lignin with maleic anhydride in which no homopolymer was formed [116]. It was reported that a significant number of lignin hydroxyl groups were lost, such that esterification by reaction of the lignin phenolic hydroxyl groups with maleic anhydride must have been the first step in generating active grafting sites.

The double effect of esterification and radical polymerization of the double bonds was also reported by Feldman *et al.* [94]. Meister *et al.* [121, 122] followed the similar approach of generating active sites on the lignin molecule using calcium chloride and a hydroperoxide. This initiation system preferentially attacks lignin repeat units, effectively creating active polymer chain growth sites. Other grafting techniques using ionic chain polymerization, the addition of complete polymer chains, and chemo-enzymatic grafting have also been reported. For a detailed description, the reader is referred to Meister [118] or the Ph.D. thesis of Kelley [123] (ionic chain polymerization), the work of Oliviera and Glasser [124, 126] (polymer chain grafting), and the work of Mai *et al.* [127, 128] (chemo-enzymatic grafting).

16.3 LIGNIN MODIFICATION

16.3.1 MODIFICATION REACTIONS

Chemical modification of lignin is another area of significant scientific work. It is largely based on the knowledge of lignin modifications used to dissolve lignin in organic solvents and applied in the determination of lignin functional groups. This dissolution is needed for various characterization techniques, so various chemical modifications, therefore, have already been well established [118, 120]. It can easily be seen that chemical modification of lignin can be used to improve the polymer–lignin compatibility and to introduce reactive sites. The available hydroxyl groups on the lignin molecule are reactive and plentiful and can act as local centers of high polarity capable of hydrogen bonding [20]. The modification of these reactive nuclei results in an effective alteration of the lignin solubility behavior [50, 118]. Specific modification of aromatic and aliphatic hydroxyl groups allows for indirect determination of the relative ratios of these alcohol groups. Esterification of the hydroxyl groups and acetylation are the commonly used methods for indirect determination of hydroxyl group content [129]. (Direct determination would use methods such as nuclear magnetic resonance spectroscopy, which does not rely on chemical modification.)

Lignin is reacted with carboxylic acids, carboxylic acid halides, or anhydrides [50]. Various acetylation procedures were described in the literature [129–134]. Acetylation also renders lignin soluble in a variety of organic solvents such as acetone, tetrahydrofuran, and chloroform, used in several characterization techniques [135, 136]. Propionation and esterification with larger chain derivatives have also been successful but have so far received relatively little attention [130, 134, 136, 137]. The addition of double-bond functionality with the addition of methacrylate [134, 138, 139], maleic anhydride [134], and styrenyl [139] groups has also been reported. Etherification of the lignin hydroxyl groups also improves the lignin solubility in organic solvents significantly [140]. The lignin hydroxyl groups were reacted with propylene oxide, ethylene oxide, and chloroacetic acid. Etherification has also been shown to reduce the lignin glass transition temperature [140].

Other modification reactions for kraft lignins include sulfonation, sulfomethylation, amination, halogenation, and nitration [118, 120]. Sulfonation and methylsulfonation add, respectively, sulfonate ($-SO_3^-$) and methylene sulfonate ($-CH_2SO_3^-$) groups to the lignin molecule. Sulfonation happens readily and quickly by reaction with sulfuric acid at specific functional groups and positions on the PPU: hydroxyl groups on the α-carbon, which is further promoted by free aromatic hydroxyl groups; ether linkages of the α-carbon and hydroxyl groups on the carbon in the γ-position to a phenyl-carbonyl pair. Dimethylsulfonation is only used if an increased content of sulfonate groups is desired. The sulfonate groups act as hydrophilic polar ends and, therefore, lignosulfonates find use as solid dispersants in water. Their applicability to polymers is limited and studies are generally focused on polar biodegradable plastics such as soy protein plastics where their polarity is beneficial [141]. However, nonionic lignosulfonate esters were shown to be water insoluble and form thermoplastics with unique properties [142]. Amination introduces $-NR_1R_2$ groups with R_i being hydrogen or organic carbon chains (aromatic or aliphatic). This is generally done by an acid-catalyzed addition of methanol and a primary or secondary amine in aqueous suspension.

Aminated lignins are water soluble. The industrial use of these lignins is very limited and applicability as a polymer additive or copolymer has not been reported. Lignin halogenation occurs readily by bubbling bromine or chlorine gas through a lignin solution, resulting in the addition of the halogens to the aromatic ring. Halogenated lignins are water insoluble and were used as fire retardant additives [118]. However, environmental concerns with halogenated aromatics make applications for these halolignins highly unlikely. Nitrated lignins are obtained by reaction of lignin with nitric acid and other concentrated acids. It introduces nitro groups ($-NO_2$) of which the majority are attached to the aromatic ring. Nitroso groups ($-NO$) occur as well but in very small amounts. Under excess nitric acid or in acidic aqueous conditions, nitrate esters ($-ONO_2$) are also formed. Nitrolignins have found no commercial applications but have been studied as additives to polyurethane with significant improvements on the mechanical properties and thermal stability with less than 3% nitrolignin addition [143]. This was attributed to the introduction of cross-links with lignin addition.

A vast amount of research was done to elucidate the effects of lignin addition to polymers, to modify lignin in order to improve interactions, and to introduce functional groups. Some of the chemical modifications described here will be used later in this work as a basis to obtain lignin derivatives soluble in unsaturated thermosetting resins.

16.3.2 MODIFIED LIGNIN IN POLYMERS

Copolymerization of etherified lignins with methyl methacrylate yielded both thermoplastic and cross-linked polymers, depending on the lignin functionality. However, samples were not tested for their mechanical properties. Ether derivatives are formed by reaction of the lignin hydroxyl groups with alkyl aryl halides, or with alcohols and acid catalysts. They can also be formed by reaction with acrylonitrile, chloroacetic acid, and alkyl chlorides among others [50]. Etherified lignin (hydroxypropyl lignin) was used in blends with polyethylene, poly(methyl methacrylate), poly(vinyl alcohol), and ethylene vinyl acetate [144]. The effect on mechanical and thermomechanical properties depends on the lignin modification and the polymer blend that was used. Improved compatibility between lignin and polymer by esterification was used in blends of lignin derivatives with cellulose acetate butyrate [145, 146], starch-caprolactone copolymer/blend [145], polyhydroxybutyrate [145], and polyvinyl chloride [147]. Methacrylated lignin was found to form cross-linked networks when copolymerized with methyl methacrylate [138]. Properties of these polymers, however, were not tested.

All of the previously described copolymers and blends with lignin have the common character that their properties deteriorate at lignin concentrations higher than 25–40% (w/w). Recently however, thermoplastic polymers were made with 85% kraft lignin content [148] and even 100% alkylated lignin [149, 150]. The 85 wt% kraft lignin polymer was a blend of underivatized industrial kraft lignin with poly(vinyl acetate) (90,000 g/mol) and plasticizers diethyleneglycol dibenzoate and indene in a 16:2:1 (w/w/w) ratio. The polymer was cast from an aqueous 82 vol% pyrrolidine solution. The degree of association between lignin molecules and average molecular weight of the blend was found to influence tensile behavior with tensile strength and modulus extending to 25 MPa and 1.5 GPa, respectively. Alkylated lignin (both methanate and ethylate) was obtained by etherification of kraft lignin [150]. The alkylated lignin was solvent cast with and without plasticizing miscible aliphatic polyesters from dimethyl sulfoxide. The 100% ethylated methylated kraft lignin exhibited tensile strength and modulus of 37 MPa and 1.9 GPa, respectively [149]. The properties of these high lignin content thermoplastic polymers are similar to currently existing petroleum-based polymers (see Table 16.2).

16.4 UNMODIFIED LIGNIN ADDITION TO SOYBEAN OIL MONOMERS

Applications of kraft lignin (KL) in unsaturated thermosetting polymers are virtually nonexistent [53]. This can be largely attributed to the insolubility of lignin in industrial polyester and vinyl ester resins as well as the ability of lignin to inhibit free-radical polymerization by forming quinonic structures, stabilized by resonance over the whole molecule [100–106]. To understand

TABLE 16.2 Comparison of properties of high-content kraft lignin polymeric samples with common petroleum-based thermoplastics.

Polymer	Tensile Strength (MPa)	Tensile Modulus (GPa)
Polyethylene (LDPE)	14	0.22
Polystyrene (HIPS)	28	2.1
Polypropylene	35	1.4
85% Kraft lignin	25	1.5
100% Ethylated methylated kraft lignin	37	1.9

Source: [149].

the dispersion/solution behavior of KL better, different industrially available KLs were added to various resins with different functionalities. Four different KLs were added to four different resin systems. Three of these resin systems are soybean-oil derived, while the fourth is an industrial vinyl ester (VE) produced by Dow Chemical Company: Dow Derakane 411–C50.

The cross-linking molecule in Derakane vinyl esters is based on a methacrylated diglycidyl ether of Bisphenol A (DGEBA) [151], which is blended with styrene to reduce the viscosity (see Figure 16.7 for chemical structures). The specified amount of styrene in Derakane 411-C50 is 50% [152]. The styrene content of 50 wt% was confirmed by H-NMR and a value for *n* of 2.645 was obtained. The three functionalized soybean oil resins are acrylated epoxidized soybean oil (AESO), hydroxylated soybean oil (HSO), and soybean oil monoglyceride (SOMG). A detailed description of the reaction pathways is given in Chapter 4. Table 16.3 summarizes the functionalized soybean oils used with their respective reactions.

The four kraft lignins examined were obtained from MeadWestvaco. They were pine, hardwood, ethoxylated, and maleated kraft lignin. They were free from any hemicellulose contamination. Their trade names were Indulin AT and PC-1369, respectively. The ethoxylated kraft lignin, sold under trade name Reax 825E, is an ethoxylated sodium salt of a highly sulfonated kraft lignin. Maleated lignin was obtained by heating pine kraft lignin (Indulin

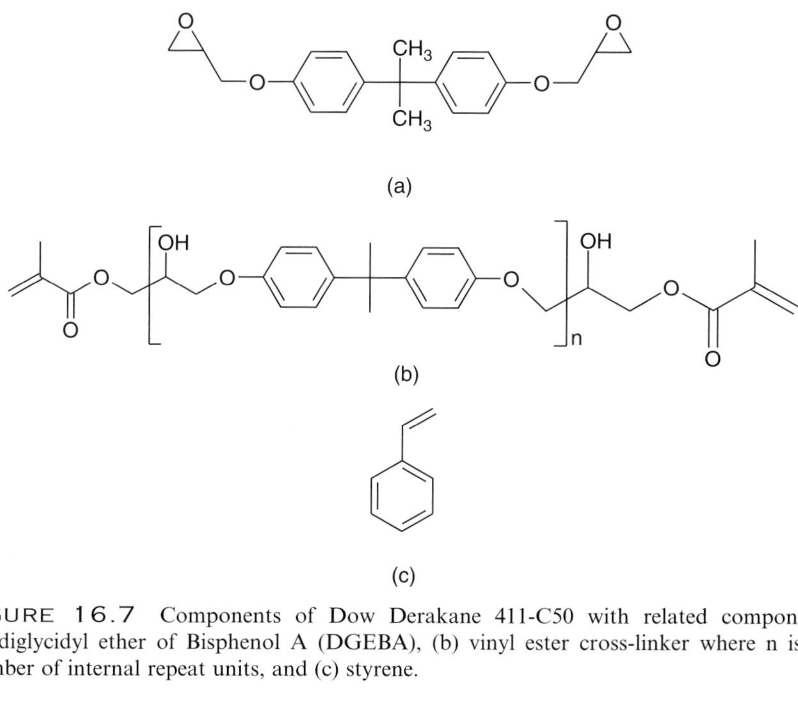

FIGURE 16.7 Components of Dow Derakane 411-C50 with related components: (a) diglycidyl ether of Bisphenol A (DGEBA), (b) vinyl ester cross-linker where n is the number of internal repeat units, and (c) styrene.

TABLE 16.3 Relevant soybean oil functionalization.

Acronym	Product	Reaction Number	Reaction
ESO	Epoxidized soybean oil	1	Epoxidation
AESO	Acrylated epoxidized soybean oil	2	Acrylation
HSO	Hydroxylated soybean oil	3	Hydroxylation
SOMG	Soybean oil monoglyceride	4	Glycerolysis

AT) and maleic anhydride. All lignins were in powder form, resembling freeze-dried instant coffee. Resin–lignin mixtures with 5% weight fraction lignin were made. They were mixed manually until complete dispersion was achieved. Samples were then left to settle for several days. The results are summarized in Table 16.4. The reason for dispersion was attributed to either the high viscosity of the resin or affinity between the resin and lignin. However, these dispersion effects are only important for the smaller particles, because gravitational forces for the bigger particles were too strong to be outweighed by either resin–lignin affinity or even resin viscosity. Note that only dispersed or dissolved lignin was accounted for.

Sedimentation to some extent was seen for all samples. It is notable that there is a small lignin fraction soluble in the low-polarity VE resin. The largest fraction, however, sedimented. The lignin solubility behavior matches the relative polarity of the respective resins: highest solubility/dispersibility in SOMG, followed by HSO and VE. AESO is too viscous to be included. These results are consistent with the notion that lignin solubility increases with the polarity of the solvent.

16.4.1 PINE KRAFT LIGNIN–AESO COMPOSITES

The highest lignin fraction that stayed dispersed was found in AESO where almost no lignin sedimented due to the high resin viscosity. HSO and SOMG need further modification of the alcohol groups to allow for free-radical polymerization. To investigate the effect of lignin as a filler, pine kraft lignin (PKL) was dispersed in the AESO resin and polymerized to obtain

TABLE 16.4 Observations of dispersion of lignin in different resins.

	Pine Kraft Lignin	Hardwood Kraft Lignin	Ethoxylated Kraft Lignin	Maleated Kraft Lignin
AESO	Viscosity	Viscosity	Viscosity	Viscosity
HSO	Affinity	No dispersion	Affinity	No dispersion
SOMG	Affinity	Affinity	Affinity	No dispersion
VE	Partially soluble	Partially soluble	No dispersion	Partially soluble

samples for dynamic mechanical analysis (DMA). AESO was mixed with varying amounts of styrene. PKL was added to this mixture in varying amounts and mixed to obtain complete dispersion. Lignin was seen to concentrate both at the top and bottom of the cured sample, believed to be due to lignin–styrene incompatibility. Therefore, only samples with 0, 5, and 10 wt% styrene were made.

A representative DMA spectrum is shown in Figure 16.8. Local maxima can be seen in the loss modulus and tan δ, representative of the existence of different phases. The storage modulus, E', and glass transition temperature, T_g, of the samples as a function of PKL and styrene content are shown in Figures 16.9 and 16.10, respectively. Expected changes were seen for the 0 and 10 wt% styrene samples. The T_g was found to increase as could be expected with a lignin T_g of about 142°C [137]. However, the increase is less than expected from simple mixing rules. For example, for the maximum 15 wt% lignin, T_g values of only 36.2° and 46.8°C were recorded for the 0 and 10 wt% styrene samples, respectively. Because lignin does not contain double bonds reactive toward free-radical polymerization, the storage modulus will decrease with increasing PKL content. The introduction of PKL plasticizes the polymer.

The incompatibility of kraft lignin with styrene, a reactive nonpolar diluent used in a large number of unsaturated thermosets, is a major problem. Kraft lignin could be incorporated in a soybean oil-based thermoset with low styrene content due to the high viscosity of the monomer mixture before polymerization. However, lignin incorporation resulted in detrimental reduction of

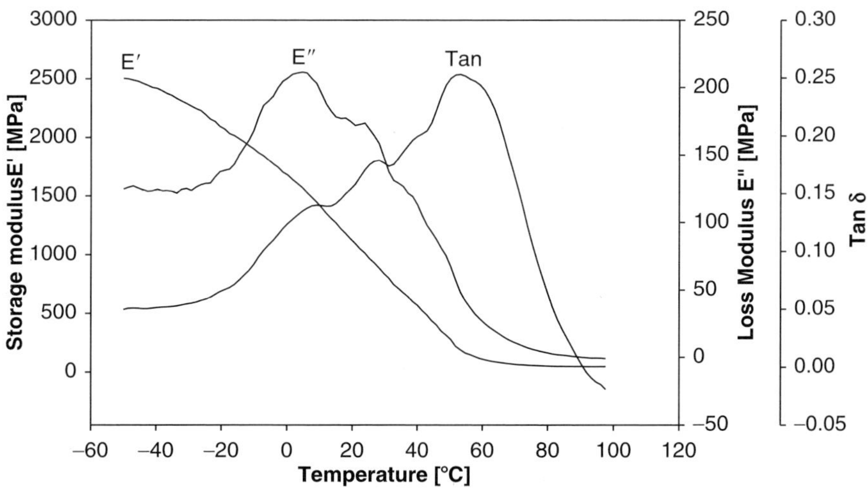

FIGURE 16.8 Representative thermomechanical behavior of the AESO/styrene/PKL samples as recorded by DMA at constant frequency in three-point bend mode.

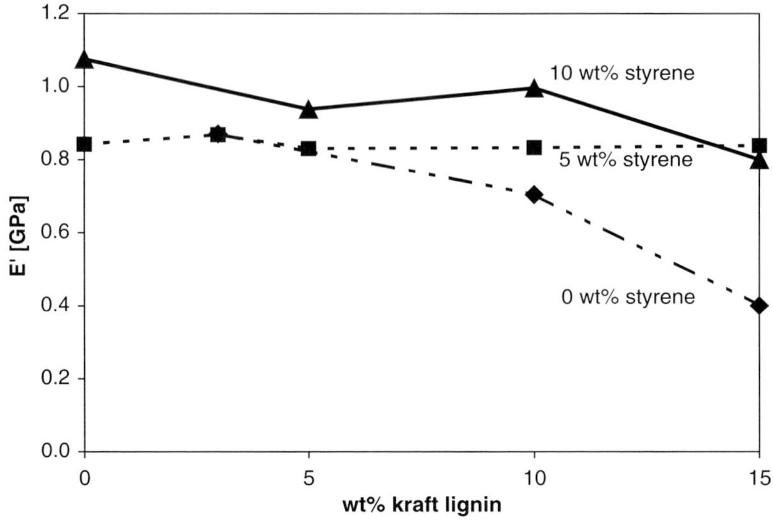

FIGURE 16.9 The storage modulus of the AESO/styrene/PKL samples at 20°C.

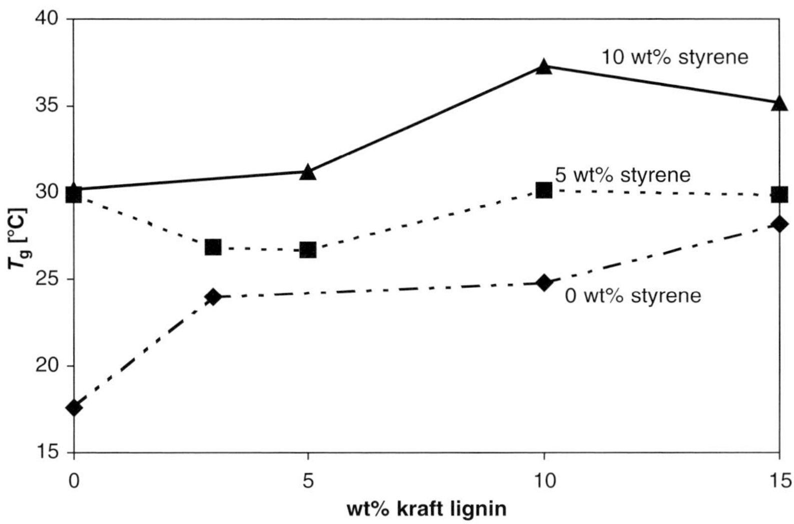

FIGURE 16.10 The glass transition temperature (T_g) of the AESO/styrene/PKL samples.

properties due to incompatibility of lignin with the resin. The study of these composite samples was kept to a minimum due to the negative trends seen with lignin addition. However, it makes a very strong case for the need to modify lignin to make it soluble in styrene and styrene-containing unsaturated thermosets such as polyesters and vinyl esters. This chemical modification could also include the incorporation of double-bond functionality, reactive toward free-radical polymerization. Improved lignin compatibility and copolymerization of lignin with the monomer mixture should improve the mechanical properties of these thermosets. Modification reactions of lignin, and its effect on lignin solubility, are presented in the next section.

Ethoxylated sulfonated lignin was not studied further due to the large number of sulfonate groups present. Sulfonation of lignin improves the water solubility of ethoxylated sulfonated lignin. Incorporation of sulfonated lignin will increase the – water uptake. This generally causes an unwanted degradation of polymer properties over time.

16.5 CHEMICAL MODIFICATION OF KRAFT LIGNIN

The results from the previous section emphasized the importance of chemical modification of kraft lignin before incorporation in an unsaturated thermosetting polymer. These chemical modifications should accomplish two effects: (1) solubilization of KL in styrene and the unsaturated thermosetting resins used, and (2) blocking of the inhibiting capability of the lignin aromatic hydroxyl groups. The inhibiting capability of the lignin aromatic hydroxyl groups has been well established [100–106]. Although inhibition of the free-radical polymerization may not be detrimental to the final polymer properties, if a sufficient amount of initiator were added, reduction in the polymerization propagation rate increases the processing time. If possible, this should be avoided, because it will increase the production cost of the polymer part. Knowing the chemical structure of pine KL (see Figure 16.6), chemical modification of the lignin hydroxyl groups is the most desirable route: It will reduce the polarity of KL significantly by reducing the amount of hydroxyl groups and their hydrogen-bonding capability, and it will block lignin's ability to form the quinonic structures for free-radical inhibition. First, the functional groups present in the two KLs used in this work will be discussed. Solubilization of lignin in styrene and the monomer mixtures used in this work will then be explained by calculation of the effect of the modification on atomic charges and with the use of solubility theory.

16.5.1 LIGNIN CHEMICAL STRUCTURE

Two KLs were used: a softwood, pine KL, sold under the trade name Indulin AT; and a hardwood KL, which was an experimental sample classified as PC-1369. Both were obtained from MeadWestvaco. Indulin AT was studied extensively, and its chemical functionality has already been identified. The analytical composition of its PPU can be represented by the following formula [20]:

$$C_9H_{7.9}O_{2.1}S_{0.1}(OCH_3)_{0.82},$$

with a molecular weight of 178 g/mol. The number and weight average molecular weight, M_n and M_w, are 1600 and 3500 g/mol, respectively [20]. The deduced polydispersity of 2.2 is typical for products of random depolymerization processes. Thus, on average there are 20 PPU per pine KL (Indulin AT) molecule. Because Indulin AT is a softwood KL, it contains largely guaiacyl units. The mole ratio of *p*-hydroxyphenol:guaiacyl:syringyl units was reported to be 14:84:2 [151]. Its chemical functional groups important to this work are combined in Table 16.5 for easy comparison with the hardwood KL PC-1369.

Data on hardwood lignin is much more varied because it is generally obtained from a blend of varying wood species. Hardwood KL from MeadWestvaco was characterized by Marton [20] to exhibit an analytical composition of

$$C_9H_{7.2}O_{1.8}S_{0.1}(OCH_3)_{1.15}$$

with a molecular weight of 183 g/mol. A polydispersity of 2.8 was measured with values for M_n and M_w of 1050 and 2900 g/mol, respectively. This amounts to an average of 15–16 PPU per hardwood KL molecule. Functional groups important to this work are combined in Table 16.5. The aromatic and aliphatic hydroxyl groups were obtained from Ref. [104]. However, Lenz also determined the pine KL functional groups, which differed from data published more recently. The hardwood KL data were therefore corrected according to the correction needed for pine KL.

Values in Table 16.5 were confirmed by ^{1}H NMR spectroscopy of the respective KLs. Some of the differences between reported hardwood KL and the values obtained in this work can easily be attributed to differences in the hardwoods from which lignin was extracted. Therefore, the values between parentheses for hardwood KL will be considered the correct values for the hardwood KL used in this work.

TABLE 16.5 Functional groups present in KL used in this work.[a]

	Softwood KL Indulin AT	Hardwood KL PC-1369
Aromatic hydroxyl (ArOH)	0.64–0.643 (0.64)	0.70 (0.73)
Aliphatic hydroxyl (AlOH)	0.439 (0.44)	0.35 (0.37)
Carboxylic acid (–COOH)	0.11 (0.11)	— (0.06)
Methoxyl (–OCH_3)	0.76–0.786	1.15
Aromatic hydrogen (ArH)	2.5 (2.5)	2.3 (2.31)
Aldehyde (–CHO)	<0.02 (0.02)	— (0.01)

Sources: Data compiled from Refs. [20, 53].

[a] Amounts shown are per PPU. Hydroxyl values for hardwood lignin from Ref. [53] were corrected proportional to corrections needed to conform softwood lignin values of Ref. [53]. Numbers in parentheses are values obtained experimentally by ^{1}H NMR spectroscopy.

16.5.2 ESTERIFICATION REACTIONS

To accomplish the conversion of alcohol groups to less polar side groups, the esterification pathway was chosen. The three most widely used reactants for alcohol esterification are carboxylic acids, acid anhydrides, and acid halides [153, 154]. They are ranked according to their relative speed of reaction with carboxylic acids the slowest of the three.

Other potential reactants are ketenes (R–C=C=O) [154, 155]. The high reactivity of ketenes with alcohols makes them powerful esterification agents. They were used in the acetylation of a variety of compounds [156]. The production of ketenes, however, requires high temperatures and specialized reactors and solid catalysts [157, 158], making this pathway less prevalent than others.

The carboxylic acid–alcohol reaction is relatively slow, reversible, and acid catalyzed and is only applicable to primary and aromatic alcohols [154]. Because kraft lignin also contains secondary alcohols, this pathway cannot be followed. Acid anhydrides and acid halides react irreversibly with all alcohol groups. The anhydride–alcohol reaction is catalyzed by tertiary organic bases such as pyridine [155] and imidazoles [159]. Primary and secondary amines are much better nucleophiles than water such that alcohol–acid anhydride reactions can even progress in aqueous media without significant hydrolysis of the anhydride [155]. The reaction product is the ester and a carboxylic acid. The acid halide–alcohol reaction can proceed uncatalyzed at room temperature. However, the formation of a hydrogen halide requires a neutralizing additive, commonly ammonia or pyridine [154, 155]. Additionally, acid chlorides are more expensive than anhydrides and very susceptible to hydrolysis. The reaction mixture has to be anhydrous to avoid the conversion to regular carboxylic acids.

16.5.3 LIGNIN ESTERIFICATION

The large variety of acid anhydrides and acid chlorides available allow for the addition of different side chains to accomplish the solubilization of lignin in styrene. Esterification of lignin was shown to improve the solubility in various solvents [119, 158]. The most common example is the acetylation of kraft lignin. This reaction is used to obtain a tetrahydrofuran-soluble lignin for gel permeation chromatography [161] and for determination of hydroxyl groups [128]. The standard reaction mixture consists of equal parts of pyridine and the acid anhydride [132].

Although pyridine is generally used as the catalyst for lignin esterification, other, more effective catalysts exist. Dimethylaminopyridine (DMAP) [160], 1- or *N*-methylimidazole (1MIM) [159, 163–165], 2-methylimidazole (2MIM) [166], and imidazole (IM) [167] have all been used. The catalytic activity of 1MIM was found to be 400 times greater than pyridine for alcohol–acetic anhydride reactions [159], while DMAP is about 10,000 times more active [162]. However, DMAP still requires the presence of pyridine to act as a proton scavenger. For the lignin esterification in this work, 1MIM was chosen. It was the most active of the liquid catalysts considered and was shown to be a very effective catalyst for acetylation of hydroxyl-terminated polymers [165] and cell wall carbohydrates [164]. DMAP and 2MIM are solid at room temperature. Because lignin has to be dissolved as well, the use of a liquid catalyst is advantageous. The reaction scheme of 1MIM catalyzed esterification has been described by Connors and Pandit [159].

16.5.4 LIGNIN SOLUBILIZATION

Solubilization of KL in styrene and the monomer mixtures used in this work (AESO/styrene and vinyl ester) was obtained by increasing the aliphatic carbon chain of the ester attached to the KL hydroxyl groups. Increasing the length of this chain can be expected to reduce the polarity of the KL and improve its solubility in nonpolar solvents. Acetylation, propionation, butyration, etc., of the KL hydroxyl groups increases the carbon chain length of the ester in steps of one carbon. The chemical compounds used are shown in Table 16.6 and the resulting esters are shown in Figure 16.11.

16.5.5 INTRODUCTION OF DOUBLE-BOND FUNCTIONALITY

Introduction of double-bond functionality to KL allows the modified KL to copolymerize with the monomers in the resin. One can expect this to be beneficial for the mechanical properties of the polymer. Two different anhydrides were used for this purpose: maleic anhydride and methacrylic anhydride. It was found that the aromatic hydroxyl groups of lignin did not

TABLE 16.6 Chemical compounds used in the chemical modification of kraft lignin and subsequent purification steps.

Name	Purity	Supplier
Acetic anhydride	99+%	Fisher Scientific
Propionic anhydride	97%	Aldrich Chemicals
n-Butyric anhydride	99%	Acros Organics
Maleic anhydride	99%	Aldrich Chemicals
Methacrylic anhydride	94%	Aldrich Chemicals
1-Methylimidazole	99%	Acros Organics
Cyclohexane	99+%	Acros Organics
1,4-dioxane	99+%	Acros Organics
Ethyl ether	Anhydrous	Fisher Scientific
Dichloromethane	99.6%	Aldrich Chemicals

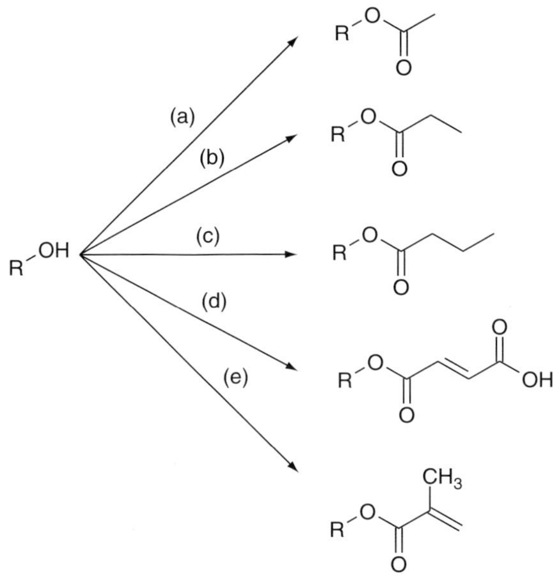

FIGURE 16.11 Conversion of hydroxyl groups to esters: (a) acetylation, (b) propionation, (c) butyration, (d) maleation, and (e) methacrylation.

react with maleic anhydride. This result is consistent with work by Cohen and Fong [168]. Maleated KL (only the aliphatic hydroxyl groups) was found to be insoluble in styrene. By varying the reaction time, KLs with varying amounts of maleates were obtained. Butyration of the remaining hydroxyl groups did not result in a styrene-soluble lignin for any significant extent of maleation.

Lignin methacrylation offers a better chance to introduce double-bond functionality reactive toward free-radical polymerization. The conversion of a hydroxyl group to a methacrylate group (see Figure 16.11) can be expected to reduce the polarity, in contrast to the formation of maleate groups. Complete methacrylation of KL was obtained by reacting KL with methacrylic anhydride in 1,4-dioxane, catalyzed with 1 MIM at 50°C. Completely methacrylated pine KL and hardwood KL were both insoluble in styrene.

Methacrylated pine KL was made with varying degrees of substitution. The remaining unreacted hydroxyl groups were subsequently reacted with butyric anhydride. Their solubility in styrene and the AESO/styrene (70/30 ratio by weight) resin system was subsequently tested. It was found that the maximum level of methacrylation for solubility in the resin system was 0.3 methacrylates per PPU. A higher degree of substitution resulted in lignin sedimentation from the resin mixture. For pine KL, this level of methacrylation allows for the addition of six methacrylate groups per average molecule, presenting a significant potential for copolymerization.

16.6 FLORY–HUGGINS THEORY

To evaluate the solubility of modified lignins in various solvents, several approaches were examined by Thielemans [169]. The Flory–Huggins theory is widely used to describe the thermodynamic behavior of polymers [169]. This theory, based on a lattice description of contact points for thermodynamic interactions, expresses the Gibbs free energy of mixing, ΔG_m, for a noncrystalline polymer by

$$\Delta G_m = RT[n_S \ln \phi_S + n_P \ln \phi_P + \chi_{PS} n_S \phi_P], \tag{16.1}$$

where n_i is the number of moles of i and ϕ_i is the volume fraction of i. The S and P stand for solvent and polymer, respectively, and χ_{PS} is the Flory–Huggins interaction parameter between polymer and solvent. The first two terms are merely the description of the configurational entropy contribution, while the last term expresses the enthalpic contribution to the free energy. The critical value for polymer solubility was derived by Flory to be a function of x, the ratios of molar volumes of polymer and solvent [170]:

$$\chi_{PS,C} = \frac{(1 + \sqrt{x})^2}{2x}. \tag{16.2}$$

To predict the Flory–Huggins interaction parameter, χ_{PS}, solubility theory was used. This theory, introduced by Hildebrand *et al.* [171], relates the enthalpy of mixing to the solubility parameters δ, of the interacting species. The solubility parameters are expressions of the cohesive energy density of the species and thus depend on the chemical structure of the compounds. Refinement of their theory led to the separation of the individual δ parameter into three different contributions [8, 172]: a dispersive (δ_d), a polar (δ_p), and a hydrogen bonding (δ_h) contribution. The Flory–Huggins interaction parameter, χ_{PS}, can then be expressed as follows [171]:

$$\chi_{\mathbf{PS}} = \frac{v_S}{RT}\left[(\delta_{dS} - \delta_{dP})^2 + (\delta_{p^S} - \delta_{p^P})^2 + (\delta_{hS} - \delta_{hP})^2\right], \tag{16.3}$$

where v_S is the solvent molar volume. By comparing χ_{PS} calculated from Eq. (16.3) with $\chi_{PS,C}$ from Eq. (16.2), we should be able to predict KL solubility. Solubility theory has already been applied to describe the solubility of various lignins in organic solvents [173, 174] and to describe the interactions between lignin, cellulose, and hemicellulose [6].

Various models exist for the prediction of the aforementioned solubility parameters [8, 172, 175, 176]. Most of these models calculate the solubility parameters based on a group contribution method, in which the contributions were obtained from homologous series studies. The model developed by Hoy [172, 176] was chosen since it is one of the more accurate available, has contributions for a large amount of functional groups, and accounts for a variety of structural features such as ring formation and isomerism. Interestingly, it predicts a hydrogen-bonding capability for aromatic rings. Although this is still controversial, the model yields good predictions and it is not the only model attributing this capability to aromatic rings [8]. An explanation for this prediction could be found in the large negative cloud formed by the aromatic π-orbital electrons.

The solubility parameters for all of the KLs discussed here were calculated based on this group contribution method. The solubility parameters for hardwood KL are shown in Figure 16.12. Increasing ester carbon chain length is seen to decrease the polar and hydrogen contributions. Surprisingly, δ_h decreases from butyrates to methacrylates while δ_p increases. The dispersive contribution varies more randomly. The styrene values were also calculated using the Hoy model for consistency:

$$\delta_d = 17.076\frac{J^{1/2}}{cm^{3/2}}, \quad \delta_p = 9.33\frac{J^{1/2}}{cm^{3/2}}, \quad \delta_h = 6.45\frac{J^{1/2}}{cm^{3/2}}.$$

To determine x, the ratio of molar volumes of solvent and polymer, one needs the actual molar volume of both. The molar volume of styrene is 114.6 cm^3/mol, calculated from its molecular weight (104.2 g/mol) and density (0.909 g/cm^3) [177]. The molar volume of lignin and its derivatized varieties in solution is not known and is impossible to determine for the insoluble KL species. It has, however, been established by an extensive amount of available data that for noncrystalline polymers the ratio of molar volume over van der Waals volume is 1.6 [174]. The van der Waals volume obeys additivity rules and can be calculated directly from group contributions. The group contributions for functional groups encountered were measured and tabulated [174].

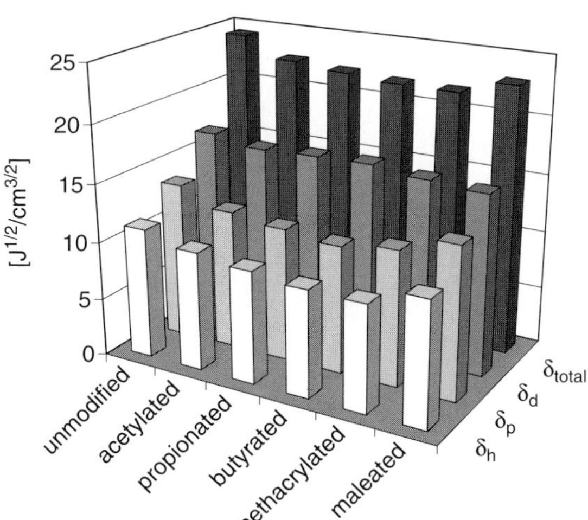

FIGURE 16.12 Solubility parameters calculated using the Hoy model [176] for unmodified and various modified hardwood KL.

The ratio of molar volumes, x, was calculated for all KL studied here and varied between 25 (unmodified) and 40.8 (butyrated) for pine KL, and between 20.2 (unmodified) and 30 (butyrated) for hardwood KL.

16.6.1 SOLUBILITY MODEL RESULTS

Figure 16.13 shows the comparison of the calculated Flory–Huggins interaction parameters χ_{PS} for all the KLs studied here with the critical value

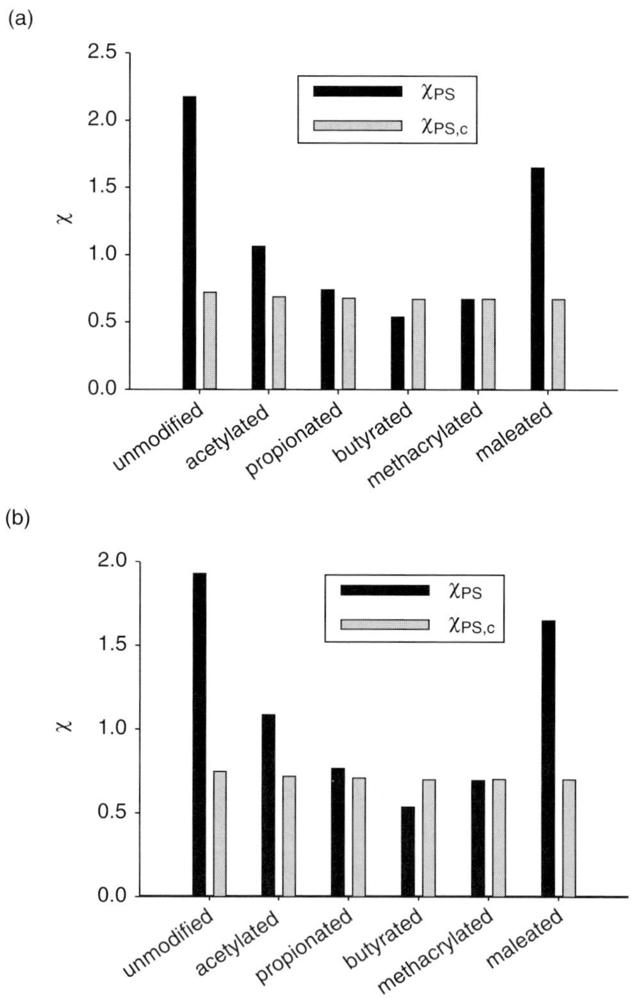

FIGURE 16.13 Comparison of calculated values for the Flory–Huggins interaction parameter, χ_{PS}, for various modified KLs in styrene and the critical Flory–Huggins parameter, $\chi_{PS,c}$, for styrene solubility: (a) pine KL and (b) hardwood KL.

calculated using Eq. 16.3 [134]. The calculations capture the experimental findings quite well: The only completely soluble KL is butyrated KL. The decrease in χ_{PS} with increasing ester carbon length is also well captured. Methacrylated KL is on the solubility boundary and is predicted to be more soluble than propionated KL, which is in contradiction with experimental observations. The partial solubility seen for propionated KL experimentally is also captured by a small difference between χ_{PS} and $\chi_{PS,C}$, pointing toward potential solubility for fractions of smaller molecular weight for which x, and thus $\chi_{PS,C}$, will be higher. The overall predictions are quite good and could be used to study the potential of future modifications on solubility of KL.

The results from the previous description are quite consistent with the experimental observations, especially considering the predictive nature of the solubility parameters and the incomplete knowledge of the exact chemical structures of pine and hardwood KL. The only inaccuracy was found with the solubility prediction of methacrylated KL, found to be on the border of styrene solubility. It was found, however, to be less soluble in styrene than propionated KL experimentally. Their small difference in χ_{PS} values (less than 10%), however, should be a source of caution and places the predictions in the "gray" area to be verified experimentally.

A cause of inaccuracy of the previous predictions might also be found in the averaging character of solubility parameters. The contributions from all functional groups are added and averaged over the whole molecule. Although this method may work well for small molecules and synthetic polymer, which are made up of constantly recurring monomer units, it may display reduced accuracy for natural polymers. Natural polymers generally exhibit a larger amount of variation of functional groups, linkages, and monomer units. Averaging small effects of these functional groups is likely to reduce the overall contribution of their local effects. To get a better understanding of local effects of the esterification reactions, the partial charges of the individual atoms in the lignin molecule, due to their differing electronegativity, were calculated by Thielemans [134, 169] and these differences were resolved. He found that the maximum charge difference, as well as the charge difference between ending carbons in the ester chain, was larger for methacrylated lignin. The increased charge difference can thus be expected to be the cause for the lower styrene solubility of methacrylated KL compared to propionated KL, which was not predicted by the solubility parameter approach.

16.6.2 SUMMARY OF LIGNIN SOLUBILTY

A variety of esterification reactions were used to alter the solubility behavior of KL. Complete butyration of KL and partial methacrylation, followed by butyration of the remaining hydroxyl groups, was found to result in styrene-soluble KL. This styrene-soluble KL was also found to be soluble in the resin system consisting of a mixture of styrene and AESO. The solubility behavior of KL and its ester derivatives could be described by the Flory–Huggins theory, combined with the prediction of solubility parameters to describe the enthalpic contribution to the free energy. The only inconsistency found in this description could be explained by small and local differences in atomic charges. These differences are averaged over the whole molecule when calculating solubility parameters, explaining the inconsistency in solubility predictions for methacrylated and propionated KL.

16.7 BUTYRATED KRAFT LIGNIN IN THERMOSETTING POLYMERS

The mechanical, thermomechanical, and fracture properties of butyrated lignin dissolved in AESO/styrene and Dow Derakane 411-C50 were investigated. For the AESO/styrene samples, the lignin was dissolved in styrene before AESO addition. Butyrated KL was directly added to the Dow Derakane 411-C50 vinyl ester. The flexural stress versus strain curves are shown in Figure 16.14. The flexural modulus and strength as a function of butyrated lignin content are shown in Figure 16.15. Both the modulus and strength are seen to go through a maximum. The modulus maximum is very pronounced. The flexural strength increases significantly with lignin addition until the

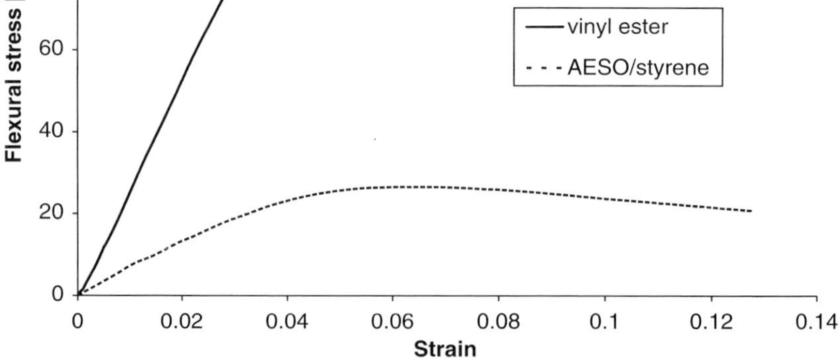

FIGURE 16.14 Typical stress–strain curve obtained from flexural testing in three-point bend mode.

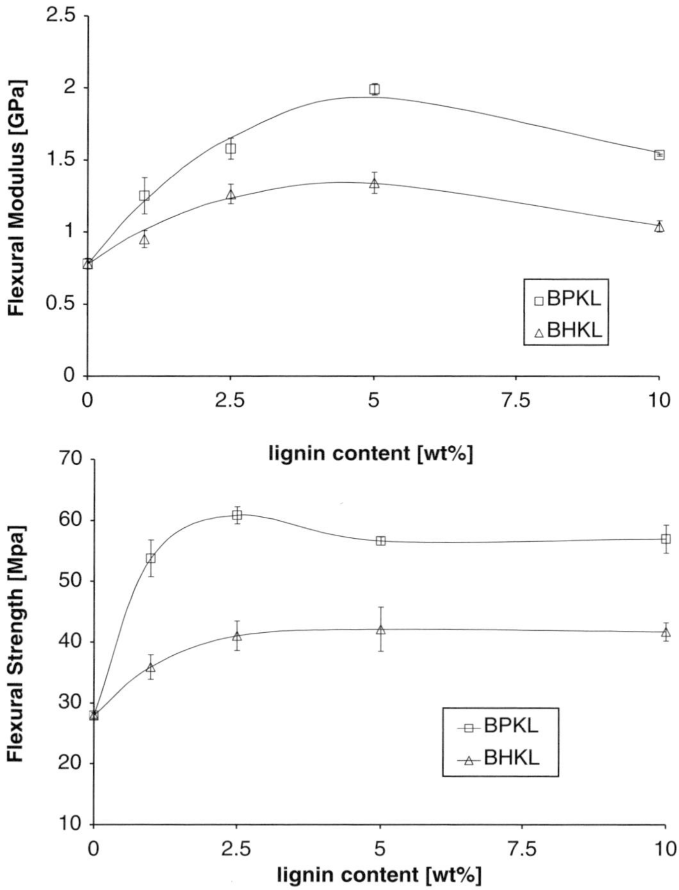

FIGURE 16.15 Flexural modulus and strength of AESO/styrene polymers as a function of butyrated KL lignin content.

maximum is reached, after which it decreases. The maximum for butyrated hardwood KL (BHKL) is much less pronounced than for the larger butyrated pine KL (BPKL).

The modulus and strength behaviors for both pine and hardwood KL are very similar, with a more pronounced effect for pine KL. The maximum relative improvements are combined in Table 16.7. The size/molecular weight of the lignin appears to be an important factor, with larger lignin size having the larger effect.

The flexural modulus and strength for the vinyl ester thermosets as a function of the butyrated KL content are shown in Figure 16.16. The underlying trend is a steady decrease in modulus and strength with increasing lignin content. At low lignin contents, however, butyrated pine KL increases the

TABLE 16.7 Maximum flexural property improvements with butyrated KL addition determined experimentally.

	Flexural Modulus		Flexural Strength	
	Maximum Increase (%)	Lignin Content (wt%)	Maximum Increase (%)	Lignin Content (wt%)
BPKL	155	5	117	2.5
BHKL	72	5	51	5

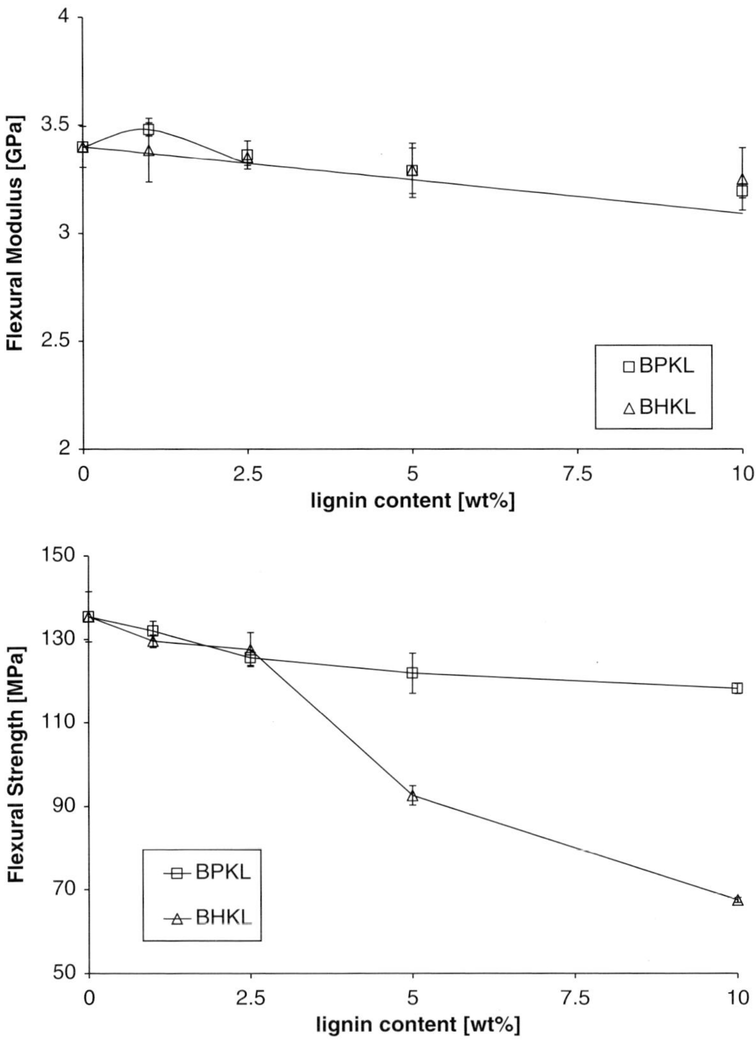

FIGURE 16.16 Flexural modulus and strength of vinyl ester polymers as a function of butyrated KL content.

vinyl ester modulus slightly. The smaller butyrated hardwood KL does not show this modulus increase. It could be hypothesized that its size is too small to affect the vinyl ester matrix significantly and is simply not noticed. DMA experiments will be used to investigate this in more detail. The flexural strength is seen to decrease quite substantially, especially at higher lignin contents. The much sharper decrease in flexural strength at high BHKL contents, compared to BPKL, is quite remarkable and cannot be explained by the different size of the lignin molecules. The better inhibition of styrene during polymerization by BHKL appears to be the major factor here. The increase in styrene concentration in the inter-microgel regions weakens these regions more than for BPKL, resulting in a more significant decrease in strength.

16.7.1 DYNAMIC MECHANICAL ANALYSIS

A typical DMA response curve for the AESO/styrene/butyrated KL thermosets is shown in Figure 16.17(a). The first, most notable result is the absence of any thermal transitions aside from the matrix material (AESO/styrene) transitions, for all lignin contents. Butyrated KL is thus completely dissolved in the matrix material and did not phase separate into aggregates. The storage modulus at 35 °C was found to behave in a similar fashion as the flexural modulus, as expected. The cross-link density, ν, calculated from the storage modulus above the T_g is found to decrease linearly with increasing lignin volume fraction, ϕ_L. The linear decrease depends on the KL and is given by:

$$\text{BPKL} \quad \nu = 2290 - 6160\phi_L \left[\frac{\text{mol}}{cm^3}\right] \tag{16.4}$$

and

$$\text{BHKL} \quad \nu = 2290 - 2586\phi_L \left[\frac{\text{mol}}{cm^3}\right]. \tag{16.5}$$

The larger BPKL does have a much more significant effect on the cross-link density than does the BHKL. Additionally, for both lignins, the cross-link density decreases more significantly than a simple volume fraction dependence. Lignin addition may also disrupt physical cross-links, because they are included in the determination of ν [170]. The decrease in cross-link density results in plasticization of the matrix.

The glass transition temperature increases with increasing butyrated KL content and goes through a maximum at higher lignin contents [Figure 16.18(a)]. The effects however are small. The value of T_g for lignin was taken to be 440 K [137]. As can be seen directly from the breadth of the damping factor peak, tan δ, shown in Figure 16.18(b) for varying BPKL contents, lignin addition does not alter the glass transition significantly.

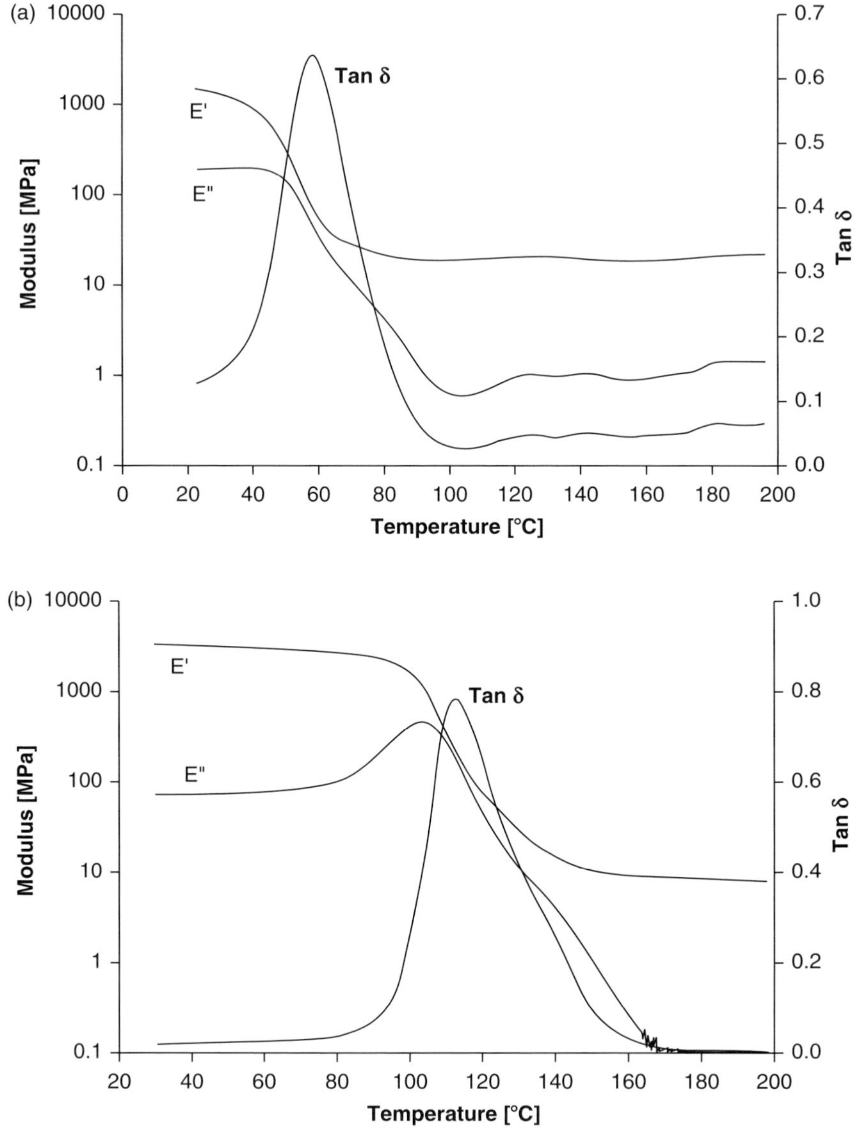

FIGURE 16.17 Typical DMA response curve for (a) AESO/styrene and (b) vinyl ester thermoset.

A typical DMA response for the vinyl ester–lignin thermosets is given in Figure 16.17(b). As with the AESO/styrene thermosets, no extra thermal transitions are see with lignin addition. The resin-soluble butyrated KL stays soluble during polymerization and is incorporated on a molecular level. The storage modulus at 35°C behaves in a similar fashion as the

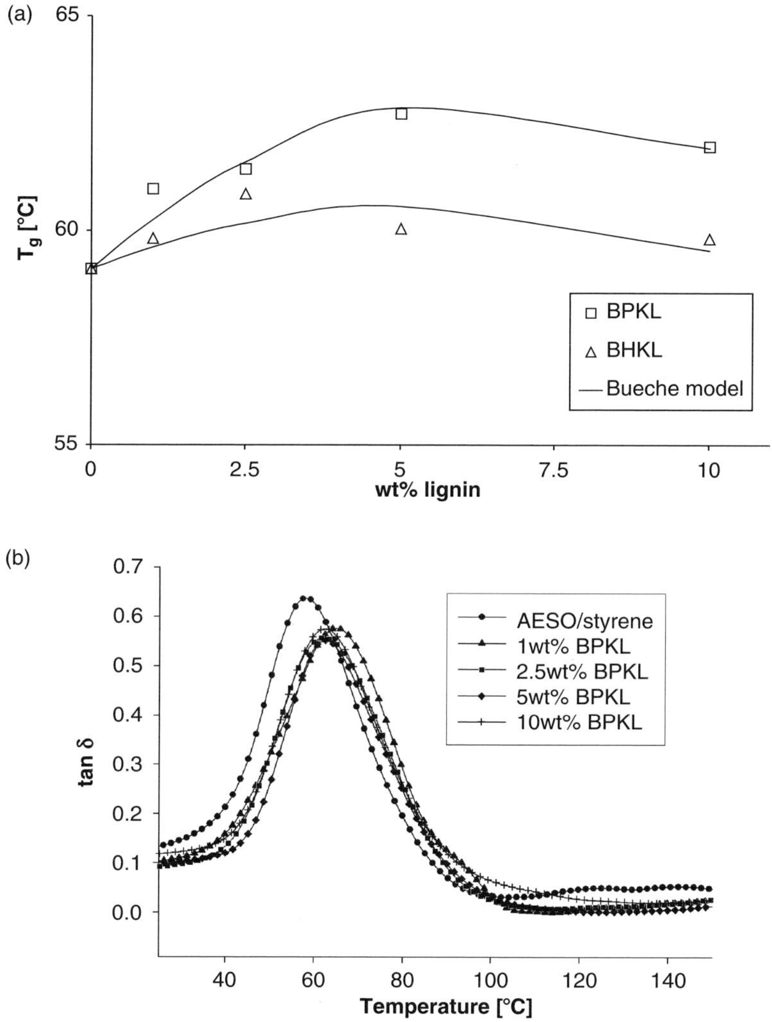

FIGURE 16.18 (a) Dependence of the glass transition temperature T_g on BPKL and BHKL content and fit of the experimental data with the modified Bueche model. (b) The variation of tan δ with BPKL content.

flexural modulus, as expected. The glass transition temperature, T_g, measured as the maximum of tan δ, behaves differently for both butyrated KLs. A slight increase is seen for hardwood lignin, while the T_g goes through a minimum for pine lignin [Figure 16.19(a)]. The cross-link density, ν, displays a similar peculiar behavior [Figure 16.19(b)]. The butyrated hardwood KL samples show a linear decrease with lignin content [Eq. (16.6)], whereas butyrated pine KL samples show a significant initial decrease, which subsides at higher lignin contents [see Figure 16.19(b)]:

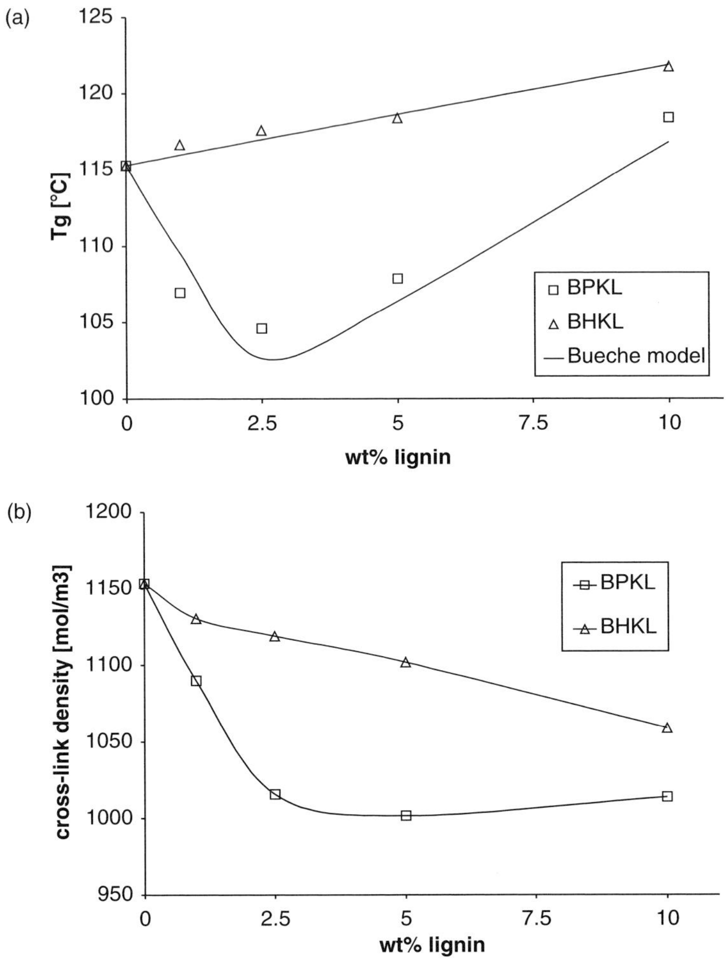

FIGURE 16.19 (a) T_g and (b) cross-link density variation for the vinyl ester thermoset as a function of butyrated KL content. Lines in part (a) are the fits of the modified Bueche model.

$$\text{BHKL } v = 1150 - 945\phi_L \left[\frac{\text{mol}}{\text{m}^3}\right]. \tag{16.6}$$

The linear decrease in cross-link density for butyrated hardwood KL is close to the volume fraction decrease. This is expected for additives that do not affect the cross-linking of the matrix. One can thus expect that at these lignin concentrations, interaction between the embedded butyrated lignin and the matrix is minimal. This is qualitatively seen for the flexural modulus data, where a linear decrease with lignin content is recorded. The next section addresses quantitative modeling of the flexural data with these findings.

Comparison of the cross-link density evolution with the T_g evolution as a function of butyrated KL content, shown in Figure 16.19, explains the discrepancy between the behavior for both KLs. The major initial decrease in cross-link density with BPKL addition is translated in a decrease in T_g. As the cross-link density decrease subsides, T_g increases due to the addition of a higher T_g component. BHKL, with a smaller cross-link density decrease, does not show this trend and the T_g increases linearly.

16.7.2 FRACTURE PROPERTIES OF BUTYRATED LIGNIN THERMOSETS

The DMA results show a plasticization effect with the addition of butyrated KL to both the AESO/styrene and vinyl ester thermosetting polymers. Plasticization is known to have a beneficial effect on the fracture properties of thermosetting polymers. It is common practice to add plasticizers to these polymers to improve their fracture properties. The fracture properties (three-point bend) of both AESO/styrene and vinyl ester thermosets were determined. The plain-strain fracture toughness, K_{1c}, is determined from the maximum in the load-deflection curve, $P_{\max}$, using [178]:

$$K_{1c} = \frac{P_{\max}}{b\sqrt{d}} f(x), \tag{16.7}$$

where $0 < f(x) < 1$ and

$$\boldsymbol{f(x) = 6x^{\frac{1}{2}}} \frac{\left[1.99 - x(1-x)\left(2.15 - 3.93x + 2.7x^2\right)\right]}{(1+2x)(1-x)^{\frac{3}{2}}}, \tag{16.8}$$

with $\boldsymbol{x} = \frac{a}{d}$ in which d is the sample thickness, b is the sample width, and a is the crack length. The strain energy release rate, G_{1c}, is calculated as

$$\boldsymbol{G_{1c} = \frac{U_1 - U_0}{bd\phi}}, \tag{16.9}$$

where ϕ is a tabulated value dependent on x, or it can be calculated using an expression given in the referenced ASTM standard [178]. The term U_1 represents the area under the load-deflection curve of the sample and U_0 is the area for an unnotched sample.

Figure 16.20 shows the relative increase in the fracture toughness, K_{1c}, and the plane-strain energy release rate, G_{1c}, of AESO-lignin thermosets. Both properties increase continuously with increasing lignin content. The enhancement is more significant for butyrated pine KL, with maximum enhancements of 86% and 180% for K_{1c} and G_{1c}, respectively. These values are comparable to epoxies toughened with rigid thermoplastics, such as poly (sulfone) and poly(ether sulfone) [179]. However, the latter thermoplastics phase separate during polymerization to form particulate or co-continuous

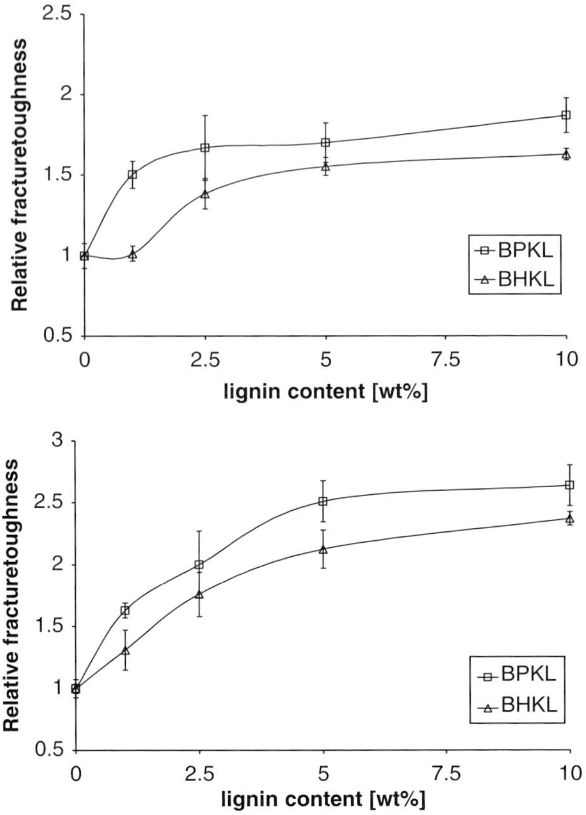

FIGURE 16.20 Relative fracture toughness ($K_{1c}/K_{1c,0}$) and plane-strain energy release rate ($G_{1c}/G_{1c,0}$) for AESO/styrene thermosets as a function of butyrated KL content.

composites and a small decrease in modulus and strength are commonly recorded.

The polymer yield strength and the cross-link density will both influence the fracture properties [180–182]. The flexural strength was found to increase at low lignin concentrations, with a linear decrease in cross-link density. This is the region with the most significant increase in fracture toughness and energy release rate because both effects work together [180, 181]. Because the glass transition temperature hardly changes with the addition of butyrated KL, the effect of shifting the test temperature closer to T_g [181] does not influence the fracture properties. At higher lignin contents, the polymer strength was found to decrease, such that fracture property improvement tapers off, due to both effects working in opposite directions [180]. From the curve trends, it appears that 10 wt% lignin is close to the lignin content limit for maximum improvement. As a comparison, rubber toughening is found to have maximum toughening improvement around 21% [183]. They do not directly affect the matrix structure, however, because they form a separate phase during polymerization [184]. The bigger improvement for larger lignin molecules is very noteworthy, because they also give better mechanical properties at the same time. Rubber toughened systems generally show a decreasing modulus and strength with increased toughening [179, 185]. On the other hand, their toughening behavior is superior with energy release rate improvements of 2 orders of magnitude reported for epoxy resins [186].

The relative fracture properties, $K_{1c}/K_{1c,0}$ and $G_{1c}/G_{1c,0}$, of the vinyl ester resin as a function of lignin content are shown in Figure 16.21. The resulting trends are in line with the findings for the AESO/styrene samples: an initial large increase, tapering off for larger butyrated KL contents, and a larger improvement for BPKL. The improvements for BHKL are still quite remarkable, considering its small effect on the cross-link density. At high BHKL content, the energy release rate is seen to decrease while the fracture toughness flattens out. This is due to the sharp decrease in yield strength, attributed

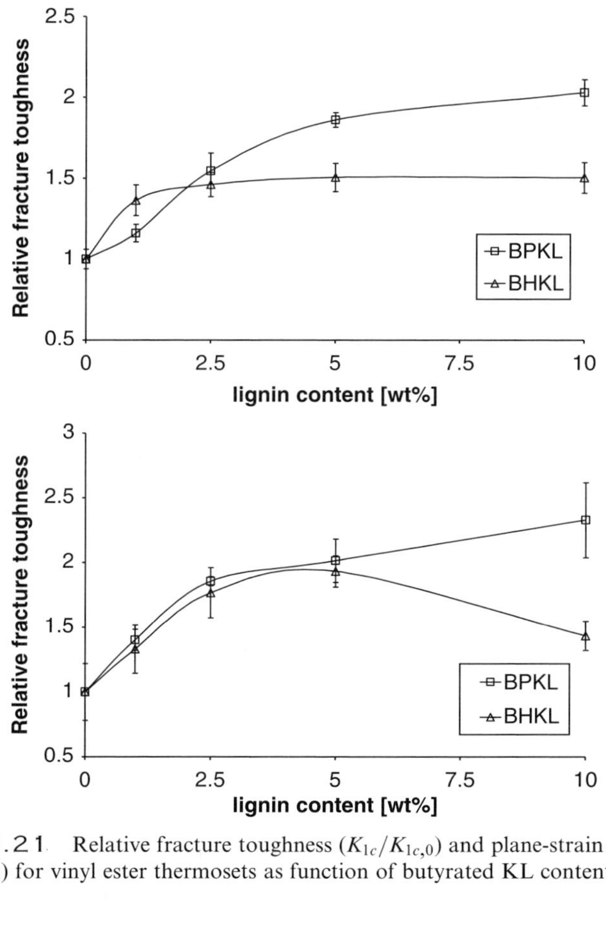

FIGURE 16.21 Relative fracture toughness ($K_{1c}/K_{1c,0}$) and plane-strain energy release rate ($G_{1c}/G_{1c,0}$) for vinyl ester thermosets as function of butyrated KL content.

to the effective styrene inhibition of BHKL during polymerization. The energy release rate has a stronger dependence on the strength than does the fracture toughness [183], such that this behavior is physically possible. Flattening out of the fracture property improvements is seen around 10 wt% BPKL, as was the case for the AESO/styrene resin. Butyrated lignin can thus be used to toughen vinyl ester to a limited extent with a relatively low loss of mechanical properties.

Figure 16.22 shows a series of scanning electron micrograph (SEM) images of the AESO/styrene fracture surface with increasing butyrated hardwood KL content. The amount of plastic deformation, in the form of "tips" pointing in the direction of crack growth, increases with lignin content. The result is an increase in fracture toughness and energy release rate, as measured. The number of deformation tips per unit area was determined from the micrographs. Figure 16.23 shows the comparison of the relative fracture toughness $(K_{1c}/K_{1c,0})$ and the relative plane-strain energy release rate

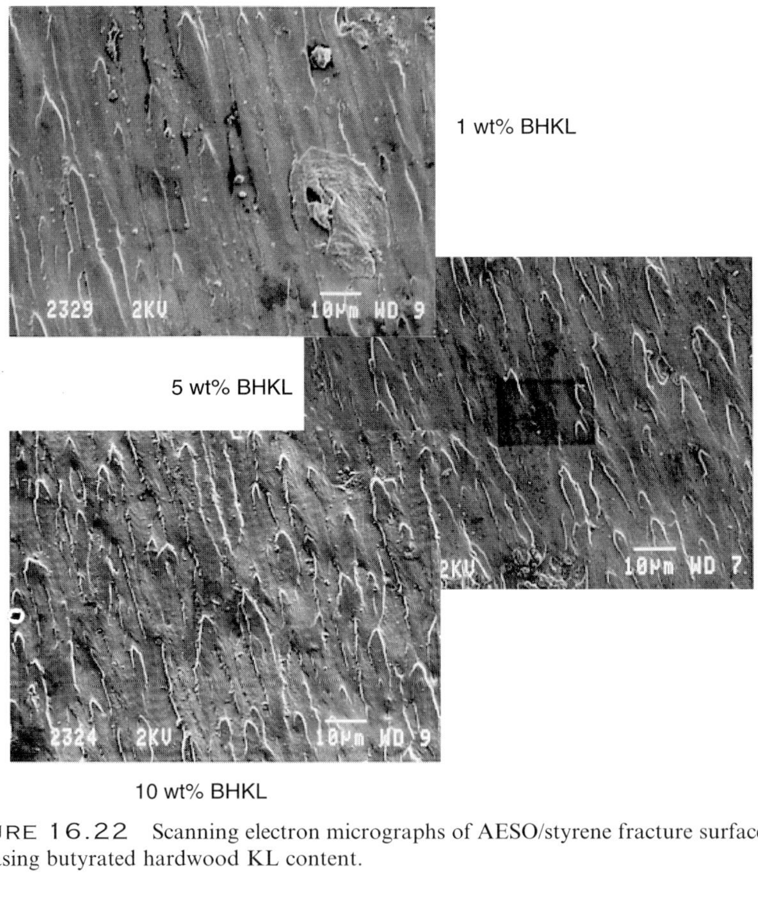

FIGURE 16.22 Scanning electron micrographs of AESO/styrene fracture surfaces with increasing butyrated hardwood KL content.

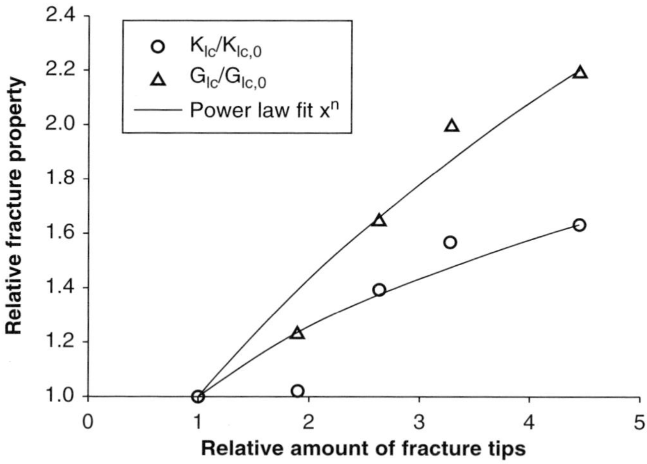

FIGURE 16.23 Relative fracture toughness and plane-strain energy release rate per fracture surface plastic deformations per unit area for butyrated hardwood–AESO–styrene thermosets.

$(G_{1c}/G_{1c,0})$ with the relative increase in crack tips per unit area for BHKL. A power dependence around 0.33 and 0.53 was found for K_{1c} and G_{1c}, respectively. The VE fracture surfaces exhibited fewer plastic deformations than the AESO/styrene thermosets. The lower plastic deformation is in agreement with the lower G_{1c} value. An increase in plastic deformation increases the energy consumed during crack propagation and consequently increases the energy release rate G_{1c}.

Methacrylated/butyrated pine KL (M-BPKL) with methacrylate substitution between 0.1 and 0.3 per PPU (2 to 6 per average pine KL molecule) was derivatized. All samples showed visible phase separation during polymerization, with the number of aggregates in the AESO/styrene resin more numerous and larger than in the vinyl ester. The AESO/styrene and vinyl ester samples with varying amounts of M-BPKL (0.2 methacrylates per PPU) were tested in three-point bend mode to determine flexural properties (Figure 16.24). The results generally mirror the trends for the butyrated lignin shown previously in Figures 16.15 and 16.16 for AESO and vinyl esters, respectively. Overall, the properties for BPKL were lower than for BPKL. This is to be expected since phase separation introduces weak regions, thus decreasing the modulus and strength of the material. However, the flexural strength of vinyl ester improved for lower lignin fractions. This strength increase, despite aggregates, can only be attributed to the methacrylates on lignin, which are expected to incorporate lignin into the matrix and provide for ideal stress transfer through covalent bonds. The incorporation of modified lignin into the polymer network offers extraordinary potential.

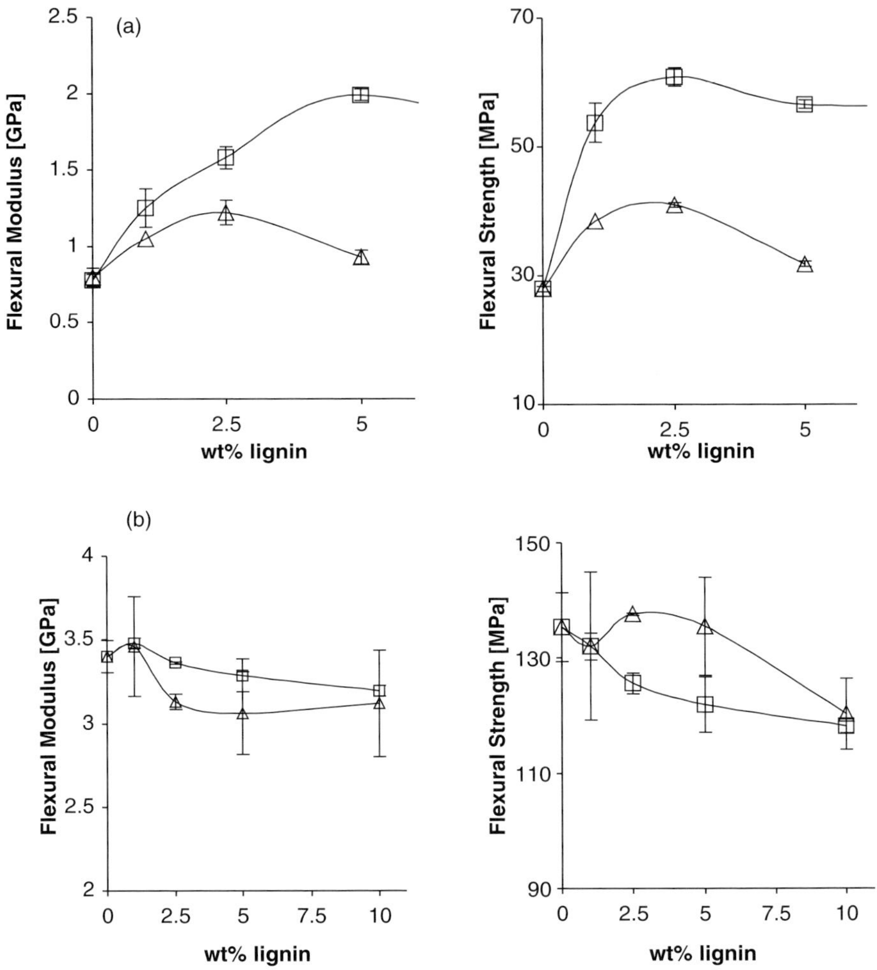

FIGURE 16.24 Flexural properties of M-BPKL (0.2 methacrylates and 0.88 butyrates per PPU) compared to the properties of BPKL: (a) AESE/styrene and (b) vinyl ester. Squares are for BPKL, triangles for M-BPKL.

The use of lignin in natural fiber composites [69] has shown some exciting potential. Treatment of natural fibers with kraft lignin decreased the contact angle and thus improved fiber wettability. This resulted in improved mechanical properties when comparing the treated to untreated fiber composites. Fracture surface SEM micrographs showed the failure to occur at the lignin–fiber interface rather than the lignin–polymer interface, vindicating the hypothesis that lignin can be used to improve the reinforcement–polymer interface. Addition of butyrated kraft lignin to the resin showed some very interesting results. No improved fiber wetting was determined by contact angle measurements. However, the SEM analysis of the fracture surface showed polymer–fiber adhesion even though significant fiber pullout did still occur. Improvements were seen in flexural properties whereas tensile properties went through a maximum.

16.7.3 SUMMARY OF BUTYRATED KRAFT LIGNIN

Butyrated and methacrylated-butyrated KL were incorporated into AESO/styrene and vinyl ester thermosets. Butyrated KL gave significant improvements in both flexural and fracture properties, with the extent of improvements dependent on the lignin size. The flexural property improvements go through a maximum due to a reduction in cross-link density with lignin addition. The reduction in cross-link density and resulting plasticization were also verified by fractography. This work shows that butyrated KL can be used to improve the properties of thermosetting polymers, as long as the cross-link density is high enough for the lignin to be *noticed* by the polymer network.

It is paramount for lignin not to have a detrimental effect on the polymer network formation as seen for high BHKL content in vinyl ester. One of the most interesting findings is that butyrated KL can improve both mechanical and fracture properties at the same time. Additionally, the amount of renewable material in the resin is increased with the use of lignin. For the AESO/styrene resin, with 5 wt% butyrated lignin, a polymer with approximately 72% renewable material is obtained. The addition of lignin also brings the mechanical properties of the AESO/styrene into the average range of values for unsaturated thermosets. Overall, lignin adds considerable value to bio-based and petroleum-based thermoset polymers, as well as making them more compatible with natural fiber composites [167]. We have barely scratched the surface of potential future applications of these green materials.

16.8 COMMENT ON BIOREFINERY

In the near future, the derivation of biofuels, bioenergy, and bioproducts as discussed herein will be treated as a biorefinery issue, similar to the petroleum refinery industry [187]. As shown in Figure 16.25, both agricultural and forestry biomass feedstock streams can be utilized in the biorefinery. The evolution of the biorefinery concept will most likely parallel that of the petroleum refinery, in which significant changes in the technology and approach can be expected. For example, petroleum refineries have evolved beyond simple separation and distillation to more complex thermal and catalytic cracking processes and reforming processes. Energy efficiency in the petroleum industry has also increased over the years. For example, the

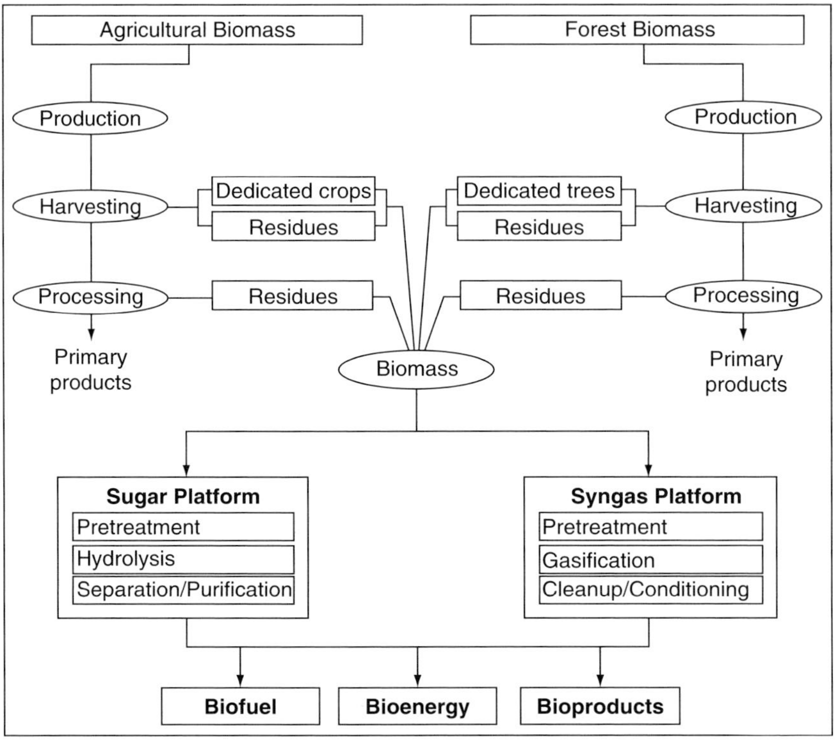

FIGURE 16.25 The biorefinery concept (IEA Report, 2004).

energy required to make polyethylene has decreased by a factor of about 7 in the past 50 years, and similar improvements are expected for future generations of biorefineries.

Biorefineries are being developed in Europe and the United States with an emphasis on the starch-based sugar platform. Lignocellulosic refining is more challenging, as it requires the hydrolysis of five sugars (glucose, galactose, mannose, xylose, and arabinose), whereas starch processing requires the hydrolysis of a single sugar (glucose). However, considerable progress is being made in the development of new low-cost enzymes to efficiently process lignocellulosics.

Chemical co-products from biomass processing have the most economic potential and are the subject of the greatest interest. The potential market for chemical bioproducts is very large. It includes the general categories of platform chemicals, adhesives and resins, plastics, paints, inks, soaps, coatings, cleaning compounds, lubricants and hydraulic fluids, greases, pesticides, toiletries, fragrances, and cosmetics. Ethanol can be readily derived as a liquid fuel or, when dehydrated to ethylene, can be utilized as the low-cost monomer for polyethylene and related plastics. To distinguish between petroleum-derived and biomass-derived products, C^{14} content will be used to determine a new product's biocontent. This method is based on the fact that the C^{14} content of very old carbon molecules in petroleum differs considerably from the C^{14} content of freshly derived biomass. Thus, in the future, we will most likely see the traditional petroleum-based plastics with a C^{14} content consistent with freshly grown biomass rather than the much lower C^{14} content derived from petroleum. The promise of the biorefinery is that we can continue to have the fuels, energy, and chemical feedstocks to support our modern lifestyle, which unfortunately is not very energy efficient or supportive of green engineering principles.

The derivation of ethanol and diesel fuels from biomass is considered a valuable but temporary solution to oil shortages generated by excess energy consumption and expansion of third world economies as they approach those of the Western nations. Critical shortages of petroleum will be noted around 2050, if not before then, and petroleum-based feedstocks for polymers and composites will continue to rise in cost. Bio-based hydrocarbon fuels will have little impact on global warming, and our future energy needs can be met in a globally sustainable manner only by solar, fusion, or fission sources. Ideally, solar or fission-fusion energy would be used to obtain hydrogen fuel by hydrolysis of water. However, the biorefineries that were initially developed for fuel and bioenergy needs will continue to have considerable impact by providing the basic renewable feedstocks for the broad range of bio-based materials and related chemical products discussed in this book.

REFERENCES

1. Goring, D. A. I. In *Lignins: Occurrence, Formation, Structure and Reactions*, Sarkanen, K. V.; Ludwig, C. H., Eds.; Wiley-Interscience, New York; 1971, pp. 695–768.
2. Batra, S. K. In *Handbook of Fiber Chemistry,* 2nd ed., Lewin, M.; Pearce, E. M., Eds.; Marcel Dekker, New York; 1998, pp. 505–575.
3. Krassig, H. A. *Polymer Monographs*, Vol. II, Elsevier, New York; 1993.
4. Gilbert, R. D.; Kadla, J. F. In *Polymer Modification: Principles, Techniques and Applications*, Meister, J. J., Ed.; Marcel Dekker, New York; 2000, pp. 21–65.
5. Timell, T. E. *Carbohydr. Chem.* 1965, **20**, 409–483.
6. Hansen, C. M.; Bjorkman, A. *Holzforschung* 1998, **52**, 335–344.
7. Hildebrand, J. H.; Scott, R. L. *The Solubility of Non-Electrolytes*, 3rd ed.; Reinhold, New York; 1949.
8. Hansen, C. M. *Ind. Eng. Chem. Prod. Res. Dev*. 1969, **8**, 2–11.
9. Sarkanen, K. V.; Hergert, H. L. In *Lignins: Occurrence, Formation, Structure and Reactions*, Sarkanen, K. V.; Ludwig, C. H., Eds.; Wiley-Interscience, New York; 1971, pp. 43–90.
10. Puls, J.; Schuseil, J. In *Hemicellulose and Hemicellulases*, Coughlan, M. P.; Hazlewood, G. P., Eds.; Portland Press, Chapel Hill, NC; 1993, pp. 1–27.
11. Timell, T. E. *Wood Sci. Technol.* 1967, **1**, 45–70.
12. Rowell, R. M.; Stout, H. P. In *Handbook of Fiber Chemistry,* 2nd ed., Lewin, M.; Pearce, E. M., Eds.; Marcel Dekker, New York; 1998, pp. 465–505.

13. Sarkanen, K. V.; Ludwig, C. H. In *Lignins: Occurrence, Formation, Structure and Reactions*, Sarkanen, K. V.; Ludwig, C. H., Eds.; Wiley-Interscience, New York; 1971, pp. 1–18.
14. Payen, A. *Compt. Rend.* 1838, **7,** 1052.
15. Payen, A. *Compt. Rend.* 1838, **8**, 52.
16. Schulze, F. *Chem. Zentr*. 1857, 321.
17. Nimtz, H. H. *Tappi J.* **56**, 124. 1973,
18. Gibbs, R. D. In *The Physiology of Forest Trees*, Thimann E., Ed.; John Wiley and Sons, New York; 1958, pp. 269–312.
19. Wardrop, A. B. In *Lignins: Occurrence, Formation, Structure and Reactions*, Sarkanen, K. V.; Ludwig, C. H., Eds.; Wiley-Interscience, New York; 1971, pp. 19–41.
20. Marton, W. In *Lignins: Occurrence, Formation, Structure and Reactions*, Sarkanen, K. V.; Ludwig, C. H., Eds.; Wiley-Interscience, New York; 1971, pp. 639–694.
21. Donofrio, C. P. In *Pulp and Paper Science and Technology,* Vol. 1, Pulp, Libby, C. E., Ed.; McGraw-Hill, New York; 1962, pp. 133–140.
22. Lehman, D. F. In *Handbook of Pulp and Paper Technology,* 2nd ed., Britt, K. W., Ed.; Van Nostrand Reinhold Company, New York; 1970, pp. 225–238.
23. Swartz, J. N.; MacDonald, R. C. In *Pulp and Paper Science and Technology,* Vol. 1, Pulp, Libby, C. E., Ed.; McGraw-Hill, New York; 1962, pp. 160–239.
24. Ericsson, E. O.; McCarthy, J. L.; Pearson, D. A. In *Pulp and Paper Science and Technology,* Vol. 1, Pulp, Libby, C. E., Ed.; McGraw-Hill, New York; 1962, pp. 240–280.
25. Nolan, W. J. In *Handbook of Pulp and Paper Technology,* 2nd ed., Britt, K. W., Ed.; Van Nostrand Reinhold Company, New York; 1970, pp. 135–143.
26. Gordon, L. J. In *Handbook of Pulp and Paper Technology,* 2nd ed., Britt, K. W., Ed.; Van Nostrand Reinhold Company, New York; 1970, pp. 159–177.
27. Stockburger, P. *Tappi J.* 1993, **76**, 71–74.
28. Saake, B., Lummitsch, S., Moranee, R., Lehnen, R., Nimz H. H. *Papier* **49**, V1–V7. 1995, *In German*
29. Puls, J.; Saake, B.; Horn, H.; Knuth, L. U.; Vila, C.; Parajo, J. C. 2003, *Afinidad* **60**, 233–238.
30. Poppiuslevin, K.; Mustonen, R.; Huovila, T.; Sundquist, J. *Pap. Puu – Pap. Tim.* 1991, **73**, 154–158.
31. McDonough, T. J. *Tappi J.* 1993, **76**, 186–193.
32. Sarkanen, K. V. In *Progress in Biomass Conversion,* Vol. 2, Sarkanen, K. V.; Tillman, D. A., Eds.; Academic Press, New York; 1971, pp. 127–144.
33. Ibrahim, M.; Glasser, W. G. *Bioresource Technol.* 1999, **70**, 181–192.
34. Focher, B.; Marzetti, A.; Crescenci, V., Eds., *Steam Explosion Techniques, Fundamentals and Industrial Applications,* Gordon and Breach Publishers, Philadelphia; 1991.
35. Avellar, B. K.; Glasser, W. G. *Biomass Bioenergy* 1998, **14**, 201–218.
36. Yan, D. B.; Krishnagopalan, G. A. *Appita J.* 2003, **56**, 391–396.
37. Argyropoulos, D. S. *J. Pulp Pap. Sci.* 2003, **29**, 308–313.
38. Carvalho, M. G. V.; Martins, A. A.; Figueiredo, M. M. L. *Appita J.* 2003, **56**, 267–274.
39. Andersson, N.; Wilson, D. I.; Germgard, U. *Nord. Pulp Pap. Res. J.* 2003, **18**, 200–209.
40. Luonsi, A.; Lento, S.; Halttunen, S.; Ala-Kaila, K. *Water Sci. Technol.* 2003, **47**, 61–69.
41. Ban, W. P.; Singh, J.; Lucia, L. A. *Holzforschung* 2003, **57**, 275–281.
42. Watson, P.; Hussein, A.; Reath, S.; Gee, W.; Hatton, J.; Drummond, J. *Tappi J.* 2003, **2**, 26–32.
43. Adler, E.; Marton, J.; Falkehag, I. *Acta Chim. Scand.* 1962, **18**, 1311–1312.
44. Gierer, J.; Lenz, B.; Noren, I.; Soederberg, S. *Tappi* 1964, **47**, 233–239.
45. Gierer, J.; Kunze, I. *Acta Chem. Scand.* 1961, **15**, 803–807.
46. Gierer, J.; Noren I. *Acta Chem. Scand.* 1962, **16**, 1976–1988.
47. Gierer, J.; Noren, I. *Acta Chim. Scand.* 1962, **16**, 1713–1729.
48. Gierer, J.; Lenz, B.; Wallin, N.-H. *Acta Chem. Scand.* 1964, **18**, 1469–1476.
49. Adler, E.; Falkehag, I.; Marton, J.; Halvarson, H. *Acta Chem. Scand.* 1964, **18**, 1311–1312.

50. Hoyt, C. H.; Goheen, D. W. In *Lignins: Occurrence, Formation, Structure and Reactions*, Sarkanen, K. V.; Ludwig, C. H., Eds.; Wiley-Interscience, New York; 1971, pp. 833–865.
51. Dutta, S.; Garver, T. M., Jr.; Sarkanen, S. In *Lignin: Properties and Materials*, Glasser, W. G.; Sarkanen, S., Eds.; ACS Symp. Ser. 397; American Chemical Society, Washington, DC; 1989, pp. 155–176.
52. Dong, D.; Fricke, A. L. *Polymer* 1995, **36**, 2075–2078.
53. (a) McCarthy, J. L.; Islam, A. In *Lignin: Historical, Biological, and Materials Perspectives*, Glasser, W. G., Northey, R. A.; Schultz, T. P., Eds.; ACS Symp. Ser. 742; American Chemical Society, Washington, DC; 2000, pp. 2–99.
(b) Tiainen, E., Drakenberg, T., Tamminen, T., Holzforschunk, 53, 529 (1999)
(c) Lenz, B.L., Tappi, 51, 511 (1968)
54. Lin, S. Y.; Zhong, X. J. *China Pulp Paper* 1990, **9**, 45–53.
55. Gargulak, J. D.; Lebo, S. E. In *Lignin: Historical, Biological, and Materials Perspectives*, Glasser, W. G.; Northey, R. A.; Schultz, T. P., Eds.; ACS Symp. Ser. 742; American Chemical Society, Washington, DC; 2000, pp. 304–320.
56. Knodt, C. B. U.S. Patent 3,035,920; 1963.
57. Toda, S. Japan Patent 02,262,524.
58. Sakagami, H.; Kawazoe, Y.; Konnom, K. Japan Patent 03,206,043.
59. Clough, T. J. U.S. Patent 5,575,334; 1996.
60. Nakanno, J.; Yamada, M. In *Proc. Pulp and Paper Research Conference,* Vol. 6; 1995, pp. 24–27.
61. Falkehag, S. I. *Appl. Polym. Symp*. 1975, **28**, 247–257.
62. Lora, J. H.; Glasser, W. G. 2002, *J. Polym. Environ*. **10**, 39–48.
63. Kharade, A. Y.; Kale, D. D. *J. Appl. Polym. Sci*. 1999, **72**, 1321–1326.
64. Alexy, P.; Kosikova, B.; Podstanska, G. *Polymer* 2000, **41**, 4901–4908.
65. Rodrigues, P. C.; Muraro, M.; Garcia, C. M.; Souza, G. P.; Abbate, M.; Schreiner, W. H.; Gomes, M. A. B. *Eur. Polym. J.* 2001, **37**, 2217–2223.
66. Toriz, G.; Denes, F.; Young, R. A. *Polym. Compos*. 2002, **23**, 806–813.
67. Li, J. C.; He, Y.; Inoue, Y. *Polym. Int*. 2003, **52**, 949–955.
68. Paauw, M.; Pizzi, A. *J. Appl. Polym. Sci*. 1993, **50**, 1287–1293.
69. Thielemans, W.; Can, E.; Morye, S. S.; Wool, R. P. *J. Appl. Polym. Sci*. 2002, 83, 323–331.
70. Setua, D. K.; Shukla, M. K.; Nigam, V.; Singh, H.; Mathur, G. N. *Polym Compos*. 2000, **21**, 988–995.
71. Li, Y.; Mlynar, J.; Sarkanen, S. *J. Polym. Sci. Polym. Phys. Ed*. 1997, **35**, 1899–1910.
72. Feldman, D.; Banu, D.; Campanelli, J.; Zhu, H. *J. Appl. Polym. Sci*. 2001, **81**, 861–874.
73. Otani, S.; Fukuoka, Y.; Igarashi, B.; Sasaki, K. U.S. Patent 3,461,082; 1969.
74. Fukuoka, Y. *Jpn. Chem. Q*. 1969, **5**, 63–66.
75. Sudo, K.; Shimizu, K. *J. Appl. Polym. Sci*. 1992, **44**, 127–134.
76. Sudo, K.; Shimizu, K.; Nakashima, N.; Yokoyama, A. *J. Appl. Polym. Sci*. 1993, **48**, 1485–1491.
77. Uraki, Y.; Kubo, S.; Nigo, N.; Sano, Y.; Sasaya, T. *Holzforschung* 1995, **49**, 343–350.
78. Kubo, S.; Uraki, Y.; Sano, Y. *Carbon* 1998, **36**, 1119–1124.
79. Kadla, J. F.; Kubo, S.; Venditti, R. A.; Gilbert, R. D.; Compere, A. L.; Griffith, W. *Carbon* 2002, **40**, 2913–2920.
80. Nada, A. M. A.; El-Saied, H.; Ibrahem, A. A.; Yousef, M. A. *J. Appl. Polym. Sci*. 1987, **33**, 2915–2924.
81. Barry, A. O.; Peng, W.; Riedl, B. *Holzforschung* 1993, **47**, 247–252.
82. Roffael, E.; Dix, B. *Holz Roh Werkstoffe* 1991, **49**, 199–205.
83. Kovrizhnykh, L. P.; Leonovich, A. A. *Russ. J. Appl. Chem*. 1998, **71**, 510–513.
84. Ball, F. J.; Moorer, H. H., Jr.; Dougherty, W. K. Canada Patent 654,728; 1962.
85. Feldman, D.; Banu, D.; Natansohn, A.; Wang, J. *J. Appl. Polym. Sci*. 1991, **42**, 1537–1550.

86. Nonaka, Y.; Tomita, B.; Hatano, Y. *Holzforschung* 1997, **51**, 183–187.
87. Hsu, O. H.-H.; Glasser, W. G. *Appl. Polym. Symp.* 1975, **28**, 297–307.
88. Saraf, V. P.; Glasser, W. G. *J. Appl. Polym. Sci.* 1984, **29**, 1831–1841.
89. Yoshida, H.; Mörck, R.; Kringstad, K. P.; Hatakeyama, H. *J. Appl. Polym. Sci.* 1987, **34**, 1187–1198.
90. Hirose, S.; Yano, S.; Hatakeyama, T.; Hatakeyama, H. In *Lignin: Properties and Materials*, Glasser, W. G.; Sarkanen, S., Eds.; ACS Symp. Ser. 397; American Chemical Society, Washington, DC; 1989, pp. 382–389.
91. Mörck, R.; Riemann, A.; Kringstad, K. P. In *Lignin: Properties and Materials*, Glasser, W. G.; Sarkanen, S., Eds.; ACS Symp. Ser. 397; American Chemical Society, Washington, DC; 1989, pp. 390–401.
92. Yoshida, H.; Mörck, R.; Kringstad, K. P.; Hatakeyama, H. *J. Appl. Polym. Sci.* 1990, **40**, 1819–1832.
93. Reimann, A.; Mörck, R.; Yoshida, H.; Hatakeyama, H.; Kringstad, K. P. *J. Appl. Polym. Sci.* 1990, **41**, 39–50.
94. Feldman, D.; Luchian, C.; Banu, D.; Lacasse, M. *Cell. Chem. Technol.* 1991, **25**, 163–180.
95. Hatakeyama, H.; Hirose, S.; Hatakeyama, T.; Nakamura, K.; Kobashiga, K.; Morohoshi, N. *J. Macromol. Sci. Pure* 1995, **A32**, 743–750.
96. Nakano, J.; Izuta, Y.; Orita, T.; Hatakeyama, H.; Kobashigawa, K.; Teruya, K.; Hirose, S. *Sen.-I. Gakkaishi* 1997, **53**, 416–422.
97. Guo, Z. X.; Gandini, A. *Eur. Polym. J.* 1991, **27**, 1177–1180.
98. Guo, Z. X., Gandini, A.; Pla, F. *Polym. Int.* 1992, **27**, 17–22.
99. Phillips, R. B.; Brown, W.; Stannet, V. T. *J. Appl. Polym. Sci.* 1972, **16**, 1–14.
100. Wang, H. X. *M.S. Thesis*, Virginia Tech, Blacksburg, VA; 1987.
101. Rizk, N. A.; Nagata, A.; Mansour, O. Y. *Acta Polym.* 1984, **35**, 61–64.
102. DePaoli, M. A.; Furlan, L. T. *Polym. Degrad. Stabil.* 1985, **11**, 327–337.
103. DePaoli, M. A.; Furlan, L. T. *Polym. Degrad. Stabil.* 1985, **13**, 129–138.
104. Furlan, L. T.; Rodrigues, M. A.; DePaoli, M. A. *Polym. Degrad. Stabil.* 1985, **13**, 337–350.
105. Barclay, L. R. C.; Xi, F.; Norris, J. Q. *J. Wood Chem. Technol.* 1997, **17**, 73–90.
106. Lu, F. J.; Chu, L. H.; Gau, R. J. *Nutr. Cancer* 1998, **30**, 31–38.
107. Pouteau, C.; Dole, P.; Cathala, B.; Averous, L.; Boquillon, N. *Polym. Degrad. Stabil.* 2003, **81**, 9–18.
108. Koshijima, T.; Muraki, E. *J. Jpn. Wood Res. Soc.* 1964, **10**, 110–113.
109. Koshijima, T.; Muraki, E. *J. Jpn. Wood Res. Soc.* 1964, **10**, 116–119.
110. Koshijima, T.; Muraki, E. *J. Jpn. Wood Res. Soc.* 1966, **12**, 139–144.
111. Koshijima, T.; Muraki, E. *J. Polym. Sci. A1* 1968, **6**, 1431–1440.
112. Phillips, R. B.; Brown, W.; Stannett, V. T. *J. Appl. Polym. Sci.* 1971, **15**, 2929–2940.
113. Phillips, R. B.; Brown, W.; Stannett, V. T. *J. Appl. Polym. Sci.* 1973, **17**, 443–451.
114. Young, R. A.; Achmadi, S. *J. Wood Chem. Technol.* 1992, **12**, 485–510.
115. Moze, A.; Zule J. *Cell. Chem. Technol.* 1999, **33**, 415–422.
116. Huawu, L.; Jiaxiang, C. *Cell. Chem. Technol.* 1991, **25**, 79–84.
117. Area, M. C.; Valade, J. L. *Cell. Chem. Technol.* 1998, **32**, 225–240.
118. Meister, J. J. In *Polymer Modification: Principles, Techniques and Applications*, Meister, J. J., Ed.; Marcel Dekker, New York; 2000, pp. 67–144.
119. Lawton, J. E.; Bellamy, W. D.; Hungate, R. E.; Bryant, M. P. *Science* 1951, **113**, 380–382.
120. Allan, G. G. In *Lignins: Occurrence, Formation, Structure and Reactions*, Sarkanen, K. V.; Ludwig, C. H., Eds.; Wiley-Interscience, New York; 1971, pp. 511–573.
121. Meister, J. J.; Patil, D. R. *J. Appl. Polym. Sci.* 1984, **29**, 3457–3477.
122. Meister, J. J.; Patil, D. R.; Augustin, C.; Lai, J. Z. In *Lignin: Properties and Materials*, Glasser, W. G.; Sarkanen, S., Eds.; ACS Symp. Ser. 397, American Chemical Society, Washington, DC; 1989, pp. 294–310.

123. Kelley, S. S. *Incorporation of Lignin Copolymers in Polyurethane Materials*, Ph.D. Thesis, Virginia Polytechnic Institute and State University; 1987.
124. Oliviera, W.; Glasser, W. G. *Macromolecules* 1994, **27**, 5–11.
125. Oliviera, W.; Glasser, W. G. *J. Appl. Polym. Sci.* 1994, **51**, 563–571.
126. Oliviera, W.; Glasser, W. G. *Polymer* **3**1994, **5**, 1977–1985.
127. Mai, C.; Milstein, O.; Hütterman, A. *J. Biotechnol.* 2000, **79**, 173–183.
128. Mai, C.; Schormann, W.; Hütterman, A.; Kappl, R.; Hütterman, J. *Enz. Microb. Technol.* 2002, **30**, 66–72.
129. Chen, C.-L. In *Methods in Lignin Chemistry*, Lin, S. Y.; Dence, C. W., Eds.; Springer Verlag, Berlin; 1992, pp. 409–422.
130. Lewis, H. F.; Brauns, F. E.; Buchanan, M. A.; Brookbank, M. A. *Ind. Eng. Chem.* 1943, **35**, 1113–1117.
131. Brauns, F. E.; Lewis, H. F.; Brookbank, E. B. *Ind. Eng. Chem.* 1945, **37**, 70–73.
132. Mansson, P. *Holzforschung* 1983, **37**, 143–146.
133. Glasser, W. G.; Jain, R. K. *Holzforschung* 1993, **47**, 225–233.
134. Thielemans, W.; Wool, R. P. *Biomacromolecules* (submitted).(2005)
135. Glasser, W. G.; Dave, V.; Frazier, C. E. *J. Wood Chem. Technol.* 1993, **13**, 545–559.
136. Li, S. M.; Lundquist, K. *J. Wood Chem. Technol.* 1997, **17**, 391–397.
137. Ralph, J.; Lu, F. C. *J. Agr. Food Chem.* 1998, **46**, 4616–4619.
138. Naveau, H. P. *Cell. Chem. Technol.* 1975, **9**, 71–77.
139. Da Cunha, C.; Deffieux, A.; Fontanille, M. *J. Appl. Polym. Sci.* 1993, **48**, 819–831.
140. Glasser, W. G.; Jain, R. K. *Holzforschung* 1993, **47**, 325–332.
141. Huang, J.; Zhang, L. N.; Chen, F. G. *J. Appl. Polym. Sci.* 2003, **88**, 3284–3290.
142. Dhara, K.; Jain, R. K.; Glasser, W. G. *Holzforschung* 1993, **47**, 403–411.
143. Zhang, L.; Huang, J. *J. Appl. Polym. Sci.* 2001, **80**, 1213–1219.
144. Ciemniecki, S. L.; Glasser, W. G. In *Lignin: Properties and Materials*, ACS Symp. Ser. 397; American Chemical Society, Washington, DC; 1989, pp. 452–463.
145. Ghosh, I.; Jain, R. K.; Glasser, W. G. In *Lignin: Historical, Biological and Materials Perspectives*, Glasser, W. G.; Northey, R. A.; Schultz, T. P., Eds.; ACS Symp. Ser. 742; American Chemical Society, Washington, DC; 2000, pp. 331–350.
146. Ghosh, I.; Jain, R. K.; Glasser, W. G. *J. Appl. Polym. Sci.* 1999, **74**, 448–457.
147. Feldman, D.; Banu, D.; Manley, R. S. J.; Zhu, H. *J. Appl. Polym. Sci.* 2003, **89**, 2000–2010.
148. Li, Y.; Mlynar, J.; Sarkanen, S. *J. Polym. Sci. Polym. Phys.* 1997, **35**, 1899–1910.
149. Li, Y.; Sarkanen, S. In *Lignin: Historical, Biological and Materials Perspectives*, Glasser, W. G.; Northey, R. A.; Schultz, T. P., Eds.; ACS Symp. Ser. **742**; American Chemical Society, Washington, DC; 2000, pp. 351–366.
150. Li, Y.; Sarkanen, S. *Macromolecules* 2002, **35**, 9707–9715.
151. Ziaee, S.; Palmese, G. R. *J. Polym. Sci. Polym. Phys.* 1999, **37**, 725–744.
152. *Material Safety Data Sheet Dow Derakane 411-C50,* Dow Chemical Company; 2002.
153. Glasser, W. G.; Barnett, C. A.; Sano, Y. *J. Appl. Polym. Sci.: Appl. Polym. Sympos.* 1983, **37**, 441–460.
154. Wagner, R. B.; Zook, H. D. *Synthetic Organic Chemistry,* 5th printing; John Wiley & Sons, New York; 1965.
155. Noller, C. R. *Chemistry of Organic Compounds,* 3rd ed., W. B. Saunders Company, Philadelphia; 1966.
156. Patai, S. *The Chemistry of Functional Groups, Part 1: The Chemistry of Ketenes, Allenes, and Related Compounds*, John Wiley & Sons, New York; 1980.
157. Libby, M. C.; Watson, P. C.; Barteau, M. A. *Ind. Eng. Chem. Res.* 1994, **33**, 2904–2912.
158. Martinez, R.; Huff, M. C.; Barteau, M. A. *Appl. Catal. A-Gen.* 2000, **200,** 79–88.
159. Connors, K. A.; Pandit, N. K. *Anal. Chem.* 1978, **50**, 1542–1545.
160. Clark, J. C.; Brauns, F. E. *Paper Trade J.* 1944, **119**, 33–36.

161. Gellerstedt, G. In *Methods in Lignin Chemistry*, Lin, S. Y.; Dence, C. W., Eds.; Springer-Verlag, New York; 1992, pp. 485–497.
162. Rowe, E. L.; Machkovech, S. M. *J. Pharm. Sci.* 1977, **66**, 273–275.
163. Wachowiak, R.; Connors, K. A. *Anal. Chem.* 1979, **51**, 27–30.
164. Bittner, A. S.; Harris, L. E.; Campbell, W. F. *J. Agric. Food Chem.* 1980, **28**, 1242–1245.
165. Dee, L. A.; Biggers, B. L.; Fiske, M. A. *Anal. Chem.* 1980, **52**, 572–573.
166. Can, E.; Kusefoglu, S.; Wool, R. P. *J. Appl. Polym. Sci.* 2001, **81**, 69–77.
167. Wellens, S. L.; Carey, M. A.; Elder, D. K. *Anal. Chem.* 1980, **52**, 1374–1376.
168. Cohen, J. L.; Fong, G. P. *Anal. Chem.* 1975, **47**, 313–316.
169. Thielemans, W. *Ph.D. Dissertation,* University of Delaware, Department of Chemical Engineering; 2004.
170. Flory, P. J. *Principles of Polymer Chemistry*, Cornell University Press, Ithaca, NY; 1966.
171. Hildebrand, J. H., Prausnitz, J. M.; Scott, R. L., *Regular and Related Solutions,* Van Nostrand-Reinhold, Princeton, NJ; 1970.
172. Van Krevelen, D. W. *Properties of Polymers: Their Correlation with Chemical Structure; Their Numerical Estimation and Prediction from Additive Group Contributions*, Elsevier Science, Amsterdam; 1997.
173. Schuerch C. *J. Am. Chem. Soc.* 1952, **74**, 5061–5067.
174. Lindberg, J. J. *Suomen Kemistilehti B* 1967, **40**, 225–228.
175. Askadskii, A. A. *Physical Properties of Polymers: Prediction and Control,* Gordon and Breach Publishers, Amsterdam; 1996.
176. Hoy, K. L. *J. Coat. Fabr.* 1989, **19**, 53–67.
177. *Materials Safety Data Sheet*, Styrene, Aldrich Chemicals; 2000.
178. *Standard Test Methods for Plane-Strain Fracture Toughness and Strain Energy Release Rate of Plastic Materials*, ASTM D5045-93; ASTM Committee on International, West Conshohocken, PA; 1993.
179. Pearson, R. A. In *Toughened Plastics I: Science and Engineering*, Riew, C. K.; Kinloch, A. J., Eds.; American Chemical Society, Washington, DC; 1993, pp. 405–425.
180. Kinloch, A. J. In *Rubber Toughened Plastics*, Riew, C. K., Ed.; American Chemical Society, Washington, DC; 1989, pp. 67–91.
181. Bradley, W. L.; Schultz, W.; Corleto, C.; Komatsu, S. In *Toughened Plastics I: Science and Engineering*, Riew, C. K.; Kinloch, A. J., Eds.; American Chemical Society, Washington, DC; 1993, pp. 317–334.
182. Williams, J. G. *Fracture Mechanics of Polymers*, Ellis Horwood Limited, Chichester, England; 1984.
183. Wool, R. P. *Polymer Interfaces: Structure and Strength*, Hanser/Gardner Publications Inc., Cincinnati, OH; 1995.
184. Seymour, R. B. In *Rubber Toughened Plastics*, Riew, C. K., Ed.; American Chemical Society, Washington, DC; 1989, pp. 3–13.
185. Montarnal, S.; Pascault, J.-P.; Sautereau, H. In *Rubber Toughened Plastics*, Riew, C. K., Ed.; American Chemical Society, Washington, DC; 1989, pp. 193–223.
186. Garg, A. C.; Mai, Y.-M. *Compos. Sci. Technol.* 1988, **31**, 179–223.
187. Mabee, W.; Gregg, D.; Saddler, J. Ethanol from Lignocellulosics: A Report to the IEA Bio-Energy Task Force; 2004.

INDEX